TRANSFORMERS

ANALYSIS, DESIGN, AND MEASUREMENT

EDITED BY

Xose M. López-Fernández

H. Bülent Ertan • Janusz Turowski

Indian Institute of Technology, Bombay

CRC Press
Taylor & Francis Group
Boca Raton London New York

CRC Press is an imprint of the
Taylor & Francis Group, an **informa** business

CRC Press
Taylor & Francis Group
6000 Broken Sound Parkway NW, Suite 300
Boca Raton, FL 33487-2742

Version Date: 20120411

International Standard Book Number: 978-1-4665-0824-8 (Hardback)

Library of Congress Cataloging-in-Publication Data

Transformers : analysis, design, and measurement / editors, Xose M. López-Fernández, H. Bülent Ertan, and Janusz Turowski.
p. cm.
Includes bibliographical references and index.
ISBN 978-1-4665-0824-8 (hardback)
1. Electric transformers. I. López-Fernández, Xose M. II. Ertan, H. Bülent. III. Turowski, J.

TK2551.T775 2012
621.31'4--dc23 2012006500

Visit the Taylor & Francis Web site at
http://www.taylorandfrancis.com

and the CRC Press Web site at
http://www.crcpress.com

Contents

Preface

Historically, transformers have been designed and widely used to transfer energy, and for measurement, protection, electrical isolation, and signal coupling. Most transformers were designed for low-frequency operations, and despite a long history of low-frequency transformer theory and design, many stray field problems have yet to be resolved. On the other hand, there has been a significant increase in the operating frequency of many switched mode power supplies to several hundreds of kilohertz and, in some cases, even up to a few megahertz. However, in the megahertz region, several problems arise related to both the windings and the standard core materials. Therefore, the recent use for some applications of coreless transformers is an alternative to transformers with a core.

In the case of power transformers, which are key and costly components in networks, the failure of a single transformer can have a major impact on the safety, reliability, and cost of electrical supply. Their maintenance or, more generally, managing their life cycle, along with improved reliability and reduction of scheduled outages, assumes more and more importance due to economic compulsions. Both users and manufacturers play an important role in the life cycle management of power transformers, and hence close cooperation between them is essential. Recent catastrophic blackouts have exposed major vulnerabilities in existing generation, transmission, and distribution systems. Severe underinvestment, aging technology, and a conservative approach to innovation are being blamed for the current sorry state of affairs. Resources need to be directed into technologies that have the potential to improve the integrity of the system. High-temperature superconductivity has such potential; it also has a positive environmental impact as it reduces the losses as well as the size and weight of the power devices significantly.

Interest in increasing operational frequencies remains alive since there are two important trends influencing the development of transformers: miniaturization and an increase in transmitted power density. Therefore, power applications of high-frequency products comprise transformers in switch-mode power supplies, power chokes, and electronic lamp ballast devices. The latter includes transformers for wind power plants, contact-less chargers for electrical cars (safety!), and contact-less energy transmission to robotic arms (unlimited twisting in all directions).

The demand for power materials running under higher operating frequencies can now be met by highly advanced materials such as ferrites. Metallic powders and amorphous metals are also becoming increasingly important for use in this area.

Solutions for measuring currents and voltages needed for protection and monitoring in power systems have been achieved using instrument transformers. New solutions such as electronic instrument transformers (sensors) are also being successfully employed. Because of their superior technical performance, current and voltage electronic transformers (sensors) are expected to play an important role in MV systems in the future. In this area of instrument transformers, problems of propagation of disturbances to secondary circuits occur depending on the design features. While the principle of operation of instrument transformers is the same as that of power transformers, the problems concerned with the design are quite different.

The advances in computational tools are promoting the development of sophisticated modeling software for the solution of 3D fields which is supporting the design of transformers in the last few decades. However, these advances have not always been accompanied

Editors

Xose M. López-Fernández received his MSc in electrical engineering and his European PhD (with first class honors) in electrical engineering from Vigo University, Vigo, Spain, in 1992 and 1997, respectively. He is currently a professor in the Department of Electrical Engineering, University of Vigo, and was a visiting professor at the University of Artois, Béthune, France. Professor López-Fernández is the founder and general chairman of the International Advanced Research Workshop on Transformers (ARWtr). He has currently been entrusted with the responsibility of leading research projects for Spanish and Portuguese utilities and power transformer manufacturers. His research interests include the design aspects of electrical machines. Dr. López-Fernández received the Alfons Hoffmann's Medal from the Polish Power Engineering Society in 2004. He is a member of the IEEE-PES Transformers Committee as well as CIGRE.

H. Bülent Ertan received his BS and MS in electrical and electronics engineering (EEE) in 1971 and 1973, respectively, from the Middle East Technical University (METU) in Ankara, Turkey, and his PhD from the University of Leeds, United Kingdom, in 1977. He then joined METU as a staff member in 1977.

Professor Ertan is currently a staff member in the Department of Electrical and Electronics Engineering at METU. His research interest is in the area of electrical machine design and electrical drives.

Dr. Ertan has served as consultant to a number of companies manufacturing electrical motors and drives in Turkey. He has directed more than 20 industry-financed research projects and many other projects financed by the Turkish Scientific and Technological Research Institute (TUBITAK). He has served as the assistant director of TUBITAK Information Technologies and Electronics Research Institute since 1999 and led the Intelligent Energy Conversion Systems Research Group at TUBITAK-SPACE Institute for 8 years.

Dr. Ertan is one of the editors of the book entitled *Modern Electrical Drives*, published in 2000 by Kluwer Academic Publishers. He has published more than 100 articles in technical journals and conferences. Currently, he holds four international patents.

Dr. Ertan is a member of IEEE and IET, and is the founder of the International Conference on Electrical Machines and Power Electronics. He is also a member of the international steering committee of the International Conference on Electrical Machines. He is an editorial board member of *Electromotion Journal* and *Advanced Technology of Electrical Engineering and Energy* (China). He also served as the chairman of the board of trustees of Mustafa Parlar Education and Research Foundation and is currently a board member.

Janusz Turowski, PhD, DSc electrical engineering, has served as full professor (retired in 2003) in electrical machines and applied electromagnetics at the Institute of Electrical Machines and Transformers (now Institute of Mechatronics and Information Systems), Technical University of Lodz (TUL), Poland. From 1999 to 2010, he was a full professor and dean at the Academy of Humanities and Economics at TUL. He received an honorary degree from the University of Pavia, Italy, in 1998. He is also a full member of the International Academy of Electrotechnical Sciences, a member of CIGRE, and a senior member of IEEE. He served as the director of IEMT from 1973 to 1992. He has authored and coauthored over 300 scientific publications, including 20 books, cited over 980 times. He has also supervised 18 PhD theses.

Dr. Turowski served as a consultant of Polish ministers and transformer works in Poland, India, China, Australia, and Canada. He is the past president and an honorary member of the Polish Association of Theoretical and Applied Electrotechnical Sciences (PTETiS). He is also a member (1978–) and vice-chairman (1999–2003) of the Electrical Committee of the Polish Academy of Sciences and the chairman (1979–2001) and honorary chairman (2001–) of the International Symposium on Electromagnetic Fields in Electrical Engineering (ISEF). He was awarded the 2004 Silver Medal by the International Academy of Electrotechnical Sciences "For a prominent contribution to the development of the world science, education technology and international co-operation."

Contributors

Erdal Bizkevelci
Information Technologies and Electronics Research Institute
Ankara, Turkey

Frederick E. Bott
Noctiluca Ltd.
Bournemouth, United Kingdom

A.J. Marques Cardoso
Department of Electromechanical Engineering
University of Beira Interior
Covilhã, Portugal

and

Departamento de Engenharia Electrotécnica e de Computadores
Instituto de Telecomunicações
Coimbra, Portugal

Arcan F. Dericioglu
Department of Metallurgical and Materials Engineering
Middle East Technical University
Ankara, Turkey

Philip Devine
Department of Engineering
University of Leicester
Leicester, United Kingdom

Jesús Doval-Gandoy
Department of Electronics Technology
University of Vigo
Vigo, Spain

H. Bülent Ertan
Department of Electrical and Electronics Engineering
Middle East Technical University
Ankara, Turkey

John Fothergill
Department of Engineering
University of Leicester
Leicester, United Kingdom

Ivanka Höhlein
Siemens Power Transformer
and
Power Transformer Factory Nuremberg
Nuremberg, Germany

A. Ibero
Electrotécnica Arteche Hermanos
Mungia, Spain

Wieslaw Jalmuzny
Department of Applied Electrical Engineering and Instrument Transformers
Technical University of Lodz
Lodz, Poland

Adolf J. Kachler
Transformer Consultant
Quality MM, Testing and Diagnostics
Nuremberg, Germany

Marceli Kaźmierski
Transformer Division
Institute of Power Engineering
Lodz, Poland

Andrzej Koszmider
Department of Applied Electrical Engineering and Instrument Transformers
Technical University of Lodz
Lodz, Poland

S.V. Kulkarni
Department of Electrical Engineering
Indian Institute of Technology Bombay
Mumbai, India

Paul Lefley
Department of Engineering
University of Leicester
Leicester, United Kingdom

Elzbieta Lesniewska
Department of Instrument Transformers and Electromagnetic Compatibility
Institute of Electrical Power Engineering
Technical University of Lodz
Lodz, Poland

Andrey K. Lokhanin
Russian Electrotechnical Institute
Moscow, Russia

Xose M. López-Fernández
Department of Electrical Engineering
Universidade de Vigo
Vigo, Spain

Ralph Lucke
Fit-Ceramics
Miesbach, Germany

Ryszard Malewski
Malewski Electric, Inc.
Montreal, Quebec, Canada

Radoslav Miltchev
Department of Computer Systems and Informatics
University of Forestry
Sofia, Bulgaria

Luís M.R. Oliveira
Departamento de Engenharia Electrotécnica e de Computadores
Instituto de Telecomunicações
Coimbra, Portugal

and

Instituto Superior de Engenharia
Universidade do Algarve
Faro, Portugal

Moisés Pereira Martínez
Department of Electronics Technology
University of Vigo
Vigo, Spain

Władysław Pewca
Transformer Division
Institute of Power Engineering
Lodz, Poland

Jeewan Puri
Transformer Solution Inc.
Matthews, North Carolina

Kjetil Ryen
Multiconsult AS
Ski, Norway

Ryszard Sobocki
Polish Consultants Society
Zabrze, Poland

Jan K. Sykulski
School of Electronics and Computer Science
University of Southampton
Southampton, United Kingdom

Janusz Turowski
Institute of Mechatronics and Information Systems
Technical University of Lodz
Lodz, Poland

Levent B. Yalçiner
Information Technologies and Electronics Research Institute
Ankara, Turkey

Ivan Yatchev
Department of Electrical Apparatus
Technical University of Sofia
Sofia, Bulgaria

Part I

Power Transformers

Marceli Kaźmierski and Janusz Turowski

Introduction

The history of power transformers is part of the history of all phenomena and devices based on electromagnetic induction. An important step forward was the moment, when the transformer production has entered on the energy industry market.

It all started in 1883 with one of the first implementations to the transmission line a single-phase 15 kVA, 1500/300 V transformer, by K. Zipernowski, M. Deri, O. Blathy (Ganz, Budapest) and others.

The first real industrial three-phase 15 kVA, 1500/300 V power transformer was built and patented in 1891 by Michal Doliwo-Dobrowolski, was Polish descent, born in Petersburg (Russia) and worked at in AEG Berlin. The history of contemporary transformer industry and science originated then with Doliwo-Dobrowolski 121 years ago.

The material presented in Part I is probably the most competent report on the state of the art in this field and will hopefully become an indispensable manual for practicing engineers in industry and in the field of power generation and distribution of electric energy. The Part I is divided into 24 chapters and deals with design, diagnostics and maintenance, economy and on-site transformer management.

Most of the problems discussed are closely linked and overlapping; hence this subdivision is rather formal, flexible, and not very precise.

This book, prepared as post-conference work, provides one of the most contemporary treatments of modern transformers. The scientists, who work at the same time as manufacturers of large transformers and consultants of most outstanding transformer works and power utilities, have also contributed to this monograph. The contribution are provided mainly from researchers and practicing, experienced engineers and should be useful to users of both very large power transformers and very small HF units. It can also be considered as a convenient practical handbook, which deals with many difficult problems that have been resolved successfully.

Fifteen outstanding international experts have contributed to this part of the book. They are in alphabetic order as follows:

A. J. Cardoso (with Luis M. R. Oliveira) on fault diagnotics by Park's Vector Approach. Adolf J. Kachler on Transformer reliability: a key issue for both manufacturers and users and (jointly with Ivanka Höhlein) on Functional and component-related diagnostics for power transformers, a basis for successful "transformer life management; Marceli Kazmierski on Life management of transformers; Andrey K. Lokhanin on Insulation problems of HV power transformers; Xose M. Lopez-Fernandez on large-shell-type transformers, Malewski; on Power transformer acceptance tests, Wladyslaw Pewca on Selected problems of transformers short circuit withstand capability; Jeewan Puri on sources, measurement and mitigation of sound levels in transformers, Kjetil Ryen on Economics in transformer management: focus on life cycle cost, design review, and the use of simple Bayesian decision methods to manage risk; Ryszard Sobocki on Transformer design review: link between design and maintenance stages; S. V. Kulkarni on Challenges and strategies in transformer design; Jan Sykulski on Superconducting transformers; and Janusz Turowski on Stray losses, screening, and local excessive heating hazard in large power transformers.

The editors of Part I would like to express their gratitude to the Authors, Workshops Organizers and to the Publisher for their excellent work.

1

Selected Problems of Transformers' Capability to Withstand Short Circuits

Władysław Pewca

CONTENTS

1.1 Introduction

Standard CEI IEC 60076-5 *Power Transformers Part 5: Ability to Withstand Short-Circuit* provides a basis for assessing short circuit withstand capability by calculation methods that are considered to be as reliable as the dynamic tests. The currently available Annex A to the standard provides information on the theoretical evaluation of a transformer's ability to withstand the dynamic effects of short circuit and serves as a guidebook that provides methodologies for evaluating short circuit withstand ability. It can be concluded from Annex A that the analysis can be carried out using simple methods such as calculating stresses and confronting them with their permissible values, regardless of deformations and changes that take place during short circuit forces [1].

The criteria for the assessment of short circuit strength, in this case, are of a contractual nature and, as a general rule, apply only to the technological and constructional solutions used by a manufacturer, taking into account a specific power range. Advanced computer-assisted methods, which are available now, permit more precise calculations and assessments, taking into account the deformities of the windings, their essential physical properties and constructional solutions, as well as some aspects of production process. It should be expected that the development of these methods will subsequently lead to the creation of a more generalized calculation method of verifying short circuit withstand capability than the one in Annex A.

This chapter presents selected computer-assisted methods of analysis of short circuit withstand capability that take into account deformations, the structure of the windings, and the physical properties of the winding materials. The results obtained using these methods were compared with the available test results and, in some cases, also with the results that could be obtained by using simplified methods.

1.2 Radial Forces, Stresses, and Deformations

1.2.1 Analytical Method of Calculation

The starting point in the analysis of withstand capability of windings against radial short circuit forces is the coil of outer winding subjected to tensile forces, the calculation model of which is shown in Figure 1.1. Generally, in addition to the discussed winding, there can also be other parts such as regulating winding or bandage that increase the transformer's strength to withstand radial forces. Provided the coil was correctly produced and there are no gaps between its conductors and the outer bandage, stress σ_k and deformations ε_k for its particular conductors (coils) can be calculated with (2n – 2) set of equations connected with the conditions of static equilibrium [2]:

$$\sigma_k = \frac{(q_k + p_{k-1} - p_k) \cdot R_k}{S_c} \tag{1.1a}$$

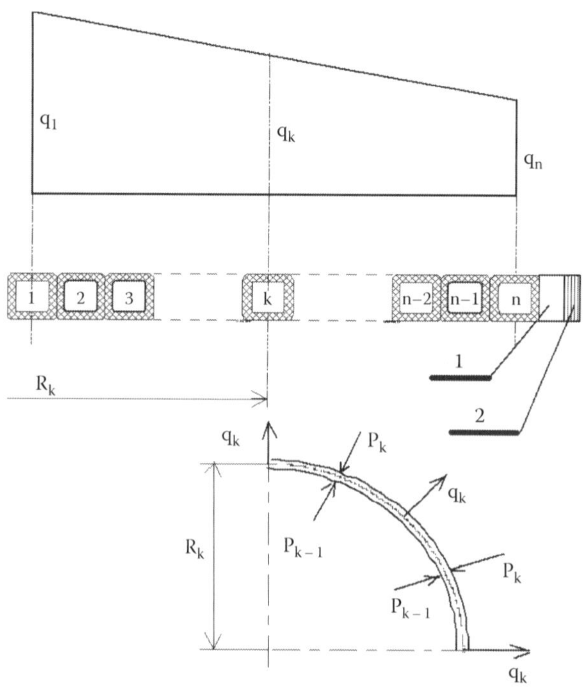

FIGURE 1.1
Electrodynamic conductor load q_k and interaction on its adjacent terns and their insulation for stretched coils: 1, axial spacers; 2, bandage.

$$\varepsilon_k \cdot R_k = \varepsilon_{k+1} \cdot R_{k+1} - \mu_k \cdot \delta_k \tag{1.1b}$$

where

q_k represents the unitary radial electrodynamics force acting on k-th conductor, p_k
p_{k-1} represents the unitary reaction forces coming from the insulation k and (k−1)
R_k represents the mean radius of k-th conductor, ε_k
ε_{k+1} represents the relative deformation of conductors k and (k+1)
μ_k, δ_k represents the relative deformation and the thickness of k-th insulation

In addition, Equation 1.1a is supplemented by the boundary conditions of the outer conductors applying to stresses on the insulation of the outer conductor p_0 and outer bandage p_n. This is usually based on the assumption that the conditions are as follows:

$$p_0 = p_n = 0 \tag{1.2}$$

Provided the mechanical characteristics (stresses σ in the function of deformity ε) are linear, the distribution of stresses on conductors and isolations can be calculated analytically; however, this solution has no practical value because constructional materials of windings demonstrate substantial physical nonlinearity (Figure 1.2). Furthermore, economic considerations as well as dimensional limitations of transformers make it necessary to apply stresses that are exerted also in the nonlinear parts of such materials.

For practical reasons, the analysis has to take into account stresses and deformities of the nonlinear character, as well as the phenomenon of accumulation of deformities, which can occur in exploitation or during dynamic tests after consecutive short circuits that can lead to the transformer's failure or its negative test result. For plastic materials such as Cu, Al the phenomenon of accumulation is widely known and can be described using the following relationship, which is also illustrated in Figure 1.3 [2]:

$$\sum_{i=2}^{I} \Delta\varepsilon_i = A \cdot (I-1)^B \tag{1.3}$$

where

$\Delta\varepsilon_i$ represents permanent conductor set as a result of succeeding stress σ_{max}
I represents multiple of the conductor loading
A, B represents empirical coefficients

Comprehensive investigations show that the coefficients depend largely on the relationship $\sigma_{max}/\sigma_{0,2}$ and also on the initial hardness of the conductors, loading (dynamic or static), and, to a lesser extent, on the temperature. Coefficient A, monotonically, but sharply, increases with σ_{max} while $B \leq 1$ and has for $\sigma_{max} \approx \sigma_{0.2}$ a relatively flat maximum. A few approximate values of A and B (listed in Table 1.1 [2]) obtained in the dynamic load tests, all of which had 20 impulses, with σ_{max} = const. and 10–20 ms duration, illustrate quantitative relationships that we have to deal with. The specified values indicate that a short circuit resulting in the average radial stress close to $\sigma_{0.2}$ creates, when repeated, a risk of unacceptable level of permanent set of the winding.

Figure 1.4 explains the method of using formula 1.1a for computations. The coil conductors interact during a short circuit, as described earlier. As a result, the post–short circuit

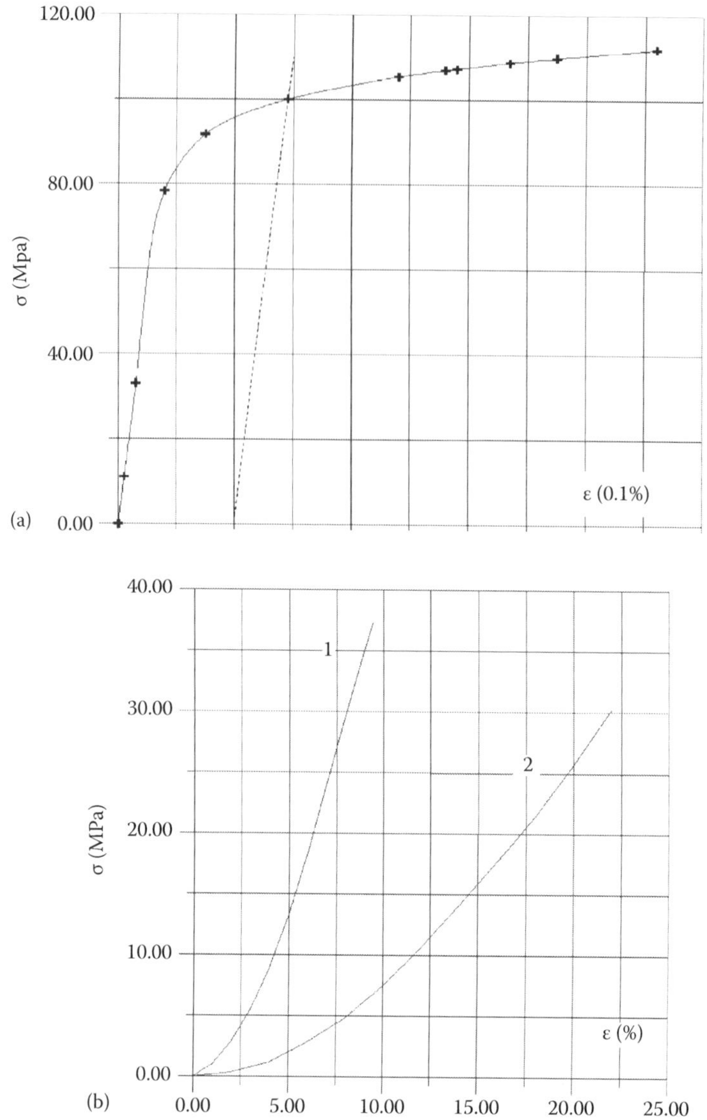

FIGURE 1.2
Strain/stress curves for (a) copper conductor, proof stress $\sigma_{0.2} = 100\,\text{MPa}$, and (b) transformer—1, board; 2, paper.

stresses lead to (see Figure 1.6) a permanent set in a coil, which needs to be included when calculating the cumulating effect in a winding. The points 1′ and 2′ correspond to the mechanical state of a selected coil conductor after the first (1′) or second (2′) short circuit, whereas σ_{12} and σ_{22} represent the related stresses in the coil conductors.

The chosen results of calculations obtained by using the program, the essential algorithm elements of which have been presented earlier, are shown in Figures 1.5 through 1.7. The calculations were made for the stretched coil of winding HV of a GSU transformer of great power, consisting of 24 paper-insulated Cu conductors, and the coefficient of filling the coil with copper was 40%, when there were axial cooling ducts, or 50%, when there

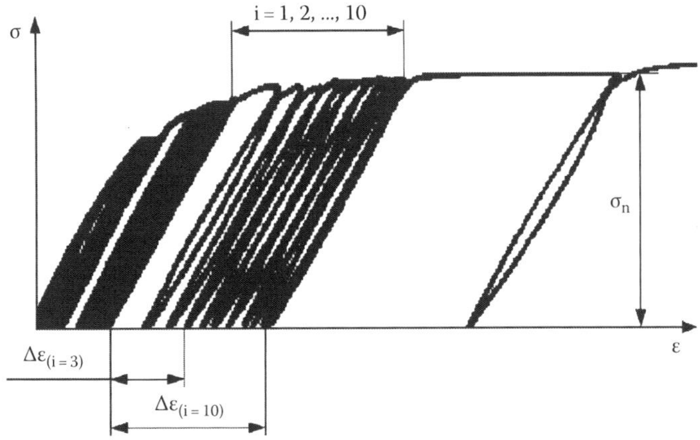

FIGURE 1.3
Variation of the conductor permanent set at repeated tensile stresses of same maximum value. Also effect of increased stresses.

TABLE 1.1

Coefficients Related to Formula (1.3) for a Selected Case

$\sigma_{max}/\sigma_{0.2}$	A	B
0.7	0.025	0.5
1.0	0.41	1.0
1.1	1.0	0.9

Source: Kozłowski, M. and Pewca, W., *J. Eur. Trans. Electr. Power (ETAP)*, 6(4), 259, 1996.

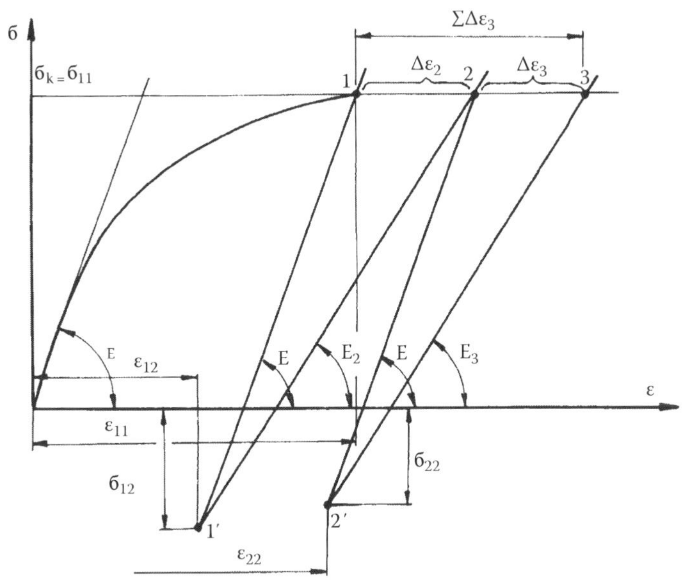

FIGURE 1.4
Approximation of cumulating effect of permanent deformations used in calculations for k-conductor.

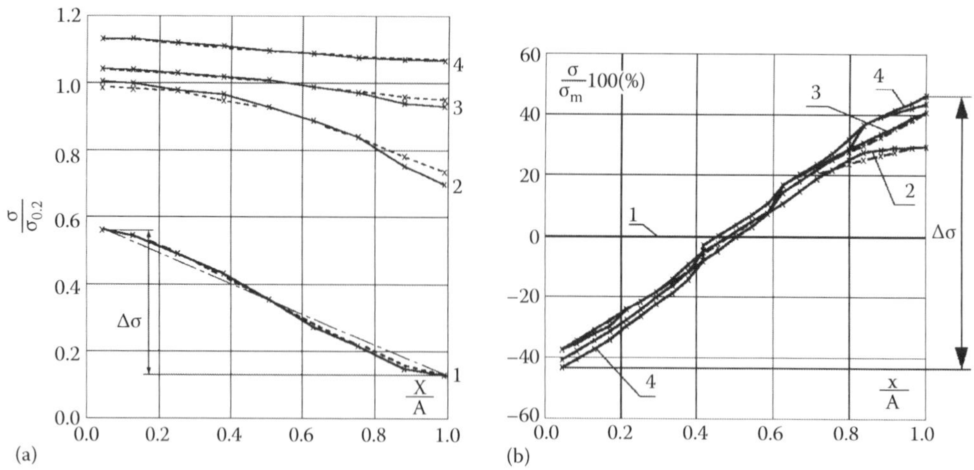

FIGURE 1.5
Distribution, along the width of the coil, of stresses in the coil conductors for different ratios $\sigma_m/\sigma_{0.2}$: 1, 0.353; 2, 0.9; 3, 1.0; 4, 1.1 (a) at short circuit state and (b) after short circuit state: *solid line*, coil with inner axial ducts; *dashed line*, without axial ducts.

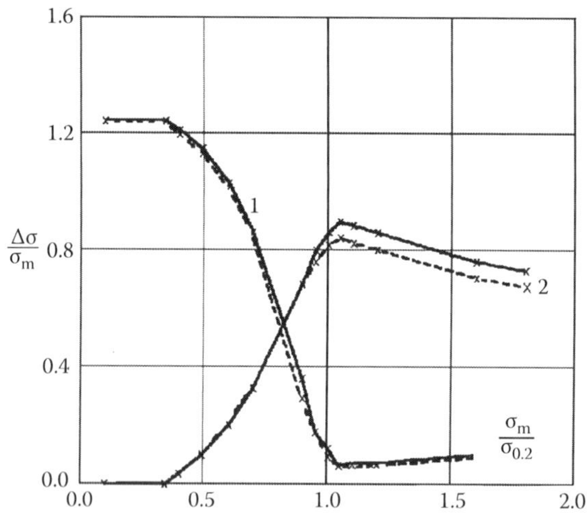

FIGURE 1.6
Effect of the $\sigma_m/\sigma_{0.2}$ ratio on the coil conductor stress difference $\Delta\sigma/\sigma_m$ at surge current (1) and after the short circuit ceases (2). (*Solid line*, coil with inner axial ducts; *dashed line*, without axial ducts).

were no such ducts. In more general terms, the results of calculation were shown in relative units, which in relation to stresses means referring them to the value of proof stresses $\sigma_{0,2}$ of conductor material or average stresses σ_m of the coil, calculated in a classic way, that is, from the relation given as follows:

$$\sigma_m = \frac{q_0 \cdot R_m}{2 \cdot s_c} \tag{1.4}$$

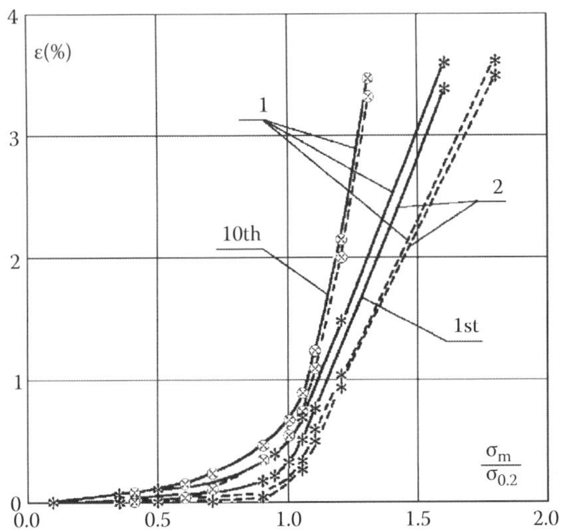

FIGURE 1.7
Strains at short circuit (*solid lines*) and permanent sets (*dotted lines*) of inner (1) and outer (2) coil conductors shown for the 1st (reference designation *) and 10th short circuits (reference designation ⊗), as a function of the $\sigma_m/\sigma_{0.2}$ ratio.

where

- q_0 is the maximum, unitary radial load for the inner conductor of the coil
- R_m is the average radius of the coil
- s_c is the cross section of the conductor

The increase of the ratio of mean tensile stress (σ_m) in the coil conductor to the apparent proof stress of the Cu conductor ($\sigma_{0.2}$) results in a more even distribution of stresses (σ) along the width of the coil. In consequence, the maximum stress rises more slowly than the mean one. It is accompanied by decreasing increments of pressure on the wire insulation (p) of maximum value (about 2% of σ_m) appearing in the middle of the width of the coil. Local pressure increases appear at axial spacers, which determine the width of the cooling duct and are caused by limited filling of the coil circumference. The calculations show the insignificant effect of inner cooling ducts on the stress distribution.

The mechanical condition of the coil, following the first short circuit fault, is illustrated in Figures 1.5 and 1.6. After a short circuit of low $\sigma_m/\sigma_{0.2}$ ratio—here 0.353, curve 1—the stress in the coil disappears. Despite uneven distribution of the stress σ under the short circuit (Figure 1.5a, curve 1), the maximum stress, appearing in an inner conductor, results in no plastic strain of the copper conductor and leaves no permanent deformations of the coil conductors.

As can be seen in Figure 1.5, with higher values of the $\sigma_m/\sigma_{0.2}$ ratio, the plastic strains appear in a inner conductors of the coil and straighten it, leading to a more even profile of the stress along the width of the coil. However, this is accompanied by permanent sets in plastically strained conductors. In consequence, the short circuit leaves new, differently distributed stresses: Inner conductors, plastically strained during the short circuit fault, are compressed by outer conductors, elastically strained and also stretched, tending to restore their original dimensions. Stress difference $\Delta\sigma$ of coil conductors, already left after the first fault, is significant (Figure 1.5).

Figure 1.6 is a somewhat different presentation of the phenomena and allows their quantitative evaluation. For $\sigma_m/\sigma_{0.2} \leq 0.35$ the short circuit leaves no permanent sets in coil conductors. For $0.35 < \sigma_m/\sigma_{0.2} < 1.05$ a certain number of coil conductors are plastically strained. This is accompanied by permanent deformations increasing with the $\sigma_m/\sigma_{0.2}$ ratio and subject to post-fault stresses in the coil. For $\sigma_m/\sigma_{0.2} > 1.05$ all conductors are plastically strained in the short circuit, resulting in practically even distribution of the stresses ($\Delta\sigma/\sigma_m$ is close to zero) and high, almost independent from the $\sigma_m/\sigma_{0.2}$ ratio, post-fault values of $\Delta\sigma/\sigma_m$ (Figure 1.6).

Figure 1.7 is a quantitative presentation of the coil deformations as a function of the $\sigma_m/\sigma_{0.2}$ ratio, conductor position (inner, outer) and the number of short circuits. It allows evaluating the relation of maximum dynamic strains in inner and outer wires and the relation of permanent sets of these conductors after the first fault.

If the $\sigma_m/\sigma_{0.2}$ is higher than 1, the differences for both types of deformations are insignificant (plastic deformation are for copper).

In addition, the relationship of dynamic strain to the permanent set of a conductor, illustrated in Figure 1.7, shows how the $\sigma_m/\sigma_{0.2}$ ratio affects the coil deformation and allows evaluating its magnitude and character. Figure 1.7 also presents the effect of the permanent set cumulation on an inner-conductor deformation: during 10 successive short circuits of the same maximum current, which is sufficient to have $\sigma_m/\sigma_{0.2} > 1$, both the dynamic strain and the permanent set still increase and become higher, when compared with the first short circuit. The difference between the dynamic strain and the permanent set decreases in the course of consecutive short circuits: The permanent set increases faster and differs slightly only after the 10th short circuit dynamic strain. Due to the cumulating effect, the increase of permanent set is bigger after the 10th short circuit than the increase of dynamic strain after the first short circuit.

The coil deformation (illustrated in Figure 1.7) is a decisive factor for the transformer short circuit withstand capability. The appearance of the deformation leads to the following two kinds of danger: disruption of wire weld connection (due to dynamic strain), and an inadmissible decrease of the major insulation dielectric strength due to significant widening of the inner oil duct at the outer winding (permanent set). Weld failures can be expected at 1% strain, so at σ_m only slightly exceeding $\sigma_{0.2}$ (Figure 1.7). Permanent sets of this kind are not permitted because of the hazard to the dielectric strength of the insulation system. Such a permanent set level in large power transformers (240–420 kV) means widening of the first cooling duct, the one that is close to the winding, by as much as 6–8 mm, up to 14–16 mm. This would result in a decrease of the dielectric strength to 85% of the original value. Such a transformer should be also rejected after a short circuit test: Its posttest impedance voltage increase would be between 4% and 5% [2,3]. This means that the permanent set of its outer winding is limited by the required dielectric strength of major insulation. Acceptable lowering by 5% of dielectric strength corresponds to a permanent set of about 0.2%, which, considering the cumulating effect, requires the $\sigma_m/\sigma_{0.2}$ ratio to be below 0.7.

Summing up the results of the analysis presented, one can come to the following conclusions:

- The existence of plastic deformations of inner coil conductors, under tensile short circuit forces, should be considered as a rule for large power transformers.
- Dynamic plastic strains result in permanent sets of conductors and leave post-fault stresses in the coils of the winding.

- The limit of deformations of a winding stretched under a short circuit is determined by an acceptable decrease in the dielectric strength of major insulation. The decrease is caused by the oil axial duct at the outer winding becoming dangerously wide due to the wire permanent set. The permissible value can be estimated to be 0.2%. With deformations of this kind, the ability of wire welds to withstand a short circuit is still sufficient, provided they were properly made.
- If the permanent set of a wire is to stay within permissible limits, the $\sigma_m/\sigma_{0.2}$ has to be below 0.7 (provided the winding was properly made). Insufficient compactness and rigidity of conductor paper insulation and insufficient winding process means uncontrolled decrease of this limit.
- The apparent elastic limit $\sigma_{0.2}$ of a wire depends on the conditions in which the winding was made. It is not possible to estimate the winding strength without knowing the ultimate value of $\sigma_{0.2}$.

The analytical calculation method presented here is used in the assessment of withstand capability of outer windings subjected to radial tensile forces and, after appropriate modifications, can be extended to inner winding compressed by short circuit forces. The experience of IEnOT shows that because of the complexity of phenomena taking place in the compressed windings, the numerical method proves to be a more effective method of analysis in this case. It is also a more general method, applicable not only to compressed windings but also to stretched ones, regardless of the shape of the winding.

1.2.2 Numerical Method of Calculation

In the numerical method applied by IEnOT, the coil of the winding in the core transformer is discretized in a flat configuration with finished elements in a manner as shown in Figure 1.8.

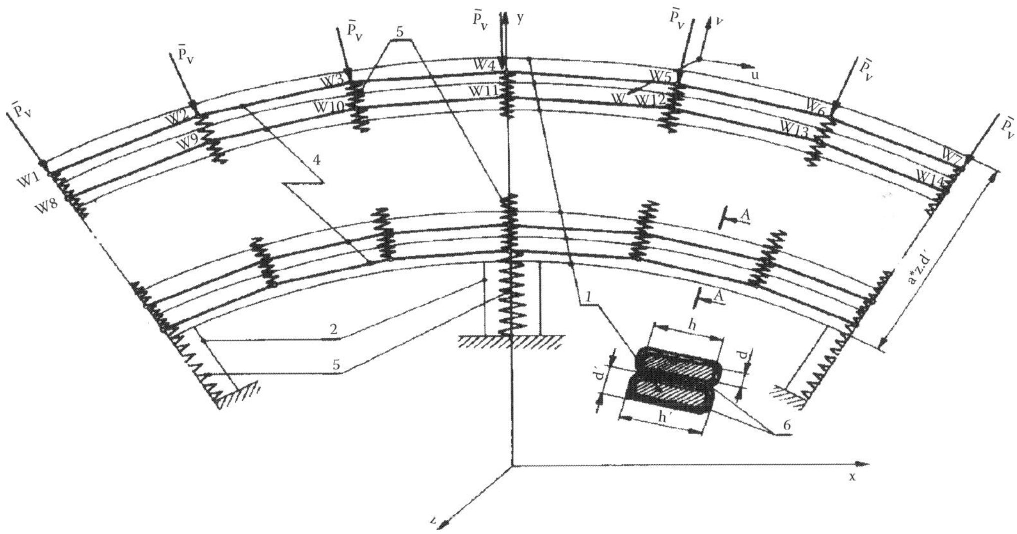

FIGURE 1.8
Discrete model of coil sector: 1, coil conductors; 2, support strips; 3, curvilinear elements representing coil conductor; 4, springy elements modeling insulation; 5, dimensions of coil conductors; 6.

It is possible to analyze only a chosen segment of the coil limited by the axis of symmetry—as demonstrated earlier—or its whole circumference, regardless of the shape of the coil itself. The conductors are represented by curvilinear plane elements and the insulation of the coil (distancing axial spacers, bundle insulation on the conductors) with springy elements. These elements are described with parameters representing their real dimensions and nonlinear mechanical characteristics (compare Figure 1.2) of the materials from which they were made. The radial load of the coil can be continuous or concentrated, and its distribution along the circumference and on particular conductors in the radial direction can be of any value or direction (stretching, compressing, circumferential tension). It is also possible to load the coil with displacements (the so-called kinematics loading), which is sometimes useful in defining mechanical characteristics of the coils or glued conductors CTC. The algorithm of the program takes into account for such generally defined loading conditions the possibility of bending of the coil conductors occurring between support strips, which makes it necessary to split the conductors into layers (usually about 10, with the radial dimension of the conductor amounting to a few millimeter), for which the distribution of stresses during bending in each layer can already be considered to be invariable.

One of the most essential characteristics of the discussed software, which bring the algorithm of calculation close to real conditions, is the introduction of springy elements of unilateral constraints (realized by the appropriate adjustment of their characteristics, for example, by compressing and stretching) and the possibility of calculating permanent sets after the first short circuit (Figure 1.4—the loading cycle 0-1-1′). Modeling the springy elements according to the unilateral constraints also makes it possible to include the following fact in the analysis of withstand capability: Insulation—in case of typical production technology, without using gluing substances—can transfer only radial compressive forces, and in case of stretching forces, it is subject to layering. Of course, in the discussed algorithm, computer calculations are carried out by applying the iteration method with an appropriately adjusted step of load increment Δq. For each calculation step, the linearization of incremental equilibrium equation is performed, which is presented in the following matrix form [4]:

$$[K_{ep}^{(i)} + K_{\sigma}^{(i)} \cdot (\sigma)] \cdot \Delta f^{(i+1)} = \Delta q^{(i+1)} \quad (1.5a)$$

where

$K_{ep}^{(i)}$ is the elastic-plastic stiffness matrix

$K_{\sigma}^{(i)}(\sigma)$ is the stress stiffness matrix

$\Delta f^{(i+1)}$ is an increment of generalized displacements vector

$\Delta q^{(i+1)}$ is an increment of loads vector

Matrix $K_{ep}^{(i)}$ and $K_{\sigma}^{(i)}$ are calculated on the basis of displacements and stresses, which are calculated at the beginning of the $(i+1)$-th step.

The incremental equation system is solved according to the variable stiffness method. The total displacement vector Δf(i) and loads vector Δq(i) are determined by the accumulation of respective vectors' increment according to the relationship:

$$f^{(i)} = \sum_{j=1}^{i} \Delta f^{(j)} \quad (1.5b)$$

$$q^{(i)} = \sum_{j=1}^{i} \Delta q^{(j)} \tag{1.5c}$$

The algorithm of the computer calculation method helps to establish, in a known way, the critical load q_{cr} occurring during a loss of stability. If the so-called proportional load q is used for the i-th step of calculation and critical load can be defined with the relationship,

$$q_{cr}^{(i)} = \lambda \cdot q^{(i)} \tag{1.6a}$$

where λ represents a multiplier of the comparative load, the problem of initial stability on i-th step of calculation can be expressed as follows:

$$[K_{(ep)}^{(i)} - \lambda \cdot K_{\sigma}^{(i)}(\sigma)] \cdot v = 0 \tag{1.6b}$$

where v is the pattern of stability loss. The Equation 1.6 can be solved by Stodel's or Lanczo's methods, and the limiting load q_{cr} is obtained from formula:

$$q_{cr} = \sum_{j=1}^{n} q_{cr}^{(j)} \tag{1.6c}$$

where n is the step of calculation, at which the singularity of matrix is not investigated and no solution for Equation 1.5 is obtained. There are also other known methods of calculating the critical load occurring during the loss of stability; however, the author's experience indicates that the best results can be achieved with the so-called deflection method, the practical use of which will be presented in Section 1.2.3.

The examples of uses of the described numerical method for calculating stresses and deformations of the windings subjected to radial short circuit forces are shown in Figures 1.9 through 1.11.

The results of calculations presented in Figures 1.10 and 1.11 also seem interesting. These calculations were carried out for the Working Group WG19.12 [5], in connection with their debate concerning calculating stresses in the conductors of layer windings, which would be a reliable basis for the assessment of strength of radial short circuit forces in this construction. There was still concern that in this type of winding, especially for units of the largest power, the existing cooling ducts between the layers may significantly impede the transfer of radial forces between them and it may not be appropriate to rely on mean stresses in the assessment of withstand capability (calculated with the formula 1.4 for such a construction).

Calculations were carried out for a certain step-up unit, the dimension of which are compatible with the power of approx. 600 MVA, provided that representatives of the producer specializing in constructions of this kind accept the following parameters for the winding HV:

- The number of layers in the winding—4, each wound with CTC conductor with dimensions 21 × (8.4 × 1.5) mm, with bundle insulation (paper) 0.56 mm/side. Proof stress for the conductor equals $\sigma_{0.2} = 210$ MPa
- The inner diameter of the winding is 1696 mm, and the outer 1847 mm. The thickness of the layer (the same as radial dimension of insulated conductor CTC) amounts to 19.7 mm and the axial cooling duct up to 6 mm

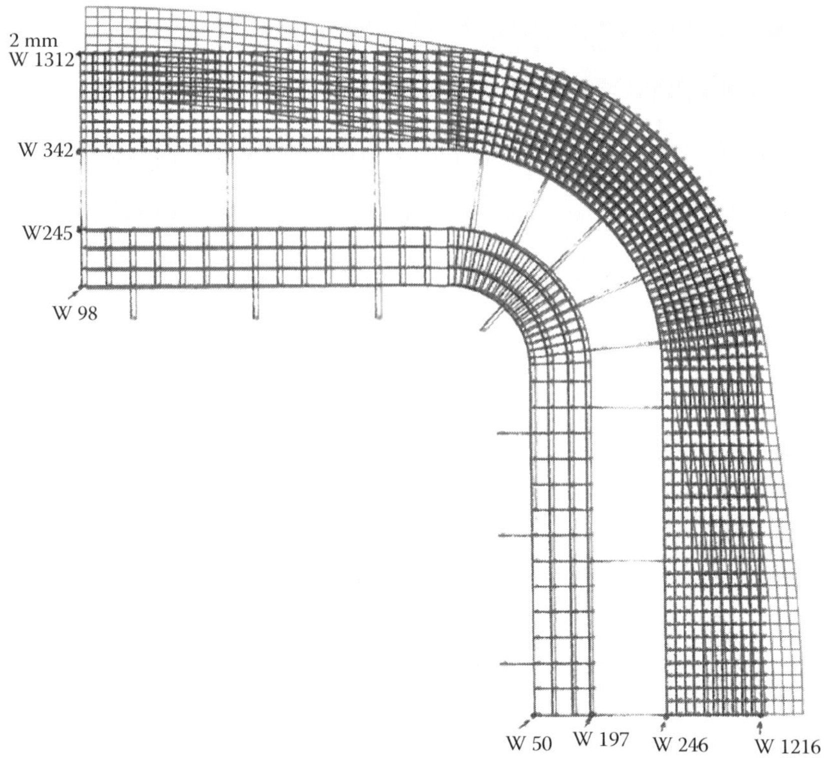

FIGURE 1.9
Winding deformation of the distribution transformer with rectangular legs at three-phase fault.

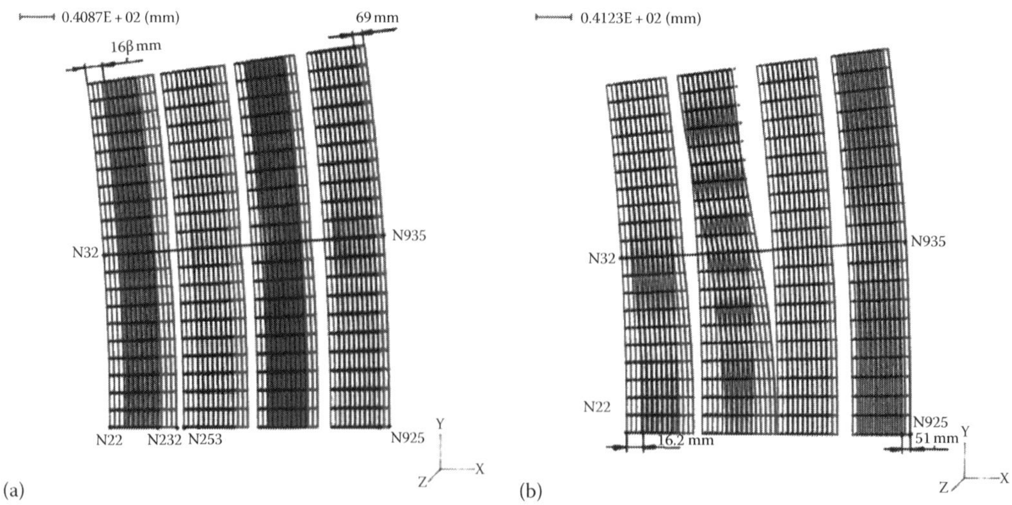

FIGURE 1.10
HV layer winding deformation of large power GSU Transformer, caused by short circuit tensile radial forces (*black line*, the winding without deformation; *gray line*, deformed winding), (a) at the amplitude of short circuit current and (b) after short circuit state.

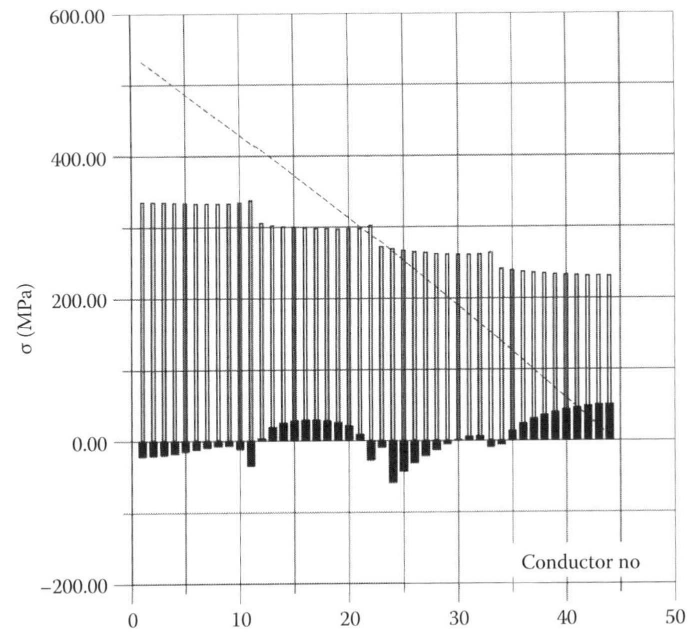

FIGURE 1.11
Stresses in HV winding conductors of large power GSU Transformer, caused by short circuit tensile radial forces; □, the amplitude of short circuit current; ■, after short circuit state result of simplified calculation, whose deformation is ignored.

- Mean tensile stresses for the whole winding equal $\sigma_m = 282\,\text{MPa}$. These data are characterized by a high value of mean tensile stresses σ_m in relation to proof stress $\sigma_{0.2}$ of the coil conductor material (relation $\sigma_m/\sigma_{0.2} \cong 1.34$), which basically contradicts the recommendations presented in the summary of Section 1.2.1. Such relations of stresses were intended to exaggerate the phenomenon of transferring loads by the neighboring layers of the winding, in order to obtain clear conclusions of calculations.

Figure 1.10 demonstrates that considerable radial deformations of layers take place in the winding, and in particular inner layer (approx. 16.8 mm in short circuit condition and approx. 16.2 mm after the first short circuit), which refers both to dynamic deformation taking place when impact short circuit current flows through the winding and permanent (plastic) sets remaining in the winding when the short circuit current disappears. Moreover—as Figure 1.10b shows—when the short circuit current disappears, two inner layers of the winding concerned are buckled. This is caused by strongly deformed outer layers pressing on these two layers. Such big deformations are unacceptable because of the stress exerted on the main insulation system [2,3] and because they can lead to a mechanical damage of the insulation arrangement itself. Due to considerable plastic sets occurring after the first circuit, one can expect—taking into account the relation (Equation 1.3)—that equally large, quickly increasing permanent sets will occur after consecutive short circuits. In the light of these results, based on design data and physical properties of the windings, one can unmistakably ascertain that the winding concerned does not have sufficient withstand capability against radial short circuit forces. The obtained result confirms the conclusions and recommendations presented in the summary of Section 1.2.1. It would not be possible to draw such explicit conclusions, if the assessment of withstand capability of the winding was based only on the comparison between stresses σ_m and $\sigma_{0.2}$ and do not take in account deplacements. Although stresses in the discussed case significantly exceed stresses $\sigma_{0.2}$, there are still no grounds to draw conclusions on the scale of risks, and it is not possible to establish the maximum stresses in the inner layer of the winding, at the dissipating gap.

The results shown in Figure 1.11 give an answer to the problem presented by the Working Group. One can conclude on the basis of their findings that in short circuit condition, a considerable equalization of stress distribution on the layers occurs in the analyzed winding, which in light of the observations presented in Section 1.2.1 confirms that plastic sets take place, as illustrated in Figure 1.10; for such a large assumed relation $\sigma_m/\sigma_{0.2}$ (approx. 1.34) the maximum stresses in the conductor of the inner layer are only about 17% higher than the average stresses σ_m approved by the designer and much lower than the maximum stresses (approx. 550 MPa) that occur in a nondeformed coil. Relatively low values of stresses, which occur after the short circuit current disappears in relation to the results presented in Figure 1.5b for a disc coil winding, can be explained by a buckling of two inner layers as well as probably big plastic sets of the layers.

In connection with the aforementioned numbers concerning the distribution of stresses, one can also jump into a misleading conclusion that since such considerable equalization of stresses took place in the coil and the theoretical maximum stresses are nearly twice as great as the calculated ones, withstand capability against radial forces of the analyzed winding seems not to be at risk. The danger of such erroneous interpretation can only occur when the assessment of withstand capability is based merely on the values of stresses in a simplified way, not taking into account the deformation of windings in the short circuit condition and when that short circuit disappears.

1.2.3 Buckling Stresses

According to the author's experience, the best results in calculating buckling stresses—which were mentioned earlier—were achieved with the deflection method. Generally, this method is based on the analyses of coil deformation (solving only statics, Equation 1.5) in the function of increasing load. The purpose of the method is to find the calculation step from which the deformations of the winding become asymmetrical and quickly increase, and in case of further, even insignificant load increment, the shape of the deformation changes. As numerical tests demonstrated, this process has its periodicity, and after the first critical load and when the shape of the coil changes, the coil becomes more fixed, and a further load increment makes it possible to calculate the second buckling pattern and the corresponding second critical load. Theoretically, the process can be continued, but because of the finite exactness of calculations and the numerical dimensions of the calculation model of the coil, sensible results are usually obtained only in the phase between the second and the third critical load. For practical reasons, of course, only the first critical load and its corresponding buckling pattern are essential.

The described algorithm of calculating critical loads is, therefore, entirely consistent with the general theory describing the phenomenon of the loss of stability, as published in many textbooks on the stability of constructions. However, the deflection method can be successfully applied only when, just as it was shown in the nonlinear calculation model of the coil presented earlier, it uses, among other things, curvilinear plane elements, taking into account nonlinear physical properties of materials. As it is known, in a typical linear calculation model based on rectilinear rod and springy elements, in which the materials have linear characteristics, the load increment does not permit achieving the aforementioned condition of the loss of stability. In such a model, deformations always remain linearly dependent on the load.

The results obtained with use of the deflection method, which are presented in the following, for coils whose models in scale 1:1 were subjected to dynamic tests in order to experimentally establish radial critical loads at a loss of stability. However, the results for the unsupported circular ring of rectangular cross section will be shown first. The shape of the ring resembles a single turn of a coil and its stability is theoretically worked out, whereas the critical load σ_{cr} is defined with the relation:

$$\sigma_{cr} = \frac{E}{4} \cdot \left(\frac{h}{R}\right)^2 \tag{1.7}$$

where

E is the modulus of elasticity of copper
h is the radial dimension of the conductor
R is the mean ring radius

Confronting the results of computer calculations with the results obtained using the Equation 1.7, taking into account the buckling patterns, which for rings are commonly known, will permit assessing the possibilities and accuracy of the method applied to calculate the critical load. It will also be possible to assess the effect of simplifications in the representation of the coil (1/4 of the circumference or the whole circumference) on the accuracy of calculations. For analytic and computer calculations, the following equation was used: $E = 1.103 \times 10^5$ MPa, h = 3.68 mm, R = 171.5 mm; the results of calculations are shown in Figures 1.12 and 1.13 and Table 1.2.

The obtained results of the numerical test for unsupported ring confirm the correctness of the presented method of calculating buckling stresses, both in relation to the

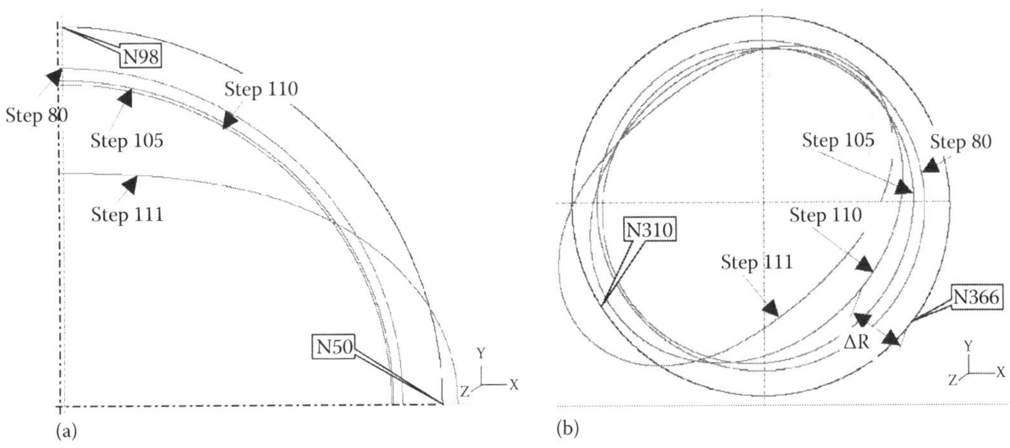

FIGURE 1.12
Deformations of the ring for different steps of radial load (*gray line*: step 80 – $\sigma_m = 8.8$ MPa, step105 – $\sigma_m = 11.55$ MPa, step 110 – $\sigma_m = 12.1$ MPa, step 111 – $\sigma_m = 12.21$ MPa; *black line*: unsupported and unloaded ring). (a) model representing a quarter of the circumference of the ring and (b) model representing the whole circumference of the ring.

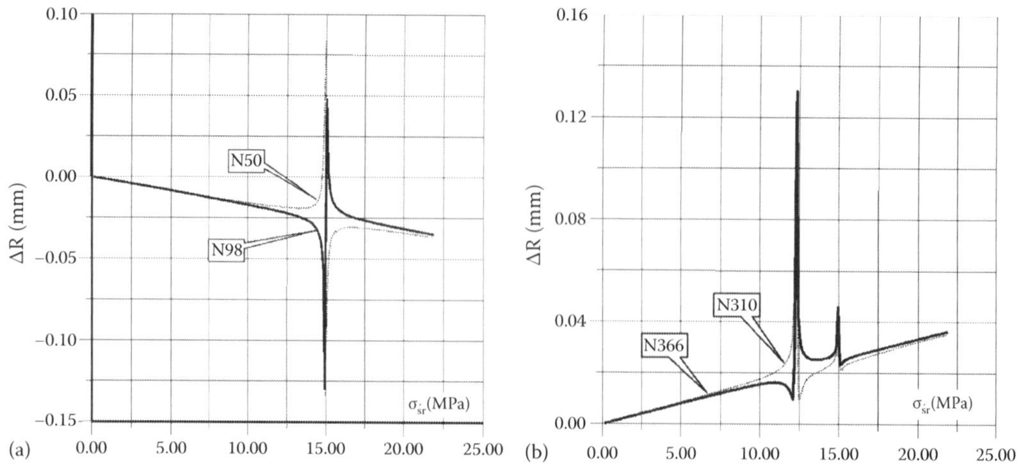

FIGURE 1.13
Radial displacement of chosen knots of the calculation ring model in function of loading, defined with average compressive stresses σ_m (position of knots on the ring is shown in Figure 1.12). (a) model representing a quarter of the circumference of the ring and (b) model representing the whole circumference of the ring.

TABLE 1.2
Critical Load σ_{kr} for Unsupported Ring

	Value of Critical Load σ_{crr} (MPa) Calculated According to		
Pattern of Buckling	**Deflection Method Representing (¼ of the Circumference)**	**The Whole Circumference**	**Formula (1.7)**
1	14.96	11.55	11.0
2	15.18	13.31	11.0
3	No data	15.07	No data

value of critical stresses themselves and to the buckling pattern (usually presented for the first two patterns). For understandable reasons, the greatest exactness was achieved for the first buckling pattern and when the coil was represented with its full circumference. Representing the coil with ¼ of the circumference, which was sometimes necessary, diminishes the size of the calculation model and the size of collections with recorded results, and also reduces the time of calculations, which makes the calculations less reliable and somewhat overstates the calculated critical loads.

The aforementioned numerical test is supplemented by the results shown in Table 1.3 and in Figures 1.14 and 1.15, referring to the real coils, for which dynamic tests [3,6,7] were carried out. Because different publications often refer to the Fischer's formula [8] when discussing the phenomenon of the loss of stability, one form of which is defined by Equation 1.8, it seems appropriate to include this simplified method in the comparisons:

$$\sigma_{cr} = \frac{E}{48} \cdot z^2 \cdot \left(\frac{h}{R}\right)^2 \tag{1.8}$$

where z is the number of supports on winding circumference.

TABLE 1.3

Value of Buckling Stresses for Selected Models

Symbol of Models	Test Results in (MPa)	Calculated Results in (MPa), According to	
		Deflection Method	Fischer's Rule [2]
MR32	74.4	70.4	31.7
M16B[a]	50.5	56.6	40.4
Del Vecc.1[b]	88.4	109.0	123.3
Del Vecc.2[b]	90.4	99.0	69.4

Source: Pewca, W. (in Polish), *Energetyka*, 1, 25, 2002.
[a] Model was made of ctc conductor [6,7].
[b] Del Vecchio's model tests [12].

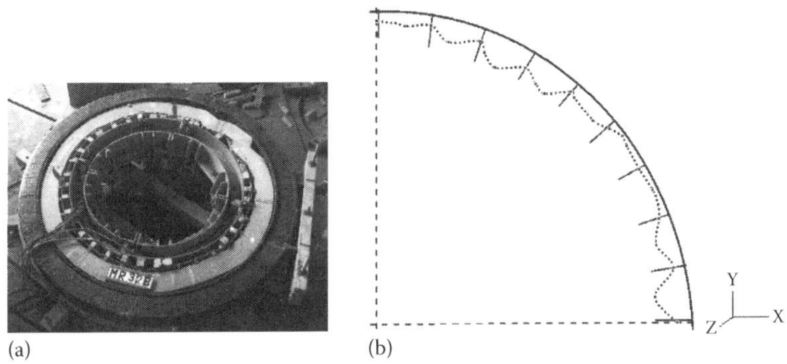

FIGURE 1.14
Test and calculation results of model MR32. (From Pewca, W. [in Polish], *Energetyka*, 1, 25, 2002.) (a) Model after destructive dynamic test and (b) calculated instability form (*dotted line*) for inner turn of coil.

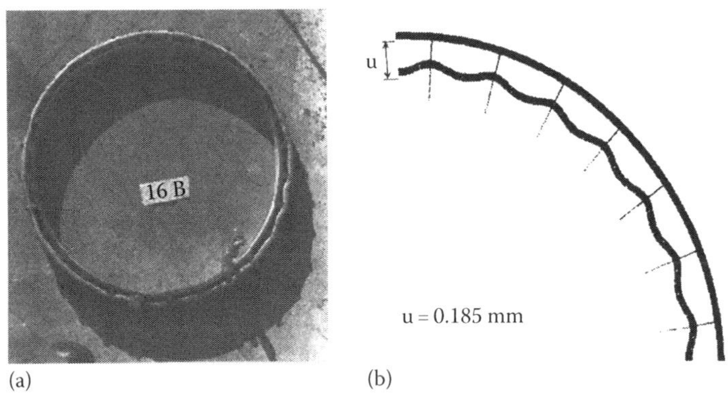

FIGURE 1.15
Test and calculation results of model M16B. (From Pewca, W. [in Polish], *Energetyka*, 1, 25, 2002; Kozłowski, M. et al., Selected short-circuit strength problems in power transformer, CIGRE Report 12-05, 1–8, 1980.) (a) model after destructive dynamic test and (b) calculated instability form for turn of coil.

Summing up the results presented, one can say that the numerical method of analyzing the strength of windings to radial short circuit forces allows bringing the calculation model of the coil (winding) close to real solutions, by taking into account the nonlinear characteristics of construction materials and specific character of insulation elements transferring not only compressive forces but also curvatures of conductors. The algorithm of method presented is universal and can be used both for the windings compressed by radial short circuit forces and for the stretched ones. It allows calculating deformations of the winding subjected to short circuit condition and after the condition disappears, which is especially important for the proper assessment of withstand capability of windings.

It should be emphasized that the exactness of calculating buckling stress is considerable, especially when the analyzed coil of the winding is represented with the full circumference. When simplified models are applied (i.e., representing ¼ of the circumference) the exactness of calculation is theoretically worse, yet it is satisfactory for practical solutions, as is confirmed with short circuit tests.

1.3 Axial Forces, Stresses, and Deformation

1.3.1 Axial Winding Vibration

Reduced stiffness of disk windings of core-type transformers for axial direction of short circuit forces, combined with the risk of the occurrence of resonance phenomena, warrants accurate methods to be applied for analyzing short circuit withstand capability. Additional reasons that support this view include the nonlinearity of the characteristics of materials and time variability of electrodynamic forces acting on the coils of windings. A typical time variability of electrodynamic axial component force acting on the end coil of the winding LV of a unit of large power is shown in Figure 1.16a [9]. This time variability is characterized by the existence of two harmonics: 50 Hz—existing in a period of time when a unidirectional component of short circuit current flows through the winding and harmonic of 100 Hz occurring at steady state of short circuit current. Figure 1.16b, on the other hand, shows the distribution of amplitudes of these axial forces acting on particular coils of this winding.

The simplified methods of analyzing withstand capability of the winding against axial forces take into account only the first maximum amplitude of electrodynamic forces for each coil, approaching the problem in accordance with the rules of static method, and considering the winding as not deformed. Stresses in insulation established in this way are not sufficiently objective data for a reliable assessment of real short circuit risks.

Time variability of axial short circuit forces and reduced stiffness of the windings (cable paper and pressboard take a significant part in the height of these windings) show that it is not possible to ignore the vibrations of the coils, which are not synchronous in every point of the winding. An accurate calculation method must necessarily include in the process of vibration also the forces that are different from electrodynamic forces acting on the coil (see Figure 1.17b) and comprising static compressive forces.

The computer program developed in IEnOT for analyzing the dynamic strength of winding to axial short circuit forces takes into account all components of the forces as well as the earlier-discussed quality of insulation (see Section 1.2.2) permitting transferring

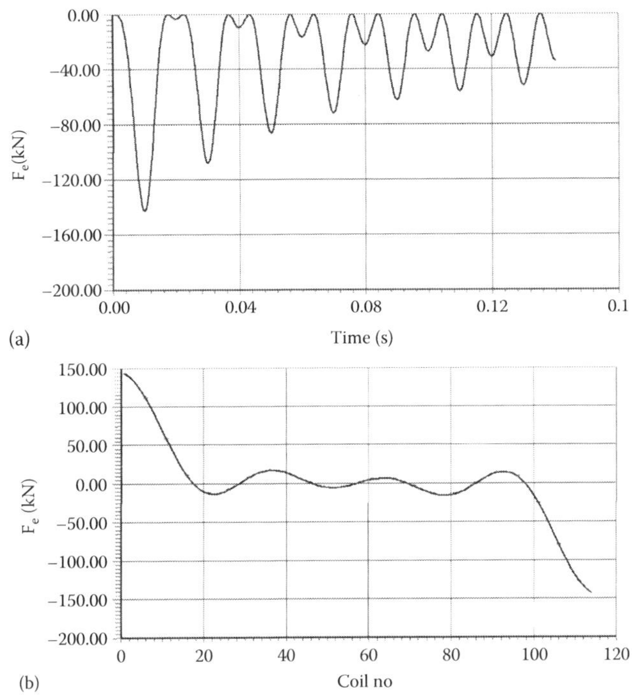

FIGURE 1.16
Axial electrodynamic forces acting on the coils of winding LV of large power GSU transformer. (From Pewca, W. [in Polish], *Energetyka*, 11, 492, 1998.) (a) time characteristic of a force acting on the end coil and (b) distribution of amplitudes of axial force per the winding coils.

only compressive forces [9]. The algorithm of this program is based on the winding model presented in Figure 1.17 and on the assumption that the electrodynamic forces will be analyzed in a flat configuration, which allows obtaining the exactness of calculation of axial forces that is satisfactory for practical reasons [9,10]. In the assumed model radial spacers are represented with nonlinear springy elements and the coils themselves with point-concentrated masses.

Calculations realized with the program, the algorithm of which is presented in this chapter, begin with estimating the static compressive forces and their distributions on every winding. To achieve this aim, the equation of the static balance [10] is used, linearized for particular calculation steps:

$$K^i(\delta)\cdot \Delta Z^{(i+1)} = F_{pr}^{(i+1)} - (S^i + T^i) \tag{1.9a}$$

where

$K^i(\delta)$ is the matrix of stiffness dependent on coil displacements δ
$\Delta Z^{(i+1)}$ is the increment of displacement vector
$F_{pr}^{(i+1)}$ is the vector of pressing forces
S^i is the vector of flexibility forces
T^i is the vector of friction forces

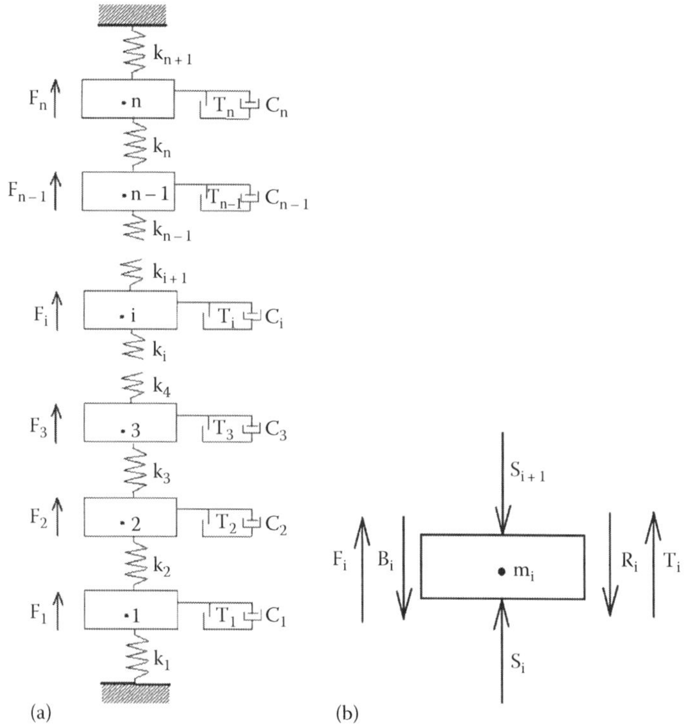

FIGURE 1.17
Winding model for dynamic analysis. (From Pewca, W [in Polish], *Energetyka*, 11, 492, 1998; Borkowski, W. et al., Non-linear analyse of dynamic withstandability of transformer winding under axial short-circuit forces, *Report of Eighth International, Symposium on Short-Circuit Currents in Power Systems*, pp. 149–154, 1998.) (a) base diagram of coil arrangement and (b) calculated forces acting on i-coil end insulation. m_i, mass of i-coil; μ_i, coefficient of "dry" friction; c_i, viscous friction coefficient; k_i, stiffness of insulation element; F_i, axial electrodynamics force; T_i, "dry" friction force; R_i, viscous friction force; B_i, force of inertia; S_i, S_{i+1}; reaction force of i- and (i+1)-insulation elements.

The input data establishing the distribution of a force pressing on particular coils comprise: a) the assumed characteristics of compressing pressboard, from which radial spacers were made and characteristics of bundle insulation on the conductors of the coil (as shown in Figure 1.2b), and b) the percentage of compression of the winding insulation on the height, as assumed by the designer. Example stress distributions in insulation coming from these forces for a transformer, which was taken into consideration earlier, were shown in Figure 1.18. These calculations were realized for two significantly different technological alternative version, namely [9]:

1. High static pressing degree of the insulation—9% of its height, by using rigid supports at the ends of winding and wound coils with low tension (force pressing conductor to the axial distance strips about 2.6 kN)
2. Limited static pressing degree of the insulation—about 7% of its height, less rigid ends supports, and force pressing the conductor to the axial strips at about 25.5 kN

The fact that compressive static forces are not directly assumed in the analysis presented, but are calculated from constructional data of the winding and data of materials and

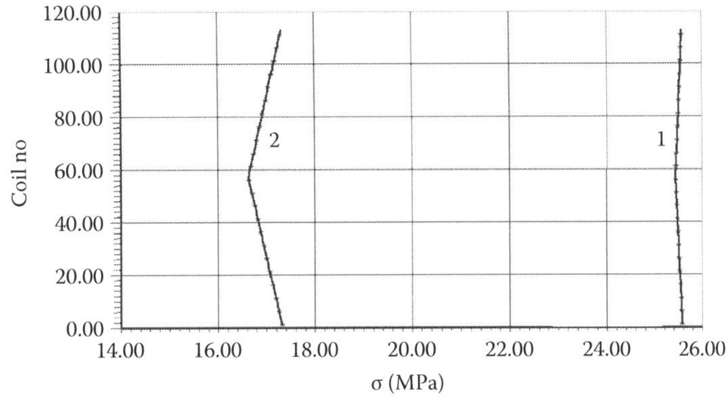

FIGURE 1.18
Stress distribution in the insulation of the winding at static pressing, the technology covering version (1) and (2). (From Pewca, W. [in Polish], *Energetyka*, 11, 492, 1998.)

production process, is an important element of the calculation methods discussed, bringing the algorithm close to real solutions. Another essential element of the algorithm is the inclusion of the distribution of compressive forces on the coils in the calculations. Experience indicates that too small static compressive force (or its diminishing during service) in relation to axial electrodynamic forces is one of the most frequent causes of transformer short circuit damages. Physically, they are caused in such cases by breaking contact between the coil and the adjacent radial spacers in the process of their vibrations. This initiates the so-called Hammer Effect, deepening the axial gaps in the winding and initiating the process of dynamic destruction of the winding structures (among other things, the radial displacement of radial spacers can happen). Identifying this calculation process is an objective element of the assessment of short circuit withstand capability of the winding against axial short circuit forces. To confront this, the assessment of short circuit strength of the winding only on the basis of values of stresses in insulation (mechanical withstand of the pressboard and cable paper to compressive stresses is considerable), which is used in simplified models, does not seem to be convincing.

In order to identify the moment of detachment of coils from radial spacers, for every calculation step and for every springy element the program analyses the direction of force S_i (see the markings from Figure 1.17b), and even if one springy element a stretching force occurs, that is [10],

$$S_i < 0 \tag{1.9b}$$

The calculations are interrupted, for example, when there is a signal that a destruction process is beginning in the winding and the compressive force resulting from design data and assumed production process is too small. If this force can be further increased for the winding (i.e., by improving with technological ways the stiffness of the insulation or increasing the percentage of compression of the winding) and the process of calculations can be brought to an end in the assumed time period of short circuit, withstand capability of the winding, from the point of view of the static compressive force, is appropriate.

Figure 1.19 illustrates time characteristic of the forces in the end-yoke insulation for the considered unit and the assumed earlier two versions of its production process. For version 1, in which the winding is strongly compressed (see Figure 1.18), there is no risk of

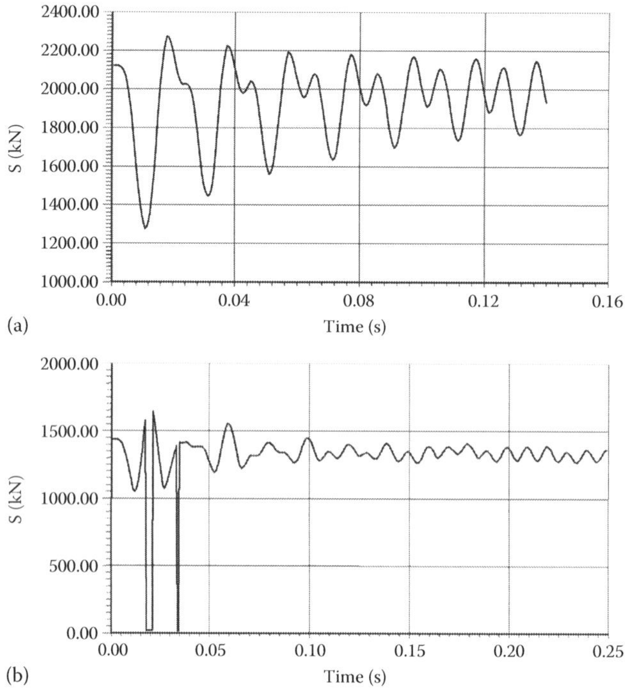

FIGURE 1.19
Forces in end insulation of the winding as functions of time for versions 1 (a) and 2 (b).

detachment of the end insulation from the winding in the whole analyzed period of time. Such a risk could occur in case of variant 2, in the initial phase of a short circuit. However, the assumed for the calculations increased friction forces on axial strips stopped the movement of the end coil downward and only because of this it was possible to avoid its detachment from the yoke insulation. The presented example illustrates how significant is the influence of technologies on short circuit withstand capability of transformers, which in quality relations is well known.

Other essential parameters that, apart from stresses in insulation, are taken into account in the assessment of short circuit strength include displacements of the coils, in particular the end ones, cooperating with the end compressive insulation. The algorithm of the discussed program establishes these parameters on the basis of known calculations in mechanics, linearized for particular calculation steps, matrix equations [10]:

$$M \cdot Z'' + C \cdot Z' + K(\delta) \cdot \Delta Z = F_{ed} - (S + T) \tag{1.10}$$

where

M is the matrix of inertia
C is the matrix of damping
F_{ed} is the vector of electrodynamic forces
Z, Z′, Z″ is the vector of displacements and its first and second derivatives

Example results of calculating displacements of end coils and stresses in insulation, which were obtained with the discussed method for the assumed earlier versions of technological solutions, are shown in Figures 1.20 and 1.21.

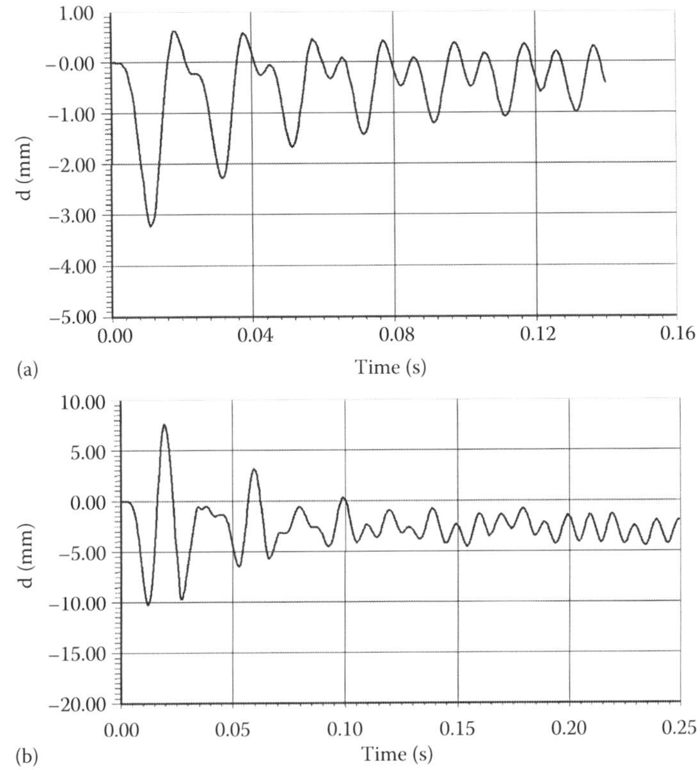

FIGURE 1.20
Axial displacement of end coils for version 1 (a) and 2 (b). (From Pewca, W. [in Polish], *Energetyka*, 11, 492, 1998.)

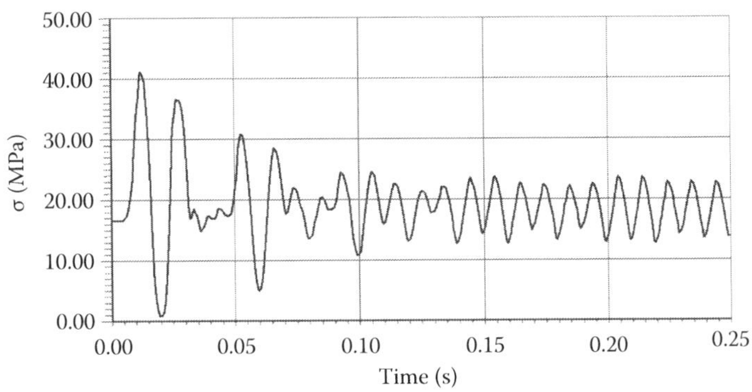

FIGURE 1.21
Pressure in radial spacers at half of the winding height, version 2. (From Pewca, W. [in Polish], *Energetyka*, 11, 492, 1998.)

Calculation results presented in the previous figure for end coils also confirm the essential influence of production process on values of coil displacements in short circuit condition. During a strong static compression of the winding (version 1) the displacement of these coils are slight, and it should be predicted that they will not have a destructive effect on the end insulation of the analyzed winding. For technological version 2, in which nearly one-third of the compressive forces were diminished (comp. Figure 1.18), over three-fold increase of the amplitude of vibrations in the initial phase of the short circuit occurs. Although the absolute value of displacement of the end coils does not seem to be critical, the ultimate assessment of short circuit strength should take into account the consequences of such displacements in the design review of end insulation. In the same variant, increased stress on radial spacers should also be expected in the middle of the height of the winding, which is documented in Figure 1.21.

Summing up the presented basic elements of the algorithm of the program for the assessment of withstand capability against axial short circuit forces and obtained with this method results, one can draw the following conclusions:

- Static forces compressing the winding and stiffness of radial insulation are—apart from the values of electrodynamics forces—basic parameters, which have an effect on the short circuit withstand strength of the winding.
- Maintaining constant mechanical contact between the coils of the winding and the adjacent radial spacers and the value of the amplitudes of coil vibrations, in particular of the end coils, can be accepted as appropriate criteria for the assessment of short circuit withstand capability of coil windings.

For stresses in radial spacers, which are affected by the aforementioned factors, it is difficult to establish precise critical values based on the physical properties of materials. In fact, practice shows that even in dynamically damaged windings, radial spacers do not demonstrate significant signs of mechanical spoil (including cases when they were damaged by arcing). However, the values of these stresses are useful in checking whether there is no risk of the coil loss stability for axial direction, which is connected with the tilting effect.

1.3.2 Critical Axial Forces

Axial loading of a winding, due to short circuit forces, can be destructive for the coil, especially because of their unacceptable tilting effect. Critical axial load at which the effect for the coil occurs is difficult to determine analytically and as a rule requires experimental investigation on models. Figure 1.22 illustrates a special arrangement for determining this critical axial load, which was used in Institute of Power Engineering [3].

Tests were performed on winding models with different coil parameters, and their results permit determining empirical coefficients in a formula for the tilting critical force F_{cr} referred to in literature [11] and also to propose a new formula [3]:

$$F_{cr} \approx \frac{z}{D} \cdot \sqrt{m} \cdot b^{2.7} \cdot A^{0.3} \qquad (1.11)$$

where

z is the number of terns (conductors) in a coil
D is the coil mean diameter
b is the conductor radial dimension
A is the radial spacer width
m is the number of spacers on the coil girt

FIGURE 1.22
Arrangement with hydraulic press for static evaluation of critical axial forces.

FIGURE 1.23
Arrangement for dynamic evaluation of critical axial forces. Supply coils and model after testing.

The static tests were followed by dynamic tests on coils of identical design and dimensions. Specially designed arrangements presented on Figure 1.23 were used in these tests. It was demonstrated that the critical axial forces, determined statically and dynamically, are similar. The difference was below 6% and only in one case reached 15%.

1.4 Conclusions

1. The results of the analysis presented proved that, regardless of the type of the winding (coil, layer) of the core-type transformer, the average stresses in the coil conductors σ_m and proof stress $\sigma_{0.2}$ of the material from which they were made

are the most essential parameters determining the withstand capability of these windings against radial short circuit forces.

2. In case $\sigma_m \geq \sigma_{0.2}$, a phenomenon of cumulating of deformations after consecutive stresses can take place, indicating insufficient strength of the winding to withstand radial tensile short circuit forces. The relation of stresses which should be considered as completely safe is $\sigma_m/\sigma_{0.2} \leq 0.7$.
3. Simplified methods of calculating buckling stresses for windings subjected to radial compressive forces do not permit achieving the accuracy of calculations that is satisfactory for design purposes. Much better results are obtained with numerical methods, which take into account deformations and nonlinear physical and geometrical properties of the windings.
4. The criterion of permissible stresses in insulation, especially when they are calculated with simplified methods, is not an objective and sufficiently precise parameter for assessing withstand capability of the windings against axial short circuit forces.
5. The analysis of winding strength to withstand the axial component of short circuit forces should be carried out taking into account the dynamics of the phenomena connected with time variability of electrodynamic force, nonlinearity of physical materials, and the occurring deformations. The basic criteria of the assessment of withstand capability include the following: (1) the correctness of adjusting the static compressive force eliminating the "hammer effect"; (2) value of maximum amplitude of axial deformation of the winding, which can be accepted for a given system of compressing the winding; and (3) stresses in insulation radial spacers are smaller than the critical ones causing tilting effect.
6. Applying appropriate numerical calculation methods in the analysis permits making a correct assessment of short circuit withstand capability of the windings based on the aforementioned criteria and taking into account essential design and technological solutions.

References

1. CIGRE WG, 12.06 (1982) Final Report of WG 06 of Study Committee 12 (Transformers), *Electra*, No 82.
2. Kozłowski M, Pewca W (1996) Short-circuit performance of a stretched transformer winding with regard to its actual mechanical characteristics. *Journal of European Transactions on Electrical Power (ETAP)*, 6(4), 259–265.
3. Kozłowski M, Pewca W (1994) Improvement in estimation criteria of transformer short-circuit withstand capability. *Archiwum Elektrotechniki*, XII(3), 577–590.
4. Borkowski W, Pewca W, Szymczyk E, Nieczorek M (1996) Numerical analysis of short-circuit withstandability of transformer windings under radial forces. *Seventh International Symposium on Short-Circuit Currents in Power Systems*, Warsaw, Poland, Report 2.12, pp. 1–6.
5. Fyvie J (2002) The short circuit performance of power transformer. The Final Report of WG12.19, *Electra* 25, 205.
6. Pewca W (in Polish, 2002) Loss of stability and buckling stresses for transformer inner windings. *Energetyka*, 1, 25–31.
7. Kozłowski M, Pewca W, Marciniak M, Weretynski W (1980) Selected short-circuit strength problems in power transformer. CIGRE Report 12-05, pp. 1–8.

8. Fischer E (1952) Die Festichkeit der inneren Roehre von Transformatorenwicklungen. *Elektrotechnische Zeitschrift*, 73.
9. Pewca W (in Polish, 1998) Influence the clamping forces and winding support stiffness on winding short-circuit axial force results. *Energetyka*, 11, 492–496.
10. Borkowski W, Pewca W (1998) Non-linear analyse of dynamic withstandability of transformer winding under axial short-circuit forces. *Report of Eighth International Symposium on Short-Circuit Currents in Power Systems*, Brussels, Belgium, pp. 149–154.
11. Hiraiishi K (1971) Mechanical strength of transformer windings under short-circuit conditions. *IEEE Transactions*, Pas-90, 2390–2391.
12. Del Vecchio RM, Poulin B, Anuja R (1999) Radial buckling strength calculation and test comparison for core form transformers. *CIGRE Transformer Colloquium*, Budapest, Hungary, pp. 1–5.

2

Insulation Problems of HV Power Transformers

Andrey K. Lokhanin

CONTENTS

2.1 Introduction

Insulation of power transformers represents a complex system consisting of various design components. The insulation may be divided into two parts: internal insulation and external insulation.

The external insulation is the insulation along the cover of bushings, air gaps between bushings of the winding, and of the earthed parts.

Internal (oil-filled, gas, cast) transformer insulation is divided into main and longitudinal insulation of windings, insulation of bushing installation, insulation of taps, tap changer, and so on.

The main insulation of windings is insulation from the given winding up to earthed parts of core, yoke, tank, and other windings (including those of the other phases).

Longitudinal insulation is the insulation between various points of one winding: between turns, layers, and coils.

The following types of internal insulation are used in oil-filled transformers:

- Solid (as rule, cellulose) insulation is the insulation between insulated conductors located closely to each other, coils or taps.
- Oil insulation gaps between a winding and the tank, a bushing screen and the tank, between taps, and so on.
- Combined oil-barrier insulation of oil gaps subdivided by barriers—insulation between windings, insulation between phases, insulation between a winding and the core, yoke, and so on.

Insulation of transformers in service is exposed to continuous actions of operating voltage and transient overvoltages. Coordination of internal insulation demands to provide withstanding of all voltage stresses during the lifetime of the transformer.

Inspection of the fulfillment insulation coordination requirements is provided by high-voltage tests, including the following voltage tests:

- Standard short-duration (1 min) power frequency voltage
- Standard long-duration (1 h) power frequency voltage with measurement of intensity of the partial discharges under $(1.3–1.5)V_m/\sqrt{3}$
- Standard switching impulse (250/2500 μs)
- Standard lightning impulse (1.2/50 μs)
- Standard chopping lightning impulse (chop time 2–5 μs)

At the present time, these voltage tests determine the basic sizes of insulation gaps. However, in connection with the development of UHV systems and deep limitations of overvoltages in EHV systems, the operating voltage can become decisive for the choice of internal insulation of transformers.

In most cases, the insulation applied consists of a combination of liquid (oil) and solid (cellulose) materials. In such combined insulation under the application of power frequency and impulse voltages, the stresses on oil channels are higher than on solid insulation. As the dielectric strength of oil is also lower than the solid insulation, in most cases the dielectric strength of transformer insulation is determined by the dielectric strength of the most stressed oil channel.

The main task of the design insulation system procedure is the choice of the criteria determining its dielectric strength.

Up to now, there is no physical theory of breakdown of the technically clean transformer oil and, hence, there is no opportunity to approach only theoretically the criteria of oil gaps' dielectric strength.

Therefore, choosing the criteria of electric strength of transformer insulation is based on the experimental results that take into account the main factors influencing the beginning and development of the insulation breakdown process.

There are several physical theories that use macroscopic mechanisms for explaining the breakdown of liquids [1,2]. These theories are based on experiments with technically clean liquids at rather large currents of conductivity and at long-duration voltages. Similar conditions take place in most cases of dielectric liquid in transformers. Results of these works are taken into account when choosing a dielectric-strength criterion.

Initial physical processes in the macroscopic mechanism of breakdown create conditions for the development of the discharge. The most significant aspects of this problem are the sources of free electrons in liquids. Their mechanisms of salvation and movement, by the influence of a molecule structure, are conditions of formation and development of electron avalanches, amounts of the energy created by electron movement. The development of these processes up to a level, which can be found out and registered by the measuring device, establishes a point of their transformation from elementary physical processes into the macroscopic phenomenon, measured in the usual manner. The point at which this transformation of elementary physical processes into a macroscopic stage occurs is accepted as the criterion for determining the electric strength. With reference to the internal insulation of transformers, this criterion is quantitatively determined by two parameters: of the electric field stress, at which point there is an initial macroscopic stage of the discharge (partial discharge), and the intensity of this partial discharge.

This criterion is acceptable, as it determines the dielectric strength for the main oil-barrier insulation intensity of electric field in the oil channel, at which point there are partial discharges, with an apparent charge of 10^{-8}–10^{-7} Cl causing irreversible damages on the surface of a barrier.

Intensity of the discharge is the main, but not the single, factor that influences the beginning and development of this process; the process is influenced by a several other factors as well, including the following ones that need attention.

2.1.1 Chemical Structure of Oil

The chemical structure of oil has a certain influence on the beginning and development of initial breakdown processes. For example, the quantity of aromatic hydrocarbons in a molecule of oil determines whether the liquid will absorb or allocate gas by acting on the electric field stresses. The chemical structure of oil influences the development of molecule dissociation processes and the course of secondary reactions with formation of gas components [3]. Presence and quantity of aromatic hydrocarbons influence the electric strength of transformer oil, especially in a nonuniform electric field. The catalytic hydrocarbon autooxidation can have an effect on the chemical compounds of oil and reduce their electric strength during aging.

2.1.2 Movement of Liquid

The electrohydrodynamical forces acting on insulating liquid under electric field stresses cause the motion of liquid. Moreover, in power transformers liquid moves with the help of operating cooling pumps and also due to the difference of temperatures between its various layers. Under some conditions, liquid movement can influence its dielectric strength, due to the movement of particles; this activity causes cavitations or promotes generation and distribution of the charges formed as a result of the flow electrification.

2.1.3 Mechanical Impurity

Harmful influence of the particles was already noticed long time ago, and it is well known that careful clearing of transformer oil can increase its dielectric strength considerably.

2.1.4 Moisture

Electric strength of oil is influenced strongly relative to the amount of water in it, especially in combination with particles. Relative humidity of oil is important for the dielectric strength. The latter depends on the chemical structure and temperature of oil and, to a degree, on oil aging.

2.1.5 Temperature

Many factors influencing the dielectric strength of transformer oil depend on the temperature. For example, essential dependence of viscosity and surface tension of oil on temperature means that cavitation mechanisms connected among themselves and liquid movements are substantially determined by temperature and oil.

Similarly, changes of temperature influence the relative oil humidity. In fact, from test results when an increase in the transformer oil electric strength proportionate to the growth of temperature is observed, it makes it evident that the water concentration in oil exists to a level that is inadmissible.

In composite insulation (oil–cellulose) the estimation of the temperature's influence on oil dielectric strength is essentially complicated by the moisture migration process, which has a significant time to achieve an equilibrium condition between the contents of water in oil and cellulose insulation.

All aforementioned factors should determine conditions at which the chosen criterion of dielectric strength can be applied.

For the estimation of dielectric strength of real designs, the basic geometrical factor (the size) that determines this parameter should be defined. However, it is still impossible to completely prove which of the two geometrical factors—area or volume—should be adopted as the "size" of a design. Most experts prefer the stressed volume, that is, the volume of oil limited to an electrode surface and by 80% or 90% equi-gradient surface [4,5]. Corresponding to an increase of the stress volume, the probability of "a weak link" occurrence, which can initiate breakdown in high-stress electric field, is higher. For example, if a breakdown is caused by particles, then a greater volume is a source of a greater number of particles that can get into an area with high electric field intensity and initiate a breakdown.

If a barrier design is applied, the opportunity of free movement of particles in the full volume of insulation becomes complicated and the electric strength of a design increases.

For the design with an identical area of electrodes in homogeneous or quasi-homogeneous fields, the width of the oil gap between the electrode and the barrier is the geometrical parameter determining its dielectric strength.

2.2 Choice of Insulation between Windings of the Transformer

Nowadays in practice in most countries, the main insulation of power transformers is made by oil barrier (Figure 2.1). This type of main insulation is reliable; it has been proven by long-term experience.

In the oil-barrier insulation, the most stressed gaps are gaps joined to windings because it is necessary to expect substantial growth of the electric field intensity due to the heterogeneity brought in by elements of the winding design (intercool channels, racks, insulation cylinders, etc.).

This hypothesis has been assumed as the basis for the development of the method of dielectric strength estimation of the main oil barrier insulation and is approved by the test of oil-barrier insulation model.

The influence of the basic design factors on the electric strength of the first oil channel (gaps), namely, the width of the channel, the design of elements between coils and barriers, the size of the channel between coils, the thickness of wire insulation, the presence of axial field of a winding, and the shape and duration of voltage have been investigated.

Results of researches [6] have shown essential dependence of the dielectric strength of the channel design between the winding and its nearest insulating barrier from its design and width (design A, B—Figure 2.2).

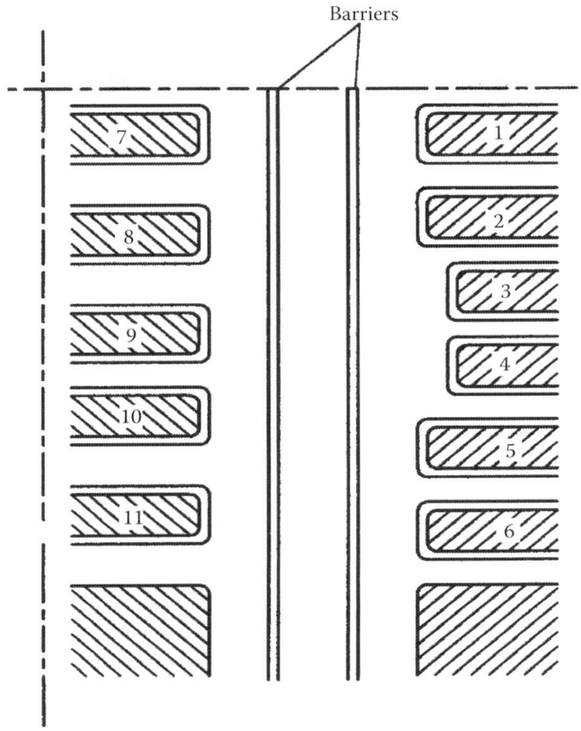

FIGURE 2.1
Oil-barriers insulation in the middle part of a transformer.

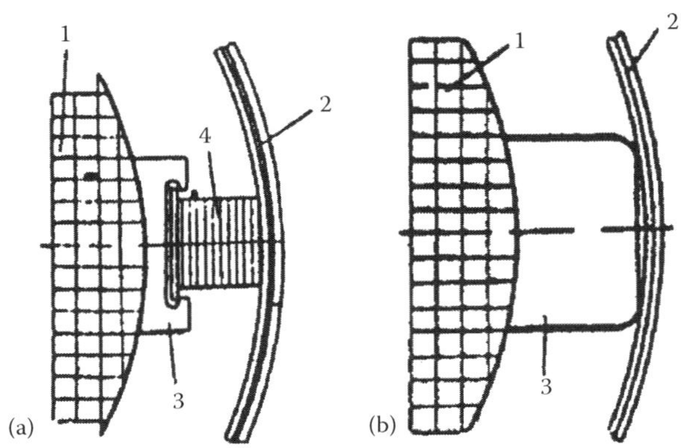

FIGURE 2.2
Design of insulation between winding and the nearest pressboard cylinder: 1, winding; 2, barrier; 3, gasket; and 4, rack. (a) With rack, and (b) without rack.

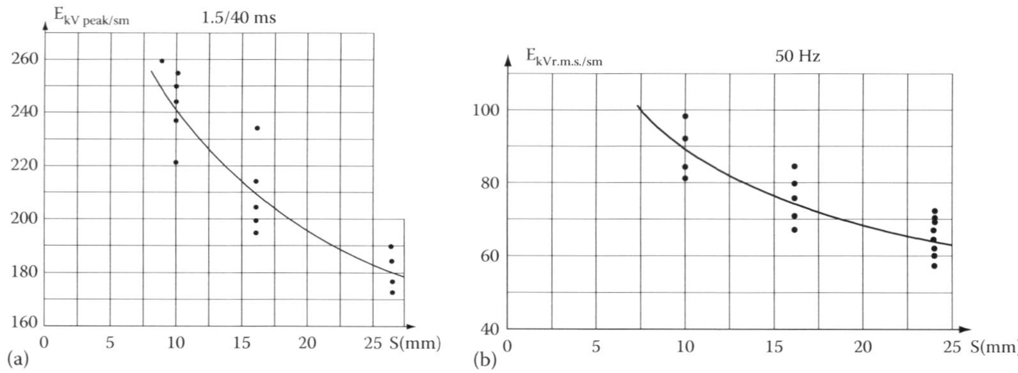

FIGURE 2.3
Dependence of the breakdown strength of oil close to a winding on its middle part. Each point represents one model; curve is plotted through mean values: (a) impulse 1.5/40 μs; (b) alternating voltage, 1 min-long application.

Dependences of oil channel dielectric strength for short-duration power frequency voltage can be expressed as follows (Figure 2.3b):

- For a design "A," providing absence of a gasket contact with a surface insulation cylinder,

$$E_{min} = 15 + \frac{67}{\sqrt{S}} \text{ kV/cm} \tag{2.1}$$

- For a design "B," irrespective of their shape

$$E_{min} = 7 + \frac{62}{\sqrt{S}} \text{ kV/cm} \tag{2.2}$$

where S is the width of the oil channel, close to a winding, cm, $6 \geq S \geq 0.3$.

The results of experiments have shown that essential reduction in radial channel dielectric strength does not occur when changing the channels between coils. So corresponding to an increase in the size of the axial channel from about 10 to 30 mm, the reduction in the radial channel dielectric strength is not more than 10%. The majority of investigations of oil channel dielectric strength dependence have been carried out with coils, the turn wire insulation of which was 0.7 mm on one side. The increase in thickness of the wire turn insulation up to 2.2 mm results in an increase of oil channel dielectric strength to 10%. The influence is caused, apparently, by the increase in the field uniformity degree in the oil channel between coils.

Researches on models have also shown that the dielectric strength of oil channels, located between adjacent cylinders, also depends on their width, but it is higher at least by 1.5 times than the electric strength of the oil channels of the same sizes but located near a winding. This result gives the basis to draw the following practical conclusion: If the oil channel in a winding at the design of oil-barrier insulation is suitable for implementing the lower, being limited only by technology requirements and conditions of a winding cooling, then the oil channel between cylinders should be of larger size and should be a proof that only design reasons are possible.

As a result of lightning impulse tests (standard impulse 1.2/50 µs), dependences of dielectric strength on the width of oil channel have been noticed (Figure 2.3a). Analytically, these dependences can be expressed as follows:

- For the design of type "A"

$$E_{\min br} = 82 + \frac{141}{\sqrt{S}} \text{ kV/cm} \quad (2.3)$$

- For the design of type "B"

$$E_{\min br} = 48 + \frac{153}{\sqrt{S}} \text{ kV/cm} \quad (2.4)$$

where S is the width of the oil channel, close to a winding, $6\,\text{cm} \geq S \geq 0.3\,\text{cm}$.

The thickness of wire insulation at a lightning impulse in the range of 0.7–2.2 mm affects the dielectric strength of the channel at less than 1 min power frequency; that is, dielectric strength increases to 5%.

Comparing the dependences $E_{\min br} = f(S)$, noticed at 1 min power frequency during an impulse test, it is easy to state that at a lightning impulse influence of the channel width close to a winding, its dielectric strength is a little bit lower than at the influence of the power frequency. So by changing the channel width from 24 to 10 mm, the electric strength at lightning impulse increases to 30% but power frequency voltage to 40%.

The distinction between the oil channel dielectric strength for the designs "A" and "B" under lightning impulse is a little bit less, especially for small channels. If for the short-duration power frequency voltage of 50 Hz, a ratio on the minimal breakdown stresses between various designs at a width of the channel in a winding of 10 mm is equal to 1.18, then at a lightning impulse the same ratio is equal to 1.10.

Researches of the main insulation dielectric strength at switching impulses have carried out two kinds of shapes: a unipolar impulse 600/1600 µs, which is equivalent to the influence of the basic peak switching over voltages with a frequency of about 250 Hz, and the voltage of the power frequency linearly increasing from 0 up to a maximum value in 0.4 s.

For transient voltage stresses, the dependence of the oil channel minimal breakdown intensity from its width can be submitted by the formula

$$E_{\min br} = A + \frac{B}{\sqrt{S}} \text{ kV/cm} \quad (2.5)$$

where S is the width of the channel ($6\,\text{cm} \geq S \geq 0.3\,\text{cm}$).

Under the stresses of the aperiodic impulse 600/1600 ms factors in a design "A," accept the following values A = 56, B = 115; under the power frequency voltage increasing linearly up to maximum for 0.4 s, A = 41, B = 107.

These dependences are a basis for dielectric strength estimation of the main insulation in the middle part of a winding under the stresses of a switching impulse.

In Tables 2.1 and 2.2 the influence in the oil channel width nearest to the winding for design "A" and "B" on dielectric strength and impulse factors are reflected by various shapes of impulses.

TABLE 2.1

Influence of the Oil Channel Width and Design of the Oil Channel Nearest the Winding on the Electric Strength of Insulation

Stresses	1 min Power Frequency	f = 50 Hz, 0.4 s	Aperiodic Impulse, 600/1600 μs	Aperiodic Impulse, 1.2/50 μs
Increase in dielectric strength at reduction of the channel width from 24 up to 10 mm (%)	41	34	31	29
Increase in dielectric strength of the 10 mm channel width at the transition from system "B" to "A" (%)	18	14	14	10

TABLE 2.2

Ratio of the Dielectric Strength Impulse to 1 min Power Frequency One (K_1)

Oil Channel Width (mm)	K_1		
	1.2/50 μs	600/1600 μs	f = 50 Hz, 0.4 s
10	1.9	1.48	1.8
24	2.1	1.58	1.34

The dielectric strength of the edge insulation (Figure 2.4) is determined by a number of factors: the type of main insulation, a ratio of distances between one winding and another and between the winding and the yoke (a pressing ring), and the design of the oil channel between the capacitor ring and its nearest barrier.

Comparison of the electric field stresses in the oil channel near a capacitor ring in the edge insulation of 110–330 kV transformers with corresponding field stresses in radial oil channel nearest to a winding in the middle part shows that degrees of heterogeneity of these fields are close enough, that is, ratio between average and maximum strength within the limits of 0.5–0.7 in both cases.

However, there is also some concern that the area of the increased field intensity in the radial oil channel in the middle part is more limited than in the oil channel at the edge of a winding, that is, the value of the stressed volume at the edge of a winding is more than that occurs in the middle part. This circumstance shows a higher dielectric strength of the oil channel at the edge of the winding than in the middle.

The results of edge insulation large-scale models test have confirmed this assumption. For all kinds of voltage stresses, dielectric strength at the edge of the winding is a little bit lower than the dielectric strength of the oil channel of the same width in the middle part of the winding. So the reduction at 1 min power frequency voltage is about 15% and at impulse about 10%.

While designing the edge insulation, it is necessary to take into account the following:

- The width of the oil gap near a capacitor ring should be as small as possible for design and technological reasons. The reduction of the channel near the capacitor ring can be achieved by using angular washers with a shape identical to the capacitor ring.

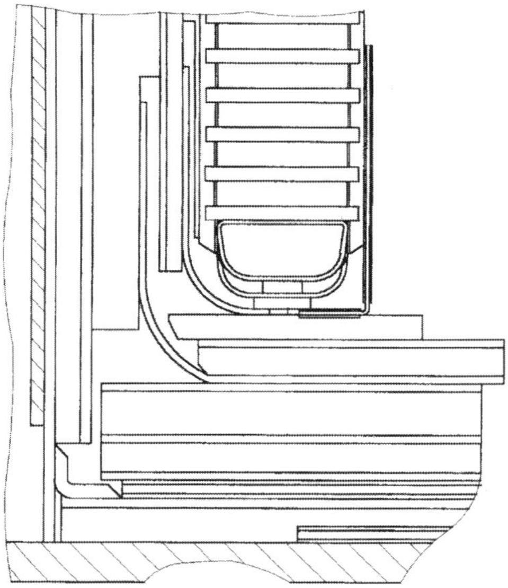

FIGURE 2.4
Edge oil-barrier insulation.

- The shape of a capacitor ring should be such that it can provide the maximal increase in a uniformity degree of a field in the oil gap between capacitor ring and washer; thus it is necessary to ensure that the oil gap width nearest to the winding does not increase.

At the establishment of permissible field intensity values, the following circumstances should be taken into account:

- Dependences $E_{min\ br} = f(S)$ are determined by values of a damaging voltage at a rather small amount of experiences; therefore, at transition to the permissible value of the field intensity, the correction factor that is taking into account this circumstance should be entered.
- Models of oil channel insulation using which researches have been carried to assess the dielectric strength of the channel at a winding have small volumes; therefore, at transition to great volumes of oil, it also should be specified by the introduction of the correction factor. This factor also takes into account the reduction of the dielectric strength because of the unforeseen deviations in the transformer.

In view of all the factors stipulated earlier the safety margin between the value of permissible intensity and its minimal breakdown value is established equal to 15%, that is,

$$E_{per} = 0.85 \cdot E_{min\,br} \tag{2.6}$$

At the estimation of the safety margin of dielectric strength under lightning impulse, it is necessary to pay special attention to the correct voltage distribution influence along a winding on the resulting intensity of a field in the oil channel.

The intensity along the surface racks of a winding in an area of the maximal stresses of electric field should also be compared with the permissible value of intensity on a surface of insulation.

Using models of main insulation, researches of main insulation dielectric strength made at lightning and switching impulses and also at 1 min power frequency voltage have been carried out, in addition to tests at different durations of power frequency till 1000 h. As a result of these tests, the voltage-second characteristic in this range was determined.

These researches have shown that dielectric strength of the main insulation is also determined at short-term stresses by dielectric strength, adjoined to a winding of the oil channel, which in turn depends on the value of average field intensity in the channel width and its design. The dielectric strength was determined as a breakdown of the oil gap, which is registered as PD with intensity 10^{-8}–10^{-7} Cl. Before the breakdown of the gap occurs, a growth of PD intensity above the noise level ~10^{-11} Cl is observed. It means that processes of ionization are occurring in small volumes of oil in places where the increased field intensity does not render influence on the oil channel dielectric strength and does not participate in the preparation of the breakdown; that is, the voltage-second characteristic in this range is determined by the statistical phenomena.

This conclusion is extremely important. If the major factor determining the dependence of the dielectric strength versus time is the statistical phenomena and during the application of a voltage on insulation there is no accumulation of any disintegration products, which results in the change of breakdown mechanisms, the dependence of breakdown intensity on time should be as follows:

$$t = A \cdot E_{br}^{-b} \tag{2.7}$$

where

t is the time before occurrence of breakdown

E_{br} is the breakdown intensity of the electric field A

b is the constants

For oil-barrier insulation b = 55/75 [7].

In Figure 2.5, the voltage-second characteristic of oil-barrier insulation is presented (from 1 min till 1000 h).

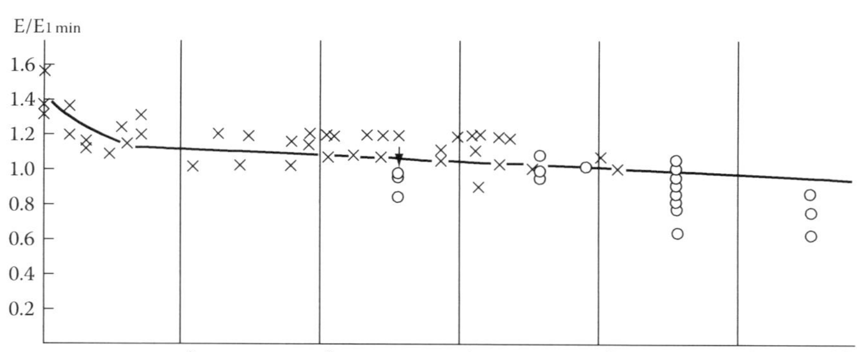

FIGURE 2.5
Volt-second characteristic of oil-barrier insulation: X, breakdown stresses; O, withstand stresses.

The character of dependence $E_{br} = f(t)$ is the same for both designs of the main insulation.

Extrapolating the voltage-second characteristic for a period of time equal to service life of the transformer (25 years) and accepting the same safety margin for 1 min power frequency voltage, the operating strength of oil-barrier insulation has been taken corresponding to 0.8 one-minute power frequency voltage strength for all designs of insulation.

However, the value of field intensity gotten by such a way should be corrected on the basis of studying physical and chemical aging processes of oil–paper insulation, and the oil irreversibly changes, resulting in the occurrence of oil and solid insulation and their influence on the insulation dielectric strength.

Special researches of oil–paper insulation have shown that power frequency voltage with an intensity at 5 kV/mm within 1000 h does not create any products of aging; that is, the cumulative effect is absent.

At the field intensity of 6.0–6.5 kV/mm in 500 h, there is a reduction in breakdown voltage of oil gap and there is a marked increase in the concentration of gases dissolved in oil.

Proceeding from these results, maximum permissible operating field intensity in oil-barrier insulation should be limited to a value of 5 kV/mm, irrespective of oil channel width.

Thus, for the definition of oil-barrier long-duration insulation, $E_{perm} = 0.8E_{perm{\cdot}1\,min}$ is used. The application of this ratio is possible only when the intensity received does not increase beyond 5 kV/mm.

It is necessary to note that in power transformers with the channel in a winding of 6 mm at operating intensity of 4.8 kV/mm is approved; there are about 15 such transformers already working for more than 20 years.

The size of insulation and oil gap between a winding and a tap is the same, as well as the oil gap "tap–wall of a tank" at the equal value of test voltages.

Such approach has been based on the similarity of electric field distribution in the aforementioned insulation gaps in transformer design.

The size of a necessary oil gap between a tap and a winding gets out on the greatest potential difference between them, taking into account the diameter and the length of tap and the thickness of insulation. Figure 2.6 shows the dependence of minimum oil gaps between tap and plane under different tap insulation (length of tap equal 3 m, R = 10 mm) from value 1 min power frequency voltage.

2.3 Longitudinal Insulation

Longitudinal insulation of high-voltage power transformers will consist of two basic elements: turn insulation (insulation between two next, close to each other turns of one coil), and coil insulation (insulation between conductors of two next coils divided by the oil channel).

Longitudinal insulation concerns also the insulation between layers of windings. In Russian practice, the winding layers are applied basically in distribution transformers of voltage classes of 6–35 kV.

Interlayer insulation can be carried out as only oil–paper and with the oil channel.

The sizes of longitudinal insulation in most cases are determined by impulse voltage, which creates nonlinear distribution of voltage along the winding.

With the purpose of reduction of stresses on longitudinal insulation of various types, so-called internal protection creating distribution of voltage on partial capacities of the

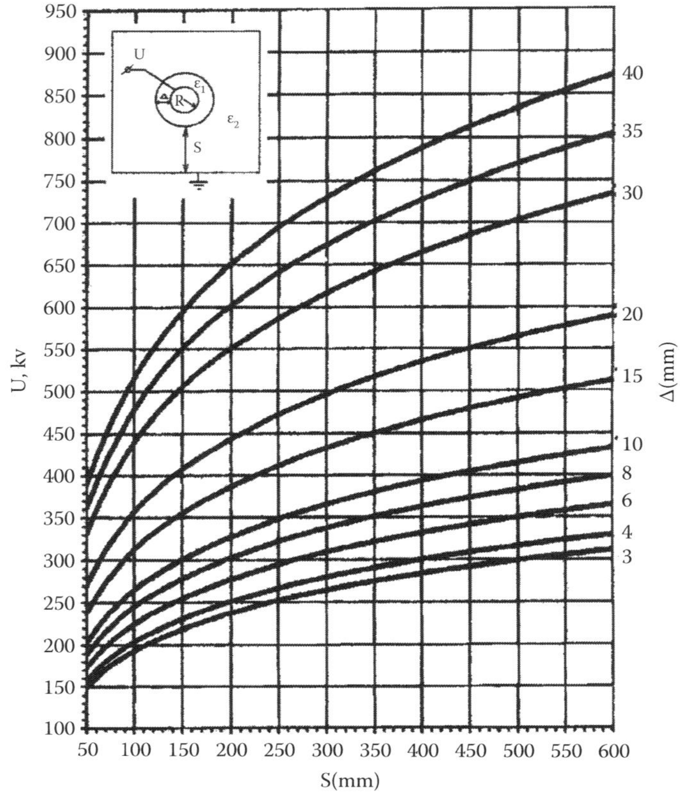

FIGURE 2.6
Dependence of the dielectric strength between top and plane from oil gap.

winding, close to being uniform, is applied. Now the basic type of "internal protection" windings of the transformer is an application of interleaving windings, full or partial that results in essential increase in longitudinal capacity of a winding and a decrease in impulse stresses between coils.

Practically, it is not considered expedient to achieve full alignment of capacitor distribution and full suppression of oscillating in a winding. Only partial improvement of capacitor distribution is sufficient. Intercoil stresses have the shape of oscillation with a rather strong attenuation. The choice of longitudinal insulation is done on the basis of experimental data received on the insulation models under aperiodic impulses stresses of standard shape.

Presently, there are no theoretical researches that would allow for finding a precise equivalent between aperiodic and oscillation stresses. Therefore, an equivalent is established on the basis of the experimental researches.

Practically, the main influence factor is the basic peak voltage amplitude and duration of impulse.

The duration of stresses between coils (ΔU_c) depends on the design of winding (interleaving or continuous) and under a full-lightning impulse lays in the limits from ten up to tens microseconds. In this case, the influence can be counted equivalent to the influence of aperiodic impulse at 1.2/50 or 1.2/20 μs. The estimation of strength should be made on

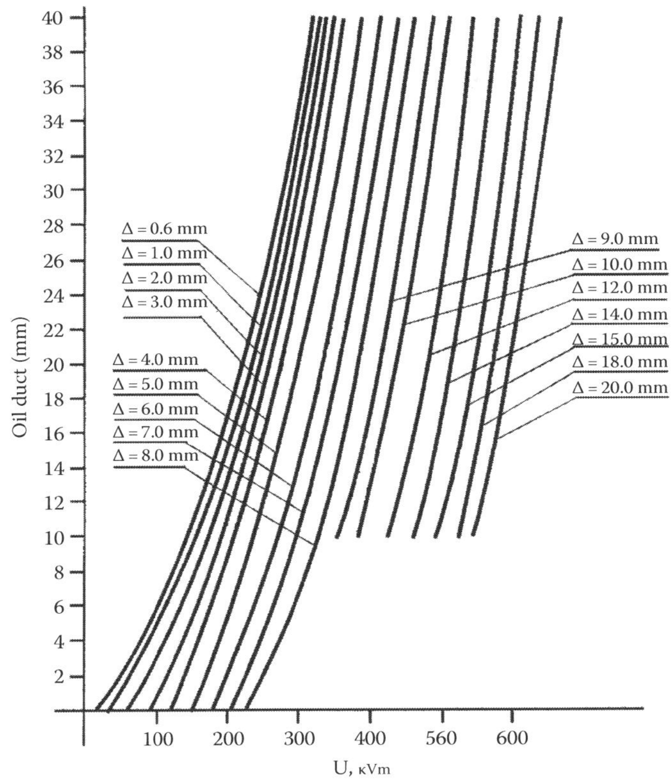

FIGURE 2.7
Dependence of the dielectric strength of intercoil insulation from the oil gap between coils.

the basis of the experimental data on the dielectric strength of intercoil insulation models under such impulse durations.

For the definition of necessary turn insulation thickness, it is necessary to know the value of the impulse voltage and its duration between adjacent turns. Curve dependences of turn insulation electric strength are determined on the basis of results of research on insulation models with different wire insulation thickness (Figure 2.7). Because of these researches the insulation cable paper with thickness of 0.12 mm and density 0.8 g/cm^3 is used. At the application of more density (1.1 g/cm^3) to the cable paper with thickness of 0.08 mm, the electric strength raises by at least 10%.

The chosen thickness turn insulation should be verified on keeping the operating voltage; thus, the permissible operating intensity is 2 kV/mm.

2.4 Conclusions

1. The analysis of power transformer oil-barrier insulation in experimental investigations shows that the criterion of this insulation's dielectric strength is the oil (channel dielectric strength of the oil) channel nearest to the winding.

2. The dependence of the dielectric strength (E) of the oil channel nearest to the winding from its width may be presented by the formula:

$$E_{\min br} = A + \frac{B}{\sqrt{S}}\,kV/cm \tag{2.5}$$

3. The voltage, second characteristic of the oil-barrier insulation under a long-duration voltage may be presented by the formula:

$$t = A \cdot E_{br}^{-b} \tag{2.7}$$

4. When determining the permissible dielectric strength, it is necessary to take into account the influence of a lot of factors such as a chemical structure of oil, the movement of liquid, impurities, moisture, and temperature.

References

1. Adamchevsky I (1972) Conductivity of the liquid dielectrics (in Russian), *Energy*.
2. Hebnerand RE et al. (1985) Observation of pre breakdown phenomena in liquid hydrocarbons, *IEEE Transactions on Electrical Insulation*, 20(2): 281–292.
3. Semionov N (1985) *About Some Problems Chemical Kinetic and Reactionary Ability* (in Russian), Academy of Sciences of the USSR, Moscow, Russia.
4. Ikeda M, Inou T (1979) Statistical approach to breakdown stress of transformer insulation, *Third International Symposium on High Voltage Engineering*, Milan, Italy, pp. 23–26.
5. Nelson JK (1989) An assessment of the physical basis for the application of design criteria for dielectric structures, *IEEE Transactions on Electrical Insulation*, 24(5): 835–847.
6. Lokhanin AK, Morozova TI, Voevodin ID, Beletsky ZM, Kuchinsky GS, Kaplan DA (1970) Problems of coordination of dielectric strength of extra high–voltage power transformer major insulation, CIGRE'70, Report 12-06.
7. Morozova TI (1976) *The Dielectric Strength of Transformer Internal Insulation Under Long Duration Voltage* (in Russian), Electrotechnica, No. 4, Moscow, Russia.

3

Stray Losses, Screening, and Local Excessive Heating Hazard in Large Power Transformers

Janusz Turowski

CONTENTS

3.1 Introduction

Power losses, hotspots, and screening in large power transformers are critical issues that need to be monitored, and sufficient precautionary and fail-safe mechanisms are to be put in place because they can affect not only active parts of the transformer but can also destroy the whole transformer unit. Correct assessment and calculation of these such phenomena can result in significant energy output, economy, and improved reliability. Extensive research during the past half a century focusing on a variety of topics, including functional specifics, three-dimensionality, nonlinear theory, and rapid computation technology, is reflected in the list of author's papers and books, starting from development of bushings and cover theory [1] through monographs [2,3], including recent publications on mechatronic impact on design and economy policy [4–6] of electrical machines and transformers. This methodology has been developed parallel to the theory and practice of active components. Nevertheless, what we need at this time a new approach based on principles of modern mechatronics, backed up by key technology features such as computer-driven real-time and artificial intelligence applications used in machines. Such approach [3] change results in better market competitiveness for the product, in this case, the transformer (Figure 3.1).

3.1.1 Principles of Mechatronics

1. System approach [8].
2. Rapid design methods* (Figure 3.6).
3. Substitution of concurrent engineering approach with mechatronic engineering.
4. Collective work (*immediate contribution of experts to a joint final program*).
5. Simple, low-priced, rapid, and easy-to-understand methods based on deep theory (*only an expert can teach easily on the subject and build such programs*).
6. Exactness of modeling vis-à-vis the need, and rejection of needless details.
7. Analytical methods, wherever possible.

* Calculation cost with the FEM ~ N^4, where $2N$ number of algebraic equations, N—half of matrix band (Z. Trzaska: Modeling and computer simulation of electric systems (in Polish). Wyd. Politechnika Warszawska, 1993).

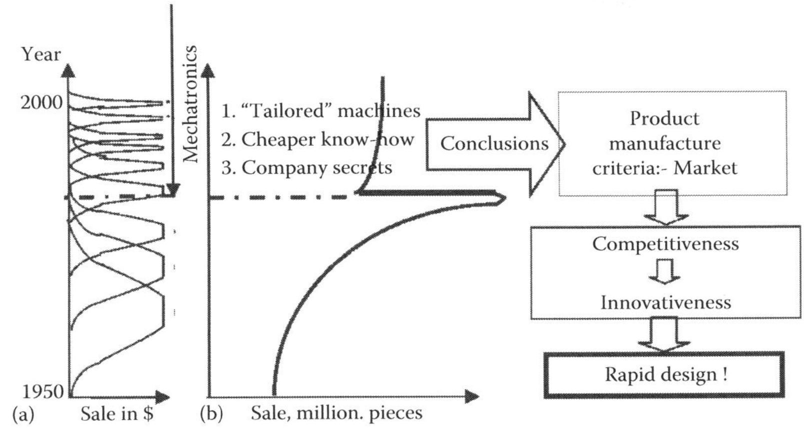

FIGURE 3.1
Impact of Mechatronics upon [4]: (a) "time to market" of introduction new products; (b) break-down market sale of small catalogue machines in the United Kingdom (By courtesy Wood, W.S., Economic importance of small El. motors, *Proceedings of 3x3x3 Seminar on Electrical Machines and Control*, Lodz, Poland, 1990.) due to mechatronic entrance 1, 2, 3.

8. Analytical approximation and linearization of nonlinear magnetic (Rosenberg, Nejman, Agarwal, Turowski 1993, [2]) and thermal parameters.
9. Expert systems:
   a. Building—quasi-static (Figure 3.2a)
   b. Motion—service or control in real time [6] (*the greater the input into the knowledge- and data-base, the simpler and faster the program*)
10. Interactive design cycle of duration in seconds (Figure 3.2b).
11. Simplification of modeling and design data for
    a. Building of elements
    b. Motion and control (Figure 3.4)
12. Structural optimization of systems and mechanisms is often more important than particular components (Koppikar, Kulkarni, Turowski [9]) (Figures 3.15 through 3.29).
13. ISO 9000. Responsibility for product quality is distributed to any work place (*one for all*).
14. Simple machines with sophisticated control systems.
15. Simple tools of design based on sophisticated, comprehensive fundamental research (Figures 3.19 through 3.21).
16. Outsourcing, that is, translocation of part or whole of component production to a specialized subcontractor or a contract manufacturer.
17. Reliability, safety, and economy aspects.

Rapid design and innovations, therefore, are the main benefits of applying such methods if one wishes to meet the market competition, according to principles of modern mechatronics [4]. Some of the spectacular examples in using innovative approaches include the feedback

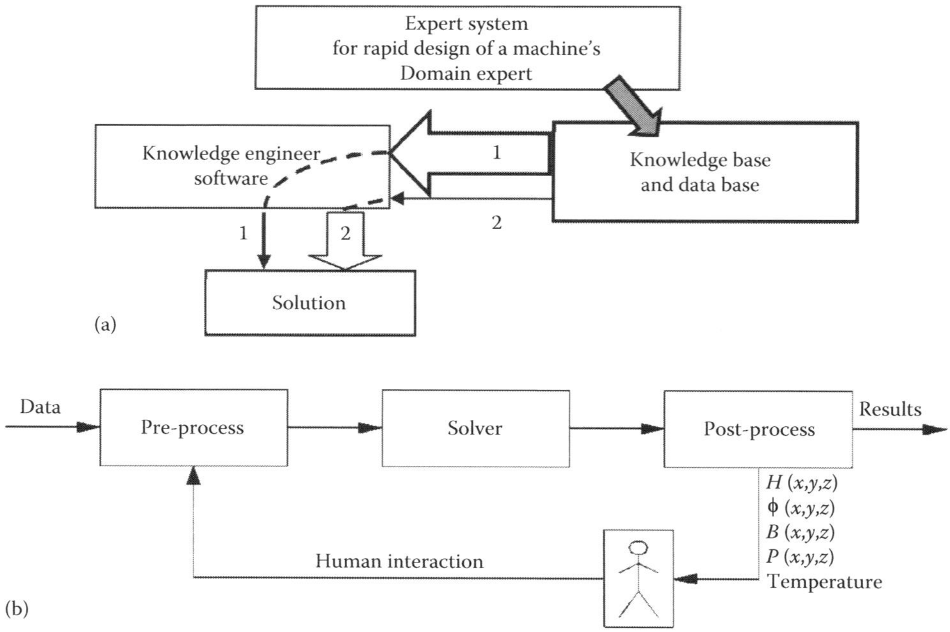

FIGURE 3.2
Block diagram of expert system type (a) design of a machines: 1, Large portion of introduced knowledge and experience—simple, inexpensive, and rapid solution, for example, 1 s; 2, small portion of knowledge and experience—difficult, expensive, labor-consuming solution, for example, 3 months. (b) Block diagram of interactive design. Interactive cycle of solution on one design variant with solver "MSR-1100," Java (Figure 3.27), and so on should be not longer than 1 to few seconds, like RNM-3D (Figure 3.19) with its ability to simulate in less than 1 s duration.

from the market of small, general-purpose machines (Figure 3.1b). Though paradoxical, Negoponte's* remark is quite relevant in the context of technological developments: "permanent improvements are the biggest enemy of innovativeness." Only innovativeness can give remarkable success.

Some of the most recommended tools used in rapid design development include (1) the expert system (Figure 3.2a) and (2) the interactive design (Figure 3.2b).

In expert system, a main role is played by the domain expert, who prepares a general model, preprocessor for interactive operation (Figure 3.2b), and scientifically based parameters (Section 3.4). Next comes the knowledge engineer, who knows how to transform instructions of the domain expert into practical and faster software applications. The third one is the solver, a separate program (e.g., MSR-1100 in Figure 3.19, Java in Figure 3.28, etc.) used for resolving a multinode 3D reluctance circuit. And the last box involves the postprocessor, which derives from the solver the following integral values: (1) flux density, $B_m(x,y,z)$, responsible for forces in windings; (2) magnetic field intensity, $H_{ms}(x,y,z)$, responsible for stray losses and local heating; (3) power density, $P_1(x,y,z)$, responsible for machine efficiency, heating, and reliability of whole unit. These results are later analyzed by the designer using interactive applications. One cycle of such analysis should be *not longer than few seconds or less*, like in RNM-3D with its <1 s.

The final program should be as easy and simple as Ohm and Kirchhoff laws.

* Negroponte N., MIT Media Lab, Cambridge, MA.

Therefore, the whole sophisticated knowledge derived from monograph [2], and so on should be hidden inside program. The regular application or the use of RNM-3D should be as easy and rapid as, for example, driving a modern, intelligent automobile, in which sophisticated automechanical parts are positioned unobtrusively inside the vehicle and do not disturb a driver in anyway.

3.2 Rapid Design and Mechatronic Impact in Transformers

In addition to applying the 17 principles of mechatronics, one also has to resolve a specific design problem [2]:

18. Forces and eddy current (additional) losses in windings.
19. Losses and excessive heating hazard in tank wall and core clampings.
20. Losses and overheating hazard in tank covers and bushing turrets.
21. Structural and material problems of electromagnetic or magnetic (shunts) screening.
22. Nonlinear dependence of load losses on current and temperatures.
23. Influence of "critical distance" of tank walls and its influence on forces and losses (3.33 and 3.34).
24. Measurements and experimental separation of stray losses from basic I^2R and total loss.
25. Experimental measurement of per-unit loss distribution in structural parts (thermometric method).
26. Effectiveness of stray flux collectors over and under winding edges (Figure 3.23).
27. Possibility of capital cost reduction, thanks to proper screening.

To satisfy the rapid design needs, it is necessary to apply the specific theory and create special scientifically based methods, addressing in turn every constructional element as shown in Figure 3.3. Such a design process can include both the classical, analytical

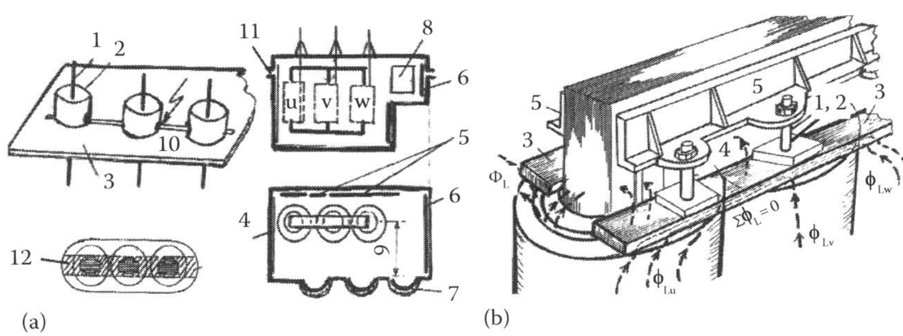

FIGURE 3.3
Steel parts in 3D field, at risk of excessive stray loss and heating [10]: (a) 1, bushing; 2, turret; 3, cover; 4, tank; 5, Fe screen (shunts); 6, Cu screens; 7, oil pockets; 8, tap changer; 9, tank asymmetry; 10, hot-spot; 11, bolted joints; 12, acceptable region of 2D calculation. (From Kazmierski, M., *Rozprawy Elektrotechniczne*, 16(1), 3, 1970.) (b) 1,2, Bolt & Al/Cu screen; 3, Stray flux collectors (Fe shunts); 4, Flat wise fields enter; 5, yoke beams.

approach, and sophisticated mesh methods. As it is seen from the simple formula for power loss density W/m^2 [2] in solid metal, basic methods of modeling and computation of almost all main parts of power transformers have been well known for many years. They were usually calculated with the 2D field models. However, as Kazmierski [11] confirmed, the 2D models are acceptable only in a limited region 1 (Figure 3.3—*12*) of transformer. These 2D methods are not acceptable for calculating eddy current losses in solid constructional parts of the core window.

Past decades have witnessed enormous progress in the development of sophisticated software for providing a variety of solutions, including complicated, nonlinear, 3D fields. However, such developments are not always in the form of inexpensive, simple to use, and rapid engineering methods, which are easy for regular, industrial design use [11].

3.3 Reliability, Safety, and Economy Aspects

Large power transformers with their optimally utilized materials and limited transportation outlines, are complicated, delicate, and extremely efficient electromagnetic structures. Electrodynamics of transformers plays an essential role in their design, reliability, and power supply.

The biggest power failure in the history was that of Great North-East Blackout of 1965. There was a complete shutdown of electric service throughout New York, eight Northeastern States, and a major portion of Canada. In the province of Ontario, some 30 million people were without power for as long as 13 h,* which confirmed the importance of transformers' reliability. It happened again in New York (1977), France (1978), USA Pacific Region (1996)† and recently in Canada, New York‡ (August 13, 2003), France, and North Italy (September 27, 2003). This cascade of tripping was mostly due to the overloading of transformers and is sufficient reason not to ignore its relationship (3.6) to overloading, especially.

Simultaneously, capitalized load-loss costs, depending on the time and place of operation, is from 3,000 to 10,000 US$/kW [5] and no-load-loss even more. Åke Carlson§ from ABB Transformer Division, Ludvika, presented a more detailed analysis of this problem. As per that analysis, the cost of capitalized load losses for generator transformers is little above 6000 US$/kW and it will be about 3000 US$/kW for network transformers.

However, he has not considered stray losses and the internal structure of transformer. He supposed probably that the transformers considered will have an optimal structure. One who knows the traditional factory methods can guess that the problem needs more in-depth 3D electromagnetic and thermal research. For instance, a proper 3D design of screening system can make it possible to eliminate about 40–50 kW stray loss or even more [12]. It means there will be savings up to US$ 0.5 millions per one large transformer. Possibilities of overheating hazards can also be eliminated. Furthermore, unplanned automatic shutdowns of a power plant cost more than US$0.5 million per day [13]. There exists opinion that, the cost for replacing a power transformer on account of unplanned automatic shutdowns of a power plant can rise from US$0.5 million to even US$0.8 million or more per day.

* IEEE PES, January 1991.
† Spectrum IEEE, April 1997.
‡ "Time," August 25, 2003, pp. 30–30.
§ Lazarz M.: *Konstruktive Massnahmen zur Verringerung der Verlustleistung in Kerntransformatoren.* XI. Internationale Tagung der Elektriker ITE. Berlin, 7-13.11.1965. Nr 3.1. Now ABB.

It was additionally shown in Refs. [9,12] that neglecting even a small 3D tank asymmetry could cause additional errors leading to losses up to 35 kW or more (Figure 3.22). These errors can be avoided by those transformer works that have rapid asymmetric packages like *RNM-3Dasm.exe* or similar packages and use them correctly.

Unfortunately, currently available professional computer programs like the most popular FEM-3D, are expensive and laborious in use. After Coulson et al. [14], FEM-3D is still "not useful for regular design use."

In spite of serious increase (since 1985) of computational capability, our recent experience shows still the same. FEM-3D needs several, if not a few hundreds, hours of work involved with using qualified specialist and expensive softwares.

At the same time, the accuracy of FEM-3D is not better than much simpler programs, for instance, the much-recommended RNM-3D.

Nevertheless, even well-known transformer works still use 2D models and semiempirical formulae, for example, Mieczyslaw Lazarz's* formula (ASEA) in W:

$$P_{ASEA} = k_2\left(f\Phi_{ms}\right)^{\alpha} \tag{3.1}$$

where

$\alpha = 1.5/1.9$

$\Phi_{ms} = \sum_1^n (H_{m,aver} \cdot t \cdot l_{average}) \cdot n$

$H_{m,aver}$ is the average in space magnetic field intensity on the width of windings, including gap

$l_{average}$ is the average winding length

Such simplified statistical formulae are still used, even though everybody knows that field lines are axial only in small region 1 (Figure 3.3a—*12*). In most parts of a three-phase transformer, the lines have significant and even dominating circumferential components. Unfortunately, such 2D formulae cannot be properly justified and are often very controversial ([2], p. 355). They are not general and are limited to a specific group of transformers. Nonetheless, with the introduction of semiempirical corrections, 2D formulae can deliver, in some range, results near to real ones. Some of them are discussed in Refs. [2,12]. However, in general, such approach can lead designers to incorrect results. Thus, it raises [5] a provocative question: *Is it worth to ignore the substantial cost savings of about US$0.5 million?*

It concerns producers, purchasers, and ordinary consumers of electrical energy, who finally have to pay for the electricity consumption.

This question follows from a comparison (Figure 3.23) between stray loss values calculated with the help of different methods, including the most primitive 2D and scientifically based rapid 3D methods that provide experimental measurements at the testing station [12].

As one can see from Figure 3.23, differences between the so-called real stray losses (i.e., measured at testing station) and those calculated three-dimensionally using a program such as RNM-3D [15] can range from 10 kW (for smaller units) to 80 kW (for large transformers with complicated, asymmetric structures). However, in Figure 3.23, the first

* Lazarz M.: *Konstruktive Massnahmen zur Verringerung der Verlustleistung in Kerntransformatoren.* XI. Internationale Tagung der Elektriker ITE. Berlin, 7-13.11.1965. Nr 3.1. Now ABB.

approach of RNM-3D computation was carried out for transformers of quarterly symmetric structures and for elliptic tanks with walls ideally screened magnetically.

For a long time, designers were almost helpless in deciding whether they wanted to do rapid checking using a 3D program focusing on field distribution, power loss density, and of hotspots localization on the internal surfaces of inactive constructional parts. Now, when a simple and user-friendly program such as RNM-3D is available, manufacturers and users can easily examine (Table 3.3) the loss density distribution, $P_1 = f(x,y)$, and its effects on the surfaces of parts integral to their transformers.

From a commercial point of view, in order to achieve cost savings, it is essential to create programs or production lines that eliminate *all components that are not absolutely necessary* and could potentially delay the design process.

It can be realized with the help of expert system approach [10,16], which consists of a strong knowledge and data-base component.

Porgram based on expert system can be faster and be of lower cost depending entirely on the amount of critical information fed into knowledge- and data-base that in turn ensures a simpler solver process.

RNM-3D is such rapid tool. FEM-3D, which is most popular nowadays, is "not useful for regular design use." It was stated by Coulson et al. from GEC-England [14] and later confirmed by us. Only rapid, expert-based packages like RNM-3D [16] give the chance to overcome [12] these problems (Figure 3.23). In fact, the differences between rapid RNM-3D and the cumbersome FEM-3D can be as large as few seconds to a several hundreds of work hours (Figure 3.7) involving highly qualified specialists [10].

Last but not the least, each 1 kWh of electrical energy needs to burn up to 0.4 kg of coal, which again is of no little ecological consequence.

3.4 Basic Theoretical Tools

3.4.1 Main Methods of Calculation

For the design of electromagnetic and electromechanical systems, two basic streams of approach (Figure 3.4) are recommended. They are as follows:

- *Maxwell's* and others field [2] theories for machine design
- *Hamilton's* principle of least action with its *Euler–Lagrange Equation* [6,17]—for the system motion and control

3.4.2 Stray Fields, Losses, and Forces

Stray fields, losses, forces, temperatures, and so on are calculated usually for short circuit, quasi-static condition of transformer. In that case, one can disregard no-load current and assume that a sinusoidal form of excitation current $I = Ie^{j\omega t}$, $I = |I|e^{j\Psi_i}$ exists.

In the first approach, for the sake of simplification, constant iron permeability $\mu(H) = const$ is assumed. It gives the basic *Maxwell's* equations for metals in the form:

$$curl(H_m) = \gamma E_m \text{ and } curl(E_m) = -j\omega\mu H_m \tag{3.2}$$

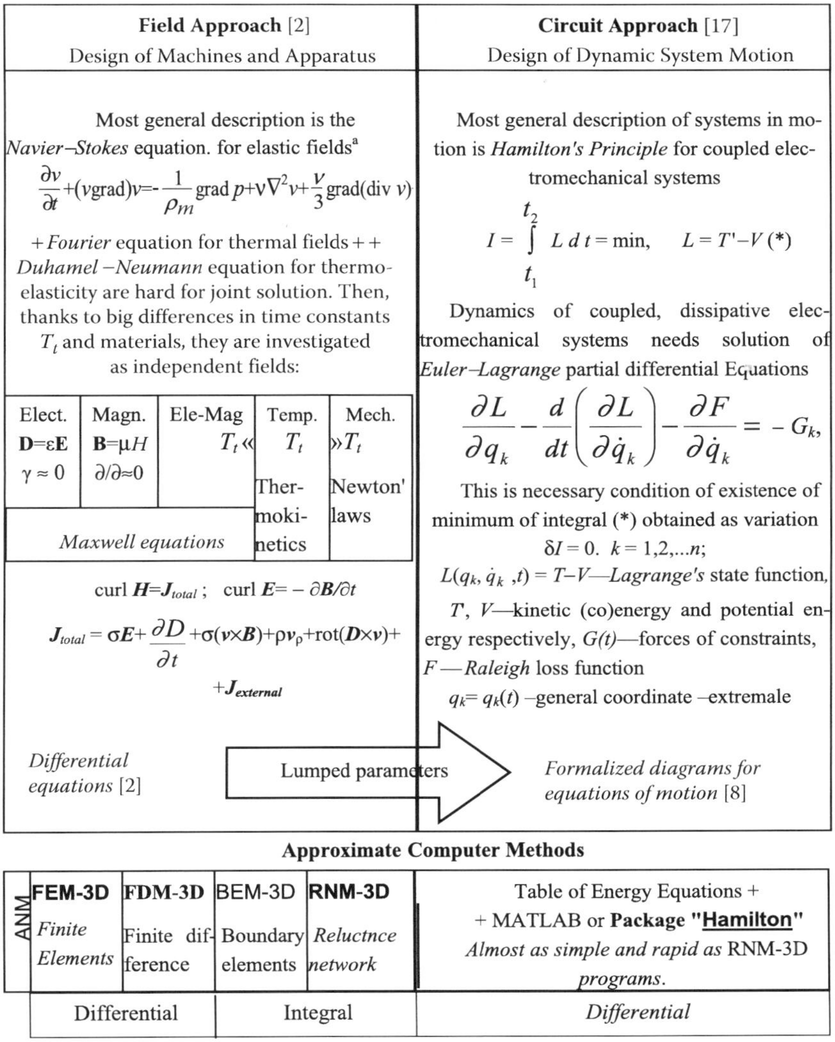

[a] This is the equation of heat exchange between fluid and surface of solid body and fluid (plasma) speed v, where $\rho_m=\rho_m(x,t)$, fluid density; $p=p(x,t)$, hydrodynamic pressure; υ, coefficient of kinetic viscosity of incompressible liquid; $\upsilon\Delta v$, density of force of internal friction; P, force acting on mass unit, for example, *Lorentz*'s force density $f_L=J\times B$, gravitational field, etc.

FIGURE 3.4
Two complementary domains of rapid design of electromagnetic and electromechanical coupled systems.

Then *Helmholtz* equation of plane-wave penetration inward (z axis) of solid metal $\partial^2 H_m/\partial z^2=\alpha^2 H_m$ and its solution in complex, exponential form is

$$H_m = H_{ms}e^{-\alpha z}, \quad E_m = \frac{\alpha}{\gamma}E_{ms}e^{-\alpha z}, \quad J_m = \gamma E_{ms} = \alpha H_{ms}e^{-\alpha z} \tag{3.3}$$

where

$$\alpha = \sqrt{j\omega\mu\gamma} = (1+j)k, k = \sqrt{\frac{\omega\mu\gamma}{2}} = \frac{1}{\delta}, \quad \lambda = 2\pi\delta \tag{3.4}$$

propagation, *attenuation*, and *phase constants* of electromagnetic wave are represented as follows: $\delta = 1/k$ is the *equivalent depth of wave penetration* of solid metal, and λ is length of electromagnetic wave in solid metal.

From Ref. [2] (p. 136) we know that at the depth $z=\lambda$ from metal surface the ratio of amplitudes of electromagnetic field, $H_{ms}(z=\lambda)$, to its surface value, $H_{ms}(z=0)$, equals $e^{-k\lambda} = e^{-2\pi} = 0.00185$ and at the depth, $z=\lambda/2$, it equals $e^{\pi} = 0.0432$.

It means that at such depths electromagnetic wave practically disappears and internal wave reflections like that in Figure 3.7 do not occur. Such a steel plate can be considered as half space, infinitely thick, and there is no influence of wall thickness d.

In conclusion, active loss dissipated in metal half – space obtained from *Poynting's* vector $P_1 = 1/2E_{mz}H_{my}^*$ in W/m² [2] is

$$P_1 = a_p\sqrt{\frac{\omega\mu_s}{2\gamma}}\frac{|H_{ms}|^2}{2} = \frac{\omega}{a_p}\sqrt{\frac{\omega\gamma}{2\mu_s}}\frac{|\Phi_{m1}|^2}{2} = \frac{\omega}{2\sqrt{2}}|H_{ms}|\cdot|\Phi_{m1}| \tag{3.5a,b,c}$$

Equation 3.5 shows the author's ([2], p. 345) basic formula for total power losses in solid and screened steel walls, in W:

$$P = \frac{1}{2}\sqrt{\frac{\omega\mu}{2\gamma}}\left[p_e\iint_{A_e}|H_{ms}|^2 dA_e + p_m\iint_{A_m}|H_{ms}|^2 dA_m + a_p\iint_{A_{St}}\sqrt{\mu_{rs}}\,|H_{ms}|^{2x} dA_{St}\right] = I^2(a+bI) \tag{3.6}$$

where

- $p_e \ll 1$ (3.13), $p_m \ll 1$ (3.14) are screening coefficients, considering (3.7)
- A_e, A_m, A_{St} are surfaces covered with corresponding (electromagnetic or magnetic) screens or not screened (St)
- H_{ms} denotes magnetic field strength on a metal surface

The value of x varies for different transformers but is typically between 1.1 and 1.14 [18].

It is a simplified formula, but there is no technical sense to endeavor for an accuracy level higher than that of a natural dispersion (Figure 3.5a) of magnetic $\mu(H)$ and thermal $\gamma(T)$ steel parameters.

3.4.3 Magnetic Nonlinearity Inward of Solid Iron

Magnetic nonlinearity, $\mu(H, z) = var$, inward (z axis) of solid iron was considered with the help of average linearization coefficients, a_p, a_q ([2], p. 322). Starting from Rosenberg's* idea, and from that of Neiman† and Turowski [1,2] J. Lasocinski, T. Janowski, and K. Zakrzewski, J. Sykulski developed a ([2], pp. 318–328) practical rapid engineering tool.

* Rosenberg E., Wirbelströme in massiven Eisen. E. u M. Vol. 41, 1923, pp. 317–325.

† Neiman L.R., Zaitsev I.A., Experimental research of skin—Effect (in Russian). Electrichestvo 2/1950, pp. 3–8.

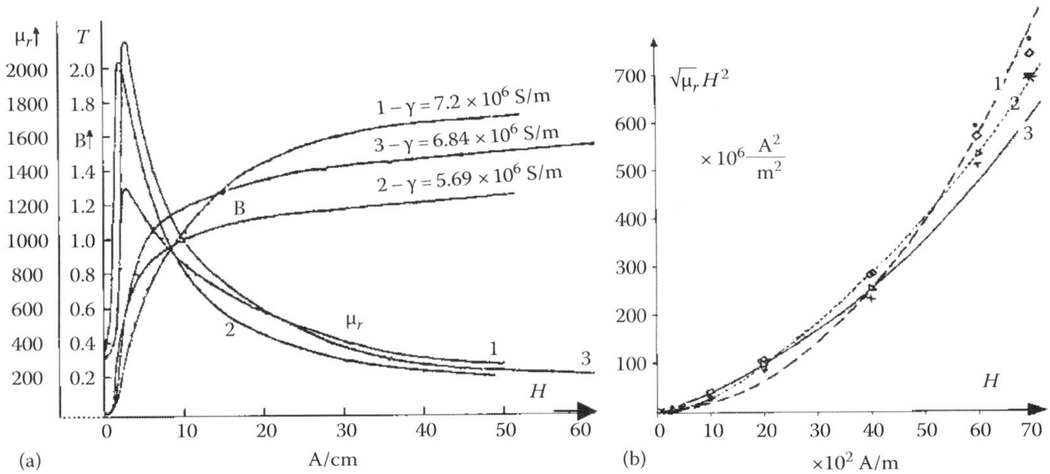

FIGURE 3.5
Nonlinear steel permeability versus *H*: (a) Discrepancy of magnetization curves (From Kazmierski, M., *Rozprawy Elektrotechniczne*, 16(1), 3, 1970.) of the samples of St3s steel used for transformers tanks. (b) Analytical approximations. (1) $1.1c_2H^{2.07}$; (2) $650H^{1.565}$; (3) $c_1H+c_2H^2$, $c_1=310\times10^2$, $c_2=7.9$. (From Turowski, J., Reluctance networks—Chapter 4, pp. 145-178, in *Computational Magnetics*, Sykulski, J. (ed.) Chapman & Hall, London, U.K., 1995; Translation from Polish of the book Turowski, J. (ed.), Analiza i synteza pol elektromagnetyeznych, Polish Academy of Science, "Ossolineum," Wroclaw, Poland, 1990.)

Shortly, for different sorts of iron, at $H_{ms}>5$ A/cm, they are $a_p=1.3$–$1.5\approx1.4$, for active power, and $a_q=0.8$–$0.9\approx0.85$, for reactive power. At lower fields, $H_{ms}\le5$ A/cm or at $\mu_s=const$, we can adopt $a_p=a_q=1$. Though simplified, they proved to be very handy and satisfactory for most engineering applications.

For instance, equivalent depth of penetration $\delta_{\mu(H)}=(1/a_p)\delta_{\mu=const}$; active reluctance $R_{\mu(H)}=0.37R_{\mu=const}$; and reactive reluctance $R_{\mu(H)r}=0.61R_{\mu=const}$, and so on ([2], pp. 323–324).

3.4.4 Magnetic Nonlinearity along the Steel Surface

Magnetic nonlinearity, $\mu(H, x, y)=var.$ along the steel surface (X, Y) was taken into account with the help of author's specific analytical approximations (Figure 3.5b) ([2], p. 316):

$$\sqrt{\mu_s}H^2=c_1H+c_2H^2;\quad \frac{1}{\sqrt{\mu}}=A_1+A_2\sqrt{\mu}H;\quad \left(\sqrt{\mu_r}H\right)^n\approx H$$

$$\sqrt{\mu_r}H^2\approx cH^b \tag{3.7a,b,c,d}$$

and corresponding exponent coefficient x in Equation 3.6.

3.4.5 Dependence of Stray Losses on Current and Temperature

Equations 3.6 and 3.7 and Figure 3.5b confirm a very important conclusion of the author ([2,19], p. 329) for load loss measurements and for overloading hazard, that *in steel elements of transformers, stray losses can increase locally faster than with I^2.*

Since after Figure 3.5b

$$H_{ms} = C\left(\sqrt{\mu_r}\,|H_{ms}|\right)^n = C\left(\sqrt{\frac{\omega\gamma}{2\mu_0}}\,|\Phi_{m1}|\right)^n \tag{3.8a}$$

where $C = 1.05 \times 10^{-5}$ $(A/m)^{1-n}$, $n = 1.77$, one can conclude that power losses in steel elements have different proportionality from current I, frequency f, and steel conductivity σ as follows:

- When field on the steel surface $H_{ms} = cI$—like in transformer coverplate

$$P_1 \sim I^{1.6} f^{0.5} \sigma^{0.5} \tag{3.8b}$$

- When flux $\Phi_{m1} = cI$—like at interwinding stray flux entering the steel wall

$$P_1 \sim I^{2.8} f^{1.9} \sigma^{0.9} \tag{3.8c}$$

This is why short circuit test should not be conducted at very small currents.

Preferably, the short circuit test should be carried out at current I bigger than rated current. It is especially recommended when we wish to be sure that transformer is resistant to overloading.

3.4.6 Choice of Calculation Methods

Current literature on electromagnetics offers at least 15 basic methods of field computation and much more combined and hybrid ones. All of them are equivalent in a theoretical sense but not equivalent in terms of practical efficiency during regular design use. Most of them need highly qualified personnel (Figure 3.6) and expensive professional commercial packages and computers. Such general commercial programs are good for education and fundamental research. They have good graphics and general capabilities. However, often they need a long time for computation (even many days) [14] and give nice lines of field distribution but without total losses.

Industrial application needs rather simple and rapid solutions, user-friendly managements, low price, with immediate engineering applicability, simple-to-understand integral output solution on PC, and practical as well as economical effectiveness.

One of such tools is the *Reluctance Network Method*—RNM-3D (Figure 3.6) with its integral form of (3.2) *Ohm* and *Kirchhoff* laws.

$$\sum_{j=1}^{m} u_{mj} = 0, \quad \sum_{i=1}^{n} \Phi_i = 0$$

Nevertheless, most accurate are the analytical models and their computer tools, that is, Analytical Numerical Methods (ANM-3D). They should be used everywhere, if possible. Unfortunately, technical analytical methods need simple structure. One of such approaches was applied effectively to the problem of transformer cover plates.

Calculation of stray fields and losses in large power transformers

RNM-3D	*Criteria*	FEM-3D
Low, secondary school level	◀ User grade level ▶	High, university level
Regular PC	⇐ Hardware ⇒	High quality
Ohm's and Kirchhoff's laws	⇐ Theory ⇒	Variation calculus
Small, smoothed by integral method	◀ Errors generation ▶	Large due to differential method
Low price	⇐ Software ⇒	Expensive and difficult
Half to few hours	Time to construct ⇐ a model ⇒	5 – 12 months
Less than 1 s	CPU time of computation of one ⇐ design variant ⇒	30 – 300 h

FIGURE 3.6
Dependence of effectiveness of modeling and simulation of electromagnetic fields, eddy current losses, and forces in windings on the computation method selected [4,11], with its corresponding set of RNM-3D packages [16,18] specially designed for fast, dedicated industrial design [20] of complex 3D, three-phase, and nonlinear electromagnetic structures.

They can include both classical analytical approach and sophisticated mesh methods.

3.4.7 Electromagnetic Screening

For rapid modeling and calculation of penetrable metal wall, an original (Figure 3.7) wave method in Ref. [2] (pp. 176–194) was developed.

From incident Z_{inc} or refraction Z_{refr} wave impedances for different metals

$$Z_{met} = \frac{\alpha}{\gamma} = (1+j)\sqrt{\frac{\omega\mu}{2\gamma}} \tag{3.9}$$

For example, $|Z_{Cu}| = 2.7\times10^{-6}\ \Omega$; $|Z_{Steel}| = 2.4\times10^{-4}\ \Omega$, at $\mu_r = 1000$ and dielectric $Z_{diel} = \sqrt{\mu_0/\varepsilon_0} = 377\,\Omega \gg |Z_{Steel}| \gg |Z_{Cu}|$, we have simple wave reflection coefficients

$$M_{in\text{-}refr} = -\frac{H_{m\,refl}}{H_{m\,inc}} = \frac{E_{m\,refl}}{E_{m\,inc}} = \frac{Z_{refr} - Z_{inc}}{Z_{refr} + Z_{inc}} \approx \pm 1 \tag{3.10}$$

from which all necessary formulae were obtained, much faster than with the help of classic analytical or FEM approach.

Particularly, the active and reactive power coefficients in penetrable metal wall (Figure 3.7) are determined as follows.

Active power P_1 in W/m^2 and reactive Q_1 in var/m^2 powers (3.6) consumed by metal wall are correspondingly:

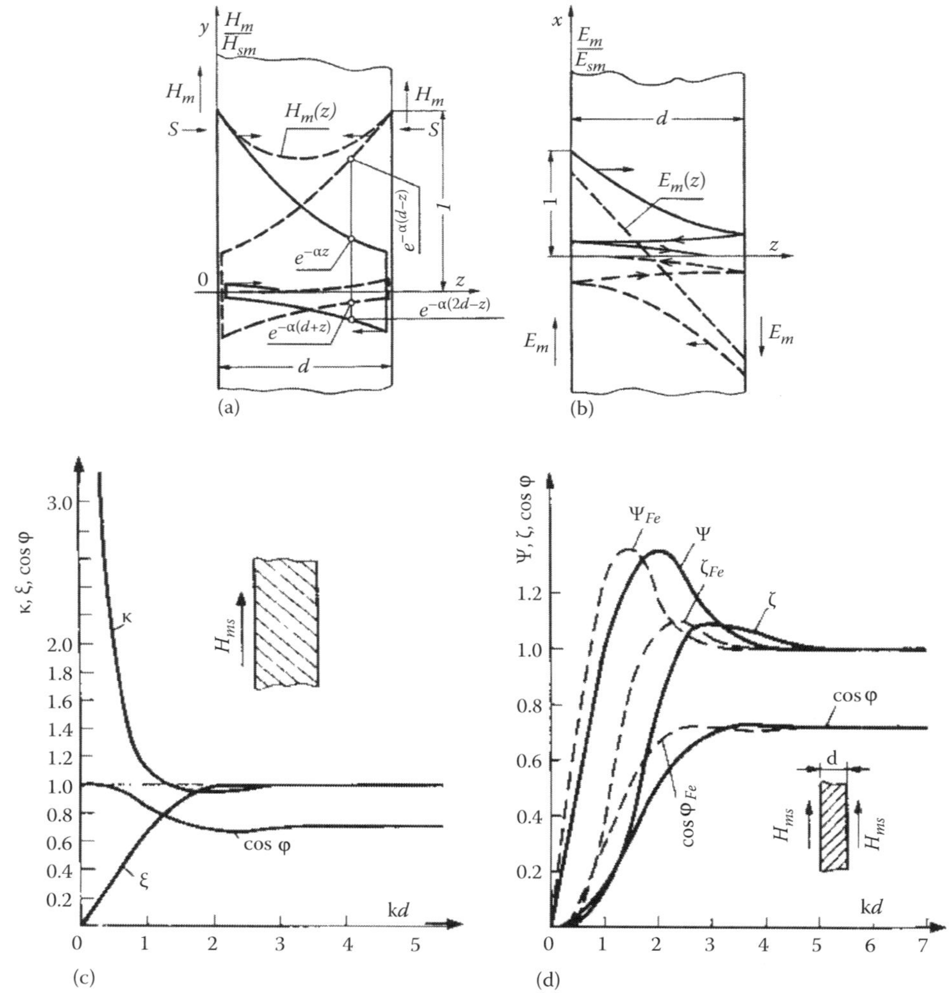

FIGURE 3.7
Internal interference of electromagnetic wave in penetrable metal wall: (a, b) Wave incident from both sides. (c) Coefficients of active κ, cos φ and reactive ξ power at one side incident. (d) Coefficients of active ζ, cos φ and reactive ψ power at both-sided symmetric-wave incident. (From Turowski, J., *Elektrodynamika Techniczna* (in Polish), WNT, Warsaw, Poland, 1993; *Tiekhniceskaja Elektrodinamika* (in Russian), Moscow, Russia, "Energia," 1974, pp. 177–191.)

$$P_1 = p_e \sqrt{\frac{\omega\mu_2}{2\gamma}} \frac{|H_{ms1}|^2}{2} \text{ and } Q_1 = r_e \sqrt{\frac{\omega\mu_2}{2\gamma}} \frac{|H_{ms1}|^2}{2} \tag{3.11a,b}$$

where

p_e represents κ, ζ (Figure 3.7c and d)

r_e represents ξ, ψ

$$\kappa = \frac{\text{sh}2kd + \sin 2kd}{\text{ch}2kd - \cos 2kd} \leq 1, \quad \zeta = \frac{\text{sh}\,kd - \sin kd}{\text{ch}\,kd + \cos kd} \leq 1 \tag{3.12}$$

Screening coefficient of electromagnetic Cu or Al screens of solid steel wall at AC field ([2], Equation 4.67) in its simplest form is

$$p_e \approx \frac{2\times 10^{-4}\,\mathrm{m}}{d} k_H \qquad (3.13)$$

where

d is screen thickness (usually d_{Cu} = 4–5 mm, d_{Al} = 7–8 mm)

$k_H \approx (H_{mse}/H_{ms})^2 \geq 1$ is coefficient of increase of magnetic field H_{ms} on metal surface after employment of electromagnetic screen; at three-phase load $k_{H3} \approx$ 1–3 ([2], Equation 4.68), but at the zero – sequence and/or third harmonic flux it could be even $k_{H0} \approx$ 15–20 ([2], Equation 4.69)

One has to remember, however, that electromagnetic screens, due to the skin effect, can generate field concentration and hotspots at their edge (Figure 3.10) ([2], p. 204).

3.4.8 Magnetic Screening (Shunting)

Nowadays, tank walls are mostly screened magnetically with the help of laminated iron shunts. However, if screen packages are situated at a wrong place, tank losses may even increase [12]. Correct packets of magnetic screen should be situated along the flux lines in tank wall (Figure 3.8—4).

But evaluation of these lines is not possible without rapid 3D solution. Therefore, producers, which do not use RNM-3D program, do not have a chance to design correct magnetic

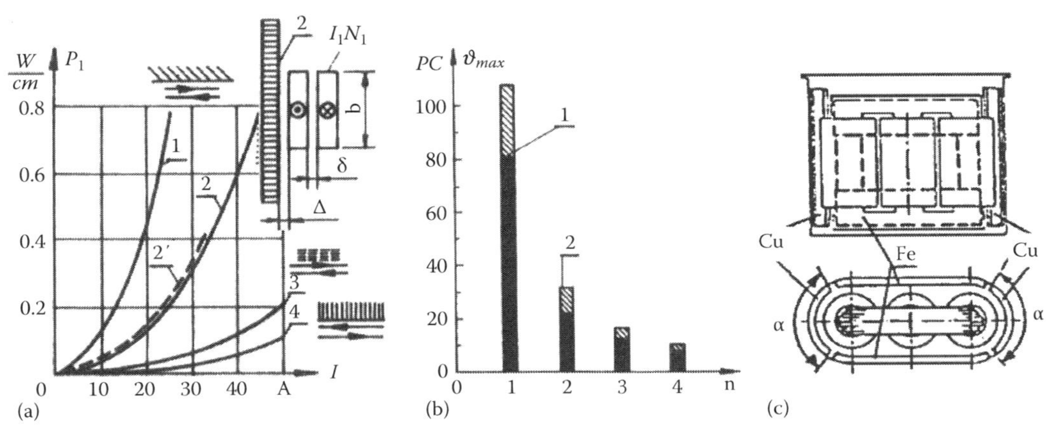

FIGURE 3.8
Effectiveness of laminated magnetic screens: (a) Strip configuration on the solid iron wall (From Turowski, J., *Elektrodynamika Techniczna* (in Polish), WNT, Warsaw, Poland, 1993; *Tiekhniceskaja Elektrodinamika* (in Russian) Moscow, Russia, "Energia," 1974, p. 405.). (b) Subdivision of external core packet into n narrower packets 3 as in (a). (From Kazmierski, M. et al., Hot spot identification and overheating hazard preventing when designing a large transformer. CIGRE 1984, Report 12-12, 1984.) (c) Combined Cu & Fe screens. (From Turowski, J., Reluctance networks—Chapter 4, pp. 145–178, in *Computational Magnetics*, Sykulski, J. (ed.), Chapman & Hall, London, U.K., 1995; Translation from Polish of the book Turowski, J. (ed.), Analiza i synteza pol elektromagnetyeznych, Polish Academy of Science, "Ossolineum," Wroclaw, Poland, 1990, p. 170.)

screening and often produce unconsciously large excessive load losses, like the ones mentioned earlier.

The magnetic, laminated iron screens (shunts) are less risky ([2], pp. 204–205) compared to electromagnetic screens. The simplest but sufficient method of calculation can be based on a comparison between the values of parallel reluctances (magnetic resistances) of both the layers ([2], p. 173). From this relationship, it follows that in comparison with dielectric and solid metal elements, screening coefficient, in first approach ([2], p. 198) will be

$$p_m = K\frac{18}{(4.2+n)^2+n^2}k_\Phi \tag{3.14}$$

where
- n is the number of 0.35 mm sheets in flat magnetic screen
- $K = 1.3$–1.5 security factor for saturation 3D field, and so on.
- k_Φ is the field excitation factor, for instance, at zero-sequence field $k_\Phi > 1$

To avoid possible saturation of screens, the number of flat sheets should be taken as bigger from the two ([2], Equation 4.65) following values:

$$n = \frac{a_c}{d_o(a_t+a_c)}\frac{\sqrt{2}IN\mu_o\delta'}{B_{1000}\,\mathrm{h}} \quad \text{or} \quad n = \frac{l_t}{l_{scr}}\sqrt{\frac{9K}{p_m}} - 4.4 - 2.1 \tag{3.15}$$

where
- a_t, a_c represent distances of gap between tank and core surface
- $B_{1000} = 1.4\,\mathrm{T}$ (isotropic iron)
- l_t, l_{scr} represent circumferential length of tank and screen
- $d_o = 0.35 \times 10^{-3}\,\mathrm{m}$ sheet thickness
- $K = 1.3$–1.5 security factor

The best but expensive are continuous screens of tank walls, which are made of upright iron strips 4 (Figure 3.8), along with flux lines. Often however, magnetic screens are partial with flat strips 3.

3.4.8.1 Surface Reluctances of Metal Walls

Skin effect and eddy current reaction are analogous to electric EMF self-inductance and impedance, which have been taken into account as the complex reluctances:

$$\underline{R}_s = R_{\mu 1} + jR_{\mu 1r} = (a_a + ja_r)\sqrt{\frac{\omega\gamma}{\mu_s}} \tag{3.16}$$

- *At nonmagnetic metals* or $\mu \approx$ *const* $a_a = a_r = 1$. Reluctance of Cu screen is

$$R_{Cu} = \frac{\alpha_2 \sin \mathrm{h}\alpha_2 d}{\mu(\cos \mathrm{h}\alpha_2 d - 1)} \cong \frac{2}{\mu_0 d} \tag{3.17}$$

At a first approach, in comparison with iron reluctance, one can adopt $R_{Cu} \to \infty$.

- *At magnetic metals*, with $\mu = \mu(H)$, $a_a \approx 0.37$, and $a_r \approx 0.61$ ([2], p.139).

Reluctance of solid iron, therefore, is

$$\underline{R}_{Fe} = R_{\mu 1} + jR_{\mu 1r} \approx (0.52 + j0.86)\sqrt{\frac{\omega\gamma}{2\mu_s}} \tag{3.18a}$$

- *At laminated* magnetic screens (shunts), in comparison with dielectric and solid metal elements, in a first approach one can opt for reluctance of laminated magnetic screens:

$$R_{Fe} \approx 0 \tag{3.18b}$$

This analysis resulted in the synthesis formula (3.6). It is a simplified formula, but there is no technical sense to endeavor here after achieving a level of accuracy that is higher than that of the natural dispersion of magnetic $\mu(H)$ and thermal $\gamma(T)$ steel parameters.

3.4.9 Excessive Heating Hazard

Overheating hazard of inactive parts for large transformers is one of the key factors of reliability for the unit and even for the whole power supplies system. From Equations 3.1 and 3.6, it follows that value H_{ms} on the metal surface is responsible for the loss density and, therefore, for a local heating.

From a thermal equilibrium equation $P_1 = \alpha'\Theta$, we can find ([2], p. 380) a permissible tangential field component (Figure 3.9) on the steel surface

$$H_{ms,perm} = 1962\left(\sqrt{1 + 3.29 \times 10^{-8} c} - 1\right) \tag{3.19}$$

where

$$c \approx 2\alpha_0' \frac{\Theta_{perm}^{1.25}}{\Theta_o^{0.25}} \sqrt{\frac{\gamma_o}{\omega\mu_o} \frac{192}{172 + t_{perm}}}$$

for permissible temperature t_{perm} (Figure 3.9), $\Theta = t - t_o$.

Above this value (3.19), excessive local heating hazard of solid steel elements (Figure 3.10), due to induced eddy currents, is possible.

In that case, additional loss reduction or cooling means are necessary.

Equation 3.19 was the basis of author's *electromagnetic overheating criteria* proposed at CIGRE'72 Plenary Session* and used until now for rapid hotspot localization (*Zhou Ke Ding* ISEF'97).

Theory of Section 3.4 resulted in the synthetic formula (3.6) and other constructional directions. These are simplified formulae, but there is no technical sense to endeavor here

* Kozlowski M., Turowski J.: Stray losses and local overheating hazard in transformers. CIGRE 1972, Report 12–10.

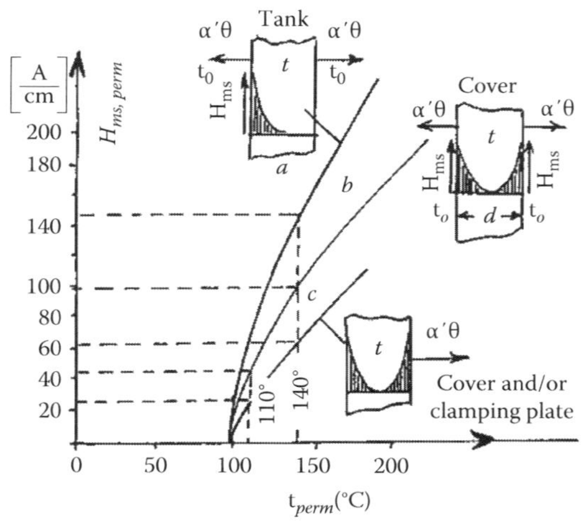

FIGURE 3.9
Permissible magnetic field strength $H_{ms,perm}$ on the solid steel surface for permitted temperature t_{perm}. (From Turowski, J., *Elektrodynamika Techniczna* (in Polish), WNT, Warsaw, Poland, 1993; *Tiekhniceskaja Elektrodinamika* (in Russian), Moscow, Russia, "Energia," 1974.)

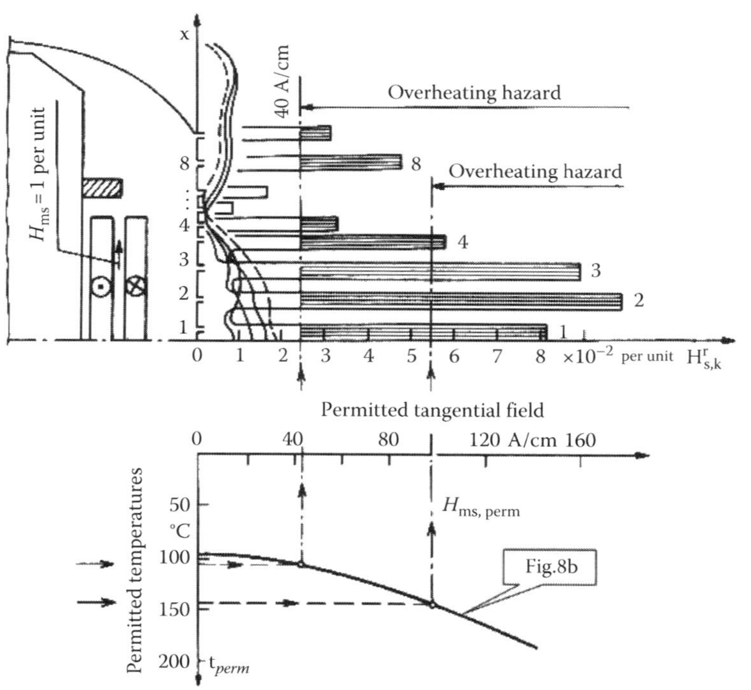

FIGURE 3.10
Localizations overheating hazard in flanged bolt joints. (From Turowski, J., *Proceedings of International Symposium on Electromagnetic Fields in Electrical Engineering–ISEF'85*, September 26–28, Warozawa, Poland, pp. 271–274, 1985.)

after achieving a level of accuracy that is higher than that of the natural dispersion of magnetic μ (Figure 3.5a) and thermal γ(*T*) parameters of constructional steel.

3.5 Losses and Overheating Hazard in Transformer Coverplates

3.5.1 Losses in Flat Cover

Problems with steel covers and bushings, including their heating, are also important to be considered from the reliability point of view (Figure 3.11).

They have a long history [1,2,22–26]. Joints *10* in Figure 3.1a are quite dangerous in the case of distribution [24] to largest [23] transformers. The problem appears especially at

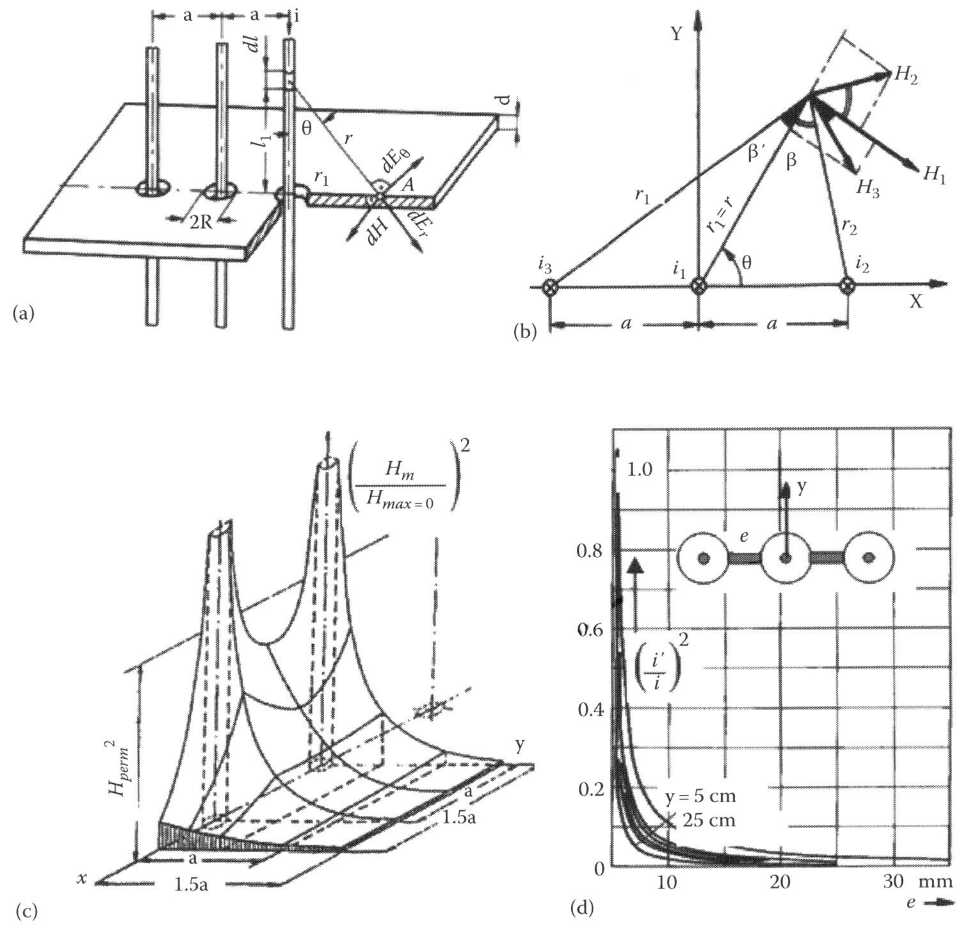

FIGURE 3.11
Analytical calculation of losses in cover (a) model, (b) Biot–Savart law application, (c) loss density distribution, (d) loss reduction with the help of nonmagnetic inserts of width *e*; $y = 5–25$ cm, distance from gap axis; *i*′, equivalent calculating current. (From Turowski, J., *Elektrodynamika Techniczna* (in Polish), WNT, Warsaw, Poland, 1993; *Tiekhniceskaja Elektrodinamika* (in Russian), Moscow, Russia, "Energia," 1974, p. 287.)

very heavy currents and bushing turrets [23]. After examination of several approaches, the best one appeared to be the analytical numerical method, ANM-3D, based on *Biot–Savart* law (Figure 3.11b) with approximate analytical [1] and numerical [22] integration of per-unit loss (3.6) ([2], p. 195):

$$P = \zeta \frac{a_p}{2} \sqrt{\frac{\omega\mu_o}{2\gamma}} \iint_A \sqrt{\mu_{rs}(x,y)} \left| H_{ms} \right|^{2x} dx \cdot dy \tag{3.20}$$

Finally, after consecutive simplifications and approximation, it was possible to obtain ([2], p. 331) a handy and technically satisfactory formula for losses in steel transformer covers in W:

$$P \approx m \cdot 5.5 \times 10^{-3} I \cdot a \left(1 + \frac{0.0056}{c} \frac{I}{a} \right) \tag{3.21}$$

where

$m = 1$ for single-phase and $m = \sqrt{3}$ for three-phase transformers, respectively
I represents bushing current in A
$c = R/a$

Formula (3.21) is valid for $I/a > 15$ A/cm in single-phase and $I/a > 28$ A/cm in three-phase transformers, respectively. In these limits, all practical structures are covered, in which the loss value is important. Figure 3.11d shows that power loss decreases rapidly with an increase of nonmagnetic insert "*e*" from width 0 to (3–5) mm and between 10 and 15 mm it becomes relatively stable because of flux expulsion.

In result ([2], p. 288), approximate formulae for losses in W in cover with nonmagnetic inserts "*e*" are

- For three-phase transformer in W:

$$P \approx 3.15 \times 10^{-2} I^2 \sqrt{\frac{\omega\mu}{\gamma}} \left(0.74 + 6 \ln \frac{a}{R} \right) \tag{3.22}$$

- For single-phase transformer in W:

$$P \approx 3.15 \times 10^{-2} I^2 \sqrt{\frac{\omega\mu}{\gamma}} \ln \frac{a}{R} \tag{3.23}$$

Formulae (3.22) and (3.23) were verified experimentally by the author, and other Polish, Macedonian, and Chinese specialists.

Another reduction of power loss and hotspots in steel coverplates, except for nonmagnetic inserts *e* (Figure 3.11d) and nonmagnetic steel, can be due to their thickness reduction and/or saturation (Figure 3.12).

In region B one can evaluate this reduction with the help of coefficient ζ (Figure 3.7d). However, it can be rather considered as a safety margin when using a basic design tool.

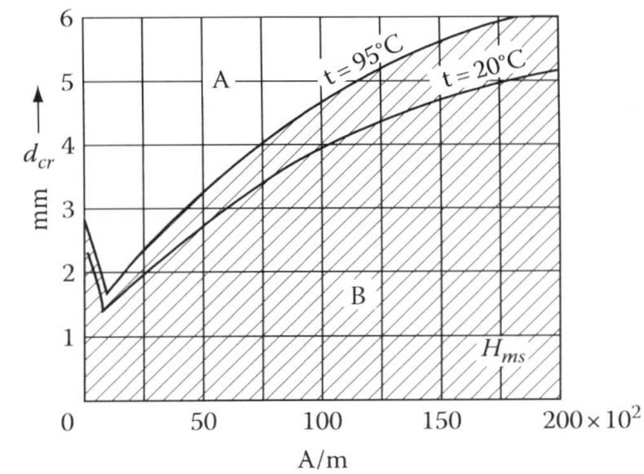

FIGURE 3.12
"Critical" thickness d_{cr} of wall, made of constructional steel, beneath which power losses depend on the plate thickness and magnetic field intensity H_{ms} on its surface due to saturation. (From Turowski, J., *Elektrodynamika Techniczna* (in Polish), WNT, Warsaw, Poland, 1993; *Tiekhniceskaja Elektrodinamika* (in Russian), Moscow, Russia, "Energia," 1974, p. 332.) A, region where losses do not depend on wall thickness; B, region where losses depend on wall thickness.

3.5.2 Method of Prediction and Elimination of Cover Overheating

Loss distribution, approximated analytically as $P(y) = Ae^{-By}$, and temperature rise, $\Theta = f(y, t)$ (Figure 3.13) on cover, have such a steep distribution that heat flow from hot to cold regions can significantly reduce the overheating hazard.

From a well-known ([2], pp. 387–389) conductivity equation as shown in Figure 3.13,

$$c\rho \frac{\partial \Theta(y,t)}{\partial t} = \lambda \frac{\partial^2 \Theta(y,t)}{\partial t^2} + q \quad (3.24)$$

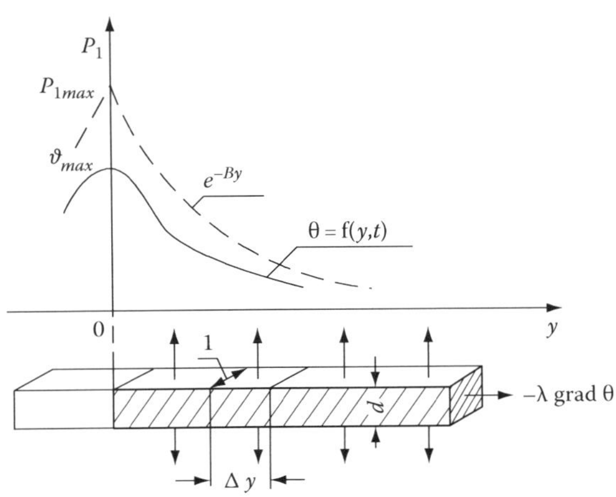

FIGURE 3.13
Loss P_1 (y,t) and temperature Θ (y,t) distribution on cover surface. (From Turowski, J., *Elektrodynamika Techniczna* (in Polish), WNT, Warsaw, Poland, 1993; *Tiekhniceskaja Elektrodinamika* (in Russian), Moscow, Russia, "Energia," 1974.)

we have maximum temperature rising over ambient temperature

$$\Theta_{max} = \frac{2A}{B^2\lambda d}\left(\frac{B}{\sqrt{2\alpha'/\lambda d}} - 1\right) = \frac{A}{\alpha'}K \tag{3.25}$$

and *unevenness heat distribution factor*

$$K \approx \frac{1}{cB\sqrt{d}+1} \leq 1 \tag{3.26}$$

on the cover surface, $K=0.5$–0.7.

Therefore, permitted maximum field (3.19) in Figure 3.9 can be increased to a value

$$H_{ms,permK} \approx \frac{H_{ms,perm}}{\sqrt{K}} \tag{3.27}$$

For instance, for locally permitted temperature of 140°C (Figure 3.9c), permitted field is $H_{ms,permK} \approx (60\ \text{A/cm}/\sqrt{0.6}) = 77\ \text{A/cm}$.

Therefore, K factor (3.26) can be considered as a safety factor in hotspot calculation.

From the same analysis ([2], p. 391), permitted bushing current can be given as

$$I_{perm} < 24.6\times10^2 a\left[\sqrt{1+3.9\left(1+\frac{2.4\sqrt{d}}{a}\right)} - 1\right] \tag{3.28}$$

3.5.3 Analytical Numerical Method of Computer Calculation of Power Losses and Heating in Transformer Covers

On the basis of theory and Equation 3.20, mentioned previously, a method for rapid modeling and calculation of cover loss with the help of Analytically Numerical Method ANM-3D was developed and published [22]. It makes it possible to carry out a fast, interactive analysis of field (Figures 3.14 and 3.15) and loss distribution as well as optimization of the structure, taking into account various parameters such as materials, geometry, excitation currents, iron saturation, skin effect, and local overheating hazard.

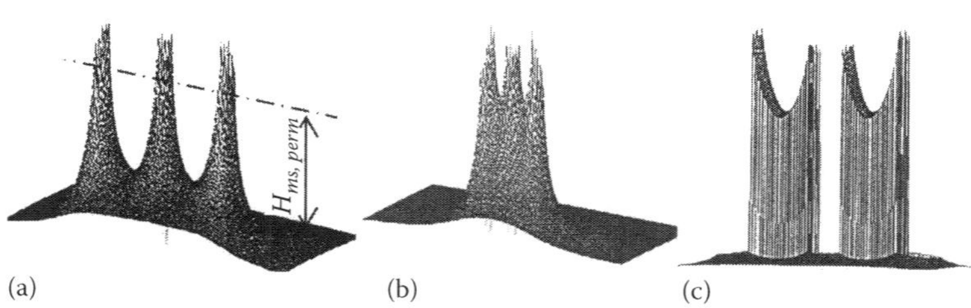

FIGURE 3.14
Distribution of the tangential field $H_{ms}(x, y)$ on the surface of flat cover made of constructional steel: (a, b) Influence of distance between bushing axes, (c) Magnetic field concentration in nonmagnetic insert $e=47$ mm as in Figure 3.11d [22] and field and loss reduction on iron surface; $H_{ms,perm}$ hot-spot localization as in Figure 3.10.

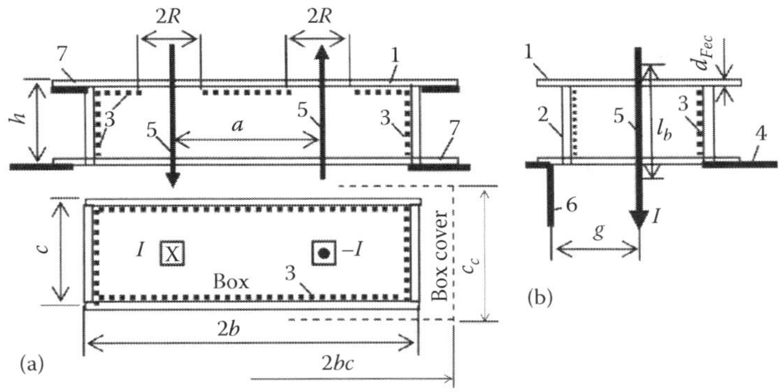

FIGURE 3.15
Model of single-phase high-current bushing turret: (a) main view, (b) end view.

All calculations from this chapter, both analytical approximate formulae and computer program, have been confirmed by many detailed measurements derived from using practical, industrial models. By using the field distribution (Figures 3.11 and 3.14), we can easily find the hotspot of excessive temperature on the cover and/or on turrets' surface.

3.5.4 Evaluation of Eddy Losses in the Cover of Distribution Transformers

Formulae 3.21 through 3.27 were developed mainly for medium and distribution transformers. In the work [24] of an international group, eddy losses in bushing mounting plates were calculated by four different methods: analytical formulation, three-dimensional Finite Element Method (FEM-3D), from measured values of initial temperature rise, and finally from measured values of steady-state temperature rise.

There is close agreement between the loss values obtained from all these four methods.

The analysis has resulted into an understanding of the loss pattern and temperature-rise phenomenon in the bushing mounting plates. The study has helped to standardize the type and material of the bushing mounting plate for various current ratings in transformers rated from 500 kVA up to 2 MVA. The scope of this work also included analysis of tank plates of small pad-mounted distribution transformers.

It is shown that judicious use of nonmagnetic stainless steel material can result into considerable energy savings for pad-mounted transformers, which are manufactured in large numbers.

FEM simulations have been carried out to find out cost-effective materials for tank plates of pad-mounted transformers. T-shaped stainless steel material is found to be effective in reducing the load loss appreciably. The results of simulations have been verified on a 225 kVA pad-mounted transformer.

The thermometric method ([2], pp. 406–411) of measured values α_0 of initial temperature rise

$$P_1 = c\rho_n \left(\frac{\partial \Theta}{\partial t} \right)_{t=0} = c\rho_m tg\alpha_0 \tag{3.29}$$

was used effectively. In (3.29) "c" is specific heat in W s/(kg K), ρ_m is mass density in kg/m^3, and $\alpha_0 = (\partial\Theta/\partial t)_{t=0}$ is initial angle of heating curve in given point of body. The loss and heat distribution analysis inside the metal body gives a chance to improve the accuracy of this method.

3.5.5 Rapid Evaluation of Excessive Local Heating Hazard in Complex High-Current Bushing Transformer Turrets

In the work [23], an analytical algorithm with numerical industrial example is presented for calculation of losses in turrets (Figure 3.15) for very large current bushings of one of the biggest single-phase transformers.

- All dimensions in Figure 3.15 are in meter.
- In computational model, $\gamma_{Fe} = 7 \times 10^6$ S/m, $\mu_r = 700 = const.$

The *following constructional and computational components are assumed*:

- Cover of bushing box of thickness $d_{FeC} > 15$ mm
- Steel walls of bushing box of thickness $d_{Fe} = 15$ mm
- Copper screens of thickness $d_{Cu} = 5$ mm
- Cover of tank of thickness $d_{Fet} > 15$ mm
- Copper loss in the bushing conductors
- Length of bushing bus participating in box heating l_b
- Losses in tank walls induced by transformer windings (Superposition)
- Eddy current loss induced in tank by leads
- Flange bolted joints

Each of these components has its own particular specifications. In spite of the seemingly simple structure they have, they do create complex problems, which should be resolved as rapidly as possible. Loss evaluation and its reduction are more important than achieving high accuracy. Theoretically, it is possible to resolve this entire problem by applying a fully sophisticated theory such as Maxwell's theory. However, practically it cannot be resolved within reasonable time and costs.

The solution should give information such as follows:

1. Influence of thickness d of the metal plates (Figure 3.7) and electromagnetic screens
2. Electromagnetic screening effectiveness (Figure 3.7)
3. Electromagnetic screening coefficients (Figure 3.7)
4. Estimation of power losses in the cover of bushing box (Figures 3.11 and 3.12)
5. Estimation of power losses and hotspots in walls (3.7)
6. Heating and hotspot in the bushing box (Figures 3.9 and 3.10)
7. Estimation of power losses in the cover of tank (Figure 3.7)
8. Estimation of power losses in tank wall from leads (Figure 3.13a and b)
9. Estimation of power losses in tank wall from leakage field of windings (Figures 3.6 and 3.19)

Thickness d in Figure 3.7 sometimes has no influence, and sometimes it plays a decisive role in power loss reduction or even in their increase.

In normal mild steel and at 50 Hz frequency, we have (3.4): $k_{Fe}=983\,m^{-1}$, $k_{Fe}d_{Fe}=18.7$, $\delta_{Fe}=1\,mm$, and $\lambda_{Fe}=6.28\,mm<d_{Fe}$; hence, no internal wave reflections.

In nonmagnetic steel ($\mu_{nr}\approx 1.5$) $k_{nFe}=17.20\,m^{-1}$. Then, for box wall layers, $d_{Fe,n}=12$–$19\,mm$, made of nonmagnetic steel, $k_{nFe}d_{Fe,n}=0.206$–0.327, $\delta_{nFe}=58\,mm$, $\lambda_{nFe}=365\,mm \gg d_{Fe,n}$. It means that there exists significant internal wave interference like in Figure 3.7, which should be examined.

In copper layer ($d_{Cu}=5\,mm$, $\gamma_{Cu}=58\times10^6\,S/m$, $\mu_r=1$) $k_{Cu}\approx 100\,m^{-1}$, $k_{Cu}d_{Cu}=0.5$, $\delta_{Cu}=10\,mm$, $\lambda_{Cu}=63\,mm>d_{Cu}=5\,mm$. Therefore, there also exists significant internal wave interference, which should be considered. At very strong fields, except for nonmagnetic steel, a Cu or Al screening is necessary.

Screening coefficients and their effectiveness depends mainly on wave impedances, $Z=E_m/H_m$, of both layers: copper and steel wall. For vacuum and air, it is $Z_0=\sqrt{\mu_0/\varepsilon_0}=377\,\Omega$, and for solid metal it is complex, $Z_{met}=\alpha/\gamma=(1+j)\sqrt{\omega\mu/2\gamma}$.

At 50 Hz, $|Z_{Cu}|=2.7\times10^{-6}\ \Omega$, $|Z_{St,\mu}=1000|\approx 2.4\times10^{-4}\ \Omega$ for nonmagnetic steel $|Z_{nSt}|=24\times10^{-6}\,\Omega$. Therefore, $Z_{diel}\ggg|Z_{St}|\gg|Z_{nSt}|\approx 8|Z_{Cu}|$.

Hence, electromagnetic wave reflection/refraction coefficients for "diel-steel-Cu layers" (Figure 3.7) are $M_{inc\text{-}refr}\approx\pm1$ as in (11).

For nonmagnetic layers (3.10) $M_{inc\text{-}rfr}=0.8\approx1$. Here returning internal reflected waves in Cu-screen are only a little weaker than at single metal wall (Figure 3.7c). For the losses and heating of metal are responsible for both components H_{ms} and E_{ms} on the metal surface as product (3.5a).

This wave interference does not change H_{ms} on Cu surface and increases a little, 1.1: $E_{ms}\approx1.1E_{ms1}$ (Figure 3.7).

At the boundary of layers $H_m(d_{Cu})\approx0$, therefore $S_z(d_{Cu})\approx0$. From (11) it follows that there is an inverse effect in the external steel layer. Here, $H_{ms}\approx1.1H_{ms1}$ and $E_m(d_{Cu})\approx0$ and $S_z(d_{Cu})\approx0$.

Consequently, power losses in each of the two layers can be approximately calculated like in the case of a single layer of thickness d_{Cu}, d_{nFe}, or d_{Fe}, respectively.

3.5.5.1 Electromagnetic Screening Coefficients

In structure considered, there are two typical screening layouts:

- *One-side* wave incident (Figure 3.7c) with power coefficient κ. In this case, power consumed by the metal wall ([2], p. 186, 187) is (3.11) with power coefficients $p_e=\kappa$ and $r_e=\xi$. At thick screens with $kd\geq1$, one can adopt $\kappa\approx1$. At thin screens ($kd \ll 1$), $\kappa\approx1/kd$. By the way, after $(kd)_{lim}\approx\sqrt{2}(2.4\times10^{-4}/377)\approx10^{-6}$, it drops to 0.
- *Both-sides* wave incident (Figure 3.7d). Now, power consumed by the metal wall ([2], p. 191) is (3.11) with power coefficients $p_e=\zeta$ and $r_e=\psi$. At $(kd)_{Fe}\geq1.5$ and $kd\geq 2$ $\zeta\approx1$. For instance, at the $k_{Fe}d_{Fe}=18.7$ ζ should be replaced by $a_p/2\approx0.7$. At $kd\ll1$, $\zeta\approx2(kd)^2/3$ and is approaching 0.

As one can see from Table 3.1 power coefficients play a crucial role in loss and hotspot calculation.

TABLE 3.1

Power Coefficients κ and ζ

	Metal Plate Layer of Thickness d	Relative Thickness kd (k/δ)	Both Sides Excitation $\zeta \approx 2(kd)^3/3$	One Side Excitation $\kappa \approx 1/kd$
1	Steel layer of box cover $d_{FeC} = 19\,mm$	$k_{nFe}d_{FeC} = 0.327$ (p. 14) $17.20\,\frac{1}{m}$ / 58 mm	$\frac{2 \cdot 0.3268^3}{3} = 0.0233$	$\kappa_{nFeC} = 1.98$[a] < 3
2	Steel layer of box walls $d_{Fe} = 15\,mm$	$k_{nFe}d_{Fe} = 0.258$ (2–3a) $17.20\,\frac{1}{m}$ /58 mm [27]	$\frac{2 \cdot 0.258^3}{3} = 0.0115$	$\kappa_{nFe} < 3.9$ Let's assume by analogy 2.6
3	Mild steel tank walls $d_{FeT} = 12.7\,mm$	$k_{Fe}d_{Fe} = 12.48$ $983\,\frac{1}{m}$ / 1 mm	$\frac{2 \cdot 0.218^3}{3} = 0.0069$	$\kappa_{nFeT} < 4.6$
4	Copper layers $d_{Cu} = 5\,mm$	$k_{Cu}d_{Cu} = 0.5$ (p. 14) $100\,\frac{1}{m}$ / 10 mm	$\frac{2 \cdot 0.5^3}{3} = 0.0833$	κ_{Cu}[b] < 2 Let us assume by analogy 1.6
5	Single mild steel layer $d_{Fe} = 19\,mm$	$k_{Fe}d_{Fe} = 18.68$ [27] $983\,\frac{1}{m}$ / 1 mm	≈ 1 $(kd)_{Fe} \geq 1.5$	≈ 1 $(kd)_{nFe\ or\ Cu} \geq 2$
6	Equivalent nFe & Cu layer $d_{eq} = 24\,mm$	$k_{eq}\,d_{eq} = 2.65$ (p. 15) $1110.6\,\frac{1}{m}$ / 9 mm	≈ 1 $(kd)_{eq} > 2.3$	≈ 1 $(kd)_{eq} > 1.5$

[a] More accurate calculation ([2], Equation 4.33) gives

$$\kappa_{nFeC} = \frac{1.92 - 0.64 \cdot 0.52 + 1.6 \cdot 0.608 - 1 + 0.64}{1.92 + 0.64 \cdot 0.52 - 2 \cdot 0.8 \cdot 0.794} = 1.98$$

[b] $$\kappa_{Cu} = \frac{e - 0.64 \cdot 0.368 + 1.6 \cdot 0.84 - 1 + 0.64}{e + 0.64 \cdot 0.368 - 2 \cdot 0.8 \cdot 0.540} \approx 1.9 \text{ (Figure 3.7c)}, \quad 2(kd)_{Cu} = 1 \text{ ([2], Equation 4.33)}$$

3.5.5.2 Power Losses in Box Cover 1

Box cover is made of nonmagnetic 19 mm steel plate, with Cu 5 mm screen 3 from internal surface and with no gap "e" between holes (Figure 3.11d). Owing to the strong concentration of main losses close to bushing, it is possible to adopt assumptions $2b_C \to \infty$ and $c_c \to \infty$ (Figure 3.15).

3.5.5.3 Equivalent Metal Layer

For rapid simulations let us apply for the loss computation one an equivalent cover layer of thickness $d_{eq} = 19 + 5 = 24\,mm$ instead of the double "nFe & Cu" cover. Conductance relation of both layers is

$(\gamma d)_{nFe}/(\gamma d)_{Cu} = (1.19)/(58.5) = 0.0655$. Then approximate equivalent conductivity of such single layer is

$$(\gamma d)_{eq} = (\gamma d)_{nFe} + (\gamma d)_{Cu} = 1.0655(\gamma d)_{Cu} = 1.0655 \times 58 \approx 62 \times 10^6\,S/m$$

Then $k_{eq} = 110.61^{1/m}$ and $k_{eq}d_{eq} = 2.65 > 2.3$.

Now box cover is made of single equivalent metal layer, excited symmetrically from both surfaces.

Since $(kd)_{eq} > 2.3$, from ([2], Equation 6.29b) power loss in the bushing box cover, the assumed bush current $I = 17$ kA is

$$P_{eqC} \approx 0.44 I^2 \sqrt{\frac{\omega\mu}{\gamma}} \ln \frac{a}{R} \approx 500 \text{ W}$$

3.5.5.4 Power Loss in Box Walls

Equation 3.6 shows that the main task of investigation is to find tangential component H_{ms1} on the surface of investigated element.

It is responsible for both the losses and heating. Owing to the symmetry of cross-section of the bushing box (Figure 3.15a) one can apply the equivalent mirror image of current I in the $x = 0$ axis with negative image $I = MI$ coefficient $M = -1$ ([2], p. 236).

Now we can apply to Figure 3.16b the approach taken in the case of a steel cavity ([2], p. 386).

Maximal (in time) value, $H_{t,ms}(x)$, of tangential component of magnetic field intensity and normal component, $B_{n,ms}(x)$, of flux density on the steel surface in Figure 3.16b have approximately a linear distribution. An important role is played by the power coefficients, depending on thickness d, geometry, and material parameters (Table 3.1).

$$B_{n,ms}(x) = B_{n,ms}\left(1 - \frac{x}{b}\right) \text{ within } 0 \le |x| \le b, \quad H_{t,ms}(x)$$

$$= H_{t,ms}\left(\frac{x}{b}\right) \text{ within } 0 \le |x| \le b, \text{ and } H_{t,ms}(x) = H_{t,ms} = const \text{ within } b \le |x| \le b + c$$

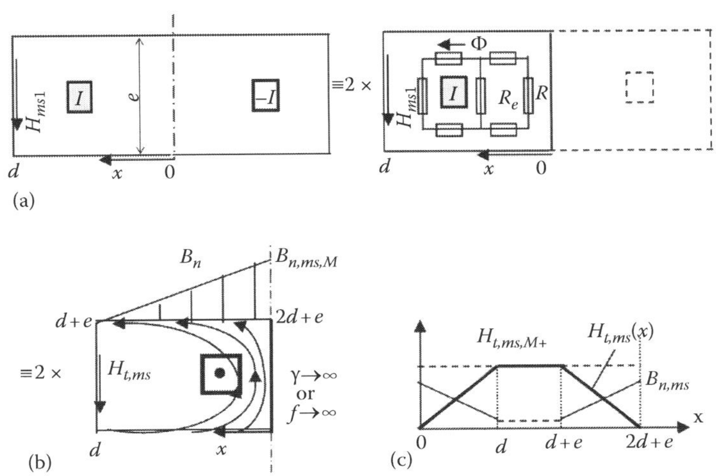

FIGURE 3.16
Simulation of bushing box (turret): (a) cross section with simplified RNM model to evaluate boundary value H_{ms1} on the internal surface of box, where $R_e = ¾R = 0.75R$; $\sqrt{2}\, I = \Phi(R + 2.75)RH_{ms1} \cdot c = \Phi R = \sqrt{2}\, I(R/(3.75R))$; (b) equivalent mirror image; and (c) field distribution on internal surface of box.

3.5.5.5 Mild-Steel Cavity Model

In analogy with the deep slot of induction motors ([2], p. 386), at no screen

$$H_{t,ms}\left(x=b+\frac{c}{2}\right)=H_{t,ms}$$

Then

$$H_{t,ms}\left(b+\frac{c}{2}\right)=\mu_0\frac{b}{2c}\sqrt{\frac{\omega\gamma}{\mu_s}}\left(\sqrt{2}I\right)=3.41\times10^5\ \text{A/m}=3410\ \text{A/cm}$$

This value is responsible for the loss density and, therefore, for local heating.

3.5.5.6 Cavity Model with Internal Cu Screen

In this case, the model of box will be similar. Only the H_{ms} on the cavity bottom is different. It can be evaluated in a simple and rapid way with the help of RNM (Reluctance Network Method) [18]. Full analysis can be made with the help of "MSR-1100" [28] package. However here, given the importance of time and economy factors, a simplest model was used to estimate an approximate boundary value H_{ms1} and other conditions of Figure 3.15.

Equivalent series reluctance for right loop (Figure 3.16a) $R_e=3/4=0.75R$, MMF is $\sqrt{2}I=\Phi(R+2.75)R$, $H_{ms1}c=\Phi R=\sqrt{2}I(R/3.75R)$, and field

$$H_{ms1}=\frac{\sqrt{2}I}{3.75\times c}=\frac{\sqrt{2}17000}{3.75\times0.3455}=18.6\times10^3\ \text{A/m}$$

In this case, we can calculate the wall losses as superposition of that in any layer separately from the formula (3.11a). Due to internal wave interference (Figure 3.7), surface value $H_{msFe}\approx1.1\cdot H_{msCu}$ is analogical to E_{ms}.

Power loss in the nonmagnetic wall layer of bushing box

$$P_{nFe}=P_{cu}\sqrt{\frac{\gamma_{Cu}\mu_{nFe}}{\gamma_{nFe}\mu_{Cu}}}\left(\frac{H_{msFe}}{H_{msCu}}\right)^2=11.29P_{Cu}=10,097\ \text{W} \tag{3.30}$$

where $\kappa_{Cu}\approx1.98$ and $\kappa_{nFe}=2.6$ (Table 3.1, line 2).

Maximum (in space) power loss density in Cu layer on smaller "c" walls of the bushing box

$$P_{c1Cu}=\kappa_{Cu}\sqrt{\frac{\omega\mu}{2\gamma}}\frac{|H_{ms}|^2}{2}=1.6\sqrt{\frac{2\pi50\times1\times0\times4\pi\times10^{-6}}{2\times58\times10^6}}\frac{|18.6\times10^3|^2}{2}=510\ \text{W/m}^2$$

where

$$\kappa_{Cu}=\frac{sh2kd+\sin2kd}{ch2kd-\cos2kd}=\frac{sh2.36-\sin2.36}{ch2.36+\cos2.36}=1.6$$

Now, we can integrate loss density in Cu layers. Then, power loss in Cu layers of bushing box walls:

$$P_{Cu} = 2h\left[2\int_0^b P_1\left(\frac{x}{b}\right)dx + P_{c1Cu}c\right] = 2h\left[2P_{c1Cu}\int_0^b\left(\frac{x}{b}\right)^2 dx + P_{c1Cu}c\right] = 2hP_{c1Cu}\left[\frac{2}{3}(b+c)\right]$$

$$= 20.634510\left(\frac{2}{3}(1.038+0.691)\right) = 894\,\text{W}$$

Total losses in Cu + nFe box:

$$P_C = 12.29P_{Cu} = 12.29\times 894 = 10987\text{ W walls} + P_{eqC} \approx 500\text{ W cover} \approx 11487\text{ W}$$

3.5.5.7 Estimation of Power Losses in Cover of Tank

From [2] (pp. 6–35) for mild-steel cover of tank outside bushing box (Figure 3.15) loss was calculated as shown in Figure 3.11, from formula (3.23) (Figure 3.17)

$$P_{T\,\text{cov}} \approx 3.5\times 10^{-2} I^2\sqrt{\frac{\omega\mu}{\gamma}}\ln\frac{a}{R} = 3.5\times 10^{-2}\times 17{,}000^2\sqrt{\frac{2\pi 50\times 700\times 0.4\pi\times 10^{-6}}{7\times 10^6}}\ln\frac{1372}{290} = 1836\,\text{W}$$

3.5.5.8 Estimation of Power Losses in Tank Wall from Leads*

This is another complicated screening, saturation, and configuration problem.

It could be resolved in the simplest way using the mirror image method ([2], pp. 236–252) of real I and fictitious image MI currents (Figure 3.18). However, image coefficient M (Figure 3.18) and equivalent quasi-permeability μ_{Qr} depend firmly on eddy current reaction, iron saturation, and shunt lamination.

It can be expressed approximately as

$$M_Q \approx \frac{\mu_{Qr}-1}{\mu_{Qr}+1}$$

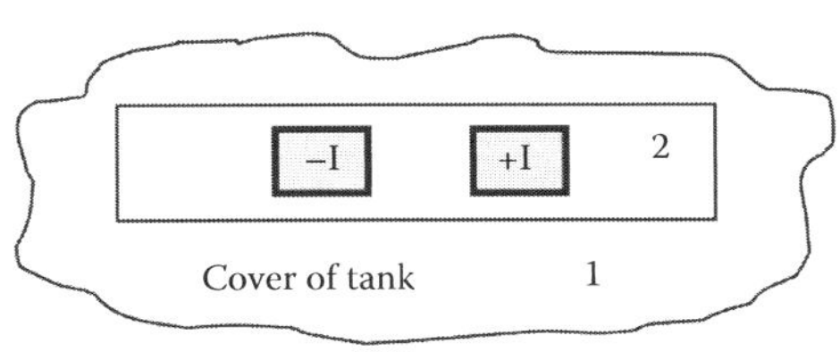

FIGURE 3.17
Cover 1 of tank outside bushing box 2.

* Author expresses his gratitude to Mr K. Kulasek from Power Transformers Division ABB Inc, Varennes, QC, Canada, for cooperation and acceptance to publish the shortened report at "XVI International Conference on Electrical Machines," Sept. 5–8, 2004, Cracow Poland [23]. Here, it was given only in a short, abstract from [23,27], along with the presentation of the author's methodology.

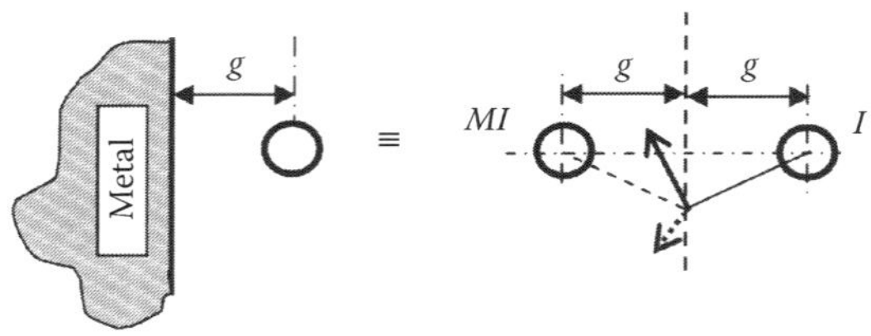

FIGURE 3.18
Estimation of field H_{ms} on metal surface with the method of mirror images. (From Turowski, J., *Elektrodynamika Techniczna* (in Polish), WNT, Warsaw, Poland, 1993; *Tiekhniceskaja Elektrodinamika* (in Russian), Moscow, Russia, "Energia," 1974, p. 236.)

and equals ([2], pp. 240–241):

- $M_Q = -1$ for superconductor or $f \to \infty$
- $M_Q = 0$ for nonmagnetic materials
- $M_Q = (\mu_r - 1/\mu_r + 1)$ for ferromagnetic with no eddy currents
- $M_Q = 1$ for ideal ferromagnetic $\mu_r \to \infty$
- $M_Q = 0.4$–0.6 for solid steel 50 Hz, $H_{ms,n}$ (for normal components)
- $M_Q = -0.9$ to -0.8 for solid Cu, 50 Hz, $H_{ms,t}$ (for tangential components)

Interaction of high current leads with metal wall, including the tank wall screening, is discussed in Ref. [2] (pp. 302–311). Analytical algorithms and computer programs, which consider magnetic nonlinearity, along with the author's analytical approximation (3.7b), are described in [2] (pp. 333–345).

Comprehensive programs were elaborated by Kazmierski [16] ([2], Fig. 6.34).

Less reliable, but very simple, experimental formulae were given by Deuring [29], which is as follows:

- For one lead $g = 17.4$ cm at 60 Hz: $P_{Lead60} = 111 \times I^{1.84} = 111 \times 17^{1.84} = 111 \times 183.66 = 20{,}387$ W
- For two leads $P_{2leads60} = 2 \times 20{,}387 = 40{,}387$ W

If the wall is screened with Al sheet 6.4 mm, for $g = 22.5$ cm and 60 Hz

- For single lead $P_{Lead60Al} = 6.7\ I^{1.96} = 6.7\ 17^{1.96} = 6.7 \times 258 = 1729$ W
- For two leads $P_{2leads60} = 2 \times 1729 = 3458$ W

Using the similar approximate approach, the losses were calculated [23] for the rest of the components:

- Losses in the nonmagnetic layer of box walls
- Power loss in the Cu layer of box walls
- Maximum (in space) power loss density on the smaller "c" walls of the bushing box
- Total power loss in the Cu layers of bushing box walls
- Total losses in box for both "Cu & *n*Fe" layers
- Total loss in the bushing box $P_{bpx} \approx 11487$ W

TABLE 3.2

Loss and Temperature Balance in Transformer Cover

Power Loss in Component Elements	Power Loss, W	Average Temperature Rise of	in K
Cover of bushing box	500	Box walls without oil cooling (at hot oil)	208
Nonmagnetic wall layers of bushing box	10,097	Bushing box walls considering oil cooling	29.7
Cu layers of bushing box walls	894	Bushing box walls and box cover	18.17
Cu and nonmagnetic steel box walls	10,987	Hot-Spot on bushing box	59.7
Total loss in bushing box	11,487		
In mild steel cover of tank outside bushing	1,836		
In copper bushing–bus–practically 0	0.52		

Source: Turowski, J., Excessive local heating hazard of bushing turrets in large power transformers. *XVI International Conference on Electrical Machines, ICEM'04*, September 5–8, 2004, Cracow, Poland, pp. 286/1–6.

First, electromagnetic wave interference (Figure 3.7) analysis in double layers, Cu–Fe wall of turret, the application of ordinary mild steel has given an intolerably huge loss of power: $P_{1Cu+nS,\max t} = \zeta\sqrt{\omega\mu\gamma/2}\,|H_{ms}|^2/2 \approx 300\,\text{kW}$ and way too high hotspot.

After the application of better materials, satisfactory (Table 3.2) results were obtained.

Average temperature rise of box walls without oil cooling (hot oil), calculated according to Ref. [2] (pp. 386–392 and 406–411), was $\Theta'_C = (P_C/\alpha_a h(4b+2c) = 208\,\text{K}$.

After considering oil cooling, this temperature dropped to an *acceptable value of 60°C.*

Average temperature rise of box walls and cover, $\Theta_{box} = 18.17\,\text{K}$, and average temperature, $t = 30 + 18.17 = 48°\text{C}$.

Hotspot temperature of bushing box, expected at the shorter "c" wall, appeared as $t_{HotSp} \approx 90°\text{C}$, *which is quite acceptable.*

Obtained results were approximately confirmed experimentally by industrial thermography in ABB—Canada Transformer Works. It is important that they were evaluated with the help of simplified but rapid methods [23].

3.6 Equivalent Reluctance Network Method Three-Dimensional "RNM-3D"

3.6.1 Existing RNM-3D Programs and Their Application

Equivalent Reluctance Method RNM (Table 3.3) is one of the simplest and fastest methods of modeling and computation. It is based on the simple, easily understandable *Ohm's* and *Kirchhoffs'* laws. RNM-3D is the 3D version (Figure 3.19). It is extremely competitive when compared with its popular counterpart, FEM-3D (Figure 3.6). The program "RNM-2Dexe" is intended mainly for the rapid design of symmetric three-leg structures (Figure 3.19). However, it can be also successfully applied to five-leg and asymmetric transformers. At asymmetry, it is only necessary to repeat calculation four times for

TABLE 3.3

Experimental Validation and Successful Industrial Implementation of the RNM-3D Was Done by the Following TW (Transformer Works) and TU (Technical Universities)

1	TW in Zychlin, Poland. Power loss and overheating of the cover. JT PhD thesis (1957)
2	TW MTZ in Moscow, Russia (1960). Hot-spot in remote transformers 160 MVA in Siberia
3	TW SKODA-Plzen. Czech Rp. (1969). *J. Kopecek, J. Klesa. Industrial Verification*
4	TW ELTA, Lodz, Poland (1987–1988)
5	TW SHENYANG, China (1991–1992)
6	TW GEC-ALSTOM OF INDIA, Allahabad, India (1991–1992)
7	TU of Sydney, Department of Electrical Engineering., Australia (1992)
8	TW BAODING, China (1993), and other universities and works
9	TW CROMPTON GREAVES, Bombay, India (1993)
10	TW of Power Board ENERGOSERWIS, Lubliniec, Poland (1993)
11	TW NGEF-ABB, Bangalore, India (1994)
12	TW CROMPTON GREAVES, Bombay, India (1994). Source code
14	TW WILSON TRANSFORMER, Australia, (1994). Testing Calculations
15	TW NORTH AMERICAN TRANSFORMERS, Milpitas, California (1995)
16	TW IRAN-TRANSFO (SIEMENS), Teheran (1995)
17	TW HACKBRIDGE-HEWITTIC AND EASUN, Madras, India (1995)
18	TU Monash. Department of Electrical Engineering, Melbourne, Australia (1995)
19	TW TRANSFORMERS AND ELECTRICALS KERALA (HITACHI), Kerala, India (1996)
20	TW BHARAT HEAVY ELECTRICALS, Bhopal, India (1996)
21	TW CROMPTON GREAVES, Bombay Program "Bolt heatings" (1996)
22	TW CROMPTON GREAVES, Bombay Program "Clampings" (1998)
23	TW CROMPTON GREAVES, Bombay Program "Turret"
24	TW CROMPTON GREAVES, Bombay Program RNM-3 Dasm (1999)
25	TW EMCO TRANSFORMERS LIMITED. Bombay, India (1997)
26	Power Institute, Warszawa-Mory, Dr. J. Kulikowski
27	TU of Palermo, Italy, Department of Electrical Engineering, Italy (1998)
28	TU of Pavia, Italy, Department of Electrical Engineering, Italy (1998)
29	TW of Power Board ZREW SA, Janów, Poland (1998)
30	TW Pouwels Trafo Belgium, Mechelen (1999), RNM-3Dexe
31	TW Ingenieria y Desarrolo Transformadores de Potencia, Tlalnepantla, Mexico Testing (2000)
32	TW Delta Star Inc. San Carlos, California. Testing (2000)
33	TW ABB-ELTA Lodz (2000)
34	TW Pouwels Trafo Belgium, Mechelen (2000). RNM-3Dasm, Source code
35	TW GE-PROLEC, Power Division, Monterrey, Mexico, Testing (2000)
36	TW Delta Star Inc., San Carlos, California. Package RNM-3Dexe (2001)
37	TW ABB Lodz. Counting more than 30 power transformer units up to 330 MVA (I, Kraj 2001)
38	TW ABB Lodz, Automatisation of stray loss modeling and calculation (M. Swiatkowski-202)
39	TW ABB Inc., Varennes, Quebec, Canada, J3X 1S4 overheating hazard in high current bushing turrets in 440 MVA 1 phase trafo (July 2003)
40	TU of Vigo, Spain
41	TW EFACEC Porto, Portugal (2006)
42	TU of Lodz, Poland—for PhD and MSc thesis. Since 1950 till now
43	TW EFACEC Savannah, Georgia (2009)
44	TW WEG Equipamentos Elétricos, SA, Brazil (2010), RNM-3Dexe

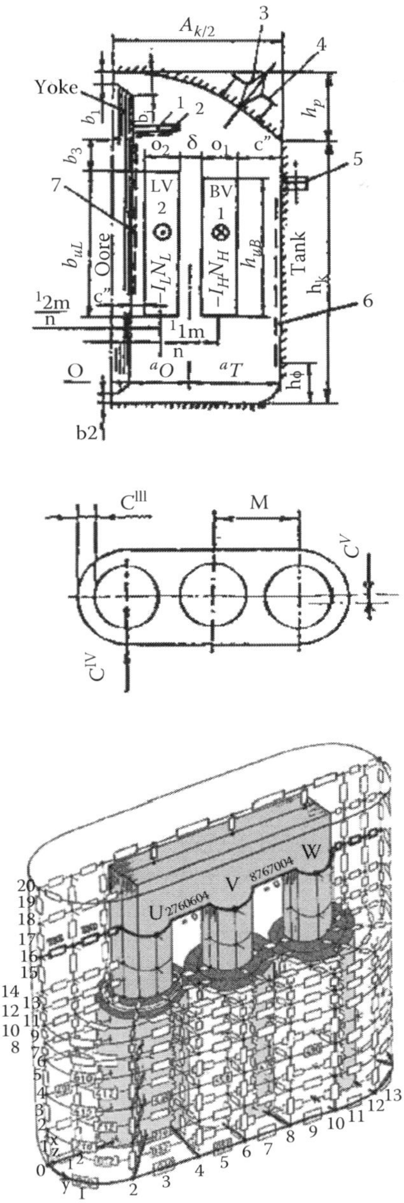

FIGURE 3.19
RNM-3Dexe model and program. (From Turowski, J. et al., *IEEE Trans. Magn.*, 26(5), 2911, 1990.)

each quarter (Figure 3.20a). This method is described in the paper "Fast 3-Dimensional Interactive Computation of Stray Field and Losses in Asymmetric Transformers" in IEE Proceedings [9].

However, for transformers with more complicated structures (Figure 3.3a), rectangular tanks 4, with oil pockets 7, tap changers 8, and so on, a special program "RNM-3Dasm,exe" was developed [20,30,31].

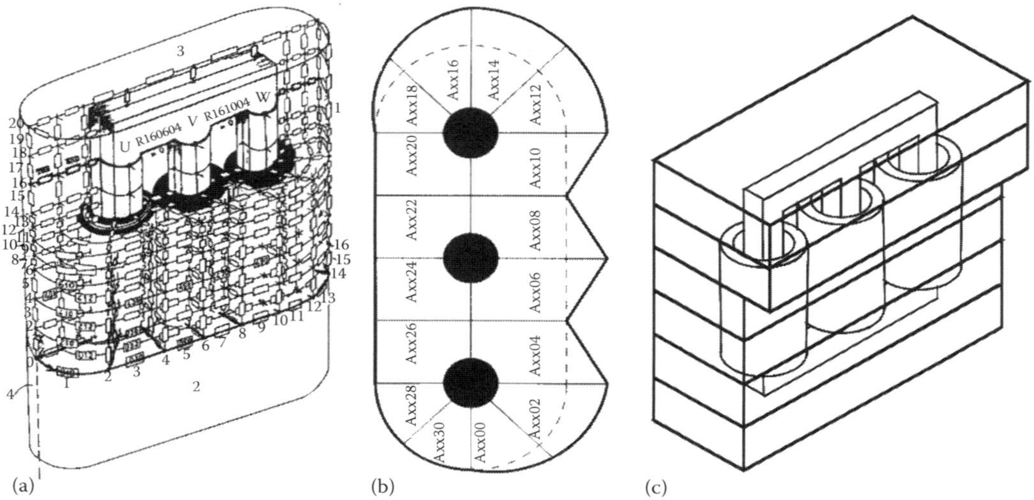

FIGURE 3.20
Rapid design of asymmetric transformer on the basis of simple symmetric "RNM-3Dexe" [16] reluctance model (Figure 3.19): (a) Symmetrization of quarterly asymmetric model. (From Koppikar, D.A. et al., *IEE Proc. Generat. Transm. Distrib.*, 147(4), 197, 2000.) (b) Deformation of symmetric (dashed line) into the asymmetric model (continuous line). (From Turowski, J. and Zwolinski, G., Hybrid model for fast computation of 3-D leakage field and eddy current losses in highly asymmetric structures, *Proceedings of Fifth Polish, Japanese Joint Seminar on Electromagnetics in Science and Technology, Gdansk'97*, May 19–21, Poland, pp. 165–168, 1997.) (c) Other deformation by sliced model (New program RNM-3Dasm,exe). (From Zwolinski, G. et al., Method of fast computation of 3-D stray field and losses in highly asymmetric transformers, *Proceedings of Fifth International Conference on Transformers, Trafotech'98*, January 23–24, Mumbai, India, pp. I-51–I-58, 1998; Zwolinski, G. and Turowski, J., Stray losses in multiwinding transformers on load. *International Conference on Power Transformers, Transformer'03*, May 19–20, pp. 63–67, Pieczyska, Poland, 2003.)

It is an extension of the basic RNM-3Dexe package (Figures 3.20b,c and 3.21a).

If one wishes to model and calculate other electromagnetic structures (reactors, shell-type transformers, isolation systems, currents, heat, and oil or air flows), he should prepare new pre- and post-processor (Figures 3.2b and 3.27) phases with corresponding parameters (reluctances). In this case, it is necessary to use one of the source code programs, "RNM-3Dsc" (symmetric) or "RNM-3Dasm,sc" (asymmetric).

In all programs of RNM-3D class, the same "MSR-1100" solver of multinode network is applied. This rapid solver, based on "Nodal Potentials Method," "Sparse Matrix," and idea of Chua and Pen-Min Lin [33] for electronic systems, was elaborated by Turowski and Kopec [28].

They were developed then in cooperation with Zwolinski [30] for the highly asymmetric transformer (Figures 3.19b,c and 3.21a). Symmetric packages were checked industrially in extended cooperations with *Crompton Greaves* (Mumbai) [9], ABB (Figure 3.23) [12], and many other transformer manufacturers (Table 3.3).

QUESTIONARY

For a rapid 3D design process with the RNM-3D software.

Electrical data: Rated power S_N ……… MVA

Frequency f ……… Hz

Rated phase currents HV/LV …….. / …….. A

Number of series turns per phase …… / …….

Rated phase voltage HV/LV …….. / ……. kV

Dimensions in mm:

h_{uHV} = ………. h_{uLV} = ………………..
a_1 = ……… a_2 = ……….. δ = …………
a_C = ………. a_T = ……… A_k = ………..
h_k = ………… h_p = ………. h_d = ……….
b_1 = ………. b_2 = ………… b'_1 = …………
l_{1m} = …….. l_{2m} = ………. b_3 = ……..
c' = …….. c'' = ……. c''' = ……. D = ……
c^{IV} = ………. M = ………. c^{V} = ………….

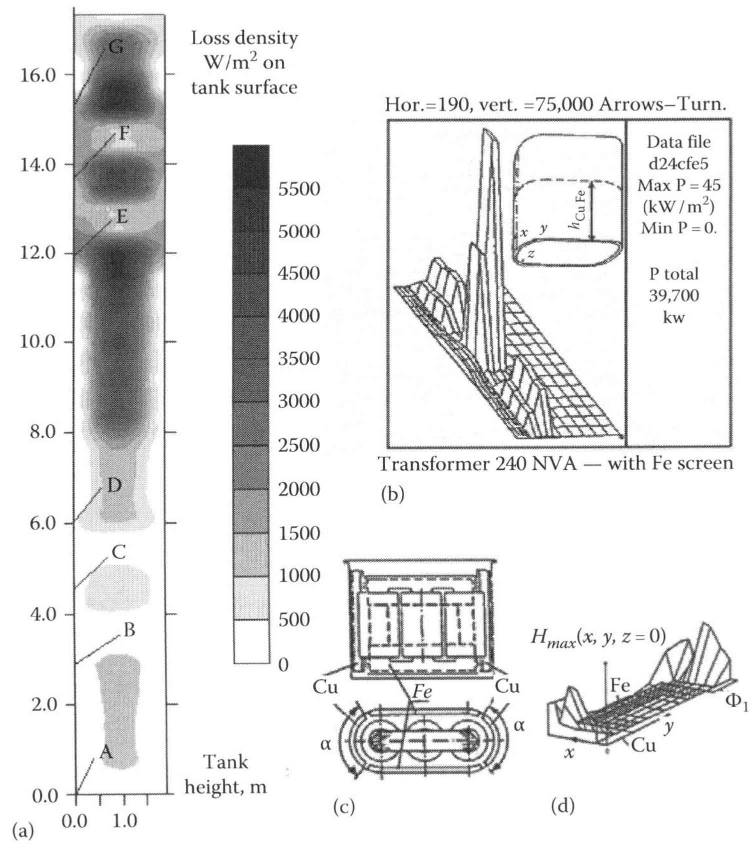

FIGURE 3.21
Examples of interactive modeling and computation of stray field and losses as well as screening optimization: (a) model of autotransformer 100 MVA, 220/132/11 kV with no screens; total stray loss 7.77 kW. (From Zwolinski, G., Stray field and losses in asymmetric transformer structures (in Polish), PhD thesis, Supervisor Turowski, J., Technical University of Lodz, Lodz, Poland, 1999.) (b) typical printout from calculation of density and total losses in five-limb transformer with partial magnetic screen; (c) design of combined Cu–Fe screening. (From Turowski, J. et al., Fast 3-D analysis of combined Cu & Fe screens in three-phase transformers, *Electromagnetic Fields in Electrical Engineering, COMPEL*, James & James, London, U.K., pp. 113–116, 1990.) (d) design and analysis of influence of frame magnetic screens (shunts) on internal surface of tank, tangential field H_{ms} distribution, responsible for loss and excessive heating hazard (Figures 3.9 and 3.11).

Other proposed elements (reluctances) material:

Copper screened, NoNo ....
Aluminum screened, NoNo ....
Screened with laminated iron (shunts),
NoNo ....
Made of solid steel, NoNo ....
Made of nonmagnetic solid steel, NoNo....
Other features, NoNo ....
Insulation or air gap, NoNo ....
Other components 1, 2, 3, 4, 5, 6, 7 in Figure 1 to calculate.,
Date .... Signature ....

For a rapid design process, it is necessary to clearly define the parameters of electric, magnetic, and geometric materials.

Figure 3.19 represents the collection of data for the PC-driven RNM-3Dexe model and program for the rapid three-dimensional (3D) simulation and calculation of stray field, forces in HV and LV windings, eddy current losses, and hotspots in tank walls and other solid metal parts, electromagnetic or magnetic (shunts) screening and optimization in three phase, and 3- (Figure 3.19) or 5- (Figure 3.21c) –leg core for quarterly symmetric structure of power transformers, with oval tanks [15,16,18]. One of the key features in this system is that one design variant (Figure 3.20a) can be calculated within less than 1 s in case of "RNM-3Dexe" and about 6 s in the case when asymmetric (Figures 3.20a and b and 3.21a) source code "RNM-3Dasm,sc" is employed.

Results of computation are shown in the form of loss density distribution and as total losses (Figure 3.21a,b and d), including that with asymmetric tank (Figure 3.21a). All principles described above were utilized to prepare effective industrial packages with the help of equivalent *Reluctance Network Method* and set of RNM-3D packages especially designed for fast analysis of complex 3D, three-phase, nonlinear electromagnetic structures.

Figure 3.22 is a good example where stray losses of more than 30 main transformers of ABB-Elta Transformer Works in Lodz (Poland) were calculated with "RNM-3D.exe" package in short time by a postgraduate student in his diploma work, under supervision of the author of this chapter.

At the same time, it was shown what errors were made by manufacturers and utilities in past because of lack of a rapid 3D computer program. As a result, the analysis system of screening was effectively improved.

3.6.2 Influence of Flux Collectors

For a long time, the role of the so-called top and bottom flux collectors and their influence on stray losses and electrodynamic forces in transformers have been the subject matter of conflicting opinions (Figure 3.3b—3). In some works, it has been suggested [35] that such collectors in the form of laminated shunts can straighten and redirect the stray magnetic lines toward the collectors as the three-phase magnetic node.

Then they vanish according to the rule $\underline{\Phi}_U+\underline{\Phi}_V+\underline{\Phi}_W=0$. Although theoretically correct, this effect has been sometimes negated by other specialists.

The answer is in the different features of transformers, which depend on the insulation clearances, critical distances of tank walls [2,36], screening systems, and so on. In order to give a full, quantitative answer on these conflicting opinions a detailed 3D analysis

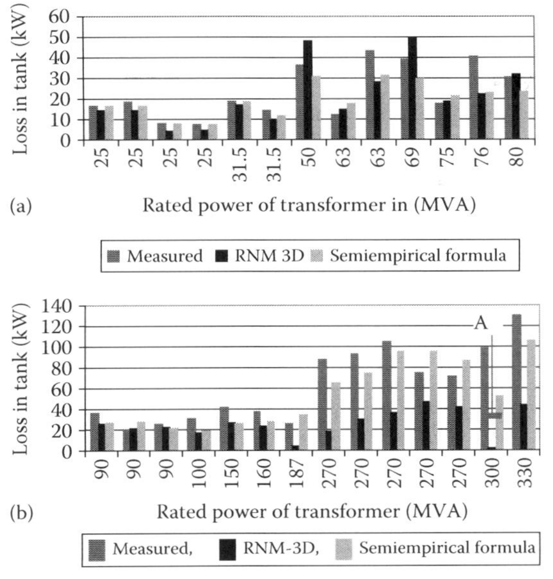

FIGURE 3.22
Stray losses in ABB-Elta transformers (a) with no screens and (b) with magnetic screens (shunts). (From Turowski, J. et al., Industrial verification of rapid design methods in power transformers. *International Conference Transformer'01*, October 5–6, Bydgoszcz, Poland, pp. 1–7, 2001.) RNM-3D-losses calculated by program *RNM-3Dexe* for quarterly symmetric transformers with oval tanks, completely screened magnetically; *A* corrected by program "RNM-3Dasm.exe."

[37] was carried out with the help of rapid RNM-3Dexe package. In RNM-3D models, a whole sophisticated theory, a complicated geometry, and many physical phenomena (magnetic nonlinearity, solid iron electromagnetic processes, skin effects, eddy current reaction, electromagnetic and magnetic screening, laminated iron shunt effects, etc.) are hidden within the analytical representation of reluctances and special composition of the program source code.

Thanks to the RNM-3D, the programs are very simple and work faster in carrying out 3D analyses. The answer was presented by the graph (Figure 3.23), and it confirmed that there is no room for any controversy.

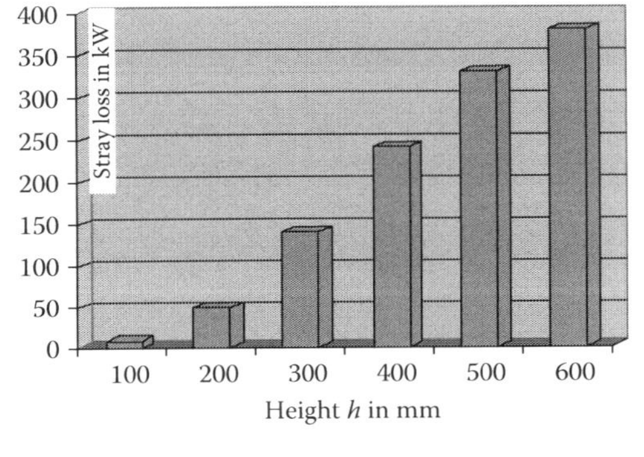

FIGURE 3.23
Influence of flux collector distance h. (From Rizzo, M. et al., *IEEE Trans. Magn.*, 36(4), 1915, 2000.)

The effect of flux collectors is then clear, and it depends on their clearance h from the winding edges.

For instance, for transformer 240 MVA, at collectors clearances $h=0$, 116, 232, and 464 mm, power losses in tank wall were $\Delta P=0$, 6.9, 78, and 307 kW correspondingly. Zero clearance of course never occurs due to insulation requirements, but it is helpful for interpolation. The ΔP values depend additionally on whole 3D configuration of screens and can be different. But generally the dependence $\Delta P=f(h)$ is almost linear.

3.6.3 Electrodynamic Crushing Forces

Photos of windings, which were destroyed by axial short circuit forces, presented during the International Conference "Transformer'01" (Bydgoszcz 2001), ARWtr'04 (Vigo 2004) and others, have shown that destructions occurred not inside but outside the core windows, that is, in HV windings.

It is well known that axial forces are products of current density $\boldsymbol{J}$ by radial component B_r of flux density in the winding. Inside the core window, due to surrounding iron, B_r is smaller than that in 3D outside field. Magnetic laminated shunting of tank walls can even increase this phenomenon a little, whereas electromagnetic Cu screens can play some positive role (see Section 3.8).

3.6.3.1 Role of 3D Calculation

Again, companies, which do not have rapid 3D program, still calculate forces with wrong, inaccurate 2D programs with intuitive correction coefficients. Calculation of forces should be, however, carried out for the outside part of windings (Figure 3.25) as 3D and not like those inside windows.

3.6.3.2 Role of Tank Screens

It is also necessary to consider the decisive influence (on B_r) of tank screens, tank and stray flux collectors [37]. Value of B_r strongly depends on the distance, kind, and arrangement of screens. "Critical distance" a_T (Figure 3.24) of tank wall [2,36] plays here an important role.

- *In magnetic (Fe) screens* (shunts), the smaller the distance a_T, the stronger the forces in the external windings.
- *In electromagnetic (Al or Cu) screens,* on the contrary, the smaller the distance a_T, the stronger the forces are in internal windings.

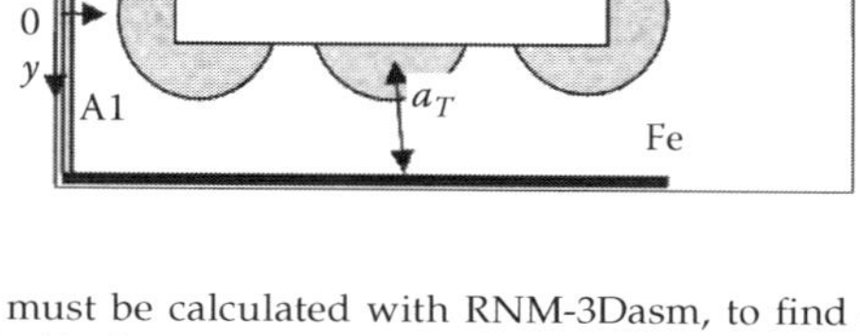

FIGURE 3.24
Winding area (shaded) which must be calculated with RNM-3Dasm, to find distribution of axial forces. Fe, laminated iron screens (shunts); Al, electromagnetic screen.

3.6.3.3 Role of Flux Collectors and Tank Critical Distance

The value and distribution of forces in windings depends also on stray flux collectors and so on. The effect of flux collectors consequently depends strongly on 3D geometry, tank asymmetry [20,30,31], and so on. All these necessitate the use of the 3Dpackage RNM-3Dasm [30] designed for asymmetric transformers and rapid interactive simulation.

For example, at a reduction of distance a_T from about 400 mm to approximately 200 mm causes increases of maximum force in HV and LV winding (at $y=0$, outside window) by 30%. Other example: At $a_T=200$ mm (Italian transformer 240 MVA), introduction of magnetic screen (shunt) increases the maximum force in HV winding (at $x=0$, outside window) by 13%. In LV winding, this force remains practically the same. Instead, the introduction of electromagnetic Cu screen decreases the maximum force in HV winding by 8%, whereas in LV winding this force remains practically the same.

3.6.3.4 General Method of Dynamic Force Calculation

Going out from general *Hamilton's Principle* and *Euler-Lagrange's* partial differential equations for coupled electromechanical systems (Figure 3.4), we have general expression [8] for electromagnetic force (Figure 3.25), where in (3.31) M_{jk} (x_k) are inductances between the particular parts of winding. These inductances are to be calculated with the help of Maxwell's theory [3]. x_k are coordinates in space and time (Figure 3.25).

$$F_x(\varphi) = f_e(x_k) = \frac{1}{2}\sum_j \sum_k i_j \cdot i_k \frac{\partial}{\partial x_k} M_{jk}(x_k) \tag{3.31}$$

where $x_k=x, y, z, t$ or r, θ, φ, t.

Then, the resultant axial force distribution around one leg is

$$F_x(\varphi) = \int_L f_e(x, \varphi)\,dx \tag{3.32}$$

Our study [38] gives evidence that the destructive forces occurring in the cylindrical windings of high-power transformers can be investigated at reasonable cost and time, with the help of RNM-3Dexe program.

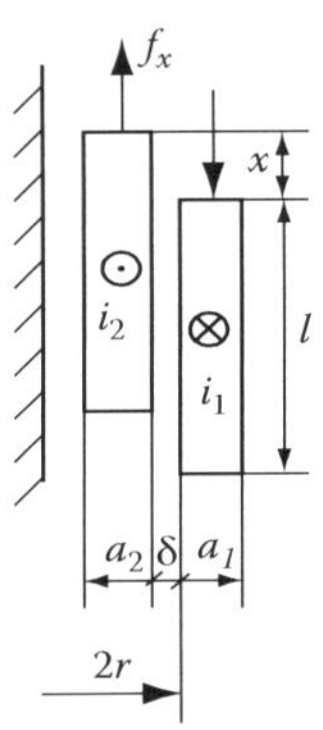

FIGURE 3.25
Axial force $F_x(\varphi)=f_e(x_k)$ between windings.

3.7 Experimental and Industrial Verification of the RNM-3D

It is of great importance and satisfaction to say that RNM-3D programs have been widely used (Table 3.3) and verified industrially in many countries, companies, and research institutions. Author received several enthusiastic opinions about the effectiveness of RNM-3D package from his collaborators from Polish and foreign (India, China, United States, United Kingdom, and others) transformer designers and manufacturers as well as from universities (Italy, Australia, Spain, etc.). They are especially important because they are a direct practical confirmation of this complicated theory.

Here are some users and experts opinions, for which author is very appreciative for these encouraging confirmations.

> ... Dear Prof. J. Turowski, ... We are using your RNM-3D package extensively and for most of the jobs stray loss figure is closely matching with the tested value. The package is very much helping to understand the dependence of stray losses on various parameters of transformer design and shunt/screen arrangement.
>
> **(Fax of October 2, 1993 from Mr G. S. Gulwadi, Chief—Design Engineer Transformer Division, Crompton Greaves LTD, Bombay)**

> ... Dear Prof. J. Turowski, ... we received the modified version of computer programme, which gave us good estimation of stray losses and I simulated provision of shunts on tank, clamp and got good results ...
>
> **(Letter of March 23, 1995 from Mr R. K. Ahuja, North American Transformer, Inc. Reliance Electric, Milpitas, CA)**

> ... Dear Prof. J. Turowski, ... The method that you use for studying the leakage field appeals to me greatly and you have now fashioned it into a reliable engineering tool. I am convinced that this method will be widely used in industry and will become the preferred method of calculation"
>
> **(Letter of 4.9.1986 from Professor P. Hammond, Head of Department of Electrical Engineering, The University of Southampton, UK)**

> ... Monsieur le Professor Turowski ... Cher Monsieur, Votre notice sur le logiciel RNM – 3D m'est bien parvenue. Toutes mes félicitations pour le beau succe du travail que vous lui avez consacré.
>
> **(Letter of 30.5.1994 from Professor M. Poloujadoff, Laboratoire D'Electrotechnique. Universities de Paris VI et XI)**

> ... Dear Prof. J. Turowski, ... I had used RNM3D very extensively and found it very useful & reliable ...
>
> **(A. K. Mathur, Design & Development Engineer. Wilson Transformers. Melbourne, Australia. From E-mail to Professor J. Turowski, dated February 8, 1999)**

> ... Dear Prof. J. Turowski, ... My name is Maria Evelina Mognaschi, I worked with Prof. Savini and Prof. Di Barba at the paper about screen design. I would like to say you that your program is very useful and complete and I liked very much work with it. I am very grateful for the program that you gave us and the opportunity to work with it"
>
> **(From: "eve.79" <eve.79@libero.it>. To: turowski@p.lodz.pl, Sent: June 30, 2003)**

> Respected Prof. Turowski, Reluctance Network Method can fulfill the requirements of very fast estimation and control of the tank stray loss
>
> **(Professor S. V. Kulkarni, Electrical Engineering Department, Indian Institute of Technology – Bombay. Powai, Mumbai, 400 076, India. [svk@ee.iitb.ac.in] February 29, 2004). Regards S. V. Kulkarni**

Design practice has shown that it is more convenient to use strictly dedicated programs for particular class of jobs rather than using very general packages.

Presently, there exist the following softwares of the RNM-3D class:

1. *RNM-3Dexe.* Reluctance Network Method Three-Dimensional. Program for fast, interactive 3D modeling and design of leakage zone and stray loss computation in large power transformers.
2. *MSR-1100.* Universal Solver for rapid, interactive solution of any kind of extended multinode models, including electromagnetic, thermal, mechanical networks, for tens of thousands of nodes.
3. *RNM-3Dexe.* Reluctance Network Method Three-Dimensional. Program for fast, interactive 3D modeling and design of leakage zone and stray loss computation in large power transformers.
4. *MSR-1100.* Universal Solver for rapid, interactive solution of any kind of extended multinode models, including electromagnetic, thermal, mechanical networks, for tens of thousands of nodes.
5. *RNM-3Dsc.* Reluctance Network Method Three-Dimensional Source code. This is the source code for RNM-3Dexe package from Item 1.
6. *RNM-3Dasm,exe.* Reluctance Network Method Three-dimensional for analysis and computation of leakage field and stray losses in power transformers with highly asymmetrical structure.
7. *RNM-3Dasm,sc.* This is the source code for RNM-3Dasm.exe package from Item 4.
8. *Turret.* Program for the fast analysis of cover and turrets losses in power transformers.
9. *Bolted Joints 3D.* Program for localization and 3D calculation of excessive heating hazard of flange bolted joints in transformers versus a position of joint.
10. *Clampings 3D.* Program for fast, interactive 3D modeling and design of optimal position and form of yoke beams, clamping plates, bolts, and other inactive structural elements of transformers.
11. *Method and Algorithms of* "Rapid Solution of Complicated Electromagnetic and Thermal Field Problem."
12. *RNM-3Dshell.* "Design of mathematical and physical model and algorithm for calculation of eddy current losses and excessive heating hazard in single phase bushing box of high currents."
13. *RNM-3Dshell.* "Design of mathematical and physical model, calculation of eddy current losses and excessive heating in Shell-type transformers" (in collaboration with Vigo University and EFACEC-Porto transformer Works).

    Overall, there is evidence that in some cases even small changes in the relationship between screen dimensions and material can reduce the power losses in large kW and increase the reliability of the transformer.

The improved screening is of evident interest if one realizes that the cost of one capitalized kW of load loss ranges from US$3,000 to US$10,000, depending on *e* and place of operation.

In this way, a manufacturer and electric power utility *can expect to save costs as much as US$1 million in one large transformer and many tons of transformer iron by the optimization of magnetic screenings (shunts).*

Additionally, it was designed similar to the simple programs used for simulation and calculation of dynamics of electromechanical systems, called

14. *Hamilton*. For rapid modeling and simulation of motion in dynamic electromechanical systems on the basis of Hamilton's principle.

3.8 "Critical Distance" of the Tank Wall

The "Critical," author has called [2] it so in 1965, is a ratio $a_{cr} = a_T/a_C$ of distance $a_T = (c'' + a_1 + \delta/2)$ of tank wall on axis of inter-winding gap δ (Figure 3.19) to the distance $a_C = (c' + a_2 + \delta/2)$ of core surface on the same axis.

At distances higher than the critical, tank walls made of solid steel have no influence on the leakage flux distribution between the core and the tank.

It means that pulling out a transformer's removable part from the tank will not cause remarkable change of stray flux penetrating the core and the windings.

It is an important factor to be considered for many leakage field phenomena. For instance, it allows separating (tare) experimentally, during short circuit test, the stray losses from total short circuit losses. In large transformers, ratio $(2c_1)/h$ and $(2c_2)/h$ is of the order 0.1–0.2. Therefore, we can evaluate simplified criteria formula ([2], p. 435) as

$$\left(\frac{a_T}{a_C}\right)_{CR} \approx \frac{\pi}{8(a_c / h)} - 1 \tag{3.33}$$

At magnetic laminated screens (shunts), relationships are approximately the same. However, at electromagnetic Cu or Al screening, this ratio is much higher and amounts approximately (see [2], p. 436, or [36]).

$$\left(\frac{a_T}{a_C}\right)_{CR} \approx \frac{h}{a_c} \tag{3.34}$$

Since in large power transformers, usually $h/a_C \approx 10$–12, and $(a_T/a_C)_{CR}$ is also 10–12. In practice, however, ratio a_T/a_C is of the order of 1–5. We can conclude, therefore, that electromagnetic (Cu or Al) screening always has an effect on the change in stray fields distribution in leakage region. Therefore, as result of Cu or Al screens on tank walls, there is always an increase in core component, Φ_C and decrease in Φ_T, that of tank component. This is accompanied by an increase in eddy current losses and axial forces in the internal LV winding and a decrease in them in the external HV winding. This is because usually external HV windings are more endangered by electrodynamic crashing forces (see Section 3.6.3).

3.9 Stray Losses Control in Shell-Type Transformers*†

All theory and practical experience presented above concerns the rather more popular core-type transformers. However, there are a variety of different kinds of, and special, transformers. One of them includes the biggest and important class of power transformers, that is, shell-type power transformer, which provides many advantages (Figure 3.26); such transformers are usually manufactured as a single-phase and later joined in three-phase bank. One of the most important manufactures of such transformers are Transformer Works in Porto-Portugal.

Although the single-phase structure seems simpler, it is in fact even more complicated and needs a completely different approach including the following: new preprocessor, postprocessor, solver, and so on. Nevertheless, after examining different programs, thus far, the RNM-3D method (Figure 3.19) has proven to be the best in such design applications.

The international team of authors: A. Soto, D. Souto, D. Couto (Portugal), J. Turowski (Poland), and XM, and Lopez-Fernandez (Spain) [8] have successfully resolved this problem as follows.

Problem was resolved using the mechatronic approach [4,8], that is, interactive design (Figure 3.27). The first step is to know the machine data, like in Figure 3.19. It is necessary,

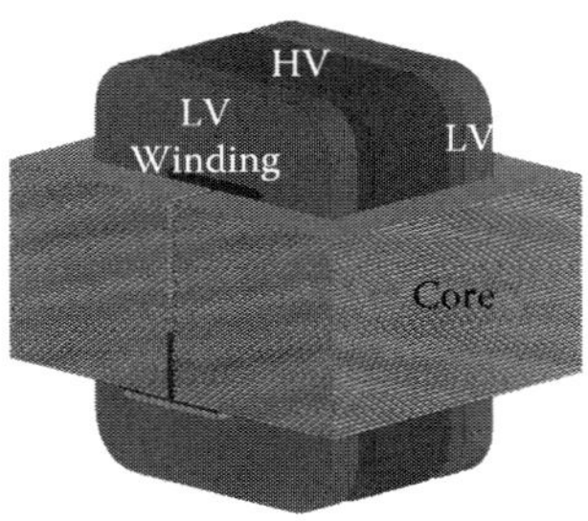

FIGURE 3.26
Single phase shell-type transformer.

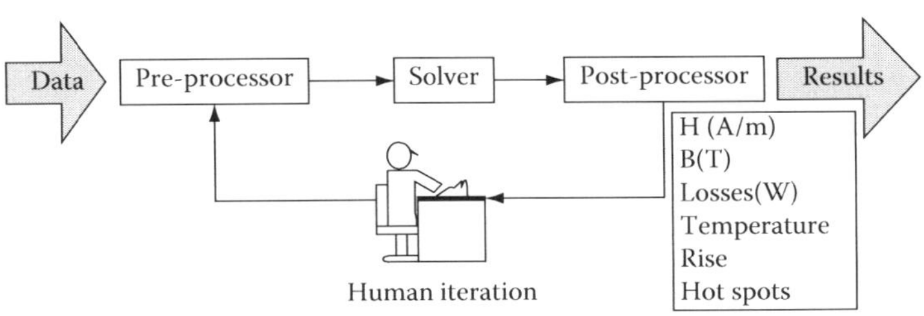

FIGURE 3.27
Rapid, interactive design and optimization of complex 3D electromagnetic structure. (From Soto, A. et al., Software for fast interactive three-dimensional modeling of electromagnetic leakage field and magnetic shunts design in shell type transformers. *International Conference on Electrical Machines ICEM'08*. Paper ID 539, Vilamoura, Portugal (Plenary lecture), 2008.)

---

* A. Soto, D. Souto, J. Turowski, X.M. Lopez-Fernandez, D. Couto [8].

† This work has been carried out with the support of Universidade de Vigo and EFACEC Project2006/2008, for which authors express their gratitude.

for example, to determine key parameters such as geometry, dimensions, electric data, materials, shielding system, and so on.

Then comes the solver phase, which is the part that requires less human interaction, because during this phase, the computer does all the work. Here, the expert system (Figure 3.2) is applied. And finally, in the postprocessor (Figure 3.27) stage, the results can be made available for analysis.

Then, designer takes a decision whether to repeat the analysis, if required, or whether the results are acceptable. The final program should be as easy and simple as Ohm and Kirchhoff laws.

The main objective of the work [8] is to implement such rapid "RNM-3Dshell" package (Figure 3.28) for the design of a ca. 600 kV, 1000 MVA shell-type transformer, for example, one manufactured by *EFACEC Energy*, in Porto-Portugal.

The "RNM-3D shell" calculates the magnetic flux through each reluctance, and from it, the rest of interest quantities can be obtained, and in each reluctance as well. The RNM-3D shell was implemented into friendly interface (Figure 3.29), which represents the main work of visual environment in *EFACEC Energía* factory where the EFACEC integrated design process is used regularly.

From RNM-3D shell interfaces it is possible to select a particular shell-type transformer in order to design and compute the main magnitudes related to stray field control. After a few seconds, by means of a set of windows, one can analyze the result magnitude and element and/or place or vice-versa (Figure 3.29), to assess the level of load from the design point of view.

Between places concerning control of the stray field and the optimization of magnetic collector exist the following: the tank walls the T-Beam [Part I]. The RNM-3D shell computes separately shunts of tank bottom, of parallel walls (inferior, superior, horizontal, and vertical in each case), of perpendicular wall (inferior and superior individually), and of T-Beam (inferior and superior again). In the last case, they happen to be stepped shunts. So the influence of the number and dimension of these steps are carefully modeled. Other parts, including "passive" elements of the transformer, are also taken into account.

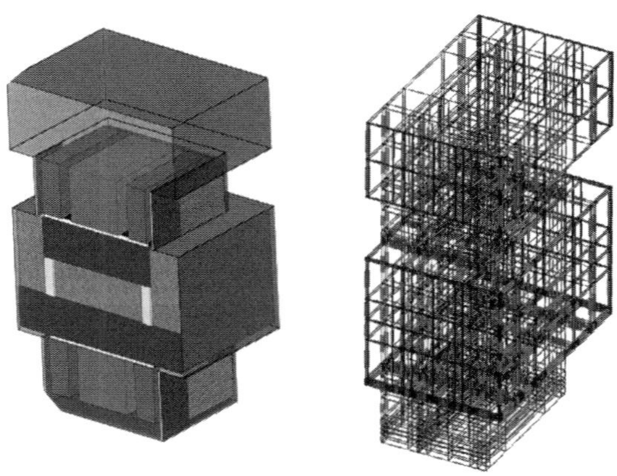

FIGURE 3.28
Basic structure of three-dimensional equivalent reluctance "RNM-3Dshell" network model created to the study half of one phase region of transformer from Figure 3.25.

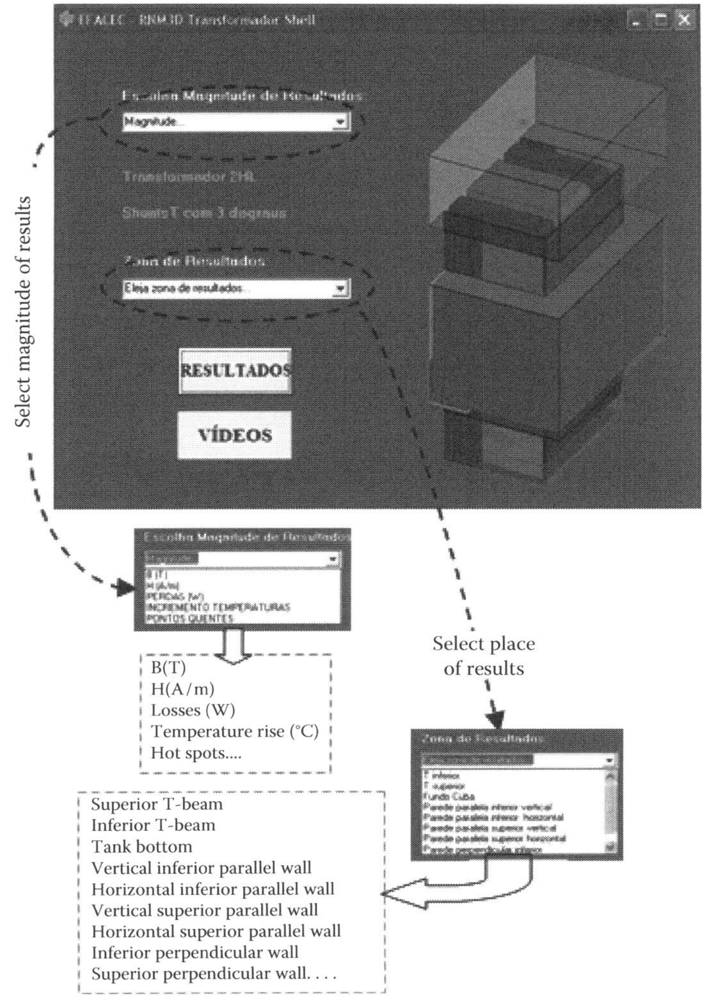

FIGURE 3.29
Interface of RNM-3D shell within EFACEC design process.

Some example of the results of possible local overheating hazard of laminated iron shunts according to the capabilities of EFACEC *RNM-3D shell* interface are shown in Figure 3.30. These results were obtained from a similar calculation of eddy current loss density distribution on the surface of the investigated part.

The RNM software presented shortly in Section 3.9 on the basis of industrial report [8] is revealed to be an important support for computing the 3D leakage field during the stage of engineering calculation and the design of shell-type power transformers. Shell-RNM-3D promises to be a fast and easy-to-use tool for overcoming the new challenges in a more demanding market each day.

Experimental measurements carried out by fiber-optic system previously discussed will soon be available to test Shell-RNM(3D) numerical results.

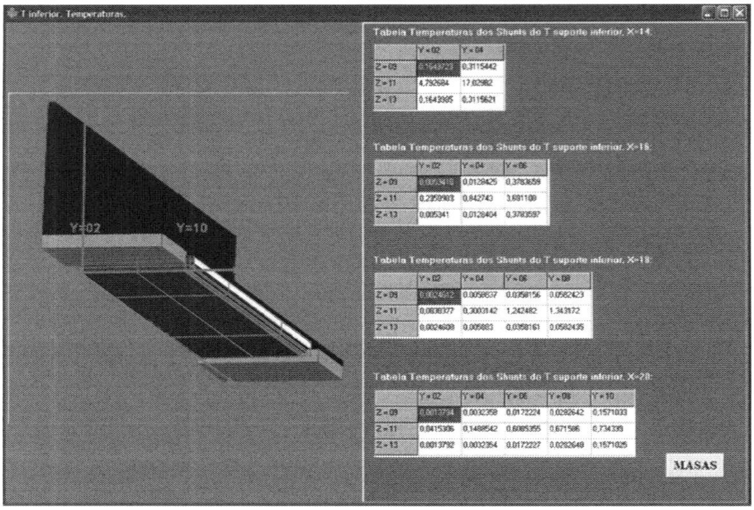

FIGURE 3.30
Tables of temperature rise in stepped shunts of T-Beam. (From Soto, A. et al., Software for fast interactive three-dimensional modeling of electromagnetic leakage field and magnetic shunts design in shell type transformers. *International Conference on Electrical Machines ICEM'08*. Paper ID 539, Vilamoura, Portugal (Plenary lecture), 2008.)

3.10 Conclusions

Highly asymmetrical 3D electromagnetic structure, with superposition of various fields, provides real challenges to design engineers. Magnetic and thermal nonlinearity, multiphase excitation, destructive electromagnetic forces, overloading, overfluxing, excessive hotspot hazard, losses in clamping and screws of particular core legs, complex 3D structure [18], with high and complex asymmetry [9], multiphaseness, and so on, are the complexities that need to be analyzed further.

All this is the burden of rapid design—a basic imperative of modern industry and economy.

Some attempts to calculate it with commercial FEM-3D software disappointed designers [14], due to the burdensome activities and the longer time required for computation. Therefore, many of the publications, unfortunately, including also professional packages [12], avoid complexity and are limited to idealized examples with too simplified geometries and at the same time too sophisticated, "prestigious" mathematics and computer techniques, not often applicable to real design tasks.

C.F. Bohren is right when he is cited in the book [39]: "... Those who are interested in application cannot avoid complexity..." and more sarcastic: "... mathematics is too important to be left to mathematicians ..."

Author's appeal to his students of design specialization: "... Mathematicians resolve what is resolvable, whereas engineers ... have to resolve everything ..." has similar meaning.

Engineering solutions should be rapid, simple, of low cost, and confirmed experimentally—at the first phase of preparing and checking reliability of methods, at the second phase of the final test, and third—at manufacturing and then at the market.

A gap between "pure" academic science and engineering jobs still exists.

This review of author's methodology delivers, it seems a most effective tools of such solutions. The applied approach is based on the expert system principle; thus, the more information and data added to the knowledge- and the data-base, the faster and simpler is the program developed on the basis of this principle. This database is created based on extensive research experiences of the author and his collaborators.

In addition to the material presented above, the following transformer problems were also resolved:

- Distribution of zero-sequence fluxes in three-phase transformers [40].
- Model investigation of stray loss in transformers [41].
- Power measurement at very low power factor or low voltage [2,42].
- Parametric formula for the calculation of losses in nonscreened tanks of large single-phase transformers [2,43].
- Criterion for selection of measurement method of power loss in transformer tanks [2,44].
- Choosing of optimal thickness of electrotechnical iron sheets [2,45].
- Problems of the extrapolation of the supplementary and short circuit losses [2,46].
- Measurement of loss density in constructional parts with the help of thermometric method [2,47].
- A new thermometric method of measuring power losses in solid metal elements [2,48].
- Additional losses in foil- and bar-wound transformers [2,49].
- The experimental prediction of power-loss distribution in metal elements of electrical machines [2,50].
- Overheating hazard in flanged bolt joints of transformers [2,51].
- Computer analysis of critical distance of tank wall in power transformers [2,52].
- Stray losses in multiwinding transformers on load [2,53].
- Method, generalized algorithm and programme "*Behakom 1.3*" for rapid modeling and simulation of motion and forces in electromechanical systems, on the basis of Hamilton principle [8,54].
- Capitalized models of stray fields and short circuit forces in large transformers (in Polish) [2,55].
- Next step involved the implementation of RNM-3D packages in the design process associated with shell-type transformers, which is carried out with success recently (Section 3.9) in cooperation with the University of Vigo-Spain (*Professor X. Lopez-Fernandez*) and Transformer Works EFACEC in Porto (Portugal) [8].

All these problems play an important role in the investigation, design, and, more important, in the reliability of large transformer units. Some of them need continued investigation and development.

Acknowledgments

The author wishes to express his sincere gratitude to all partners from many Polish and foreign transformer works, universities, and research institutions and to his PhD and MSc students who collaborated and discussed with him these interesting problems of electrodynamics of transformers and electrical machines. He would especially like to mention his own Institute of El. Machines and Transformers (now IM&IS) of TU-Lodz; Italian Professors M. Rizzo (Palermo) and A. Savini (Pavia), who collaborated during the development of RNM-3D scientific applications; Professor S.V. Kulkarni and Crompton Greaves Co., Mumbai, India, for industrial testing; his son, Dr. Marek Turowski (now director of Micro/Nano Electronics, CFDRC, United States) for permanent assistance at RNM-3D software & solver development; and M. Kopec, MSc (director of TUL Computer Centre), Dr. G. Zwolinski IM&IS TUL at asymmetric modeling, Professor X. Lopez-Fernandez for the recent cooperation within "Advanced Research Workshop on Modern Transformers—ARWtr05" in Vigo University, Spain, and plenty of other transformer manufacturers cited in Table 3.3 and many other colleagues and institutions for extending fruitful collaboration.

References

1. Turowski J (December 1957) Electromagnetic field and power loss in transformer covers (in Polish), PhD thesis, Technical University of Lodz, Lodz, Poland.
2. Turowski J (1993) *Elektrodynamika Techniczna* (in Polish), WNT, Warsaw, Poland and (1974) *Tiekhniceskaja Elektrodinamika* (in Russian), Moscow, Russia, "Energia."
3. Turowski J (2008) *Fundamentals of Mechatronice* (in Polish), Wydawnictwo WSHE, Lodz, Poland.
4. Turowski J (2004) Mechatronics impact upon electrical machines and drives, *Proceedings of International Aegean Conference on Electrical Machines and Power Electronics—ACEMP'04*, May 26–28, Istanbul, Turkey, pp. 65–70.
5. Turowski J (2002) 3-D models of capitalised cost of stray losses and short circuit forces in transformers (in Polish), *Proceedings of Fourth Conference on Power and Special Transformers*, October 25–27, Kazimierz Dolny, Poland, pp. 69–75.
6. Turowski J (2004) Innovative challenges in technology management, *Transition Economies in the European Research and Innovation Area: New Challenges for their Science and Technology*, H Jasinski (ed.) Wydawnicwo Naukowe Wydziału Zarządzania Uniwersytetu Warszawskiego, Warsaw, Poland.
7. Wood WS (1990) Economic importance of small EI, *Proceedings of 3x3x3 Seminar on Electrical Machines and Control*, Lodz, Poland. December 7, 1990.
8. Soto A, Souto D, Turowski J, Lopez-Fernandez XM, Couto D (September 2008) Software for fast interactive three-dimensional modelling of electromagnetic leakage field and magnetic shunts design in shell type transformers. *International Conference on Electrical Machines ICEM'08*. Paper ID 539, Vilamoura, Portugal (Plenary lectures) September 06–09.
9. Koppikar DA, Kulkarni SV, Turowski J (July 2000) Fast 3-dimensional interactive computation of stray field and losses in asymmetric transformers, *IEE Proceedings Generation, Transmission, Distribution*, 147(4), 197–201.
10. Turowski J (2002) Expert system in transformer design (in Polish), *Przeglad Elektrotechniczny*, 78(3), 63–69.
11. Kazmierski M (1970) Estimating approximate methods in the analysis of electromagnetic fields in transformers (in Polish), *Rozprawy Elektrotechniczne*, 16(1), 3–26.
12. Turowski J, Kraj I, Kulasek K (2001) Industrial verification of rapid design methods in power transformers. *International Conference Transformer'01*, October 5–6, Bydgoszcz, Poland.
13. Rickley AL (1989) 1979–1989 A decade of nuclear progress, *Power Engineering Review*, 8, No. 5, 25.
14. Coulson MA, Preston TW, Reece AB (1985) 3-Dimensional finite element solvers for the design of electrical equipment. *IEEE Transaction on Magnetics*, Vol. MAG=Z1, No. 6, November, pp. 2476-2479.

15. Turowski J (1999) Application of reluctance network to estimate stray losses in transformer tank, *International Conference on Power Transformers, TRANSFORMER'99*, SEP (ed.), April 27–30, Kolobrzeg, Poland, pp. 19–26, Invited plenary lecture.
16. Turowski J, Turowski M, Kopec M (September 1990) Method of three-dimensional network solution of leakage field of three-phase transformers, *IEEE Transactions on Magnetics*, 26(5), 2911–2919.
17. Turowski J (1984) *Theory of Electrical Machines. AC Machines* (in Polish), 3rd edn., Technical University of Lodz, Lodz, Poland.
18. Turowski J (1995) Reluctance networks—Chapter 4, pp. 145–178, in *Computational Magnetics*, Sykulski, J. (ed.) Chapman & Hall, London, U.K. Translation from Polish of the book Turowski J. (ed.) (1990) Analiza i synteza pol elektromagnetyeznych, Polish Academy of Science, "Ossolineum," Wroclaw, Poland.
19. Jezierski E, Turowski J (1964) Dependence of stray losses on current and temperature. CIGRE 1964, Report 102.
20. Zwolinski G, Turowski J, Kulkarni SV, Koppikar DA (1998) Method of fast computation of 3-D stray field and losses in highly asymmetric transformers, *Proceedings of Fifth International Conference on Transformers, Trafotech'98*, January 23–24, Mumbai, India, pp. I-51–I-58.
21. Kazmierski M, Kozlowski M, Lasocinski J, Pinkiewicz I, Turowski J (1984) Hot spot identification and overheating hazard preventing when designing a large transformer. CIGRE 1984, Report 12-12.
22. Turowski J, Pelikant A (November 1997) Eddy current losses and hot-spot evaluation in cover plates of power transformers, *IEE Proceedings, Generation, Transmission and Distribution*, 144(6), 435–440.
23. Turowski J (September 2004) Excessive local heating hazard of bushing turrets in large power transformers. *XVI International Conference on Electrical Machines, ICEM'04*, September 5–8, 2004, Cracow, Poland, pp. 286/1–6.
24. Kulkarni SV, Olivares JC, Escarela-Perez R, Lakhiani VK, Turowski J (September 2004) Evaluation of eddy losses in cover plates of distribution transformers, *IEE Proceedings—Science, Measurement and Technology*, 151(5), 313–318.
25. Kim D, Hashn S, Kim S (September 1999) Improved design of cover plates of power transformers for low eddy current losses, *IEEE Transactions on Magnetics*, 35(5), 3529–3531.
26. Koppikar DA, Kulkarni SV, Srinivas PN, Khaparde SA, Jain R (July 1999) Evaluation of flitch plate losses in power transformers, *IEEE Transactions on Power Delivery*, 14(3), 996–1001.
27. Turowski J, Przytula A (1961) Rapid solution of complicated electromagnetic and thermal field problem—Design of mathematical and physical model. Calculation of eddy current losses and excessive heating hazard in single phase bushing box of high currents (Part I) 4th version, IMSI—ABB Internal Report 28.7.03. Zeszyty Nauk. Politechniki Lodzkiej, "*Elektryka*," No. 8, pp. 91–114.
28. Turowski M, Kopec M (1986) Dedicated network program for interactive design of screens and field distribution in electromagnetic devices, *Rozprawy Elektrotechniczne*, 32, No. 3, 835–843.
29. Deuring WG (June 1957) Induced losses in steel plates in the presence of an alternating current, *AIEE Transactions*, 75, 166–173.
30. Turowski J, Zwolinski G (1997) Hybrid model for fast computation of 3-D leakage field and eddy current losses in highly asymmetric structures, *Proceedings of Fifth Polish, Japanese Joint Seminar on Electromagnetics in Science and Technology, Gdansk'97*, May 19–21, Poland, pp. 165–168.
31. Zwolinski G (April 1999) Stray field and losses in asymmetric transformer structures (in Polish), PhD thesis, Supervisor Turowski J, Technical University of Lodz, Lodz, Poland, pp. 63–67.
32. Zwolinski G, Turowski J (2003) Stray losses in multiwinding transformers on load. *International Conference on Power Transformers, Transformer'03*, May 19–20, Pieczyska, Poland.
33. Chua LO, Lin, P-M (1981) *Komputerowa Analiza Ukladow Elektronicznych* (in Polish), WNT, Warsaw, Poland, pp. 19–193.
34. Turowski J, Kopec M, Turowski M (1990) Fast 3-D analysis of combined Cu & Fe screens in three-phase transformers, *Electromagnetic Fields in Electrical Engineering, COMPEL*, James & James, London, U.K., vol. 11, pp. 113–116.

35. Darley V (1992) The practical application of FEM in transformer design & development, *COMPEL*, 11, 125–128.
36. Turowski J, Kopec M, Rizzo M, Savini A (1991) 3-D analysis of critical distance of tank wall in power transformers, *Proceedings of International Symposium on Electromagnetic Fields in Electrical Engineering, ISEF'91*, September 18–20, Southampton, U.K., pp. 229–232.
37. Rizzo M, Savini A, Turowski J (July 2000) Influence of flux collectors on stray losses in transformers, *IEEE Transactions on Magnetics*, 36(4), 1915–1918.
38. Rizzo M, Savini A, Turowski J (2002) Destructive forces in large power transformers, *Proceedings of the International IGTE Symposium on Numerical Field Calculation in Electrical Engineering*, September 16–18, Graz, Austria.
39. Lakhtakia A (ed.) (1993) *Essays of Formal Aspects of Electromagnetic Theory*, World Scientific, Singapore.
40. Turowski J, Pawlowski J, Pinkiewicz I, Zeszyty N (1963) Model investigation of stray loss in transformers (In polish), *Politechniki Lodzkiej Elektryka*, 52, No. 12, 95–115.
41. Turowski J (1965) Power measurement at very low power factor or low voltage, *Proceedings IEE*, 112(4), 740–741.
42. Turowski J (1969) Formula for the calculation of stray losses in nonshielded tanks of single-phase large transformers (In polish), *Rozprawy Elektrotechniczne*, 15, No. 1, 149–176.
43. Janowski T, Turowski J (1970) Criterion of choice of the method of power loss measurement in transformer tank (In polish), *Rozprawy Elektrotechniczne*, 16, No. 1, 205–226.
44. Turowski J (1964) Choice of optimum thickness of electrical sheet from point of view of core quality (In polish), *Przegląd elektrotechniczny*, 60, No. 8, 361–368.
45. Turowski J (1964) Discussion, *Proceedings of the supplementary CIGRE, 20th Plenary Session*, 1–10 June, 1964, Paris, Group 12 (Transformers), Vol. 1, pp. 279 and *Proceedings of IEE*, 112, N0. 4, April 1965, pp. 740–741.
46. Turowski J, Kazmierski M, Ketner A (1964) Measurement of stray losses in construction elements by thermometric method (In polish), *Przeglad Elektrotechniczny*, 60, No. 10, 439–444.
47. Niewierowicz N, Turowski J (1972) New thermometric method of measuring power losses in solid metal elements, *Proceedings IEE*, 119(5), 626–636.
48. Turowski J (Jan 1976) Additional losses in foil- and bar-wounded transformers *IEEE Transactions on Power Application and Systems*, and (1976) *Winter Power Meeting and Tesla Symposium*, New York, Paper A-76151. Catalog Power Eng Soc, No. 76, CH 1075-1-PWR, pp. 1–7.
49. Komeza K, Turowski J, Wiak S (1984) *International Conference Reliability and Lifetime of Electrical Rotating Machines*, October 3–5, Budapest, Hungary, No. 14, pp. 72–76.
50. Turowski J (1985) *Proceedings of International Symposium on Electromagnetic Fields in Electrical Engineering—ISEF'85*, September 26–28, Warszawa, Poland, No. 63, pp. 271–274.
51. Savini A, Turowski J (1988) *Electromagnetic Fields in Electrical Engineering*, Plenum Press, New York, pp. 119–127.
52. Zwolinski G, Turowski J (2003) *International Conference on Power Transformers, Transformer 03*, May 19–20, Pieczyska, Poland.
53. Turowski J, Bracha M, Haik J, Kosiorowski P, Merdala G (2004) WSHE, Lodz, Poland, December 7, 1990.
54. Turowski J (2002) *Proceedings of Fourth Konferencja N-T, pt Transformatory Energetyczne i Specjalne*, September 25–27, Kazimierz Dolny, Poland, pp. 69–75.
55. Turowski J (1982) *Obliczenia elektromagnetyczne elementów maszyn i urządzeń elektrycznych* (in Polish), WNT, Warsaw, Poland, and (1986) *Elektromagnitnyje rasczoty elemientow elektriczeskich maszin* (in Russian) "*Energoatomizdat*," Moscow, Russia.

4

Superconducting Transformers

Jan K. Sykulski

CONTENTS

4.1 Introduction

Recent catastrophic blackouts in New York, London, and Italy (August and September 2003) have exposed major vulnerabilities in existing generation, transmission, and distribution systems. Severe underinvestment, aging technology, and conservatism in the approach to innovation are being blamed and have created a situation where reliability of the entire system is under question. Resources need to be directed into technologies that have the potential to improve the integrity of the system; high temperature superconductivity has such potential. It also has a positive environmental impact by significantly reducing the losses as well as size and weight of the power devices.

The advent of high-temperature superconducting (HTS) materials has renewed interest in the possibilities for superconducting power apparatus offering real economic benefit, within power ratings typical of present system practice. Previously developed low-temperature superconductors (LTS) required cooling by liquid helium to about 4.2 K, with advanced cryogenic technology that is expensive both in terms of cost and of refrigeration power expended per unit of heat power removed from the cryostat. The technology for the new materials, which may be based on liquid nitrogen (LN2) at temperatures up to about 78 K, is simpler and cheaper, and the ratio of refrigeration power used to heat removed is reduced from over 1000 to about 25. There is significant activity around the world regarding applications of HTS materials to cables, motors, generators, fault current limiters, energy storage devices, and transformers. This review builds on the material published previously by the author [1–9] and reflects on the experiences with HTS technology in electrical power applications at the University of Southampton, United Kingdom. More information may also be found on a dedicated web page [10].

4.2 Properties of HTS Materials

The announcement in April 1986, by Muller and Bednorz (IBM) of superconductivity in the perovskite structure lanthanum-barium-copper oxide at 30 K, was an important step toward a wider application of superconductivity [11]. This was followed by the discovery of Wu and coworkers in January 1988, of $Y_1Ba_2Cu_3O_{7-8}$ (YBCO or 123), with a transition temperature of 93 K, bringing superconductivity above the boiling point of liquid nitrogen (77.4 K at 1 atm), a cheap and widely available cryogen. There has since been much effort on the search for new materials and the optimization of processes for production of thin films (<1 μm), thick films (10–100 μm), bulk materials, wires, and tapes (Figure 4.1) in single or multifilament composites. Many practical problems remain to be solved, but the potential for engineering application is clear.

Unlike LTS materials, HTS superconducting ceramics have highly anisotropic electronic structure, which causes critical current and critical field to have different values on two perpendicular planes. Randomly oriented polycrystalline HTS superconductors have low transport critical current, due not to intrinsic material properties but to misalignment between crystallites and weak superconducting regions at grain boundaries. Alignment or texturing of the material during the process of crystal growth improves properties and can be achieved by various means: temperature gradients, zone melting, 2D configurations, etc.

The most promising known HTS materials for high current applications are given as follows, with their common numerical reference names in brackets:

1. Yttrium compounds (YBCO)
   a. $Y_1Ba_2Cu_3O_{7-x}$ (123) $T_c = 92\,K$
   b. $Y_2Ba_4Cu_7O_{15-y}$ (247) $T_c = 95\,K$
2. Bismuth compounds (BISCCO)
   a. $Bi_2Sr_2Ca_1Cu_2O_y$ (2212) $T_c = 80\,K$
   b. $Bi_{2-x}Pb_xSr_2Ca_2Cu_3O_y$ (2223) $T_c = 110\,K$
3. Thallium compounds
   a. $(TlPb)_1Sr_2Ca_2Cu_3O_9$ (1223) $T_c = 120\,K$
   b. $Tl_2Ba_2Ca_2Cu_3O_{10}$ (2223) $T_c = 125\,K$
4. Mercury compounds
   a. $Hg_1Ba_2Ca_2Cu_3O_{10}$ (1223) $T_c = 153\,K$

4.3 Modeling and Simulation

HTS materials exhibit strong flux creep effects [12]. When modeling power loss mechanism due to a flow of alternating current, it is therefore not sufficient to use a critical state

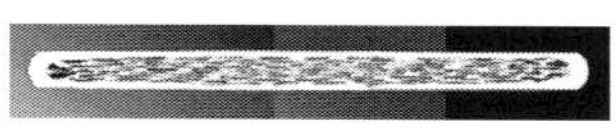

FIGURE 4.1
Multifilament HTS tape.

model [13], which has proved successful when dealing with LTS, and it is necessary to consider the flux creep *E–J* characteristic. A flux creep region is well described by the Anderson–Kim model:

$$E = K_1\sqrt{H}\sinh\left(K_2\sqrt{H}J\right) \tag{4.1}$$

where K_1 and K_2 are parameters related to temperature and the properties of the material [12], or a simpler relationship suggested by Rhyner [14], where

$$E = E_c\left(\frac{J}{J_c}\right)^{\alpha}, \quad \rho = \rho_c\left(\frac{J}{J_c}\right)^{\alpha-1} \tag{4.2}$$

The critical current density J_c corresponds to an electric field E_c of $100\,\mu\text{Vm}^{-1}$, and $\rho_c = E_c/J_c$. The power law (4.2) contains the linear and critical state extremes ($\alpha = 1$ and $\alpha \to \infty$, respectively). It has been found that for practical HTS materials, $\alpha \approx 20$ and thus the system is very nonlinear.

Using the Rhyner model, the governing field equation takes the following form in a 2D space:

$$\frac{\partial^2 E}{\partial x^2} + \frac{\partial^2 E}{\partial y^2} = \mu_0\frac{\partial}{\partial t}\left\{\sigma_c\left|E\right|^{(1/\alpha)-1}E\right\} \tag{4.3}$$

where $E = E_z/E_c$ for brevity and $\sigma_c = J_c/E_c$ [6]. Using a rectangular space mesh $\Delta x \times \Delta y$, a finite difference scheme may be built, which yields

$$\left|E_{ij}^{(k+1)}\right|^{(1/\alpha)-1}E_{ij}^{(k+1)} = \left|E_{ij}^{(k)}\right|^{(1/\alpha)-1}E_{ij}^{(k)} + \Delta t\cdot C_{ij} = K_{ij} \tag{4.4}$$

$$E_{ij}^{(k+1)} = \left|K_{ij}\right|^{\alpha-1}K_{ij} \tag{4.5}$$

where

$$C = \{\mu_0\sigma_c(\Delta x)^2\}^{-1}\{(E_{i+1,j}^k + E_{i-1,j}^k) + R^2(E_{i,j+1}^k + E_{i,j-1}^k) - 2(R^2+1)E_{i,j}^k\} \tag{4.6}$$

and $R = \Delta x/\Delta y$. The indices i and j denote the nodal addresses in the (x,y) space.

The loss over a cycle can then be calculated using numerical integration as

$$\int_0^T\int_V (J\cdot E)\,dV\,dt \tag{4.7}$$

The stability requirements for the numerical solution are severe, and in order to make the process as fast as possible, the time steps are adjusted at each step using the criterion

$$\Delta t \le \frac{\sigma_c\mu_0\left|E\right|^{(1/\alpha)-1}E_c^{1-(1/\alpha)}}{2\alpha((1/\Delta x^2)+(1/\Delta y^2))} \tag{4.8}$$

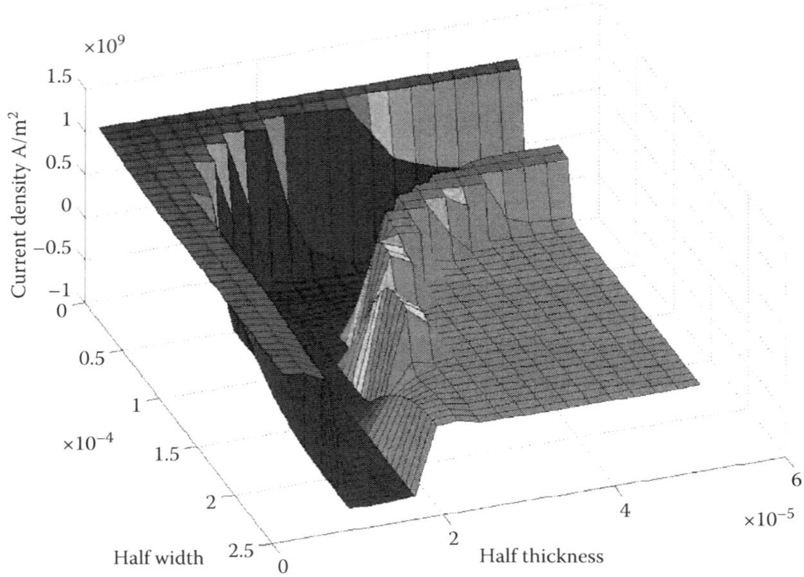

FIGURE 4.2
Current density distribution at the instant of a quarter of the maximum current.

It has been shown in Ref. [6] that field penetration in HTS tapes can be modeled accurately as a highly nonlinear diffusion process described earlier. Knowing AC losses is of paramount importance in design of HTS power devices as these losses are released at low temperature. Figures 4.2 through 4.4 show typical results of simulations.

The effects of nonlinearity have been found to be much stronger than experienced in linear conductors, and, thus, for larger currents (deeper field penetration), the 2D analysis is essential, despite a large aspect ratio of the tape. The analogy with the linear case only works well for smaller currents where a simpler 1D model is quite acceptable.

4.4 240 MVA Grid Autotransformer

A design feasibility study was conducted for a 240 MVA HTS grid autotransformer [1]. The principal feature of the design is the removal of the copper windings and their replacement by HTS equivalents. These are only a fraction (<10%) of the bulk of the conventional windings. The inevitable result is windings of reduced mechanical strength, which will stand neither the radial bursting forces nor the axial compressive forces that occur during through-fault conditions without special strengthening structural features. The tap winding is kept outside the cryostat to avoid the heat, which could otherwise flow into liquid nitrogen through a large number of connections. With the bulk of the ohmic losses (in common and series windings) removed, it is possible to cope with the core loss and remaining ohmic loss (in the tap windings) by forced gas cooling, leading to an oil-less design of transformer. This has a great advantage of reducing fire risk and environmental

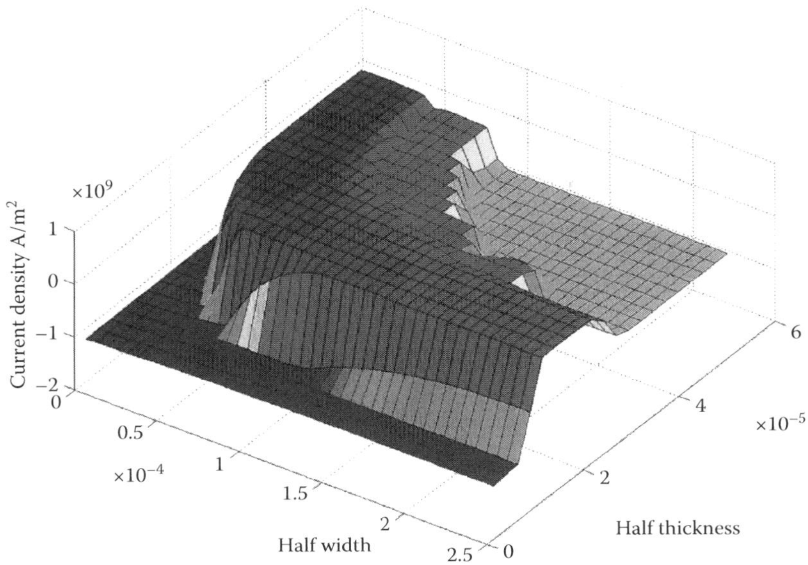

FIGURE 4.3
Current density distribution at the instant of max current.

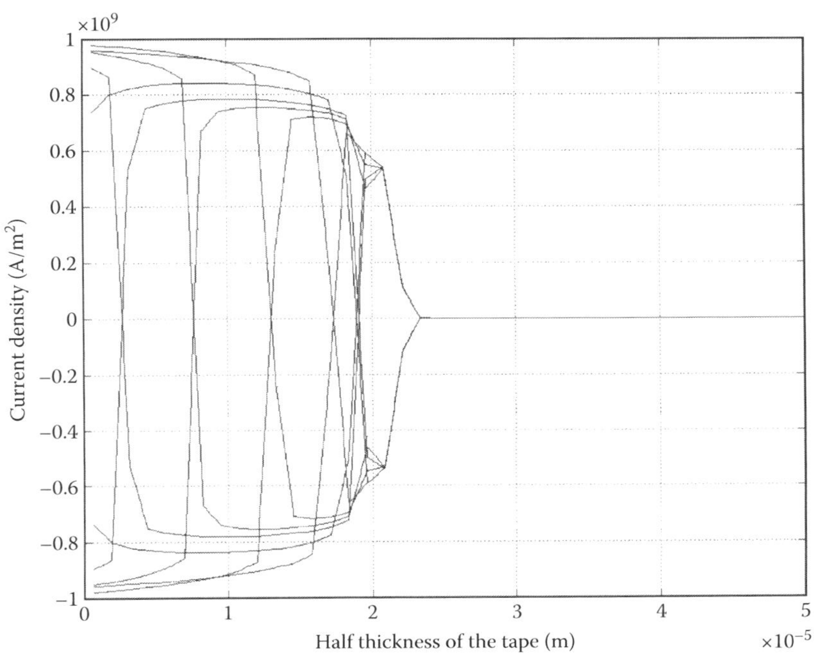

FIGURE 4.4
Current density profiles through the tape thickness.

hazard from oil spillage. Furthermore, the need for an explosion-proof outer steel tank is removed—though some form of enclosure must be provided for weatherproofing and acoustic noise reduction.

The superconducting windings have small thermal mass and are cryogenically stable (i.e., capable of returning to normal operation after a period of abnormal heating without disconnecting and cooling down) over only a small range of temperature rise. In consequence, the HTS design has very little capability to recover from a through fault without disconnection, in contrast to a conventional transformer. This is the chief weakness of the HTS design; however, the HTS design can *survive* a through fault, though it subsequently needs disconnecting for a period, and there is also good overload capability.

The principal parameters of the suggested transformer design are listed as follows:

1. kVA: 240,000
2. Normal volts: 400/132 kV
3. Tappings: 132 kV + 15% − 5% in 14 steps
4. Line current at normal volts: 346/1054 A
5. Diagram no.: Yy0 Auto
6. Guaranteed reactance: 20%
7. Rated current densities:
   a. Series winding* = 39.1 A/mm^2
   b. Common winding* = 36.9 A/mm^2
   c. Tap winding = 3.0 A/mm^2 (conventional)

A short summary of the principal features is provided as follows (based on [1]):

Core: Similar to a conventional design, but with leg length and window spans reduced and consequent saving in core weight and overall size.

Tap change arrangements: Copper tap windings retained, adjacent to core leg, outside the cryostat, connected to a tap changer, both of conventional design.

Cooling of core and tap winding: Forced-convection gas cooling, probably nitrogen. Gas is forced through axial ducts to cool the core legs and tap windings. Top and bottom yokes fitted with fiberglass cowling to contain the fanned gas and direct it over yoke surfaces.

Common and series windings: Constructed of rectangular section composite conductor comprising about 33% superconducting fibers and 67% matrix metal. The interturn insulating tape is applied to one face of the conductor. Individual tapes making up the conductor will be transposed in a normal way. Disks at the high-voltage ends of the series windings interleaved in pairs.

Winding reinforcement: Outer diameter of series (high voltage) winding reinforced with fiberglass hoops (possibly prestressed) or continuous cylinder. Substantial inner cylinder with multiple spacing sticks supports inside diameter of common winding. Annular clamping plates top and bottom of complete in-cryostat winding assembly, pulled together by eight through-bolts or studs, all constructed of insulating material.

* Average over composite conductor section, comprising both superconducting and matrix materials.

Cryostats: One cryostat per leg, each cryostat comprising a vacuum vessel constructed of double-skinned fiberglass, with the vacuum continuously pumped. Cryostat pressure will be slightly in excess of 1 bar, containing nitrogen at 78 K. The intermediate-voltage leads pass through the top lid. High-voltage lead passes centrally through cryostat wall.

Fault and overload capability: For any substantial through fault, disconnection is required and a period of minutes is needed before reconnecting. Transformer can survive the most severe through fault for about 170 ms, within which time disconnection must be achieved. Internal faults sensed by inbuilt monitors or terminal voltage and current sensors, which initiate disconnection. Overload capability of 100% is expected.

Housing: Oil-less design obviates need for a tank, which is replaced by a steel structure carrying load exerted by bushings. External housing required for weatherproofing, gas-seal for the forced-convection nitrogen coolant, and acoustic noise reduction.

Table 4.1 summarizes the total losses of the HTS transformer design and compares them with the corresponding figures for a conventional "reference" design. Losses are expressed in percentage form with the total loss in conventional design taken as 100%. Table 4.2 shows all the significant global features, covering size and construction as well as performance.

Table 4.3 shows estimates of saving/expenditure components based on continuous operation in rated conditions. It is clear that an expenditure on extra equipment and materials is offset by the enormous value of the saved losses (taken over a 10 year period and discounted at 9.5% per year). However, because of the redundancy built into the system for security, the load factor of a grid transformer is remarkably low and may be taken as 0.23 average and 0.26 rms. For such load conditions, a pessimistic estimate suggests that the total equivalent first-cost saving may now become a net increase of first-cost equivalent expenditure of about 20%.

On the other hand, a common practice is to have two transformers fully rated normally connected in parallel. It is thus worth considering an arrangement of an HTS transformer normally connected, in parallel with a conventional transformer normally disconnected but capable of being switched on quickly when required (e.g., during through fault). Hence, savings on the losses of the conventional transformer will be very significant. This arrangement is discussed in the next section.

TABLE 4.1

Loss Analysis (Total Loss of Conventional Design = 100%)

	HTS	Conventional
Core loss	8	9
Clamp stray loss	5	5
Tank loss	—	7
Total copper loss	<1 (tap)	79
Refrigerator power	7	—
Gas-cooling fan loss	2	—
Estimated total loss	23	100

TABLE 4.2

Comparison of Technical Features

Parameter	HTS	Conventional
Core length[a]	88.5	100
Core height[a]	82.4	100
Core thickness[a]	100	100
Window, height[a] × width[a]	70 × 78.5	100 × 100
Core weight[a]	80	100
Winding weight[a] (common and series)	6.3	100
Tap winding weight[a]	100	100
Cooling of core and tap winding	Forced N_2 gas	ONAN/OFAF
Cooling of common and series windings	Liquid N_2	ONAN/OFAF
Guaranteed % reactance	20	20
B in core, T	1.67	1.67
J rated (average of C and S), rms, A/mm^2	38	2.83
Rated loss, total[a]	23	100
Overload capability	2 pu, many hours	1.3 pu (6 h), 1.5 pu (30 min)
Through-fault capability (+ doubling transient)	2 pu, 64 ms	5 pu, 3 s
Survival time at 5 pu (+ doubling transient)	166 ms	Seconds (>3)

[a] Values for the conventional design taken as 100 for comparison purposes.

TABLE 4.3

Cost Savings on Continuous Full Load

Savings/(Expenditure)	%[a]
Saving on core plate	1
Saving on continuously transposed copper	7
Saving on copper losses, discount over 10 years	65
Cost of refrigeration plant	−21
First-cost equivalent expenditure on refrigerator drive power, discount over 10 years	−6
Cost of AC conductor, total of 7371 amp-km	−10
Total equivalent first-cost saving	36

[a] All values are referred to the overall first cost of the conventional transformer, taken as 100%.

4.5 Parallel Operation

Figure 4.5 shows a typical arrangement of parallel connected transformers. Each transformer is usually rated to take the maximum load on its own, and because that maximum load rarely occurs, the load factor for a typical National Grid transformer in the United Kingdom is low, around 23%. A typical probability density function for such a transformer is shown in Figure 4.6.

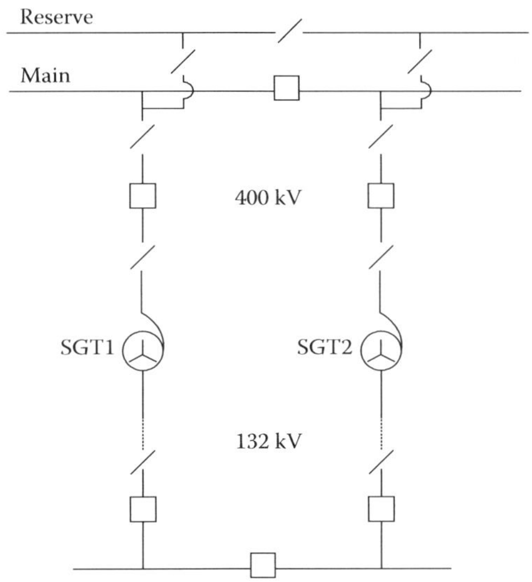

FIGURE 4.5
Conventional arrangement of parallel transformers.

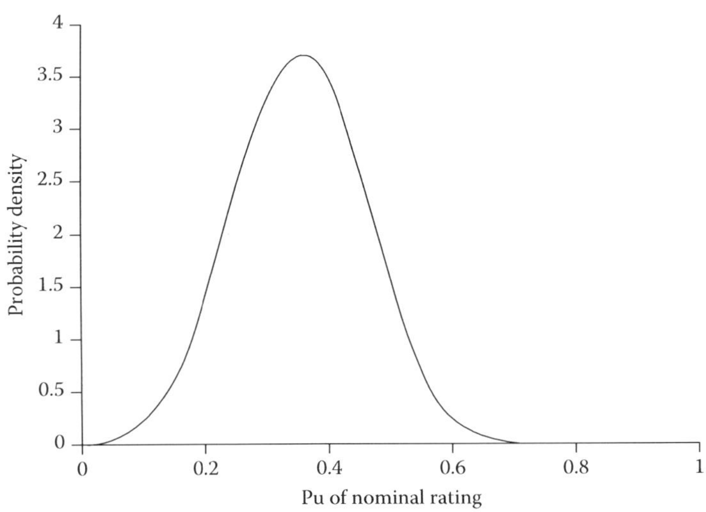

FIGURE 4.6
Probability density of load for a typical grid transformer.

A suggested scenario is therefore to replace one of the conventional transformers (say, SGT1 in Figure 4.5) with a HTS "equivalent." In normal operation, the SGT2's breaker is open, and the load passes through SGT1 (HTS). In the event of a through fault, SGT2's breaker is rapidly closed and SGT1's opened. This may require faster than usual breakers and associated protection, but there is no reason to think that this may not be achieved within the 200 ms required. Thus, in this configuration, it should hardly be necessary to ever use SGT2 (only during faults or maintenance outages), and so it may be possible to derate this transformer. The modified probability density distribution is shown in Figure 4.7.

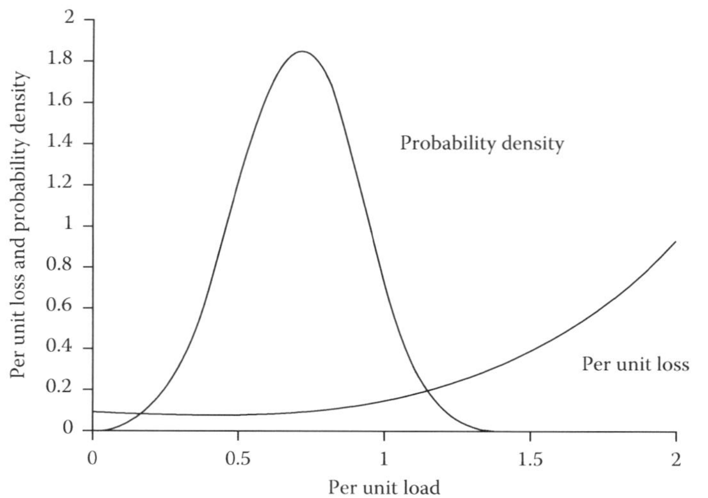

FIGURE 4.7
Load probability density and loss as a function of load for a HTS transformer in parallel with a (normally) unconnected conventional unit. The mean load is around 0.7 pu.

Following the argument put forward in Ref. [5], assuming a capital cost of a 240 MVA transformer of £1.0 M, per unit loss to be 600 kW and cost of losses over the transformer life to be £3000/kW, we can assess the costs as follows:

Costs (£k)	Superconducting + Conventional	2 × Conventional
Transformer capital	1000 + 1230	2000
Losses	0.105 × 600 × 3	0.426 × 600 × 3
Total	2419	2768

This simplistic analysis shows that the combination of HTS and conventional transformer is a lower cost option. Moreover, by inspecting Figure 4.7, it is clearly possible to optimize this further by reducing the nominal rating of the HTS transformer thus reducing capital cost. It can be seen that the load would "hardly ever" go much above 1.5 pu so the HTS transformer could be of a smaller rating.

4.6 HTS Transformer Demonstrator

A small single-phase 10 kVA HTS demonstrator transformer was designed, built, and tested at Southampton University [8,9]. In order to limit the material cost (predominantly the cost of the superconducting tape), it was decided that the nominal rating at 78 K should be 10 kVA and only one winding, the secondary, should be superconducting; this also had the benefit of allowing direct comparison of performance between conventional and superconducting windings. The secondary current at this load is 40 A. Since the large space

FIGURE 4.8
Field plots with and without flux diverters.

required for thermal insulation between copper and superconducting windings increases the radial flux densities and leakage reactance, a three-limb construction was used with the two windings on the center limb. This arrangement minimizes both of these problems and also simplifies the design and construction. It is essential to reduce the radial component of leakage flux in the superconducting winding, as being perpendicular to the face of the tape, it is most detrimental to its performance. For example, to carry the peak current of 9.5 A per tape, the perpendicular component of peak leakage flux density must be less than 15 mT, compared with 110 mT for the parallel (axial) component. Field plots with and without the flux diverters are shown in Figure 4.8.

Flux diverters are placed close to the ends of the superconducting winding in the cryostat and are constructed from low-loss materials to minimize the heat that must be removed from the cryostat. The flux diverters are made from powdered iron epoxy composite. A small test ring of this material produced a relative permeability of 6, and the 50 Hz iron loss at 78 K with a peak density of 40 mT is less than 1500 W/m^3. This loss is about 25% higher than that at room temperature due to the lower resistivity (the eddy currents are resistance limited). A more expensive alternative, using flux diverters constructed of segments of ferrite with a relative permeability in excess of 100, was considered. However, finite element analysis showed that, although the higher permeability did reduce the radial flux density in the end coils, circulating currents would be more difficult to control. The expected increase in the maximum capacity of the transformer is therefore very small and cannot justify the use of ferrite rings.

Flux diverters have proved to be a very effective means of controlling local field distribution and reducing the winding losses. Figure 4.9 demonstrates such losses as a function of secondary current. The graph also validates the numerical procedures developed for estimating AC loss in HTS windings described in Ref. [7]. The most important observation resulting from these results is that the design requirements differ quite significantly from conventional (i.e., nonsuperconducting) transformers where the local shape of leakage field would not have mattered, whereas for the HTS winding, a twofold reduction of loss has been achieved. This emphasizes the importance of very careful field modeling using finite element or similar software.

The epoxy-filled fiberglass cryostat (Figures 4.10 and 4.11) consists of an inner annulus—which contains the HTS winding and liquid nitrogen—fitting inside the outer annulus. The cavity between the two halves is filled with superinsulation and evacuated to form the thermal insulation around the inner cryostat. The whole assembly is threaded by the

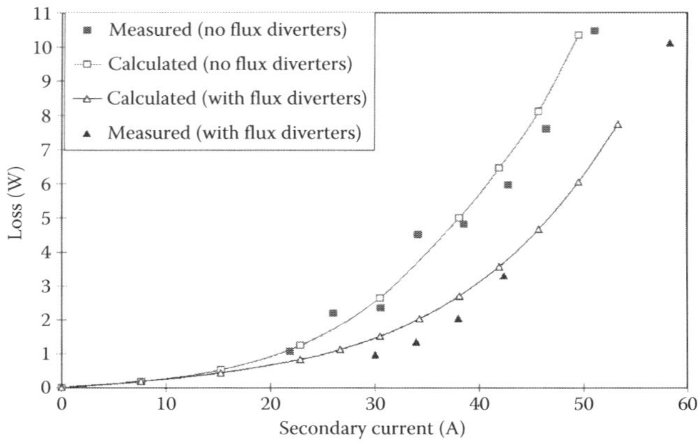

FIGURE 4.9
Measured and calculated losses with and without flux diverters.

FIGURE 4.10
HTS winding of the 10 kVA demonstrator.

center limb of the transformer laminated core. The complicated shape of the cryostat is necessary because it is not feasible to operate with a "cold" core in view of the loss of nitrogen that would be caused by the iron losses. Neither can the cryostat be made of electrically conducting material because it would act as a short-circuited turn.

4.7 Conclusions

High-temperature superconductivity has great potential in electric power applications (generators, motors, fault current limiters, transformers, flywheels, cables, etc.) as losses and sizes of devices are significantly reduced. The technology is now mature, and prototypes are considered. The ability to predict and reduce all "cold" losses is crucial to show economic advantages of HTS designs.

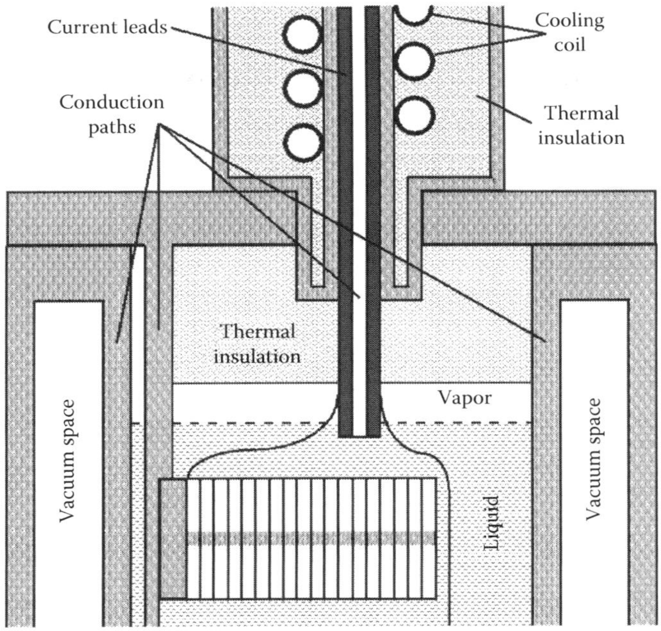

FIGURE 4.11
Cryostat and current leads.

References

1. Sykulski JK et al. (January 1999) High temperature superconducting power transformers: Conclusions from a design study. *IEE Proceedings: Electrical Power Applications*, 146(1), 41–52.
2. Ship KS, Goddard KF, Sykulski JK (2002) Field optimisation in a synchronous generator with high temperature superconducting field winding and magnetic core. *IEE Proceedings: Science, Measurement and Technology*, 149(5), 194–198.
3. Ship KS, Sykulski JK (2004) Field modelling and optimisation of a high temperature superconducting synchronous generator with a coreless rotor. *Fifth IEE International Conference on Computation in Electromagnetics*, Stratford-upon-Avon, U.K., April 19–22, 2004, pp. 109–110, ISSN 0537-9989.
4. Sykulski JK (2003) Applications of high temperature superconductivity in electrical power devices: The Southampton perspective. *Proceedings of Superconductivity UK*, IEE Savoy Place, London, U.K.
5. Sykulski JK et al. (2000) The design, construction and operation of high temperature superconducting transformers: Practical considerations. *Proceedings of 38th Session of the International Conference on Large High Voltage Electric Systems (CIGRE)*, Paris, France, No 12-203, August 27 to September 1, 2000.
6. Sykulski JK et al. (2000) 2D modelling of field diffusion and AC losses in high temperature superconducting tapes. *IEEE Transactions on Magnetic*, 36(4), 1178–1182.
7. Sykulski JK, Goddard KF, Stoll RL (2000) A method of estimating the total AC loss in a high-temperature superconducting transformer winding. *IEEE Transactions on Magnetic*, 36(4), 1183–1187.

8. Sykulski JK et al. (1999) Design of a HTS demonstrator transformer. *International Cryogenic Engineering Conference*, Institute of Physics Publishing, Bournemouth, U.K., pp. 571–574.
9. Sykulski JK et al. (1999) High temperature superconducting demonstrator transformer: Design considerations and first test results. *IEEE Transactions on Magnetics*, 35(5), 3559–3561.
10. http://www.superconductors.org/. Online.
11. Muller KA, Bednorz JG (1986) Possible high Tc superconductivity in the Ba-La-Cu-O system. *Journal of Physics B*, 64, 405–407.
12. Anderson PW (1962) Theory of flux creep in hard super-conductors. *Physical Review Letters*, 9, 309–311.
13. Bean CP (1964) Magnetization of high-field superconductors. *Reviews of Modern Physics*, 36, 31–39.
14. Rhyner J (1993) Magnetic properties and ac losses of superconductors with power law current-voltage characteristics. *Physica C*, 212, 292–300.

5

Challenges and Strategies in Transformer Design

S.V. Kulkarni

CONTENTS

5.1 Introduction

Significant advances have been made in design, analysis, and diagnostic techniques for transformers in the last decade. Competitive market conditions, improvements in computational facilities, and rapid advances in instrumentation are the main enabling factors for all-round improvements in transformer engineering. The ever-increasing growth and complexity of power systems make the task of designing reliable and cost-effective transformers more difficult. A transformer as a system consists of several components; it is essential that integrity of these components, individually and collectively, is ensured.

This chapter summarizes challenges in design of transformers in today's competitive market conditions. Advanced computational techniques are being used to accurately predict performance parameters of transformers. A transformer is a complex three-dimensional (3-D) electromagnetic structure, and it is subjected to variety of stresses, viz., dielectric, thermal, electrodynamic, etc. In-depth understanding of various phenomena occurring inside the transformer is necessary. Most of these can now be simulated on computers so that suitable changes can be made at the design stage to eliminate potential problems. The design and manufacturing practices and processes have significant impact on the performance parameters of the transformer, and the same have been identified in this chapter at appropriate places.

The design of a transformer mainly constitutes the analysis of the following parts: magnetic circuit, windings, insulation, cooling system, and tank/structural parts. Impedance characteristics, eddy and stray losses, hot spots, short-circuit stresses, dielectric stresses during high-voltage surges, structural stresses (including those like seismic stresses if specified by the user), power system interaction aspects, etc., are the principal considerations and influencing factors for the design. There is continued interest in confronting these design issues as is evident from the recent books [1,2] and numerous papers on various aspects of transformer design.

With the great advances that have taken place in computational tools, routine design calculations can be efficiently programmed along with the calculation of performance parameters. Optimal design can be worked out with due considerations of design and manufacturing constraints. With the ever-increasing competition in the global marketplace, there are continuous efforts to optimize the material cost of transformers. In most of the contracts, the transformers have to be delivered in a short period of time, and, hence, the speed of design and manufacture of the transformers is a key issue. It should be ensured that the process of optimization does not lead to nonstandard designs, which not only increase engineering efforts but may also lead to quality issues during manufacturing. It is a common practice of leading transformer manufacturers to develop their own customized finite element method (FEM) programs for optimization and reliability enhancement of transformers. The two-dimensional (2-D) FEM analysis, which is widely used for stray loss estimation/control, winding temperature rise calculations, short-circuit force calculations, etc., can be integrated into the main electrical design optimization program.

The transformer is a complex 3-D structure having materials with nonlinear characteristics and anisotropic properties. Superimposition of various physical fields, viz., electrostatic, electrodynamic, thermal, etc., poses a real challenge while designing the transformer and optimizing its material cost. Many times, design requirements for these fields are in conflict with each other. This chapter has identified such complex design issues and commented on strategies to tackle them.*

Recent trends in the area of computation are also enumerated. Coupled field computations are increasingly being done for accurate computation of performance parameters of transformers under steady-state and transient operating conditions.

5.2 Magnetic Circuit

Due to continuous research and development efforts by steel manufacturers, core materials with improved characteristics are being made available for use to transformer manufacturers. In the last few decades, the trend has been also to use thinner laminations to take advantage of lower eddy losses. The lowest thickness currently available is 0.23 mm. However, with a lower thickness of laminations, the core building time may increase since the number of laminations for a given core area increases.

5.2.1 Building Factor

Expected loss reduction is not always obtained with better core material grades. One of the reasons for this anomaly is the inaccurate assessment of the core building factor that is defined as the ratio of *built transformer core loss* to *material Epstein loss*. The building factor is generally found to increase as the grade of the material improves from the conventional cold rolled grain oriented (CRGO) material (e.g., M4) to Hi-B material (e.g., M0H) to scribed (domain refined) material. This is due to the fact that the flux is not along the grain orientation at the corner joints, and the increase in W/kg due to deviation from the direction of grain orientation is higher for a better grade material. Hence, building factors based on experimental/test data should be used. The advantage gained in losses for better grades should be checked vis-à-vis their higher cost, for the proper selection of material grade.

The building factor is also a function of various design and manufacturing practices and processes, viz., core construction, type of core joint, number of laminations per lay, overlap length, angle of overlap, gap at joints, and operating flux density. Single-phase two-limb transformers give significantly better performances than three-phase cores [3]. Step-lap joints, which consist of a group of laminations (commonly 5–7), stacked with a staggered joint, have excellent performance figures (lower losses, lower noise, etc.) as compared to mitered joints [4]. Both the types of joints are shown in Figure 5.1. The step-lap joint can give noise level reduction of about 4–5 dB.

* *Note*: A few of the discussions in Sections 5.2 through 5.6 are reporduced with permission from S.V. Kulkarni and S.A. Khaparde, *Transformer Engineering: Design and Practice*. © 2004 Marcel Dekker, CRC Press, Taylor & Francis Group.

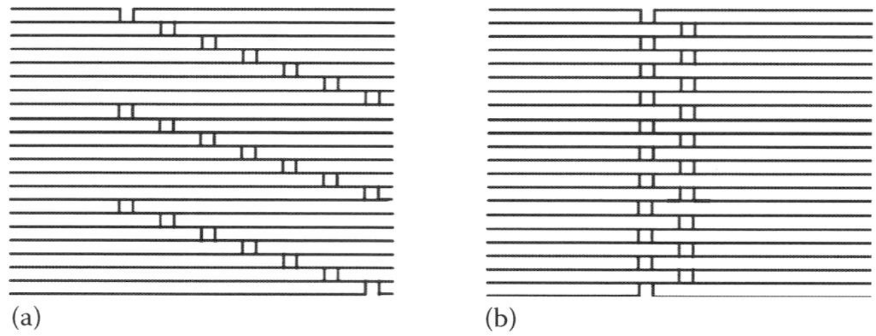

FIGURE 5.1
Core joints. (a) Step-lap joint and (b) mitered joint.

The gap at the joints should be kept at minimum possible value. A smaller overlapping length at the joints gives better core performance. The lesser the laminations per lay, the lower the core loss is. As the number of laminations per lay reduces, the manufacturing time for core building increases, and, hence, core building with two laminations per lay is most popularly used. The building factor is strongly influenced by the operating flux density; it may deteriorate more rapidly for better grade materials with the increase in the operating flux density. The higher the corner weight proportion, the higher the core loss is, since greater loss occurs in corner joints of limbs and yokes due to cross-fluxing and crowding of flux lines in them. The building factor is low for higher ratios of window height to width due to comparatively lower core weight at the corners in such cases.

5.2.2 Types of Core Construction

The type of core construction adopted is mainly decided by user's specifications, manufacturing limitations, and transport restrictions. Although it is economical to have all the windings of three phases in one core frame (three-phase three-limb construction), spare unit considerations and manufacturing/transport constraints may favor single-phase three-limb option. In large transformers, three-phase five-limb construction is used in order to reduce their height for transportability. In this construction, the effect of nonsinusoidal flux density distribution in the yokes (due to lower reluctance of the main yoke as compared to that of magnetic path represented by the end yoke and end limb) is generally compensated by providing higher yoke area. In some cases, a particular requirement of zero-sequence impedance characteristics may decide the choice of the core construction. Zero-sequence impedance is equal to positive-sequence impedance for a bank of single-phase transformers. Zero-sequence impedance is much higher for the three-phase five-limb construction than the three-limb type due to low reluctance path of yokes and end limbs available to the in-phase zero-sequence fluxes, and its value is equal to the corresponding value of positive-sequence impedance (the assumption made here is that the applied voltage during the zero-sequence test is small enough so that the yokes and end limbs are not saturated). An elaborate discussion on open-circuit and short-circuit zero-sequence impedances is available in Ref. [1]. For large transformers, if short-circuit forces are critical, one can design the core with two limbs per phase (e.g., bank of single-phase two-limb transformers). In this type of construction, magnitudes of short-circuit forces are lower because of lower ampere-turns/height. The bolted yoke construction leads to local flux distortions at the position of holes, and the core losses increase. The boltless yoke construction results in better core material utilization and loss reduction. In this construction, suitable epoxy resin can be applied at the edges of yoke laminations to strengthen the bonding of laminations, which results in reduction of noise level.

5.2.3 Overexcitation Conditions

The magnetic circuit of a transformer, which is often subjected to overexcitation conditions, deserves special design considerations. Although there has been considerable improvement in losses and magnetizing volt-amperes with the development and use of better core grades, the saturation flux density has remained almost the same (about 2.0 T). The peak operating flux density generally gets limited by the overexcitation conditions specified by users. For the 10% continuous overexcitation specification, a peak operating flux density of 1.73 T could be the upper limit. For the 5% continuous overexcitation specification (power system with well-maintained voltage profile), the flux density of 1.8 T may be used as long as the core temperature and noise levels are within permissible limits (these limits are generally achievable with the step-lap core construction) [1]. Under the overexcitation conditions, the spill-over flux (from the core) and exciting current are in the form of pulses having high harmonic content that increases the eddy losses and temperature rise in windings and structural parts. The provision of extra yoke area may improve the performance under overexcitation conditions since eddy losses in structural parts, due to flux leaking out of core due to its saturation under overexcitation conditions, are reduced to some extent [5].

5.2.4 Core Temperature Rise

The core area (net iron area) needs to be maximized to get an optimum transformer design. The cooling ducts reduce the useful core area, and, hence, their number should be as low as necessary. An accurate determination of temperature profile of the core is required for deciding the number and positions of cooling ducts. The ducts reduce both the surface temperature rise of the core relative to that of oil and the temperature rise of the interior of the core relative to that at the surface. A typical temperature rise (with respect to surrounding oil) profile of a core section between two cooling ducts, obtained by 2-D FEM thermal analysis, is shown in Figure 5.2. The core has three cooling ducts out of which one is along its center line.

Actually, the calculation of temperature distribution in the transformer core is a complex 3-D problem with nonuniform heat generation. Also, the thermal properties of the core are anisotropic in the sense that the thermal conductivity along the plane of laminations is quite different from its value across them. Hence, most accurate temperature rise distribution can be obtained by using 3-D FEM thermal simulation with the anisotropic thermal material properties taken into account. The limit on the temperature of core surface, which is usually

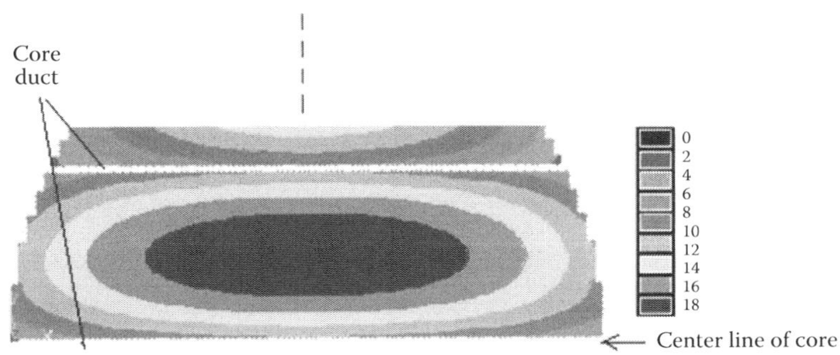

FIGURE 5.2
Temperature rise of core by FEM analysis.

in contact with the insulation (between core and frame), is the same as that for the windings. For the interior portions of the core, which are in contact with oil (film), the limit is 140°C.

A number of studies have been reported in the literature, which have analyzed various factors affecting core performance. Numerical techniques (finite difference method, FEM, etc.) have been used to calculate spatial flux distribution and core losses. With the rapid development in computational facilities, it is now possible to simulate core performance using 3-D models that take into account magnetic anisotropy and nonlinearity [6].

5.3 Windings

The design of windings is the most critical constituent of the transformer design. It is influenced by electrostatic, electrodynamic, and thermal aspects. These aspects often have conflicting design requirements. Hence, while optimizing the winding design, a number of performance parameters need to be accurately calculated, viz., winding stray losses (eddy current and circulating current losses), thermal gradients, short-circuit stresses and strengths, dielectric stresses and strengths, etc.

5.3.1 Choice of Winding Conductor

The choice of winding conductor type, viz., strip, bunch, or continuously transposed cable (CTC) conductor, gets decided by the considerations of material cost optimization, eddy loss control, winding space factor, and ease of winding process. These three types of commonly used conductors are shown in Figure 5.3.

The winding conductor is generally subdivided into a number of parallel conductors to reduce eddy loss since it is proportional to square of the conductor thickness. However, the dimensions of conductors have to be enough to withstand short-circuit forces. If the width to thickness ratio of a rectangular conductor is more than 7, there could be difficulty in winding it. Also, the winding space factor gets impaired in the radial direction due to the subdivision of the conductor since each individual parallel conductor has to be insulated. The bunch conductor, in which usually two or three parallel conductors are bunched in a common paper covering, improves the space factor. CTC conductor, in which it is possible to use quite a small thickness and width, is popularly used in large power transformers for minimizing eddy loss and for winding ease. The higher cost of the CTC conductor gets generally compensated by the advantage gained due to reduction of losses in cases of transformer specifications with higher value of load loss capitalization (dollars per kW of load loss). The low voltage winding in large generator transformers is usually designed with the CTC conductor to minimize the eddy loss and improve productivity (manual

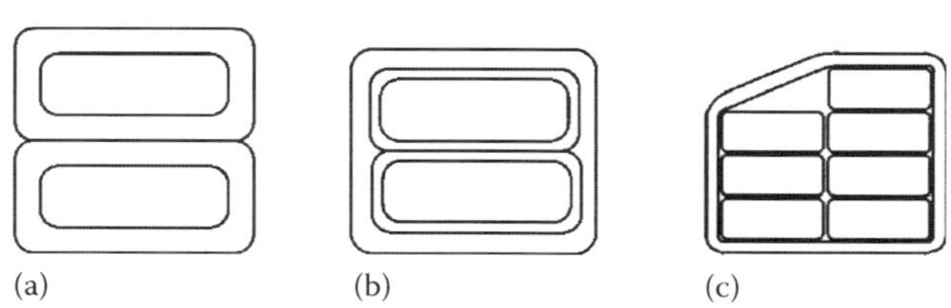

FIGURE 5.3
Winding conductors. (a) Strip, (b) bunch, and (C) CTC.

transpositions at the winding stage are avoided due to transpositions of parallel conductors at regular intervals within the CTC conductor).

5.3.2 Eddy Loss Evaluation and Optimization

The formulae for computation of eddy loss in thin and thick winding conductors are given in Ref. [1] along with their derivations. The eddy loss in a thin winding conductor (i.e., dimension of the winding conductor is less than or comparable to its depth of penetration) per unit volume is given as

$$P_E = \frac{\omega^2 B^2 t^2}{24\rho} \tag{5.1}$$

where
- ω is frequency (rad/s)
- B is peak value of tangential leakage flux density on conductor surface (T)
- t is dimension of the conductor perpendicular to field (m)
- ρ is resistivity of the conductor material (Ω-m)

If the eddy loss is determined by means of 2-D field calculations, the eddy loss due to axial and radial leakage fields can be obtained separately by using the previous equation; t is thickness (or width) of the conductor if loss due to axial (or radial) leakage field is being calculated. Analytical methods (Roth's method [7], Rabin's method [8]) or 2-D FEM can be used to calculate the eddy loss due to axial and radial leakage fields. FEM analysis is quite popularly used for the eddy loss calculations. It is assumed (for thin conductors) that the eddy currents do not have any influence on the leakage field. The values of axial and radial leakage field components are assumed to be constant over any conductor and equal to their values at its center. The total eddy loss for each winding is calculated by the integration of the axial and radial eddy loss components of all conductors in the winding cross section. With the quantum improvement in the speed and power of computational tools, the accurate analytical/numerical methods for eddy loss calculation can be integrated into the main design optimization program. 2-D methods give reasonably accurate eddy loss values. Three-dimensional magnetic field calculations have also been used to get very accurate results for large transformers [9]. The three components of the flux density (B_x, B_y, and B_z) obtained from 3-D field solution are resolved into two components, viz., the axial and radial components, which then enables the use of Equation 5.1 for the eddy loss evaluation. The higher computational efforts involved in 3-D computations may get justified for large power transformers since the accuracy of computations could be critical for them.

If conductor dimensions are reduced to decrease eddy loss in windings, the DC resistance (I^2R) losses in windings increase. Hence, optimization of the total winding losses (I^2R and eddy losses) should be done. For the minimization of eddy loss (and also axial short-circuit forces) due to radial leakage field, it is essential to minimize the radial flux along the height of the windings. For a winding with tappings within its body, a high value of radial flux density may cause excessive loss and temperature rise. Balancing of ampere-turns per unit height of LV and HV windings should be done (for various sections along their height) at the average tap position to minimize the effects of the radial field. The conductor dimensions in the tap zone can be judiciously chosen different to minimize the risk of hot spots. The hot-spot consideration limits the width of the winding conductor. For a frequency of 50 Hz, the maximum width that can be used in high radial leakage field zones is usually in the range of 12–14 mm, whereas for 60 Hz it is of the order of 10–12 mm [1]. For calculating the temperature rise of a disk/turn, its eddy loss (calculated accurately by FEM analysis) and I^2R loss should be added. The guidelines given in Ref. [10] are useful for choosing the conductor width for eliminating hot spots in windings. A combination of strip and CTC conductors can be used to minimize winding losses; the CTC conductor is used only for a few top and bottom disks minimizing the eddy loss. For this purpose, however, a proper method must be established for making the joint of the CTC and strip conductors.

5.3.3 Winding Temperature Rise

Some of the thermal design requirements for the windings are in conflict with the dielectric and short-circuit strength considerations. For example, the lower the paper insulation on the conductor, the lower the winding thermal gradient over oil is. On the contrary, lower paper insulation may not satisfy the insulation strength consideration. Similarly, the first winding duct (formed due to axial insulation spacers at the inside diameter of the winding) should be as small as possible to have higher dielectric withstand (kV/mm) to counter higher stresses at the conductor corners. However, the duct width lower than 6 mm is generally not allowed in large transformers from the thermal considerations. Narrow ducts or manufacturing deficiencies result in higher flow resistance and corresponding higher temperature rise within the winding. The first duct is usually of 6–8 mm size. Another example of such conflicting requirements is the design of radial spacers. The radial spacers between disks/turns cover about 30%–40% of the winding surface, making the covered area ineffective for the convective cooling. Thus, although higher spacer width is desirable from the short-circuit withstand considerations, it is counterproductive for cooling. All these examples underline the need of judicious choice of various components of the transformer to satisfy different engineering aspects of its design. Checking of such conflicting design criteria should be integrated into design optimization program to avoid quality issues/redesigning at a later stage.

The cooling is improved with the increase in the thickness of radial spacers (contrary to their width). Although the radial spacers may not be required from the dielectric considerations in low voltage windings, they are required for providing cooling ducts. For a given radial depth of the winding, a certain minimum thickness of radial spacers is required for effective cooling (otherwise oil may largely flow through the axial ducts at inside and outside diameters of the winding due to higher flow resistance of thinner radial ducts, resulting in higher temperature rise at the middle portion of the radial depth). When the radial depth of a winding is quite high, an additional axial cooling duct in the middle of the radial depth (apart from the two axial ducts at the inside and outside diameters) may be provided. This allows provision of thinner radial spacers. The arrangement, although improves the axial space factor of the winding, worsens the radial space factor. The short-circuit strength is also weakened in the radial direction since a perfect load transfer between the sections separated by the duct is not possible. This discussion shows that the design and dimensioning of the axial and radial spacers need to be judiciously done, which may also depend on the manufacturing practices.

Guided oil flow arrangement is commonly used in power transformers for the effective cooling of the windings. The oil guiding is usually achieved by use of washers, as shown in Figure 5.4.

The washers have to be accurately sized to have a proper sealing at desired locations for eliminating oil leakages. The location and number of these oil guiding washers have to be

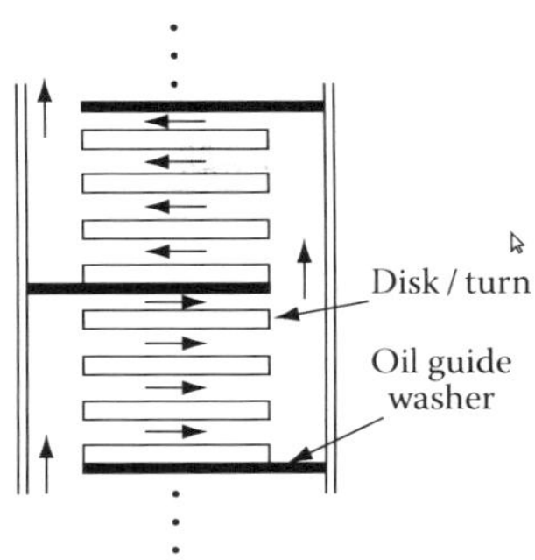

FIGURE 5.4
Directed oil flow.

appropriately selected. The value of hot-spot temperature depends on the number of radial ducts between two consecutive oil guiding washers, width of radial ducts, and width of axial ducts [11].

Static electrification has been identified as the cause of failure of a few power transformers with directed-flow forced oil cooling. This phenomenon occurs due to the friction between the oil and solid insulation components. High levels of localized electrostatic charges (due to charge separation) can get generated leading to a very high voltage inside the transformer depending on the type of oil and its velocity. The methods of avoiding/suppressing the static electrification are now well known and practiced by manufacturers [1]. The oil flow rate inside windings should be calculated to check that it is within the acceptable limits from the electrification considerations (oil pump specification may thereby get influenced).

Traditionally, the temperature rise calculations for transformer windings have been heavily relying on empirical factors. There have been continuous efforts for determining the temperatures through accurate formulations. It is reported in Ref. [12] that the variation in temperature with winding height is close to the linear distribution with forced oil cooling, whereas for naturally oil-cooled transformers, it can be quite nonlinear.

Users are increasingly specifying the direct winding hot-spot measurement through fiber-optic sensor. It is essential to accurately calculate the loss densities and the corresponding gradients for predicting locations of hot spots in windings. The hot-spot temperatures measured at these locations should be reasonably close to the calculated values. The resistivity of conductor varies along the winding height with temperature. The I^2R and eddy losses vary directly and inversely with resistivity (and therefore with temperature), respectively. Also, there is variation of the leakage field and corresponding winding eddy losses along the winding radial depth. These aspects should be taken into account during the hot-spot calculations.

5.3.4 Types of Windings

The windings of power transformers can be broadly classified into two types: layer- (spiral) and disk-type windings. The layer-type winding is mostly used for LV windings of distribution and power transformers. There are a number of design variants for disk windings. Simple continuous disk winding is used for HV windings up to 66 kV. The type of HV winding for power transformers is predominantly decided by the impulse stresses and the corresponding strength considerations. It is well known that the initial impulse distribution inside a winding is a function of distribution constant $\alpha = \sqrt{C_G/C_S}$, where C_G and C_S

are the total ground and series capacitances of the winding, respectively. The higher the value of α, the greater is the deviation between the initial nonlinear distribution (decided by winding capacitances) and the final linear distribution (decided by winding inductances); the amplitudes of oscillations that occur between these two distributions are also higher. The initial voltage distribution of the winding can be improved and made as close to the ideal linear distribution ($\alpha = 0$) by increasing its series capacitance and/or reducing its capacitance to ground. Since methods for reducing ground capacitances result into uneconomical designs, the series capacitance of the winding is increased by using different techniques.

For a conventional continuous disk winding shown in Figure 5.5, the value of α is generally higher and in the range of 5–30. This is due to the fact that the series equivalent of all the turn-to-turn and disk-to-disk capacitances comes out to be a low value. With an increase in the voltage rating, the insulations between turns and between disks have to be increased, further worsening the total series capacitance. The inherent disadvantage of low series capacitance of the continuous winding was overcome by electrostatic shielding in earlier days of transformer technology till the advent of interleaved windings. In an interleaved winding, as shown in Figure 5.6, with eight turns per disk, two consecutive electrical turns are separated physically by a turn that is electrically much farther along the winding. It does not require any additional space as in the case of electrostatic shielding (which was a very cumbersome method). The continuous disk winding with local electrostatic shields in a close vicinity of line end disks can be used up to 220 kV; here the

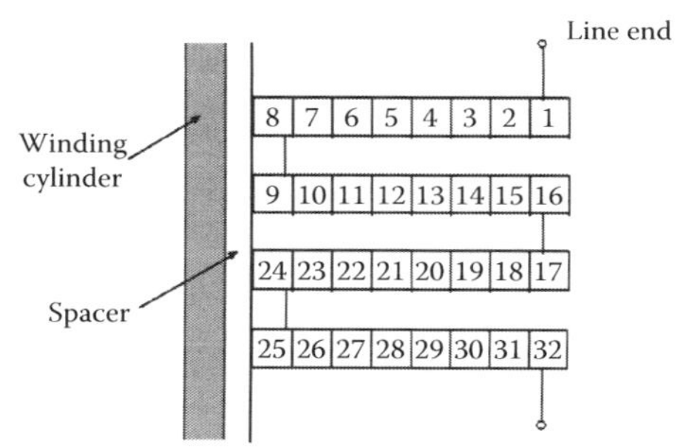

FIGURE 5.5
Continuous disk winding.

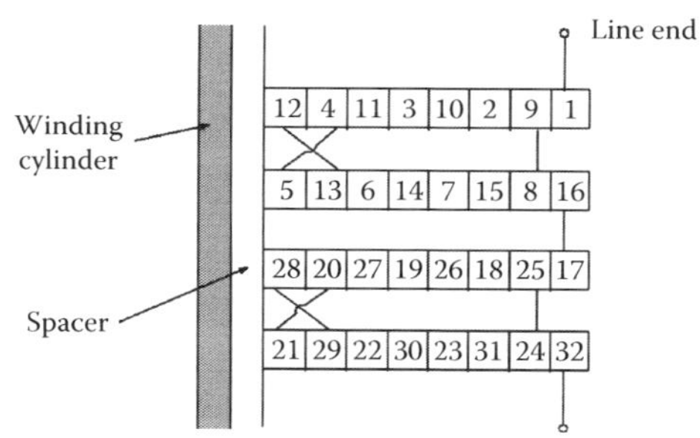

FIGURE 5.6
Interleaved winding.

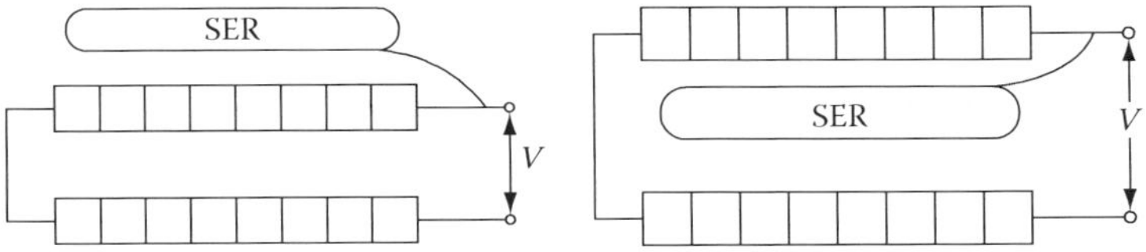

FIGURE 5.7
Continuous disk windings with SER and SR.

concept of electrostatic shielding is used in a limited way. The continuous disk windings with a static end ring (SER) at line end and a static ring (SR) between disks are shown in Figure 5.7. These electrostatic shields not only reduce local dielectric stresses by virtue of providing a large equipotential surface with a good corner radius but they also improve the effective series capacitance locally.

Interleaving requires higher paper covering on conductors from the working stress considerations. The interleaved windings are more effective in designs with higher turns per disk, and, hence, they may not be attractive for use in high-voltage high-rating transformers that are naturally designed with larger core diameters (and corresponding smaller winding turns). Also, as the rating increases, the current carried by the winding increases necessitating the use of a number of parallel conductors. The interleaved winding with more than two or three parallel conductors is not attractive from productivity point of view. For such cases, shielded-conductor (wound-in-shields) winding is commonly used nowadays for high-voltage windings of large power transformers. It results in a reasonable increase in the series capacitance in comparison with the continuous disk winding. The other major advantage is that the CTC conductor, which is well suited for windings with high current rating in large transformers, can be used with this shielded-conductor winding technology. A typical disk pair of shielded-conductor winding is shown in Figure 5.8. The first k turns are shielded in the two disks. This type of winding requires dropping of turns in the shielded disks, and the consequent change in ampere-turns/mm should be taken into account from the short-circuit force considerations (i.e., mismatch of ampere-turns between LV and HV windings should be checked and minimized). Here, the shield conductors are assumed to be floating. However, in another design, the first shield conductor is tied to a fixed potential by connecting it to the turn at outside diameter of the winding.

The series capacitances of various types of windings can be calculated by the method based on the stored energy [1,13,14]. A rigorous analytical method is presented in Ref. [15] to calculate the equivalent series capacitance of windings, which can be also

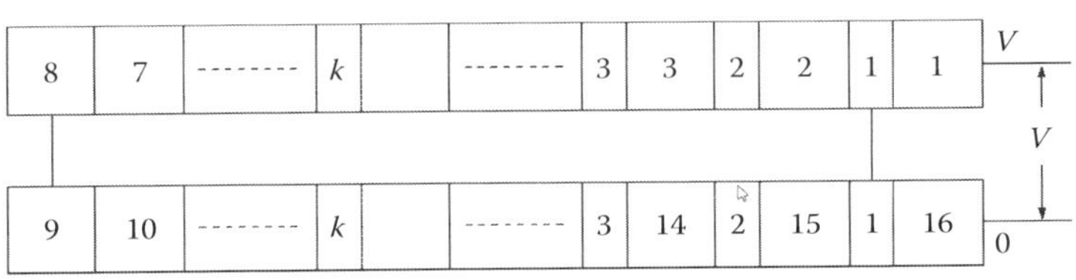

FIGURE 5.8
Shielded-conductor winding.

used to determine their natural frequencies and internal oscillations. Fringing effects and corresponding stray capacitances cannot be accurately taken into account in these analytical methods, and, hence, the numerical methods like FEM can be effectively used for the purpose [1].

Equivalent circuit representation is commonly used for finding response of transformer windings to high-voltage surges. The inductances and capacitances are distributed along the windings. The calculation of self and mutual inductances is straightforward [16,17]. Thus, if windings are represented by lumped inductances and capacitances, partial differential equations get replaced by ordinary simultaneous differential equations that can be solved by numerical analysis using computers. Transient response can be calculated using the trapezoidal rule of integration through the companion network approach [18] in which mutually coupled elements are replaced by equivalent uncoupled elements making the analysis simpler. The impulse distribution can also be calculated conveniently by using state variable method [1]. Using such methods, the disk voltages can be determined enabling the design of inter-disk insulation.

A number of failures of large transformers have been attributed to the part winding resonance phenomenon. If the frequency of an oscillating voltage, generated, for example, by a line fault, coincides with one of the fundamental natural frequencies of a winding or a part of winding, resonant overvoltages occur. Natural frequencies of core-type transformers normally lie between 5 kHz and few hundred kHz [19], and there could always be a chance that the frequency of an external oscillating disturbance is close to any of the natural frequencies of the winding. If exact natural frequencies of the network can be determined, it may be possible to change the winding type to avoid the possibility of resonance conditions (for which a closer cooperation between users and manufacturers of transformers is desirable). It has been reported in Ref. [20] that certain aperiodic overvoltages (whose waveshapes are much different than that of the standard test wave shapes), such as fast-front long-tail switching overvoltages, can lead to high internal voltage stresses in transformer windings. External countermeasures, viz., connection of external arrester to winding, use of shunt capacitors, and connection of nonlinear resistors in parallel with tap winding, have been adopted. The winding design can also be suitably modified. The surge performance of a power transformer having taps on HV winding has been analyzed in Ref. [21] for two cases, viz., interleaved winding and noninterleaved continuous disk winding. A fully interleaved winding may not always be a right solution; a winding with a graded series capacitance may sometimes prove to be a better option. For some designs, the series capacitance of the main winding can be gradually reduced in two or three steps (by the change in the degree of interleaving [1]), and its neutral end part (electrically adjacent to the tap winding) can be a continuous disk winding. In such a case, the tap winding can also be of continuous disk type.

Very fast transient overvoltages can get generated by switching operations and fault conditions in gas-insulated substations. In a worst-case scenario, a transient with a rise time of 10 ns and amplitude of 2.5 per-unit is possible. This steep fronted section of the wave is often followed by an oscillatory component having frequency in the range of 1–10 MHz [22]. In addition to producing severe intersection/inter-turn voltages, it may also lead to a part-winding resonance. For deciding countermeasures, the knowledge of inter-turn voltage distribution is essential, for which it is necessary to represent winding turns individually in the simulation models so that the very high frequency performance of the winding is assessed accurately [23].

5.4 Insulation

Sound design practices, use of appropriate insulating materials, controlled manufacturing processes, and good housekeeping ensure quality and reliability of transformers. Due to severe competition and constraints on space and weight, margins between withstand levels and stress levels are reducing. This requires greater efforts from designers for accurate calculation of dielectric stress levels at various critical electrode configurations inside the transformer under various test voltage levels and different test conditions. Advanced computational tools (e.g., FEM) are being used for accurate calculation of stress levels while designing minor (within winding) and major (between windings, between winding and core, and between leads and ground) insulations.

5.4.1 Design Insulation Level

There are basically four different types of tests, viz., lightning impulse test, switching impulse test, short-duration power frequency test, and long-duration power frequency test with partial discharge measurement, for which the insulation has to be designed. The first three tests check the insulation's overvoltage withstand capability, and the purpose of the long-duration test is mainly to check the insulation behavior under the working voltage stress. While the inter-disk spacings are decided by impulse stresses, the end clearances between windings and yoke may get predominantly decided by the power frequency test voltages. The conductor insulation is decided by either the impulse stress or working voltage stress [1].

After having calculated the voltage magnitudes inside the windings, at their ends, at the surface of leads, etc., for the aforementioned four tests, the values are generally converted to one equivalent test level called as Design Insulation Level (DIL), which is generally short-duration power frequency level. Thus, at a given point inside a transformer, there is only one DIL, which is highest of the equivalent 1 min power frequency voltage levels corresponding to the four tests. The multiplying factors for converting the estimated voltages under various tests to 1 min power frequency level are given in Ref. [24]. The basis for using a certain value or range of these factors has been explained in Ref. [1].

5.4.2 Design of Oil Ducts

It is well known that the gap between LV and HV windings is subdivided into many oil ducts by means of solid-insulating cylinders called barriers. The dielectric strength of the solid insulation is much higher than that of the oil. Since electric stress is inversely proportional to permittivity, the stress in the oil is higher than that in the solid insulation. The permittivity of oil is about half that of the solid insulation resulting in the oil stress, which is twice the solid insulation stress in uniform fields. Hence, the barriers should be as thin as possible subject to mechanical strength consideration; this gives more space for the oil ducts. Thus, in an oil-paper-pressboard insulation system, the strength of the insulation arrangement is predominantly decided by the strength of oil ducts. The lower the oil duct width, the higher the withstand stress (kV/mm) is. Widths of oil ducts away from winding corners can be higher (but generally not greater than 14 mm) as compared to that of the first duct. Insulating barriers do the additional function of acting as barriers against the propagation of discharge streamers in the oil. The HV–LV gap value is usually based on the average stress, which is generally in the range of 5.5–6 kV_{rms}/mm [1,25]. The average stress is calculated as DIL divided by the effective oil gap (after converting solid insulation thickness into equivalent oil duct). For a winding facing core (end limb clearance in single-phase three-limb construction or innermost winding to core clearance), a lower value of average stress is used due to the presence of sharp corners of core. An electrostatic shield may be required around the core/end limb in critical cases.

5.4.3 Design of End Insulation

While assessing the end insulation design, the strengths of oil gaps and solid–oil interfaces need to be evaluated. The electric stress distributions along various contours between windings and between yoke and windings are commonly analyzed by using FEM for the design of oil gaps. High-voltage leads of inner windings need to be properly taken out from the end insulation region. The cumulative stress distribution in oil gaps can be determined by the procedure given in Ref. [26]. For each oil gap, the cumulative oil stress values are compared with a reference strength curve described by a formula such as [27]

$$E_{oil} = 18 d_1^{-0.38} \text{ kV}_{rms}/\text{mm} \tag{5.2}$$

where d_1 is the oil gap distance in mm between covered electrodes oil is considered without gases

If the minimum margin between withstand and stress curves is approximately same for all the oil ducts, an optimum insulation arrangement is obtained. The barriers/angle rings in the end insulation region should be adjusted in such a way that the minimum margin in each resulting oil duct is greater than some fixed safety margin value and that these minimum margins are in the same range of values [1]. The minimum safety margin should be decided by insulation designers depending on the quality of oil–solid insulation system and technology of drying and oil-impregnation processes. The placement of angle rings/caps is a critical aspect of the end insulation design. These should be placed along the equipotential lines; otherwise, higher creepage stress along their surface may reduce the dielectric strength. The calculated cumulative creepage stress is compared with the withstand given by [28]

$$E_{creep} = 15 d_2^{-0.37} \text{ kV}_{rms}/\text{mm} \tag{5.3}$$

where d_2 is the creepage path length in mm. Ideally as one goes away from the winding corner, the radius of angle ring/cap should be more to minimize the creepage stress. The first angle ring/cap should be as close to the winding as possible for having smaller first oil duct width. Stresses at the winding corners are reduced by placing SER at winding ends, which should be placed as close to winding end disk/turn as possible all along the circumference.

5.4.4 Design of Internal Insulation

All the insulation components within the winding, viz., conductor paper covering, insulation between layers in the radial direction, insulation between turns or disks in the axial direction, and special insulating components that are placed close to the insulated conductors, are categorized as internal/minor insulation. The conductor insulation has to withstand a continuous working voltage stress in addition to withstanding short- and long-duration overvoltages during tests at manufacturer's works as well as during operation at site. The conductor paper covering thickness is commonly expressed as a function of stressed paper area, which decides the allowable working stress [1]. This criterion is particularly important for interleaved main and tap windings in which physically adjacent turns are electrically many turns apart, resulting in significantly higher working stress. To counter primarily the lightning impulse stresses, special insulating components are used closely hugging the paper-insulated winding conductors. These special insulation components, called as disk angle rings or disk angle caps, are usually placed at line-end disks or turns having high stresses. The disk or turn insulation should not be increased, at the places where stresses are high, without detailed calculations. This is because the increase in strength due to extra insulation may not be commensurate with the increase in stress due to a reduced series capacitance.

For optimization and reliability enhancement, it is important to quantify the effects of various factors on the stress levels at the electrode corners. A systematic statistical analysis called as *Analysis of Variance* can be done by using *Orthogonal Array Design of Experiments* to do parameter design (optimum combination of factors) and tolerance design (identify factors whose manufacturing variation needs to be controlled). One such analysis, done for optimization of insulation in a 400 kV class transformer, is reported in Ref. [29].

5.4.5 Design of Lead Clearances

In high-voltage transformers, lead to ground clearances need to be judiciously selected. The maximum electric stress in paper insulation and oil can be calculated accurately by FEM analysis. Strength of oil duct can be calculated based on stressed oil volume (SOV) concept. The SOV between maximum stress and contour corresponding to 90% of maximum stress is first calculated by FEM analysis. The 50% power frequency breakdown probability stress can be calculated from SOV by the formula [30]

$$E_{50} = 11.5(\text{SOV})^{-1/9.5} + 2.5\,\text{kV/mm} \tag{5.4}$$

where SOV is in cm^3. The safe withstand value is lower than E_{50} value and can be decided by the transformer manufacturer depending on the acceptable value of % breakdown probability and the quality of manufacturing processes [30,31]. Working tolerances need to be added. For lower voltages, mechanical tolerances and lowering clearances as well as magnetic clearances may dominate over dielectric considerations. Two distinct approaches for determining withstand for oil ducts have been discussed in the literature, viz., one based on distance and other based on SOV. The details of these two approaches are given in Ref. [1], and the consistency of strength values, given by them, has been elaborated in Ref. [27].

When grounded electrode is not smooth and has sharp corners, it becomes difficult to standardize the clearances. In any case, a minimum of 2 mm radius should be provided at corners. In such cases, in the absence of detailed analysis, 1 mm of oil gap per kV of DIL seems to be a safe design guideline [1]. The earlier discussion is applicable for jump clearances in oil between lead and ground. When leads are supported by solid insulation from a grounded part, as shown in Figure 5.9 with the corresponding equipotential lines obtained from FEM analysis, cumulative creepage stresses and strengths need to be computed.

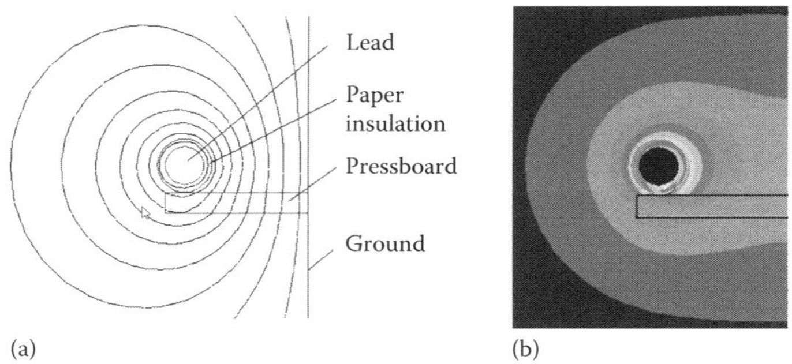

FIGURE 5.9
FEM analysis for optimum lead clearances. (a) Equipotential plot and (b) stress plot.

5.5 Stray Loss Control

Stray losses in a power transformer may form a large part of total load loss if not evaluated and controlled properly. The uncontrolled stray field will result into excessive losses in the tank and various clamping structures leading to hot spots. Computation of stray losses in a transformer is complicated because of its highly asymmetrical and 3-D structure. Magnetic nonlinearity, limitations of experimental verification methods, inability in isolating exact stray loss components from tested load loss values, etc., are the additional hurdles.

5.5.1 Factors Affecting Stray Losses

5.5.1.1 Type of Surface Excitation

There can be mainly two types of surface excitation for structural components in transformers. Incident field can be either tangential (e.g., bushing mounting plate) or normal (e.g., tank). If hysteresis and nonlinearity of magnetic characteristics are not taken into account, the expression for the eddy loss per unit surface area of a plate, subjected to (on one of its surfaces) a magnetic field of rms value (H_{rms}), is [1]

$$P_e = \sqrt{\frac{\omega\mu}{2\sigma}} H_{\text{rms}}^2 \tag{5.5}$$

where
σ is conductivity
μ is permeability of the plate material

Surfaces of structural steel parts are often saturated due to the skin effect necessitating the use of appropriate linearization coefficient in which case the total power loss for a steel plate with a permeability μ_s can be given in terms of the peak value of the field (H_0) as

$$P = a_1 \iint_S \sqrt{\frac{\omega\mu_s}{2\sigma}} \frac{H_0^2}{2} ds \tag{5.6}$$

The term a_l in the previous equation is the linearization coefficient. The equation is applicable to a plate excited by a tangential field on one of its sides; the plate thickness being assumed sufficiently larger than the depth of penetration so that it becomes a case of infinite half space. For magnetic steel used in transformers, the linearization coefficient has been taken as 1.4 in Ref. [32].

The analytical calculation of stray loss in terms of the normally incident field is not straightforward. In many analytical formulations, the two orthogonal tangential field components are first calculated from the normal component using Maxwell's equations. The resultant tangential component can be used to calculate the losses as per Equation 5.6.

5.5.1.2 Material Properties

The stray loss behavior in a structural component is also a function of its material. Loss variation with the thickness of structural plate shows distinct patterns for aluminum, magnetic steel, and stainless steel materials [1,13]. When a plate made of nonmagnetic and highly conductive material (e.g., aluminum and copper) is used in the vicinity of field due to high currents or leakage field from windings, it should have sufficient thickness (at least comparable to its depth of penetration) to avoid its overheating and to be able to minimize the stray loss in the structural component shielded by it. For a plate made of magnetic steel, since its depth of penetration is usually much less than the thickness required from mechanical design considerations, the eddy loss in it cannot be controlled by changing its thickness. Hence, either magnetic shunts or eddy current shields are used to minimize the stray losses in magnetic structural components in medium and large transformers. When a structural component is made of (nonmagnetic high resistivity) stainless steel, its thickness should be small as possible (permitted from mechanical design considerations) in order to get a lower loss value since its depth of penetration is much higher. It should be appreciated that losses in magnetic steel (mild steel) and nonmagnetic steel plates could be comparable for their thickness of around 10–13 mm [1,13]. More elaborate discussions on the earlier two factors and the other factors (frequency, load, temperature) are available in Ref. [1].

5.5.2 Stray Loss Components

5.5.2.1 Core Edge Loss

It occurs due to flux incident radially (normally) on the core laminations. The loss could be substantial in large power transformers, particularly for generator transformers. The first step of the core is split into two or three parts to reduce the loss. The use of laminated flitch plate is recommended in large transformers since it acts as magnetic shunt reducing the core edge loss.

5.5.2.2 Flitch Plate Loss

Although the losses occurring in a flitch plate may not form a significant part of the total load loss of a transformer, the local temperature rise can be much higher due to high value of incident flux density and poorer cooling conditions. There are mainly four different designs of flitch plate, viz., plain type without slots, with slots at two ends, with slots of full length, and laminated. The plain type is used for small transformers, whereas designs with slots are used for medium ratings. The slots are made in the regions where the flux is incident on the plate. The material of the plate in the first three designs can be either

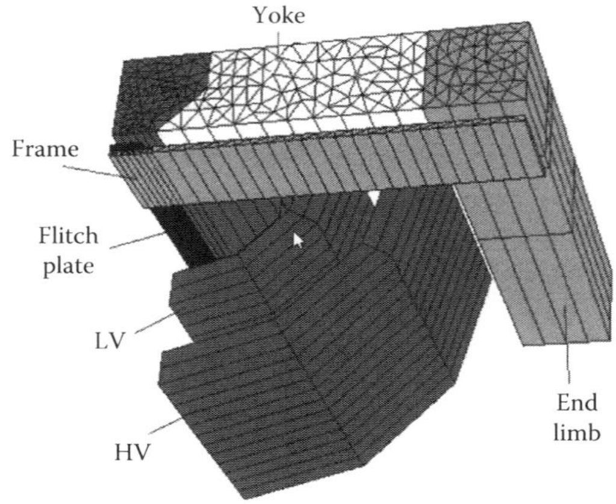

FIGURE 5.10
3-D FEM model for evaluation of stray loss in flitch plate. (Reprinted with permission from Koppikar, D.A., Kulkarni, S.V., Srinivas, P.N., Khaparde, S.A., and Jain, R., Evaluation of flitch plate losses in power transformers, *IEEE Transactions on Power Delivery* © 1999 IEEE.)

magnetic or nonmagnetic. The laminated flitch plate design is preferred for large transformers (especially generator transformers) since it not only results into least losses in the plate but it also reduces the core edge loss as mentioned earlier. The eddy current patterns and losses in these various types of flitch plates are analyzed in Ref. [33] by using 3-D FEM analysis. The 3-D model is shown in Figure 5.10. It is shown in the paper that slots are more effective in reducing losses in nonmagnetic stainless steel plate as compared to magnetic mild steel plate.

5.5.2.3 Frame Loss

Although hot spots seldom develop in frames (since cooling conditions are better), the stray loss in them can be appreciable if their considerable surface area is exposed to leakage field and high current field. It has been shown in Ref. [34] that losses in the frame and tank have mutual effect on each other. Nonmagnetic steel is not recommended as a material for frames since it is expensive and difficult to machine, and stray losses will be lower only if its thickness is sufficiently small. The loss in frames due to leakage field can be reduced by the use of either aluminum shields or nonmetallic platforms (for supporting windings). The loss due to high current field is reduced by having *go-and-return* arrangement of LV winding leads passing by the frames.

5.5.2.4 Stray Loss in Tank

This is generally the largest component of total stray loss due to exposure of tank's large surface area to leakage field/high current field. The stray loss is controlled by either magnetic or eddy current shielding. The magnetic shunts (made up of CRGO/CRNGO laminations stacked together) can effectively control stray losses in the tank. They offer low reluctance path to the incident leakage field constraining its path in a predictable fashion. There are two designs of magnetic shunts, viz., width-wise and edge-wise, as shown in Figure 5.11. The width-wise shunt is less effective due to nonmagnetic gaps, between laminations, encountered by the flux at the entry point. However, the manufacturing

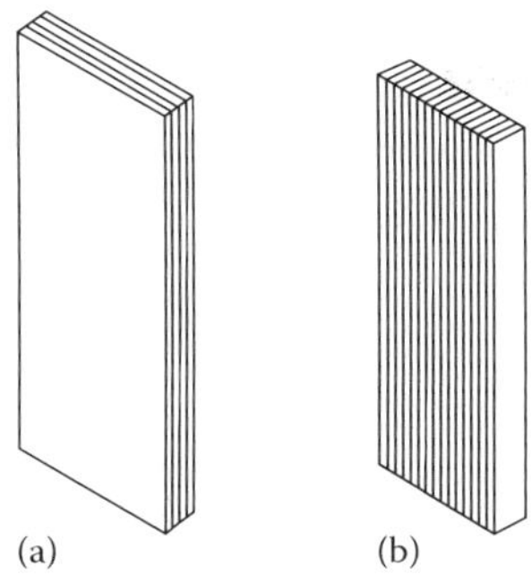

FIGURE 5.11
Tank shunts. (a) Width-wise and (b) edge-wise.

process of the edge-wise shunt is more elaborate, and the decision for using it should be based on the analysis of loss advantage vis-à-vis its higher manufacturing time/cost. Magnetic shunts, which are placed parallel to the yoke at the top and bottom ends of the windings, are called as yoke shunts. These not only can guide the leakage field safely back to the core minimizing stray losses in the tank and other structural components but they also tend to make the leakage field in the windings axial, thus minimizing the winding eddy loss due to the radial field at the winding ends. The gap between the shunt and yoke must be kept sufficiently small for the effective control of the leakage field. The yoke shunts are more effective and convenient for use in three-phase five-limb transformers as compared to three-phase three-limb transformers [1]. The eddy current shields (of aluminum or copper) have the advantage that they can be fitted on odd shapes of the tank. For shielding a tank plate from the high current field, the eddy current shields are invariably used due to ineffectiveness of magnetic shunts on account of gaps between them. However, the flux repelled by eddy current shields should be analyzed since it may cause losses in nearby magnetic structures. As discussed earlier, the thickness of these shields should be adequate for their effectiveness and for reducing the loss and temperature rise in them.

5.5.2.5 Stray Loss in Bushing Mounting Plate

The bushing mounting plates are made of either mild steel or stainless steel material. With the increase in currents carried by bushings, the eddy current loss and the related heating effects increase. Hence, for higher currents, nonmagnetic stainless steel plates are used. For intermediate currents, mild steel plates with nonmagnetic inserts, as shown by hashed lines in Figure 5.12, are used. Analysis of losses in bushing mounting plates is reported

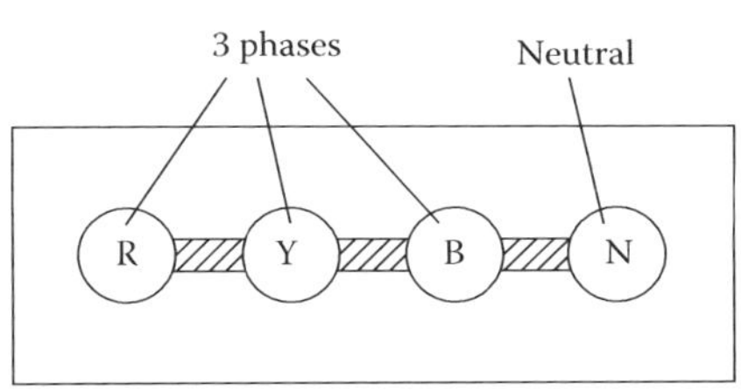

FIGURE 5.12
Bushing mounting plate.

in Ref. [35]. The losses calculated by an analytical method [36,37] and 3-D FEM have been compared with those calculated from steady-state temperature rise and initial temperature rise measured on a typical plate.

5.5.3 Methods for Stray Loss Estimation

2-D analytical/numerical methods and 3-D analytical methods were used earlier for calculation of stray losses in various structural components. With great advances in computational power and tools, application of numerical methods such as FEM, finite difference method, boundary element method, etc., is now possible for the calculation of 3-D fields inside a transformer and for accurate estimation of stray losses in structural components. For rapid and reasonably accurate calculations, reluctance network method [32] has also been used, which is a hybrid analytical-numerical method in which the analytical approach is used for the portion of the geometry involving eddy currents. In another method, the problem of modeling small skin depths in FEM analysis has been resolved by using the concept of surface impedance [38]. The review of all such methods is given in Ref. [39].

5.6 Short-Circuit Withstand

A steady increase in unit ratings of transformers and simultaneous growth of short-circuit capacities of networks have made *short-circuit withstand* as one of the most important aspects of the power transformer design. The short-circuit test failure rate is high for large transformers. The test data from high power test laboratories around the world indicate that on an average practically one transformer out of four has failed during the short-circuit test, and the failure rate is above 40% for transformers above 100 MVA rating [40].

5.6.1 Radial and Axial Forces

The leakage flux density at any point in the winding can be resolved into two components, viz., one in the radial direction that produces axial force and the other in the axial direction that is responsible for radial force. The radial forces act outward on the outer winding tending to stretch the winding conductor, producing a tensile stress (hoop stress), whereas the inner winding experiences radial forces acting inward tending to collapse or crush it, producing a compressive stress. For layer windings having multiple layers, the average radial stresses in various layers are different, the stress being highest in the layer next to the LV–HV gap. This happens due to the fact that the layers do not firmly support each other and there is no transfer of load between them.

Mismatch of ampere-turn distribution along the height of LV and HV windings, axial asymmetry between the two windings, taps in a winding, unaccounted shrinkage of insulation, etc., are reasons for higher radial field and corresponding axial forces. A small axial displacement of windings or misalignment of magnetic centers of windings can eventually cause enormous axial forces since the resulting axial forces are in such a direction that the asymmetry and the end thrusts on the clamping structures increase further.

5.6.2 Computational Methods

Over the years, calculation of forces has been studied from static considerations. Different methods for calculation of forces are well documented in 1979 by a CIGRE working group [41]. Static forces can be calculated by any one of the following established methods, viz., Roth's method, Rabin's method, method of images, and FEM. Withstand is checked for the first peak of asymmetrical short-circuit current.

Methods for assessing dynamic performance under short circuits are not yet perfected due to lack of precise knowledge of dynamic characteristics of various materials used in transformers. Transformer winding, made up of large number of conductors separated by insulating material, can be represented by an elastic column with distributed mass and spring parameters, restrained by end springs representing insulation between windings and yoke [42]. When a force is applied to an elastic structure, displacements and stresses depend not only on magnitude of force and its variation with time but also upon the natural frequencies of the structure. In the radial direction, due to the larger ratio of stiffness to mass for the winding conductor, the natural frequency is much higher than 50/60 and 100/120 Hz (the fundamental frequency and twice the fundamental frequency of the excitation force). Hence, the possibility of increase in displacements by resonance effects under the action of radial forces is usually remote.

In case of nonavailability of accurate dynamic characteristics of the materials, natural frequencies are calculated, and it is ensured that they are reasonably away from the power frequency or twice the power frequency. Hence, established static calculations along with natural frequency determination could form a basis of short-circuit strength calculations [43,44] until the dynamic analysis is perfected and standardized. If natural frequency is close to either 50 or 100 Hz (60 or 120 Hz), it can be altered by using different prestress value or by altering mode of vibration by suitable subdivision of windings [45].

5.6.3 Failure Modes and Short-Circuit Withstand

5.6.3.1 Radial Forces

The strength of outer windings subjected to the outward forces depends on the tensile strength of the conductor. If the stress exceeds the yield strength of the conductor, a failure occurs (e.g., damage to conductor insulation, local bulging of winding, and breaking of conductor due to improper joints). Use of work-hardened conductor, design of winding with lower current density, and use of a conductor with a certain minimum 0.2% proof strength are some of the countermeasures for improving the strength.

The strength of inner windings subjected to the inward forces depends on their support structure. The radial collapse of inner windings is more common as compared to outer windings whose outward bursting usually does not take place. Conductors of inner windings, which are subjected to the radial compressive load, may fail due to forced buckling (bending between supports) or free buckling, as shown in Figure 5.13. The former failure occurs when the inner supporting structure (axial spacers, winding cylinder, and core) as a whole has stiffness higher than that of conductors. The free buckling is an unsupported buckling mode in which the span of the conductor buckle bears no relation to the span of axial supporting spacers. There are many factors that may lead to the buckling phenomenon, viz., winding looseness, no firm support to winding from inside, eccentricities in windings, lower stiffness of supporting structures as compared to that of the conductor, etc. Use of thicker insulating cylinders for supporting inner windings, proper alignment

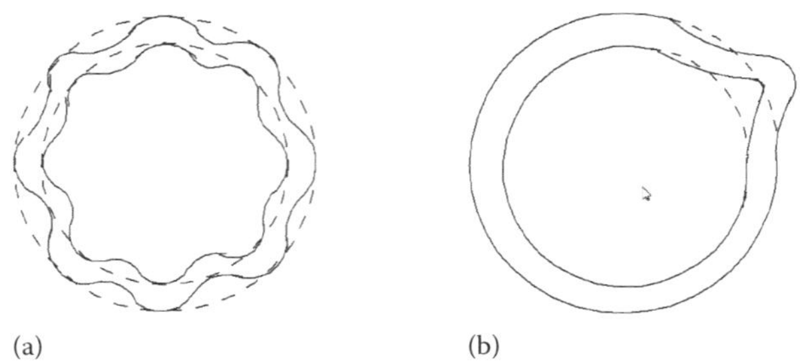

FIGURE 5.13
Buckling phenomena. (a) Forced buckling and (b) free buckling.

and positioning of axially placed spacers to give adequate support to the inner windings, and placement of tight-fitting wooden dowels on the core in close contact with the insulating cylinders are some of the ways to improve the withstand. In one of the design philosophies [46], the strength provided by the inner supporting structures is completely ignored, and the windings are designed to be completely self-supporting. The current density used is lower in such designs increasing the cost of the transformer, which can be easily justified for large and critical transformers.

5.6.3.2 Axial Forces

There are two principal types of failures due to axial forces, viz., bending between radial spacers and tilting. Under the action of axial forces, the winding conductor can bend between the radially placed insulation spacers. The higher the conductor dimensions (and hence the moment of inertia) and the lower the span between two radial supports, the lower the bending stress is. The conductor bending can result into a damage of its insulation. In tilting, there is turning of cross section of conductors around the perpendicular axis of symmetry. The conductor material and frictional forces oppose the conductor tilting. These two resisting forces are usually considered separately to arrive at critical stress and load [45]. The tilting strength due to conductor material decreases inversely as the square of winding radius, suggesting that the large windings should be carefully designed. Two modes of failures in CTC conductor, viz., cable-wise tilting and strand-wise tilting, are analyzed in Ref. [47]. The epoxy-bonding of CTC conductor greatly enhances its strength since the epoxy coating effectively bonds the strands increasing the resistance against the strand-wise tilting (as well as buckling). Some of the other countermeasures against axial forces are, viz., estimation of natural frequencies of windings and ensuring that there is no excited resonance, strict sizing/dimension control during processing and assembling of windings so that they get symmetrically placed, minimization of radial flux for designs with taps in body of main winding, judicious selection of prestress (clamping pressure applied on the windings after the completion of core-winding assembly) value, etc.

Revision has been made in IEC 60076-5 standard, second edition 2000–2007, reducing the limit of change in impedance from 2% to 1% for category III (above 100 MVA rating) transformers. This revision has far reaching implications for transformer manufacturers because a much stricter control on the variations in materials and manufacturing processes will have to be exercised to avoid looseness and winding movements. General guidelines and precautions that can be taken at the specification, design, and production stages of transformers for improving the short-circuit strength are summarized in Ref. [1].

5.7 Recent Trends in Computations

The recent research trends show that many of the complex design problems in transformers, involving more than one physical field (electrostatic, electromagnetic, structural, thermal, etc.), are increasingly being solved by using coupled field formulations [48]. Coupled field treatment is required for problems in which the involved fields (e.g., electromagnetic and thermal) interact either strongly or weakly. The problems are either strongly or weakly coupled depending on the degree of nonlinearity and the relative time constants of the involved fields. A weakly coupled problem is solved using cascaded algorithm; the coupling is performed by applying the results from the first analysis (involving only one field) as the loads for the second analysis (which involves the other field) greatly simplifying the formulation. For a strongly coupled problem, the governing equations representing the coupled fields are solved simultaneously with all the necessary variables taken into account. This approach is essential when the coupled-field interaction is highly nonlinear and the involved fields have comparable time constants.

5.7.1 Thermomagnetic Problems

In transformers, the problem of estimation of temperature rise of conducting parts due to induced eddy currents is generally solved as a weakly coupled problem since the thermal time constant is much higher than the electromagnetic time constant. Transient solution of a thermomagnetic problem is computationally very expensive due to large difference between the time constants of the magnetic and thermal fields. When nonsinusoidal voltages and/or currents are involved, the magnetic field will also contain harmonics. The coupled thermomagnetic analysis of a transformer subjected to current harmonics is outlined in Ref. [49], which uses a mixed frequency and time domain approach. In this approach, the time-dependent magnetic vector potential (A) is represented by the sum of its harmonic components as

$$A(t) = \sum_{h=1}^{N} A_h(t)e^{jh\omega t} \tag{5.7}$$

where N is the number of harmonics.

5.7.2 Transient Analysis

To model various transient phenomena in transformers involving coupled fields, which are categorized as coupled circuit-field problems, the current density in windings must be known. If a transformer is fed by voltage sources through some electrical circuitry, the currents in the conductors are not known in advance. The time transients, nonlinear materials, nonlinear circuit components, etc., are real challenges while simulating the field-circuit couplings that need to be treated as strongly coupled problems.

For analyzing behavior of transformers under a transient condition, such as short-circuit or inrush phenomenon, a 3-D formulation is essential. The complete transient simulation of a power transformer using a 3-D model with ***A*-*V*-*A*** formulation coupled with circuit equations is described in Ref. [50]:

$$\nabla \times \frac{1}{\mu} \times \boldsymbol{A} - \nabla \frac{1}{\mu} \nabla \cdot \boldsymbol{A} + \sigma \nabla V + \sigma \frac{\partial \boldsymbol{A}}{\partial t} = 0 \tag{5.8}$$

and

$$\nabla \cdot \left(-\sigma \nabla V - \sigma \frac{\partial \boldsymbol{A}}{\partial t} \right) = 0 \quad \text{in } S_1 \tag{5.9}$$

$$\nabla \times \frac{1}{\mu} \times \boldsymbol{A} - \nabla \frac{1}{\mu} \nabla \cdot \boldsymbol{A} = \boldsymbol{J} \quad \text{in } S_2 \tag{5.10}$$

where
- σ is conductivity
- μ is permeability
- V is electric potential
- $\boldsymbol{J}$ is current density vector
- S_1 is the eddy current region
- S_2 is the region without eddy currents

The current density magnitude can be expressed in terms of instantaneous current as

$$J = \frac{N_c}{A_c} i(t) \tag{5.11}$$

where
- N_c is the number of turns
- A_c is the total cross-sectional area of the windings

Analysis of transient state of a power transformer, taking field and connected circuit into account with complete 3-D description of the winding and magnetic core, introduces large number of unknowns in the case of the magnetic vector potential formulation (***A***-*V*-***A***). In order to reduce the number of unknowns, ***T***-Ω formulation is used. In this case, the magnetic scalar potential (Ω) is considered in whole domain, while the electric vector potential (***T***) is considered only in the conducting domain.

5.7.3 Noise Level Prediction

Although various methods for noise level reduction in transformers are well known and practiced [1], the prediction of noise level is still a complex and challenging problem. No-load magnetostrictive strain, load-controlled noise caused by the Lorentz forces, and noise produced by fans and oil pumps are the main sources of the noise in transformers. To estimate noise radiations from a transformer due to load current, the electromagnetic, structural, fluid-flow, and acoustic fields must be taken into account with appropriate couplings [51]. In the case of loaded power transformer, the main coupling between the mechanical and magnetic fields is due to the magnetic volume force $\boldsymbol{f}_v$ resulting from the Lorentz forces that cause the winding vibrations,

$$f_v = \boldsymbol{J} \times \boldsymbol{B} = \left(\boldsymbol{J}_0 - \sigma \nabla V - \sigma \frac{\partial \boldsymbol{A}}{\partial t} + \sigma \boldsymbol{v} \times (\nabla \times \boldsymbol{A}) \right) \times (\nabla \times \boldsymbol{A}) \tag{5.12}$$

where
$\boldsymbol{B}$ is the flux density
$\boldsymbol{J}_0$ is the source current density
$\boldsymbol{v}$ is the time derivative of mechanical displacement

Magnetic and mechanical fields are coupled strongly. On the contrary, the mechanical and acoustic fields are coupled weakly for which the calculated tank surface displacements are applied as inputs to the acoustic field model to calculate the radiated transformer noise.

5.8 Conclusions

In today's very competitive and demanding market conditions, it is a very difficult task to design a cost-effective power transformer that meets high quality and reliability standards. This chapter has addressed many of these challenges and has suggested strategies to tackle them. It has encompassed all the important aspects of the transformer design technology. The chapter has summarized briefly some of the important design considerations elaborated in Ref. [1].

It has been emphasized that the accurate estimation of core building factor is of paramount importance since the expected loss reduction for better grades may not be obtained in the absence of analysis of test/experimental data. The operating flux density value in the core has to be decided after due considerations to overexcitation conditions. The effect of various design and manufacturing practices on the core performance in the normal and overexcitation conditions can be analyzed using 3-D numerical techniques that take into account magnetic anisotropy and nonlinearity.

The design of windings is influenced by electrostatic, electrodynamic, and thermal aspects that often have conflicting requirements. Hence, it is required to judiciously choose winding parameters for satisfying different design criteria. In order to minimize the effects of radial leakage fields, the conductor width has to be properly selected. The CTC conductor is commonly used for large transformers to reduce eddy losses and improve productivity. Loss densities in various parts of windings need to be accurately calculated to predict the possible hot-spot locations. This is particularly important when fiber-optic sensors are installed to directly measure the hot spots. The type of high-voltage winding used depends mainly on impulse withstand considerations. Interleaved windings, used for increasing series capacitance of windings, should be selectively used. In certain cases, winding with a graded capacitance design proves to be a better option. Shielded-conductor windings are suitable when turns per disk are lower (as in large power rating transformers). A greater cooperation between manufacturers and users of transformers is desirable to eliminate the possibilities of part winding resonance and to mitigate the effects of very fast transient overvoltages.

For higher-voltage rating transformers, comprehensive insulation design philosophy, use of appropriate insulating materials, and controlled manufacturing processes are essential. Due to severe and competitive market conditions, insulation arrangements may be used for which margins between strengths and stresses could be critical. It is necessary to calculate the stresses at various electrodes inside a transformer by techniques such as FEM analysis. The design of gaps between windings, between leads and ground, and between windings and core amounts to designing of the resulting oil ducts optimally. The bulk oil withstand and creepage strength considerations are the deciding factors.

Stray losses in large power transformers can become considerable part of the load loss if they are not evaluated and controlled properly. The evaluation and control of stray losses in core, flitch plates, frames, tank, and bushing mounting plate have been discussed. For the structural components, judicious use of magnetic and nonmagnetic materials is required. Adequate dimensioning and proper placements of shields are necessary for the effective control of the stray losses.

Failure of transformers due to short circuits is a major concern for transformer users. The success rate during actual short-circuit tests is far from satisfactory. The static force and withstand calculations are well established. Efforts are being made to standardize and improve the dynamic short-circuit calculations. Failure modes due to the axial and radial forces have been highlighted in the chapter. Quality of manufacturing processes has a significant impact on the short-circuit withstand. A number of precautions that can be taken at the design and manufacturing stages of transformers for improvement in short-circuit withstand have been elaborated.

Finally, the recent advances in the area of analysis and computation of various phenomena in transformers are enumerated. The complex design problems involving more than one physical field are solved by coupled field formulations. Estimation of temperature rise of conducting parts due to induced eddy currents, analysis of transformers under transient conditions, prediction of noise level, etc., are the problems for which the coupled field treatment is generally essential.

This chapter has thus highlighted a number of design and manufacturing issues that continue to attract the attention of researchers. It has encompassed all the important aspects of transformer engineering including the recent advances in research and development activities. The ongoing developments in computational tools are expected to provide better solutions to overcome many of the challenges identified in the chapter.

References

1. Kulkarni SV, Khaparde SA (2004) *Transformer Engineering: Design and Practice*. Marcel Dekker, CRC Press, Taylor & Francis Group, New York.
2. Del Vecchio RM, Poulin B, Feghali PT, Shah DM, Ahuja R (2001) *Power Transformer Design Principles: With Applications to Core-Form Power Transformers*. Gordon and Breach Science Publishers, Canada.
3. Girgis RS, te Nijenhuis EG, Gramm K, Wrethag JE (1998) Experimental investigations on effect of core production attributes on transformer core loss performance. *IEEE Trans Power Deliv*, 13(2), 526–531.

4. Mechler GF, Girgis RS (2000) Magnetic flux distributions in transformer core joints. *IEEE Trans Power Deliv*, 15(1), 198–203.
5. Koppikar DA, Kulkarni SV, Khaparde SA (1998) Overfluxing simulation of transformer by 3D FEM analysis. *Fourth Conference on EHV Technology*, IISc Bangalore, India, pp. 69–71.
6. Silva VC, Meunier G, Foggia A (1995) A 3-D finite element computation of eddy current and losses in laminated iron cores allowing for electric and magnetic anisotropy. *IEEE Trans Magn*, 31(3), 2139–2141.
7. Boyajian A (1954) Leakage reactance of irregular distributions of transformer windings by method of double Fourier series. *AIEE Trans Power Appar Syst*, 73(Pt III-B), 1078–1086.
8. Rabins L (1956) Transformer reactance calculations with digital computers. *AIEE Trans Commun Electron*, 75(Pt I), 261–267.
9. Girgis RS, Scott DJ, Yannucci DA, Templeton JB (1987) Calculation of winding losses in shell form transformers for improved accuracy and reliability—Part I: Calculation procedure and program description. *IEEE Trans Power Deliv*, PWRD-2(2), 398–410.
10. Kozlowski M, Turowski J (1972) Stray losses and local overheating hazard in transformers. *CIGRE*, Paris, France, Paper No 12-10.
11. Kamath RV, Bhat G (1998) Numerical simulation of oil flow through cooling ducts of large transformer winding. *International Conference on Transformers, TRAFOTECH*, Mumbai, India, pp. I1–I5.
12. Pierce LW (1992) An investigation of the thermal performance of oil filled transformer winding. *IEEE Trans Power Deliv*, 7(3), 1347–1358.
13. Karsai K, Kerenyi D, Kiss L (1987) *Large Power Transformers*. Elsevier Publication, Amsterdam, The Netherlands.
14. Del Vecchio RM, Poulin B, Ahuja R (1998) Calculation and measurement of winding disk capacitances with wound-in-shields. *IEEE Trans Power Deliv*, 13(2), 503–509.
15. Chowdhuri P (1987) Calculation of series capacitance for transient analysis of windings. *IEEE Trans Power Deliv*, PWRD-2(1), 133–139.
16. Grover FW (1947) *Inductance Calculations: Working Formulae and Tables*. Van Nostrand Company, Inc, Princeton, NJ.
17. Miki A, Hosoya T, Okuyama K (1978) A calculation method for impulse voltage distribution and transferred voltage in transformer windings. *IEEE Trans Power Appar Syst*, PAS-97(3), 930–939.
18. Kasturi R, Murty GRK (1979) Computation of impulse voltage stresses in transformer windings. *Proc IEE*, 126(5), 397–400.
19. Preininger G (1993) Resonant overvoltages and their impact on transformer design, protection and operation. *International Summer School of Transformers, ISST'93*, Technical University of Lodz, Lodz, Poland, Paper No 11.
20. Musil RJ, Preininger G, Schopper E, Wenger S (1981) Voltage stresses produced by aperiodic and oscillating system overvoltages in transformer windings. *IEEE Trans Power Appar Syst*, PAS-100(1), 431–441.
21. De A, Chatterjee N (2000) Part winding resonance: Demerit of interleaved high-voltage transformer winding. *Proc IEE Electr Power Appl*, 147(3), 167–174.
22. Cornick K, Filliat B, Kieny C, Muller W (1992) Distribution of very fast transient overvoltages in transformer windings. *CIGRE*, Paris, France, Paper No 12-204.
23. De Leon F, Semlyen (1994) A complete transformer model for electromagnetic transients. *IEEE Trans Power Deliv*, 9(1), 231–239.
24. Dahinden V, Schultz K, Kuchler A (1998) Function of solid insulation in transformers. *Transform 98*, Munich, Germany, pp. 41–54.
25. Lakhiani VK, Kulkarni SV (2002) Insulation design of EHV transformers—A review. *International Insulation Conference, Insucon 2002*, Berlin, Germany, pp. 283–287.
26. Nelson JK (1994) Some steps toward the automation of the design of composite dielectric structures. *IEEE Trans Dielectr Electr Insulation*, 1(4), 663–671.

27. Nelson JK (1989) An assessment of the physical basis for the application of design criteria for dielectric structures. *IEEE Trans Electr Insulation*, 24(5), 835–847.
28. Derler F, Kirch HJ, Krause C, Schneider E (1991) Development of a design method for insulating structures exposed to electric stress in long oil gaps and along oil/transformer board surfaces. *International Symposium on High Voltage Engineering, ISH'91*, Dresden, Germany, pp. 1–4.
29. Koppikar DA, Kulkarni SV, Dubey AK (1997) Optimization of EHV transformer insulation by statistical analysis. *International Symposium on High Voltage Engineering, ISH'97*, Montreal, Quebec, Canada, Vol. 6, pp. 289–292.
30. Ikeda M, Teranishi T, Honda M, Yanari T (1981) Breakdown characteristics of moving transformer oil. *IEEE Trans Power Appar Syst*, PAS-100(2), 921–928.
31. Ikeda M, Menju S (1979) Breakdown probability distribution on equi-probabilistic V-T characteristics of transformer oil. *IEEE Trans Power Appar Syst*, PAS-98(4), 1430–1437.
32. Turowski J, Turowski M, Kopec M (1990) Method of three-dimensional network solution of leakage field of three-phase transformers. *IEEE Trans Magn*, 26(5), 2911–2919.
33. Koppikar DA, Kulkarni SV, Srinivas PN, Khaparde SA, Jain R (1999) Evaluation of flitch plate losses in power transformers. *IEEE Trans Power Deliv*, 14(3), 996–1001.
34. Kerenyi D (1988) Stray load losses in yoke-beams of transformers. *Electromagnetic Fields in Electrical Engineering*, Savini A and Turowski J (eds), Plenum Press, New York, pp. 113–118.
35. Kulkarni SV, Olivares JC, Escarela-Perez R, Lakhiani VK, Turowski J (2004) Evaluation of eddy losses in cover plates of distribution transformers. *IEE Proc Sci Meas Technol*, 151(5), 313–318.
36. Turowski J (1993) *Technical Electrodynamics* (in Polish), WNT, Warszawa, Poland.
37. Turowski J, Pelikant A (1997) Eddy current losses and hot spot evaluation in cover plates of power transformers. *Proc IEE Electr Power Appl*, 144(6), 435–440.
38. Holland SA, O'Connell GP, Haydock L (1992) Calculating stray losses in power transformers using surface impedance with finite elements. *IEEE Trans Magn*, 28(2), 1355–1358.
39. Kulkarni SV, Khaparde SA (2000) Stray loss evaluation in power transformers: A review. *IEEE PES Winter Meet*, Singapore, Paper No 0-7803-5938-0/00.
40. Bergonzi L, Bertagnolli G, Cannavale G, Caprio G, Iliceto F, Dilli B, Gulyesil O (2000) Power transmission reliability, technical and economic issues relating to the short circuit performance of power transformers. *CIGRE*, Paris, France, Paper No 12-207.
41. Sollergren B (1979) Calculation of short circuit forces in transformers. *Electra*, Report No 67, pp. 29–75.
42. Patel MR (1973) Dynamic response of power transformers under axial short circuit forces, Part I: Winding and clamp as individual components. *IEEE Trans Power Appar Syst*, PAS-92, 1558–1566.
43. Macor P, Robert G, Girardot D, Riboud JC, Ngnegueu T, Arthaud JP, Chemin E (2000) The short circuit resistance of transformers: The feedback in France based on tests, service and calculation approaches. *CIGRE*, Paris, France, Paper No 12-102.
44. Lakhiani VK, Kulkarni SV (2002) Short circuit withstand of transformers: A perspective. *International Conference on Transformers, TRAFOTECH*, Mumbai, India, pp. 34–38.
45. Waters M (1966) *The Short Circuit Strength of Power Transformers*. Macdonald and Co Ltd, London, U.K.
46. Bertagnolli G (1996) *Short Circuit Duty of Power Transformers. The ABB Approach*. Golinelli Industrie Grafiche, Legnano, Italy.
47. Patel MR (2002) Instability of the continuously transposed cable under axial short circuit forces in transformers. *IEEE Trans Power Deliv*, 17(1), 149–154.
48. Kulkarni SV, Kumbhar GB, Nabi M (2003) Current trends in coupled field formulations in electrical machinery. *6th International Symposium on Electromagnetic Fields*, Aachen, Germany, pp. 287–291.
49. Driesen J, Belmans R, Hameyer K (2000) The computation of the effects of harmonic currents on transformers using a coupled electromagnetic-thermal FEM approach. *Proceedings of Ninth International Conference on Harmonics and Quality of Power*, Orlando, FL, Vol. 2, pp. 720–725.

50. Renyuan T, Shenghui W, Yan L, Xiulian W, Xiang C (2000) Transient simulation of power transformers using 3D finite element model coupled to electric circuit equations. *IEEE Trans Magn*, 36(4), 1417–1420.
51. Rausch M, Kaltenbacher M, Landes H, Lerch R, Anger J, Gerth J, Boss P (2002) Combination of finite and boundary element methods in investigation and prediction of load-controlled noise of power transformers. *J Sound Vib*, 250(2), 323–338.

6

Large-Shell-Type Transformers: Aspects of Design

Xose M. López-Fernández

CONTENTS

6.1 Introduction

A transformer is a device for stepping up, isolating, or stepping down the voltage of an alternating electric signal, and it is widely used for transferring energy from an alternating current in the primary winding to that in one or more secondary windings.

The importance of transformers is due to the vital role that they play in the electrical power systems, which has made possible the power generated at low voltages (LVs) to be stepped up to extra-high voltages (EHVs) for transmission over long distances and then transformed to LVs for utilization at proper load centers. So power transformers represent the largest portion of capital investment in transmission and distribution substations. In addition, power transformer outages have a considerable economic impact on the operation of an electrical network. More than that, it has its own manufacturing problems associated with insulation, dimensions, and weights because of demands for ever rising voltages and capacities.

There are two main different types of power transformer construction: the core form type and the shell form type. The core form unit is adaptable to a wide range of design parameters and economical to manufacture. However, shell form transformers have a high *kVA*-to-weight ratio and find favor on EHVs and high *MVA* applications. Shell forms have better short-circuit strength characteristics and are more immune to transit damage but have a more laborious intensive manufacturing process.

While core form transformer predominates all over the world, shell form large power transformers are widely used in North America. Therefore, this chapter tries to show the main different aspects of design and their influence in the performance of shell form transformers. A picture of 675 MVA 352/22 kV one-phase GSU shell-type transformer is shown in Figure 6.1.

6.2 Core Form and Shell Form Transformers

As an example of the former type of classification, one has generator step-up (*GSU*) transformers, which are connected directly to the generator and raise the voltage up to the line transmission level or power distribution transformers (*DTs*), which are the final step in

FIGURE 6.1
Shipping of 675 MVA 352/22 kV one-phase GSU shell-type transformer.

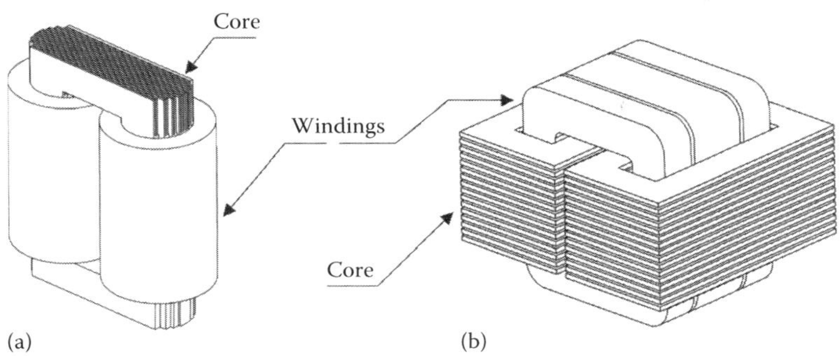

FIGURE 6.2
Single-phase transformer: (a) core form and (b) shell form.

a power system, transferring single-phase power directly to the household or customer. One late type of classification, perhaps the most important is the distinction between the following:

- Core-form-type transformers
- Shell-form-type transformers

Conventionally, in the core form construction, the two or more vertically arranged laminated steel core limbs have two or more windings concentrically arranged around each core limb. In its simplest form, the windings are commonly separated into LV and high-voltage (HV) winding sections. In an alternative construction, the LV and HV coils are interleaved vertically for shell form construction. In any case, the coils are separated from each other by a dielectric (insulating) material.

The basic difference between a core form and shell form transformer is illustrated in Figure 6.2. Both of them refer to the arrangement of the steel core with reference to the windings. In the core-type transformer, the winding surrounds the laminated steel core. In the shell-type transformer, the steel core surrounds the windings. The choice depends on the economics, both labor and material involved in construction and installation.

Each of these types of construction has its own advantages and disadvantages. Perhaps the ultimate determination between the two comes down to a question of cost.

6.3 Transformer Basic Construction

Although transformers are primarily classified according to their function in a power system, they also have subsidiary classifications according to how they are constructed.

The basic design of a transformer consists of two or more electrical circuits comprising primary and secondary windings, each made of multi-turn coils of conductors with one or more magnetic cores coupling the coils by transferring magnetic flux between them. Therefore, in its simplest form, a transformer consists of a laminated iron core around which are wound two or more sets of windings. Voltage is applied to one set of windings of the primary, and the load is connected to the secondary.

The core can be constructed in different ways with reference to the windings.

6.3.1 Different Magnetic Circuit Construction (Core Structures)

A transformer core is made of magnetic material, serving as a path for magnetic flux. Core construction can be either stacked or wound. Core and shell forms are the two basic types of core construction used in power and DTs. In a core-type transformer, a single core loop (path of magnetic circuit) links two identical winding coils. The core consists of limbs and yokes. A limb is the part of the core surrounded by windings. The remaining parts of the core, which are not surrounded by windings and are used to connect the limbs, are called yokes. In a shell-type transformer, the laminations constituting the iron core surround the greater part of the windings.

A single-phase core-type transformer is illustrated in Figure 6.3. The windings on the two limbs are identical consisting of primary and secondary windings. Figure 6.3a shows a scheme of a single-phase core, with the dashed lines indicating the magnetic path. Since the core has very high permeability, the magnetic flux is considered to be inside the core only.

A single-phase shell-type transformer is illustrated in Figure 6.3b. In that core construction, a single winding links two core loops (paths of magnetic circuit).

The vast majority of three-phase transformers are found in core-type construction. Their principal magnetic circuit design for a three-phase transformer is the three-limb

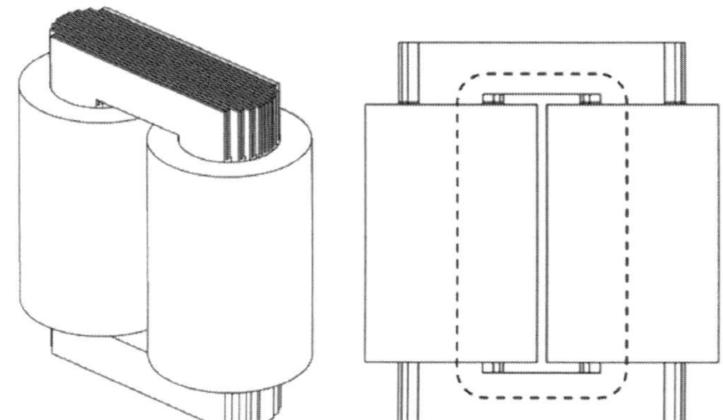
(a)

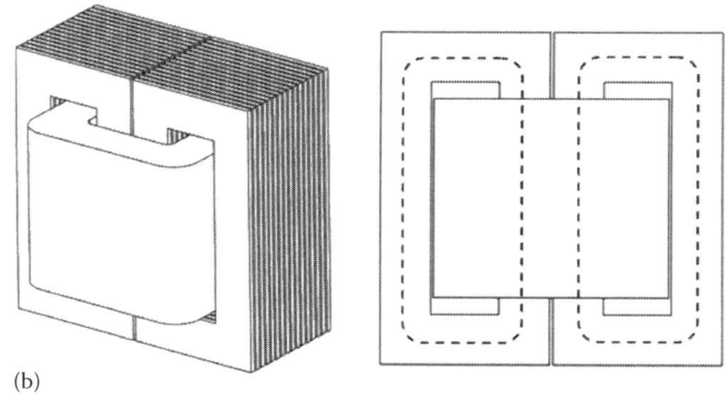
(b)

FIGURE 6.3
Single-phase transformers and their magnetic paths: (a) core form and (b) shell form.

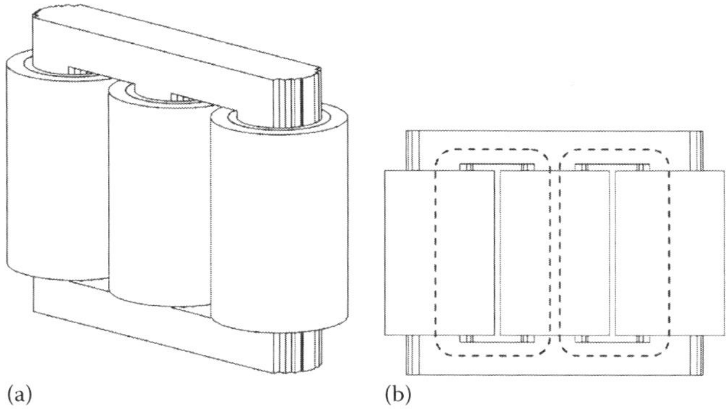

FIGURE 6.4
Three-phase transformer: (a) three-limb core type and (b) three-limb magnetic circuit.

core form, as illustrated in Figure 6.4. Three parallel, vertical limbs are connected at the top and bottom by horizontal yokes. The primary and secondary windings of one phase are wound around one core limb, which is different from a single-phase core-type transformer.

The five-limb, core form magnetic circuit (see Figure 6.5) has three limbs with windings and two unwounded side limbs of lesser cross section. The yokes connecting all five limbs also have a reduced cross section in comparison with the wounded limbs.

The shell-type design is different from the three-phase core-type design. In a shell-type transformer, the windings are usually constructed from stacked coil spirals like pancakes. The three-phase variants of shell-type transformers are illustrated in Figures 6.6 and 6.7. The first shell-type-transformer variant is the conventional one, in which the three phases are mounted on a frame with three horizontal cores having a common central line. The core limbs inside the windings have essentially a rectangular cross section and the adjoining parts of the magnetic circuit surround the windings like a shell. Meanwhile, the second variant is mounted on a core with seven limbs, in which the wounded limbs are oriented in a different way (see Figure 6.7).

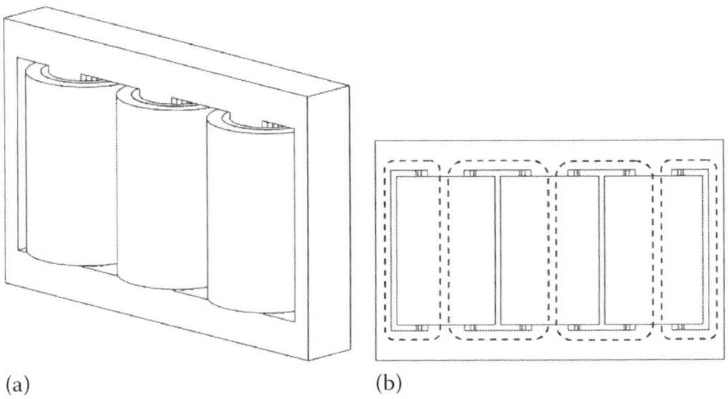

FIGURE 6.5
Three-phase transformer: (a) five-limb core type and (b) five-limb magnetic circuit.

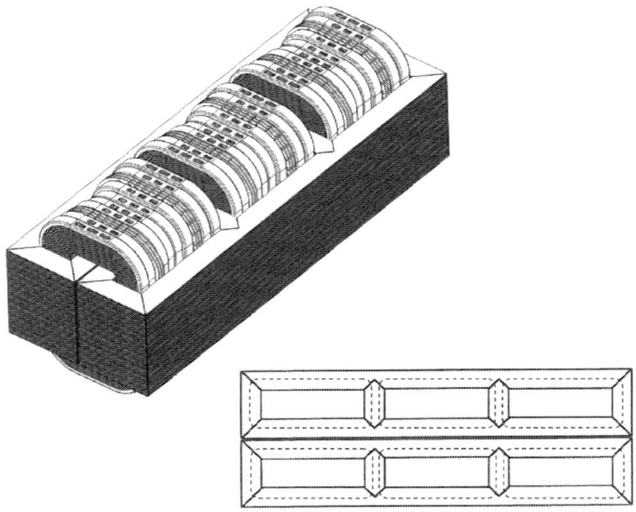

FIGURE 6.6
Shell-type transformer: three-phase conventional shell form magnetic circuit.

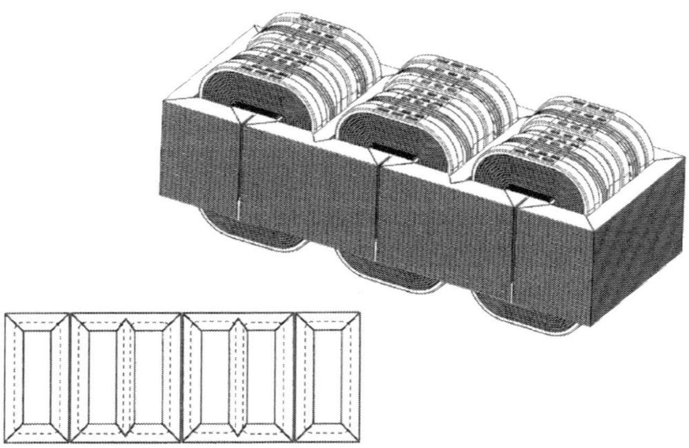

FIGURE 6.7
Shell-type transformer: three-phase seven-limb shell form magnetic circuit.

6.3.2 Core

6.3.2.1 Core Construction

The power transformer core, which provides the magnetic path to channel the flux, consists of thin sheets of high-grade steel, called laminations, which are electrically separated by a thin coating of insulating material. The sheet can be stacked, with the windings built separately and assembled around the core sections. The core steel can be hot or cold rolled, grain oriented or non-grain oriented, and even laser scribed for additional performance. Thickness ranges vary from 0.23 up to 0.36 mm.

Usually, a computer-controlled automatic machine cuts grain-oriented silicon steel sheets with high accuracy and free of burrs, so that magnetic characteristics of the grain-oriented silicon steel remain unimpaired. Then, silicon steel sheets are stacked. Each core limb is fitted with tie plates on its front and rear side, with resin-impregnated glass tape wound around the outer circumference. Also, sturdy clamps applied to front and rear side of the upper and lower yokes are bound together with glass tape. Limbs and yokes are jointed at an angle of 45° to utilize the magnetic flux directional characteristic at the corner of core made of steel sheets.

The core cross section can be circular or rectangular. Rectangular cores use a single width of steel sheets, while circular cores use a combination of different sheets widths to approximate a circular cross section commonly referred to as cruciform construction. The type of steel and arrangement depends on the transformer rating related to cost factors such as labor and performance.

Just like other components in the transformer, the heat generated by the core must be adequately dissipated. While the steel and coating may be capable of withstanding higher temperatures, they will come in contact with insulating materials with limited temperature capabilities. In larger units, cooling ducts are used inside the core for additional convective surface area, and sections of lamination may be split to reduce localized losses.

The core is held together by, but insulated from, mechanical structures and is grounded to a single point in order to avoid the concentration of electrostatic charge. The core ground location is usually some readily accessible point inside the tank, but it can also be brought through a bushing on the tank wall or top for external access. Multiple core grounds, such as a case whereby the core is inadvertently making contact with otherwise grounded internal metallic mechanical structures, can provide a path for circulating currents induced by the main flux as well as a leakage flux, thus creating concentrations of losses that can result in localized heating.

The maximum flux density of the steel core is normally designed as close to the knee of the saturation curve as practical, accounting for required overexcitations and tolerances that exist due to materials and manufacturing processes. For power transformers, the flux density is typically between 1.3 and 1.8 T, with the saturation point for magnetic steel being around 2.03–2.05 T.

In core form construction, there is a single path for the magnetic circuit. Figures 6.3a and 6.8a show a sketch of a single-phase core, with the dashed lines showing the magnetic flux path. For single-phase applications, the windings are typically divided on both core limbs as shown. In three-phase applications, three-phase core form transformers usually employ a three-limb core and the windings of each phase are typically on the same core limb, as illustrated in Figures 6.4 and 6.8b. Windings are separately constructed of the core and placed on their respective core limbs during core assembly. The last kind of core form, showed in Figure 6.5, is related with power station gigantic transformers, which employ five-limb core to prevent leakage flux, minimize vibration, increase tank strength, and effectively use space inside the tank. But they are usually transported by ship and freed from restrictions on in-land transport. The sectional areas of the yoke and side limb are 50% of that of the main limb; thus, the core height can be reduced to a large extent compared with the two-limb core.

In shell form construction, the core provides multiple paths for the magnetic circuit. Figure 6.3b is a sketch of a single-phase shell form core, with the two magnetic flux paths illustrated. The core is typically stacked directly around the windings, which are usually "pancake-type" windings (see Figure 6.9b), although some applications are such that the core and windings are similarly assembled to core form. Due to the advantages in short-circuit and transient voltage performance, shell forms tend to be used more frequently in the largest transformers, where conditions can be more severe mechanically. Variations of three-phase shell form construction include five- and seven-limbed cores, depending on size and application as shown in Figures 6.6 and 6.7.

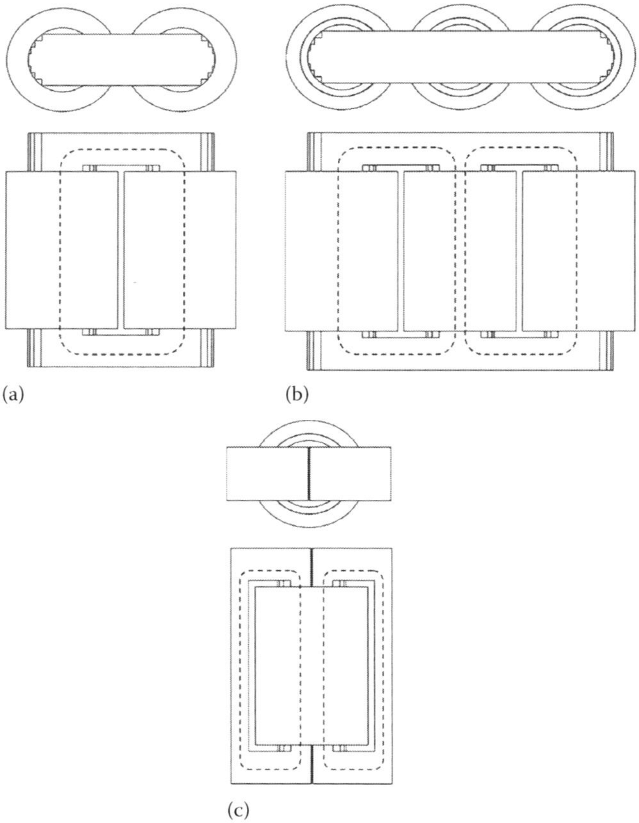

FIGURE 6.8
Sketch of construction. (a) Single-phase core-type, (b) three-phase core-type, and (c) single-phase shell-type constructions.

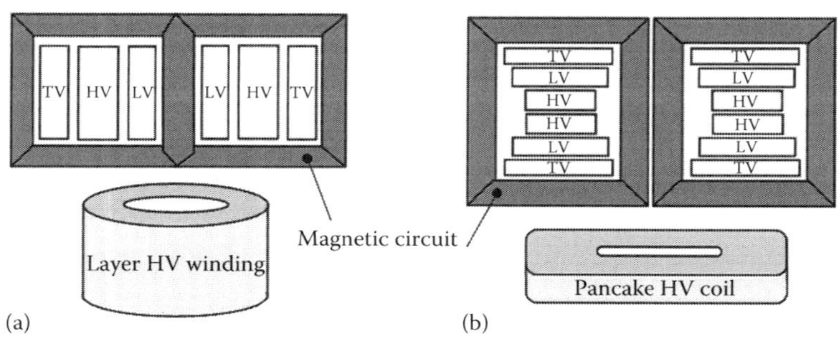

FIGURE 6.9
Schematic cut from transformer. (a) Core-type transformer and (b) shell type.

In the core form, the windings are wrapped around the core, and the only return path for the flux is through the center of the core. Since the core is located entirely inside the windings, it contributes to keep the structural integrity of the transformer frame. Core construction plays a key role when compactness is a major requirement. Connected by a core limb tie plate fore and hind clamps by connecting bars, the core is constructed with silicon strip held in a sturdy frame, which resists both mechanical force during hoisting the core-and-coil assembly and as well forces during short circuits. Although some core-type units are built for medium- and high-capacity use, it is the shell type that offers a construction to be used in larger transformers actually. It is in fact due to the shell form transformer is completely enclosed the windings surrounded by the iron circuit. So the core acts as a structural member, reducing the amount of external clamping and bracing required, capable to withstand large forces created by the leakage flux.

6.3.3 Winding

The power transformer windings are typically stranded with a rectangular cross section, although some transformers at the lowest ratings may use sheet or foil conductors. Multiple strands can be wound in parallel and joined together at the ends of the winding, in which case, it is necessary to transpose the strands at various points throughout the winding to prevent circulating currents around the loop(s) created by joining the strands at the ends.

Individual strands may be subjected to differences in the alternating flux field due to their relative positions within the winding, which create differences in voltages between the strands and drive circulating currents through the conductor loops. Proper transposition of the strands cancels out these voltage differences and eliminates or greatly reduces the circulating currents that are sources of stray losses, such as those mentioned in Section 6.5.6. A variation of this technique, involving many rectangular conductor strands combined into a cable, is called continuously transposed cable (CTC), as shown in Figure 6.26.

The type of winding depends on the transformer rating as well as the core construction. There are three basic types of windings: circular; rectangular, and elliptical. When considering concentric windings, it is generally understood that circular windings have inherent higher mechanical strength than rectangular windings, whereas rectangular coils can have lower associated material and labor costs. Rectangular windings permit a more efficient use of space, but their use is limited to small power transformers and the lower range of medium power transformers, where the internal forces are not extremely high. As the rating increases, the forces significantly increase, and they need additional strength in the windings, so circular coils or shell form construction are used. In some special cases, elliptically shaped windings are selected.

Coils can be wound in an upright, vertical orientation, as it is necessary with larger, heavier coils, or they can be wound horizontally and placed upright upon completion.

In core form transformers, the windings are usually arranged concentrically around the core limb, as illustrated in Figure 6.9a, which shows a winding being wounded over another winding already wound on the core limb. This lends itself to cylindrical shaped coils. Generally HV and LV coils are wound concentrically, with the LV coil inside the HV one. This ensures better magnetic coupling between the coils.

In the shell form design, the core is wrapped or stacked around the coils. This leads itself to flat oval-shaped coils called "pancake" coils, with the HV and LV windings stacked on the top of each other, generally in more than one layer each in an alternating way. In such interleaved arrangement, as shown in Figure 6.9b, individual coils

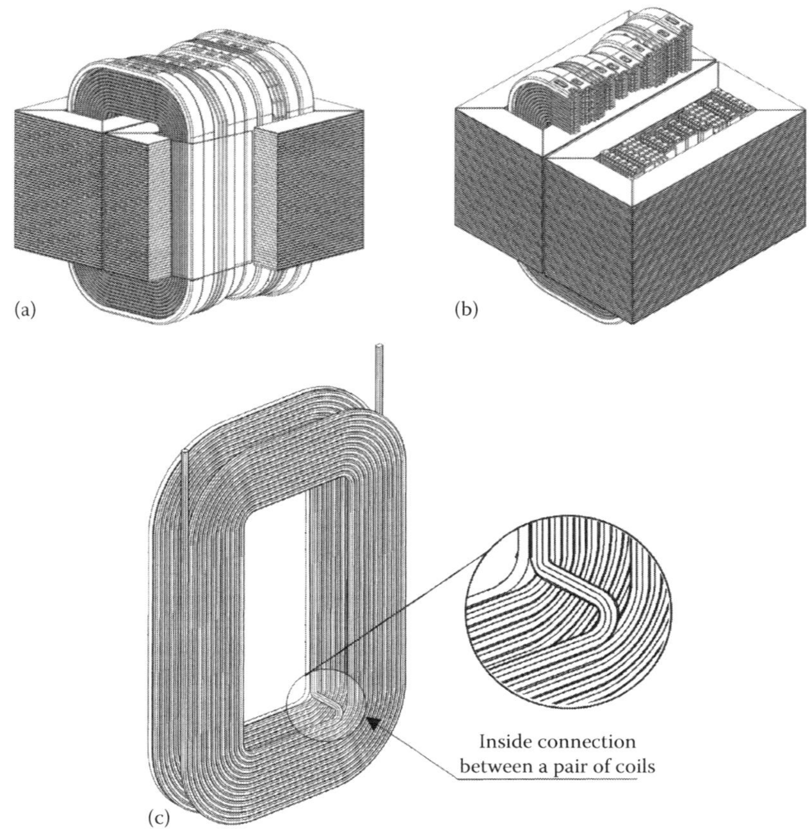

FIGURE 6.10
Configuration of active parts of shell-type transformer. (a) Shell core section (a quarter at the right corner), (b) shell form winding sections (vertical and window sections at the right side), and (c) two coil winding groups with inside connection between them in a continuous way.

are stacked, separated by insulating barriers and cooling ducts. The coils are typically connected with the inside of one coil of an adjacent coil (see Figure 6.10c). Similarly, the outside of one coil is connected to the outside of an adjacent coil. Sets of coils are assembled into groups, which then form the primary or secondary winding. In the shell-type transformer "pancake" construction, the HV and LV coils are alternately placed around the core. Such arrangement of coils reduces the reactance between the coils and improves the operation of the transformer, particularly in large-size transformers where heavy currents are experienced. The shell-type transformers also provides arrangements for simpler oil cooling paths.

6.4 Shell-Form-Type Transformer Construction

The shell form transformer technology was developed by Westinghouse Electric Corporation and was licensed to a number of manufactures. Nowadays, there are very few manufactures that still make both single-phase and three-phase shell form transformers, despite the fact that this design has definite advantages in certain applications. It is composed of a magnetic core enclosing the windings, which provides robustness to short-circuit and transportation efforts and compactness of the design to match transportation and hauling restrictions as shown in Figure 6.14.

One particular feature of the power shell form transformers is that windings are constructed from a number of pancake multi-turn coils, such as those shown in Figure 6.10. Individual coils are connected in series to form groups, and the groups are connected in series to form packs. A winding has one or more sections in parallel. Each turn of a coil is made from a number of individually taped copper strands in parallel as shown in Figure 6.15. The turn itself contains several layers of conductors.

The coils are arranged in alternating groups of HV and LV windings referred to as an interleaved arrangement (see Figure 6.13). This allows the short-circuit forces of the HV and LV windings to act in opposite directions as indicated in Figure 6.31, thus partially canceling each other out and further increasing the short-circuit withstand capability of the transformer. For a shell form transformer, as the transformer capacity increases, the size of each coil is kept similar, but more HV and LV coil groups are added (see Figure 6.32), which reduces the ampere-turns per winding group. This keeps the short-circuit force inside the transformer relatively constant even for very large-capacity units. Core form transformer designs cannot employ this technique as the mean turn length of the coils would become prohibitively large due to the winding arrangement.

Figure 6.10a represents a section within the iron core. Figure 6.10b shows the upper quarter section of the windings. A vertical section show the cut of core, winding, core clamping, T-beam supports, and tank bottom of a single-phase transformer (see Figure 6.11).

In general, the number of coils in a shell form transformer is relatively low and the surface area of each coil is large. That creates a large capacitance between the coils and a low

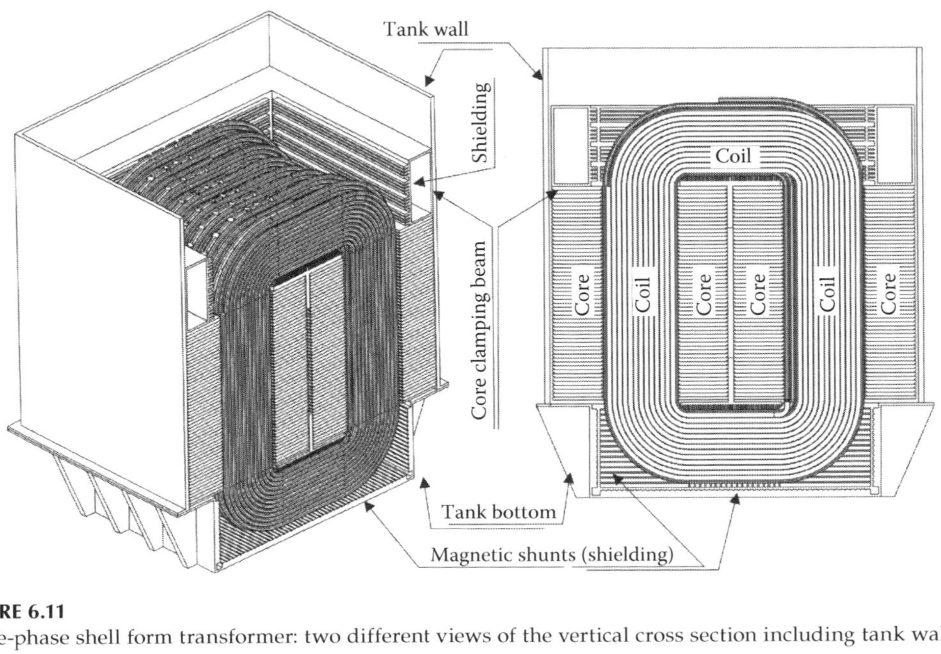

FIGURE 6.11
Single-phase shell form transformer: two different views of the vertical cross section including tank wall and bottom.

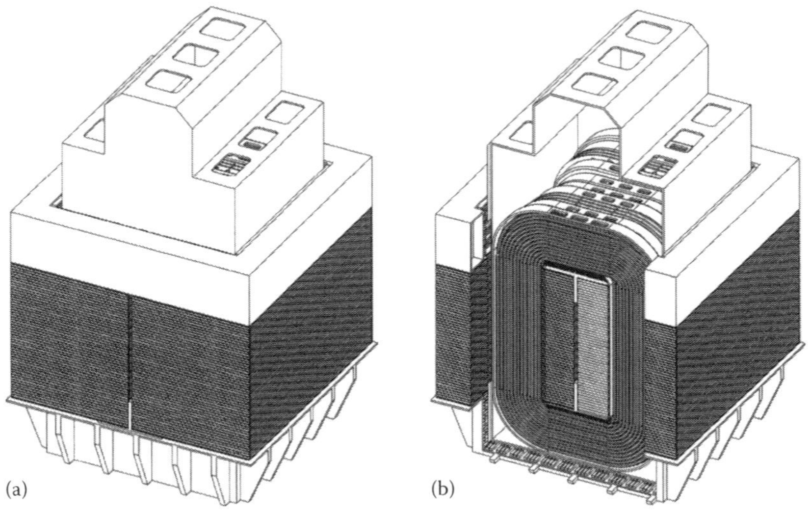

FIGURE 6.12
Schematic view of single-phase shell form transformer pack with insulation cover package. (a) Active part including tank bottom and core clamping beam. (b) Front section of core, clamping beam, up T-beam, low T-beam, tank bottom, and phase winding pack.

capacitance to ground. This allows for the surge voltage distribution across the winding to be almost uniform, with no oscillations (Figure 6.12). The static plate located on the end line of each winding also enhances the surge voltage distribution and plays an important role in smoothing the steep front of the impulse wave (see Figures 6.13 and 6.33). The shell form winding design can achieve a very high dielectric strength against impulse voltage due to the mentioned design of the windings and insulation structure.

As many large-capacity transformers are likely to invite increased leakage flux, wood and nonmagnetic steel are used or slits are provided in steel members to reduce the width for preventing stray loss from increasing on metal parts used to clamp the core and for preventing local overheating. The core interior is provided with many cooling oil ducts parallel to the lamination to which a part of the oil flow forced by an oil pump is introduced to achieve forced cooling as shown in Figure 6.34.

6.4.1 Design Concept and Requirements

The desire of the purchaser is to obtain a transformer at a reasonable price that will achieve the required performance for an extended period of time. The desire of the manufacturer is to construct and sell the product, at a profit, that meets the customer goals. The specification and purchase contract are the document that combines both purchaser requirements and manufacturer's commitments in a legal format. The specification will typically address the transformer service condition, ratings, general construction, control and protection, design and performance review, testing requirements, and transportation and handling. Since it is impossible to address all issues in a specification, the industry uses standards that are acceptable to purchaser and supplier.

From successful design point of view, a commercial transformer requires the selection of a simple structure so that the core and coils are easy to manufacture. At the same time,

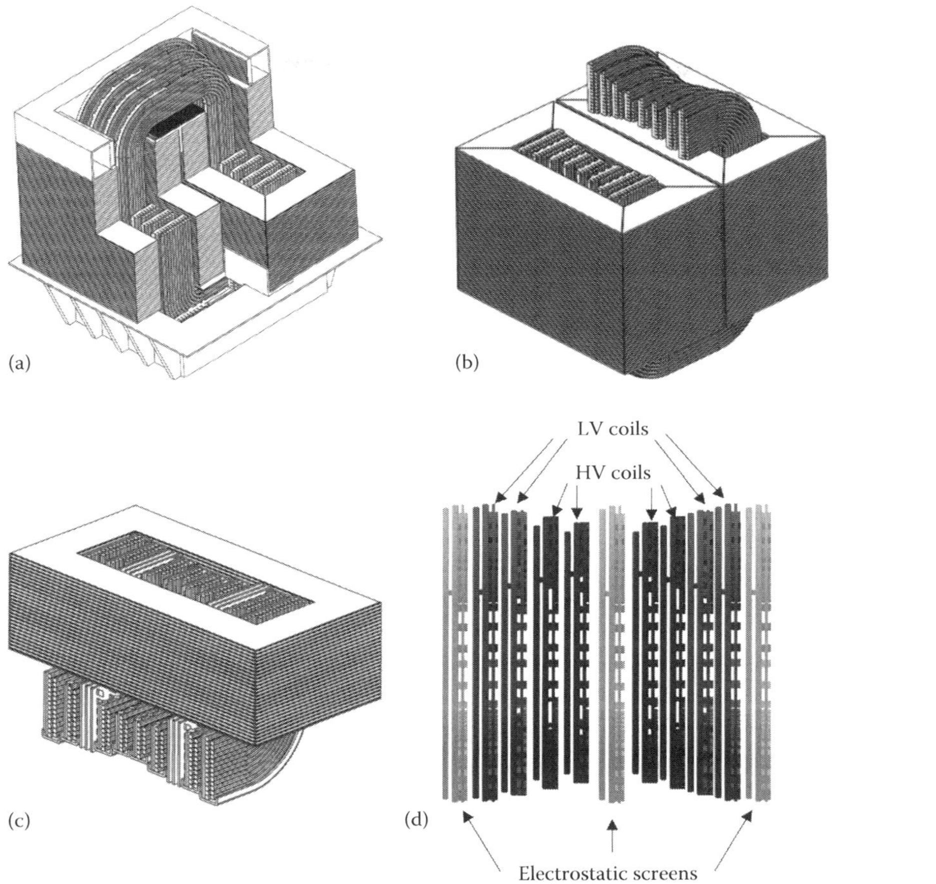

FIGURE 6.13
Coil arrangement. (a) Detailed horizontal and vertical section of active part (windings and core) without insulation between coils nor electrostatic screens. (b) Active part with horizontal section of left window windings without insulation between coils neither electrostatic screens. (c) One quarter of active part with insulation between groups of coils but without electrostatic screens. (d) Set of coils according to one-phase arrangement into the window section.

the structure should be as compact as possible to reduce required materials, shipping concerns, and footprint. The form of construction should allow convenient removal of heat, sufficient mechanical strength to withstand forces generated during system faults, acceptable noise characteristics, and an electrical insulation system that meets the system steady state and transient requirements. According to previous classification, there are two common transformer structures in use nowadays. When the magnetic circuit is encircled by two or more windings of the primary and secondary, the transformer is referred to as a core-type transformer. When the primary and secondary windings are encircled by the magnetic material, the transformer is referred to as a shell-type transformer.

In the core type, the windings are layers or concentric cylinders in a radial disposition from a core column. For the shell type, the windings are disk or rectangular pancakes symmetrically distributed from the winding half point and placed axially over a core column or beam.

In both designs, each winding has several layers or disks and electrostatics screens to improve the internal voltage distribution, which are organized in the following sequence:

- The LV winding and/or tertiary (TV). It is the winding close to the core column by the isolation level.
- Followed by an electrostatic screen. This is connected to ground and is located between the tertiary and the common winding.
- Then, the medium voltage winding (MV) or common.
- Followed by another electrostatic screen. This is connected to MV and is placed between this and the regulation winding.
- Now, the regulation winding (TAP). This is connected in series with MV winding and it allows a voltage regulation in a certain range.
- Followed by an electrostatic screen again. This is connected to the regulation winding and placed between this one and the series winding.
- The HV winding or series.
- Finally, an electrostatic screen. This is connected and is placed over the series winding.

Shell form transformer construction features a short magnetic flux path and a longer mean length of electrical turn. The shell form transformer has a larger core area and a smaller number of winding turns than the core form of the same output and performance. Additionally, the shell form generally has a larger ratio of steel to copper than an equivalently rated core form transformer. The most common winding structure for shell form winding is the primary-secondary-primary (LV-HV-LV) called as two high-low windings (2 H-L). But it is not uncommon to encounter shell form winding LV-HV-LV-HV-LV, identified as 4 H-L as illustrated in Figures 6.14 and 6.32. The winding structure for both the primary and secondary windings is normally of the pancake-type winding structure.

Shell form transformers completely enclose the windings inside the core assembly. Therefore, the core of a shell-type transformer completely surrounds the windings, provides a return path for the flux lines both through the center and around the outside of the windings. Then, the core also acts as a mechanical structural member, reducing the amount of external clamping and bracing required. This is especially important in larger and extra-large applications, where extra-large forces are created by the leakage flux.

Since shell form transformers windings are designed with coil groupings, it permits to achieve any impedance relationships between windings using various ampere-turns per coil group. Multi H-L grouping is not an economical construction for core form units due to its inherent technical limitation. As transformers get larger it becomes desirable to reduce the ampere-turns per coil group to control short-circuit forces. However, the control of such forces is not as readily accomplished with core form transformers, especially in the larger power and higher-voltage classes. Therefore, shell-type transformer has the natural capability to withstand short-circuit and transport stresses. It covers all transformer applications providing a compact and robust alternative especially for large GSU

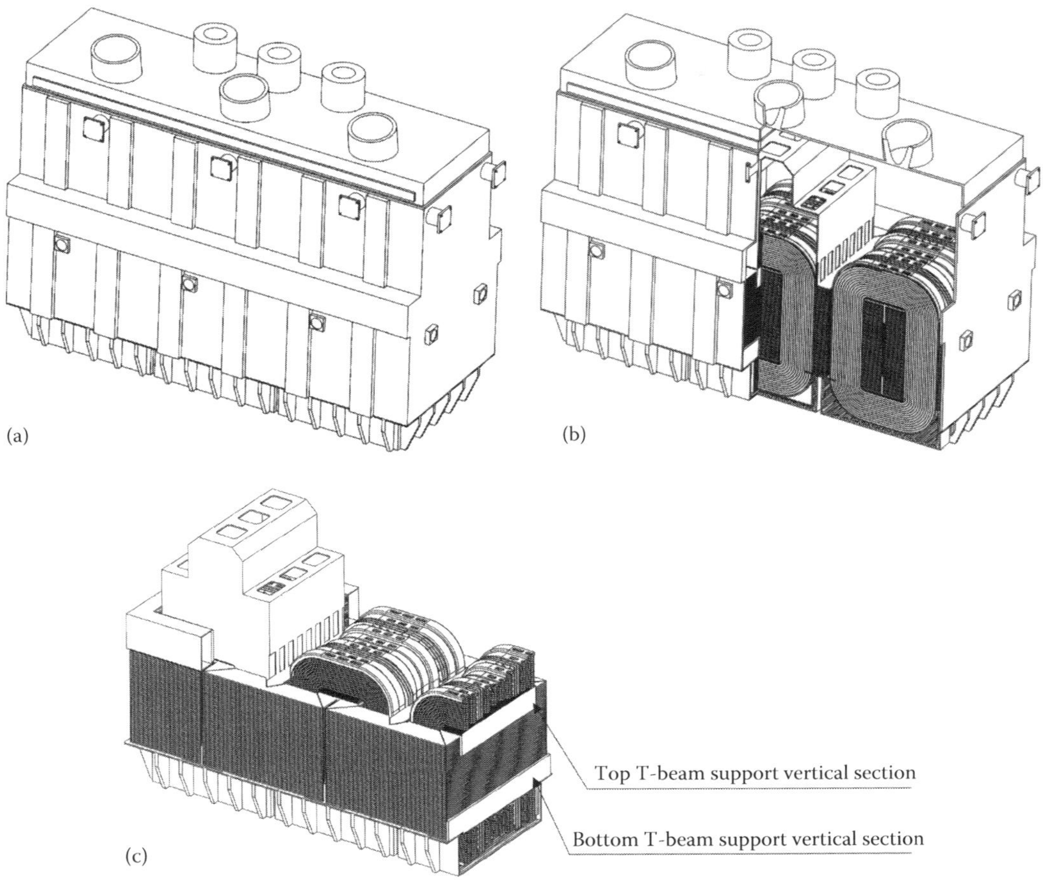

FIGURE 6.14
Schematic drawing of the three-phase shell type transformer. a) Completed three phase transformer with active part into the tank with cover mounted. b) Vertical section of the right phase without insulation cover package, and a vertical section at the middle of the central phase. c) Without cover and neither tank wall, the left phase is completed, the middle is only the active part, and on the right is a vertical section of windings, core and tank bottom, highlight the tongue supports (top and bottom T-Beams).

transformers and HV substation autotransformers with a product range up to 1100 MVA and 550 kV.

Based on the previous discussion, one can summarize the main characteristics of the shell transformer as follows:

- Compact design to meet transportation limits, including the possibility of lay down transportation
- Exceptional mechanical robustness and natural capability to withstand short-circuit events and transportation accelerations
- Flexibility to electrically and mechanically match existing units that make it the best option for replacement of transformers, meeting dimensional and connection arrangements and impedance requirements, even those with low impedances and high impedances to tertiary winding and lay down operation

6.4.2 Pancake Windings

Several types of windings are commonly referred to as "pancake" windings due to the arrangement of conductors into disks. However, the term most often refers to a coil type that is used almost exclusively in shell form transformers. The conductors are wound around a rectangular form, with the widest face of the conductor oriented either horizontally or vertically. Figure 6.15 illustrates how these coils are typically wound. This type of winding leads itself to the interleaved arrangement previously discussed in last section (Figure 6.16).

The shell form design is a completely different design from the three-phase core form design. In a shell form transformer, the windings are constructed from flat coiled spirals that are stacked together like pancakes. For this reason, the windings in a shell form design are often referred to as "pancake" windings. In simple pancake windings, the HV

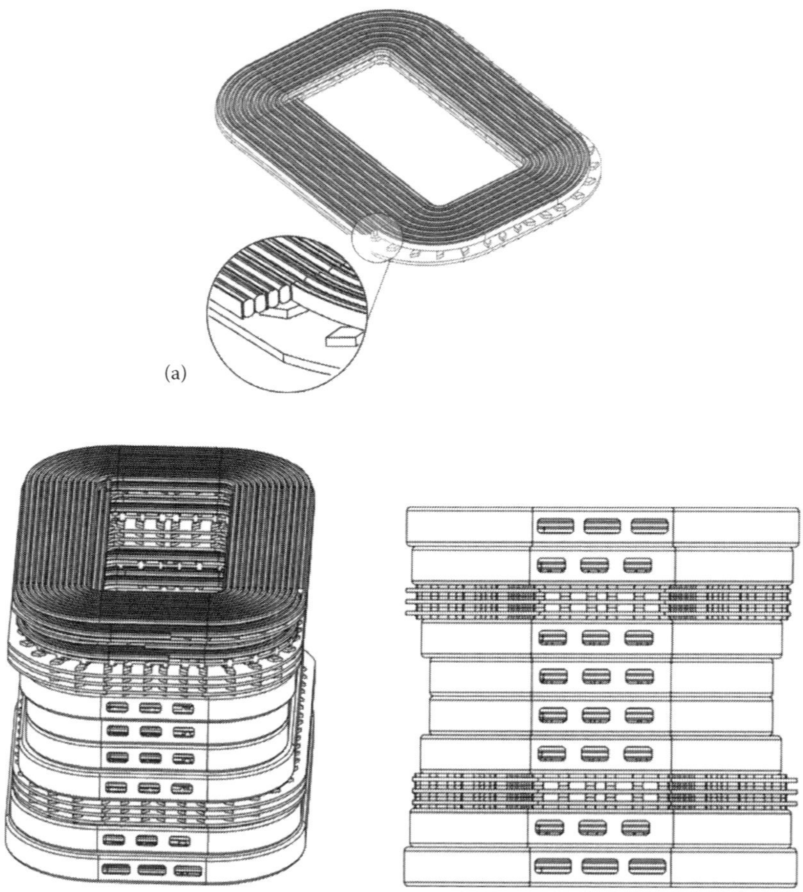

(a) (b) (c)

FIGURE 6.15
Different views of a winding package. (a) Coil made of four conductors per turn on an insulator base with radial spacer to channel the cooling fluid. Detail of the four conductors turned. (b) Stacked pancake windings with the two upper groups of LV coils without insulation around but on insulation bases and radial spacers to channel the cooling fluid. (c) One single-phase package with insulate per group of coils.

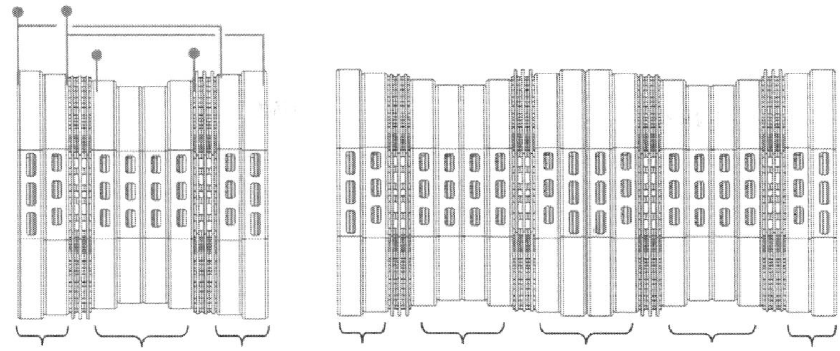

FIGURE 6.16
View of a phase pancake winding with interleaved arrangement: (a) 2 H-L, (b) 4 H-L.

winding is sandwiched between the two halves of the LV winding shown in Figure 6.16. Instead of a circular shape, the pancake coils are actually square shaped with the outer corners rounded off (see Figure 6.15).

The center of the pancake coils is hollow and square shaped. Core limbs with square cross sections pass through the centers of the pancake coils. The core limbs are laid horizontally so the coils are stacked horizontally on edge (see Figure 6.14). The return paths for the core go around the coils forming a "shell" around the windings, hence named the shell form.

6.5 Performance Aspects of Shell-Type Transformers

6.5.1 Leakage Magnetic Field

Power transformers represent the largest portion of capital investment in transmission and distribution substations. In addition, power transformer outages have a considerable economic impact on the operation of an electrical network. Today's challenges in the power transformer industry are loss evaluation, high reliability requirements, and low cost and weight restrictions. To face to these challenges manufacturers need sophisticated design techniques that provide a synergistic effect that leads to product performance improvements and optimum designs.

One of the most important aspects governing the transformer life expectancy is the hot-spot temperature value. An underestimation of stray losses in power transformer easily leads to the formation of hot spots. Therefore, the stray loss evaluation is an essential aspect to calculate hot-spot temperature. The stray losses in transformer are caused by the time variable leakage flux that induces electromotive forces (e.m.f.s), which results in both eddy currents in metallic parts of transformer like tank wall, core, clamps etc., and additional circulating currents in coils due to bridges established in strands brazed.

The previous discussion states that the calculation of leakage field is one of the most important problems in the power transformer design. The leakage field must be calculated accurately in order to analyze stray loss and electromagnetic forces. The loss evaluation often imposed by customers forces manufacturers to make every effort to develop improved methods of loss calculations and reduction. As such, their effort is related with

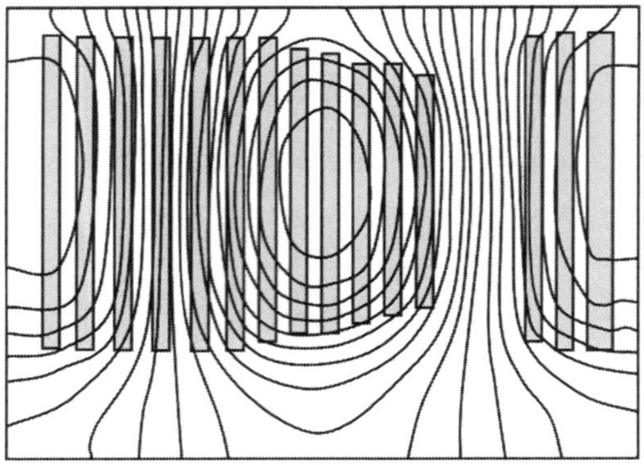

FIGURE 6.17
The 2D flux distribution in a window section at steady state.

winding eddy and circulating current losses, stray losses in the core and structural parts of the transformer, magnetic and electromagnetic shielding requirements, impedance, short-circuit forces, core loop voltage, and, in general, loss densities at potential regions of both excessive heating and local hot spots. All of these effects and aspects are primarily determined by the electromagnetic leakage field strength of the transformer windings. A typical 2D pattern plot of steady-state lines of flux is shown in Figure 6.17. Here, it is interesting to note that in any point of space the flux density has two components that act in different way. One, the radial (horizontal) component of the field is the most important in the determination of the transformer performance parameters. Another one, the axial (vertical) component of the magnetic field is important in determining radial short-circuit forces, in calculating winding eddy losses, and in determining the winding circulating currents that flow across the breadth dimension of a coil.

The losses induced by the leakage flux are a 3D phenomenon and should correspondingly be modeled by a 3D mathematical analysis.

6.5.2 Flux Distribution around a Coil

Each turn of a coil is made from a number of individually taped copper strands in parallel as shown in Figure 6.15b. Winding eddy losses are the result of eddy currents induced within individual strands and between strands of a turn by the leakage flux. An estimation of the magnitude of the eddy currents and the associated losses can be obtained once the leakage flux distribution is known.

Magnetically, a coil represents four different regions depending on their position with respect to the core and tank (see Figure 6.18). Each region is characterized by flux density distribution values. So, in Figure 6.19b, calculated values of radial and axial flux densities along a coil of a large power transformer, at both the insides and the outside edges of the coil, are presented. Here, the maximum leakage flux density values on coils appears, which is taken as reference one at 100%.

The significant variation of magnitude of the flux densities along the coil is placed at the corner region (CR). At the middle of the limb region (LR) of the same coil, the radial flux densities at the inside and outside edges of the coil are practically the same and represent about 50% of the maximum value inside along the LV coil due to axial flux as shown in Figure 6.19a. One can easily foresee the difference in eddy losses and in circulating currents losses.

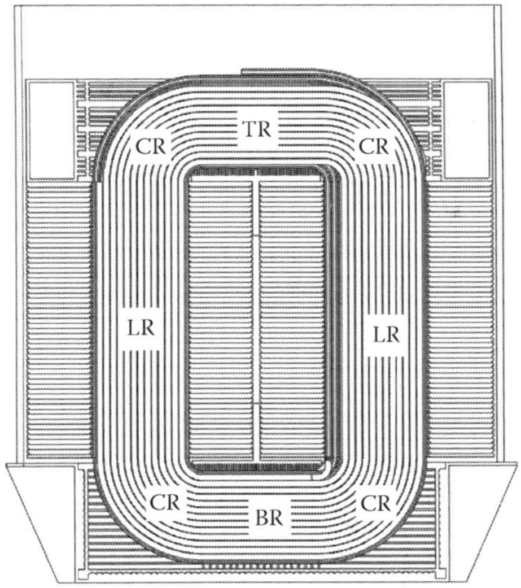

FIGURE 6.18
Coil regions characterized by the leakage flux value in shell form transformers: Top end region (TR); Bottom end region (BR); Limb region (LR); and Corner region (CR).

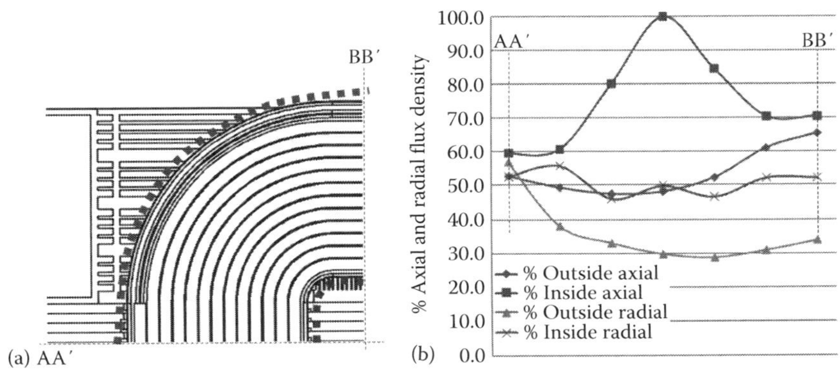

FIGURE 6.19
Leakage flux density distribution. (a) Outside and inside along the LV coil at CR. (b) Axial and radial leakage flux density distribution in percent over the maximum leakage value place at 100%.

The large differences between the flux distribution in the *LR* and that at the two ends of the phase is attributed to both the magnetic boundaries of the coils and to the larger contribution of the CR of the coil to the strength of the field at the inside edge of the two ends of the coil. This tends to drive more flux in the *coil support* (T-beam) and wall shunts. Calculations at other regions of the winding have indicated identical results regarding flux distribution characteristics. The results appear to be independent on the side of the winding at which they are obtained (tank wall side or interphase side).

6.5.3 Flux Distribution across a Coil

A sample of the flux density distribution across the breadth of a HV coil of a large power transformer is presented in Figure 6.20. That figure presents very interesting characteristics

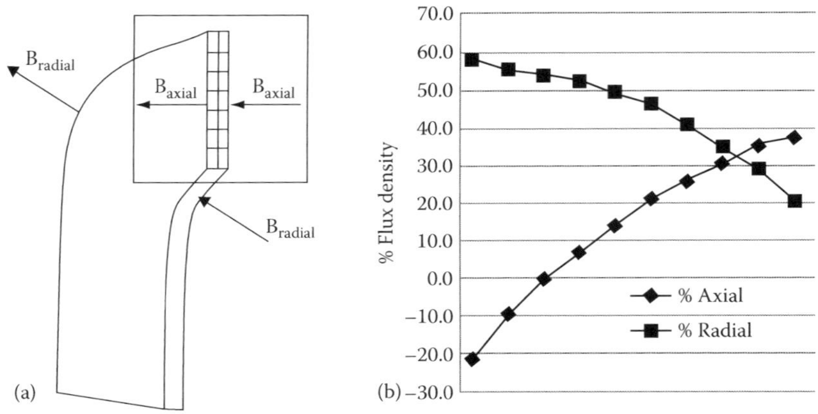

FIGURE 6.20
Flux densities distribution across a HV coil. (a) At top end region (TR) (b) axial and radial flux density distribution in percent over the maximum leakage value shown in Figure 6.19.

of both the *axial* and the *radial* components of the leakage flux. While the variation of the magnitude of the radial component across the coil breadth is almost symmetrical about the coil axis in the LR of the transformer, this is not the case at the TR and BR ends of the coil. This proves that extrapolating the distribution of this component of flux in the LR to the ends of the coils is not valid (refer to Figure 6.20). Consequently, transposition techniques based on the uniform symmetric flux distribution around a coil are not necessarily optimum for minimizing circulating currents and subsequent losses. In a typical large power transformer, an optimum location for the transposition obtained using 3D calculations actually reduced the circulating currents to half its level resulting in a sizable reduction in the load losses of the transformer.

At higher magnitude of flux at the inside coil edges will also produce short-circuit forces that are unbalanced about the axis of the coil giving rise to large moments in addition to the resultant compression forces applied to the coil surface and coil-to-coil insulation.

Regarding the distribution of the axial component of the leakage field, it is evident from Figure 6.20 that the distribution deviates significantly from the perfect asymmetric distribution, where the flux density distribution in one-half of the coil is generally assumed to be equal and opposite to that in the other half as could be obtained from 2D field analysis. Consequently, using such a distribution will result in values of losses, circulating currents, and short-circuit forces that deviate significantly from their actual values. For example, the asymmetric distribution when used to calculate the radial forces on the coils results in no net force. However, because of the dominance of the magnetic field at the inside edge of the coil, there is a net force on the coil that is not commonly recognized. The same can be well applied to the calculation of the circulating current. While the asymmetric distribution of the field across the coil breadth would hardly result in circulating currents, the actual asymmetric distribution will result in substantial circulating currents and consequently extra losses and possibly excessive local heating of parts of a strand or a number of strands in a coil depending on the cooling provided.

6.5.4 Magnetic Shielding

Turning the attention to the flux density distribution in magnetic shielding members outside the coils, for fear of local heating in the metallic structural elements, such as tank

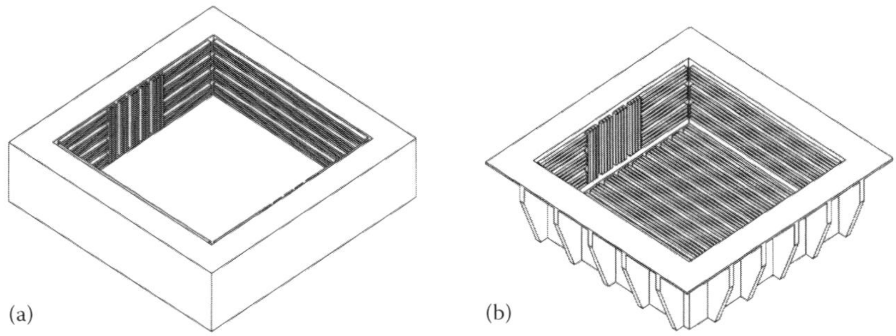

FIGURE 6.21
The bigger mechanical elements considered to be shielded by magnetic shunts to channel the leakage flux are (a) core clamping beam (upper beam) and (b) lower tank (tank bottom).

walls and T-beam supports, two alternative means can be used. The first one is that the threatened elements are usually covered with magnetic steel strips (magnetic shunts). The purpose of these magnetic shunts is to channel the leakage flux back to the main magnetic circuit. When precalculations give clues that the leakage flux at the vicinity of threatened elements would not be important enough, the latter are covered with nonmagnetic plates. The second mean are aluminum or copper plates, which are alternatively used (electromagnetic screens). The purpose of these plates is to expel, instead of channel, the leakage flux by means of induced currents.

The main mechanical elements considered to be shielded are the tank walks, core clamping, tank bottom, and T-beam supports. Figure 6.21 shows the inside of core clamping and the inner of tank bottom covered by magnetic shunts.

6.5.4.1 T-Beam Shielding

One important element to avoid the localization of hot spots is the T-beam supports. The role of those elements is to keep the winding package in vertical right position and support the core and winding weights. They are made of solid steel just placed between both bodies of the core. There are two T-beam supports, one placed on the upper face of the core and the second one under the core, resting their extremes on the wall of the tank bottom (see Figure 6.22). Therefore, the T-beam faces have to be shielded, and it is traditionally done employing magnetic shunts, as is appreciated in Figure 6.22b. There, the inaccuracies and deficient predictions of leakage flux values have to be carefully avoided. Otherwise, oversaturation at shielding could lead to overfluxing and localized hot spots. Those hot spots will be placed inside the active part where it is very difficult for cooling.

Both the amount of required shielding and optimum shielding configurations imply the use of 3D field calculations. As shown in Figure 6.23, the flux density is not uniform inside the shielding with flux driving a large part of the shielding into saturation. That figure shows the very high values inside T-beam magnetic shunts. They are in percent over maximum leakage flux collected in Figure 6.19. Even in the middle of the shielding, the 2D calculation predicts a lower value of flux density than what is really there.

6.5.5 Third Component of Flux

Certain wiring configurations of shell form transformers, because of the multiple paths available for the flux flow, are susceptible to higher core losses due to harmonic generations.

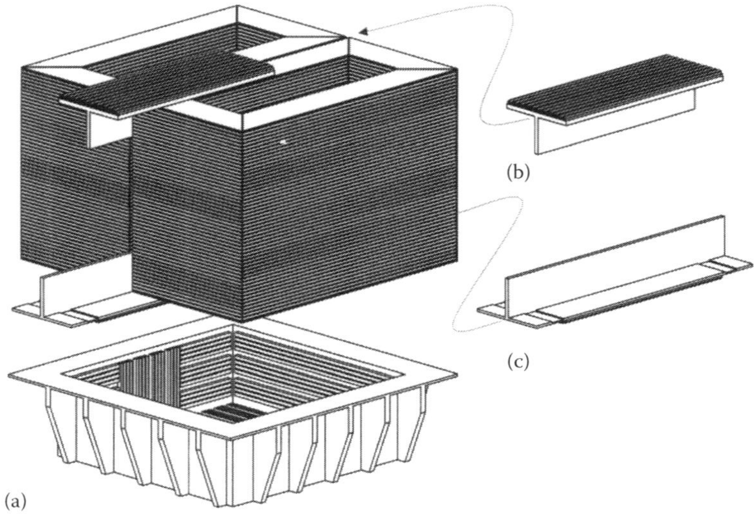

FIGURE 6.22
T-beam supports: (a) bottom tank covered by magnetic shunts under single core separated by T-beam supports, (b) top T-beam covered by magnetic shunts, and (c) bottom T-beam.

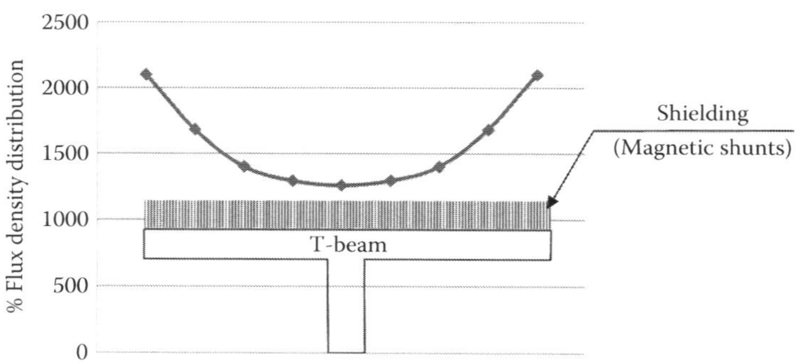

FIGURE 6.23
Magnetic shunts on T-beam support to channel the leakage flux. Calculated flux density distribution at T-beam shunts along a cross section. The values are expressed in percent over maximum leakage flux value shown in Figure 6.19.

The "third harmonic" flows primarily in the core and can triple the core losses. These losses occur primarily in Wye–Wye configured transformers. This third component of the field gives rise to additional stray losses and could contribute to excessive core heating. This component of flux cannot be calculated using the 2D field analysis. The same reasoning is true for the design of the sidewall shielding, which can only be achieved through 3D field calculations.

6.5.6 Stray Losses

The main sources of stray loss are eddy-current and hysteresis losses in tanks, structural members, leads, and shielding. The calculation of winding eddy losses requires a detailed knowledge of the winding geometry and the leakage flux distribution at specified points located in regions of the coils. Calculating the eddy losses, circulating currents and circulating current losses in each coil, one can estimate their total value in the whole transformer windings.

The field solution for the spatial flux density has to be obtained in each point of the geometry over one quadrant (Figure 6.13c) of each coil to provide the desired accuracy. The flux is more concentrated in certain points of the CRs where the field varies rapidly. There are few points where the field is relatively uniform as refer to Figure 6.19, while in the center the flux density reaches its highest value considered to be constant.

The 3D components of flux density (B_x, B_y, B_z) are converted into two components, namely, the axial and the radial components ($B_{axial,}$ B_{radial}). Only eddy-current losses and induced voltages tending to produce circulating currents as it is considered in the following text.

6.5.6.1 Calculation of Eddy-Current Losses

A coil strand is subjected to two components of flux density normal to its sides (see Figure 6.25) that produce eddy currents. These are the horizontal and the vertical components.

The vertical loss component is that caused by the radial component of flux density B_{radial} and occurs in the strand width dimension. The horizontal component is that caused by the axial component of flux density B_{axial} and occurs in the strand height dimension.

6.5.6.2 Circulating Currents

As illustrated in Figure 6.24, there are two types of circulating currents that flow in a transformer coil:

- Strand-to-strand (STS) currents flow within a group of strands that are brazed together at both ends of the coil.
- Group-to-group (GTG) currents flow between groups of strands within a turn.

Each of these types of circulating currents has two components: one is due to the radial flux (B_{radial}) and the second one is due to the axial flux (B_{axial}). The total circulating eddy current in a strand is the sum of all four components of the circulating eddy currents. To calculate circulating eddy currents, the voltages that circulate these currents within a coil

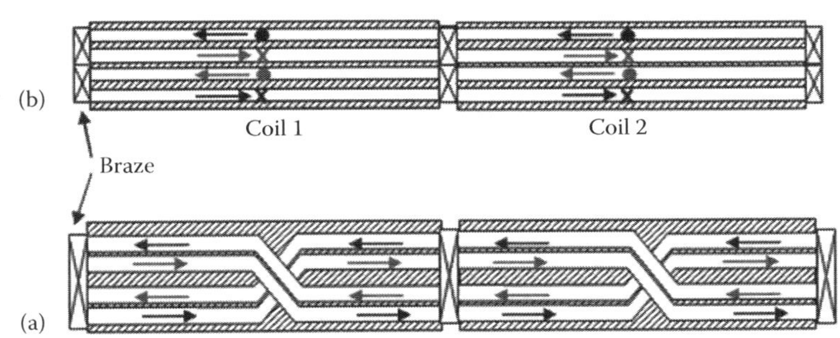

FIGURE 6.24
Paths of STS and GTG circulating currents in a coil and between coils of a winding group: (a) STS circulating current and (b) GTG circulating current.

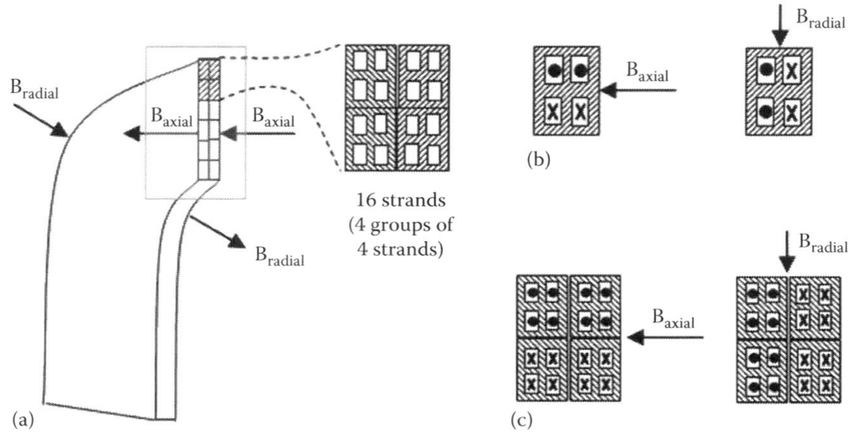

FIGURE 6.25
Standard turn showing paths of STS and GTG circulating currents within each strand. (a) Reacting circulating current along strands under axial or radial excitation. (b) Turn cross section made of 16 strands, 4 groups of 4 strands. (c) Reacting circulating current into each strands dependent on B excitation (B_{axial} or B_{radial}).

or from coil-to-coil within a winding group are first found by integrating the flux density components over the width, height, and length of the coil or coils, taking into account the transpositions that exist in the coils. Such voltage tending to cause circulating current between any two strands is the difference between the voltages being generated in each strand. Each strand group is brazed together at both ends of the coil and to the next coil; however, the individual groups are not connected together except at the two ends of the winding group.

As an example, Figure 6.25 shows a turn made of a total of 16 strands. Each of those strands is divided in four groups of four strands each, each group occupying one corner of the turn. Each of these strand groups is brazed together at both ends of the coil, and the next coil braze is said to be split into four parts, two in the turn-height direction and two in the coil-depth direction.

Several features in the shell form winding design have to be taken into account to calculate the circulating currents and make the task very complex, such that only a rigorous and carefully devised algorithm can deliver the correct values.

Different types of transpositions, strand groupings in a turn, and split coil connections can be considered in order to reduce the circulating currents. An effective alternative to cancel the circulating currents is the use of the CTC as illustrated in Figure 6.26. CTCs do not generate STS circulating currents because the short pitch of the transposition effectively eliminates any voltages between strands. Splitting CTC at the braze joints does not result in circulating currents either, as long as the coil contains only one cable. When two or more cables are used in parallel, the cables act like individual strands in a standard coil and circulating currents can exist. If CTC cables are brazed together at the joints between coils, then there will be circulating currents in the cables.

6.5.6.3 Eddy Losses due to the Horizontal Flux

At the middle of a winding group, the horizontal flux component is at a maximum, and, correspondingly, this component of loss has a peak at this location. Typical distribution

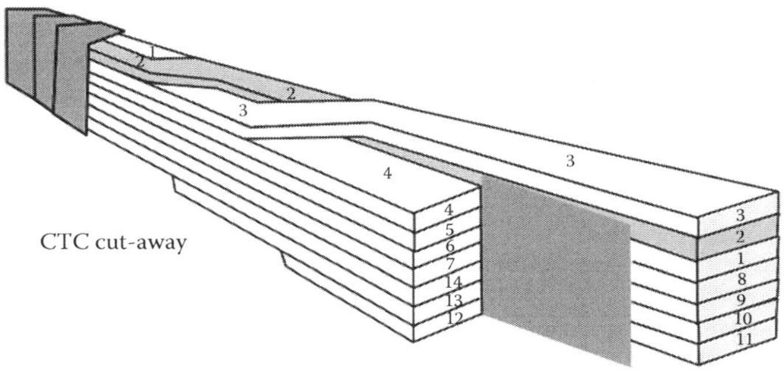

FIGURE 6.26
Continuously transposed cable.

of this loss component is plotted in Figures 6.27 and 6.28 for a 2 H-L and 4 H-L designs, respectively.

The horizontal component of flux cannot be calculated to any reasonable degree of accuracy using simplified assumptions based on the Ampere turns distribution. The experience evidences that 2D values are considerably lower than the 3D values, differing from each other up to 30%. Such deviation is considered substantial, especially for units with

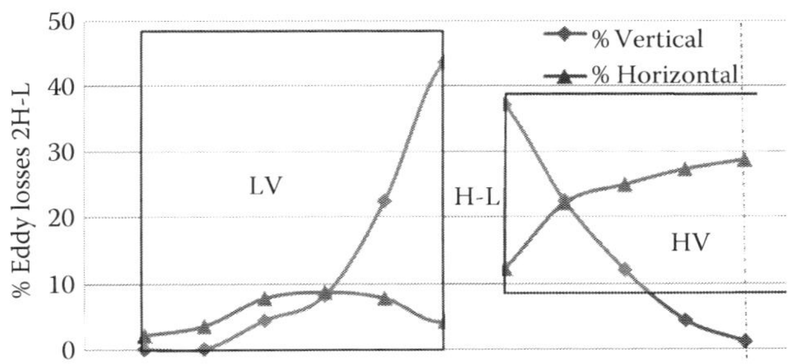

FIGURE 6.27
Typical symmetric distribution of eddy losses in pancake winding of 2 H-L designs: horizontal component distribution and vertical component distribution. The values are expressed in percent over maximum losses value shown in Figure 6.28.

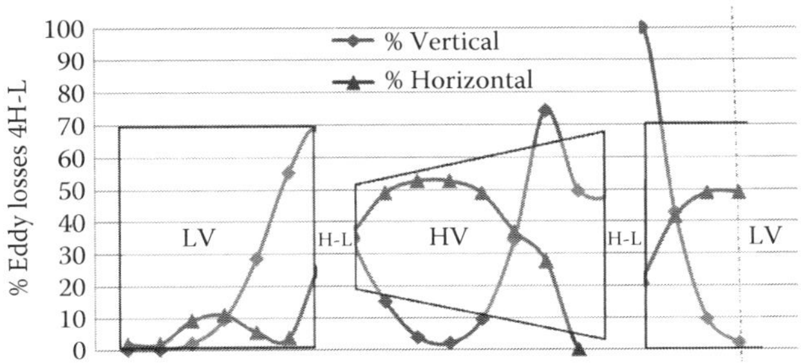

FIGURE 6.28
Typical symmetric distribution of eddy-current losses in pancake winding of 4 H-L designs: horizontal component distribution and vertical component distribution. The values are expressed in percent over maximum losses value shown in this figure.

high load loss evaluation. The 2D values will always be lower because axial flux densities around the pancake coil are (except at the outside region of the coil corners) higher than those at the middle of the LR. Loss densities of a magnitude equal to about twice those at the LR are found at the inside CRs of the coils. The significance of using 3D field methods provides insight into the actual contribution of each coil to the total value of the loss component. Such information is very valuable to designers for effective winding loss reduction.

6.5.6.4 Eddy Losses due to the Vertical Flux

In the H-L gap, the magnetic field is mostly vertical with very little horizontal component at the coil ends. The distribution of this loss component is plotted in Figure 6.27. Meanwhile, Figure 6.27b shows the typical distribution of eddy losses due to a vertical flux in pancake winding of 2 H-L design, and as in Figure 6.28b, for a 4 H-L design.

6.5.6.5 Circulating Current Losses

According to Section 6.5.6.2, the correct relative contributions of the individual coils can be given as a result from excessive circulating currents. One example of a typical distribution of circulating current losses in pancake winding of 4 H-L designs is given in Figure 6.29. Those effects can be canceled by means of its optimum location with 3D calculations and applying proper cable transpositions in the respective coils. The shape of transposition of four conductors of a cable is shown in Figure 6.30.

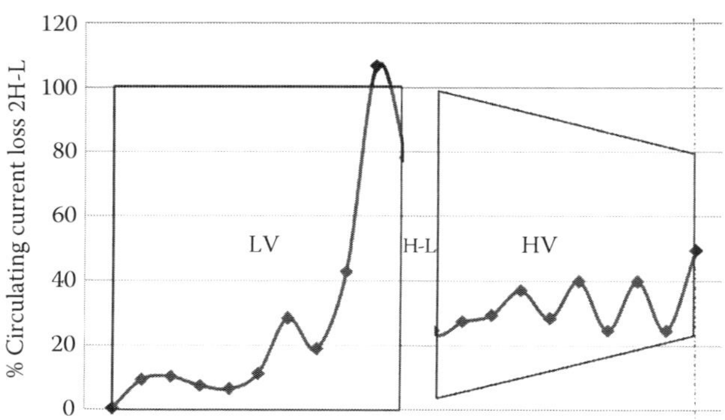

FIGURE 6.29
Typical symmetric distribution of circulating current losses in pancake winding of 2 H-L designs. The values are expressed in percent over maximum losses value shown in Figure 6.28.

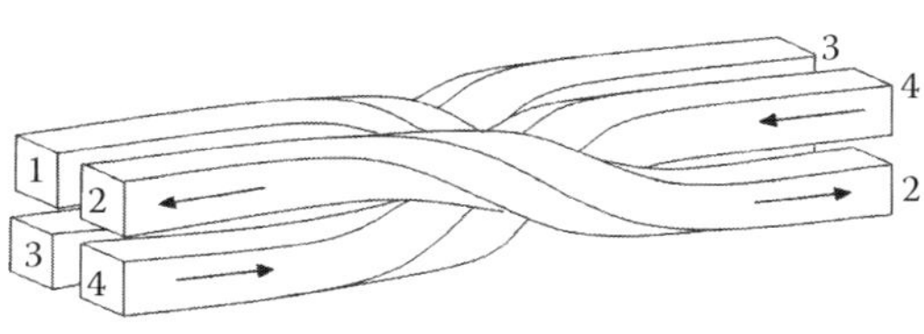

FIGURE 6.30
Shape of the transposition of four conductors of one cable.

6.5.6.6 Effect of Transposition

The magnitude of the circulating currents is very sensitive to the cable transposition and its exact location. For example, in the case of moving the location of a certain transposition, 10% of the turns in the wrong direction increased the circulating currents to about three times its original magnitude. Only the 3D calculation can lead to the optimum location of the transposition that would result in practically zero circulating currents.

6.5.7 Forces between Windings

In transformers, all coils that carry current in the same direction attract one another and coils that carry currents in opposite directions repel one another. Hence, all the coils of the primary attract one another as do all the coils of the secondary; however, primary coils repel the secondary coils and vice versa. Therefore, the flux that links the two windings of the transformer together also creates a force that tends to push the conductors apart. Under normal operating conditions, these forces are relatively small, but in case of short-circuit conditions or the carrying of very large currents, these forces can multiply to hundreds of times the normal value and may become great enough to damage the transformer if the coils do not adequately support these short-circuit stresses. When the coils are concentrically placed, the forces produced are radial, which may tend to distort the shapes of the coils.

Coils and cores must be mechanically capable of withstanding electromagnetic forces in both the radial and axial directions generated during the operation of the transformer. The entire coil and winding assembly must be firmly braced, both on the top and bottom and all around the sides. Bracing also helps to hold the coils in place during shipping.

When the electrical centers of the windings are aligned, the net force in the axial direction is zero. However, if there is a misalignment of the electrical centers, axial forces act in opposite directions in the LV and HV windings. Under short-circuit conditions, these forces might become substantial enough to cause the LV and HV windings to displace relative to each other, which can be catastrophic. Strip or sheet conductors are used in the LV winding to minimize the differences in electrical centers, and wire with a flat profile is used in the HV winding to increase the surface area for better adhesion. For example, insulation paper with epoxy coatings is used to help hold the conductors in place.

In addition to the manufacturing steps mentioned earlier, designers run simulations to predict short-circuit forces and the resulting stresses. These stresses are compared to allowable limits, which are determined by the manufacturer to provide a guideline for a safe design.

6.5.7.1 Short-Circuit Performance of Shell Form Transformers

In shell form transformers, the main component of the leakage flux is the radial (vertical) one. This develops short-circuit forces in axial (horizontal) direction shown in Figure 6.31, which tend to separate the LV windings from the HV windings. Such forces compress the pancake coils toward their magnetic centers instead of exploding the coils outward, unlike the HV coil in a typical core form transformer. Compression forces are usually easier to deal with than tensile forces, which give the shell form transformer a distinct advantage over the core form design. Since the HV coils are normally sandwiched between the LV

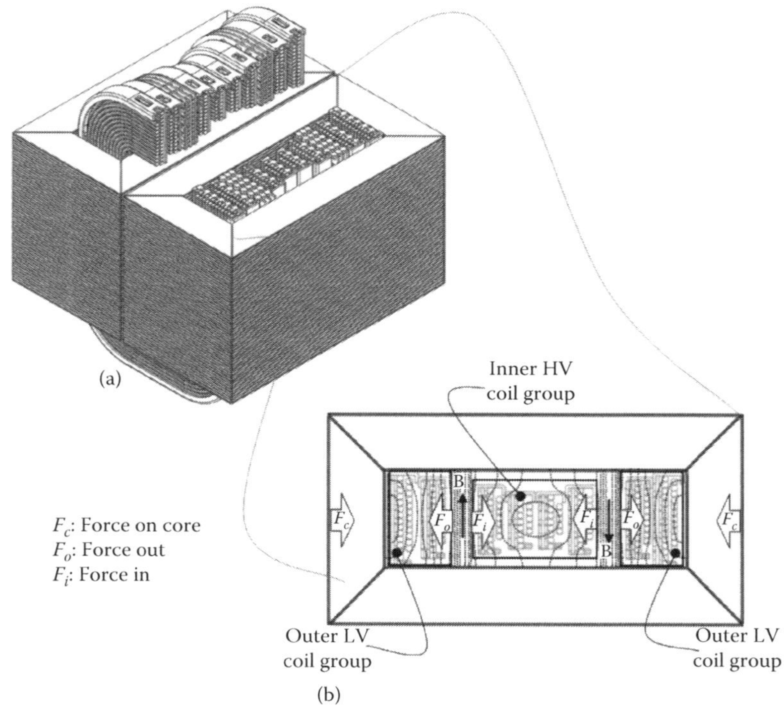

FIGURE 6.31
Short-circuit forces into shell form transformers: (a) a 3D section window, (b) acting forces (F) into a window between HV and LV gap due to a leakage flux B.

coils in a shell form transformer, the axial forces tend to compress the HV pancake coils toward the center of the winding pancake. The axial forces are restrained by fill material that is wedged between the coils, between the winding packs, and between the LV coils on the outer winding packs and the transformer tank, providing mechanical bracing. The resulting configuration is an assembly that is securely braced in all acting directions and has great mechanical short-circuit strength.

The fill material is not actually solid, but it contains numerous ducts for cooling oil to flow. Wedges at the end surfaces of the windings compress the entire coil structure as the upper tank is lowered on top of the lower tank, resulting in a structure that resists all movements.

Since the short-circuit force acts perpendicularly to the pancake coil, then the outer coil group (LV) is pressed to the core and the tank and the inner coil group (HV) is compressed within the coil group composed of both conductor and insulation materials. Thus, the width of the leakage flux path would be changed by the deformation of the winding, resulting in an impedance change.

Generally, in the core form transformer, the deformation of the concentric coil arrangement is mainly due to the buckling of the winding. This means that even when the impedance is slightly changed, it might be very critical for the core form transformer. However, in the shell form transformer, it would not result in very serious damages when the change is slight. The impedance change is generated by the following factors:

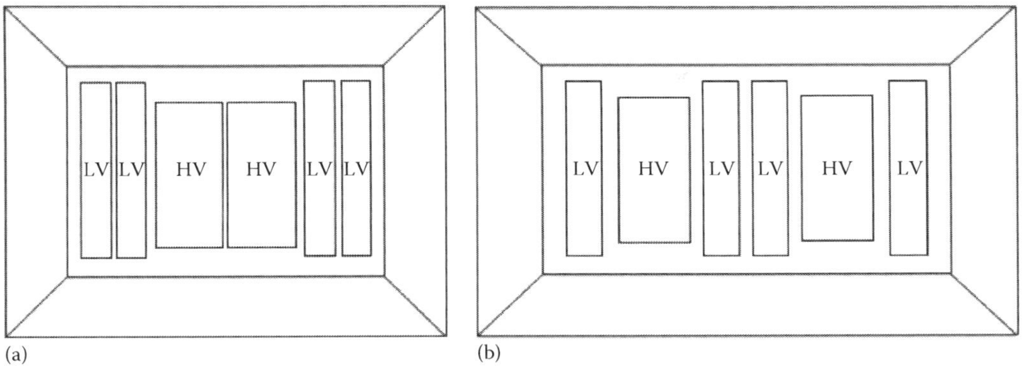

FIGURE 6.32
Coil arrangement with flexibility for increasing or decreasing the number of interleaved coil groups in shell form transformers: (a) 2 H-L designs and (b) 4 H-L designs.

- Deformation of tank
- Deformation of core
- Deformation of insulation material

In order to attain the appropriate strength against short-circuit forces, the conductor strength, the tank strength, and the core clamping force are mainly considered at the design stage. In addition, high performance in the shell form transformer can easily control the short-circuit force itself by adjusting the number of the interleaved coil groups as shown in Figure 6.32.

As a summary related with shell-type transformers, one can point out the following:

1. The short-circuit strength of the transformer is one of the most important subjects for improving the reliability of the transformer.
2. The short-circuit force can be controlled by the number of interleaved coil groups at the coil arrangement in the shell form transformer.
3. The tank should be designed to withstand the short-circuit force in the shell-type transformer.

6.5.8 3D Shell Form Transformer Performance

Years of experience and a large database coupled with regression analysis techniques allow the designer to predict the total losses in the transformer. The conventional formulas used to calculate stray and eddy losses in transformers tend to be specific to particular physical design configurations. However, presently, significant changes are being introduced to the physical configuration of the transformer, and the classical calculation procedures can generate answers that differ significantly from test results.

On the other hand, 2D techniques for electromagnetic field calculation were used by most of the major transformer manufacturers throughout the world. For some calculations, such as the determination of leakage field densities within winding configurations of cylindrical core-type transformers, these are satisfactory techniques. In this case, 2D techniques provide enough accurate field results. When calculating field densities outside of these cylindrical coils or for shell-type winding configurations outside the core or near the top and bottom of the core (see Figures 6.19 and 6.20), 2D field analyses are not accurate. In particular, the 2D analysis cannot be applied to the portions of the windings outside the core or near the top and bottom of the core. There, the calculation error could be achieved up to 30%.

In the case of the shell-type transformer, it is necessary to use 3D magnetic field analyses in order to take into account their complex geometry and to achieve the accuracy necessary to predict local losses and prevent overheating.

The manufacturer of shell-type transformer has to do an effort to assess 3D performance parameters of power transformers by means of extensive calculations and measurements of actual flux density distribution along coils, across coils, and across the T-beam shielding. From that work, the following could be determined:

1. Load losses, including winding eddy and circulating current and stray losses, which can be included as a part of the design process of commercial units.
2. Components of load loss of each coil. Such information is used for the reduction of load losses through proper choice of strand dimensions, optimum location of transpositions (see Figure 6.30), appropriate braze splitting, etc. That permits an average of 30% reduction in winding eddy and circulating current losses. Meanwhile, in one transformer, the optimum location reduces the total circulating current losses up to an 8% of total load and almost eliminates a 10% circulating current loss in the LV winding.
3. Circulating currents in each strand of each coil. Location and the circulating currents in the most heavily loaded strands can be found. The designer uses this information for appropriate choice of strand dimensions and to check for possible hot spots in the winding and avoid it if needed, either by providing special cooling or better reducing circulating currents through any of the means available to him at the design stage. This results in the enhancement of the reliability of the transformer as well as reduction of losses.
4. New designs of the T-beam shielding and magnetic shielding like sidewall shielding design. The new shielding improves the flux distribution across the shielding width so that the shielding is not saturated at any segment. This not only reduces stray losses but also prevents high localized heat generation in the shielding as well as in the portion of the core beneath the coil support member, hence enhancing the reliability of the transformer operation. That avoids the generation of hydrocarbon gases associated with paper and oil decomposition, which can be detrimental to the transformer operation.

6.5.9 Electric Insulation

The windings of the transformer must be separated (insulated) from each other as well as from the core, tank, or other grounded metallic elements. The actual insulation between the turns of each winding can usually be provided by an enamel coating or a few layers of paper. The insulation is necessary because the entire voltage drops across the windings and ground. That voltage creates an electric field intensity distribution into the volume of the transformer (see Figure 6.33). The field values have to be satisfied by the insulation system implemented during design stage. Otherwise, if the dielectric strength of insulation is exceeded, the dielectric breakdown could occur with failure consequences for the transformer.

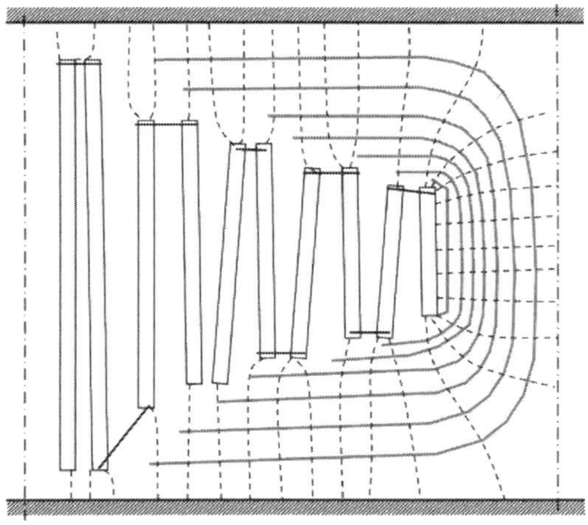

FIGURE 6.33
2D electrical field intensity distribution in a section window at steady state.

In conventional transformers, the insulation is controlled by spacers between conducting electric parts made of a variety of insulating materials, depending on required temperature classes, design, cost, and other performance related with proper requirements. Commonly used materials include cellulose fibers, paper, board, ceramic materials, pressboard, and glass fiber-filled materials such as epoxy or polyester, where the glass can be in the form of discontinuous short fibers, a glass mat, or a fabric. In fact, nonconductive materials, such as plastic, hardwood, or plywood blocks, are used to separate the windings from each other and from the tank walls. These separations in the construction allow paths for fluid or air to circulate, both adding strength to the insulation and helping to dissipate the heat, thereby cooling the windings. Allowing oil to circulate between the windings, the turn-to-turn insulating level can be appreciably increased and the amount of heat built up in the windings can be efficiently dissipated. This is especially important in large, HV transformers, where the heat builds up and the turn-to-turn separations must be controlled.

Actually, transformers are designed to withstand impulse levels several times and, in some cases, hundreds of times higher than one operating voltage. This is to provide adequate protection in the case of a lightning strike, a switching surge, or a short circuit.

A wide range of dielectric elements in the form of vertical sticks, axial, and radial spacers, called collectively spacers, must be made of electrically insulating material. The insulating material must be able to withstand heat and fluctuations of temperature. Since the coils and insulating layers are immersed in fluid, which aids in transporting heat away from coils, the insulator material should ideally be resistant to the commonly used fluids. Also, the spacers must be able to withstand the mechanical stresses developed during manufacturing and electrical/mechanical stresses during the operation of the transformer, such as during a short-circuit event.

6.5.10 Cooling

In operating transformers, windings and core are heated by copper and iron losses. In order to keep transformer temperature at a suitable range, cooling is necessary, especially for large-capacity transformers.

There are many forms of losses in a transformer, and although they have different sources, the resultant product of these losses is heat buildup within the tank. Transformer losses can be divided into two general categories: load losses and no-load losses. No-load losses are independent of the applied load and include the following:

- Hysteresis losses in the core laminations
- Eddy-current losses in the core laminations
- Joule losses due to no load exciting current
- Stray eddy-current losses in the tank core clamps, bolts, and other metallic components
- Dielectric losses

Load loses consist of the Joule copper losses across the winding electrical resistances that are produced by the applied current (I^2R) and of the stray currents in the windings that appear when the load is applied.

The result of the generation of the losses is the heating of active parts of the transformer. In order to avoid the overheating and hot spots in vital points of the transformer, it is essential to adopt cooling requirements. That is the reason for which the most large power transformers have their windings immersed in some type of fluid. That fluid not only acts as a cooler but also helps to increase the electrical insulation between coils, windings, and between those and the ground. Although larger-dry-type transformers are constantly being produced, in many new forms of construction, such as resin cast and gas filled, the most common method of insulating the windings and dissipating the heat is by submerging the active part, the windings and core, in an insulating fluid. Oil is a widely used fluid.

The use of oil permits to optimize both dielectric strength and cooling efficiency of shell form transformers also. In shell form transformers, the space between windings and core is entirely surrounded with the barrier insulation. Therefore, cooling oil has to be directed to the insulation space. From the viewpoint of cooling, oil flow is needed only at heating sources such as windings and core. Oil flow path to heat sources such as windings and core is furnished in the insulation structure. Cooling is not necessary at other insulation areas and blockade for oil flow is furnished there. However, manufacturing of barrier structure requires gap spacing between blockades, which causes some leakage of oil flow. The leakage of oil flow might result in additional source for oil flow electrification. Elimination of unnecessary leakage oil path enables efficient use of cooling oil and also leads to low electrification of insulation materials by minimizing the sources of oil flow electrification at the same time. This is due to the discharge of static electrification in the power transformer that can be caused either by electrification of pressboard or by accumulation of space charge in oil. The flow of liquid through the windings can be based solely on natural or forced convection, and its flow can be somehow controlled through the use of strategically placed barriers within the winding. When the windings are of disk type, a term which also encompasses the shell form type (pancake windings), it is known to provide axial and radial spacing through appropriate use of axial spacers and/or radial spacers. Thus, spacers provide desired dielectric distance between conductors and tank wall, and adequate flow of coolant fluid around the windings. These radial spacers are typically glued to a sheet insulation called a washer in the shell form construction. In another typical setup, the axial and radial spacers are combined to form a comb-shaped configuration

Figure 6.34 illustrates the inside structure between coil groups inserting flow channels of shell form transformers. Windings are entirely covered by the oil/paper insulation and

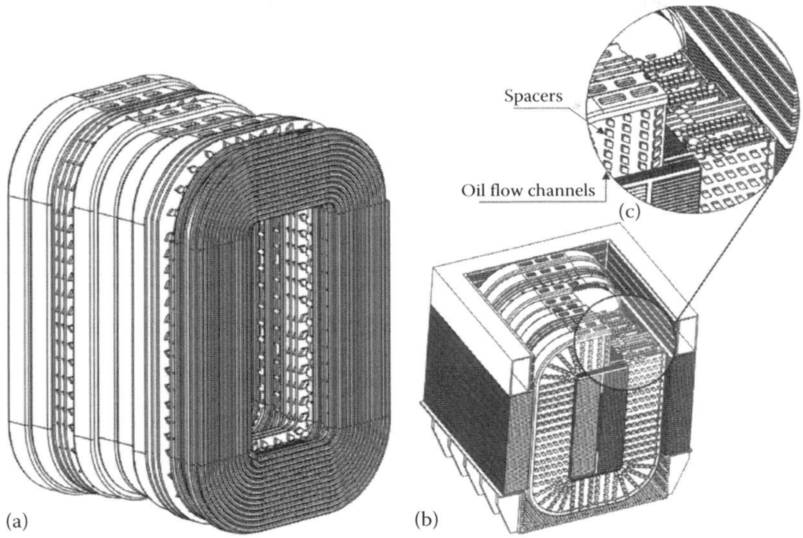

FIGURE 6.34
Inside structure between coil groups inserting oil flow channels. (a) Single-phase package. The right LV coil group is without insulation and blockade for oil flow. (b) A front view of where radial spacers are glued to sheet insulation. (c) Details of winding and spacers.

the core tightly surrounds the winding insulation. Careful attention has to be paid there where the distance between windings and iron is the smallest in the transformer, since the highest electrical stress could appear inside the core windows. Dielectric strength of oil gap is strongly influenced by dimensions of the surface area and gap length. Since the scale effect of gap length is more predominant than that of surface area, fundamental nature of the dielectric strength of oil gap is expressed by the relation between electrical field strength E (kV/rnm) and gap length d (mm) at the discharge inception (E-d characteristics).

In the shell form insulation structure, the oil flow path or blockade is not built in the pressboard materials but is formed at the construction process. At the construction of the pressboards, manufacturing adjustment is necessary to form three 3D barrier systems. This requires some gap spacing between blockades since during drying process after construction contracts pressboard, which also requires some freedom in space for contraction.

6.5.11 Remarks on Shell Form Power Transformer Construction

The main disadvantages of the shell form design are that it is much more difficult to assemble than the core form design and there are tighter electrical clearances to contend with. The major advantages of the shell form design are that it is more compact than core form transformers and it has great mechanical strength. As a remarked summary of advances of shell form, one can mention the following:

- *High short-circuit withstand capability:* The shell form transformer is a mechanically strong design for a transformer, with the coils arranged vertically and completely surrounded by the iron core. The tank of the shell form transformer is a "form-fit design" that fits just over the core, creating a very strong reinforcement against possible coil movement. This helps the transformer to withstand the high mechanical forces that are experienced during external short-circuit conditions.

- *High dielectric strength:* In shell form transformers, the pressboard materials that surround the coils subdivide the oil insulation into many small gaps. The oil-impregnated pressboard materials have a much higher dielectric strength than the oil gaps. The coils and solid insulation are also easily arranged along equipotential lines, providing a high dielectric strength against insulation breakdown. Due to this arrangement, insulation breakdown cannot occur unless the solid insulation is punctured. And from the insulation structure of shell form transformers, for a discharge to break down the insulation and cause an internal fault in the transformers, it must puncture several layers of this pressboard material. This provides a highly reliable insulation system. For these reasons, shell form transformers offer an advantage for EHV applications.
- *Efficient cooling capability:* Shell form transformers have an efficient cooling capability due to the vertically arranged pancake coils. This allows the oil to easily flow across the entire coil surface and cool the transformer. The form-fit design also directs almost all of the oil flow over the coil surfaces, eliminating oil flow to unnecessary areas inside the transformer. This efficient cooling ensures a lower hot spot temperature vs. the average winding temperature when compared with a core form design. It increase longer life of the insulation system and also provides greater ability to handle short-term overloading.
- *Excellent control of leakage magnetic flux:* The shell form transformer provides excellent control of leakage magnetic flux to prevent against local heating inside the transformer. The coil windings are almost completely surrounded by the iron core, which channels most of the leakage flux from the windings. Additional tank shield areas are added both on the top and on the bottom to further eliminate leakage flux. Also, the wire size of each coil can be selected flexibly, depending on the leakage flux in the windings, to prevent from local heating in the winding itself. Similar to the core form transformer, designs cannot freely select the wire size of each coil due to their continuous winding construction. This may lead to local heating in the winding of core form transformers.
- *High flexible design:* The shell form interleaved HV and LV winding group arrangement allows for flexible design of the transformer to meet virtually any requirement. The impedance between windings can easily be adjusted, which allows for easy parallel operation with any other transformer, even other core form transformers. It is also easy to lead out taps from any winding, due to the interleaved arrangement of the coils, which makes it easier to design and manufacture transformers with tap changers. It is also easy to build coils of high current capacity, which makes shell form designs especially suitable for transformers of ultra-large current capacity. Shell form transformers are appropriate EHV, large-capacity applications due to the inherent flexibility and advantages that the design offers for these applications.
- *Compact size:* Due to the form-fit construction of the shell form transformer, unnecessary oil volume in the transformer tank is eliminated. The oil volume of a shell form transformer is generally approximately 40% less than a comparable core form transformer. This also provides for a compact transformer with a compact footprint. Thus, for cases where an existing shell form transformer must be replaced with a new one, the new shell form transformer can easily fit to the existing foundation. Sometimes, core form transformers cannot easily fit onto existing shell

form transformer foundations, and, also, the increased oil volume of the core form unit may require the existing oil containment facilities to be increased to handle the extra oil volume, which can greatly increase the cost, duration, and complexity of the installation work.

- *Highly reliable design:* All of the advantages mentioned earlier are combined to provide a transformer design with a very high degree of reliability and robustness, even in the most severe of the operating environments. Reliability is well proven by the vast number of units in operation over many years for critical and demanding applications worldwide.

6.6 Conclusions

This chapter collects a wide range of material on large-shell-type transformers. It represents an attempt to show the specific particularities of design and performance of this type of transformer, which is not easy to find in the literature. The author thinks that the effort put in this chapter is justified since the large shell-form-type transformer clearly offers advantages compared to its main competitor core type. The particular advantages are not always well assumed for the transformer users, since the demand of shell type is representing a low percentage compared to core type. Therefore, it is hoped that the reader finds argument enough along this chapter to evaluate the convenience to consider shell-type transformers as an alternative in more applications than previously thought. The material was generated during the period of the author's collaboration with the EFACEC Energía in Porto-Portugal from 2004 to 2010, thanks to access to the Large Shell Power Transformers Division. The chapter should particularly be useful not only to undergraduate and postgraduate students, but also of utility to engineers, designers, and researchers.

Acknowledgments

The author would like to express his thanks to Patricia Penabad-Durán and Casimiro Alvarez-Mariño for their invaluable help during the preparation of this chapter. Thanks also to Ricardo Sanchez Gregorio for his precise support during the elaboration of figures on shell-type transformers. Finally, EFACEC Energia for the advice during the selection of the content of this chapter.

7

Transformer Design Review: A Link between Design and Maintenance Stages

Ryszard Sobocki

CONTENTS

7.1 Introduction

Large and high-voltage power transformers are comparatively complicated and expensive elements of a substation. Hence, purchase of such transformers will need to be based on a comprehensive technical requirement. No matter how detailed the purchase specifications are, there are still some elements that may cause misunderstanding between a purchaser and a supplier. Moreover, there is also the question of whether the technical document must contain extremely detailed information. To avoid problems, an increasing number of purchasers have decided to organize a so-called design review. The information provided in the following sections should help with carrying out the design review process.

7.2 Scope and Steps of a Design Review: Economics

7.2.1 Scope and Steps of a Design Review

As far as the purpose of a design review is concerned, the CIGRE brochure [1] in Clause 4.2 "Design Review" says:

> A design review, initiated and chaired by the Purchaser should be held for the purpose of conducting an in-depth review of the ordered power transformer and to allow the Purchaser to have a clear understanding of the overall design.

The main tasks involved in the design review process are as follows:

- Get acquainted with a project of the ordered transformer on the basis of calculations, drawings, documents (and explanations) prepared by a manufacturer.
- Discuss the project and exchange opinions to determine whether the proposed solutions will meet the requirements of a technical specification.
- Ensure compliance with the purchaser proposals and requests. Introduce changes to the proposed solutions where needed.
- Obtain necessary approvals for the selected solutions, drawings, connection diagrams, auxiliary equipment, and technical documentations.

The design review comprises meetings wherein modalities of the project are discussed, including harmonizing of documents and drawings.

The documents and the drawings prepared by the manufacturer are forwarded to the purchaser for thorough verification. Opinions and remarks are sent back to the manufacturer and are taken into consideration and discussed during the meetings at the manufacturer's end.

Typically, multiple meetings are needed to perform the design review. The first meeting is organized when a design of the ordered transformer is in an advanced stage but not completed. Given the situation that the delivery time stipulated by the purchaser is shorter these days, the first meeting should be organized without a delay, in order to enable the manufacturer execute supply of the ordered transformer on time. After the first meeting, it is advisable to organize the next meeting at the customer site where the transformer is to be installed. It may also be advisable to include a visit of the purchaser's staff at the factory while manufacturing an active part of the transformer.

Following substantial problems should be carefully discussed during the design review:

- General project of the transformer: demonstration that all substantial requirements should be met
- General arrangement of the transformer: tank, cooling equipment, bushings, conservators, control cabinet(s), substantial valves and fittings, and so on

- Design of an active part: core, windings, insulation system, electrical, and mechanical withstand
- Stray flux: screening of a tank and cover, temperatures, and local overheating
- Cooling of the transformer: cooling equipment, temperatures of windings under normal loading and overloading, controlling devices, temperature sensors, and so on
- Noise level under normal load and overloading; acoustic shield on a tank (if requested or needed)
- Selection of auxiliary equipment: tap changer, bushings, current transformers, temperature sensors, online monitoring devices (if any), and so on
- Selection of protection devices: Buchholtz relays, safety valves, and so on
- Spare parts. Training of the purchaser staff
- Manufacturing process: stabilization of windings, drying, and impregnation
- Factory tests (type and routine), after assembling tests
- Delivery on site, assembling on site
- Technical documentation for operation and maintenance purposes: drawings, connection diagrams, photos (including video) showing the assembled transformer and the assembled active part

Transportation and installation of large, heavy transformers may create a problem. It would be a good practice to make a supplier responsible for transportation on site, unloading and putting a transformer on a foundation, as well as execute both mechanical and electrical installation of the transformer. Condition on site shall be made clear to the supplier. Often a visit of the supplier staff on site is essential. Details of handling facilities need to be discussed for the ease of unloading and installation and also to facilitate, if necessary, reinstallation and transportation of the transformer to other substation or to a workshop. To monitor conditions during transportation, a railway carriage shall be equipped with an impact recorder to record any mechanical shock during the transportation, which may damage the transformer.

An after-assembly test can be performed either by the supplier or by another team (company) under the supplier's supervision. The test results shall be compared with data obtained during a factory test.

It is not necessary that the supplier representative must attend a trial run; however, he shall have the right to supervise it.

7.2.2 Design Review versus Economics

The economical aspect should not be neglected while preparing the technical specifications of the transformer, especially during the design review.

The CIGRE brochure [2], in Clause 3: "Specification and Purchase" states:

> Because of the relatively large capital expenditure involved when purchasing a transformer, most utilities are generally very well aware of the economic factors and savings that can be achieved at this stage of transformer's life cycle.

The expected life time of modern power transformers is at least 50 years, apart from some computerized devices installed in the transformer (see Subclause 7.5.10) whose life time is much shorter. Hence, during the design review the expected life time should not be forgotten. Moreover, the scope for, and cost of, maintenance should be considered. Generally, there is no problem with the exchange of some elements of auxiliary equipment. On the contrary, any work (repair) involving the active part is extremely expensive and inconvenient. A lot of attention should be concentrated on the active part from economics point of view, as it could generate substantial savings during the life time of the transformer.

7.3 System Operating Conditions and Environmental Data

As a matter of fact, it is the purchaser's responsibility to provide the manufacturer with all necessary information. The purchaser's data should, on a minimum, include the following:

- *Climate conditions—ambient air temperature*: minimum, maximum, and average; altitude above see level; ambient air pollution level; and sulfur dioxide air contamination.
- *Power system characteristic*: highest voltage for equipment, short-circuit level, earthing of system neutral(s), variations in supply voltage, seismic level, fault clearance (only if it is unexpectedly long).

Table 7.1 is a practical example of the characteristic of a power system.

If values of the short-circuit level (item 2) and ratios of impedances (items 3 and 4) are not known for a particular location, then approximate data can be taken from IEC Standard No. 60076-5.

Variation of the supply voltage refers to a rated voltage on the supply side of the transformer.

7.4 Required Parameters and Ratings

List of various required parameters and ratings may comprise even several dozen items. In case of large units, generally only a limited number of them may be difficult to understand

TABLE 7.1

Practical Example of the Characteristic of the Power System

	Winding	HV	LV	TV
1	Highest system operating voltage (kV)	245	123	17.5
2	Short circuit level	20	6GVA	Negligible small
3	Ratio X/R	14	12	N/A
4	Ratio X_0/X_1	0.8	1.5	N/A
5	Earthing of a system neutral	Solidly earthed		Isolated
6	Variation of a supply voltage	+10% −20%	—	—

and, thus, may require clarification. Some of the most important parameters to be taken into account are as follows:

- *Voltage regulation, particularly for multiwinding transformers with both on-load and off-load circuit regulation*: Apart from specifying the voltage regulation range(s) and the number of steps, it shall be stated whether the regulation will be performed at the rated power or the power will be reduced on some positions of the tap changer.
- *Operation at a voltage higher than a rated value*: If there is substantial fluctuation of a supply voltage that require deep adjustment by an on-load tap changer, it is possible that the rated tapping voltage is remarkably lower than the supply voltage. It causes an oversaturation of a core. Necessary steps are to be put in places to avoid such occurrences.
- *Insulation level*: In general, standards list various insulation levels for each high voltage supplied by the equipment (rated voltage of a system). It is not necessary to select the highest voltage level, as it may have an influence on the cost of the transformer. A set of testing voltages shall be selected with or without switching impulses. One shall remember that in the case of high-voltage transformer, surges transferred from a main winding(s) to a winding with a comparatively low voltage rate (e.g., an auxiliary winding) may surpass its insulation level. In such cases, protection (i.e., surge arresters) features shall be permanently installed on a transformer.
- *Short circuit impedance*: When a transformer has two main windings and an auxiliary winding, it seems it is a good practice not to list values of the short circuit impedances of that winding and main ones but to impose the highest short circuit symmetrical three-phase current for short circuit in a network connected to the auxiliary winding. Let the supplier select the value of the short circuit impedances. When a purchased transformer can operate in parallel with an existing unit then not only short circuit impedance on a principal tapping but also the values for extreme tapings shall be listed.
- *Sound level*: Particular attention shall be paid to the location at which the level must be low. Details of acoustic screens shall be carefully verified, as the screening may hinder installation on site.

7.5 Overview of a Transformer Design

7.5.1 General Overview of a Project and a Manufacturing Process

They invariably will be the first subject of a first design review, particularly when a design adopted by a particular supplier is less known to a purchaser.

General overview of the manufacturing process shall comprise the following:

- ISO and other quality procedures
- Manufacturing of windings, a core, and a tank (equipment and technology)
- Assembling
- Drying and impregnation

- Preparation for testing
- Preparation for transportation and transportation
- On-site assembling and filling with oil

Drying and impregnation of an insulation can influence upon whether a transformer can withstand high-voltage electricity not only during a factory test but also during a service activity. Hence, the process shall be carefully explained by the supplier. At lower voltage a typical classic vacuum oven can be sufficient. In case of higher voltage, modern vacuum oven using a kerosene vapor is preferred.

Transportation of transformers with oil is more secure; hence, it shall be preferred. However, large transformers must be often transported without oil. A tank is then filled with dry air or nitrogen. Provision shall be made to maintain sufficient overpressure to prevent wet ambient air entering the tank.

Assembling a transformer on site, particularly a large, heavy unit, can require extra area, which shall be at least temporarily prepared to facilitate operating heavy hoisting cranes. If special, less common oil treatment equipment is necessary during assembling it shall be specified.

7.5.2 Tank and Its Main Elements

Important elements:

- General construction
- Location of the main elements against the tank: conservator, bushings, cooling elements, tap changer, rolling and lifting facilities, and so on
- Arrangement of pipes, valves, and the main fittings
- Surface preparation and painting
- Connection of a cover with the tank
- Leakage flux control

In case of large transformers, it is not all the same how bushings, conservators, tap changers, and coolers (cooling cubicles) are located on the tank against each other. This is particularly important when the transformer needs to be installed on an existing substation. Local requirements and customs shall not be neglected. Arrangement of pipes, valves, and other elements is important too.

According to current practice, in many types of transformers a cover is welded to the tank. This will help avoid the leakages caused by the main gasket, often a reason for the leaks caused.

Leakage flux control typically requires installation of magnetic shields. This can have an influence on the tank design.

Tanks of large transformer are comparatively high. Hence, from maintenance point of view it is a good practice to ask for some facilities (a ladder, a stage, etc.) to safely access the top of the tank and to conveniently watch its cover and elements on it when the transformer is switched off! They enable access to the Buchholtz relay(s).

Necessity of restriction of the sound level (precisely, a sound power level) of transformers has become more and more important, especially for units with high-rate power and large dimensions. One of the methods of noise restriction is the implementation of acoustic shields. Such shields covering the tank walls are comparatively effective and inexpensive. They can lay just on walls, or be separate from it by a small distance, but still fixed to them. The shields have to be made from incombustible materials. Provisions shall be made to ensure access to necessary valves, sensors, and so on.

7.5.3 Core and Its Frame

More important elements:

- General design, method of lamination
- Losses and cooling
- Overexcitation limitation
- Insulation of the core against its fame and the tank
- Earthing system
- Clamping: method, materials
- Sensors to measure the core temperature (if any)

Magnetic steel (as all other types of steel) is able to operate at high temperatures, which can be dangerous to mineral oil and cellulose insulation. Hence, any loss on account of overheating at the core shall be produced to the customer. This is particularly important under the overexcitation condition, during which loss distribution can be extremely heterogeneous.

IEC Standard No. 60076-1 limits overexcitation to 5% above a nominal value. When a transformer is supplied with a voltage exceeding a rated tapping voltage of some taps at 5% higher than the overexcitation condition, it shall be duly specified. It is advised to require a permanent operation (or at least during >1 h) of network transformers at a voltage 10% higher than a nominal value.

A core frame is insulated against laminations, and the whole assembly is insulated against the tank. Lack of proper insulation may create closed loops in which circulating current can be induced. Insulation system shall be carefully designed and selected to enable successful operation during the whole lifetime of the transformer. One shall remember that free water accumulates on the bottom of a tank, which causes deterioration of the insulation against the tank. The core repair can be done at the factory after dismantling windings, but it is an expensive job.

Some utilities require installation sensors to measure the core temperature. If this is the case, at least two sensors shall be installed, in order to have one spare readily available.

7.5.4 Windings and Their Insulating System

Windings are one of the most important part of a transformer. Hence, a lot of attention is paid to them. Some important elements are listed below:

- General arrangement: types of windings positions against a core, type of conductors, and its insulation
- Insulation design: a sketch of an insulation layout, demonstrating how the insulation is designed to fulfill requirements
- Eddy current loses, local overheating

- Cooling of windings, cooling method, temperature distribution, hotspot factor
- Clamping of windings
- Short circuit–withstanding capacity: distribution of radial and axial forces at various short circuits and permissible values
- Leads to bushings and a tap changer
- Leakage flux control

Windings must withstand both overvoltage and short circuit forces. Hence, the manufacturer shall show the results of calculation of overvoltage distribution together with permissible values. Such data shall be supported by results of test of samples. Accordingly, distribution of both radial and axial forces gained by adequate calculation shall be produced. Permissible values for the adopted types of windings shall be clearly stated. Results of tests done on models shall support them.

A hotspot temperature stipulates thermal durability of winding insulation. Distribution of eddy current losses and temperature distribution obtained by calculation constitute a base for discussion. Values of the hotspot factors for all windings shall be proved by verified calculation methodology. It shall be emphasized that a high value of the hotspot factor (i.e., higher than a value of 1.3, as proposed by IEC Standard No. 60076-7) may reduce dramatically the overloading capacity of the transformer.

Clamping of windings and its maintenance via service is essential to ensure that windings can mechanically withstand. A significant percentage of large damages of transformers is caused by inadequate capacity of the leads to withstand electricity, as well as their fastening; therefore, design of them shall be revised in such situations.

In case of large transformers, particularly generator transformers, a problem of leakage flux control by no means shall be given overemphasis. Lack of adequate stray flux control would be a reason of local overheating, which in turn would cause permanent oil decomposition, that is, producing of combustible gases.

One of the essential factors assuring the capacity of the transformer to withstand short circuit forces is the proper mechanical stabilization of windings and their blocs before placing them on core columns. Technology of the stabilization becomes more complicated and expensive as dimensions of winding rises. Hence, adoption of simple technology that is adequate only for medium-power transformer to large units could not ensure the expected lifetime of the transformer. Detailed attention is advised here. It is a good practice to include the detailed requirement of the technology in the technical specifications.

7.5.5 Cooling System

Cooling system, even of medium size transformers, is quite complex. Improper design and improper (i.e., of poor quality) elements would cause a lot of problems in service. Some topics involved are as follows:

- Cooling method (variation with loading)
- Cooling elements: radiators, coolers, oil pumps, and fans
- Cooling system control and its cabinet
- Measurement of the oil temperature and estimation of the winding temperatures, and indication of oil flow

Loading of network transformers, varies vastly compared to that of generator transformers. In order not to overcool them at low loading and/or at low ambient temperature, cooling capacity shall be adjusted automatically. In a number of cases, even a cooling method shall vary, for example, from ON to OFAF via ONAF (or OFAN) upon loading. Moreover, the number of operating coolers (cooling cubicles) shall depend on loading, that is, oil and winding temperature.

For measurement of oil temperature and estimation of winding temperature, high-quality and durable instruments shall be selected. It is advised to use modern, microprocessor-controlled devices to ease adjustment of cooling capacity to varying loading levels.

7.5.6 Conservators and Pipes

Some essential topics:

- Main conservator and the on-load tap changer conservator
- Preservation system against moisture: only silica gel breather, rubber membrane, rubber air bags, nitrogen blanket
- Pressure/vacuum capabilities
- Adequate pipes arrangement for evacuation of gas collected during operation
- Auxiliary equipment: shut-off (cut-off) valves, oil-level indicator(s)

When the on-load tap changer is used, it shall have a separate, small conservator to prevent mixing of oils. There is a number of oil preservation systems in the conservator. For larger transformers, the system with the rubber airbag seems to be the most appropriate. However, some utilities prefer the nitrogen blanket.

While filling the transformer with oil, air pockets can occur in a tank. Pipes shall be arranged to ensure that all air pockets are evacuated. Some utilities use shut-off (cut-off) valve to prevent excessive oil flow from the conservator to the tank. It can happen when substantial oil leakage from the tank occurs, possibly due to any rupture. Large transformers shall be equipped with oil-level indicators, with a remote indicator informing at least the maximum and the minimum oil level.

7.5.7 Factory Type and Routine Tests

Relevant, important topics:

- *Lightning impulse tests*: tap positions, detection methods, and terminal connection
- *Switching impulse tests*: tap position to meet required voltage on terminals of both windings
- *Long-duration AC test with partial discharges measurement*: single- or three-phase supply, tap position, method of PD measurement, and permissible PD level
- *Heat run tests*: method of test, recalculation of the test results, estimation of overload requirements.
- Sound-level test
- *Special tests*: FRA measurement, and so on
- Some requirements for the routine tests

To enable proper assessment of a lightning impulse test two detection methods shall be used (not one as required by IEC Standard No. 60076-3). Moreover, they shall be utilized at chopped impulse as well.

Both switching impulses and AC voltages are distributed with a turn ratio. Hence, in some cases where transformers have two windings, they shall undergo such tests that prescribe the test voltages for both windings, which may not be met by merely adjusting the position of a tap changer. Such cases shall be indicated during the design review. Evidence has to be produced by the manufacturer that the insulation system would meet requirements.

According to the IEC Standard No. 60076-3 permissible partial discharge (PD) level is comparatively high (i.e., in the order of 500 pC). Manufacturers using good technology obtain much lower value. Consequently, lower PD level shall be agreed (in the order of 100 pC or so).

Results of a heat run test shall not only confirm that rise of oil temperature measured is within the permissible level and that windings are not surpassed but also furnish to the customer the data needed to estimate loading and overloading capacity of the transformer. Hence, a multiple heat run tests shall be performed.

Special tests shall include tests giving fingerprints for further measurements in a service. It is advised to include no-load test lasting several hours (at least >6 h) with overfluxing lasting, say 1 h.

7.5.8 Loading and Overloading Capacity

Power transformers, apart from generator transformers and some other minor groups work under loads varying in a wide range. Some of them are overloaded during normal or abnormal network conditions. There are standards, for example, IEC Standard No. 60076-7, discussing in detail how to calculate permissible loading and overloading of existing units at various levels of loading.

When a new transformer (or a series of them) is purchased, the problem shall be reversed. On the basis of loading either recorded at existing locations or estimated on the basis of a network calculation, the required loading capacity of the purchased transformer can be listed. Provision can be made for load variation during a calendar day. Seasonal variation of an ambient temperature shall also be taken into account. Since loading capacity depends on the transformer design (e.g., a hotspot factor), detailed discussion with a manufacturer may bring a proper estimation of the requirements of a purchaser and a specific design feature.

7.5.9 Auxiliaries and Fittings

From the current maintenance point of view, auxiliaries and fittings are particularly important. If their quality or durability falls short of the expected standard, it would invariably increase the maintenance cost and may in turn force shutting down the transformer. A large number of different types of fittings, valves, thermometers, and so on are installed on a transformer. It is no doubt that a substantial portion of time reserved for a design review will be devoted to focusing on auxiliaries and fittings. Detailed drawings of electrical installation shall be prepared by a supplier on the basis of a purchaser's general (or better, more specific) requirements. Current transformers are commonly mounted on oil part of capacitively graded bushings and on leads from windings to medium voltage bushings.

7.5.10 Online Monitoring Devices and Systems

The CIGRE brochure [3], in Clause 5.4: "On-line continuous monitoring techniques" states:

> A number of utilities are now using on-line monitoring devices to respond to the increasing emphasis on reducing unplanned outages and equipment failures, improving power quality, and deferring capital and maintenance expenditures.

These days, a wide range of various online monitoring devices are commercially available. Currently, the most common are devices that can monitor online the contents of water and inflammable gas (one, or more) soluble in insulation oil. Installation of such devices is advised on heavy-duty transformers. From the maintenance point of view, only devices with verified reliability and durability shall be selected so as not to mislead a staff or cause problems that can in turn increase the maintenance cost.

There always exists the need for online monitoring of high-voltage bushings, since their sudden failure could cause dramatic damage to the whole transformer. Essential question is, how can the device take into account permanent minor variation of a voltage on a transformer's line terminals? It is advised to use such devices that continuously monitor the voltages via voltage transformers (always installed in substations) and analogue/numerical transducers. Thus, verifying the reliability and durability of such devices is important.

In an extremely high-voltage transformer, a problem of PDs could persist. Currently, there are devices (sensor) available that operate at ultra high frequency, in the range of up to few GHz. Such sensors can be inserted into transformers via some valves, which typically are installed to filter oil. If such valves are to be used in the sensors installed now or to be installed in future, then proper valves shall be selected. Moreover, they shall be installed in such places (mostly a transformer cover) that enable pushing of sensors deep inside to "catch" signals but not to cause flashover between live elements and the sensor.

Communication outlets of the above devices shall be incorporated into online monitoring system, when foreseen. It enables remote access to the devices, for example, analogically, the microprocessor-controlled device controlling the cooling system of a transformer (see Subclause 7.5.5).

An increasing number of transformers equipped with online monitoring system are now available, particularly in the case of large, heavy-duty unit operated in remote substations.

The number of signals produced by the transformer intended for the monitoring system is comparatively big, especially when cooling system has a lot of oil pumps and fans. It seems it is a reasonable practice to connect all of them to concentrator(s), which convert them to numerical signals. Two concentrators are advised: main and standby, to ensure 100% redundancy. They shall be connected with a part of a monitoring system located in a control room by fiber-optic cables. There are various transmission protocols, and the selected one have to be indicated in the technical specification. It is proposed to use the protocol according to IEC Standard No. 61850.

The concentrators with their auxiliary elements can be installed in the control cabinet of the transformer. Assuring proper temperature and humidity inside the cabinet is essential in such cases.

The signals generated by a transformer are needed by the SCS system too. In order to avoid mistakes, a careful list of all signals shall be prepared. It should clearly indicate a recipient of each signal. An element producing the signal (contactor, temperature sensor, current transformer, etc.) shall be shown. Terminals of a strip shall be indicated too. Sometimes, especially in the case of new suppliers, preparation and harmonization of the list could be a time-consuming job.

7.6 Practical Examples

In what follows, some practical examples are shown. They are mainly focused on large network transformers.

7.6.1 Voltage Regulation

In the case of autotransformers, location of a tap changer from an electrical point of view shall be given. In order to avoid a mistake, not only a percentage value of regulation but also a range in "kV" shall be listed. When the regulation range is comparatively wide, it would be more economical not to require full power on extreme minus taps. This is to avoid overdimensioning of a winding with regulation (Table 7.2).

7.6.2 Rated Impulse Voltage Level

For three-phase transformers switching voltage between phases is 50% higher than the ones against earth. The value of a chopped wave is 10% higher than that of the full wave, according to IEC Standard No. 60076-3.

As it was mentioned earlier (see Subclause 7.5.7), not always the rated switching impulse of both windings can be met. If it happens, the rated level of the winding with the highest U_h prevails (Table 7.3).

7.6.3 Rated AC-Induced Voltage Level

Table 7.4 shows values related to the AC long-duration (ACLD) test, according to IEC Standard No. 60076-3. Moreover, a three-phase supply was selected, at which a voltage between phases is $\sqrt{3}$ times higher than against earth. If a single-phase supply is chosen, then the ratio would be 1.5.

Comparatively high value of PDs level was allowed, however, but still lower than that required by the above-mentioned standard.

TABLE 7.2

Practical Example of the Voltage Regulation

Type of Regulation	On-Load
Voltage variation	CFVV (constant flux)
Location of a tap changer	At the HV side (i.e., in a series winding)
Tapping range	±16% (from 220 kV, i.e., ±35.2 kV)
Number of taps	±16 steps (i.e., 32 positions)
Required power on tapings other than the rated tap	+16% to +11%, at the rated power (full-power taps); +10 to −10%, at the rated power (full-power taps); −16% to −11%, at the current equal to the rated tapping current of the tap −10% (reduced power taps)

TABLE 7.3

Practical Example of the Rated Impulse Voltage Level

Rated lightning impulse–withstand voltage (full wave/chopped wave) for line terminals		
1	HV winding ($U_h = 420\,kV$)	1300/1430 kV
2	LV winding ($U_h = 245\,kV$)	850/935 kV
3	TV winding ($U_h = 17.5\,kV$)	95/110 kV
Rated switching impulse–withstand voltage for HV line terminals		
1	Phase to earth	1050 kV
2	Phase to phase	1575 kV
Rated switching impulse–withstand voltage for LV line terminals		
1	Phase to earth	750 kV
2	Phase to phase	1125 kV

TABLE 7.4

Practical Example of the Rated AC–Induced Voltage Level

Rated short duration voltage U_1		
1	Phase to earth	
a	HV winding	420 kV
b	LV winding	245 kV
2	Phase to phase	
a	HV winding	$\sqrt{3}*420 = 727\,kV$
b	LV winding	$\sqrt{3}*245 = 424\,kV$
Rated 30 min duration voltage U_2 for HV line terminals		
1	Phase to earth	$1.5*420/\sqrt{3} = 364\,kV$
2	Phase to phase	$\sqrt{3}*1.5*420/\sqrt{3} = 630\,kV$
Rated 30 min duration voltage U_2 for LV line terminals		
1	Phase to earth	$1.5*245/\sqrt{3} = 212\,kV$
2	Phase to phase	$\sqrt{3}*1.5*245/\sqrt{3} = 368\,kV$
PD level at U_2		$\leq 200\,pC$

7.6.4 Rated Short Circuit Impedance

Rated values refer to the rated power. The IEC Standard No. 60076-1 stipulates tolerances for the measured values equal ±10% for the nominal tap and ±15% for other taps. In case of a parallel operation, values not only for the rated tap but also for the extreme taps shall be listed. Moreover, it is advised to reduce the tolerance for the extreme taps up to ±10%. These days calculation methodology and productive repeatability enable such restriction. In order to avoid mistakes, one shall list the range within which the measured values shall lie.

In the case of a three-winding transformer one shall be aware that not all values of the short circuits impedances of the three pairs of windings are possible to achieve (see Clause 7.4; Table 7.5).

TABLE 7.5

Practical Example of the Rated Circuit Impedance

The pair of HV and LV windings		
1	Nominal tap (i.e., 0%), taking into account a tolerance value according to Standard No. IEC 60076-1; that is, the measured value shall be in range	12.5%; from 11.25% to 13.75%
2	Maximal tap (i.e., +10%), taking into account a tolerance value according to Standard IEC 60076-1; that is, the measured value shall be in range	16.4%; from 14.76% to 18.04%
3	Minimal tap (i.e., −10%), taking into account a tolerance value according to Standard IEC 60076-1; that is, the measured value shall be in range	10.0%; from 9.00% to 11.00%
Pairs of HV-TV and LV-TV windings		
1	Short circuit impedance of these pairs of windings should be so chosen that in case of operation on nominal tap, with taking into account the network short circuit power, the three-phase short circuit current in network connected to TV winding will not exceed	40 kA

7.6.5 Loading and Overloading

This problem is much more important for network transformers than for generator transformers, which typically continuously operate at a rated load (or close to it).

IEC Loading Guide No. 60076-7 distinguishes three loading cases:

- Normal cyclic loading
- Long-time cyclic overloading
- Short-time emergency overloading

Cyclic loading can be represented by sinusoidal variation with time of a day as shown below:

$$S(t) = S_{max} \cdot \left[A + B \cdot \sin\left(\frac{t}{12} \pi \right) \right] \tag{7.1}$$

where

S_{max} is the maximal value of loading
A is the coefficient of an average loading
B is the coefficient of variation of the loading (±)
$A + B = 1$

The maximum loading S_{max} occurs at the moment $t = 6\,h$ against a reference moment $t = 0$, and the average loading ($S_{max} * A$) takes place at the moments $t = 0$ and $12\,h$ against the reference one.

It is difficult to propose a representation of the short-time emergency overloading. It depends on local conditions. We can assume that at the moment of occurrence of S_{max} equal to a rated power of a transformer of the normal cyclic loading the load jumps to a value of S_{emrg} for a period of T_{emrg}. After that period, the loading returns to the normal cyclic loading. The period T_{emrg} depends on the time needed to rearrange a system to return to the normal cyclic loading or at least to the short time emergency overloading. The value of $T_{emrg} \approx 1 \ldots 2\,h$ seems to be a reasonable approximation.

TABLE 7.6

Example Data of Criterion Values and Loading

		"A"	"B"	"C"
Maximal load current of HV winding (A)		650	700	750
Permissible value of oil temperature (°C)		105	115	115
Permissible value of windings temperature (hotspot) (°C)		120	130	140
Season, ambient air temperature				
"cold," $\Theta_a \leq +5°C$	$S_{max} \geq$ MVA	190	215	240[a]
"hot," $\Theta_a = +30°C$		170	185	220[a]

"**A**," normal cyclic loading; "**B**," long-time cyclic overloading; "**C**," short-time emergency overloading.

[a] S_{emrg}.

The proposed criterion values and loading parameters are shown in Table 7.6. It is emphasized that it is an example only!

Criterion values of temperatures were taken from the above-mentioned loading guide.

Assuming a value of a hotspot factor up to 1.3, a rated power of a typical OFAF network transformer to fulfill values from Table 7.5 is in the order of 130–140 MVA. When the value of the hotspot factor rises then the representative rated power of the transformer to fulfill loading, as given in Table 7.6, have to rise remarkably.

References

1. Working Group WG 12.22. Guidelines for conducting design review for transformers 100 MVA and 123 kV and above, CIGRE 2002, Brochure No. 204.
2. Working Group WG A2.20. Economics of transformer management, CIGRE 2004, Brochure No. 248.
3. Working Group WG A2-18. Life management techniques for power transformers, CIGRE 2003, Brochure No. 227.

8

Life Management of Transformers

Marceli Kaźmierski

CONTENTS

8.1 Introduction

Power transformers are key and costly components in networks, and loss of a transformer can have an essential impact on the safety, reliability, and cost of supply of electrical energy. Increasing over recent years, demands for high reliability of the supply on the one hand, and common policy among utilities to maximize use of existing networks at their full design capability, on the other, have resulted in a number of 20- to 30-year-old or older large power transformers being still in operation at full loading. Their maintenance, or—more generally—managing their life with improved reliability and reduction of scheduled outages, becomes more and more important.

Effective transformer life management, consisting of a high variety of aspects, can vary greatly depending on local conditions and practices.

In fact, the life-cycle management for power transformer begins with the specification for a new transformer and finishes with the "end of useful life" (technical, economic, strategic) [26]. The successive stages in the life management have been dealt with in detail by SC A2—formerly SC 12—(Transformers) of CIGRE, namely:

- Specification, tender review [35]
- Design, design review [36]
- Manufacture, factory tests
- Operation [37]

Both users and manufacturers play important roles in these stages, and, hence, close cooperation between them is essential in achieving success in life cycle management. Moreover, over recent years, each of the stage is usually assisted by an economic consideration [38].

The fundamental objective of life management has been defined simply by WG 12.18 [37] as "to get the most out of an asset" by ensuring that actions are carried out to promote the longest possible service life or minimize the lifetime operating cost, whichever is the most appropriate.

The starting point for the effective life management of a transformer is its proper condition assessment. The base for the condition assessment is created by two main factors:

1. Measurement of a number of physical parameters, realized after all without the transformer switching off.
2. Systematic knowledge, arranged in clear rules adopted to a computer processing, is after all about degradation mechanisms, failure modes, and their symptoms as well as measurement methods and their identification possibilities.

The chapter is dealing with selected elements of life management during the transformer operation.

8.2 Failure Model

The basic failure model in a concept of the transformer life management and condition assessment assumes [37] that there are a number of key functions or parameters, such as dielectric and mechanical strength, and that failure occurs when the withstand strength of the transformer with respect to one of these key properties is exceeded by operational stresses (Figure 8.1).

Operational stresses consist of stresses under normal operation and those due to intermittent events such as lightning strikes or short circuits. For proper protection, both of them should be below the withstand critical level of a new, properly designed and manufactured transformer that is checked during factory (commissioning) tests.

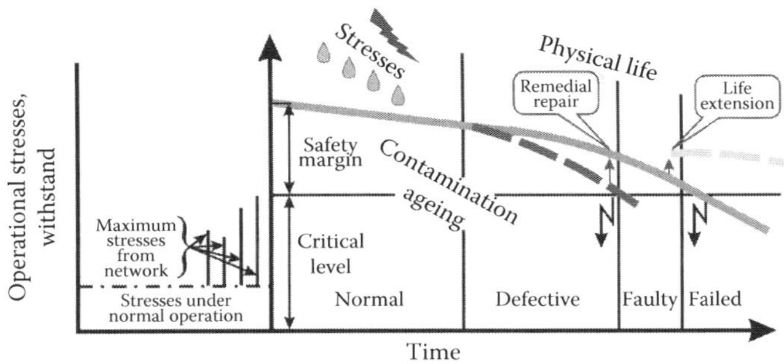

FIGURE 8.1
Conditions of a transformer in the course of its life cycle. (According to Guide for life management techniques for power transformers, CIGRE '03, Brochure No. 227, 2003.)

The withstand strengths of a transformer is naturally decreasing over its life due to various aging processes (normal aging) but may deteriorate faster than normal under the influence of agents of deterioration (e.g., moisture) or if some abnormal destructive deterioration process (e.g., tracking) occurs. The last may have a reversible character (often referred to as defects) and irreversible ones (faults).

A key element of the transformer life management is quantitative assessment of the safety margin (Figure 8.1). Usually, the margin is related to mechanical condition of windings, electrical, mechanical, and thermal wearing of solid insulation and oil, condition of OLTC, and so on.

Obviously, the main operational stresses are of random nature. Therefore, it is practically impossible to predict when the final failure will occur, but, if remnant strength and operational stress are estimated adequately, it would be possible to determine when the circumstances were such that a failure could occur, which means when the transformer enters into an area of increased unreliability.

8.3 Deterioration Factors

Multiple stresses that transformers are subjected to during operation are an origin of deterioration factors. Typically, they are thermal, dielectric, chemical, mechanical, and also environmental in nature. It is commonly assumed that the service life of transformers is determined by the life of the paper–oil insulation system.

8.3.1 Insulation Aging

Insulation aging is a 4D problem due to thermal, dielectric, chemical, and mechanical stresses, which are highly dependent on operational conditions. In the degradation of cellulose (solid insulation) and oil several mechanisms are involved [16,31,32]. The most important factors influencing the degradation are temperature and moisture.

The main mechanism for the degradation of the oil is oxidation. Oxidation causes the formation of polar degradation products, with acids and sludge as the final products. Also water is produced in the oxidation process. At the end, the production of large amounts of these breakdown products may lead to loss of the insulating properties of the oil.

Degradation processes of cellulose are assisted by shortening of molecule chains that are characterized by paper degree of depolymerization (DP), which is the indicator of mechanical strength of the insulating paper. For new cellulose, the DP value is between 1000 and 1200. The molecule length of degraded cellulose is reduced to a DP value of about 200 that correspond to 30%–40% of the initial mechanical strength of the paper. In this condition, cellulose is brittle and the durability against mechanical stresses is strongly reduced.

The degradation processes of oil and the degradation of the solid insulation have many interactions. The degradation products of one may interfere in the degradation of the other. In some cases, they may also, to a significant extent, accumulate in the other component. The net effect of such phenomena may be very difficult to predict, and the full extent of this interaction and the details of the mechanisms are still far from completely understood.

The diagram shown in Figure 8.2 illustrates schematically the different factors governing the degradation of the oil and the paper. Discussed problems are presented more detail in lecture [11].

8.3.2 Thermal Factors

The heat generated by the transformer in operation causes degradation of characteristics of component materials of the transformer and a structural deterioration of the transformer due to a stress induced by a temperature change. The deterioration of the insulation system is determined by the deterioration of insulation at hotspot that is governed by the well-known Arhennius–Dakin formula [6].

For example, due to thermal deterioration, the average DP degree of polymerization drops to about 50% in 30 years, and mechanical strength of the insulating paper is lowered accordingly [13].

Faster progress of deterioration at higher temperatures is a characteristic of overload condition. In that case, a new hazardous factor may occur—air bubbles from the insulating paper as the result of a sharp temperature rise of the windings. The temperature at which air bubbles begin to occur varies with the moisture content of insulating paper. According to experimental results presented in Ref. [13], the temperature limits are as follows:

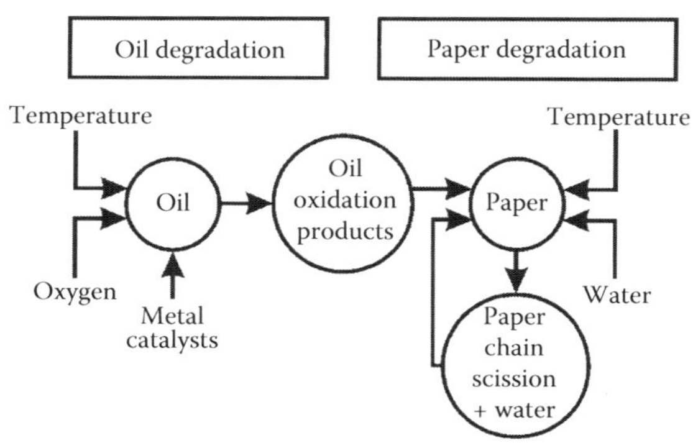

FIGURE 8.2
Schematic degradation mechanism for the paper–oil system. (According to Leemans, P. et al., Control, diagnostic and monitoring of power transformers, CIGRE '98, Report 12-213, 1998.)

- 180°C at 0.5% moisture content
- 160°C at 2% moisture content
- 135°C at 4% moisture content

According to Ref. [21] at 2% water content, the bubble evolution temperature is 140°C–150°C.

Contamination of the oil by particles and fibers has also a significant influence on the temperature at which bubbling takes place [40].

8.3.3 Dielectric Factors

The electric field, particularly in areas of its concentration, is deterioration agent for the insulation system. Electrical overstressing resulting in dielectric breakdown may be caused by external influences such as atmospheric (lightning) or switching overvoltages or internal influences such as winding resonance. Dielectric breakdown may also be a secondary effect of a range of primary causes such as mechanical overstressing, aging, or contamination.

The typical failure scenario for an initially defect-free insulation subjected to the impact of the other degradation agents (e.g., thermal aging or oil contamination) is as follows [37]:

Nondestructive partial discharges (PD) at operating voltage and reduction in impulse withstand strength ⇒ Appearance of destructive PD ⇒ Progressing surface discharges occurrence ⇒ Creeping discharge occurrence ⇒ Breakdown

8.3.4 Mechanical Factors

Mechanical overstressing may be caused by current stresses such as overloads, short circuits, overexcitation, or inrush currents that impose electromagnetic forces on the winding structure leading to displacement and possibly dielectric breakdown or/and to increasing the transformer vibration and noise. The structure is disturbed each time a current stress occurs, with each stress increasing the risk of damage. Mechanical overstressing may also arise due to vibration caused by transport shocks or resonance phenomena.

The typical scenario of a transformer failure due to the mechanical factors may have the following form [37]:

Loose clamping ⇒ Distortion of winding geometry ⇒ PD appearance ⇒ Creeping discharge progressing ⇒ Breakdown

It should be noted that the mechanical factors are extremely dangerous for old transformers with the insulating paper of DP = 450 or less.

8.3.5 Environmental Factors

These include rain (acid rain), sunshine (ultraviolet rays), corrosive gases, particles of sea salt, and so on. Mainly, the transformer tank and its accessories are affected by those factors.

8.3.6 Deterioration of Accessories

The service life of a transformer depends more or less on the life of its accessories. The most important and also failed more often than the other, are OLTC, bushings, and coolers.

8.4 Condition Classification

Because the most of the deterioration factors listed earlier are of random nature, it is not practicable to quantify withstand strengths and operational stresses for each unit in operation. Usually, their condition is used to assess the expected reliability. The classification listed in Table 8.1 connects the transformer condition with required action to be performed on the unit.

The classification is of general applicability. Nevertheless, the symptoms of failures, defects, and faults vary greatly for the different failure and aging modes/mechanisms.

8.5 Maintenance Strategies

The time-based maintenance (TBM), previously used worldwide, is recently replaced by condition-based maintenance (CBM) that is more reliable and more economic effective. The latter results, for example, from a lesser number of the transformer shut-offs needed for any maintenance operations. Moreover CBM meets requirements to enable important decisions on life management to be made by utilities. A pragmatic two-stage-condition-based methodology has been proposed by WG 12.18 (Figure 8.3) [37].

The main purpose of the first stage, which may be described as monitoring or detection, is to filter out those items of transformers that are operating normally. The techniques applied to solve the question have to be applied regularly, and need to be cheap, sensitive, and broadband so that any potential problem can be detected (but not necessarily diagnosed) at an early stage. Online techniques are preferred here.

TABLE 8.1

Transformer Condition Classification

Condition	Definition
Normal	No obvious problems, no remedial action justified
Aged; normal in service	Acceptable, but does not imply defect free
Defective	No significant impact on short-term reliability, but asset life may be adversely affected in long term unless remedial action is carried out
Faulty	Can remain in service, but short-term reliability likely to be reduced. May or may not be possible to improve condition by remedial action
Failed	Cannot remain in service. Remedial action required before equipment can be returned to service (may not be cost effective, necessitating replacement)

Source: Guide for life management techniques for power transformers, CIGRE '03, Brochure No. 227, 2003.

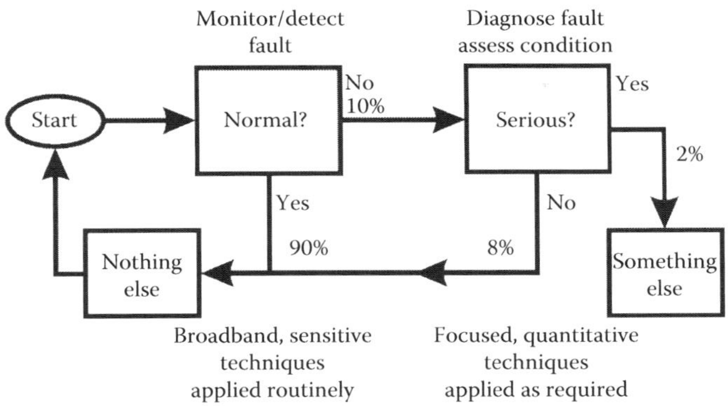

FIGURE 8.3
Two-stage-condition-based methodology. (According to Guide for life management techniques for power transformers, CIGRE '03, Brochure No. 227, 2003.)

The second-stage process—described as diagnosis or condition assessment—would normally only be carried out on those transformers—usually less than 10%—which could not definitely be classified as normal. To these transformers, more precious and more expensive tests would be applied in order to answer the question whether it is serious or whether it is reliable and so on.

Tests would be applied as required to address this question and can be more expensive and off-line but need to be focused on individual attributes in an attempt to arrive at an unambiguous diagnosis. To identify a problem the following measures are used:

- Quantification of defective condition based on the knowledge about the process of the physical mechanism of the defect development
- Comparison of the diagnostic measurement results with some references (finger printing)
- Trend analysis that illustrates the manner and the change, along with time of the defect development
- Statistical analysis that can help in understanding the variability of available data

The next advanced steps in the methodology are Risk-Based Maintenance (RBM) and Reliability-Centered Maintenance (RCM) [12,28]. RBM is based on setting of inspection frequency by specification, application, age, and other factors and is combined with simplified methodology and inspection procedures.

RCM is one approach for obtaining high reliability at lower costs. Generally, it is a computerized decision-making tool that allows to manage a maintenance program with an optimum strategy. It may be defined also as a structural decision process, based not on an attempt to estimate the reliability of each apparatus, but on the consequences of the functional failures for equipment itself.

The decision process has the following sequence:

$$\text{Failure symptoms} \Rightarrow \text{Identification} \Rightarrow \text{Development forecasting}$$
$$\Rightarrow \text{Preventive maintenance}$$

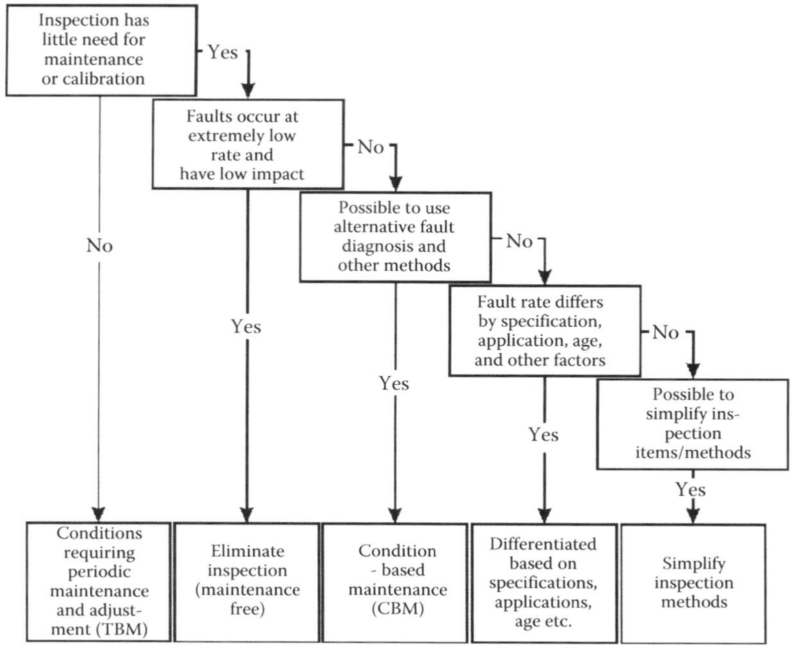

FIGURE 8.4
Transformer inspection streamlining flowchart used in Japan. (From Kawamura, T. et al., Improvement in maintenance and inspection and pursuit of economical effectiveness of transformers in Japan, CIGRE '02, Report 12-107, 2002.)

This process allows to minimize maintenance operations. Systematic failure analysis and assessment of the transformer design from reliability point of view create a feedback form.

Choosing a strategy depends on a number of local factors, such as a transformer importance in the network, the failure rate, effectiveness of used diagnostic methods, and so on. For example, in France [22] the decision about strategy used for the maintenance of the biggest units is taken by the National Transformer Diagnostic Center (CNDT). Special care is applied to generator transformers operated in nuclear power stations. In Russia [29], a special diagnostic program has been elaborated, addressed to about 20,000 transformers operating for 20–30 years. Inspection Streamlining Flowchart used in Japan, characterized by joining different maintenance strategies—according to local needs, which leads to lower the maintenance cost, is presented in Figure 8.4.

8.6 Online Monitoring Systems

Considerable attention has recently been given worldwide to improving and refinement of monitoring techniques of large power transformers and to development of new techniques, including online (in real time), computer-aided, multisensor systems that are able to store, process, and interpret large volumes of data and guide the operator in faultsignature recognition. Such systems offer new possibilities, unattainable by conventional methods, in diagnosing power transformers during their normal operation as well as in emergency conditions. It is estimated [25] that only approximately 35% of transformer failures are detectable in advance by means of the traditional off-line methods. The systems are in operation in a number of countries; for example, see Refs. [3,4,8,14,18,19,24,33].

The primary purpose of condition monitoring is to detect the first signs of an incipient fault, aging development, or other problems and monitor their evolution to enable an operator to take appropriate action and avoid major failure. On the other side, the systems make it possible to record different relevant stresses that can affect the transformer lifetime.

8.6.1 Signals/Sensors

The extent of condition monitoring depends primarily on possibility of online measurements of different inner and outer parameters of a transformer. The measurements usually cover currents and voltages, oil and insulation testing, partial discharge, temperature, mechanical vibration, pressure, condition of accessories (pumps, fans, bushings, on-load tap changer). The various sensors used are based on new microprocessor and fiber-optic techniques. In detail, the following basic measurement quantities are typically recorded:

1. Voltage of each phase, at least at the high-voltage side
2. Current of each phase, at least at the high- and medium-voltage side
3. Overvoltages coming to the transformer (through the measuring tap of the capacitor bushing)
4. Oil temperature at the top and in the inlets and outlets of each cooler; ambient temperature (usually PT100 sensors)
5. State of oil pumps and fans
6. Oil level, moisture in oil, and dissolved gas concentration
7. Operational data of OLTC (tap position, number of switching operations) and the power consumption of the motor-drive mechanism
8. Capacitance, leakage current, and tan δ of bushings; oil pressure of bushing

8.6.2 Structure of the System

Logic structure of the monitoring system is presented in Figure 8.5.

Inputs to the monitoring system include raw input data from sensors or sensor systems as well as data introduced into the system by means of a keyboard in an off-line manner, the collected history of data relevant to the evaluation of the transformer condition. Models and expert structure that enable the evaluation and interpretation of the data are also essential to the system. Finally, user priorities and constraints associated with each utility or user are required.

The "process" of monitoring (middle shaded section in Figure 8.5) is separated into five stages, namely:

1. Primary data acquisition
2. Data evaluation
3. Analysis, state evaluation

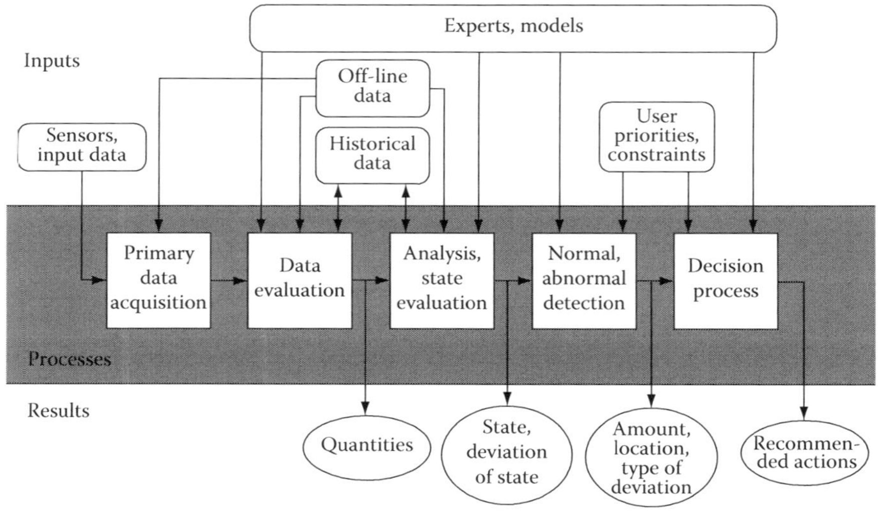

FIGURE 8.5
Generalized logic structure of the monitoring system. (An idea was taken from On behalf of Study Committees 15 and 33, Joint Working Group 15/33.08. Status report progress on high-voltage insulation monitoring systems for in-service power apparatus, CIGRE '96, Report 15/21/33-20, 1996.)

4. Normal, abnormal detection
5. Decision process

Outputs consist of quantities of basic parameters, determination of the transformer state, amount, location and type of deviation (if any), and recommended actions.

Not all monitoring systems fully employ all of these stages in a completely developed form. Simple versions may provide elementary-state detection, such as an alarm signal when certain critical parameters exceed preset limits.

In practice, the system covers functions of "short-" and "long-"term character. The first is connected with current monitoring of a transformer, warning the personnel of the excess of permissible limits or of conditions likely to end with an outage, and "prompting" remedy actions for such situations. "Short-"term function is also responsible for edition reports (on request, cycling, alarm), operational determination of optimal and permissible loading of the transformer, controlling of oil pumps, fans, and on-load tap changer. Long-term functions are essentially focused on long-term loading forecasting and overall diagnostic reports, including expected lifetime of a transformer and determination of dates for nearest inspections and repairs.

8.6.3 Software

Usually online system software consists of the following main elements:

- Online measurement module (data acquisition from sensors and status of equipment registration).
- Diagnostic/decision module (current diagnosis, operational determination of permissible loading, long-term analysis, strategic decisions).

- Database (an operational database, online and off-line databases with a long-term buffer memory, a historical database, a knowledge/rules database). The last, in fact being the basic diagnostic module, represents the knowledge of experts and is updated during the system operation.
- Automatic archiving program—increasing volume of data in large systems is estimated to be in the range of 100 MB or more per month.
- Expert system—the core of the monitoring system. It analyzes processed signals supplied by a number of local measurement subsystems, realizes the inference process, and then issues the diagnosis.
- Control (managing) program—controlling operation of the monitoring system and responsible for online operation of measurement subsystems; operations on databases, including data buffering; supervising of expert system operation, including quick access in an emergency condition; edition of alarm signals; and so on.

An example relationship between the main software elements (system implemented for PPGC) is presented in Figure 8.6.

8.6.4 Location of the System/Data Distribution

It is rather rare that a monitoring system, excluding pilot installations, is in operation as an autonomic system. Usually, the system monitors other transformers in one substation, and it has to cooperate or has to be included into other systems working on the substation, such as Substation Control System (SCS), Asset Management System (AMS), or Dispatcher Control System (DCS). From the other side, monitored and processed data by the system should be accessed by a number of authorized users in a number of different points, usually located far away. To perform these tasks the newest solutions, including Internet technology are used (Figure 8.7).

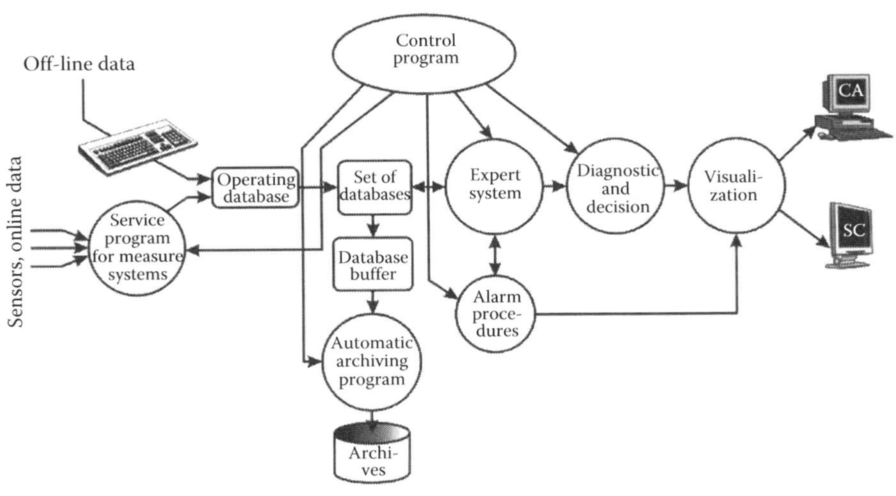

FIGURE 8.6
Schematic relations between software main elements of the monitoring system; CA, system administrator computer; SC, substation computer.

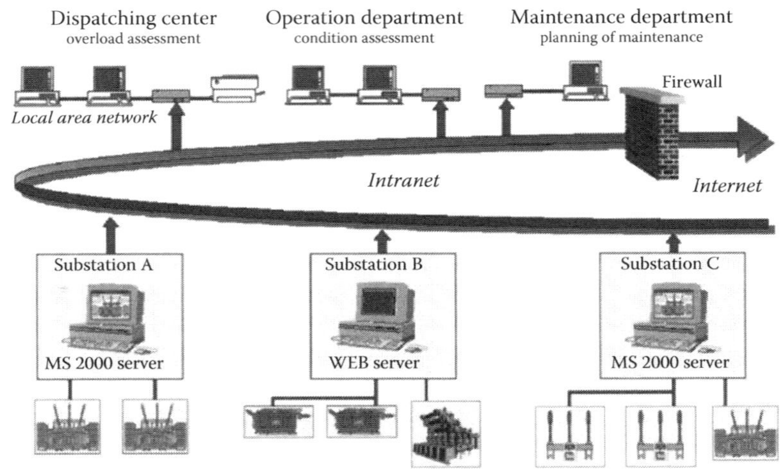

FIGURE 8.7
Data access by use of the Internet. (According to Tenbohlen, S. et al., Experienced—based evaluation of economic benefits of online monitoring systems for power transformers, CIGRE '02, Report 12-110, 2002.)

8.6.5 Benefits from Monitoring System Installation

The cost of a serial online monitoring system is still high, especially in relation to rather low transformer failure rate. Advantages compensating the cost include after all the following:

- Reduced risk of fault occurrence and consequent outage of a large power transformer with all financial implications.
- Reduction to a necessarily minimum the extent and, consequently, the cost of repairs, by preventing a serious fault occurrence inside the tank. Moreover, a planned outage can be scheduled to be carried out at the most convenient time, that is, when load is down or replacement energy costs are minimal. Outage time can be decreased by engaging a spare transformer and oil-processing equipment beforehand.
- More effective, optimum loadability, resulting in lower operation costs.
- Optimization of strategic tasks—repair, refurbishment, and replacement.
- Effective policy in regard to spare transformers.
- Possible reduction of insurance costs.

More detailed benefit calculations as a result of lowering the failure rate, associated loss, and a new investment avoidance are given, for example, in Refs. [4,5,33].

8.7 Transformer Lifetime

For operating personnel who have to make a decision regarding the repair or relocation of existing units, the questions of remaining life and how to extend the life are legitimate ones.

The lifetime of transformers is typically assumed to be 25 or 30 years, and in some cases, up to 40 years. Apart from the small number of units that suffer major failure within that time scale, actual lifetime, assuming normal loading, can be much longer, and many transformers still give satisfactory service after operating for more than 50 years, especially if they are not moved.

Life expectancy can be significantly influenced in service by the loading pattern, maintenance policy, and by the steps taken to protect the transformer against externally imposed overstressing and contamination. Moreover, assessment of the life expectancy should cover not only the transformer internal components but also the tank and its accessories (Figure 8.8). When a transformer is to be used for 40–50 years, its accessories should be subjected to planned inspection/renewal and so controlled that they do not become critical parts in the judgment of the transformer life. Transformer accessories such as cooling equipment and control and supervisory devices require little more than routine visual inspection and operational testing. Nevertheless, bushings and OLTC need more attention.

Factors that determine the remaining life of a transformer can be categorized under three headings [6]:

- Strategic
- Economic
- Technical

End of life may be dictated by any one factor or by any combination.

The strategic factor relates to the transformer ability to carry the loads, short circuit currents, network service voltages, overvoltages, and other "normal" stresses applied in service. When the load increases beyond the rating of the transformer, the user is faced with the need to decide whether to overload the unit or to replace it with a new one. The old transformer may be moved to another location or scrapped. Moving a transformer to a new site carries some risk, especially for an older unit. Transformers can continue to operate satisfactorily despite their age, provided they are not disturbed mechanically.

The economic factor includes the cost of losses and maintenance costs. The cost of maintenance outages in terms of undelivered energy or of more expensive supply arrangements may also be relevant.

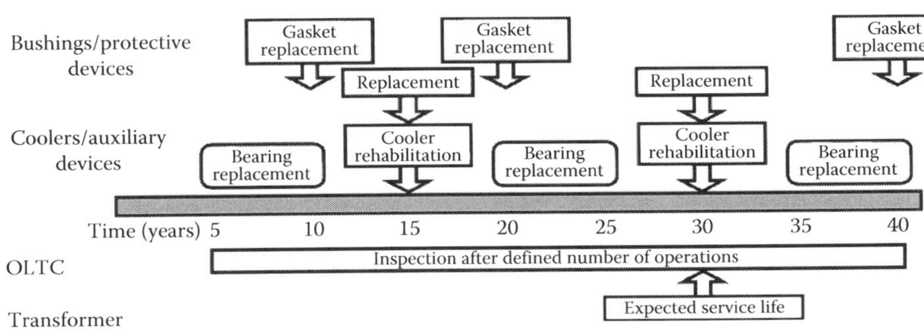

FIGURE 8.8
Time schedule of operation on transformer accessories. (According to Kawamura, T. et al., Site maintenance operations on oil-immersed transformers and the state of renewal for low-cost operations in Japan, CIGRE '04, Report A2-209, 2004.)

Cost of undelivered energy should be considered if loss of a transformer results in loss of the utility's ability to supply load. Most systems will have alternative supply possibilities, although the additional costs may be considerable, especially in the case of generator transformers in high-merit plant.

On the other hand, avoiding catastrophic technical failures with all implications—technical and economic—is estimated as having far greater significance than any other economic evaluation. Concern is needed not only for the safety of the transformer but even more so for the safety of personnel and the environment. The importance of earliest possible detection of incipient faults was stressed as being far greater in significance than any other evaluation of remaining life.

Although strategic and economic factors are relatively straightforward, technical considerations are more complex and include assessment of aging, overstressing, and contamination that can have a significant impact on remaining life, perhaps resulting in sudden catastrophic failure.

Assessing remaining life in the context of the technical factors mentioned in Section 8.3 is only possible on a statistical base because the various initiating causes are indeterminate. Therefore, the task of risk assessment is reduced to predicting when a particular type of fault will occur based on statistical data, preferably for transformers of similar type and class, and taking into account design, age, history, and so on. The factors influencing the risk of failure must also be addressed and controlled.

8.8 Life Extension/Repair/Refurbishment

A very important decision—an answer for the question as to what to do with an old transformer being in operation over a number of years—needs to be supported by technical and economic analysis. Economic factors are of great importance when making decision regarding any investment during a transformer operation (Figure 8.9). In any case, the economic estimation should include not only costs of losses and maintenance but also risk of outages with all financial implications (reinstallation, transportation, auxiliary equipment, lost sales revenue, more expensive supply arrangement, etc.), risk of environmental pollution, risk of fault occurrence during transportation, and so on.

A reasonable time horizon of the transformer life extension is about 45–60 years [2]. It may be attained substantially by means of three different ways:

1. Through better protection of the transformer against external stresses. Operations, such as ZnO overvoltage limiters installation with proper protective characteristics and superior reliability or installation of an inductor limiting short circuit currents, located in the transformer neutral point, are performed outside the transformer. Also lowering the transformer operation temperature (hotspot and oil) by means of more effective cooling system may be considered. At the end, avoiding contamination is primarily concerned with the exclusion of water and air, by effective design, monitoring, and maintenance.
2. The aging process is slowed down by properly controlling three main decisive factors: hotspot temperature, water content in oil and in solid insulation, and oxygen

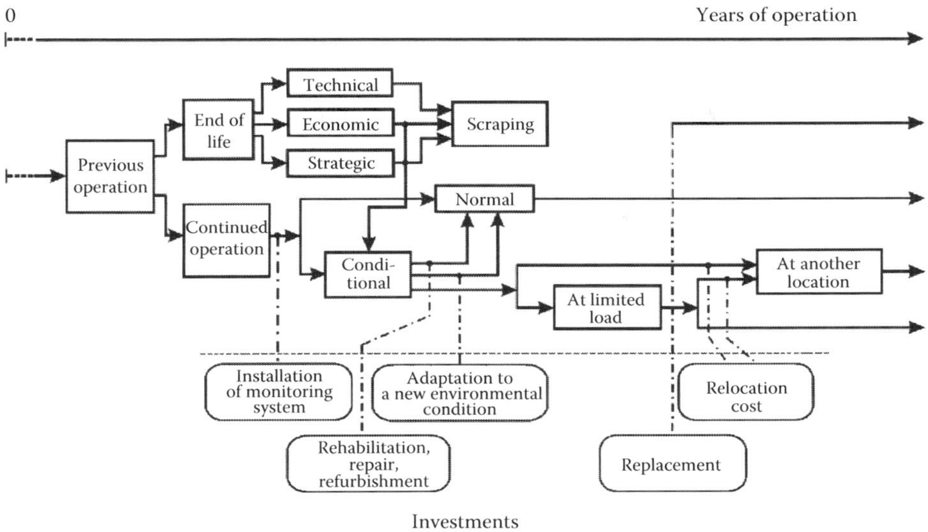

FIGURE 8.9
Exemplary possible options of maintenance of an old transformer and associated investments.

content (oxidation rate) in oil. Overloading is the main cause of accelerated aging, and extreme cyclic loading results in undesirable thermal cycling and mechanical stressing. Controlling loading so that the unit is always operated below its thermal aging threshold is the best contribution to minimized aging and longer life. The significance of adequate maintenance and monitoring of transformer condition cannot be overstated from that point of view.

3. Through proper operation inside the transformer in a course of rehabilitation, refurbishment, and repair. The operations are focused on stopping or even reversing the aging effects.

8.8.1 Rehabilitation/Repair/Refurbishment

Rehabilitation/repair/refurbishment procedures, preceded by proper and accurate evaluation of transformer technical condition and supported by economic analysis may extend the life of transformers that have either exceeded or are approaching the end of their normal lifetime.

The normal objective in repair is to correct known defects and restore the transformer to its condition prior to the failure or anticipated failure.

Rehabilitation comprises a major overhaul to restore the transformer to healthy condition. It involves correction of known defects, gasket replacement, control wiring replacement (if necessary), overhaul or replacement of supervisory devices, overhaul of cooling system, repair of leaks, corrosion treatment, and painting. Oil reconditioning or replacement will be included if necessary. Core and winding inspection may be included if a problem is suspected. Reclamping of windings may also arise. Rehabilitation does not extend the life of a transformer but rather enables it to meet normal life expectancy. On the other hand, some extension of life will almost certainly be achieved.

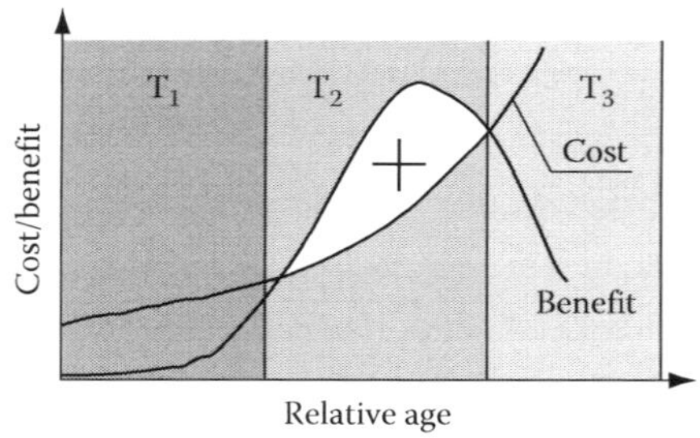

FIGURE 8.10
Rehabilitation benefit analysis during the transformer lifetime. (According to Boss, P. et al., Rehabilitation based on a technical-economical analysis, CIGRE '02, Report 12-106, 2002.)

Refurbishment is taken to involve rehabilitation as described earlier with some degree of upgrading or uprating of the transformer, possibly including replacement of bushings, tap changers, leads, and additional cooling. In some cases, new windings may be envisaged.

From an economic point of view, any decision on rehabilitation, repair, refurbishment, or replacement of a transformer must be made with reference to the results of a profit-and-loss account. The following factors should be taken into consideration:

- Technical deterioration of the transformer resulting in increased risk of fault occurrence, shorter remaining lifetime, and reduced reliability of supply in a case when no action is undertaken
- Cost of rehabilitation/repair/refurbishment and their influence on the transformer specifications, remaining lifetime, and increasing safety of the transformer, personnel, and the environment
- Transformation cost of a new transformer, replacing the old one

Figure 8.10 shows a conceptual method for determining the relative cost/benefit for a rehabilitation program as a function of the relative age. In this chart, T_1 represents the initial stage of the transformer life where modifications or rehabilitation are either not needed or not economically justified. T_2 represents the period in an older transformer where repairs or upgrades are needed and yet there is sufficient remaining life to economically justify the upgrades. In period T_3, the transformer is either too old or too obsolete to justify the cost of an upgrade.

Rehabilitation of the transformer insulation system, including also onsite, is an activity that is most frequently engaged in practice. It consists of filtering, degassing, and drying of oil, washing by hot oil, and then drying (LFH, DC methods) of the insulation system; see, for example, Refs. [1,9,10,15,20,23,34].

8.9 Refurbishment Study Case

8.9.1 Refurbishment of Generator Transformers

In the first half of the 1990s, in a few large Polish power plants, reconstruction of power units has started with planned increase of its power from 200 to 235 MW [14]. In consequence, loading of the 240 MVA generator transformers of $U_n = 123$, 245, and 420 kV rated voltage should be increased to 270 MVA. These transformers were in operation between 1974 and 1983, and the diagnostic tests indicate a good condition of their insulation systems. The following three possibilities were considered within the scope of technical and economic analysis: (1) overloading of 240 MVA transformer with additional coolers, (2) refurbishment, and (3) replacement by new 270 MVA rated units. Recalculated temperature rises (of top oil [$\Delta\theta_{or}$], and of HV winding—the hottest winding—average [$\Delta\theta_{wr}$] and hotspot [$\Delta\theta_{hr}$]) for 270 MVA loading listed in Table 8.2 indicate that replacement of windings in the 245 kV unit is necessary. The HV rises in winding temperature in 123 and 420 kV units exceed standardized values only slightly, and it would be possible to leave them unchanged without any significant risk. It has also been found that modernization of the outer core stack was necessary for all the transformers because of their elevated heating at the new rated condition.

The refurbishment extent was finally decided to include the following:

- Use of the old cores with necessary modernization of the outer core stack (all units).
- Use of the old tanks and covers with replacement of existing copper screens by magnetic shunts (all units).
- New windings (HV and LV) with new insulation system for 123 and 245 kV units and use the old windings for 420 kV transformers. For 123 kV transformer the following two options of windings were considered: (a) use of conventional wires (increase of the copper weight by 16%) and (b) use of transposed cable conductors (5% increase). Increase of the copper weight for 245 kV windings (conventional wires) was about 35%.
- Increase the number of coolers up to 7 (all units).

Relation of losses (no load loss ΔP_0 and load loss ΔP_k) and the approximate transformation cost K (sum of cost of purchasing/refurbishing of a transformer K_z, and the capitalized cost of losses) in considered versions of 270 MVA transformers are listed in Table 8.3.

8.9.2 Refurbishment of Network Transformers

Autotransformers 160 MVA/230 kV operated in Polish 220 kV network were manufactured between 1965 and 1980 (Figure 8.11) [27]. Relatively high failure rate (2–3 units per year) of the autotransformers, resulted from a weak short circuit withstand made the refurbishment of the unit necessary. The decision about the refurbishment has been supported by the economic effectiveness analysis [30].

TABLE 8.2

Temperature Rises in 240 MVA Transformers Overloaded up to 270 MVA OFAF Cooling

U_n (kV)	123		245		420
Number of coolers	5	7	5	7	7
$\Delta\theta_{or}$ (°C)	35	28	39	31	31.9
$\Delta\theta_{wr}$ (HV) (°C)	77	67	86	76	66
$\Delta\theta_{hr}$ (HV) (°C)	100	87	104	92	87

TABLE 8.3

No-Load and Load Losses and the Transformation Cost (Relative Values) and Impedance Voltages for Various Versions of 270 MVA Transformers

				Cost	
	ΔP_0	ΔP_k	u_z (%)	K_z	K
U_n (kV)			127		
Existing[a]	1.00	1.00	12.76	—	—
Refurbished[b]	1.00	0.72	12.8	0.53	0.89
New	0.91	0.70	13.0	1.00	1.00
U_n (kV)			245		
Existing[a]	1.00	1.00	15.9	—	—
Refurbished	1.00	0.77	16.5	0.62	0.93
New	0.79	0.74	15.9	1.0	1.00
U_n (kV)			420		
Existing[a]	1.00	1.00	17.0	—	—
Refurbished	0.93	0.87	17.0	0.50	0.94
New	0.53	0.82	17.0	1.0	1.00

[a] 240 MVA overloaded up to 270 MVA.
[b] For use of transposed cable conductors.

FIGURE 8.11
Active part of RTdxP 125000/220 autotransformer, version before refurbishment.

Failure analysis disclosed that most often the LV winding failed due to axial and radial forces (Figure 8.12). The failures of HV and MV windings were less frequent and might be estimated as a succession of the LV failure.

It was decided to improve a short circuit withstand by means of changing the design of windings as well as to improve operation parameters such as the loss level and cooling conditions.

FIGURE 8.12
An example of LV winding failure of RTdxP 125000/220 autotransformer, version before refurbishment.

Four options were considered:

- *Case A*. Only common (MV) and tertiary windings (LV), together with a part of an insulation system associated with them, were to be changed. LV vector connection (delta) as in the origin.
- *Case B*. As Case A, LV star connected.
- *Case C*. As Case A and additional HV rough regulating winding was to be changed.
- *Case D*. All windings and the whole insulation system were changed. It was intended also to eliminate multilead bushings on 100 kV side. New on-load tap changers were installed. A core was modernized; hence, on-load loss was reduced.

At the end it was decided to choose Case D (Figure 8.13).

Comparison of some main specification items—before and after the refurbishment—is given in Table 8.4.

8.10 Final Remarks

Demands for high reliability of large power transformers—new and those being in operation for 20–30 years—result in intensive development of diagnostic techniques. The progress development concentrates on more precise measurement techniques and application of computer-aided monitoring systems to allow on-hand estimation of the measurement results and condition diagnosis. The life management concept of transformers in operation has also changed. It should cover the all transformer life periods with specification formulation at the beginning. Closer cooperation between transformer users and manufacturers is also observed.

A key element in the transformer life management is the condition diagnostic. In that area, the diagnostic performed within the TBM is recently replaced by those performed

FIGURE 8.13
Active part of RTdxP 125000/220 autotransformer, version after refurbishment.

TABLE 8.4
Autotransformer RTdxP 125000/220, Comparison of Some Main Specification Parameters—Before and After the Refurbishment

Parameter	Before Refurbishment	After Refurbishment
Rated power (MVA)	160/160/50	160/160/50
Voltage (kV)	230 ± 12 × 1.01%/120/15.75	230 ± 10 × 1.0%/120/10.5
No load loss (kW)	84.96	64.1
Load loss (Tap "0," pair of main windings) (kW)	362.45	362.23
Leakage reactance (Tap "0," pair of main windings) (%)	9.94	10.08
Vector group	Yy0d11	Yy0d11

within the condition-based maintenance that is more reliable and more economical. Nevertheless, a failure that takes place when the transformer withstand begins to lower against stresses from the network system (Figure 8.1) is still assessed as a statistical category.

The possible benefits from the usage of the presented concept of life management of power transformers include the following recommendations:

- Maintaining a transformer minimizing de-energizing
- How to operate a defective unit
- Minimizing the remedy actions
- Comprehensive life extension program
- Priority in-field repair and online processing

The technical operations to be performed by the transformer is strongly conditioned by economic aspects that should be thoroughly assessed. For example, essentially the transformation costs should be a decisive factor in choosing between the refurbishment and the replacement. Nevertheless, the financial resources, availability of spare transformers, as well as higher—in comparison with the new transformers—risk of failure, shorter lifetime, and possibly higher running cost may also greatly influence the final decision.

On the other hand, avoiding catastrophic technical failures with all implications—technical and economic—is estimated as having far greater significance than any other economic evaluation. The economical implications are connected with the repair or replacement of the failed unit or, perhaps even more significantly, with the cost of alternative supply arrangements. Major failure may also have safety and environmental implications through explosion, fire, or oil spillage. It seems that online monitoring systems are the new powerful tools for effective life management.

New problems arise when online monitoring systems are installed on transformers. On the one hand, the equipment involved should be of the reliability not worse than the transformer reliability (also in the expected lifetime category), and on the other, the transformer design should make it possible to install the equipment.

At the end, it should be noted that existing online monitoring installations are useful tools only for comparatively slow (days, minutes, seconds) changes of the transformer condition. There is no guarantee that a failure due to quick random external events will be noticed in advance [7].

References

1. Aschwanden T, Schenk A, Kreuzer, Fuhr J, Hässig M (2004) On-site repair, refurbishment and high voltage tests of large power transformers in the transmission grid, CIGRE '04, Report A2-203.
2. Austin P (1997) Widing inter-strand insulation failure identification and location of fault, Sydney Transformer Colloquium, October 5–10, Sydney, Australia.
3. Bengtsson T, Kols H, Foata M, Léonard F (1998) Monitoring tap changer operations, CIGRE '98, Report 12-209.
4. Boss P, Lorin P, Viscardi A, Harley JW, Isecke J (2000) Economical aspects and practical experiences of power transformer on-line monitoring, CIGRE '00, Report 12-202.
5. Boss P, Horst T, Lorin P, Pfammatter K, Fazlagic A, Perkins M (2002) Rehabilitation based on a technical-economical analysis, CIGRE '02, Report 12-106.
6. Breen G (1992) Essential requirements to maintain transformers in service, CIGRE '92, Report 12-103.
7. Breckenbridge T, Harrison TH, Lapworth JA, MacKenzie E, White S (2002) The impact of economic and reliability considerations on decisions regarding the life management of power transformers, CIGRE '02, Report 12-115.
8. Burgos JC, Pagán E, García B, Anguas JI, Raqmos A, Montávez D, Perez E (2002) Experiences in managing transformers through maintenance operations and monitoring systems, CIGRE '02, Report A2-206.
9. Domżalski T, Kaźmierski M, Kozłowski M, Olech W (1994) Repair on-site of EHV transformers in the polish grid, CIGRE '94, Report 12-202.

10. Domżalski T (2004) On-site management of transformers, ARWtr Lecture, October 28–30, Vigo, Spain.
11. Kachler AJ, Höhlein I (2004) Functional and component related diagnostics for power transformers, a basis for successful "transformer life management," ARWtr Lecture, October 28–30, Vigo, Spain.
12. Kawamura T, Fushimi Y, Shimano T, Amano N, Ebisawa Y, Hosokawa N (2002) Improvement in maintenance and inspection and pursuit of economical effectiveness of transformers in Japan, CIGRE '02, Report 12-107.
13. Kawamura T, Ichikawa M, Hosokawa N, Amano N, Sampei H (2004) Site maintenance operations on oil-immersed transformers and the state of renewal for low-cost operations in Japan, CIGRE '04, Report A2-209.
14. Kaźmierski M, Sobocki R, Domżalski T, Olech W (1997) Polish experience with life management of power transformers, Sydney Transformer Colloquium, October 5–10, Sydney, Australia.
15. Koestinger P, Aronsen E, Boss P, Rindlinsbacher G (2004) Practical experience with the drying of power transformers in the field, applying the LFH technology, CIGRE '04, Report A2-205.
16. Lapworth JA, McGrail AJ, Heywood R (1999) Moisture distribution in power transformers and the successful management of moisture related problems, *Proceedings of the Sixty-Sixth Annual International Conference of Doble Clients*, Sec 5–11.
17. Leemans P, Randoux M, Even A, Dhuyvetter G, Dekinderen E (1998) Control, diagnostic and monitoring of power transformers, CIGRE '98, Report 12-213.
18. Liebfried T, Knorr W, Kosmata A, Sundermann U, Viereck K, Dohnal D, Breitenbauch B (1998) On-line monitoring of power transformers—Trends, new developments and first experiences, CIGRE '98, Report 12-211.
19. Lingdren SR, Moore HR (1997) Diagnostic and monitoring techniques for life extension of transformers, Sydney Transformer Colloquium, October 5–10, Sydney, Australia.
20. Mendes JC, Marcondes RA, Westberg J (2002) On site repair of HV power transformers, CIGRE '02, Report 12-114.
21. Oommen TV (2000) Bubble evolution from transformer overload, IEEE Insulation Life Subcommittee, October 17, Niagara Falls, Canada.
22. Patelli JP, Tanguy A, Taisne JP, Devaux F, Ryder S, Chemin E (2002) French experience with decision making for damaged transformers, CIGRE '02, Report 12-111.
23. Pinkiewicz I, Kaźmierski M, Olech W, Malinowski J, Sobocki R (2004) On-site processing of insulation system of large power transformers and hot-spot computer determination, CIGRE '04, Report A2-208.
24. Poittevin J, Tenbohlen S, Uhde D, Sundermann U, Borsi H, Werle P, Matthes H (2000) Enhanced diagnosis of power transformers using on and off-line methods: results, examples and future trends, CIGRE '00, Report 12-204.
25. Savchenko E, Sokolov V (1997) Effectiveness of life management procedures on large power transformers, Sydney Transformer Colloquium, October 5–10, Sydney, Australia.
26. Shenoy V (1997) Life management of power transformers, Sydney Transformer Colloquium, October 5–10, Sydney, Australia.
27. Sieradzki S (2003) Refurbishment and manufacturing possibilities of network autotransformers 160 MVA/220/110 kV in Energoserwis Co. Lubliniec (in polish), *XIII Power Engineering Conference on Power Engineering, Modernization and Development*, September 10–12, Kliczkow, Poland, pp. 61–68.
28. Skog JE (1994) RCM—Reliability centered maintenance, Minutes of the *Sixty-First International Conference of Doble Clients*, pp. 2–3. 1-2-3.4.
29. Smekalov VV, Dolin AP, Pershina NF (2002) Condition assessment and life time extension of power transformers, CIGRE, Report 12-102.
30. Sobocki R, Kaźmierski M, Olech W (2002) Technical and economic assessment of power transformers, the polish practice, CIGRE '02, Report 12-104.

31. Sokolov V, Vanin B, Griffin P (Jun 1998) Tutorial on deterioration and rehabilitation of transformer insulation, CIGRE WG 12.18 Colloquium and International Seminar, Lodz, Belchatow, Poland.
32. Sokolov V (Jun 1998) Experience with transformer insulation maintenance, CIGRE WG 12.18 Colloquium and International Seminar, Lodz, Belchatow, Poland.
33. Tenbohlen S, Stirl T, Bastos G, Baldauf J, Mayer P, Stach M, Breitenbauch B, Huber R (2002) Experienced-based evaluation of economic benefits of on-line monitoring systems for power transformers, CIGRE '02, Report 12-110.
34. Tenbohlen S, Stach M, Lainck T, Gunkel GWH, Bräsel E, Altmann J, Daemisch G (2004) New concepts for prevention of ageing by means of on-line degassing and drying and hermetically sealing of power transformers, CIGRE '04, Report A2-204.
35. Guide for customer specifications for transformers 100 MVA and 123 kV and above (2000) CIGRE '00, Brochure No. 156.
36. Guidelines for conducting design reviews for transformers 100 MVA and 123 kV and above (2002) CIGRE '02, Brochure No. 204.
37. Guide for life management techniques for power transformers (2003) CIGRE '03, Brochure No. 227.
38. Guide on economics of transformer management (2004), CIGRE'04, Brochure No. 248.
39. On behalf of Study Committees 15 and 33, Joint Working Group 15/33.08 (1996) Status report progress on high-voltage insulation monitoring systems for in-service power apparatus, CIGRE '96, Report 15/21/33-20.
40. Working Group 12.17 (2000) Effect of particles on transformer dielectric strength, CIGRE '00, Brochure No. 157.

9

Power Transformer Acceptance Tests

Ryszard Malewski

CONTENTS

9.1 Acceptance Tests as the Main Tool of Quality Control

Checking the assembly of transformer under test: ratio measurement, winding resistance, magnetizing current, short-circuit impedance between all windings, winding group connections, OLTC, CTs, bushings, control circuits, accessories [1].

Dielectric tests: induced voltage test with partial discharge (PD) measurement and acoustic location, impulse test with lightning full and chopped impulse, switching impulse, and applied voltage tests.

Induced voltage test: Induced voltage test has been defined to predict a reliable performance of the transformer insulation in service for 25 or 30 years. A theoretical base of such prediction was given by Cramp [2], who extrapolated the volt-time curve of an oil gap typical for transformer insulation.

This extrapolation extends from the short time of lightning-impulse stress to somewhat longer stress time induced by the switching impulse, a substantially longer stress applied during the low-frequency induced voltage test to a very long time of service at the rated voltage. With such extrapolation, the induced voltage level was set at 150% of the rated voltage U_n, and duration of the test was standardized as 1 h (2 h in certain national standards). PDs have to be recorded during this test to confirm an adequate dielectric strength of the examined insulation. Besides, a short-time (~30 s) enhancement of the test voltage level to ~175% U_n is required to simulate switching over voltages in the transmission system and to make sure that PDs ignited by such transient are extinguished and do not sustain (Figure 9.1).

9.1.1 Shielded HV Laboratory for Measurement of Partial Discharges

Ideally, the examined transformer insulation should be free from PDs. However, the background noise of the test circuit determines the acceptable level of PDs recorded by the measuring instrument.

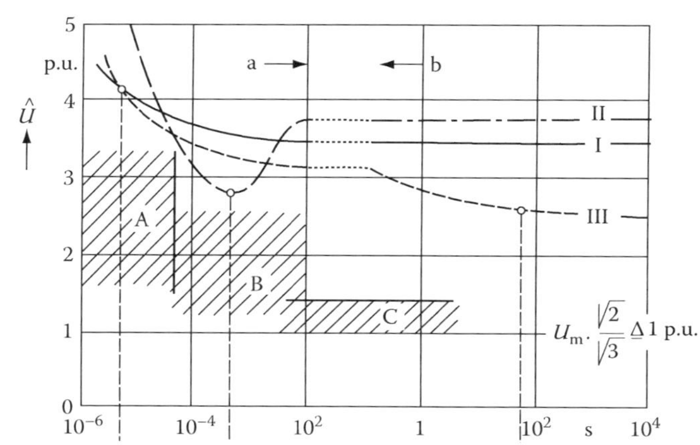

FIGURE 9.1
Volt-time curve of HV insulation: I, compressed gas; II, air; and III, paper-oil. A, lightning impulse test; B, switching impulse test; and C, low-frequency test: a, transient, and b, continuous stress. Peak value of the phase-to-ground voltage is defined as 1 pu.

Broadcasting stations, sparking of electric traction pantographs, ignition of internal-combustion engines, electric welding, etc., radiate high-frequency electromagnetic disturbances. These disturbances induce the background noise in the (PD) measuring circuit.

An efficient electromagnetic shield of the HV test laboratory can reduce this background noise. A well-designed HV test hall shall attenuate by ~75 dB the electric component of the electromagnetic field radiated by AM broadcasting stations in the frequency range of PD measurement from ~20 to ~800 kHz (Figures 9.2 and 9.3) [3].

In an industrial environment, stray currents circulate in the upper soil layers, between different grounding points of supply transformers, large motors, and other loads installed in different factory buildings. A large metal enclosure of the shielded HV test hall offers a low-impedance path for these stray currents. An on-and-off switching of such current induces a transient disturbance in the PD measuring circuit.

FIGURE 9.2
HV test hall equipped with an electromagnetic shield.

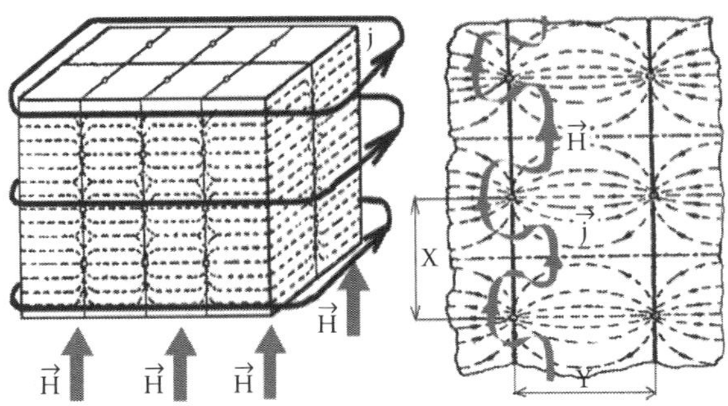

FIGURE 9.3
An external electromagnetic field induces a current that flows in the shield walls and prevents the applied field from penetration into the shielded area. The right sketch shows an exaggerated effect of the disturbance field H penetration between the spot-welded adjacent metal panels. Big doors and windows interrupt the induced current path and allow the external disturbance field to penetrate inside the shielded room. (From Malewski, R. et al., Measurement of partial discharges in an industrial HV laboratory. *CIGRE*, Paris, France, 2002, paper 33–303.)

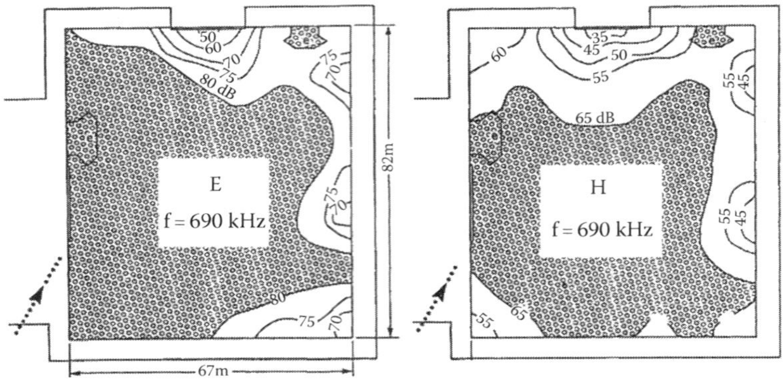

FIGURE 9.4
Attenuation of the electric (E) and magnetic (H) field component inside a large shielded HV test hall. A leakage of the disturbance field in vicinity of the large entrance door, as well as near the windows, can be identified on this graph. (From Rizk, F. et al., *IEEE Trans.*, PAS-94(6), 2077, 1975.)

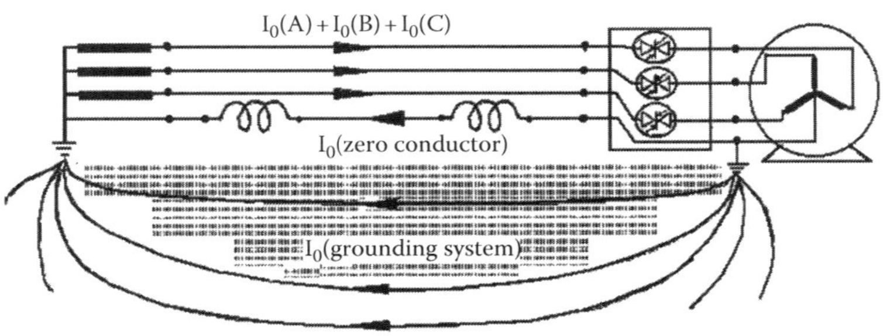

FIGURE 9.5
Thyristor-controlled loads generate higher harmonics. Upper layers of soil offer a lower-impedance path to the higher-harmonics return current than the "zero" conductor. Its inductive impedance increases with the signal frequency, whereas the soil and buried metal objects provide a lower-impedance path to harmonics and transient disturbance currents. (From Malewski, R. et al., Measurement of partial discharges in an industrial HV laboratory. *CIGRE*, Paris, France, 2002, paper 33–303.)

To reduce the background noise to ~10 pC, the grounding system shall have a low impedance (less than 1 Ω). Moreover, the electromagnetic shield has to be isolated from structural elements of the building and from the soil upper layers [4] (Figures 9.4 through 9.6).

9.1.2 Partial Discharge Measuring Instruments and Techniques

PDs shall be recorded using a broadband instrument with multichannel or multiplexed input that allows monitoring the discharge activity on the three-phase bushings of the upper- and lower-voltage windings [4,5].

Some standards and specifications allow for a narrowband measurement of PDs since such measurements are immune to electromagnetic disturbances induced in the test circuit by external sources. Although easier to perform, the narrowband measurement does not detect PDs that are located deep inside of the winding. Discharges from certain winding sections are heavily attenuated and not revealed by the narrowband instrument. Some HV test laboratories cannot perform broadband measurements of PDs since they

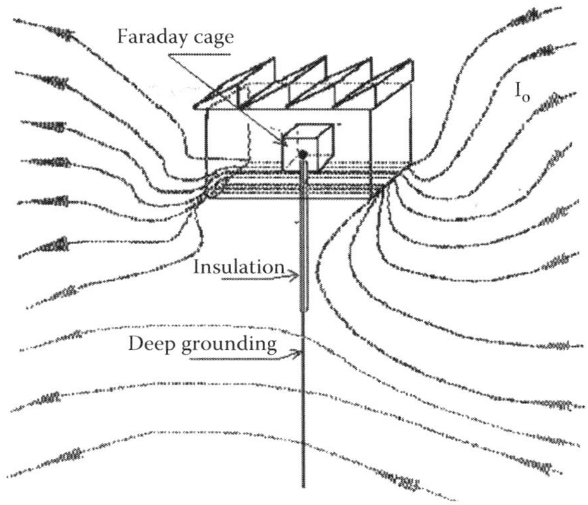

FIGURE 9.6
The building structural steel and metallic structures embedded in soil provide a low-impedance path for ground-circulating currents. The electromagnetic shield has to be grounded by a single rod insulated from the soil upper layers to prevent ingress of conducted disturbances. (From Malewski, R. et al., Measurement of partial discharges in an industrial HV laboratory. *CIGRE*, Paris, France, 2002, paper 33–303.)

are not equipped with an electromagnetic shield and out of necessity measure PDs, or radio-influence voltage (RIV), using a narrowband instrument [6]. However, many transformer owners do not accept results of such measurements.

A low and steady level of PDs recorded during the induced voltage test is considered as its successful outcome. Significant variation of PD activity and an increasing level during the second half of the test duration may require a repeated drying under vacuum or disqualify the transformer.

9.1.2.1 Advanced Processing of Partial Discharge Records

PDs can develop in small gas cavities between cellulose insulation that has not been saturated with oil, in water vapor microbubbles, and on sharp edges or protrusions of the winding or lead conductors. Metal particles, conducting objects on an undefined (floating) potential, even sharp edges of grounded objects, such as magnetic shields welded to the inner tank wall, may be a source of PDs. An advanced processing of the recorded PDs, as well as their location by microphones attached to the tank wall, is used to identify the nature and origin of such discharges.

Modern digital analyzers record each discharge with three parameters: charge, its polarity, and phase of the test voltage (point on the wave). A three-dimension plot of the number of discharges recorded during a fixed time interval (for instance 10 s) is plotted as a function of the charge and phase of the test voltage (Figures 9.7 and 9.8).

Such graph reveals the nature of some typical discharges: a protrusion on the conductor at HV will show up a high number of discharges around the maximum of the negative half-cycle of test voltage. Discharges in a cavity will be identified by their occurrence at the test-voltage descending slope. Excessive moisture results in a high number of unstable discharges distributed at random over the whole test-voltage cycle.

To assist test engineers in their interpretation of the obtained records, such typical distributions have been stored in the analyzer memory for comparison to the current test

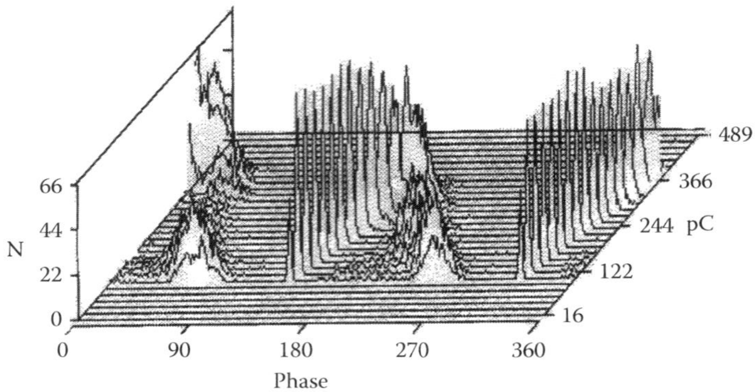

FIGURE 9.7
Number of PDs recorded on 400 kV, 125 MVA single-phase, shell-type transformer during 10 s, plotted as a function of their charge and the test-voltage phase. Characteristic sickle-form pattern of PDs indicates a cavity-type discharge located inside the insulation. This PD pattern was observed at the beginning of the test.

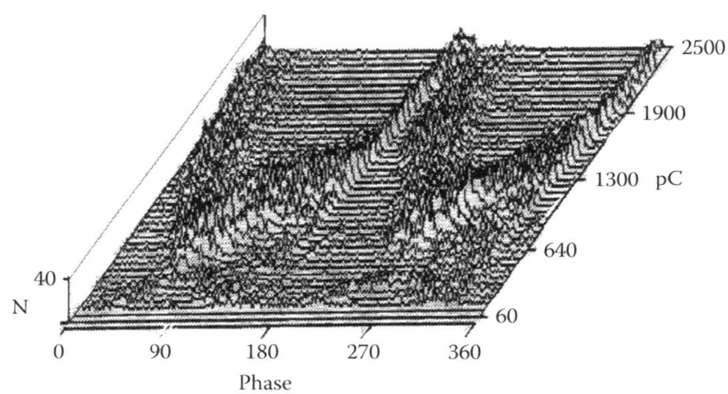

FIGURE 9.8
PDs recorded on the same transformer at the end of 1 h induced voltage test.

results. Moreover, the operator can develop his own reference distribution of PDs caused by an identified flaw of the transformer insulation. Results of such comparison are displayed as the probability of the known mechanism, such as cavity, protrusion, floating object, etc. (Figure 9.9).

9.1.2.2 Acoustic Location of Partial Discharges

To locate the PD source, a set of microphones is attached to the transformer tank, and their acoustic signals are registered on a multichannel recorder. Initially, the time of arrival of each signal is different since the acoustic signal has to pass a different path through the solid insulation and oil to each microphone attached to the tank. The microphones are then moved in such a way that the acoustic signal arrives simultaneously to each microphone. In such a case, the microphones are distributed on a circle that has its center on the axis pointing to the PD source (Figure 9.10).

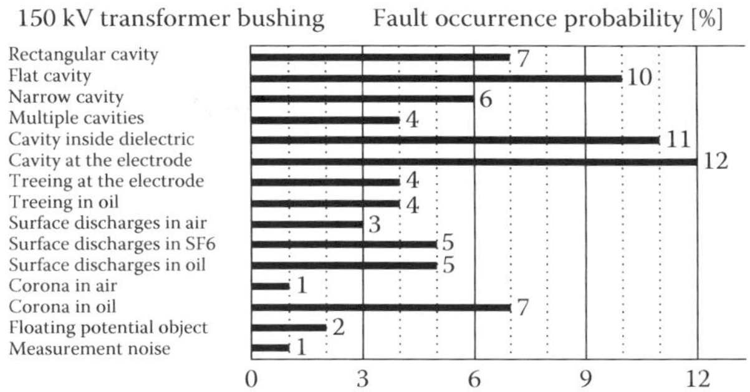

FIGURE 9.9
Probability of occurrence of the previously identified insulation faults to the PD pattern recorded on 150 kV bushing. (From Claudi, A. et al., Digital techniques for quality control and monitoring of HV apparatus. *Stockholm Power Tech Conference*, Stockholm, Sweden, 1995, Paper STP HV 01–06–0579.)

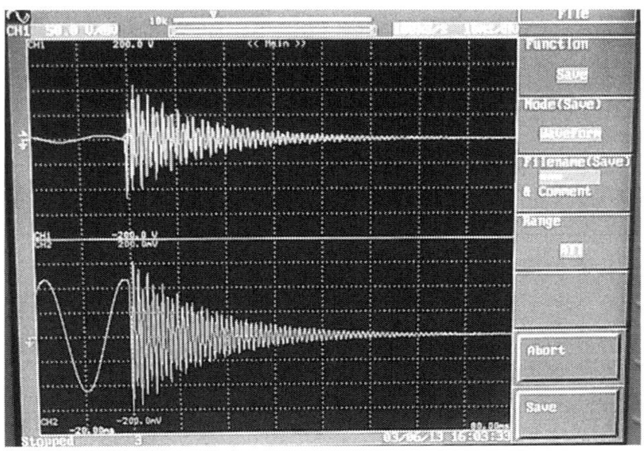

FIGURE 9.10
Acoustic signals from microphones located at the circle with an axis pointing to the PD site in the winding.

This source can be located by triangulation of such axes obtained with microphones attached to the tank walls and cover. The triangulation can also be performed by specialized software, without the need to reposition the microphones. This technique is well established in the test laboratories where the test voltage can be gradually increased to the onset of PDs in one site. However, the acoustic location of PDs in a transformer in substation is much more difficult, and often impossible, since several PD sites are active at the working voltage, and their acoustic signals are superposed.

9.1.3 Impulse Test

Historically the impulse test was considered as a check of the transformer insulation integrity under the stress resulting from the lightning stroke to the nearby section of the line. The transmission voltage was 110 kV at these times, and the main threat to the transformer insulation was the powerful lightning. Initially, the test outcome was simple: breakdown or withstand, and the number of impulses applied in the laboratory was limited, as each subsequent impulse increased the breakdown probability.

At present, 800 kV is the highest transmission-voltage class, and switching, rather than lightning transients may cause the critical stress. Our knowledge of the lightning impulse has also increased since the time of Prof. Berger who recorded lightning parameters by an impulse oscilloscope installed at San Salvadore mountain. Modern, digital recorders have recorded multiple lightings, and the high rate of rise on the subsequent stroke was revealed. Metal oxide overvoltage limiters have replaced traditional surge arresters with carborundum slabs and multiple-electrode quenching gap. These changes resulted in a different waveform of transients applied to the winding insulation and to the air insulation of three-phase bushings.

Nevertheless, the standards still specify the same form and the number of impulses to be applied to the examined transformer. As the rationale, it was said that at the high-voltage (HV) transmission (i.e., 220 up to 400 kV), the lightning waveform is not very much different from that at the lower-voltage levels. Besides, power companies operating the extra-high-voltage (EHV) transmission system (i.e., 735, 765, and until recently 1150 kV) wrote their own specifications and procedures for transformer testing. For instance, the American Electric Power Company has specified an impulse with a short front and long tail to reproduce the stress resulting from 765 kV transformer switch-on by an air-blast breaker.

A common point of all these standards and specifications is the concept of linear behavior of the examined insulation with voltage. At first, the lightning impulse is applied at the test voltage reduced usually to 50% or 66% of the basic insulation level (BIL), and the winding response is recorded. The neutral terminal current is usually taken as the response, but sometimes is the transient voltage induced in the lower-voltage winding. Then the same impulse form is generated at the full (100% of BIL) test voltage and applied to the transformer. Comparison of records obtained at the full and reduced test voltage may reveal a different form of the 100% BIL applied impulse and of the winding response. Such difference is an indicative of nonlinear behavior of the examined insulation with the increased test voltage. In practical terms, such nonlinearity means a local breakdown of coils or turns insulation or a powerful PD. According to the standards, there is no requirement of PD free behavior of the insulation subjected to the impulse test. However, even a local breakdown of the turn-to-turn insulation disqualifies the transformer. Effectively, an acceptance of say million-dollar transformer depends on an interpretation of a small glitch on the winding response record. Under the circumstances, the inspector representing the transformer maker may claim that the difference is due to a PD, but the buyer's inspector will refuse to accept such test since the record reveals an insulation breakdown.

9.1.3.1 Digital Recording and Processing of Impulse Test Results

With the digital impulse recorders, the impulse waveform and winding response can be stored, the records taken at the full and reduced voltage can be superimposed, subtracted, and the difference magnified to make it legible.

Moreover, the frequency spectrum of transients recorded in time domain can be calculated using the fast Fourier transform (FFT) [7–17]. Then the winding trans-admittance, or transfer function, can be calculated as a quotient of the winding response spectrum and the spectrum of applied impulse.

The transfer function reveals resonant frequencies that are determined by an inductance and capacitance of respective winding sections. A short circuit of such section changes its inductance and capacitance, and in consequence, the resonant frequency is changed (Figure 9.11).

A PD results in a charge leakage from the affected winding section to the grounded core or tank. On the transformer winding equivalent circuit, this can be represented as

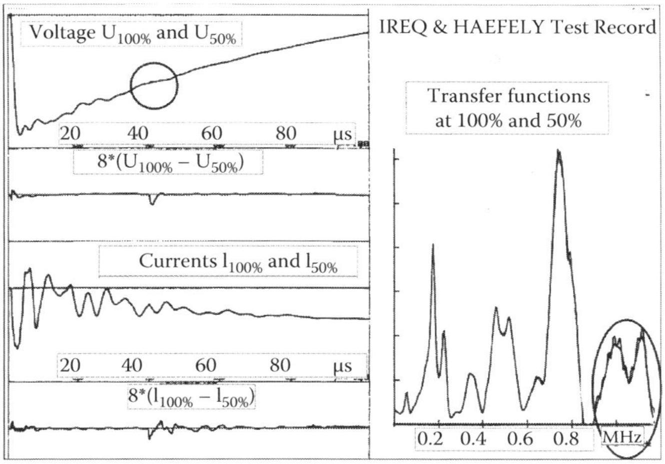

FIGURE 9.11
Impulse test report with superimposed impulse and winding response recorded at the full (100%) and reduced (50%) test-voltage level. Difference between the respective records is amplified eightfold to enhance legibility. Transfer function is derived from the compared records. A local insulation fault is revealed by new resonant frequencies between 0.8 and 1 MHz on the transfer function derived from the full-voltage level test. (From Malewski, R. et al., Checking electromagnetic compatibility of a HV impulse measuring circuit with coherence function. *ERA European Seminar of HV* Measurements, Arnhem, the Netherlands, October 1994.)

an insertion of leakage resistor. Such resistor does not change significantly the resonant frequency but increases the resonance damping. This appears as lowered resonance peak on the winding transfer function.

Using this criterion, the acceptable PD can be distinguished from a local breakdown of the insulation that disqualifies the transformer.

A complementary algorithm, referred to as coherence, is often used to assess the reliability of the transfer function derived from the digitized records of the test impulse and winding response (Figure 9.12).

These records may be affected by electromagnetic disturbances, as well as by digital conversion error. These errors are transferred to the frequency spectra by FFT and show up on the winding transfer function at a certain frequency range. This range can be determined by calculation and plotting the coherence on the transfer function graph. Normally, the coherence amounts to one but drops to a lower value where the transfer function is corrupted by disturbances or digitalization error.

The calculation of transfer function and its coherence is implemented in modern digital impulse recorders.

9.1.3.2 Disturbances of Impulse Voltage and Current Records

The impulse voltage has to be reduced from a megavolt range to a few volts to be measured by digital impulse recorder. Such signal shall be free from electromagnetic interference, despite a broadband measuring system that is by nature, susceptible to induced and conducted disturbances.

Equipotential ground plane is required to provide a stable reference potential to the voltage divider, current-measuring shunt, impulse recorder, impulse generator, chopping gap, and the transformer under test. Such reference ground plane is usually provided by an

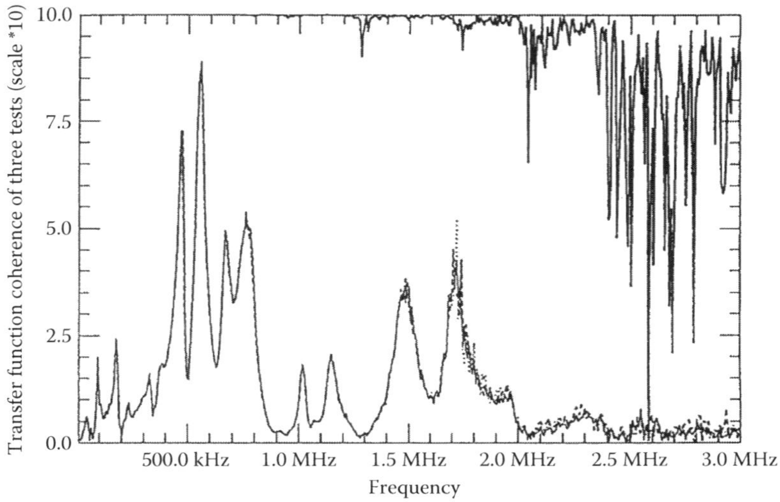

FIGURE 9.12
Transformer winding transfer function derived from three tests and superimposed. The coherence graph (upper curve, scale from 0 to 1) reveals the effect of digital conversion noise above 1.5 MHz. (From Malewski, R. et al., Checking electromagnetic compatibility of a HV impulse measuring circuit with coherence function. *ERA European Seminar of HV* Measurements, Arnhem, the Netherlands, October 1994.)

expanded copper mesh embedded in the concrete floor, and equipped with access points distributed over the impulse test area (Figure 9.13).

Transient potential difference is induced between the grounding points of the voltage divider VD, current-measuring shunt SH, transformer under test TR, and digital impulse recorder DR. Such transient potential difference shows up on the impulse voltage and current records as an irregular high-frequency oscillation superimposed at the beginning of the recorded impulses.

The current induced by such potential difference is split between the ground plane and shield of coaxial cables that bring the recorded signal from voltage divider and shunt to the impulse-recording instrument. A voltage drop on the cable shield resistance is added up to the recorded signal and masks the measured impulse waveform. To reduce this disturbance, the ground floor impedance has to be as low as possible, and the coaxial measuring cables shall be installed in steel conduits under the ground plane. The steel conduit acts as a magnetic core that increases the inductance of this current path, with respect to the expanded copper mesh (Figure 9.14).

9.1.4 Loss Measurement

The manufacturer specifies the transformer load loss and no-load loss in the bid submitted to the buyer. For the utility that will buy the transformer, its total cost is composed of two components, namely, the price and the capitalized cost of loss over say 30 years of service. Bids received from different manufacturers are compared, and the decision of who will get the contract for delivery of the power transformer may depend on the loss indicated in the manufacturer specification. At this stage, the manufacturer can only evaluate the loss by calculations since the transformer has not yet been ordered and built.

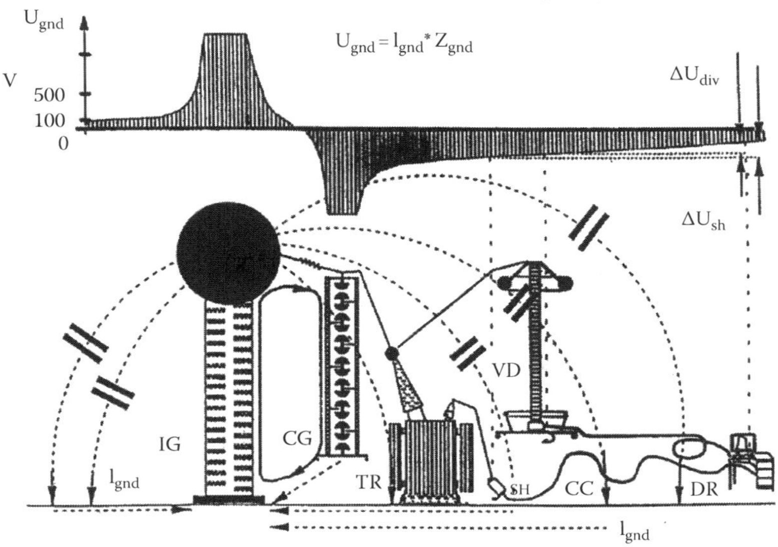

FIGURE 9.13
Potential drop U_{gnd} on the ground-plane impedance induced by the impulse current I_{gnd} circulating between the impulse generator IG and chopping gap CG. Moreover, an impulse current is induced in unshielded cables and instruments by a rapid discharge of the ground capacitance of impulse generator HV electrode. (From Malewski, R. et al., Checking electromagnetic compatibility of a HV impulse measuring circuit with coherence function. *ERA European Seminar of HV* Measurements, Arnhem, the Netherlands, October 1994.)

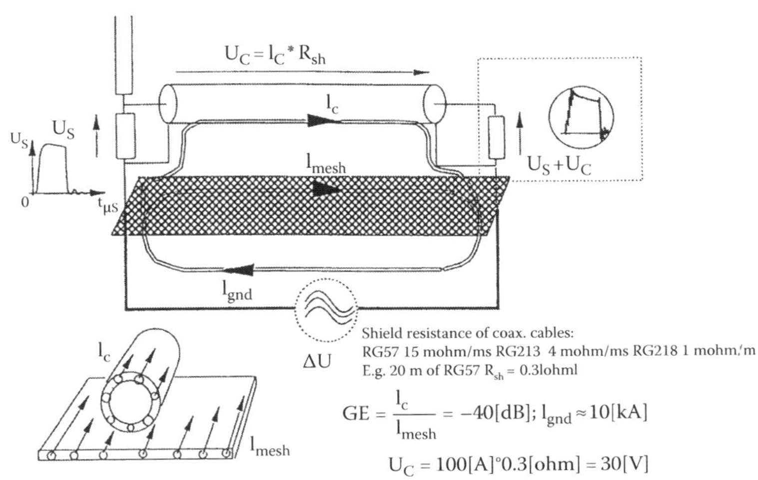

FIGURE 9.14
Voltage divider and impulse recorder grounded to two different points of the grounding mesh. Current induced between these points I_{gnd} is split between the ground plane I_{mesh} and the measuring cable shield I_c. Ratio of these two currents is often considered as a measure of the ground floor efficiency GE. (From Malewski, R. et al., Checking electromagnetic compatibility of a HV impulse measuring circuit with coherence function. *ERA European Seminar of HV* Measurements, Arnhem, the Netherlands, October 1994.)

The actual loss value is determined during the acceptance test, and if the measured loss exceeds the value specified in the bid, then a heavy penalty is imposed on the manufacturer. Typical figures are a few thousand dollars for every kilowatt of an excessive loss. For this reason, an accurate loss measurement has a direct effect on the manufacturer profit.

The measured loss is sometimes contested due to an uncertainty caused by phase angle error of the potential and current transformer, as well as by the wattmeter. There is no easy way to calibrate the loss measuring system [18]. European practice consists in certification of the individual instruments that form the loss measuring system by a reference institution, such as Physikalisch-Technisches Bundesanstalt (PTB). In America and Far East countries, the whole measuring system is calibrated in the transformer manufacturer laboratory by comparison to a reference system brought to the site by the National Research Council Canada.

9.1.4.1 Load-Loss Measurement

With the secondary winding short-circuited, the input impedance of a large power transformer is predominantly reactive. The resistive component represents the sought load loss and may by two orders of magnitude lower than the total impedance. In other words, the power factor (cos Φ) of short-circuited transformer may be as small as 1% (Figure 9.15).

The measured loss uncertainty depends on the examined transformer power factor and on the angular uncertainty of the loss measuring system. To measure this loss with say 1% uncertainty, the measuring system that sees the total input voltage and current must have uncertainty lower than 100 ppm (parts per million) (Figure 9.16).

In the case of large shunt reactors, used to compensate impedance of long transmission lines, the loss measurement is even more difficult. Such reactors are characterized by an extremely low power factor, for instance, 0.2% or less for shell-type reactors.

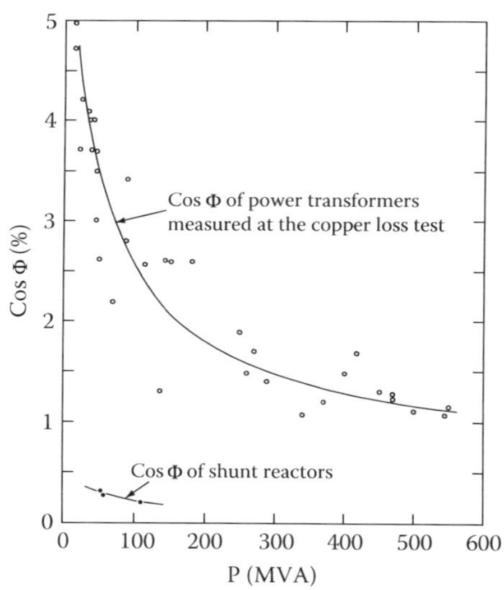

FIGURE 9.15
Power factor (cos Φ) of large transformers with short-circuited secondary winding and of HV shunt reactors plotted against their rated capacity. (From Cancino, A. et al., Testing and loss mesurement of HV shell-type shunt-reactors at very low power factor. *CIGRE*, Paris, France, 2004, paper No. D1- 303.)

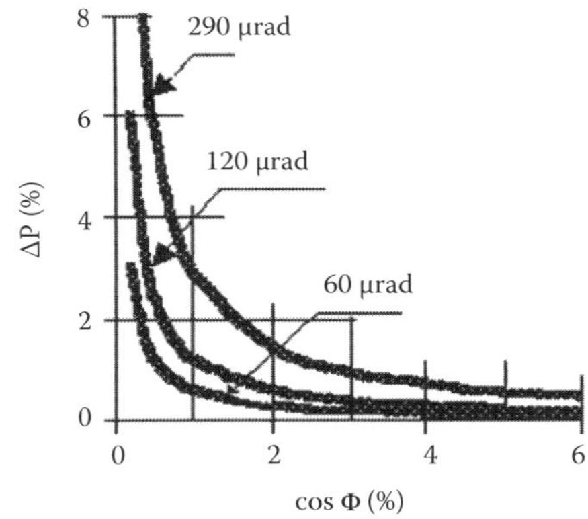

FIGURE 9.16
Loss power uncertainty (ΔP) as a function of the power factor (cos Φ) and the measuring system angular uncertainty (ΔΦ) expressed in microradians.

9.1.4.2 No-Load Loss Measurement

Utilities that transmit power from a remote hydroelectric plant to load center by long lines may operate the transformers at the voltage exceeding the rated level. This results in a nonlinear behavior of the transformer core, highly distorted magnetization current and increased loss at the fundamental and higher harmonics. To ensure that the transformer can operate at, for example, 107% U_n, a low no-load loss is required by the technical specification. To measure no-load loss at such voltage level, the HV laboratory supply system has to provide an undistorted voltage waveform to the transformer that may be driven to saturation during a part of the power frequency cycle. The regulated voltage supply available in HV laboratory cannot maintain the sine test-voltage waveform, since the transformer draws a heavy current when the core is saturated. The no-load loss measured at the distorted voltage does not correspond to the respective loss dissipated by the transformer in the HV power transmission system that maintains the sine voltage waveform.

To correct the measured no-load loss for the voltage distortion, a correction factor has been derived from the ratio of average (so-called flux voltage) and the root mean square voltage (rms.) measurement. Effectively, the mean-value and rms value voltmeters have been used to provide the correction factor. This traditional technique has been standardized and works reasonably well if the voltage waveform distortion is small. However, in the case of large (say 400 MVA single-phase) transformers, the test voltage waveform has to be maintained by a large capacitor bank or a series of harmonic filters installed in parallel with the examined transformer [19]. A digital record of the test voltage and magnetization current should be taken, and the power drawn at the fundamental and at the higher harmonics should be calculated from the frequency spectrum of the test voltage and the measured current.

9.1.5 Transformer Loss Measuring System

It is difficult to meet such requirement with conventional potential and current transformers, as well as with an electrodynamic wattmeter. Contemporary loss measuring systems use compressed gas, standard capacitors to measure voltage, and "zero-flux" current transformers.

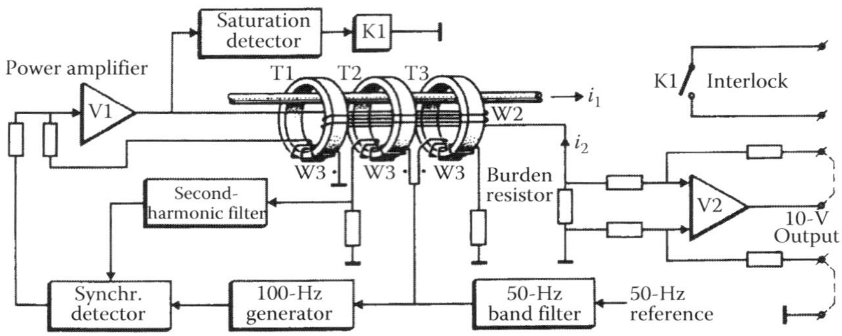

FIGURE 9.17
Zero-flux current transformer with the angular and magnitude uncertainty reduced by an operational amplifier and electronic control circuit.

The "zero-flux" current transformer contains an active circuit that feeds an auxiliary winding with a current controlled in such a way to cancel the primary magnetic flux in the core. This feature reduces the angular uncertainty and also provides an autoranging to supply the current signal at the proper level to the wattmeter (Figure 9.17).

Time division multiplier wattmeters are most often used for their ability to take accurate measurements at a very low power factor [20]. However, this high accuracy can only be achieved within a narrow dynamic range of the input: voltage and current range. The autoranging is then necessary to meet this requirement. A built-in calibrator is often provided with the complete (three-phase) loss measuring system to check the low-voltage electronic parts for drift.

The standard capacitor is equipped with an electronic low-voltage arm that includes an autoranging system. In such a way, the wattmeter receives the voltage signal at the level corresponding to the best part of the wattmeter dynamic range. In the low-voltage arm, the capacitor used in the operational amplifier feedback circuit has a low, but not negligible, loss.

This effect is compensated by a "T" filter; however, such compensation is frequency-dependent and has to be checked periodically (Figure 9.18).

9.1.5.1 Loss Measurement of HV Shunt Reactors

A specialized bridge is needed to measure shunt-reactor loss at an extremely low power factor [21]. Essentially, Schering bridge is used with some modifications. The bridge compares

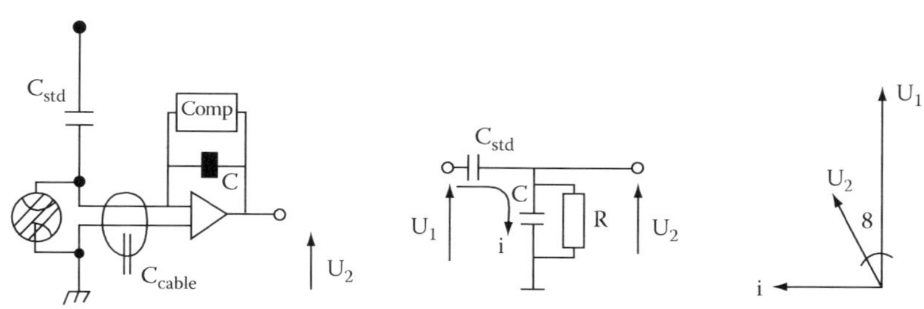

FIGURE 9.18
Compressed gas, HV standard capacitor C_{std} with a low-voltage capacitor C in the feedback loop of an operational amplifier. Angular uncertainty δ is compensated by the tuned "T" network connected in parallel to the capacitor C.

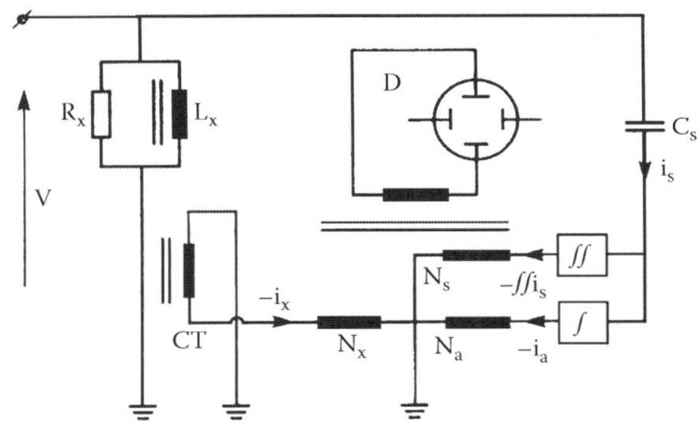

FIGURE 9.19
Schering bridge modified to measure shunt-reactor loss. A double integrator is inserted in the standard-capacitor path of the capacitive current component. This double integrator immunizes the bridge reading against power frequency fluctuations. (From Malewski, R. et al., Mesures des pertes dans les réactances shunt EHT par un pont C-tgδ insensibilisé aux fluctuations de la fréquence du réseau. *IEEE Canadian Conference*, Montréal, Quebec, Canada, 1978, Cat No. 78 CH 1373.0, Reg 7, pp. 489–490 [in French].)

the shunt reactor to the standard capacitor and reads the reactor current phase angle. Such reading changes with the fluctuation of power frequency since the current in the inductive and capacitive arm of the bridge varies in opposite direction when the pulsation $\omega = 2{\cdot}\pi{\cdot}f$ drops or increases with power frequency f.

A special compensation system using a double integrator of the standard-capacitor current is needed to correct frequency fluctuation effect and should be installed if the test voltage is obtained from a power system affected by frequency fluctuations [22]. The current flowing through the standard capacitor is expressed by the formula $i_s = U{\cdot}\omega{\cdot}C_s$ that changes to $\iint i_s = U{\cdot}C_s/\omega$ after double integration. Then the reactive component of the shunt-reactor current $i_x = U/\omega L_x$ has the same frequency dependence as the double-integrated current of the standard capacitor (Figures 9.19 and 9.20).

This technique is suitable for loss measurement of single-phase shunt reactors. However, there are a growing number of three-phase reactors with an on-load tap changer that allows for control of the reactive power in a large dynamic range. One way to measure loss of such reactors consists in subsequent measurements taken on each of the reactor three-phase windings. Alternatively, simultaneous measurement of the total three-phase loss can be taken using a high-accuracy measuring system, such as developed by "Measurement International" in cooperation with the National Research Council Canada.

9.2 Gathering of Reference Patterns on a New Transformer for Subsequent Diagnostic Procedures Onsite

The acceptance test may include records of the following:

1. Content of gas dissolved in transformer oil for dissolved gas analysis (DGA) performed during the heat run test
2. Winding frequency response to be used as reference for frequency response analysis (FRA) of transformers on-site

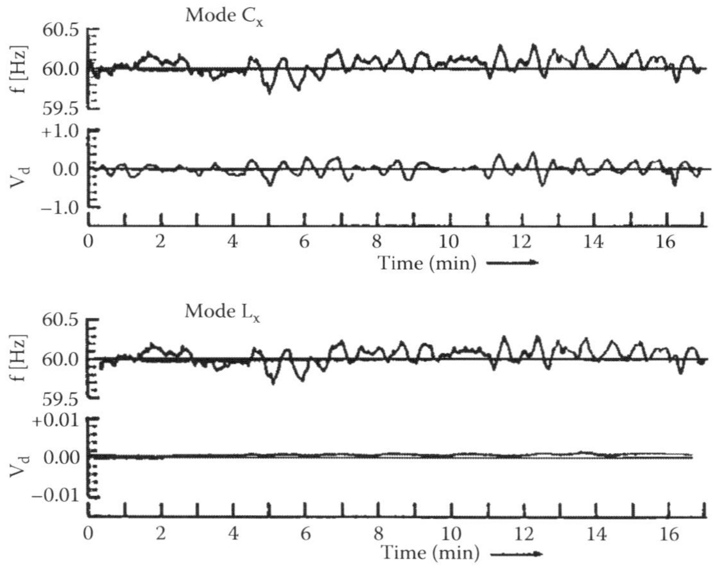

FIGURE 9.20
Imbalance signal V_d of the reactor loss measuring bridge without compensation of power frequency fluctuations (Mode C_x) and with the double integrator inserted in the reference current is circuit (Mode L_x). (From Malewski, R. et al., Mesures des pertes dans les réactances shunt EHT par un pont C- tgδ insensibilisé aux fluctuations de la fréquence du réseau. *IEEE Canadian Conference*, Montréal, Quebec, Canada, 1978, Cat No. 78 CH 1373.0, Reg 7, pp. 489–490 [in French.].)

3. PD phase-resolved patterns obtained at the rated and test voltage levels
4. Assessment of HV insulation condition based on dielectric polarization methods:
   a. Recovery voltage measurement (RVM)
   b. Frequency-domain spectroscopy (FDS) of the insulation loss factor tgδ and capacitance C
   c. Polarization-depolarization current (PDC)

9.2.1 Dissolved Gas Analysis

DGA of an oil sample taken from the transformer during the test provides the basic diagnostic information, which can be compared to blood test performed by physician to assess the health condition of the examined patient. The content of gases dissolved in oil and their composition are interpreted according to standard codes and reveal an insulation dielectric fault or accelerated degradation. Several codes have been proposed by research centers and standardized by the International Electrotechnical Commission (IEC), Institute of Electrical and Electronics Engineers (IEEE), and national standards. Outcome of DGA performed according to these codes may vary, but the gas content measured on a transformer during the acceptance tests can be used as reference for subsequent gas analyses in oil samples taken from transformer in service.

On-line monitoring of gases dissolved in oil provides an early warning signal of oncoming dielectric fault and can trigger alarms, as well as switch off the transformer to prevent an explosion. Technical specifications often require the advanced gas sensors to be installed on the transformer in the factory, and such sensors have to be checked and calibrated during acceptance tests.

9.2.2 Frequency Response Analysis

Short circuit in the network adjacent to the transformer results in the circulation of the current in the windings that can be 20 and even 40 times higher than the rated current. Dynamic force acting on the winding is proportional to the square of this current and may persist for a few cycles of power frequency until the breaker clears the fault. Windings are pressed in the factory to ensure that their mechanical strength withstands the dynamic force caused by short circuit in the network or by inrush current. However, the pressing force tends to weaken since cellulose shrinks and looses its elasticity with years of operation at high temperature. Such winding is prone to deformation or coil displacement by the high short-circuit current, and utilities try to monitor the winding integrity to prevent the transformer failure in service.

Historically, a change of winding geometry was detected by periodic measurement of transformer short-circuit reactance at power frequency and its comparison to the nameplate value. Such method may reveal a major winding displacement, but local deformations of coils or turns can only be detected by measurement of winding frequency response.

Winding trans-admittance, often referred to as transfer function, is measured in the frequency range from ~100 Hz to 1 or 2 MHz. Sometimes the transformation ratio of the high- and low-voltage windings is recorded in this frequency range. The obtained amplitude and phase characteristic constitutes a "finger print" that identifies the winding geometry. A displacement of coils or turns changes some of the winding natural resonance frequencies and can be detected by comparison to the winding frequency response recorded in the factory.

As most of the diagnostic methods, the FRA is based on comparison of the records taken in service to the reference records obtained on the new transformer. Unfortunately, older transformers do not have such reference since the winding response method has gained popularity only a few years ago. Under the circumstances, the frequency response of one winding in three-phase transformer is compared to the response of other phase winding or to the corresponding winding in a twin transformer. Successful detection of winding displacement depends on reproducibility of the frequency response records taken with different instruments and somewhat different connections between the instrument and transformer bushings. International Council on Large Electric Systems (Conseil International des Grands Réseaux Électriques [CIGRE]) has set up a working group (CIGRE A2 WG 26) to standardize characteristics of the measuring instruments and their connections to the transformer bushing, and outcome of this study has largely reduced the irreproducibility of frequency response measurements.

Another investigation was initiated to develop a model of transformer winding that can be used as reference to compare the measured frequency response. This is a rather complex and difficult task, but a certain progress has been already achieved and is expected that such model will be developed for a few typical winding styles.

9.2.3 Partial Discharge Phase-Resolved Patterns

Progress in detection and analysis of PDs in transformer insulation has enabled to discriminate the miniscule pulses produced by PD in transformer winding from the intense, broadband electromagnetic noise caused by corona on HV line conductors and other perturbations injected into the winding of transformers in service. Essentially, development of high-order digital filters and spectral analysis of PD pulses allowed retrieving these relatively weak signals buried in noise. Such schemes based on active, tunable filters are still under development, but introduction of optical signal transmission that separates the measuring system ground potential from the potential of HV bushing tap has opened the door to further improvement of the phase-resolved PD detection systems.

Besides, the ultrahigh-frequency (UHF) sensors built into transformer tank detect PD spectral components that exceed the noise spectrum. Such systems have already been implemented in power transformers in service. At present, there are problems with calibration of the measured signals in the apparent charge units (pC) universally accepted to quantify PD intensity. Problem of PD localization in the windings has not yet been resolved, although some investigations claim that it is possible to process signals from a few sensors to locate PD source.

Such techniques are oriented to detect and locate PD in insulation of transformers in service. It is believed that general acceptance of such monitoring schemes shall be preceded by their application to acceptance testing in HV laboratories in parallel with conventional PD measuring systems.

9.2.4 Assessment of HV Insulation Condition Based on Dielectric Polarization Methods

Dielectric polarization of paper-oil transformer insulation has been employed for long time to detect moisture content and degradation of cellulose. Theory of polar molecules that contribute to the insulation capacitance up to their relaxation frequency is marked by milestone studies by Debay, Dakin, Curie von Schweidler, Jonscher, and other prominent scientists.

Historically, first instruments developed to assess moisture content in transformer insulation were the mega-ohm meters (Meggers) that measured insulation resistance at 15 and at 50 s, as well as bridges that measured insulation capacitance at 2 and at 50 Hz. Ratio of these two respective readings had to stay within an experimentally determined range to qualify transformer as "wet" or "dry." The main drawback of such, somewhat primitive, techniques was the conductivity of oil, which cannot be ignored when interpreting the outcome of such measurements.

The dielectric loss factor tgδ has been introduced in Europe by Schering and in the United States by Dobble in the decade of 1920. Although initially limited to power frequency measurements, tgδ has been the reliable and universally recognized indication of moisture in transformer insulation [23].

9.2.4.1 Recovery Voltage Method

Progress in instrumentation and growing demand for condition assessment of service worn transformers stimulated development of the recovery voltage method. Once charged to a direct voltage of ~1 kV, then rapidly discharged and open-circuited, the examined transformer insulation will generate a recovery voltage that is caused by relaxation of polarized molecules. The recovery voltage will attain a maximum value that depends on the insulation moisture content, on the time of charging, and on temperature. Contemporary RVM instrument plots the maximum value of the recovery voltage as a function of the charging time and introduces a correction for temperature. The plotted characteristic passes through a maximum that can be calibrated in moisture content in oil expressed in percents. A software package assists the user in interpretation of the recorded characteristic, although rather critical remarks were published on the applicability of such method to acceptance test of power transformers [24].

9.2.4.2 Frequency Dielectric Spectroscopy of the Insulation Loss Factor tgδ and Capacitance C

Development of C-tgδ bridge, which can sweep the frequency range from 0.1 mHz to some 100 Hz, has enabled detection of moisture in paper-oil insulation, as well as cellulose degradation [25]. Technical problems caused by measurement of a very small current that flows through the insulation capacitance at the very low frequency had to be addressed and resolved. To perform such measurement, the variable-frequency test voltage has to be amplified, but not beyond a certain level, since the insulation behavior may change with voltage.

Another critical problem consisted in development of insulation samples with strictly controlled moisture content and degree of cellulose decomposition. Such samples have been necessary to calibrate the C-tgδ bridge readings in terms of the sought percentage of water content in oil and cellulose aging. A statistically significant number of the "reference samples" was needed to write the specialized software that assists interpretation of the bridge readings [26].

9.2.4.3 Polarization–Depolarization Current

Dielectric polarization can be measured in time or frequency domain. Although the physical phenomena do not depend on the measuring method, the instrumentation is distinctly different, as well as interpretation of the readings. Direct voltage is applied to the examined transformer insulation, and the polarization current is measured. Depolarization current is measured when the charged insulation is short-circuited. The measured current through the transformer insulation is very small and can be measured only by an electrometer [27].

The PDC time of decay may attain several hours, and the electrometer readings have to be taken at regular intervals to plot the time-domain characteristic. Interpretation of such characteristic consists in fitting of a few exponential curves with the time constant that correspond to the molecule relaxation and a line that reflects oil conductivity. Specialized software assists the analysis, and the interpretation is based on data derived from insulation samples with controlled moisture content and cellulose degradation.

To compare and evaluate reliability and practical applicability of these three diagnostic methods based on dielectric polarization, a research project was initiated and completed by three universities, two power companies, and a transformer repair facility [28]. Sponsored by European Union, the project REDIATOOL has contributed to the calibration of compared measuring instruments and improvement of the reading interpretation software. Indirectly, it has added credibility to the results obtained by the examined methods and instruments.

It is believed that utilities operating large and expensive transformers will perform subsequent evaluations of the paper-oil insulation condition using one of the methods based on dielectric polarization. As most of the diagnostic methods, they depend on comparison of currently obtained results to the reference measurement taken at the factory during transformer acceptance test. The test station shall be then equipped with appropriate instrumentation, and the test engineers have to familiarize themselves with intricacies of dielectric polarization measurements.

9.3 Specialized Instruments Required for Transformer Testing References

9.3.1 Frequency Response Analysis

Major instrument manufacturers have supplied their models for evaluation by the working group CIGRE A2 WG26, and summary of the WG investigations has been published as CIGRE Brochure [29]. The equipment to record the winding frequency response can be divided into time- and frequency-domain instruments.

The former apply a steep-front, low-voltage impulse to the examined winding and measure the applied impulse waveform, as well as the winding response. Digital record of these two transients is converted to frequency domain using FFT algorithm, and the transfer function is derived as quotient of the response and applied impulse spectrum.

The latter uses a variable-frequency voltage source that sweeps the frequency range from some 100 Hz to 1 or 2 MHz. This voltage is applied to the examined winding, and its response is recorded as a function of frequency.

Theoretically, both procedures are equivalent, but, in practice, there are important differences. The winding transfer function has a large dynamic range of 100 to even 120 dB. This is the ratio of the transfer function highest peak to its lowest trough. Such dynamic range can be achieved if the constant magnitude voltage is swept over the whole frequency range of interest, and the response-measuring circuit has a very high signal-to-noise (S/N) ratio.

The spectrum of steep-front impulse applied to the winding decays with frequency, and around 1 MHz, the spectral components amount only to a small fraction of the low-frequency components. The winding response is recorded by an analog-to-digital (A/D) converter of a high dynamic range and high sampling rate. The latter shall be not smaller than 10 ns/sample, and such fast digitizer can have the dynamic range of say 12 bits ($1/2^{12} = 2.4 \times 10^{-4}$) that corresponds to ~72 dB.

This is substantially lower than the required 100 (or even 120) dB and means that the transfer function troughs will be buried in the digitizer quantization noise. In conclusion, the frequency sweep type of instruments can provide a higher dynamic range than the impulse type.

Another issue is the accuracy of the response phase-angle measurement. Some sweep frequency type of instruments cannot detect correctly the phase-angle zero crossing that occurs at the transfer function resonance peaks. In most cases, the phase angle is neglected at the evaluation of FRA. However, a more advanced investigation and, in particular, modeling of transformer need the additional information hidden in the phase-angle characteristic.

The connection of FRA instrument to the examined transformer bushing affects the measured frequency response, and every effort has to be made to standardize the connection and ensure reproducibility of the measurements. One instrument maker has developed an active probe at the end of measuring cable. Such probe has high (1 MΩ) input impedance and an amplifier (Darlington) that converts it to the low (50 Ω) impedance of coaxial cable. Such probe effectively eliminates the influence of cable layout between the FRA instrument and the transformer bushing tap.

9.3.2 Partial Discharge Phase-Resolved Patterns

Detection and localization of PD in transformers in service is much more difficult than in HV laboratory during the acceptance test. In the latter case, the test voltage can be controlled to identify onset and extinction of an individual PD source, and external perturbations are largely reduced by the electromagnetic shield as well as by separation of the supply and control circuits by filters and isolation transformers.

Nevertheless, instruments have been designed to capture the miniscule PD pulses in the presence of intense broadband noise prevailing in HV stations [30]. A high-speed multichannel recorder acquires signals from couplers attached to the transformer bushing tap and stores the data on computer disc. The analog filters used in conventional PD measuring instruments have been replaced by algorithms that offer the selectable bandwidth, which can be changed to respond to varying ambient conditions. An external antenna unit collects disturbances from, for example, corona pulses. These are suppressed if they appear simultaneously in the PD measuring channel and in the external antenna channel. Cross talk between three-phase windings is used to localize PD sources using three-phase amplitude relation diagram algorithm. An optical signal transmission between the couplers and acquisition-processing unit eliminates the common mode disturbances.

UHF PD detection systems have not yet been advertised as commercial product, but research centers and universities have developed their prototypes.

9.3.3 Assessment of HV Insulation Condition Based on Dielectric Polarization Methods

Commercial instruments are available for transformer insulation condition assessment using dielectric polarization methods. The oldest RVM instrument has been offered together with the software package that assists the interpretation of the recorded characteristics [31]. It should be noted that previous software version, criticized by many experts, has been modernized and corrected. Besides, it is possible to use the direct-voltage source of RVM instrument to record the PDC by an external electrometer. A specialized PDC recording instrument is also available [32], as well as the FDS bridge [33].

All three methods, RVM, FDS, and PDC, have been used to assess the transformer insulation moisture content and aging and have their advocates as well as opponents. The method of assessment of transformer insulation condition should be indicated in the technical specification by the utility that orders the transformer, since the subsequent measurements in service shall be carried out using the same instrument and procedure.

Quality of the result interpretation software should be considered when choosing the transformer insulation assessment method. Such software is based on results derived from calibrated samples of paper-oil insulation and also from the examined transformers. A statistically significant number of such samples is required to draw conclusions that will be included in the result interpretation software.

Acknowledgments

This chapter reflects collective experience of CIGRE Working Groups 33.03 and A2-26, as well as IEEE Power System Instrumentation and Measurements Committee, which the author had privilege to serve at different positions. Numerous discussions at the meetings and exchanges with members of these professional associations are gratefully acknowledged here.

References

1. Carlson, A., Fuhr, J., Schemel, G., Wegscheider, F., *Testing of Power Transformers*. ABB Business Area Power Transformers, Zurich, Switzerland, 2003.
2. Cramp, M.G., A statistical basis for transformer oil breakdown. Rensselaer Polytechnic Institute, Troy, NY, 1959, PhD Thesis.
3. Rizk, F., Gervais, Y., Lührmann, H., Performance of electromagnetic shields in HV laboratory. *IEEE Transactions*, PAS-94(6), pp. 2077–2083, 1975.
4. Malewski, R., Mokański. W., Wierzbicki. J., Measurement of partial discharges in an industrial HV laboratory. *CIGRE*, Paris, France, 2002, Paper #33-303.
5. Rochon, F., Malewski, R., Vaillancourt, G., Gervais, Y., Acquisition et traitement des signaux de décharges partielles lors des essais diélectriques sur grands transformateurs triphasés. *IEEE Canadian Conference*, Montreal, Quebec, Canada, 1984 (in French).
6. Vaillancourt, G., Malewski, R., Train, D., Comparison of three techniques of partial discharge measurements in power transformers. *IEEE Transactions*, PAS-104(4), 900–909, 1985.
7. Malewski, R., Poulin, B., Impulse testing of power transformers using the transfer function method. *IEEE Transactions*, PWRD-3(2), 476–489, 1988.
8. Malewski, R., Göckenbach, E., Neue Möglichkeiten der Beurteilung von Stoßspannungsprüfungen and Transformatoren durch Verwendung eines digitalen Messystems. *ETZ-Archiv*, 11(6), pp. 179–186, June 1989 (in German).
9. Malewski, R., Wolf, J., Transfer function of HV windings connected directly to SF_6 breakers. CIGRE Working Group 33.03, Cairns, Queensland, Australia, 2001, Paper #44.
10. Malewski, R., Claudi, A., Josephy, Ch., Jud, S., Checking electromagnetic compatibility of a HV impulse measuring circuit with coherence function. *ERA European Seminar of HV Measurements*, Arnhem, the Netherlands, October 1994.
11. Malewski, R., Schierig, S., Shielding and grounding of HV test laboratories. *HighVolt Kolloquium '03*, Dresden, Germany, May 2003, pp. 22–23, Paper #4.3.
12. Malewski, R., Douville, J., Lavallée, L., Tschudi, D., Dielectric stress in 735 kV generator transformers under operating and test conditions. CIGRE, Paris, France, 1990, Paper 12-203.
13. Malewski, R., Dielectric tests of 800 kV class power transformers. *7th International Symposium on HV Engineering*, Dresden, Germany, IHS, 1991, Paper #25.08.
14. Malewski, R., Claudi, A., Gockenbach, E., Maier, R., Fellmann, K.-H., Five years of monitoring the impulse test of power transformers with digital recorders and the transfer function method. *CIGRE*, Paris, France, 1992, Paper #12-201.
15. Claudi, A., Malewski, R., Gulski, E., Digital techniques for quality control and monitoring of HV apparatus. *Stockholm Power Tech Conference*, Stockholm, Sweden, 1995, Paper STP HV 01-06-0579.
16. Vailles, Ch., Malewski, R., Dai-Do, X., Aubin, J., Measurements of dielectric stress EHV power transformer insulation. *IEEE Transactions*, PWRD-10(4), 1757, 1995.
17. Malewski, R., Feser, K., Claudi, A., Gulski, E., Digital techniques for quality control and in-service monitoring of HV power apparatus. *CIGRE*, Paris, France, 1996, Paper #15/21/33-03.
18. Malewski, R., Arsenau, J., So, E., Moore, W.J.M., A comparison of instrumentation for measuring the losses of large power transformers. *IEEE Transactions*, PAS-101(6), 1570–1573, 1983.
19. Malewski, R., Kosztaluk, R., Train, D., Measurements of iron loss in large power transformers. *CIGRE*, Paris, France, 1988, Paper No. 12-01.

20. Clarke, K.J., Malewski, R., Digital wattmeters for loss measurements on power transformers. *IEEE Transactions*, PWRD-2(1), 94–101.
21. Cancino, A., Ocon, R., Malewski, R., Testing and loss measurement of HV shell-type shunt-reactors at very low power factor. *CIGRE*, Paris, France, 2004, Paper No. D1-303.
22. Malewski, R., Douville, J., Mesures des pertes dans les réactances shunt EHT par un pont C-tgδ insensibilisé aux fluctuations de la fréquence du réseau. *IEEE Canadian Conference*, Montréal, Quebec, Canada, 1978, Cat No. 78 CH 1373.0, Reg 7, pp. 489–490 (in French).
23. Malewski, R., Kobi, R., Kopaczyński, D.J., Lemke, E., Werelius, P., Instruments for HV insulation testing in substations. *CIGRE*, Paris, France, 2000, Paper 12/33-06.
24. Kachler, A., Baehr, R., Zaengl, W., Breitenbauch, B., Sundermann, U., Kritische Anmerkungen zur Feuchtigkeitsbestimmung von Transformatoren mit der "Recovery-Voltage-Method." ETZ, Jg 95, 19, 1238–1245, 1996.
25. Gäfvert, U. et al., Dielectric spectroscopy in time and frequency domain applied to diagnostics of power transformers. *6th International Conference on Properties and Applications of Dielectric Materials*, Jiaotong University, Xian, China, June 21–26, 2000.
26. Blennow, J., Ekanayake, Ch., Walczak, K., Garcia, B., Gubanski, S., Field experiences with measurements of dielectric response in frequency domain for power transformer diagnostics. *IEEE Transactions*, PWRD-21, 681–688, 2006.
27. Houhanessian, V., Zaengl, W., Application of relaxation current measurements to on-site diagnosis of power transformers. *Conference on Electrical Insulation and Dielectric Phenomena*, Minneapolis, MN, October 19–22, pp. 47–51, 1997.
28. Gubanski, S., Blennow, J., Karlsson, L., Feser, K., Tenbohlen, S., Neumann, C., Moscicka-Grzesiak, H., Filipowski, A., Tatarski, L., Reliable diagnostics of HV power transformer insulation for safety assurance of power transmission system, *REDIATOOL—European Research Project. CIGRE*, Paris, France, 2006, Paper D1-207.
29. CIGRE A2 WG26 (2008) Mechanical condition assessment of transformer windings using Frequency Response Analysis (FRA). *CIGRE*, Brochure #342 ELECTRA #237, April 2008, pp. 34–45.
30. Schaper, S., Kalkner, W., Plath, R., Synchronous multi terminal on site PD measurements on power transformers. *14th Symposium on HV Engineering*, Beijing, China, 2005. Also in (2008) MPD 600, Omicron Technology pamphlet, May 2008.
31. Schlag, A., The recovery voltage method for transformer diagnosis. *Tettex Information Bulletin*, 1991. Also in (2000) The 5461 RVM automatic *Recovery Voltage Meter, Haefely Information Bulletin*, 2000.
32. Werelius, P., Measuring C-tgδ in broad frequency range. Discussion of CIGRE, 2000, Paper JS 12/33-06. Also in (2006) IDA 200 Insulation Diagnostic System, *GE-Programma Information Bulletin*, 2006.
33. Alff, J.-J., Houhanessian, V., Zaengl, W., Kachler, A., A novel compact instrument for the measurement and evaluation of relaxation currents conceived for on-site diagnosis of electrical apparatus. *IEEE International Symposium on Electrical Insulation*, Anaheim CA, April 3–5, 2000. Also in (2000) Electrical Insulation Diagnostic System PDC-Analyser-1MOD, *ALFF Engineering Technical Bulletin*, 2000.

10

Functional and Component-Related Diagnostics for Power Transformers, a Basis for Successful "Transformer Life Management"

Adolf J. Kachler and Ivanka Höhlein

CONTENTS

10.1 Introduction

Transformer life management (TLM) has recently gained an increasing interest due to economic pressure. This forces utilities to new ways of asset management and risk assessment to cut the total operating costs [1–3].

Mr. Finance demands cost reduction measures and wants to move from "time-based" to "condition-based" maintenance and risk assessment. This, however, is only possible on the basis of consequent maintenance (fingerprint, regular trend analysis) by means of a whole system of off- and online diagnostic controls [4–6].

This chapter describes most of the so far established procedures for failure/aging detection from the laboratory to on-site analysis. The comprehensive literature [1–45] contains most of the successful failure and aging condition assessment and classification routines.

It discusses the main aging factors, the state of the art of transformer diagnostics (in general and component related). The consequent application of condition assessment and of life extension techniques are the key to considerable cost reduction opportunities.

10.2 Objectives of Condition Assessment and TLM

Transformers are in general very reliable, with a life expectancy of well beyond 30–40 years.

Failures and defects occur when the withstand strength such as dielectric, thermal, or mechanical strength is exceeded by operational stresses.

The withstand strength decreases over its lifetime due to various processes:

- Normal aging
- Under destructive processes (e.g., tracking)
- Under the influence of agents of deterioration (e.g., air, moisture, and contaminations)
- Under abnormal system and transient stresses

The fundamental objective of TLM starting from condition assessment and condition classification up to life extension techniques is to provide the longest possible service life and to minimize the lifetime operating costs [4–6]. Operational stresses may vary greatly, and they are normally divided into the following:

TABLE 10.1
Definitions of Condition Classification

Condition	Definition
Normal	No obvious problems, no remedial action justified
Aged, normal in service	Acceptable, but does not imply defect-free
Defective	No significant impact on short-term reliability, but asset life may be adversely affected in long term unless remedial action is carried out
Faulty	Can remain in service, but short-term reliability likely to be reduced; may or may not be possible to improve condition by remedial action
Failed	Cannot remain in service; remedial action required before equipment can be returned to service (may not be cost effective, necessitating replacement)

Source: CIGRE, Guide for life management techniques for power transformers, Brochure No. 227, 2003.

- Dielectric conditions (continuous voltages and overvoltages/inherent and transient system/atmospheric overvoltages, frequency of occurrence and amplitude)
- The thermal conditions (loading/overloading cycles and ambient temperatures)
- The dynamic stresses due to system short circuit stresses (shrinkage of insulation [normal aging] and increase of dynamic stresses)
- Chemical changes due to the effect of agents

Most of the critical operational stresses are of random nature and not very well known; therefore, failure cannot be predicted. This is why we need condition assessment and classification. The authors of [5] have provided a very good classification system (see Table 10.1). This classification is of general applicability.

Before going into functional/component-related diagnostic tools, it is important to discuss the "aging factors" of power transformers.

10.3 Main Aging Factors

10.3.1 General Aging Aspects

The state of the insulation system determines the life expectancy of a transformer. Moisture, oxygen, and temperature lead to an inevitable decrease of the mechanical strength of the solid insulation and to a reduction of the oil insulation properties. Aging is defined as irreversible changes of the electrical insulation system. The main insulation aging factors are the following:

- Air/oxygen
- Moisture
- Temperature
- Mechanical/electrical stress
- Insulation contamination

These factors are strongly interlinked and act together.

10.3.2 Impact of Air/Oxygen

Air/oxygen ingress causes

- Aging of the oil already at service temperatures (<75°C–90°C)
- Aging of the solid insulation

10.3.2.1 Aging of the Oil due to Air/Oxygen

The oil parameters are severely influenced by the ingress of air and are also catalyzed by some transformer materials (e.g., copper). The resulting products are alcohols, aldehydes, ketones, acids, and moisture, which all deteriorate the oil parameters and lead to an accelerated aging of the solid insulation at normal operating temperatures (≥75°C–90°C).

There is a clear difference between open breathers and closed-system-type transformers. For open breathers, aging starts at less than 100°C, whereas in closed systems, aging needs >100°C.

10.3.2.2 Aging of the Solid Insulation due to Air/Oxygen

Air contributes to oxidative degradation of cellulose by production of carbonyl compounds and lots of acids.

Paper ages very slowly under ideal conditions (moisture ≤0.5% and nitrogen blanket) at ≤90°C up to 38 years. Ingress of 2% moisture and air reduces the lifetime to 2 years [8]. Furans are built from cellulose under air. In sealed transformers, very low values of furans have been found in the United States [9]. Cellulose ages much faster in air-saturated oil (Figures 10.1 and 10.2).

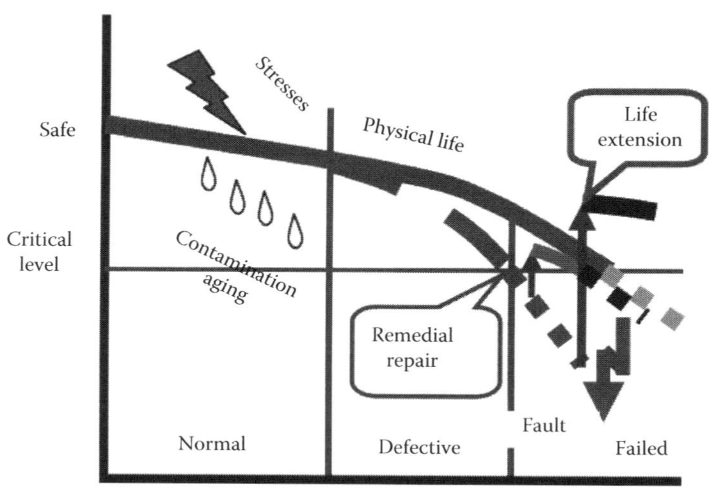

FIGURE 10.1
Conditions of a transformer in the course of its life cycle. (From CIGRE, Guide for life management techniques for power transformers, Brochure No. 227, 2003.)

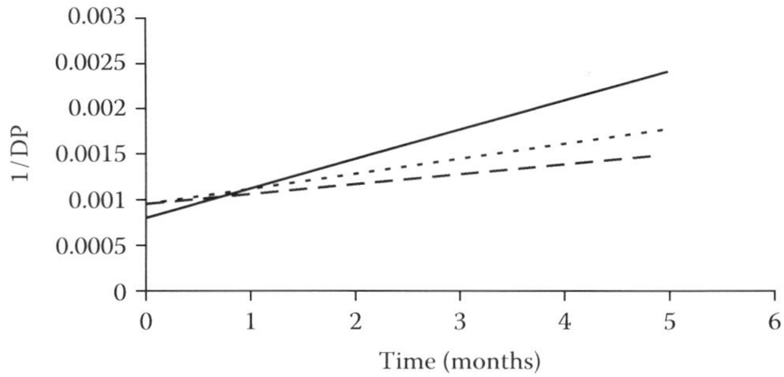

FIGURE 10.2
Rate of depolymerization at 85°C. (From Kachler, A.J. and Höhlein, I.) Thick line, air, uninhibited oil; dotted line, air, inhibited oil; and dashed line, absence of air, uninhibited oil.

10.3.3 Impact of Moisture

Water/moisture in a transformer is developed from the following:

- Aging of cellulose
- To some extent, from oil aging
- Mainly through diffusion/ingress external moisture

Once water has penetrated and distributed in oil and solid insulation, it causes

- Accelerated aging of cellulose
- Reduction of dielectric strength of oil and solid insulation
- Risk of bubble formation causing partial discharges (PDs)

The moisture paper/oil distribution curves [10] can only be used for a coarse estimation of moisture in the solid insulation. So far only the Karl Fischer titration can determine the % by weight moisture in solids. A very promising method for moisture content determination in solids is the PDC measurement [11,33–36,39].

Transformers with humid insulation and high operating temperatures age approximately 20 times faster than normal [12,13]. Figure 10.3 shows that the aging process of cellulose is severely influenced by moisture.

10.3.3.1 Reduction of the Dielectric Strength in Oil by Moisture

Water solubility in oil is very temperature and aging status dependent. Furthermore, it is dependent on the aromatic content present (Figure 10.4).

The saturation value of the oil reflects very well its dielectric margin—Figure 10.5. It is well observed that up to 20% moisture saturation does not influence significantly the breakdown voltage. A further increase in the saturation reduces the dielectric properties of the oil. A saturation of 80% means a full loss of the dielectric strength.

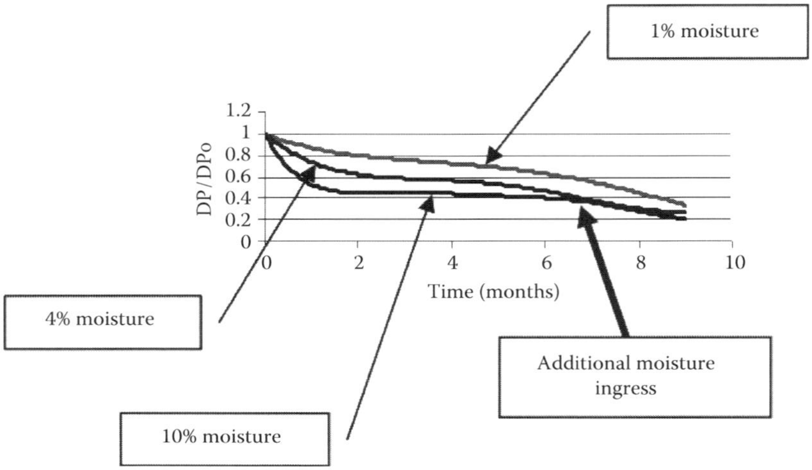

FIGURE 10.3
DP of pressboard under different moisture contents at 95°C. (From Höhlein, I. and Kachler, A.J., Progress in transformer ageing research. Impact of moisture on DP of solid insulation and furan development in oil at transformer service temperatures. CIGRE, Report D1-309, 2004.)

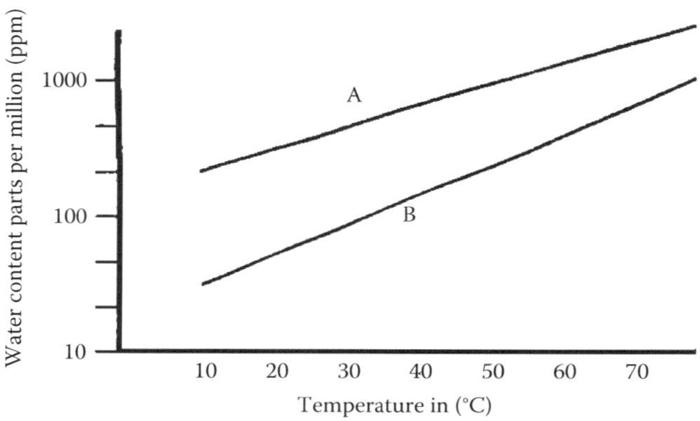

FIGURE 10.4
Water solubility in oil depending on the aromatic content. Oil A contains about 30% aromatics; oil B is a low aromatic. (From Meyers, S.D., *A Guidance to Transformer Maintenance*, Transformer Maintenance Institute, Akron, OH, 1988.)

10.3.3.2 Reduction of Mechanical and Dielectric Strength of Paper and Pressboard with Moisture

Insulation paper normally consists of coniferous wood pulp purified from lignin and contains 90% of cellulose. The so-called Kraft paper corresponds to a degree of polymerization (DP) of 1000–1200 in the new state. Figure 10.6 shows the rapid decrease of the tensile strength below a DP of 500.

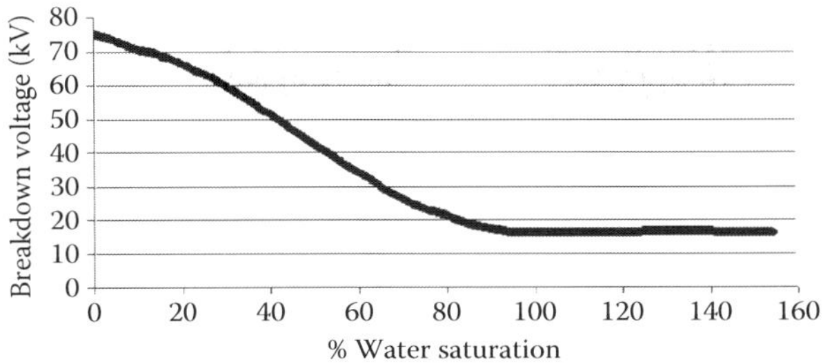

FIGURE 10.5
Breakdown voltage versus percent water saturation in a new mineral oil. (From Fonfana, I. et al., *IEEE Electr. Insul. Mag.*, 18(3), 18, 2002.)

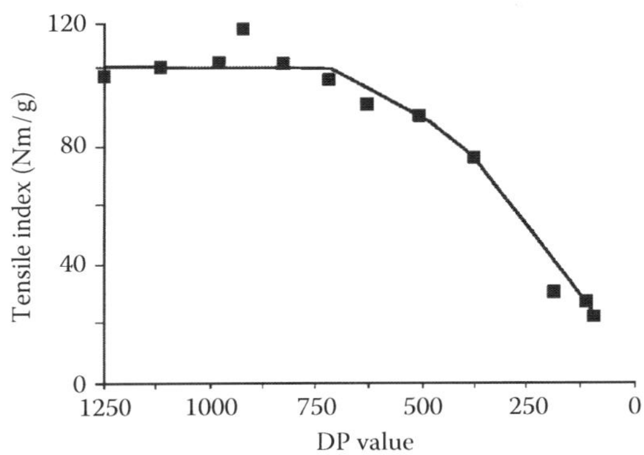

FIGURE 10.6
Relationship between mechanical strength and DP value for not thermally stabilized Kraft paper. (From Lundgaard, L., Ageing and restoration of transformer windings, Sintef energy research, Technical Report, 2001.)

10.3.3.3 Accelerated Aging of Paper with Moisture

Mechanical properties of paper are strongly reduced with moisture in paper [8,14]. The rate of thermal aging is proportional to the moisture content [15,16]. The Emsley method predicts that under the exclusion of moisture and oxygen, a DP value of 200 can only be expected for operating temperatures >90°C–100°C [12,17–20] (Figure 10.7).

Under the ingress of moisture, DP values <200 are reached even at normal operating temperatures (<90°C).

10.3.3.4 Impact of Temperature

According to the Arrhenius equation, an increase of 6°C–10°C in the range of ≥90°C–120°C leads to a duplication of the aging rate. Temperature always acts in combination with other aging factors and contributes to the thermal degradation of cellulose [8,21].

Hydrolysis, oxidative, and pyrolytic degradation are the three main processes that dominate cellulose aging. Up to 120°C hydrolysis is the key process. Table 10.2 shows the thermal classification of solid insulation materials.

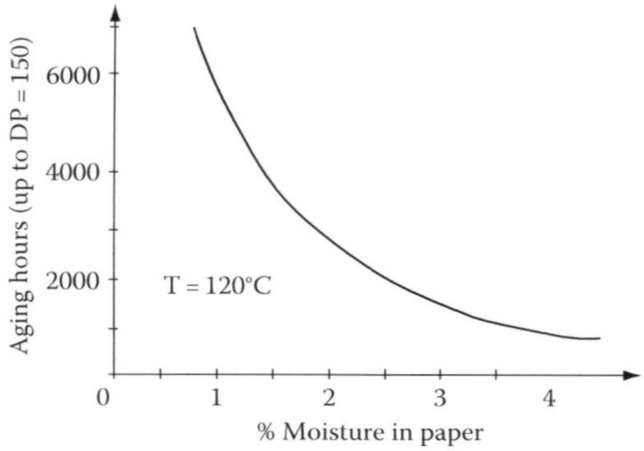

FIGURE 10.7
Dependence of the paper has achieved a value of DP of 150 from the moisture content in paper at 120°C.

TABLE 10.2
Thermal Classification of Solid Insulation Materials

Class Designation	Maximal Permissible Temperature	Material
105	105	Impregnated cellulose
120	120	Thermal upgraded cellulose

Source: IEC TC14 WG 29, Guide for the design and application of liquid immersed transformers using high temperature insulation materials, IEC 60076-14, Draft 3/2003.

TABLE 10.3
Maximal Operating Temperatures of Different Insulating Fluids

Type of Oil	Maximal Operating Temperature (°C)
Mineral insulating oil according to IEC 60296	105
Synthetic ester according to IEC 61099	130
Dimethylsilicone according to IEC 60836	155

Source: IEC TC14 WG 29, Guide for the design and application of liquid immersed transformers using high temperature insulation materials, IEC 60076-14, Draft 3/2003.

There are no tests to determine the insulation class of liquids, but there are experience values (Table 10.3).

Sludge formation, affinity to moisture absorption, and rate of oxidation all affect the thermal capability of a liquid [22].

10.3.3.5 Interaction between the Main Aging Factors

All dangerous effects caused by air, moisture, temperature, contamination, and electrical/mechanical overstresses interact according to Figure. 10.8. CIGRE WG 12.18 has developed a comprehensive aging model [5].

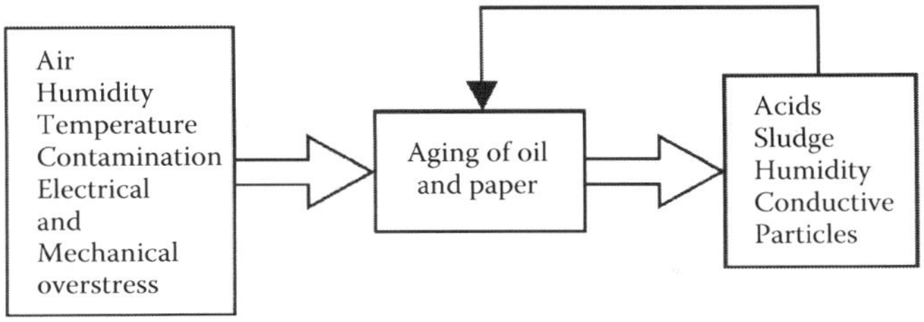

FIGURE 10.8
Interaction between the various aging factors. (From Höhlein, I. et al., Transformer life management, a customer's guideline for ageing analytics and laboratory diagnostics, service guidelines for transformers in service. Siemens Brochure, 61D7074 TV/EK 100551 A02031.)

10.4 State of the Art of Transformer Diagnostics

Utilities, manufacturers, and universities have developed a wide range of diagnostic procedures to detect normal/abnormal aging, defective or faulty conditions, and, last but not the least, the loss of life or functional/component defects that urgently require corrective actions.

TLM/condition assessment is primarily directed to transformer life extension and operational cost cutting. Most diagnostic systems (including online monitoring) serve to detect changes in the insulation system due to the following:

- Thermal
- Dielectric
- Chemical
- Mechanical aging

With the objective to detect/identify

- Hotspots
- Degradation of the insulation
- Excess moisture in the insulation
- Localized faults/defects
- PDs
- Partial rupture/mechanical defects, brittleness
- Chemical or thermal aging

None of these defects can be detected by a singular diagnostic procedure. Therefore, we need a multitude of different methods.

Furthermore, to enable trend analysis and condition assessment, we frequently need a sequence of samples to identify the rate of progressiveness and the quantity of indicated signature. With the existing diagnostics, the identification of thermal, dielectric, and chemical deficits is possible today. However, large unit performance must always be treated individually, besides the general criteria.

10.5 Functional and Component-Related Diagnostic

Condition assessment of power transformers in respect to remaining life is not yet generally available. But in the past 10–20 years, numerous diagnostic procedures have been developed—see Refs. [1,2,45].

Condition assessment to date is based on both fingerprints and off- and online monitoring and sampling of important functional aspects at regular or irregular intervals. The various electrical, mechanical, chemophysical analysis procedures are widespread and component related [4,41].

The CIGRE WG 12.18 [5] has come up with a very clear synonym approach of TLM.

It all starts with detection of normal/defective or faulty conditions.

10.5.1 Functional-Based Methodology

1. A transformer is defined by the following subsystems [5]:
   a. Electromagnetic circuit
   b. Current carrying circuit
   c. Dielectric system
   d. Mechanical structure
   e. Cooling system
   f. Bushings
   g. OLTC
   h. Oil preservation and expansion system
   i. Protection and controls/monitoring
2. The function-based methodology must sequentially check on possible defects in any one of the subsystems, which may lead to malfunction of the whole transformer.
3. This requires multistep diagnostics.

10.5.2 Key Requirements for Functional Service Ability

In respect to the functionality of the transformer subsystems, there are four key serviceability requirements:

- Electromagnetic ability and integrity to transfer energy under specified conditions
- Integrity of current carrying circuits
- Dielectric withstand strength under specified stresses and normal aging
- Mechanical withstand strength (short circuit fault current conditions)

The functional failure model, developed by [5], is clearly directed to identify

- What defects/faults can be expected in the transformer core and coil assembly?
- What is the possible defect evolution into a failure?

Some defects may be attributed to design/production deficits; therefore, any defect/failure investigations must start with both a *design* and *production* quality review. But it is also necessary to perform a *review of the effective system/operating conditions.*

10.5.3 Functional Diagnostic Procedures for Power Transformers

The technical literature [4,5,41 and many others] provides an excellent source of diagnostic procedures. But so far very little is said about the normal/abnormal/defective and faulty signatures or symptoms. For the sake of clarity, we present here a summary of

- Generally applicable diagnostics (Section 10.5.3.1)
- Dielectric failure recognition in magnetic circuits (Section 10.5.3.2)
- Dielectric failure recognition in windings/main insulation (Section 10.5.3.3)
- Thermal failure recognition (Section 10.5.3.4)
- Mechanical failure recognition (Section 10.5.3.6)
- Critical aging of oil and celluloses (Section 10.5.3.5)
- Dielectric/mechanical failure in OLTC (Section 10.5.3.6)
- Dielectric/thermal failure in bushings (Section 10.5.3.7)

10.5.3.1 Generally Applicable Diagnostic

Here we list all the well-accepted diagnosis procedures, which are or may be used irrespective of the subsystems and of the functional components (Table 10.4):

10.5.3.2 For Dielectric Failures in Magnetic Circuits (Table 10.5)

TABLE 10.4

Generally Applicable Diagnostics

Dissolved gas-in-oil analysis (sampling DGA), fingerprint, trend indicator by regular sampling, in the future also as an online early warning system
Oil analysis (BDV, H_2O, dissipation factor neutralization, etc.)
ITF
Oil aging (according to Baader or IEC/ANSI)
Dielectric response measurement PDC and FDS analysis, to determine moisture and aging status
Particle counting in oil (sampling)
Furan analysis (HPLC)—in the future, possibly a powerful aging diagnostic procedure of oil and cellulose

TABLE 10.5

Diagnostic Procedures for Dielectric Failure Recognition in Magnetic Circuits

Ratio measurement/turn to turn insulation
Magnetizing current measurement with low voltage (LV) (fingerprint)
No load loss measurement, single phase with LV (fingerprint)
For core sparking, PD measurement with acoustic localization
DGA for overheating of core and winding conductors
Winding resistance measurement
Magnetic flux distribution (with Rogowski coils on upper yoke and limbs)

10.5.3.3 For Dielectric Failures in Windings and Main Insulation (Table 10.6)

10.5.3.4 For Thermal Failure Recognition (Table 10.7)

10.5.3.5 For "Critical Aging" in Oil and Cellulose

The most complex failure is "critical aging." It is an integral process, which is highly influenced by the mode of operation (overload cycles, low load cycles with or without reduced cooling, overvoltage transients in frequency of occurrence and magnitude, etc.) and impact of aging agents.

For accelerated/critical aging detection, online monitoring of the operational conditions and particularly the temperatures of

TABLE 10.6

Diagnostic Procedures for Dielectric Failure Recognition in Windings and Main Insulation

DGA
Oil analysis (BDV, dissipation, neutralization, etc.)
ITF
Humidity in oil
Dielectric response (PDC/FDS) measurement for moisture/aging
Karl Fischer for moisture
DP
Power factor and capacitance
Winding resistance
Dew-point measurement
PD measurement and localization
Transfer function analysis
Frequency response analysis
Humidity in oil by sampling/online monitoring

TABLE 10.7

Diagnostic Procedures for Thermal Defects Recognition

DGA
Oil analysis
ITF
Furan derivate analysis
Control of winding temperatures/oil temperatures
Dielectric response measurements (PDC, FDS)—for excess moisture control
Hotspot controls by thermovision
Magnetizing current control for core and/or winding thermal defect (depending DGA in oil analysis results)

- The cooling medium
- The windings
- The ambient
- The system condition

will be very helpful to assess aging in the future. Table 10.8 shows some of the most important procedures to detect critical aging.

10.5.3.6 For Mechanical Failure Recognition

Mechanical defects due to mechanical impact during transport or due to long-lasting critical system short circuits may be very critical. For the latter also the frequency of occurrence and amplitude of short circuit SC currents are important (Table 10.9).

The dynamic impacts either produce

TABLE 10.8

Some of the Most Important Procedures for Recognition of Critical Aging due to Thermal and/or Dielectric Problems

DGA
Oil analysis
ITF
Moisture in oil
Moisture in solid insulation by Karl Fischer titration dielectric response measurement (PDC/FDS)
Power factor and capacitance
DP
Furan derivate analysis
Thermovision of hotspots in tanks or magnetic shields
Winding hotspot measurement
Online monitoring of operational conditions
PD measurement and localization

TABLE 10.9

Most Important Diagnostic Procedures for Recognition of Mechanical Defects

3-Dimensional shock recorder monitoring for excess dynamic impacts during transport
Impedance measurement with LV precision instrumentation
LV impulse injection (LVI) response measurements
DC resistance measurements/solder joint control on multistrand conductors
FRA comparison with fingerprint and phase-wise
Transfer function analysis
Clamping force control
Noise and vibration measurement
Visual inspection
Repeat of dielectric test at 80%–100% U test

- Winding distortion/displacement (radial/axial/buckling/twisting)
- Loosening of the clamping and of the lead support

The most important diagnosis procedures are as follows.

10.5.3.7 For Dielectric/Thermal/Mechanical Failure Recognition of OLTC

OLTCs are important components of power transformers. According to a CIGRE Evaluation of Power Transformer Failure Statistics from the 1980s [42], the tap changers are involved in up to 40% of the cases.

This is completely against our Siemens/MR experience. According to our experience, tap changers are in less than 16% involved.

Typical defects or faults are described in the OLTC failure models presented by [5,30].

Here we indicate some of the most effective diagnostic procedures to detect OLTC defects (Table 10.10).

10.5.3.8 For Dielectric/Thermal Aging and Failure Detection in Bushings

Also the bushings are of highest importance for successful transformer operation.

There are two major defects or faults in bushings (Table 10.11):

- Dielectric deterioration of the bushing insulation in case of overstressing and heavy outside contamination
- Thermal run away, due to dielectric overheating

TABLE 10.10

Diagnostic Procedures for Dielectric and Thermal Defects in OLTCs (Aging Defects)

DGA
DC resistance measurement
Surface resistance measurement according to Ref. [30]
Ratio measurement and coupling control
Torque measurement/monitoring for contact erosion
PD measurement and localization
Oscilloscopic control of one complete cycle of OLTC operation
Oil and gas tightness control of the diverter switch

TABLE 10.11

Diagnostic Procedures for Dielectric/Thermal Aging and Failure Detection in Bushings

Power factor and capacitance measurements as original fingerprint and consecutive checks during operation/online monitoring
Comparative C + tan delta measurements between phases
DGA of bushing according to IEC 599
PD measurements
Oil level control
Moisture control in oil
Dielectric response measurement (PDC) for moisture determination (aging)

Both these aspects are very effectively shown in the failure model for condenser-type bushings [5].

The predominant diagnostic procedures are as follows.

10.5.4 Condition Assessment

One of the most demanding engineering tasks is to derive from the symptoms of deficits the nature of the fault, the location of the faulty/defective component. But even more important is the validation of continued serviceability or of the involved risks.

- Is it normal? or is there a fault?
- Is it serious? or is it fit for service?
- What is the reliability or the risk?

These are the key questions for condition assessment.

One quite successful way of condition assessment is

- Definition of characteristics of defective conditions
- By fingerprinting from the factory
- By trend analysis

With all the diagnostic procedures given in Section 10.5.3, we now need good identification of defective symptoms—but more importantly, we need to develop guidelines in limitations of diagnostic signatures (see, e.g., Appendix A).

10.6 Most Powerful Aging/Defect/Fault Detection Procedures

As shown in Chapters 5 and 6, there are three major diagnostic procedures that are equally applicable for the detection of "dielectric, thermal, and chemical aging" of power transformer insulation:

- Dissolved gas analysis (DGA)
- Oil parameters analysis
- Furan analysis

Mechanical/dynamic deficits cannot be detected by these powerful diagnostics. They need special procedures (mainly electrical) and visual controls.

10.6.1 DGA

The DGA has a history of >35 years and has become the no. 1 procedure to detect aging and defective or faulty condition of the whole transformer insulation. IEC/IEEE/BS/VDE standards have developed the so-called

- IEC Interpretation Scheme
- Roger Interpretation Coefficients
- MSS/VDE Müller, Schliesinger, Soldner Ratios [32]

All of these interpretation schemes can identify electrical/thermal faults or a mixture of them.

Figure 10.9 provides excellent guidance for the identification and interpretation of DGA, resulting in respect to nature and criticality.

Note: Not only the ratios but also the absolute values of the individual failure gases and the rate of gas production are important for the right interpretation.

In the DGA, the thermal oil aging can be followed additionally by the development of the carbon monoxide and the solid insulation aging by the development of carbon monoxide and carbon dioxide.

10.6.2 Oil Parameters

Premature aging can be detected by measuring of the oil values:

- Neutralization value
- Interfacial tension (ITF)
- Loss factor at 90°C
- Breakdown voltage
- Water content in oil (Karl Fischer)
- Water content in the solid insulation by PDC measurement
- Change of color (not specified in IEC 60422), but this is often the first indicator of oxidation and oil aging

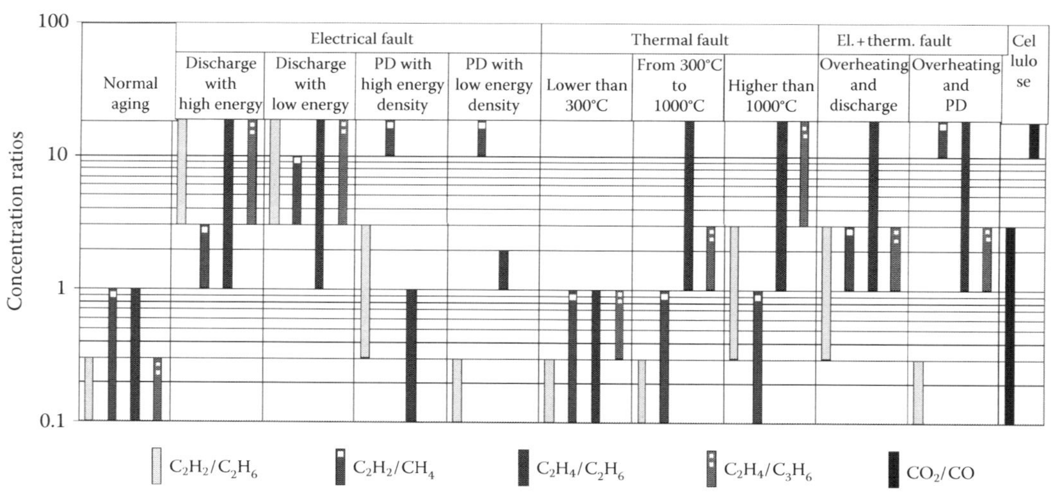

FIGURE 10.9
MSS/VDE DGA Interpretation Scheme. (From Müller, R., Gas analysis, precautionary monitoring for transformers. Presented at the 1980 Hannover Fair, transformer union information reprint: 54.5.19/6.80E, 1980.)

10.6.3 Furans

Hydrolysis, pyrolysis, and oxidation are the processes that cause decomposition of the paper.

During aging of solid insulation, a decrease of the DP takes place, and various decomposition products are built—for example, furan compounds.

The furanic compounds are soluble in oil, and their analysis is relatively simple. The main compound is 2-furfural or 2-furfuralaldehyde (2-FAL). The furans are indicators of the aging status of the solid insulation. The furanic content in oil, however, is dependent on

- Oil temperature
- Neutralization value
- Sludge content
- Ratio oil/paper
- Type of oil (inhibited/not inhibited)
- Type of paper (Kraft or thermally upgraded)
- Moisture content of paper and oil

Therefore, further research on samples from transformers is necessary. A good example for the impact of moisture is shown in Figure 10.10.

10.6.4 Correlation between DP Value and Furanic Content Furans

At present there is no exact correlation between DP values and furans. But there are approximate correlations [45]. This requires further research.

A trend analysis based on furan development is possible to characterize the thermal behavior [19,29]. Both the absolute value and the rate of furan formation are very

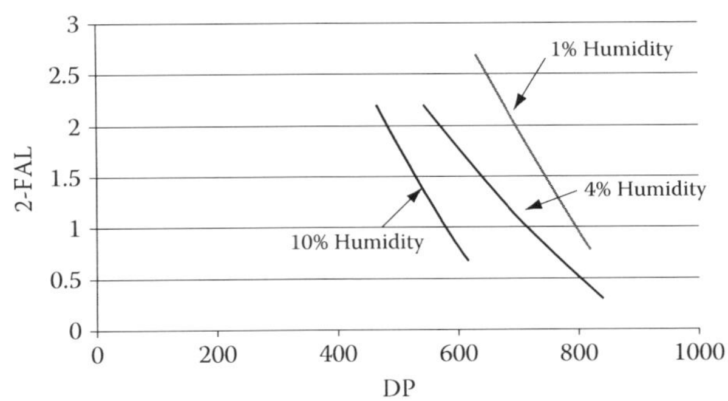

FIGURE 10.10
2-FAL/DP development from pressboard at 95°C and different starting moisture contents. (From Höhlein, I. and Kachler, A.J., Progress in transformer ageing research. Impact of moisture on DP of solid insulation and furan development in oil at transformer service temperatures, CIGRE, Report D1-309, 2004.)

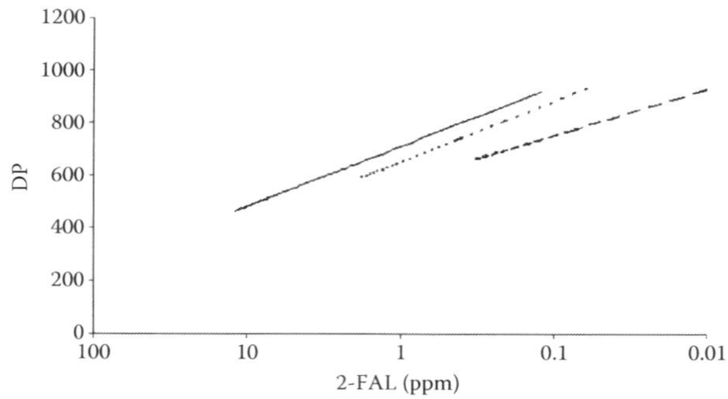

FIGURE 10.11
DP as a function of 2-FAL at 85°C, pressboard. Thick line, air, uninhibited oil; dotted line, air, inhibited oil; and dashed line, absence of air, uninhibited oil. (From Höhlein, I. and Kachler, A.J., Progress in transformer ageing research. Impact of moisture on DP of solid insulation and furan development in oil at transformer service temperatures, CIGRE, Report D1-309, 2004.)

important for the evaluation of the state of the solid insulation. Figure 10.11 clearly shows the DP/furan changes for different conditions.

10.7 Further Required Research

Despite the many successful diagnostic procedures, there is still a need for continued research to establish loss of life or to determine remaining life. The DP is the best indicator for both.

It is, however, invasive, that is, it takes sample cutting from the solid insulation (leads, windings, etc.). Most of these samples cannot be taken from the hottest area and are therefore not really representative.

Possible alternatives are as follows:

- Dielectric response analysis (PDC analysis)
- Furan derivate analysis

Both methods are noninvasive and deliver excellent information about the aging status.

However, furan analysis depends highly on temperature and moisture, and PDC results are influenced by oil conductivity; therefore, we recommend studying the various procedures in more detail to determine the following:

1. Relative moisture (RM) and age/loss of life (DP)
2. Furan analysis and age/loss of life (DP), that is, we need to work on the following:
   a. Correlation of RM, furan compounds, and DP values.
   b. In addition, research should also concentrate on more stable sensors for hydrogen (H_2) and other failure gases for online monitoring.
   c. Establishment of more guidelines for all diagnostic procedures to determine normal/defective/faulty conditions.

10.8 Conclusions

The presented diagnostic tools, both in general and functionally/component related, have proven to be quite effective. This is also supported by the examples of Appendix A. In detail we conclude the following:

1. The most powerful diagnostic tool is the DGA. It may be used to identify dielectric, thermal, and chemical aging problems.
2. Oil analysis as established in all facts is the second most powerful diagnostic tool.
3. It normally takes two or more different diagnostic approaches to really identify the nature of the defect.
4. The existing diagnostic procedures are mature and can also be used to identify defects in peripheral components (cooling plant, bushings, etc.).
5. Both off- and online diagnostics can be extremely successful to avoid catastrophic and costly outages, if applied consequently.
6. Consequent maintenance, with fingerprinting, trend analysis, and ad hoc on-site diagnostic testing as well as life extension measures are indispensable for TLM.
7. Chapter 8 suggests the future research.
8. The state of the art in transformer diagnostics allows to identify dielectric, thermal, and chemical aging and defective problems.
9. For the impact of air, moisture, and temperature, "closed system transformers" are less susceptible to aging than "open breathers."

Appendix A: Examples of Successful Diagnostics from German Utilities and Manufacturers

Case No. 1—300 MVA, 245 kV—ST, Made in 1978

Defective Insulation System

Fault symptom: Excessive water content, aged insulation system.

Detected by: PDC, Karl Fischer in oil, oil analysis, DGA, DP.

History

1998—There was a warning with a gas in the Buchholz relay. The DGA analysis did not give any indication of transformer failure. The oil analysis, however, showed an excessive water content in oil and the presence of sludge. The color, neutralization number, and the loss factor were also elevated. The sludge led to a tarnish on the tap changer contacts and elevated surface resistance. The contacts have been cleaned through switching.

2000—The oil was changed against an inhibited one. The oil numbers, color, acidity, and loss factor became better, but the increased water content was further present.

2002—An LFH drying of the transformer has been performed (approximately 100 L water was removed).

A paper sample was taken from phase V. It showed a depolymerization (DP) value of 350—indicating an advanced cellulose destruction, surely influenced by moisture. The water extracts from the LFH drying were very acidic—they contained in high amounts acetic and formic acid and various aldehydes, among which furfural was in the highest concentration.

Diagnosis: Excessive moisture in the insulation system, dielectric deterioration of the insulation system.

Action taken: LFH drying on site.

The ST is successfully operating since 2002.

Case No. 2—438 MVA, 420 kV—ST, Made in 1969

Defective Insulation System

Fault symptom: Oil analysis, indicating aged and humid insulation system.

Detected by: Oil analysis, furans.

History

This transformer was in service approximately 20 years before it was used as standby transformer for a period of about 8 years. The routine oil checks after recommencement (Table 10.12) indicated an accelerated aging of the insulating system.

In consequence of this result, the oil checks have been intensified. They clearly indicated (Table 10.13) highly aged insulation. In addition high level of moisture and extremely low breakdown voltage of the oil were detected.

TABLE 10.12

Consecutive Oil Analysis

	1994	1995	1999
Appearance	Pure	Pure	Pure
Color	2.5	3.0	3.0
Refractive index	1.4722	1.4725	1.4732
Neutralizing index (mg KOH/g oil)	0.10	0.10	0.12
Dielectric loss factor	0.035	0.069	0.083
Water according to Karl Fischer (mg/kg oil)	15	21	36
Breakdown voltage	>60	>60	>60

TABLE 10.13

Water Content and Breakdown Voltage of Oil

Date	Temperature (°C)	Water acc, Karl Fischer (mg/kg)	Breakdown Voltage (kV/2.5 mm)
October 14, 1999	58	36	36
October 27, 1999	61	45	28
November 08, 1999	55	33	31
November 11, 1999	56	38	38
November 18, 1999	48	57	34
November 22, 1999	63	45	20

TABLE 10.14

Furan Analysis

Substance	Concentration (mg/kg)
5-HMF (5-hydroxymethyl-2-furfural)	<0.05
2-FOL (2-furfural alcohol)	<0.05
2-FAL (2-furfural)	6.00
2-ACF (2-acetylfuran)	<0.05
2-MEF (5-methyl-2-furfural)	<0.05

The additional furfural analysis (Table 10.14) indicated also a severe aging of the solid insulation.

The transformer obviously had reached the end of its lifetime. It was taken out of service. The determination of water in insulation according to Karl Fischer resulted in values up to 5%.

DP—Determination of insulation resulted in an average value of 181, which was an indication of a complete cellulose degradation.

Diagnosis: Highly aged insulation and oil, end of transformer life.

Action taken: The transformer was scrapped.

Case No. 3—234 MVA, 345 kV—ST, Made in 1991

Defective Winding

Fault symptom: Elevated DC resistance.

Detected by: DGA, DC resistance measurement.

History

After 11 years of service, rising amounts of fault gases from thermal oil decomposition with small amounts of acetylene have been detected by DGA (Table 10.15). Furthermore, the ratio carbon monoxide/carbon dioxide was <3, which suggested an electrical degradation of cellulose. The transformer was inspected on site. A resistance measurement was carried out. There was a remarkable increase in the DC resistance. The transformer was sent for repair to the factory.

One broken twin conductor of two twins with burnt out insulation paper was detected. The damage was a defective solder joint at the outmost turn of the ground end disk (Figure 10.12).

TABLE 10.15

Last DGA before Taking Out of Service

Type of Gas	ppm (v/v)
Hydrogen	1060
Methane	2481
Ethane	703
Ethylene	2187
Acetylene	4
Carbon monoxide	450
Carbon dioxide	995

FIGURE 10.12
Broken twin conductor as a consequence of a defective solder joint.

Diagnosis: Defective solder joint.

Action taken: The winding was repaired and the transformer was taken in service again and is working trouble free.

Case No. 4—60 MVA, 123 kV—Compensating Reactor, Made in 1979

Defective Insulation System

Fault symptom: Sludge in oil.

Detected by: DP, furans, oil analysis.

History

After 20 years of successful operation, time-based maintenance (DGA, oil and furan analysis) indicated severe aging of oil and insulation. The reactor was brought back in the factory for refurbishment of the cooling system. DGA showed no failure indications (MSS Code 01002), but suggested a severe aging of the solid insulation. The oil analysis revealed also a strongly aged oil (color 5.5, neutralization value 0.18 mg KOH/g oil) with sludge formation. The oil sludge has plugged the cooling ducts of the radiators and led to overheating. The paper samples of the winding were covered heavily with sludge, and the DP value was only 200. The measured furans showed values of 2-FAL of 11.5 mg/kg oil. The solid insulation has reached the end of life.

Diagnosis: accelerated aging of the insulating system as a consequence of inadequate cooling.

Action taken: A complete refurbishment was carried out—winding, oil, and cooling system.

Case No. 5—60 MVA, 123 kV—ST, Made in 1984

Defective OLTC [30]

Fault symptom: High surface and DC resistance, excessive fault gases from thermal oil degradation.

Detected by: DGA, DC resistance, surface resistance measurement [30].

DGA indicated high amounts of fault gases from thermal oil degradation.

FIGURE 10.13
Eroded contacts of the tap selector with a coke layer.

History

The DC resistance measurements taken on the OLTC showed that only marginally increased contact resistance could be determined in the entire tap selector area. After a few switching operations had been carried out, they were reduced to the normal level and did not show any current dependency. However, unusually high and unstable contact resistances were determined at the change-over selector. For this case a special surface resistance measurement was developed. Even after an increased number of change-over selector operations, the resistance values did not show any improvement. This behavior indicated quite a heavy coke layer on the contacts, which could not be removed by simple switching.

The transformer was opened, the selector was inspected, and the change-over contacts were exchanged. The change-over contacts were covered with a thick carbonized layer with a significant surface erosion—see Figure 10.13. Such contact damage mainly appears on change-over selectors made of copper (movable contact) and brass (fixed contact), which are rarely operated. Caused by the slight tarnish, the contact resistance increases, and, consequently, the temperature of the contacts, too. The increased contact temperature accelerates the growth of tarnish, and the temperature is high enough to decompose the oil, and oil carbon is generated (Figure 10.13). In the course of time, heavy layers (oil carbon) will be built up.

Diagnosis: Oil-carbon layer on the copper and brass selector over contacts.

Actions taken: Exchange of the contacts, on-site repair with consequent oil degassing.

References

1. Allan DJ (2001) *Why Transformers Fail*. EurotechCom, U.K., pp. 99–112.
2. Kachler AJ (2001) Unique transformer failure statistics an important contribution to economics of power transformer management. *CIGRE*, Dublin, Ireland, Colloquium SC12.
3. Baehr R (2001) Transformer technology: State of the art and trends of future development. *Electra*, 198, 13–19.

4. Kachler AJ (1997) Diagnostic and monitoring technology for large power transformers (fingerprints, trend analysis from factory to on-site testing). *CIGRE*, Sydney, Australia, Colloquium SC12.
5. CIGRE (2003) Guide for life management techniques for power transformers. Brochure No. 227.
6. Kachler AJ (2003) Power transformer reliability: A cooperate responsibility of manufacturers and users. *CIGRE Workshop*, Florianopolis, Brazil.
7. Kachler AJ, Höhlein, I (2005) Ageing of cellulose at transformer service temperatures. Part 1. Influence of type of oil and air on the degree of polymerisation of pressboard, dissolved gases and furanic compounds in oil, *IEEE, Electrical Insulation Magazine*, 21(2), April 2005, pp. 15–21.
8. Schroff DH et al. (1985) A review of paper ageing in power transformers. *IEE Proceedings*, 132 (Pt C), 312.
9. Griffin PJ (1996) Experience in testing for furanic compounds in the USA. *CIGRE 1996, CIGRE WG 15.01*, Paris, France.
10. Du Y et al. (1999) Moisture equilibrium in transformer paper, oil systems. *IEEE Electrical Insulation Magazine*, 15, 11–20.
11. Leibfreid T et al. (2002) Ageing and moisture analysis of power transformer insulation system. CIGRE, Report 12-101.
12. Emsley M et al. (2000) Degradation of cellulosic insulation in power transformers. Effect of oxygen and water on ageing in oil. *IEE Proceedings*, 147 (3), 115.
13. Lundgaard L (2001) Ageing and restoration of transformer windings. Sintef energy research, Technical Report.
14. Gussenbauer I (1993) Grenzwerte der Isolationseigenschaften für die Betriebssicherheit von Leistungstransformatoren, M&M, 93 (10), 445.
15. Bingelli J (1964) The treatment of transformers. Quality and completion criteria of the process. CIGRE, Report 12-110.
16. Bouvier B (1970) Nouveaux criteria pour characteriser la degradation thermique dùne isolation a base de papier. Lab Central des Industries Electrique, 72, 1029.
17. Gasser HP (1999) Determining the ageing parameters of cellulosic insulation in transformers. *IEE, High Voltage Engineering Symposium*, Dublin, Ireland, p. 467.
18. Emsley AM et al. (1994) Review of chemical indicators of degradation of cellulosic electrical paper in oil filled transformers. *IEE Proceedings Science, Measurement and Technology*, 141 (5), 324.
19. Emsley AM et al. (2000) Degradation of cellulose insulation in power transformers, Part 2. *IEE Proceedings Science, Measurement and Technology*, 147 (3), 110.
20. Lundgaard L (2002) Transformer winding ageing and restoration. CIGRE, Report 15.01.
21. Höhlein I et al. (2001) Transformer life management, a customer's guideline for ageing analytics and laboratory diagnostics, service guidelines for transformers in service. Siemens Brochure, No. 61D7074 TV/EK 100551 A02031.
22. IEC TC14 WG 29, Guide for the design and application of liquid immersed transformers using high temperature insulation materials. IEC 60076-14, Draft 3/2003.
23. Sokolov V et al. (2001) Moisture equilibrium and moisture migration within transformer insulation system. *CIGRE WG 12.18*, Dublin, Ireland, June 18–21.
24. CIGRE (2000) Effects of particles on transformer dielectric strength. Brochure No. 157.
25. Ferguson R et al. (2002) Suspended particles in the liquid insulation of ageing transformers. *IEEE*, 18 (2), 17.
26. Kachler AJ et al. (2004) Transformer life management, German experience with condition assessment, *ETG, Proceedings*, Köln, Germany, pp. 317–329.
27. Gäfverts H, Frimpong G (1998) Modelling of dielectric measurements on power transformers, CIGRE, Report 15-103.
28. Leibfried T, Kachler AJ (2002) Insulation diagnostic on power transformers using polarisation and depolarisation current (PDC) analysis. *IEEE International Symposium on Electrical Insulation*, Boston, MA, pp. 170–173.
29. Höhlein I et al. (2002) Transformer ageing research on furanic compounds dissolved in insulating oil. CIGRE, Report 15-302.

30. Krämer A et al. (1994) On site determination of the condition of on, load tap changer contacts if a hot spot has been indicated by gas in oil analysis in power transformers. CIGRE, Report 12-205.
31. Tenbohlen S et al. (2000) Enhanced diagnosis of power transformers using on and off line methods: Results, examples and future trends. CIGRE, Report 12-204.
32. Müller R (1980) Gas analysis, precautionary monitoring for transformers. Presented at the 1980 Hannover Fair, transformer union information reprint: 54.5.19/6.80E.
33. Der Houhanessian V, Zaengl W (1996) Vor-Ort Diagnose für Leistungstransformatoren. Bull. SEV/VSE, issue 23/26, pp. 19–28.
34. Aschwanden T et al. (1998) Development and application of new condition assessment methods for power transformers. CIGRE, Report 12-207.
35. Kachler AJ et al. (1996) Kritische Anmerkungen zur Feuchtigkeitsbestimmung von Transformatoren mit der Recovery Voltage Methode. *ETZ*, Jg. H19, pp. 1238–1245.
36. Kachler AJ (1999) Ageing and moisture determination in power transformer insulation systems. Contradiction of RVM methodology, Effects of geometry and oil conductivity. *Transformer, International Conference on Power Transformers*, Kolobrzeg, Poland.
37. Sokolov V (1993) Effective criteria of oil condition on large power transformers, diagnostic and maintenance techniques, *Proceedings of CIGRE Symposium*, Berlin, Germany.
38. Sokolov V et al. (1999) Experience with in field assessment of water contamination of large power transformers, *Proceedings EPRI Substation Equipment Diagnostic Conference VII*, New Orleans, LA.
39. Zaengl W (2001) Dielectric spectroscopy in the time and frequency domain for power equipment/transformers, cables etc. *12th International Symposium on High Voltage Engineering*, Bangalore, India, Keynote Speech, Session 9, pp. 76–85.
40. Meyers SD (1988) *A Guidance to Transformer Maintenance*, Transformer Maintenance Institute, Akron, OH.
41. Pratt F (1986) Diagnosis methods for transformers in service. CIGRE, Report 12-06.
42. Dietrich on behalf of CIGRE WG (2005) An international survey of failures in large power transformers in service. CIGRE 12-05, *Electra*, 88, 21–48.
43. Höhlein I, Kachler AJ (2004) Progress in transformer ageing research. Impact of moisture on DP of solid insulation and furan development in oil at transformer service temperatures. CIGRE, Report D1-309.
44. Fonfana I, Wasserberg V, Borsi H, Gockenbach E (2002) Challenge of mixed insulating liquids for use in high, voltage transformers, Part 1: Investigations of mixed liquids. *IEEE Electrical Insulation Magazine*, 18 (3), 18–25.
45. Fushimi Y, Shimano T et al. (2002) Improvement in maintenance and inspection and pursuit of economical effectiveness of transformers in Japan. CIGRE, Report 12–107.

11

Sources, Measurement, and Mitigation of Sound Levels in Transformers

Jeewan Puri

CONTENTS

11.1 Introduction

In modern communities, due to increasing density of residential housing near substations and transformers, there is an increased prevalence of local ordinances specifying sound levels at commercial and residential property lines. Therefore, it is appropriate that a good understanding of sources, measurement, and the mitigation options of sound energy radiated by transformers be developed for properly specifying sound levels in transformers. A good understanding of these principles can help us minimize the environmental impact of transformer noise on the neighboring communities.

Transformer cores were recognized as major source of sound levels in transformers. However, due to the increased demand for lower sound levels in transformers, winding noise has become a significant component of the radiated sound energy. The presence of current harmonics in modern load configurations has also become an important consideration in designing low sound level transformers.

The demand for low sound levels has added new complexity to the measurement process of the radiated energy for transformers. As a result of this complexity, it has become important to understand the nature of the sound energy and how it is radiated.

11.2 Transformer Sound Level

In order to evaluate a sound source, we must understand the following basics used for the quantification of sound energy.

11.2.1 Sound Pressure Level

The main quantity used to describe sound is the size or the amplitude of the pressure fluctuations at a human ear. The weakest sound a healthy ear can detect has an amplitude of 20 millionth of a Pascal (20 μPa). This pressure deflects the human eardrum by a distance less than the diameter of a single hydrogen molecule. Amazingly, the human ear can tolerate pressures more than a million times higher. Thus, using a Pascal scale to measure sound would lead to very large to very small numbers. In order to avoid this, a decibel or dB scale is used. It expresses the measured sound pressure P as a logarithmic ratio between it and a reference sound level of 20 μPa as follows:

$$\text{Sound pressure level} = Lp\,\text{dB} = 10 * \log\left[\frac{P}{P_0}\right]^2 \tag{11.1}$$

where P_0 is the reference pressure of 20 μPa.

Decibel scale gives better approximation of the human perception of relative loudness than a Pascal scale.

11.2.2 Perceived Loudness

The ear of a healthy young person can hear frequencies ranging from 20 Hz to 20 kHz. In terms of sound pressure levels, the audible sounds range from threshold of hearing at 0 dB to threshold of pain that can be over 130 dB.

Although an increase of 6dB represents doubling of the sound pressure, in actuality, a change of 10dB subjectively appears to be twice as loud. The smallest change in sound that we can perceive is about 3dB.

The perception of loudness is determined by many complex factors. One such factor is that the human ear is not equally sensitive at all frequencies. It is most sensitive to frequencies between 2 and 5kHz and less sensitive at higher and lower frequencies.

11.2.3 Sound Power

A sound source radiates energy resulting in a sound pressure. Sound energy is the cause and pressure is the result. The sound pressure that we hear or measure with a microphone is dependent upon the distance from the sound source and the acoustic environment (or sound field) in which the sound waves travel. Therefore, by measuring sound pressure, we cannot quantify how much noise a machine makes. We have to therefore find sound power because it is unique descriptor of the noisiness of a sound source.

11.2.4 Sound Intensity Level

Sound intensity describes the rate of energy flow through a unit area and is measured in watts per square meter.

Sound energy may be transmitted in many directions. There may be energy flow in one direction and not in others. Sound intensity is a vector quantity as it has magnitude and direction. Sound pressure on the other hand is a scalar quantity and has magnitude only.

We usually measure sound intensity in a direction perpendicular to the unit area through which the sound energy is flowing. It is measured as the time averaged rate of energy flow per unit area. If there is no net energy flow in the direction of measurement, there will not be recorded intensity.

Like sound pressure, sound intensity level L_i is quantified using a dB scale and is expressed as a ratio of measured intensity I in W/m^2 and a reference intensity I_0 in W/m^2 as follows:

$$L_i = 10 * \log\left[\frac{I}{I_0}\right] \qquad (11.2)$$

where I_0 = reference level = 10^{-12} W/m^2.

11.2.5 Sound Pressure and Sound Intensity Relationship

For a free progressive wave, there is a unique relationship between the mean square sound pressure and intensity. This relationship at a particular point and in the direction of intensity may be described as follows:

$$I = \frac{p_{rms}^2}{\rho c}[\mathrm{W/m^2}] \qquad (11.3)$$

where

I is the intensity in W/m^2

p_{rms}^2 is the mean square pressure in N/m^2

ρc is the characteristic resistance of the medium in rayls (equals to 405 for air, 20°C at 0.751 m of Hg)

As described earlier in (11.2), sound intensity level

$$L_i = 10 * \log\left[\frac{p_{rms}^2/\rho c}{I_0}\right] \quad (11.4)$$

$$L_i = 10 * \log\left[\frac{p_{rms}}{p_0}\right]^2 + 10 * \log\left[\frac{p_0^2/\rho c}{I_0}\right] \quad (11.5)$$

From this expression, L_i can be defined as

$$L_i = L_p - 10 * \log K \quad (11.6)$$

where

$$K = \text{Constant} = I_0 * \frac{\rho c}{p_0^2} \quad (11.7)$$

By definition,

$$\frac{p_0^2}{I_0} = \frac{(20 * 10^{-6})^2}{10^{-12}} = 400 \quad (11.8)$$

Therefore, if $K = I$, then $10 * \log K = 0$. Therefore, under normal pressure and temperature conditions,

$$L_i = L_p \quad (11.9)$$

Therefore, in free field space, noise pressure and noise intensity measurements yield the same numerical value.

11.3 Sound Energy Measurement

Sound level of a source may be measured sound pressure or sound intensity method. Both of these measurements yield acceptable results. Sound intensity measurement procedure has now been added to the IEC Standard 60076-10 as an acceptable alternative. Similar changes are also being made to the IEEE C57.12.90 and C57.12.91 standards.

11.3.1 Sound Pressure Level Measurement

A sound level meter is designed to respond to sound approximately the same way as the human ear and give objective, reproducible measurements of sound level. A typical sound meter consists of a condenser-type microphone, a processing section, and a read-out unit.

The microphone converts the sound signal in to an equivalent electrical signal which is amplified by a preamplifier before being processed by passing through filters. These filters are designed so that their sensitivity to frequencies varies the same way as the human ear. This has resulted in three internationally recognized characteristics termed as "A," "B," and "C" weightings. Nowadays, the "A" weighting is most commonly used since it correlates the best with the human hearing perceptions.

11.3.2 Sound Intensity Measurement

Sound intensity is the time-averaged product of pressure and particle velocity. A single microphone can measure pressure, but measuring particle velocity is more complicated.

With Euler's linearized equation, the particle velocity can be related to pressure gradient (i.e., the rate at which pressure changes with distance). This equation is essentially Newton's second law of motion applied to a fluid. Newton's law relates the acceleration given to a mass to the force acting on it. If we know the force and the mass, then we can find the acceleration and then integrate it with respect to time and find the velocity.

With the knowledge of pressure gradient and the fluid density, the acceleration (or deceleration) can be calculated as follows:

$$a = -\frac{1}{\rho}\frac{\partial P}{\partial r} \tag{11.10}$$

where a is the acceleration produced in a fluid of density ρ by a pressure ∂P over a distance ∂r.

Therefore, by integrating the above with respect to time, the particle velocity u can be determined as follows:

$$u = -\int \frac{1}{\rho}\frac{\partial P}{\partial r}\,dt \tag{11.11}$$

The intensity measuring instruments use two microphones facing each other and separated by a small distance for measuring the pressure gradient by taking the difference between the pressures Pa and Pb measured by them and dividing it by the distance Δr between them, as can be seen in Figure 11.1. This signal can now be integrated and related to the particle velocity u as follows:

$$u = -\frac{1}{\rho}\int \frac{(Pa - Pb)}{\Delta r}\,dt \tag{11.12}$$

Since intensity is the time-averaged product of pressure and particle velocity,

$$I = -\frac{P}{\rho}\int \frac{(Pa - Pb)}{\Delta r}\,dt \tag{11.13}$$

where

$$P = \frac{(Pa + Pb)}{2} \tag{11.14}$$

This is the basic principle of sound intensity measurement.

As discussed earlier, in free sound field conditions like the ones in specially constructed anechoic chambers, sound pressure measurements will be equal to sound intensity measurements. In any other sound field condition, sound pressure measurements will be

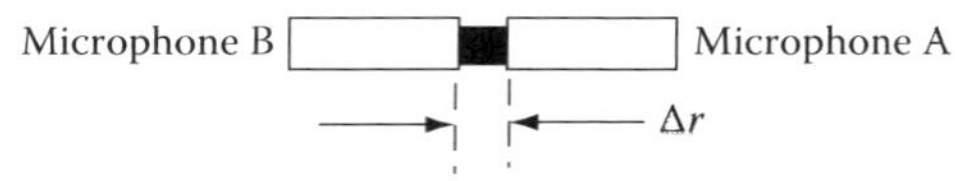

FIGURE 11.1
Intensity measuring instruments.

exaggerated by ambient noise, sound reflections from the walls, standing sound waves, and the reactive sound fields near the transformer. Therefore, sound intensity measurements will be less than sound pressure measurements. Also, semi-reverberant sound fields tend to inaccurately yield lower sound intensity levels.

Through extensive experimental measurements, it has been proven that for making sound intensity measurements of acceptable accuracy, the difference between the simultaneous sound pressure and sound intensity measurements must be less than 6 dB (A). This difference is defined as (P – I) index and is an important measure of the adequacy of the sound level measurement environment.

11.4 Sources of Sound in Transformers

Unlike fan and pump noise, sound radiated from a transformer is of tonal nature consisting of even harmonics of power frequency. The low-frequency tonal nature of noise makes it harder to mitigate than broadband noise that comes from other sources. This is because low frequencies propagate farther with less attenuation. They are perceived more than broadband noise even with high background noise levels. This combination of low attenuation and high perception makes tonal noise dominant problem in neighboring communities around transformers. The sound producing mechanisms in transformers may be characterized as follows.

11.4.1 Core Noise

When a strip of core steel is magnetized, it undergoes a small change in its dimension (usually only a few parts per million). This phenomenon is called magnetostriction. This change is independent of the direction of magnetization and therefore occurs at twice the power frequency. Magnetostriction is the predominant source of core noise in transformers. Because magnetostriction curve is nonlinear, higher even harmonics also appear at higher induction levels (above 1.4 T).

Core induction level is a complex function of induction level and the core mass. Typically, core noise and induction relationships are derived through predictive relationship based on measurement experience. The following equation is perhaps one of the best descriptors of change in sound level ΔL that occurs when the induction is changed from $B1$ to $B2$ and as a result, the core mass changes from $G1$ to $G2$:

$$\Delta L\,\text{dB} = 10\log\left[\left(\frac{B1}{B2}\right)^{8}\left(\frac{G1}{G2}\right)^{1.6}\right] \tag{11.15}$$

Compressive stresses increase magnetostriction and hence the core sound level. Core geometry, core material, and the wave form of the excitation voltage can also influence the magnitude and the frequency component of core noise. Mechanical resonance in tank walls and core mounting structure can also increase the sound level of a transformer.

11.4.2 Load Noise

Load noise is caused by vibrations in tank walls, magnetic shields, and windings due to the vibrations caused by the electromagnetic forces resulting from the leakage field produced by load currents. Load noise can be a significant factor in low sound level transformers. The loaded windings produce radial as well as axial vibrations. These vibrations are produced at twice the power frequency.

Except in very large transformers, radial vibrations do not have a significant effect on load noise. The compressive electromagnetic forces produce axial vibrations and can be a major source of sound in poorly supported windings. In some cases the natural frequency of the supporting structure may resonate with electromotive forces and greatly increase the sound level of the transformer.

Based on factory tests on 65 transformers of 60–1000 MVA rating for 50 Hz application, the following predictive relationship was developed by a European manufacturer for relating their A-weighted sound power level at their self-cooled MVA rating:

$$\text{Winding sound power level} = L_W = 39 + 18\log\left\{\frac{M}{Mo}\right\}\text{dB} \tag{11.16}$$

where
M is the MVA rating of the transformer
Mo is the reference level (equals to 1 MVA)

It is reasonable to expect that this relationship will differ slightly for different manufacturers depending upon their clamping practices and fabrication materials used in their respective factories.

Through many decades, the contribution of the load noise to the overall transformer sound level has remained moderate. However, in modern transformers designed with low induction levels and improved core materials for meeting low sound level specifications, load noise has become a significant contributor to their sound level.

11.4.3 Fan and Pump Noise

The heat generated by winding and core losses in a transformer has to be removed by fans that blow air over radiators or coolers. The noise emitted by cooling fans is normally of broad band nature. Factors that affect the total fan noise include tip speed, blade design, the number of fans used, and the arrangement of the radiators.

Cooling equipment usually influences the overall sound level of transformers of smaller MVA rating and low induction level design.

11.5 Sound Level Measurement Standards

IEC sound level measurement standard IEC60076-10 accepts sound intensity and sound pressure level measurement methods for determining transformer sound levels. IEEE

Standards C57.12.90 and C57.12.91 are also being modified to accept these measurement methods.

11.5.1 Standard Transformer Sound Levels

In the IEEE standards, it is expected that all transformers must meet the standard sound levels listed in NEMA Publication TR-1, Tables 02 through 04, for power, distribution, and dry-type transformers (see Appendix 11.A) and ST-20, Table 3-9, for dry-type lighting transformers (see Appendix 11.B).

These tables have been in publications for a long time, and their true source and logic is not easily traceable. These tables represent the maximum levels of flux density that was commonly used in commercial transformers, and consensus in the industry is that they are still valid as a minimum standard for modern transformers. The logic of these tables can be analyzed using the following formula that was discussed earlier:

$$\Delta L\,\mathrm{dB} = 10\log\left[\left(\frac{B1}{B2}\right)^{8}\left(\frac{G1}{G2}\right)^{1.6}\right] \tag{11.17}$$

where ΔL is the change in sound level that occurs when the induction is changed from $B1$ to $B2$ and as a result, the core mass changes from $G1$ to $G2$. It is also well recognized that the core weight is proportional to $(\mathrm{kVA})^{3/4}$.

Therefore, if induction is unchanged, then

$$\Delta L\,\mathrm{dB} = 10\log\left[\left(\frac{kVA_1}{kVA_2}\right)^{1.6\times 0.75}\right] \tag{11.18}$$

or

$$\Delta L\,\mathrm{dB} = 10\log\left[\left(\frac{kVA_1}{kVA_2}\right)^{1.2}\right] \tag{11.19}$$

By inspection of all the NEMA tables, the writer has noticed that by changing the exponent of this equation from 1.2 to 1.03, the sound levels listed in all the tables can be derived, as shown in Appendices 11.A through 11.D. This can be an effective way of reviewing these standards in the future.

IEC standards place no mandatory restrictions on transformer sound levels unless the customer specifies specific sound levels. They also believe that the IEEE standards are too liberal for modern low sound level transformers.

11.5.2 Sound Level Measurement Procedures

IEEE and IEC standards have been harmonized to specified similar sound level procedures whether sound intensity or sound pressure measurement methods are used (Figure 11.2).

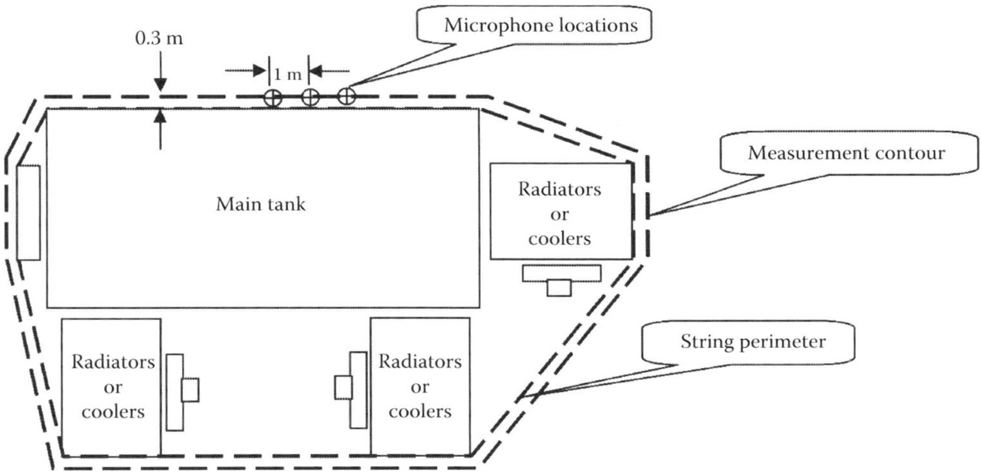

FIGURE 11.2
Sound level measurement.

Both the standards specify that sound level measurements should be made:

1. With 100% core excitation and under no load conditions.
2. Under self-cooled condition, microphones shall be placed on a measurement contour at distance of 0.3m from the string perimeter drawn around the transformer tank and the cooling equipment.
3. Under forced cooled condition, the measurement contour shall be 0.3m from the tank sides with no cooling equipment and 2m away from the sides that have cooling equipment mounted on it.
4. The transformer shall be positioned at least 3m away from any reflecting surfaces.
5. The average sound pressure on sound intensity level measurement may be calculated as follows:

$$\overline{L_{pA0}} = 10 \log_{10}\left(\frac{1}{N}\sum_{i=1}^{N} 10^{0.1L_{pAi}}\right) \quad (11.20)$$

where
N is the total number of measuring positions
$\overline{L_{pA0}}$ is the uncorrected average A-weighted sound pressure level
L_{pAi} is the A-weighted sound pressure level at the ith measuring location with the test object energized

or

$$\overline{L_{IA}} = 10 \log_{10}\left[\frac{1}{N}\left(\sum 10^{0.1L_{IAi}}\right)\right] \quad (11.21)$$

where
N is the number of measurements
$\overline{L_{IA}}$ is the average A-weighted sound intensity level
L_{IAi} is the measured A-weighted sound intensity level at location i

Both IEC and IEEE standards recognize that sound pressure level measurements may be influenced by ambient noise and reflections from the surrounding surfaces. Both the standards, therefore, allow the following:

1. The sound pressure level measurements may be corrected for reflections. This correction K may be computed as follows:

$$K = 10\log_{10}\left(1+\frac{4}{A/S}\right) \tag{11.22}$$

where $A=\alpha S_v$, being α the average acoustic absorption coefficient (see Table 1), S_v the total area of the surface of the test room, and S the measurement surface area. The maximum allowed correction is 7 dB.

2. The measured sound pressure level may be corrected for ambient noise level. The ambient noise shall be measured on the measurement contour with the transformer de-energized. These measurements shall be made before and after measuring sound level of the transformer. These measurements must not differ more than 3 dB from each other, and the higher of the two must be at least 8 dB lower than the measured sound level of the transformer. The average ambient sound pressure level shall be computed as follows:

$$\overline{L_{bgA}} = 10\log_{10}\left(\frac{1}{M}\sum_{i=1}^{M}10^{0.1L_{bgAi}}\right) \tag{11.23}$$

where
M is the total number of measuring positions
L_{bgAi} is the measured A-weighted background noise pressure level at the ith measuring position

3. The measured sound pressure level L_{pA} may be corrected as follows:

$$\overline{L_{pA}} = 10\log_{10}\left(10^{0.1\overline{L_{pA0}}} - 10^{0.1\overline{L_{bgA}}}\right) - K \tag{11.24}$$

11.5.3 Sound Power Level Determination

For demonstrating compliance with the local ordinances, it becomes necessary to calculate sound pressure level at properties located away from the transformer.

Sound power level provides a measure of the total sound energy radiated from the transformer.

Since sound pressure and sound intensity measurements yield the same numerical result, the following equations may be used to calculate sound power levels in dB form these measurements:

$$L_{WA} = L_{IA} + 10 \log (S) \text{dB} \quad (11.25)$$

or

$$L_{WA} = L_{PA} + 10 \log (S) \text{dB} \quad (11.26)$$

where S is the area of the radiating surface, which is equal to 1.25 multiplied by transformer height (in meters) and by measurement contour length (in meters).

11.5.4 Sound Power Level Calculation at Far-Field Receiver Location

From the sound power level, sound pressure level at any desired distance from the transformer can be calculated.

Given the sound pressure level requirement of L_{PD} at a distance D from the transformer, the maximum sound power level L_W for the transformer can be computed as follows:

$$L_{WA} = L_{PD} + 10 \log \left(2\Pi D^2\right) \text{ dB} \quad (11.27)$$

Therefore,

$$L_{PA} = L_{WA} - 10 \log (S) \text{dB} \quad (11.28)$$

11.6 Factors Influencing Sound Levels in Field Installation

In order to assure repeatability, sound level measurements are made in the factory under controlled conditions. The results of these measurements can be significantly influenced by the operating environment of the transformer.

11.6.1 Load Power Factor

During the factory tests, core and winding wound levels are measured separately under no-load and full-load conditions, respectively. Assuming that these vibrations are in phase with each other, the power levels of these measurements are added together to determine the overall sound level of the transformer. In the field, this assumption may not be accurate due to the nonunity load power factor.

11.6.2 Internal Regulation

The load currents cause an internal voltage drop in the transformer. This can change the core induction level and hence the sound level of the transformer.

11.6.3 Load Current Harmonics

The sound levels in the factory are measured using 60 Hz voltage and current sources. Presence of harmonics in load current can influence the sound level by adding to the sound power level of the winding and/or of the core by inducing harmonic voltages in the upstream system connected to the transformer. No simple methodology is available in the industry for quantifying the effects of load current harmonics on sound levels. More work needs to be done in this area.

The following is a simple suggested approach toward predicting the effects of load current harmonics on the core and winding sound levels:

By definition, dB sound power level

$$L\,\text{dB} = 10 \log\left\{\frac{W}{Wo}\right\}\text{dB} \tag{11.29}$$

where

W is the sound power in W
Wo is the reference power (equals to 10^{12} W which is the threshold of audibility)

Therefore, the load sound power in W is

$$Lww = 10^{(0.1*Lw)} * Wo \tag{11.30}$$

and the no-load sound power in W is

$$Lcw = 10^{(0.1*Lc)} * Wo \tag{11.31}$$

where Lw and Lc are winding and core power levels, respectively.

The winding vibrations are produced by the electromagnetic forces acting on the winding conductors exposed to stray magnetic fields produced by load currents. From the basic principles, we can see the following proportionalities:

$$F \sim B \cdot I \sim I^2 \tag{11.32}$$

where

F is the vibrational force in N
B is the stray flux density in T
I is the winding current in A

It is also recognized that the vibrational velocity and amplitude are proportional to the vibrational force. As the sound power is proportional to the square of the vibrational velocity, it can be derived that sound power is proportional to the fourth power of winding current and second power of frequency as follows:

$$W \sim \nu^2 \sim (w \cdot x)^2 \sim F^2 \sim I^4 \sim f^2 \tag{11.33}$$

where

ν is the vibrational velocity
x is the vibrational amplitude
$\omega = 2\pi f$ is the angular acoustical frequency

Knowing the winding sound power in watts at 60 Hz operation, the previous proportionalities can be used to calculate the sound power of various harmonics of other frequencies and current levels as follows:

$$Lww(f_N) = Lww(60) * \left\{ \frac{I_1}{I_N} \right\}^4 * N^2 \tag{11.34}$$

where
$Lww(60)$ is the calculated winding sound power in watts at 60 Hz load
I_1 is the fundamental (60 Hz) current magnitude
I_N is the magnitude of the Nth harmonic current

The total winding sound power "*Lww*" in watts, including the harmonic load, can be computed as follows:

$$Lww = Lww(60) + \sum_{N=13}^{N=3} Lww(f_N) \tag{11.35}$$

And the winding sound power level in dB

$$Lw\,\text{dB} = 10 \log \left\{ \frac{Lww}{10^{-12}} \right\} \tag{11.36}$$

In order to estimate the change in core noise, the effects of load current harmonics on the applied voltage profile and hence on the core flux have to be explored.

Depending upon the impedance of the upstream system, the harmonics in the load current can induce voltage harmonics that can distort the profile of the voltage applied to the transformer. This can significantly influence the peak flux density and hence the core sound level. It is well recognized that the induced voltage

$$E = K * N * \frac{d\Phi}{dt} \tag{11.37}$$

where
N is the number of turns in a winding
Φ is the flux in the core
K is constant

Therefore,

$$\Phi = \frac{1}{K * N} \int E dt \tag{11.38}$$

This means that integrating the applied voltage profile could provide a much better estimate of the peak flux density and hence the core sound level. It is therefore recommended that a representative voltage profile should be considered for obtaining a better estimate of the increase in transformer sound level due to harmonic loads.

11.6.4 Acoustical Resonance

In the case of indoor installations, for example, a room with walls of low sound absorption coefficient, the sound from the transformer will reflect back and forth between the walls resulting in a buildup of the sound pressure level in the room. Assuming a reverberant field, the number of decibels by which the sound pressure level measured around the transformer will increase may be approximated according to the following equation:

$$\text{dB buildup} = 10\log\left[1+\frac{4(1-\alpha)A_T}{\alpha A_U}\right] \tag{11.39}$$

where
- A_T is the surface area of the transformer
- A_U is the area of the reflecting surface
- α is the average absorption coefficient of the surfaces

In a room with concrete walls (with an absorption coefficient of 0.05) and with a sound reflecting surface area of four times that of the transformer ($A_U/A_T=4$), the increase in sound pressure level at the transformer will be 13 dB. However, covering the reflecting surfaces of this room with sound-absorbing material of absorption coefficient 0.3 will reduce this buildup to 5.2 dB.

Similar increases in sound levels can take place in installations with walls that partially surround the transformer; in these cases, special calculation techniques will be needed to quantify the sound level increase.

11.7 Sound Level Mitigation

The best and most effective sound mitigation results from good design methodology. However, conditions in the field can necessitate measures that can minimize the discomfort experienced by the neighboring communities due to the sound emitted by transformers.

11.7.1 Factory-Installed Sound Abatement

The following design features are most effective in lowering sound levels in transformers:

1. Adopt lower induction levels—sound level as induction to the fourth power.
2. Use lower sound level core technology—minimize saturation at core joints by adopting modern step lap joint designs and minimize joint gaps.
3. Minimize mechanical stresses in core steel—use flat lamination and adopt effective core clamping design.
4. Place core material in tension through innovative core assembly procedures.
5. Avoid resonant frequencies in tank panels and core frames.

In addition to the features mentioned earlier, the following design options may also be used for further reduction of sound levels in transformers:

1. *Double-walled tanks*: Surrounding the main transformer tank with another tank separated by air can reduce the sound level by 5–8 dB. This method is expensive and harder to maintain and increases the shipping weight of the transformer.
2. *Sound housings*: These are paneled enclosures that may be bit around the transformer with a layer of sound-absorbing material in between. A sound level reduction of 8–12 dB is possible through this arrangement. This method of sound reduction greatly reduces the cooling capabilities of the transformer tank which has to be compensated with additional cooling equipment.

11.7.2 Field-Installed Sound Abatement

Barriers are the most commonly used method for reducing the sound radiated by transformers in certain direction. The noise reduction is achieved by blocking the line of sight between the transformer and the receiver, as shown in Figure 11.3a and b showing the plan and side view of the transformer. Only the sound that is diffracted around the edges of the barrier can reach the receiver in the shadow region.

The acoustic insertion loss IL due to a barrier is defined as

$$IL = 10 \log\left(\frac{I}{I_B}\right) \tag{11.40}$$

where I and I_B are the sound intensity levels at the receiver location without and with the barrier present.

11.7.2.1 One-Sided Barrier

The acoustic insertion loss for the barrier in the shadow region shown in Figure 11.3a can be 5–20 dB depending upon the design of the barrier.

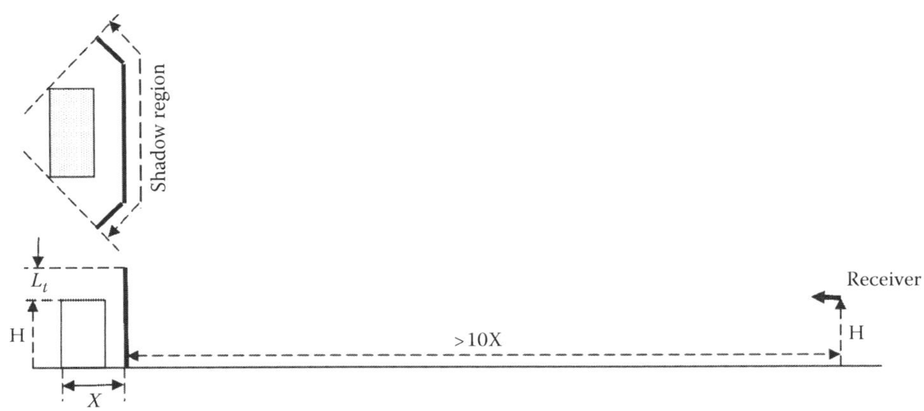

FIGURE 11.3
Noise reduction by blocking the line of sight between the transformer and the receiver.

For a barrier to be effective, it must have an effective weight per unit surface area in excess of 6 lb/ft^2 and must be free of gaps or other sound leaks. Doubling the weight of per unit area of the barrier can increase the insertion loss by 6 dB, and halving the weight can reduce the insertion loss by 6 dB.

The insertion loss of the barrier can be predicted and is a function of the sound frequency, barrier extension L_t above the transformer tank, and the distance at which the barrier is placed from the back of the transformer. It is also assumed that the receiver is located at a distance 10 times the distance X, as shown in Figure 11.3b.

11.7.2.2 Four-Sided Barriers

This type of barriers provides 360° coverage around the transformer. They normally designed to extend 3–6 ft above the cover of the transformer. Four-sided open top barriers without sound-absorbing lining can merely act as short stack and merely direct sound energy upward instead of radiating it horizontally. Such barriers may not be effective in providing insertion loss at distance beyond 500 ft.

Four-sided barriers must never be designed without sound-absorbing lining.

11.7.2.3 Four-Sided Enclosures

The insertion loss of this type of barriers, as shown in Figure 11.4, is more predictable through its design configuration. The insertion loss with this configuration can be in excess of 13 dB.

11.7.2.4 Walk-In Enclosures

These enclosures are designed to provide a continuous barrier around the transformer at distance of 5 ft or more to allow room for access to the interior. They may be provided with sound-absorbing internal lining depending upon the interior requirement. They could be made of steel plate or concrete material. Internal sound-absorbing lining influences the sound level inside the enclosure. In addition the sound dissipating qualities determine the insertion loss of the barrier:

$$IL = TL + 10\log\alpha - C \tag{11.41}$$

where

TL is the transmission loss of the barrier material
α is the absorption coefficient of the interior wall surface of the enclosure
C is a constant to account for sound energy leakage—typically 3 dB

FIGURE 11.4
Four-sided enclosure.

Appendix 11.A

TABLE 11.A.1

Dry Type (Progression Analysis—TR-1, Table 0-4)

kVA		Self-Cooled, Ventilated			Self-Cooled, Sealed		
			Delta dB			Delta dB	
From	To	dB	Pres.	Calc.	dB	Pres.	Calc.
0	50	50			50		
51	150	55	5	5	55	5	5
151	300	58	3	3	57	2	3
301	500	60	2	2	59	2	2
501	700	62	2	2	61	2	2
701	1000	64	2	2	63	2	2
1001	1500	65	1	2	64	1	2
1501	2000	66	1	1	65	1	1
2001	3000	68	2	2	66	1	2
3001	4000	70	2	1	68	2	1
4001	5000	71	1	1	69	1	1
5001	6000	72	1	1	70	1	1
6001	7500	73	1	1	71	1	1

TABLE 11.A.2

Dry Type (Proposed and Present Sound Levels—TR-1, Table 0-4)

kVA		Self-Cooled, Ventilated Sound Level, dB		Self-Cooled, Sealed Sound Level, dB	
From	To	Present	Proposed	Present	Proposed
0	50	50	51	50	51
51	150	55	56	55	56
151	300	58	59	57	59
301	500	60	61	59	61
501	700	62	63	61	63
701	1,000	64	64	63	64
1,001	1,500	65	66	64	66
1,501	2,000	66	67	65	67
2,001	3,000	68	67	65	67
3,001	4,000	70	70	68	70
4,001	5,000	71	71	69	71
5,001	6,000	72	72	70	72
6,001	7,500	73	73	71	73
7,501	10,000		74		74

Appendix 11.B

TABLE 11.B.1

Dry-Type Lighting (Progression Analysis—ST-20, Table 3-9)

kVA		Self-Cooled, Ventilated Up to 1.2 kV			Self-Cooled, Ventilated Above 1.2 kV			Self-Cooled, Sealed		
			Delta dB			Delta dB			Delta dB	
From	To	dB	Pres.	Calc.	dB	Pres.	Calc.	dB	Pres.	Calc.
0	9	40								
10	50	45	5	8	50			50		
51	150	50	5	5	55	5	5	55	5	5
151	300	55	5	3	58	3	3	57	2	3
301	500	60	5	2	60	2	2	59	2	2
501	700	62	2	2	62	2	2	61	2	2
701	1000	64	2	2	64	2	2	63	2	2
1001	1500	65	1	2	65	1	2	64	1	2
1501	2000	66	1	1	66	1	1	65	1	1
2001	3000	68	2	2	68	1	1	65	1	1
3001	4000				70	2	1	68	2	1
4001	5000				71	1	1	69	1	1
5001	6000				72	1	1	70	1	1
6001	7500				73	1	1	71	1	1

TABLE 11.B.2

Dry-Type Lighting (Proposed and Present Sound levels—ST-20, Table 3-9)

kVA		Self-Cooled, Ventilated Up to 1.2 kV		Self-Cooled, Ventilated Above 1.2 kV		Self-Cooled, Sealed	
		Sound Level, dB		Sound Level, dB		Sound Level, dB	
From	To	Present	Proposed	Present	Proposed	Present	Proposed
0	9	40	42				
10	50	45	50	50	51	50	51
51	150	50	55	55	56	55	56
151	300	55	58	58	59	57	59
301	500	60	60	60	61	59	61
501	700	62	62	62	63	61	63
701	1000	64	63	64	64	63	64
1001	1500	65	65	65	66	64	66
1501	2000	66	66	66	67	65	67
2001	3000	68	68	68	69	66	69
3001	4000			70	70	68	70
4001	5000			71	71	69	71
5001	6000			72	72	70	72
6001	7500			73	73	71	73

Appendix 11.C

TABLE 11.C.1

Liquid-Filled Distribution Transformers (Progression Analysis—TR-1, Table 0-3)

kVA		Self-Cooled, Ventilated		
			Delta dB	
From	To	dB	Pres.	Prop.
0	50	48		
51	100	51	3	3
101	300	55	4	5
301	500	56	1	2
	750	57	1	2
	1000	58	1	2
	1500	60	2	2
	2000	61	1	1
	2500	62	1	1

TABLE 11.C.2

Liquid-Filled Distribution Transformers (Proposed and Present Sound Levels—TR-1, Table 0-3)

kVA		Self-Cooled, Ventilated	
		Sound Level, dB	
From	To	Present	Proposed
0	50	48	45
51	100	51	48
101	300	55	53
301	500	56	55
	750	57	57
	1000	58	58
	1500	60	60
	2000	61	61
	2500	62	62

Appendix 11.D

TABLE 11.D.1

Audible Sound Level for Oil-Immersed Power Transformers (At OA, OW, and FOW Ratings)

	Proposed and Present Sound Levels											
	350 kV and below BIL		450 kV, 550 kV, 650 kV BIL		750 kV, 825 kV BIL		900 kV, 1050 kV BIL		1175 kV BIL		1300 kV and above BIL	
kVA	dB Pres.	dB Prop.	dB Pres.	dB Prop.	dB Pres.	dB Prop.	dB Pres.	dB Prop.	dB Pres.	dB Prop.	dB Pres.	dB Prop.
700	57	57	59	59		60		61		62		63
1,000	58	58	60	60		62		63		64		65
1,250	59	59	61	61		62		64		65		65
1,500	60	60	62	62		63		64		65		66
2,000	61	61	63	63		64		66		67		68
2,500	62	62	64	64		65		66		67		68
3,000	63	63	65	65	66	66		67		68		69
4,000	64	64	66	66	67	67		68		69		70
5,000	65	65	67	67	68	68		69		70		71
6,000	66	66	68	68	69	69		70		71		72
7,500	67	67	69	69	70	70		71		72		73
10,000	68	68	70	70	71	71		72		73		74
12,500	69	69	71	71	72	71	73	73	74	74	75	75
15,000	70	69	72	71	73	73	74	74	75	75	76	76
20,000	71	71	73	73	74	73	75	75	76	76	77	77
25,000	72	72	74	74	75	74	76	76	77	77	78	78
30,000	73	72	75	74	76	75	77	77	78	78	79	79
40,000	74	73	76	75	77	76	78	78	79	79	80	80
50,000	75	74	77	76	78	77	79	79	80	80	81	81
60,000	76	75	78	77	79	78	80	79	81	80	82	81
80,000	77	76	79	78	80	79	81	81	82	82	84	83
100,000	78	77	80	79	81	79	82	81	83	82	84	83

Note: Increase sound level at self-cooled rating as follows for computing acceptable sound levels at forced cooled ratings:

Rating	Sound Level
FA and FOA first stage	(Sound level at OA + 2 dB) but not less than 67 dB
FA and FOA second stage	(Sound level at OA + 3 dB)
FOA with coolers	(Sound level at OA + 3 dB)

References

1. IEC 60076-10, *Determination of Sound Levels*, 2005.
2. K. Karsai, D. Kerenyi, and L. Kiss, *Studies in Electrical and Electronic Engineering 25, Large Power Transformers*.
3. Westinghouse Electric Corp. and Bolt Beranek and Newman Inc., Report on power transformer abatement-sponsored by empire state electric energy research corporation, New York.
4. Bruel & Kjaer, Sound intensity theory, Technical Review Publication No.3-1982.
5. Bruel & Kjaer, Sound intensity theory, Technical Review Publication No.4-1982.
6. Bruel & Kjaer, Sound intensity, Publication No. BR 0476-14.
7. T.R. Specht, Technical Memo—Transformer Division Westinghouse Electric Corp., Noise level of indoor transformers, October 1955.
8. E. Reiplinger, Study of noise emitted by power transformers based on todays viewpoint, *Cigre International Conference on Large High Voltage Electric Systems*, Paris, France, 1988 Session, August 28 to September 3.
9. CIGRE Working Group 12.12, Transformer noise: Determination of sound power level using the sound intensity measurement method. *Electra*, 144, October 1992.

12

Power Transformers' Fault Diagnostics by Park's Vector Approach

A.J. Marques Cardoso and Luís M.R. Oliveira

CONTENTS

12.1 Introduction

The electricity market was, for a long time, dominated by national or regional monopolies. During the last decades, this picture has changed drastically, particularly in Europe and in the United States. The trend toward free competition and privatization with the related demands on return on investments results in a cost consciousness among utilities. In this context, monitoring and on-site diagnostics are seen as a possible way of optimizing existing assets. The main driving forces are to reduce maintenance costs, to prevent forced outages with the related consequential costs, and to work existing equipment harder and longer [1]. Transformers constitute the largest single component of the transmission and distribution equipment market. The global market for transformers was valued at US$14.07

billion in 2002 [2], and it is projected to exceed US$36.7 billion by 2015 [3]. Therefore it is quite obvious the need for the development of on-line diagnostic techniques that would aid in transformer maintenance. This, in turn, implies an exhaustive knowledge about the most likely failures that can occur and also about its main underlying causes.

Two major transformer reliability surveys, carried out under the auspices of CIGRE [4] and IEEE [5] in order to assemble objective data on the performance of transformers in service, provide useful information.

12.2 International Survey on Failures in Large Power Transformers in Service

The main objectives of this international survey, conducted in 1978 by the Working Group 05 (Transformer Reliability) of Study Committee 12 (Transformers) of CIGRE (International Conference on Large High Voltage Electric Systems), were to pinpoint the main causes of transformer failures and to evaluate transformer outage times [4].

The survey involved transformer units designed for networks with a highest system voltage of not less than 72 kV, without any limitations on rated power, not older than 20 years, and installed on generation, transmission, and distribution systems (including HVDC systems).

The analysis took in more than 1,000 failures,* relating to a total population of more than 47,000 unit-years, which corresponds to a general failure rate figure, irrespective of the voltage classes and function of the units, of the order of 2%. Nevertheless, if voltage classes are taken into account, it seems that the failure rate increases with voltage.

The data available were also analyzed as a function of the failure first component involved and of the presumed cause. The statistically more substantial results are those concerning substation transformers with on-load tap changers (OLTCs).†

Regarding the first component involved (Figure 12.1), it may be noted that about 33% of failures are due to the windings. The failures were also subdivided as a function of downtime (Figure 12.2) into three classes (not more than 1 day, from 1 to 30 days, and more than 30 days) in order, however roughly, to be able to correlate downtime with the components involved in the failures. From Figure 12.2, it can be seen that the longer downtimes are associated with winding faults.

Regarding the presumed causes of failures, and despite uncertainty about the reliability of the collected data, it would seem to be possible to state that failures due to design, manufacture, and materials, which often involve long outage times, represent a high percentage of failures, as shown in Figure 12.3.

12.3 Transformer Reliability Survey: Industrial Plants and Commercial Buildings

The Power Systems Reliability Subcommittee of the IEEE Industry Applications Society has been conducting surveys of the reliability of electrical equipment, transformers

* The failures with forced outage were 70% or more of the total failures.

† In order to be able to compare these results with the ones obtained in the IEEE Transformer Reliability Survey [5], the faults associated with the OLTC are not considered in this work.

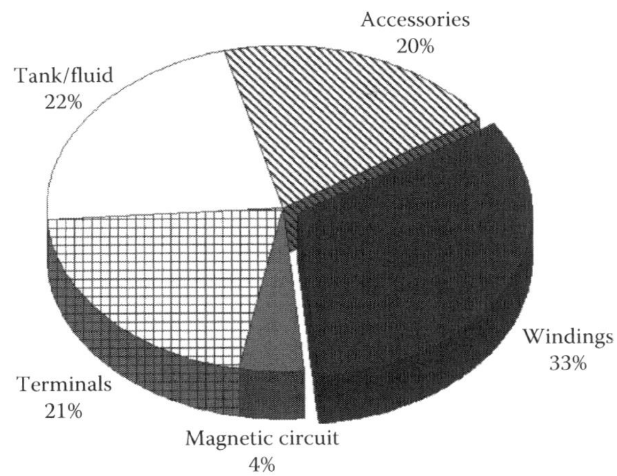

FIGURE 12.1
Typical failure distribution for substation transformers.* (From Cardoso, A.J.M. and Oliveira, L.M.R., *Int. J. COMADEM.*, 2, 5, 1999.)

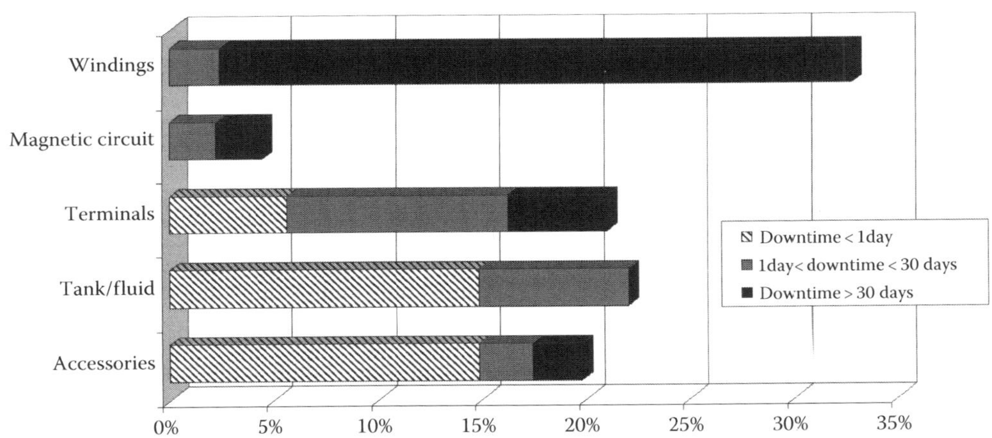

FIGURE 12.2
Typical downtime as a function of failure distribution for substation transformers.* (From Cardoso, A.J.M. and Oliveira, L.M.R., *Int. J. COMADEM.*, 2, 5, 1999.)

included, in industrial and commercial power systems [5]. The 1979 survey results, which are presented in the IEEE Standard 493-1990, reveal a failure rate of 0.62% for the liquid-filled power transformers and a failure rate of 1.9% for rectifier transformers.†

The 1979 survey limited the choices of failure type to "windings" and "others" as shown in Table 12.1 for power and rectifier transformers. Clearly, the most significant failure type occurred in power transformer windings.

* In order to be able to compare these results with the ones obtained in the IEEE Transformer Reliability Survey [5], the faults associated with the OLTC are not considered in this work.

† Relay or tap changer faults were not considered in calculation of failure rates.

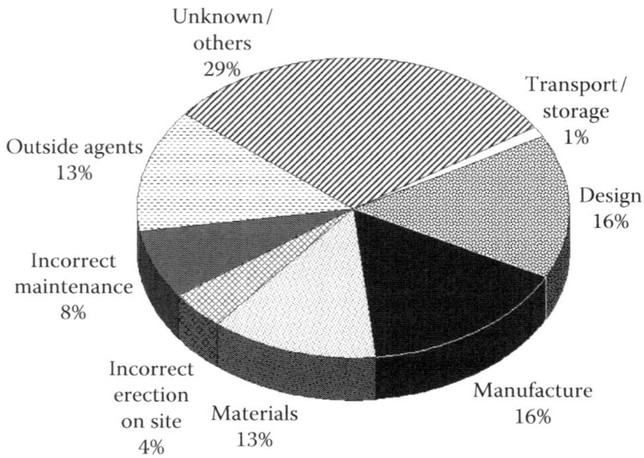

FIGURE 12.3
Presumed causes of failures for substation transformers. (From Cardoso, A.J.M. and Oliveira, L.M.R., *Int. J. COMADEM.*, 2, 5, 1999.)

TABLE 12.1
Type of Failure for Power and Rectifier Transformers (1979 Survey)

	All Power Transformers		All Rectifier Transformers	
Type of Failure	**Number of Failures**	**Percentage**	**Number of Failures**	**Percentage**
Windings	59	53	8	50
Others	53	47	8	50

Source: Cardoso, A.J.M. and Oliveira, L.M.R., *Int. J. COMADEM.*, 2, 5, 1999.

Although it is not possible to correlate downtime with the failure type, it is, however, possible to correlate downtime for all electrical equipment surveyed. From that analysis, it can be concluded that transformers' downtime per failure is, in general, the highest one.

Figure 12.4 summarizes the suspected failure responsibility for power transformer failures. The results show that manufacturer defects and inadequate maintenance are responsible for the majority of power transformer failures (i.e., 60%).

12.4 Corollary of Transformer Reliability Surveys

As a corollary, and similarly to what happens with other electrical equipment, namely, with three-phase induction motors [7], it can also be stated here that beyond the correct installation and utilization of transformers, areas where standardization can perform a very important role, a substantial reliability improvement still implies the development of reliable diagnostic techniques that would contribute not only for a more effective quality control and therefore reducing manufacturing defects but also for the implementation of adequate maintenance strategies [6].

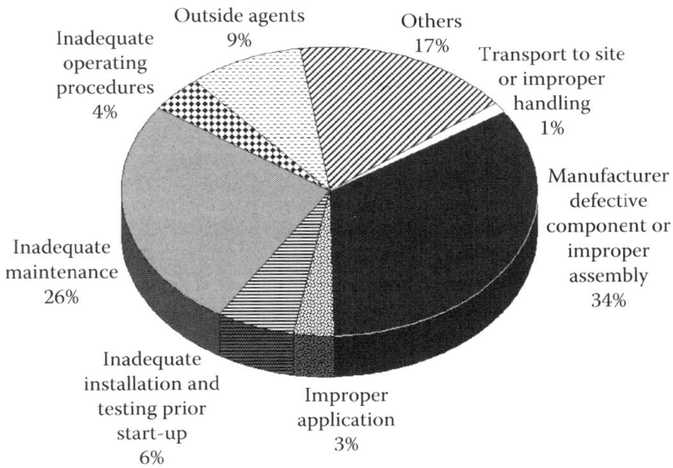

FIGURE 12.4
Suspected failure responsibility for power transformers (1979 survey). (From Cardoso, A.J.M. and Oliveira, L.M.R., *Int. J. COMADEM.*, 2, 5, 1999.)

12.5 Transformers' Diagnostic Techniques

Several approaches have been developed with the purpose of providing an accurate and complete diagnosis of power transformers.

12.5.1 Dissolved Gas Analysis

A transformer is subject to electrical and thermal stresses. These two stresses can break down the insulating materials and release gaseous decomposition products. Overheating, corona, and arcing are three primary causes of fault-related gases. Different patterns of gases are generated due to different intensities of energy dissipated by various faults. The analysis of dissolved gases is a powerful tool to diagnose developing faults in oil-filled power transformers. By sampling and examining the insulation oil of transformers, ratios of specific dissolved gas concentrations, their generation rates, and total combustible gases are often used as the attributes of classification by diverse dissolved gas analysis approaches, such as the key gas analysis, the Doernenburg and the Rogers gas ratio methods described in the ANSI/IEEE Standard C57.104-1991. These methods would find the relationship between the gases and the fault conditions, some of which are obvious and some of which may not be apparent—hidden relationships [8–10].

Expert systems, fuzzy logic, and artificial neural network approaches have been proposed to reveal some of the hidden relationships in transformer fault diagnosis [9–12].

12.5.2 Partial Discharges

It is commonly known that partial discharges in a transformer deteriorate the insulation to the point of destruction [13].

Partial discharge pulses generate at their point of origin electromagnetic waves, acoustic waves, local heating, and chemical reactions. Theoretically, these phenomena, if detectable, would constitute possible indicators of a partial discharge defect. The established, but not always successfully applied, methods for partial discharge defect location are based on acoustic techniques, electrical techniques, or a combination of both [1,14].

12.5.3 Temperature

Several approaches have been adopted to develop hot-spot sensors suitable for insertion into transformer windings. Measurement systems exist which show promise of practical application in commercial transformers. The most promising results of development seem to be devices making use of an optical fiber transmitter connected to a crystal sensor, which convert the incoming light beam into an optical signal characteristic of the sensor temperature. These devices have demonstrated their properties in factory tests and, to some extent, in service [15].

Nevertheless, some problems concerning the widespread use of this diagnostic technique remain to be solved: Temperature sensors can only be installed in the windings when the transformer is manufactured or repaired; the optimum location of the sensor is difficult to predict; and since the temperature sensors are inserted into the insulation structure under high-voltage conditions, special precaution must be applied in order to preserve the electrical and mechanical strength of the insulation system [15].

Alternatively, infrared camera measurement is worthy of mention, which is already in use for indirect indication of transformer hot spots.

12.5.4 Vibrations

Vibration signal analysis is being pursued as a likely means of achieving a dependable transformer winding mechanical integrity diagnostic tool. The vibration sensors are magnetically mounted piezoelectric accelerometers attached to the sides and top of the transformer tanks. The signals are optically isolated for transmission to a data recorder. The occurrence of winding looseness has been investigated [16]. A similar approach has been also applied as a method for diagnostics of OLTCs [1].

Vibration measurement and analysis may, however, prove to be complicated due to the various sources which cause vibration of a transformer (i.e., primary excitation, leakage flux, mechanical interaction, and on/off load switching) and the various locations where vibration signals may be taken [16,17].

12.5.5 Leakage Flux

This is a traditional method for detecting changes in the winding geometry [1]. The windings' mechanical displacement results, mainly, in modifications of the leakage flux radial component. By using search coils, conveniently installed in the transformer, it is possible to measure such modifications [18].

However, some of the problems previously mentioned regarding temperature sensors installation still apply here.

12.5.6 Frequency Response Analysis

By measuring the transfer function of the transformer, deformations of the windings can sometimes be detected, provided that a reference fingerprint of the unit is available. Deformation or changes in geometric distances of the windings lead to changes in internal capacitances, and thereby a change in the transfer function of the transformer. In practice, an impulse is injected on one side, and the Fourier spectrum is measured of both the impulse and of the response on the other side. The transfer function is calculated by dividing the two spectra [1,19].

As an alternative to the formerly used time-domain testing (low-voltage impulse method) [20], the frequency response analysis is one of the more frequently used techniques for diagnosing deformations of the transformer windings. However, this technique presents as major disadvantages the need to set the transformer out of service, and it involves great uncertainties due to the fact that the result is affected by a large number of factors [1,21].

12.5.7 Excitation Current

The single-phase excitation-current test can be used to detect undesirable conditions in single-phase or three-phase transformers. Normally the test results are analyzed by comparing currents between all three phases in a given transformer and between the similar single-phase units. Certain problems, however, can be detected on the basis of current change when the measurements are performed on different load tap changer positions. Accordingly, understanding how the load tap changer affects the current magnitude of individual phases is essential for developing proper analysis [22,23]. Moreover, this method also requires the transformer to be disconnected.

12.6 Corollary of Transformer Diagnostic Techniques

As a corollary, it can be stated here that the development of diagnostic techniques is of prime importance, which can be applied without taking the transformer out of service and which can also provide a fault severity criteria, in particular, for determining winding defects, since this is the most significant type of failure occurring in power transformers and this is also the one with an associated longer downtime [6].

12.7 Transformer Behavior under Incipient Turn-to-Turn Winding Fault Occurrence

The most difficult transformer winding fault for which to provide protection is the fault that initially involves only one turn [24]. Initially, the insulation breakdown leads to internal arcing, which results into a low-current, high-impedance fault [25]. Usually, this incipient interturn insulation failure does not draw sufficient current from the line to operate an ordinary overload circuit breaker or even more sensitive balanced protective gear [26]. This turn-to-turn fault will then progress, with random propagation speed, involving additional turns and layers, leading to a high-current, low-impedance fault [27]. The transformer will, in fact, only be disconnected from the line automatically when the fault has extended to such degree as to embrace a considerable portion of the affected winding [26].

As stated earlier, the initial turn-to-turn insulation defect leads to an arcing fault, which is a dangerous form of a short circuit that may have a low current magnitude. Arcing faults usually cause a damage that is limited to the fault area, and pose a great danger to the transformer [28]. When the voltage potential between the affected turns breaks down the insulation, a spark discharge takes place. The arc ignition and extinction depend on this threshold voltage [29,30], resulting on a fault of intermittent nature. The behavior of the transformer winding currents under the occurrence of this type of fault should be clearly understood in order to allow the detection of the failure in its incipient stage.

If the fault occurs in the primary winding, the short-circuited turns act as an autotransformer load on the winding, as shown in Figure 12.5a, where R_{sh} represents the fault impedance. However, if the fault takes place on the secondary winding, the short-circuited turns act as an ordinary double winding load, Figure 12.5b [26].

Several tests were made on custom-built transformers in order to characterize the behavior of the transformer under incipient winding fault occurrence. The transformers used have additional tapings connected to the coils, allowing for the introduction of shorted turns at several locations in the winding. In the results presented in this work, a *YNyn*0 winding connection and a balanced resistive load were used in the laboratory tests (Figure 12.6).

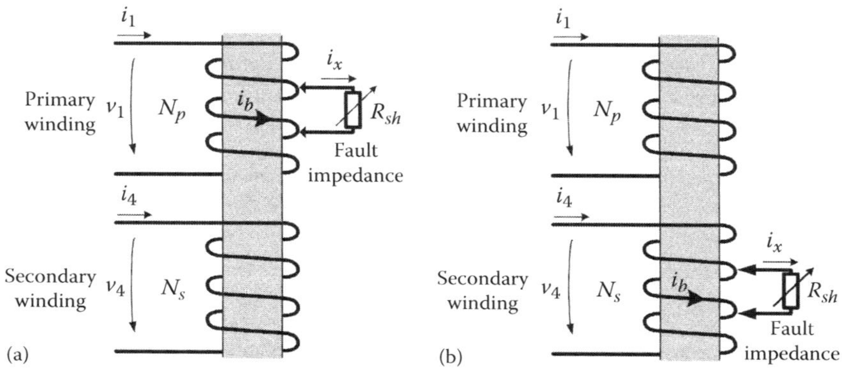

FIGURE 12.5
Equivalent circuits for a fault occurring in the (a) primary winding and (b) secondary winding (phase *R*). (From Oliveira, L.M.R., Cardoso, A.J.M., and Cruz, S.M.A., *Electr. Power. Syst. Res.* 81, 1206, 2011.)

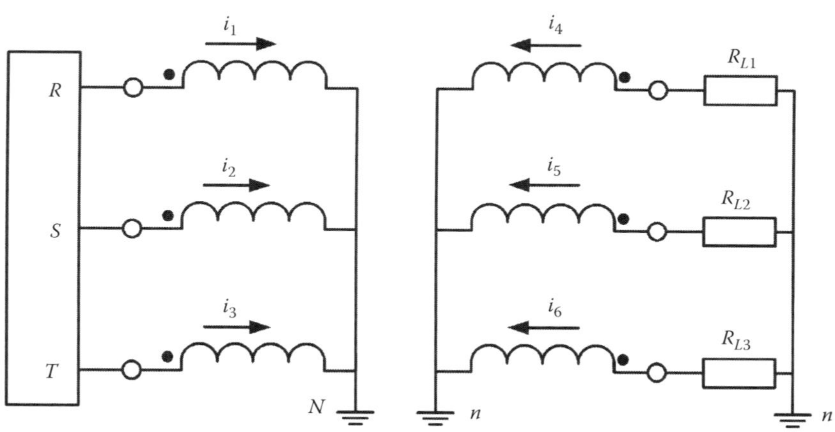

FIGURE 12.6
Transformer *YNyn*0 winding connection. (Reprinted with permission from Oliveira, L.M.R. and Cardoso, A.J.M., A permeance-based transformer model and its application to winding interturn arcing fault studies, *IEEE Trans. Power. Deliv.*, 25, 1589–1598, Copyright 2010, IEEE.)

12.7.1 Permanent Faults

The permanent faults are introduced in the transformer by connecting a shorting resistor at the terminals of the affected turns. The value of this resistor was chosen so as to create an effect strong enough to be easily visualized but simultaneously big enough to limit the short-circuit current and thus protecting the test transformer from complete failure when the short is introduced.

Figure 12.7a presents the waveforms of the primary-side line currents for the case of four permanent shorted turns in the phase *R* of the transformer primary winding (notation as per Figures 12.5 and 12.6). The occurrence of primary-side interturn short circuits leads to an increment in the magnitude of the current in the affected winding, as compared to a healthy condition, which results in an unbalanced system of primary currents. For this reason, the magnitude of the primary neutral current ($i_{n1} = i_1 + i_2 + i_3$) is also affected. In the presence of the primary winding interturn short circuits, the secondary-side currents do not present any relevant change as compared to the transformer's

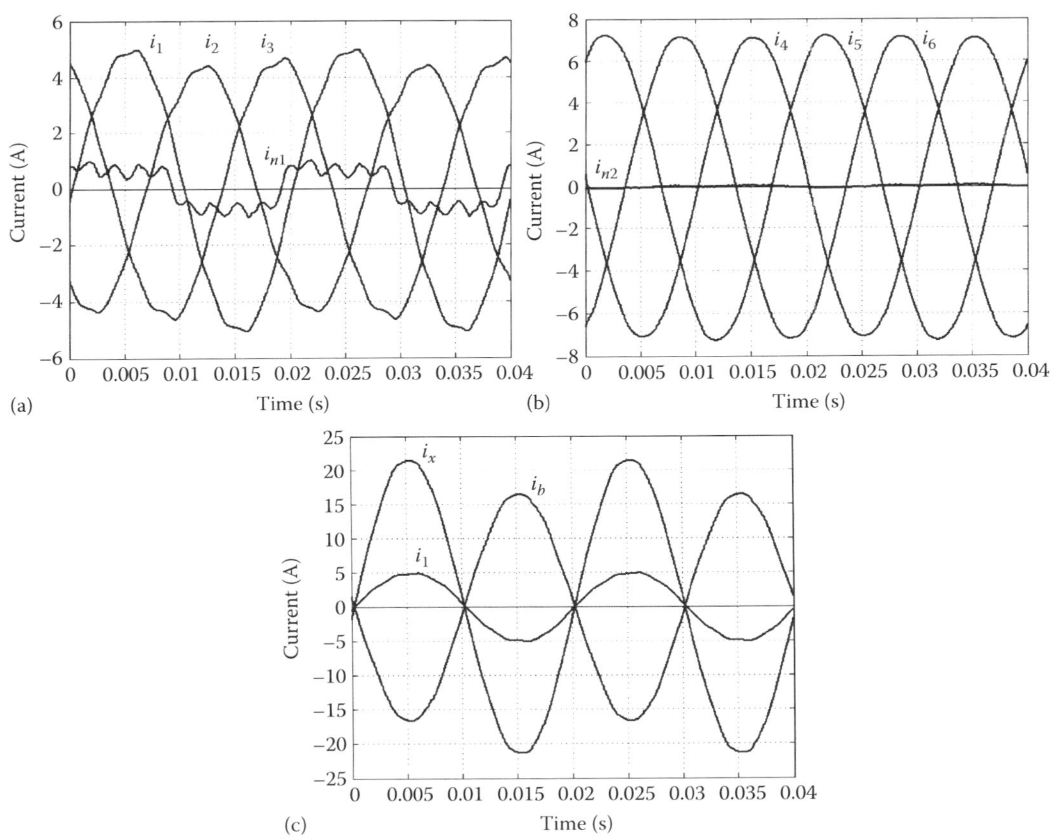

FIGURE 12.7
Experimental test results for the case of four permanent shorted turns in the primary winding (phase *R*): (a) primary-side current waveforms, (b) secondary-side current waveforms, and (c) current waveforms in the affected winding. (Reprinted with permission from Oliveira, L.M.R. and Cardoso, A.J.M., A permeance-based transformer model and its application to winding interturn arcing fault studies, *IEEE Trans. Power. Deliv.*, 25, 1589–1598, Copyright 2010, IEEE.)

healthy operation, remaining an approximately balanced three-phase system, as shown in Figure 12.7b. The current waveform in the shorted turns, i_b, and the current waveform in the fault impedance, i_x, are shown in Figure 12.7c. The current i_b is approximately in phase opposition with i_1 due to the autotransformer action of the shorted turns. The current in the short-circuit auxiliary resistor, i_x, has a higher magnitude than i_b since $i_x = i_1 - i_b$.

The input current in the affected winding can be divided into three terms:

$$i_1 = -i'_4 + i_{exc1} + i'_x \tag{12.1}$$

where

i'_4 is the secondary winding current referred to the primary side
i_{exc1} is the excitation current
i'_x is the current in fault impedance, also referred to the primary side

Obviously, the first two terms are related with the healthy operation of the transformer, being the third a result of the shorted turns. The faulty current is reflected to the primary side through the turns ratio between the number of shorted turns, N_b, and the total number of turns of the primary-side winding, N_p:

$$i'_x = i_x \times \frac{N_b}{N_p} \tag{12.2}$$

As a result, the increase in the magnitude of the primary-side winding current due to an incipient insulation defect, with only a few turns involved, is small, even if the faulty current is large, and it is very likely that the fault remains undetected by the protection devices, until it progresses to a catastrophic failure. The severity of the fault depends not only on the number of shorted turns but also on the value of the faulty current, which is limited by the fault impedance [31,32].

In the case of secondary-side winding faults, the additional load produced by the shorted turns also results in an increment in the magnitude of the correspondent primary-side winding current, as compared to a healthy condition, as shown in Figure 12.8a. Again, the line currents of the secondary side do not suffer any significant alteration with the introduction of the defect, Figure 12.8b. However, with this type of fault, the current in the shorted turns is in phase with the line current of the affected winding, as shown in Figure 12.8c, and it takes larger values than the current in short-circuit auxiliary resistor.

12.7.2 Intermittent Faults

The arcing faults are introduced in the transformer by connecting a custom built power electronics board, at the terminals of the affected turns. Basically, the power electronics circuit consists of two modules, each one with an IGBT in series with a power diode, connected in antiparallel, as shown in Figure 12.9a. The arc ignition and extinction depend on the threshold voltage, which is regulated by the firing angle and the conduction time (pulse width) of the IGBTs, Figure 12.9b. Again, an auxiliary shorting resistor was used to maintain the faulty current within safe values.

Figure 12.10a presents the arc current waveform, i_x, for the case of four intermittent shorted turns in the phase R of the transformer primary winding, maintaining the same

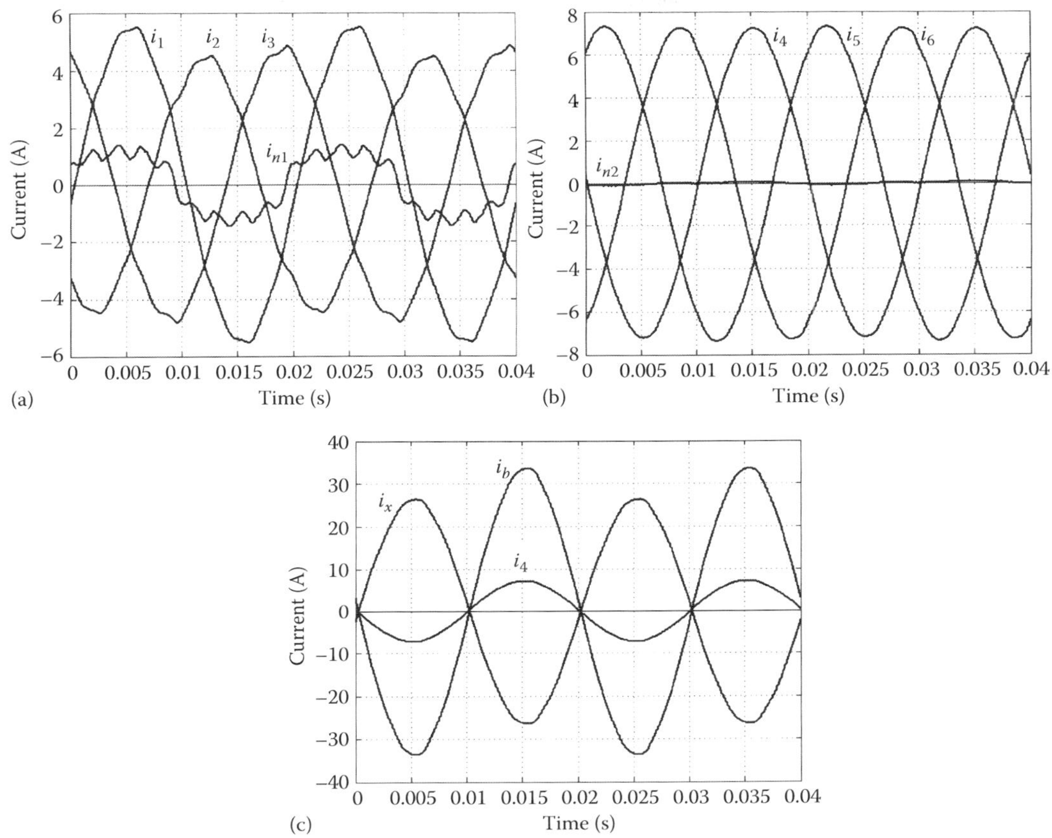

FIGURE 12.8
Experimental test results for the case of four permanent shorted turns in the secondary winding (phase *R*): (a) primary-side current waveforms, (b) secondary-side current waveforms, and (c) current waveforms in the affected winding. (Reprinted with permission from Oliveira, L.M.R. and Cardoso, A.J.M., A permeance-based transformer model and its application to winding interturn arcing fault studies, *IEEE Trans. Power. Deliv.*, 25, 1589–1598, Copyright 2010, IEEE.)

load conditions and transformer winding connections mentioned earlier. In order to clearly visualize the transient phenomena, a pulse width of approximately 800 μs was chosen.

As in the case of permanent faults, the arc current is reflected to the primary side through the turns ratio N_b/N_p, resulting in a pulse of relatively small magnitude in i_1, as shown in Figure 12.10b. Once again, the secondary-side currents do not present any significant change with the introduction of the defect.

The current waveform in the shorted turns is shown in Figure 12.10c. When i_x is zero, the current in the defected turns is equal to i_1. During the arc discharge, i_b presents a deep notch since $i_b = i_1 - i_x$.

For the same aforementioned conditions, but now for the case of four intermittent shorted turns in the phase *R* of the transformer secondary winding, the current waveforms in the shorting resistor, in the primary and secondary-side windings of the affected phase, and in the shorted turns are presented in Figure 12.11a through c, respectively. In comparison with the previous test, the only significant difference is the current waveform in the shorted turns, which is now given by $i_b = i_4 - i_x$.

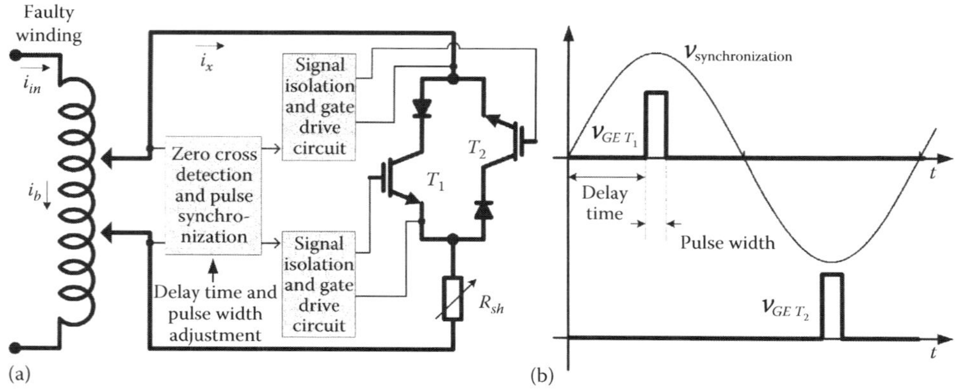

FIGURE 12.9
(a) Intermittent fault equivalent circuit with a power electronics board and (b) gate signals of the IGBTs. (Reprinted with permission from Oliveira, L.M.R. and Cardoso, A.J.M., A permeance-based transformer model and its application to winding interturn arcing fault studies, *IEEE Trans. Power. Deliv.*, 25, 1589–1598, Copyright 2010, IEEE.)

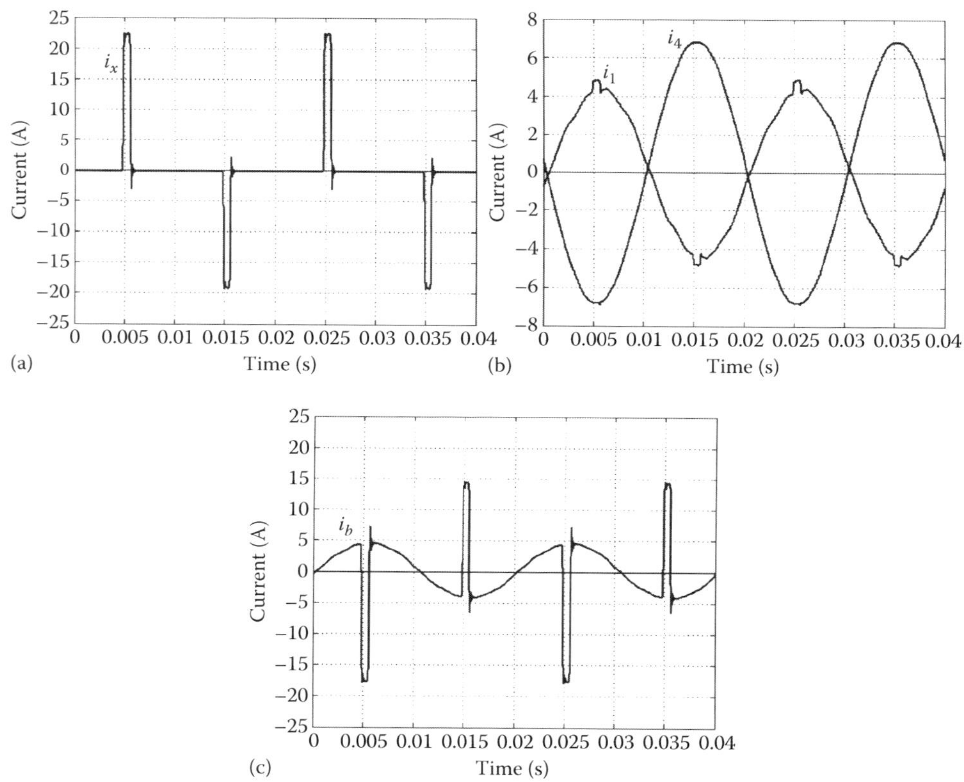

FIGURE 12.10
Experimental test results for the case of four intermittent shorted turns in the primary winding (phase *R*): (a) arc current waveform, (b) primary- and secondary-side current waveforms of the affected phase, and (c) current waveform in the shorted turns. (Reprinted with permission from Oliveira, L.M.R. and Cardoso, A.J.M., A permeance-based transformer model and its application to winding interturn arcing fault studies, *IEEE Trans. Power. Deliv.*, 25, 1589–1598, Copyright 2010, IEEE.)

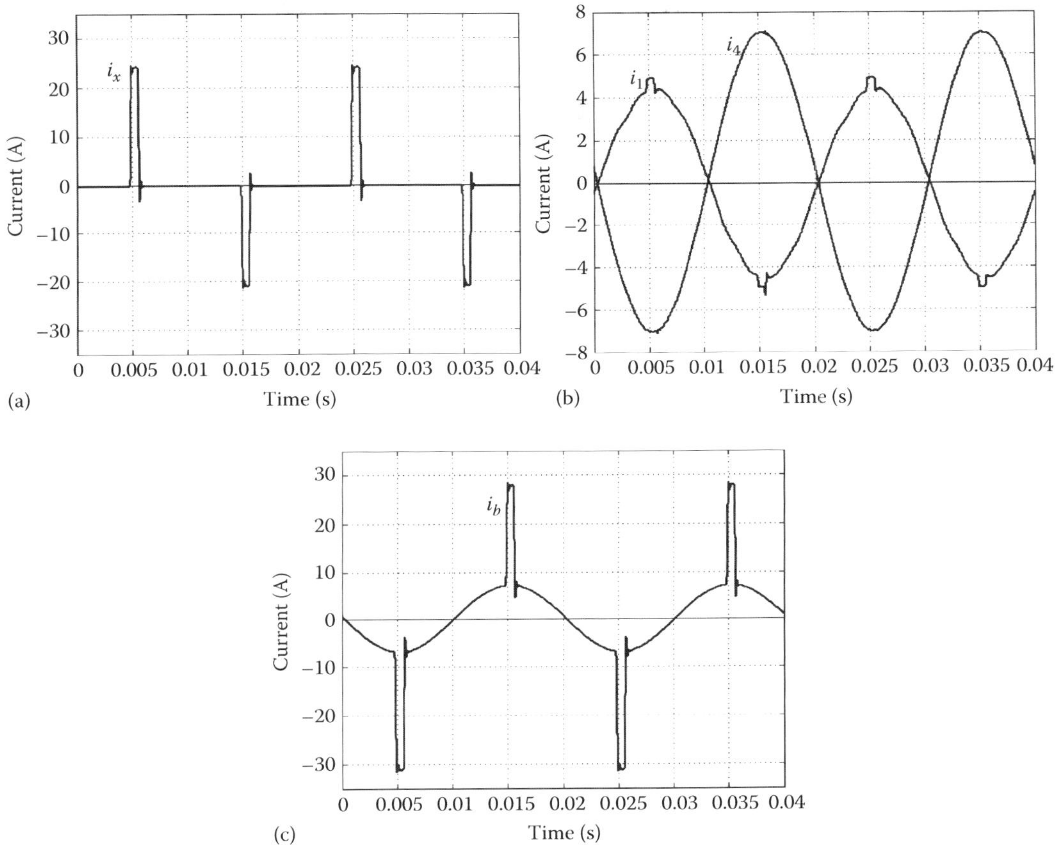

FIGURE 12.11
Experimental test results for the case of four intermittent shorted turns in the secondary winding (phase R): (a) arc current waveform, (b) primary- and secondary-side current waveforms of the affected phase, and (c) current waveform in the shorted turns. (Reprinted with permission from Oliveira, L.M.R. and Cardoso, A.J.M., A permeance-based transformer model and its application to winding interturn arcing fault studies, *IEEE Trans. Power. Deliv.*, 25, 1589–1598, Copyright 2010, IEEE.)

12.8 Park's Vector Approach

12.8.1 Supply Current Park's Vector Approach

As a function of mains phase variables (i_1, i_2, i_3), the transformer current Park's vector components (i_D, i_Q) are as follows:

$$i_D = \left(\frac{\sqrt{2}}{\sqrt{3}}\right)i_1 - \left(\frac{1}{\sqrt{6}}\right)i_2 - \left(\frac{1}{\sqrt{6}}\right)i_3 \tag{12.3}$$

$$i_Q = \left(\frac{1}{\sqrt{2}}\right)i_2 - \left(\frac{1}{\sqrt{2}}\right)i_3 \tag{12.4}$$

Under ideal conditions, the three-phase currents lead to a Park's vector with the following components:

$$i_D = \left(\frac{\sqrt{6}}{2}\right)\hat{I}_M \sin(\omega t) \quad (12.5)$$

$$i_Q = \left(\frac{\sqrt{6}}{2}\right)\hat{I}_M \sin\left(\omega t - \frac{\pi}{2}\right) \quad (12.6)$$

where
$\hat{I}_M$ is the maximum value of the supply current (A)
ω is the angular supply frequency (rad/s)
t is the time variable (s)

The corresponding representation is a circular locus centered at the origin of the coordinates. Under abnormal conditions, (12.5) and (12.6) are no longer valid, and consequently the observed picture differs from the reference pattern. The operating philosophy of the Park's vector approach is thus based on identifying unique signature patterns in the figures obtained, corresponding to the transformer current Park's vector representation.

Figure 12.12 presents the experimental primary-side phase current Park's vector patterns for several percentages of shorted turns in the primary windings and for different faulty phases. The primary-side phase current Park's vector pattern, corresponding to the healthy operation, differs slightly from the circular locus expected for ideal conditions due to, among others, the nonlinear behavior and asymmetry of the magnetic circuit.

The occurrence of primary-side interturn short circuits manifests itself in the deformation of the primary-side phase current Park's vector pattern corresponding to a healthy condition, leading to an elliptic representation, whose ellipticity increases with the severity of the fault and whose major axis orientation is associated to the faulty phase. Similar conclusions, concerning the transformer supply current Park's vector patterns, can be drawn for the occurrence of secondary interturn short circuits, under the same load conditions and winding connections [32].

The transformer secondary-side current Park's vector pattern does not provide any indication about interturn short circuits that may occur either in the primary or in the secondary side of the transformer. However, it plays a very important role for discriminating the presence of unbalanced loads.

Consider, for example, the case of an unbalanced load with $R_{L2}=R_{L3}=R_{L1}/3.5$ and, simultaneously, 5% of shorted turns in the primary winding of phase R. From the resultant primary-side current Park's vector pattern, shown in Figure 12.13a, it is difficult to detect the fault, being necessary to make use of the secondary-side current Park's vector pattern to recognize the unbalanced load condition of the transformer. Consequently, with this diagnostic technique, it is difficult to discriminate between unbalanced loads and winding faults.

12.8.2 On-Load Exciting Current Park's Vector Approach

To overcome the aforementioned difficulty, an improved diagnostic technique was implemented, which consists in the analysis of the on-load exciting current Park's vector pattern and therefore unaffected by the transformer's load conditions. The on-load exciting current waveforms are computed by adding the primary and secondary winding currents,

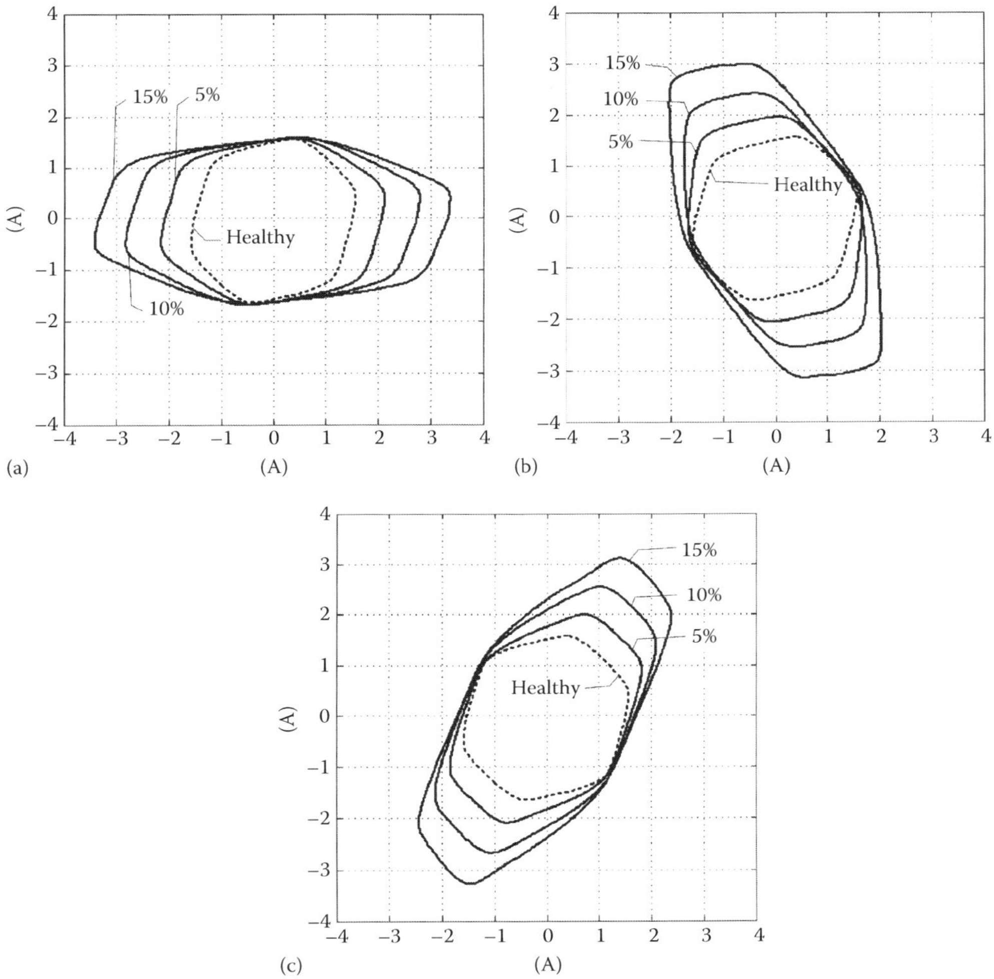

FIGURE 12.12
Experimental primary-side phase current Park's vector patterns for the case of a *YNyn*0 connection and a balanced resistive load, with several percentages of permanent shorted turns in the primary windings and for different faulty phases: (a) phase *R*, (b) phase *S*, and (c) phase *T*. (From Oliveira, L.M.R., Cardoso, A.J.M., and Cruz, S.M.A., *Electr. Power. Syst. Res.* 81, 1206, 2011.)

both referred to the primary side. For the *YNyn*0 winding connection (Figure 12.6), the on-load exciting currents are as follows:

$$i_{e1} = i_1 + i_4 \cdot \left(\frac{N_2}{N_1}\right)$$

$$i_{e2} = i_2 + i_5 \cdot \left(\frac{N_2}{N_1}\right)$$

$$i_{e3} = i_3 + i_6 \cdot \left(\frac{N_2}{N_1}\right) \tag{12.7}$$

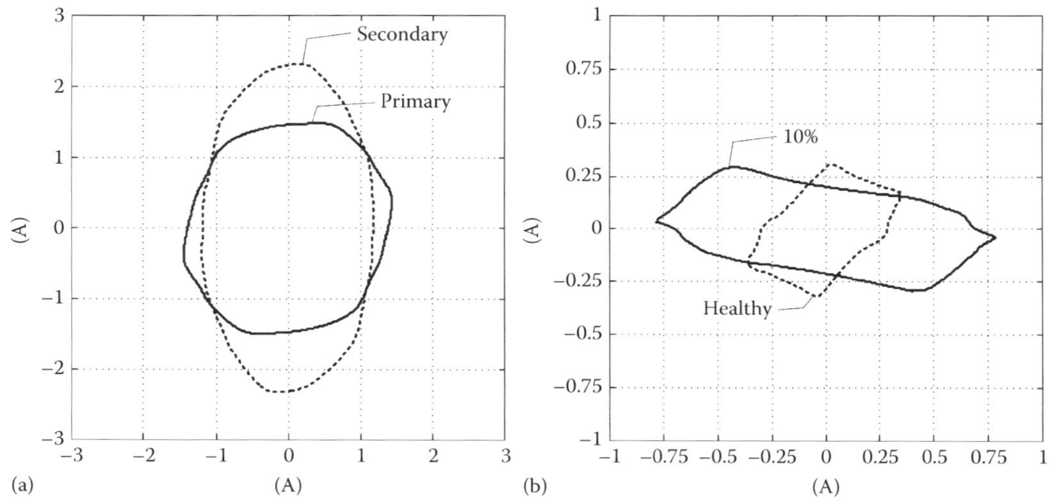

FIGURE 12.13
Experimental primary- and secondary-side winding current Park's vector patterns (a) and on-load exciting current Park's vector patterns (b), for the case of a *YNyn*0 connection, a unbalanced resistive load and 5% of shorted turns in the primary winding (phase *R*). (From Oliveira, L.M.R., Cardoso, A.J.M., and Cruz, S.M.A., *Electr. Power. Syst. Res.* 81, 1206, 2011.)

For the case of other transformer connections, the on-load exciting currents can be obtained by using the same basic principle, but with slightly different computations [33].

With this approach, the on-line characteristic of the formerly diagnostic technique is maintained.

Figure 12.14a presents the on-load exciting current Park's vector pattern for the case of a *YNyn*0 winding connection and for a healthy operation of the transformer. This pattern also

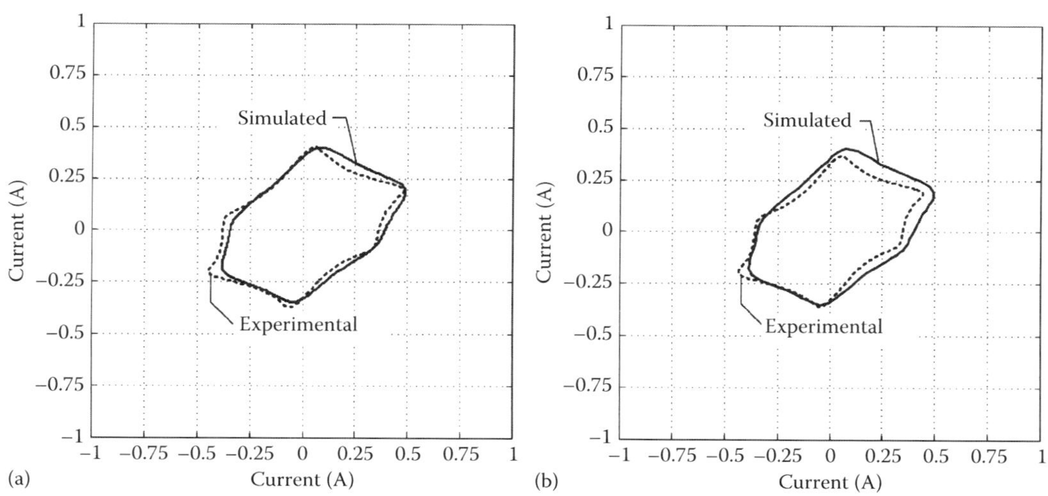

FIGURE 12.14
On-load exciting current Park's vector pattern (a) and no-load exciting current Park's vector pattern (b), for the case of a *YNyn*0 connection and healthy operating conditions. (From Oliveira, L.M.R., Cardoso, A.J.M., and Cruz, S.M.A., *Electr. Power. Syst. Res.* 81, 1206, 2011.)

differs from the circular locus expected for an ideal situation due to, among others, the nonlinear behavior and asymmetry of the magnetic circuit. This is a well-known phenomenon, which is revealed by the unbalanced and distorted nature of the exciting currents obtained from any three-phase no-load test. In fact, the exciting current Park's vector pattern, obtained at no-load conditions, presents the same characteristics, as shown in Figure 12.14b.

For the same case presented earlier, concerning the simultaneous occurrence of unbalanced loads and winding faults, the resultant on-load exciting current Park's vector pattern can be seen in Figure 12.13b, from which the fault is clearly detected. The same operating philosophy of the supply current Park's vector approach can again be applied to identify the severity and the phase location of the fault. Additionally, the on-load exciting current Park's vector approach enhances the severity of the fault, as compared to the former diagnostic technique, as shown in Figure 12.15.

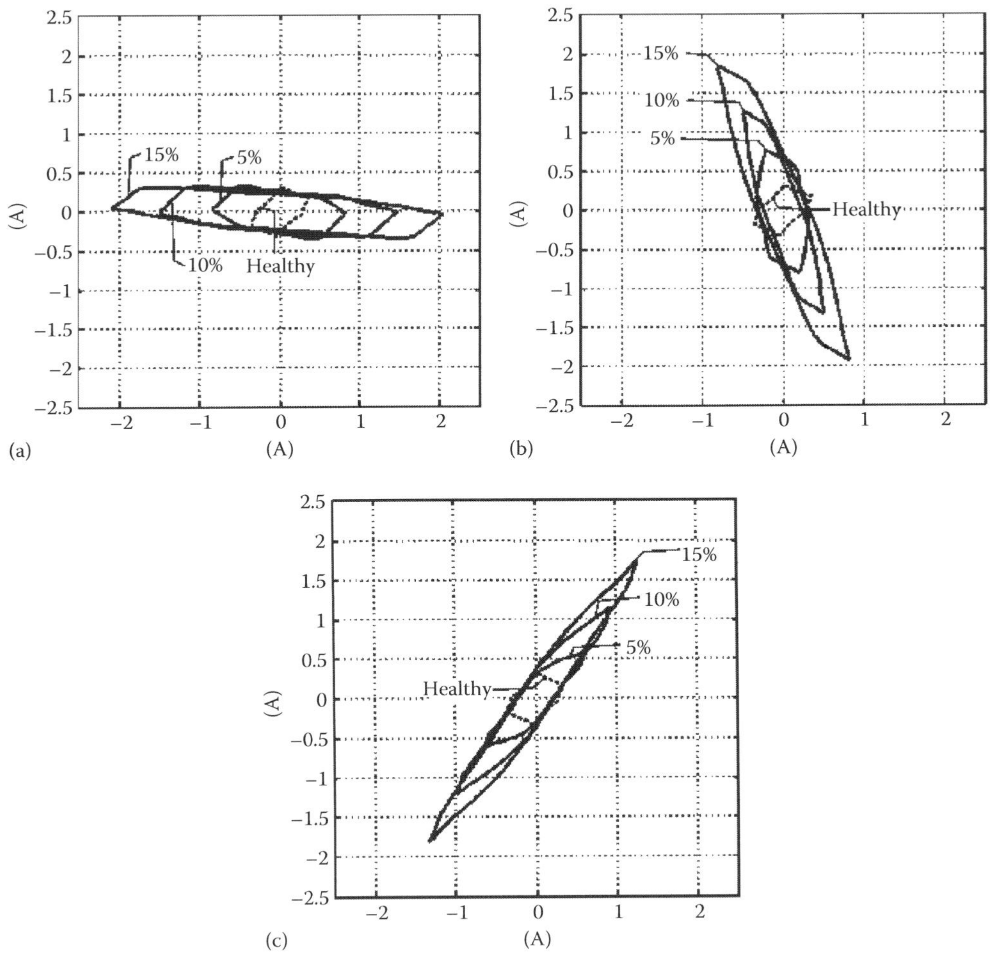

FIGURE 12.15
Experimental on-load exciting current Park's vector patterns for the case of a *YNyn0* connection and a balanced resistive load, with several percentages of permanent shorted turns in the primary windings and for different faulty phases: (a) phase *R*, (b) phase *S*, and (c) phase *T*. (From Oliveira, L.M.R., Cardoso, A.J.M., and Cruz, S.M.A., *Electr. Power. Syst. Res.* 81, 1206, 2011.)

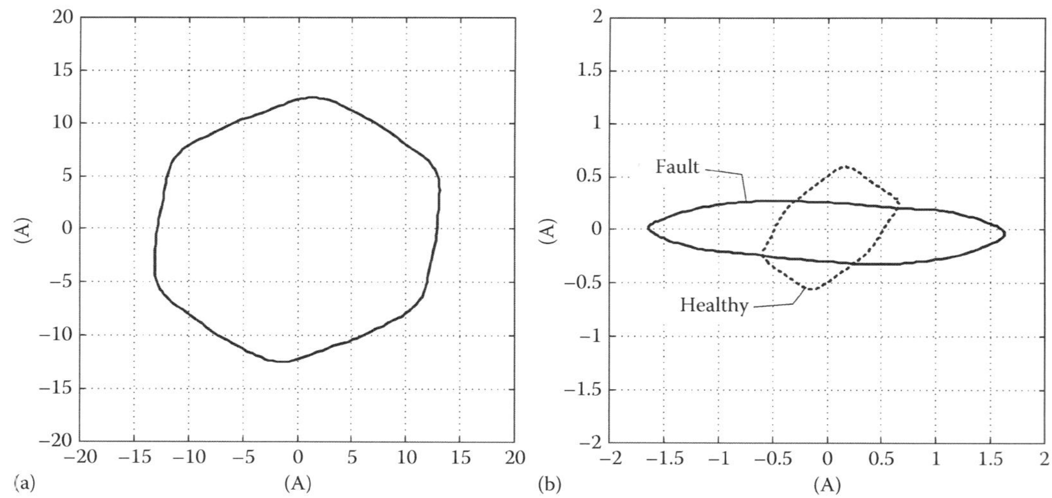

FIGURE 12.16
Simulated supply current Park's vector pattern (a) and on-load exciting current Park's vector pattern (b), for the case of a *YNyn*0 connection, a balanced resistive load and two shorted turns (≈1%) in the primary winding (phase *R*).

For the case of an incipient fault, involving only two shorted turns (≈1%), the supply current Park's vector pattern does not provide any indication about the interturn short circuit (Figure 12.16a), whereas the analysis of the on-load exciting current Park's vector approach clearly reveals the presence of the fault, Figure 12.16b. Other experimental and simulated tests carried out for different types of the transformer windings connection lead to similar conclusions to the ones presented earlier [31,32].

For the case of intermittent faults, the same operating philosophy is applied, but a spiked on-load exciting current Park's vector pattern is obtained. As shown in Figure 12.17, the faulty phase is now detected by the pulsed pattern orientation [32,34]. Figure 12.18 presents the evolution of the on-load exciting current Park's vector pattern when the magnitude of the faulty current is increased from $0.5 \times \hat{I}_{1n}$ to $1.5 \times \hat{I}_{1n}$. The severity of the fault is directly related with the magnitude of the pulsed pattern. Additionally, the results clearly indicate that the proposed diagnostic technique is sensitive to low-level faults.

The on-load exciting current Park's vector approach takes into account the currents in all the three primary and secondary phases. Too much information is thus contained in this representation, which cannot be completely extracted by the analysis of the resulting geometric pattern. Also, a fault severity factor (SF) cannot be easily defined [31,35].

Firstly applied to diagnose AC motor faults, a new technique has been introduced [36], the so-called extended Park's vector approach (EPVA), in order to allow a more in-depth characterization of the unit condition.

12.8.3 On-Load Exciting Current EPVA Signature

The EPVA is based on the spectral analysis of the AC level of the current Park's vector modulus:

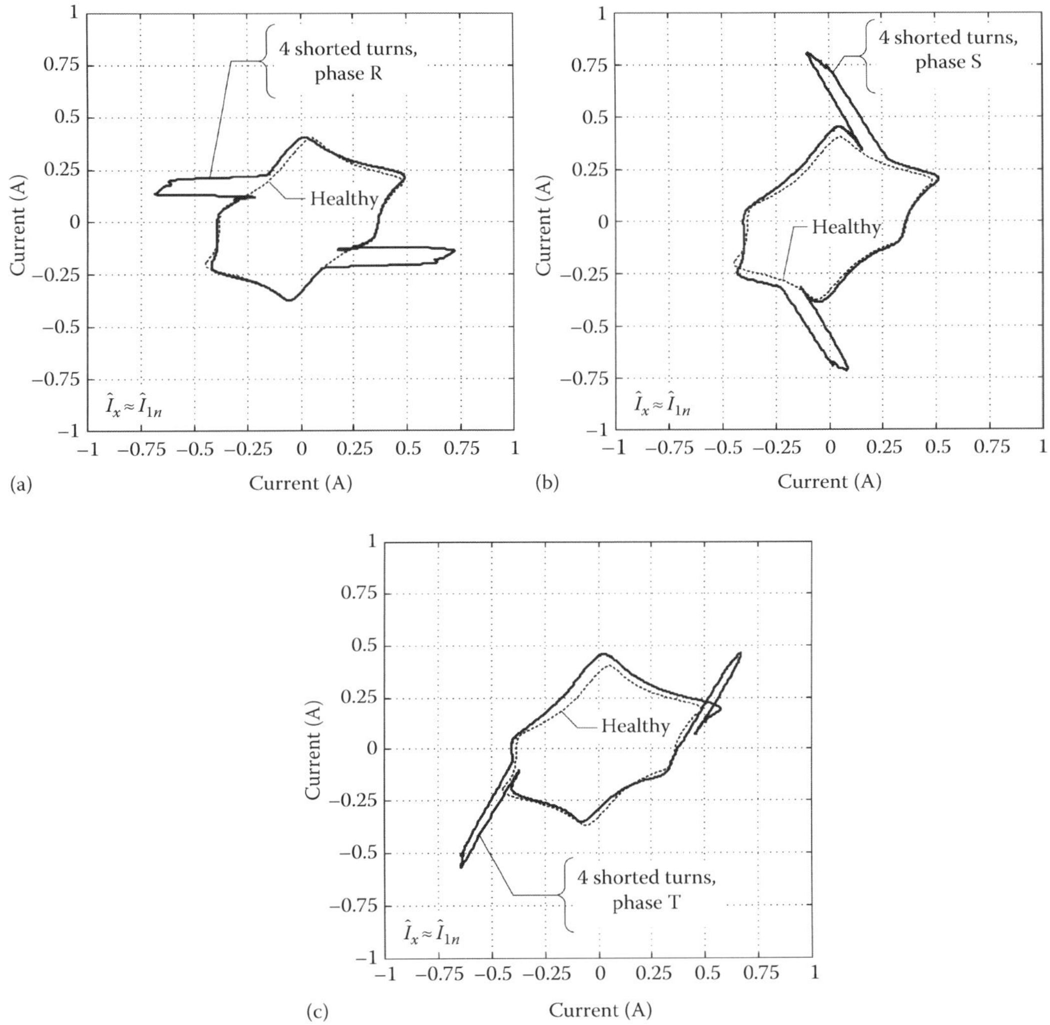

FIGURE 12.17
On-load exciting current Park's vector patterns for the case of a *YNyn*0 connection and a balanced resistive load, with four intermittent shorted turns in the primary windings and for different faulty phases: (a) phase *R*, (b) phase *S*, and (c) phase *T*. (Reprinted with permission from Oliveira, L.M.R. and Cardoso, A.J.M., A permeance-based transformer model and its application to winding interturn arcing fault studies, *IEEE Trans. Power. Deliv.*, 25, 1589–1598, Copyright 2010, IEEE.)

$$\left|i_{eD} + ji_{eQ}\right| = \sqrt{i_{eD}^2 + i_{eQ}^2} \tag{12.8}$$

Ideally, under healthy conditions, the EPVA signature will be clear from any spectral component, that is, only a DC value is present in the current Park's vector modulus.

The occurrence of winding interturn short circuits (on the primary or on the secondary side) leads to an increment in the magnitude of the on-load exciting current in the

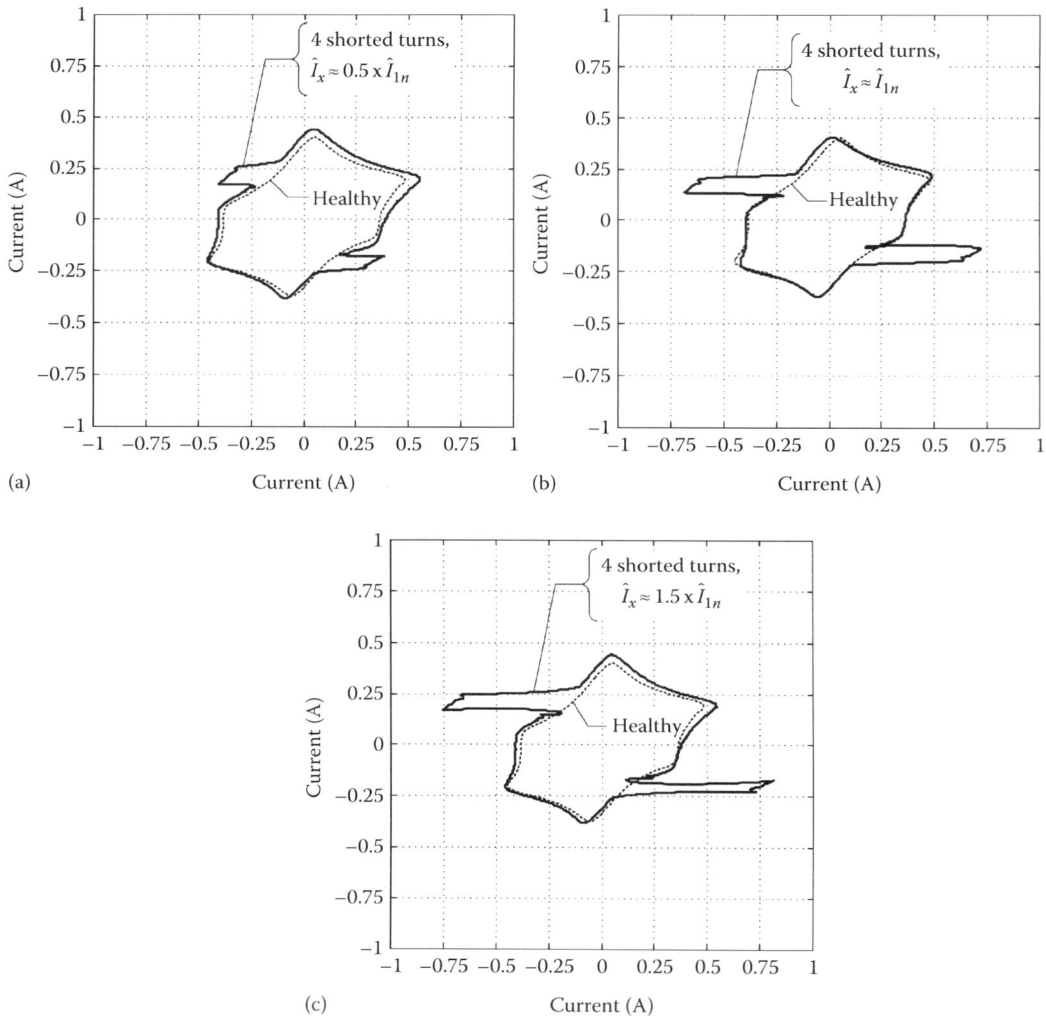

FIGURE 12.18
On-load exciting current Park's vector patterns for the case of a *YNyn*0 connection and a balanced resistive load, with four intermittent shorted turns in the primary winding of phase *R* and for different values of the magnitude of the pulsed faulty current, $\hat{I}_x$ (in terms of the rated magnitude of the current in the affected winding, $\hat{I}_{1n}$): (a) $\hat{I}_x \approx 0.5 \times \hat{I}_{1n}$, (b) $\hat{I}_x \approx \hat{I}_{1n}$, and (c) $\hat{I}_x \approx 1.5 \times \hat{I}_{1n}$. (Reprinted with permission from Oliveira, L.M.R. and Cardoso, A.J.M., A permeance-based transformer model and its application to winding interturn arcing fault studies, *IEEE Trans. Power. Deliv.*, 25, 1589–1598, Copyright 2010, IEEE.)

affected phase, as compared to a healthy situation, which results in an unbalanced system of currents. Under these conditions, the on-load exciting current Park's vector modulus will contain a dominant DC level and an AC level, at twice the supply frequency ($2f$), whose existence is directly related to the asymmetries in the transformer. The amplitude of this spectral component is directly related to the extension of the fault. In this way, an indicator of the degree of asymmetry can be defined as the ratio between the amplitude of the spectral component at the frequency of $2f$ and the DC level of the on-load exciting current Park's vector [36]. The proposed fault SF is thus expressed by the following:

$$SF = \frac{\max\left(\left[\sqrt{i_{eD}^2 + i_{eQ}^2}\right]_{(2f)\,\text{component}}\right)}{\text{average}\left(\sqrt{i_{eD}^2 + i_{eQ}^2}\right)} \tag{12.9}$$

Further details of the theoretical principles related with the EPVA can be found in [36].

Figure 12.19 presents the EPVA signatures for the case of a *YNyn*0 winding connection, rated load conditions, and for a healthy operation of the transformer. As stated earlier, under ideal conditions, the EPVA signature would be clear from any spectral component. However, in practice, a spectral component with a small amplitude, at a frequency of $2f$ (100 Hz), is present, which is originated by the same reasons responsible for the deformation of the exciting current Park's vector pattern (Figure 12.14).

The EPVA signatures, for the case of two shorted turns in the primary winding of phase *R* and for several values of the current in the faulty turns (in terms of the rated current of the affected winding, I_{1n}), are shown in Figure 12.20. It can be seen that the amplitude of the $2f$ spectral component (100 Hz) increases with the severity of the fault, representing a good indicator of an insulation defect.

Figure 12.21 presents the evolution of the SF, given by (12.9), with the current in the two shorted turns (phase *R*). The values of the SF increase monotonically with the increase of the faulty current [36].

The results clearly indicate that the proposed diagnostic technique is sensitive to low-level faults.

Figure 12.22 shows the evolution of the SF with the number of shorted turns (phase *R*). Similar conclusions to the ones presented for the case of incipient faults can be drawn.

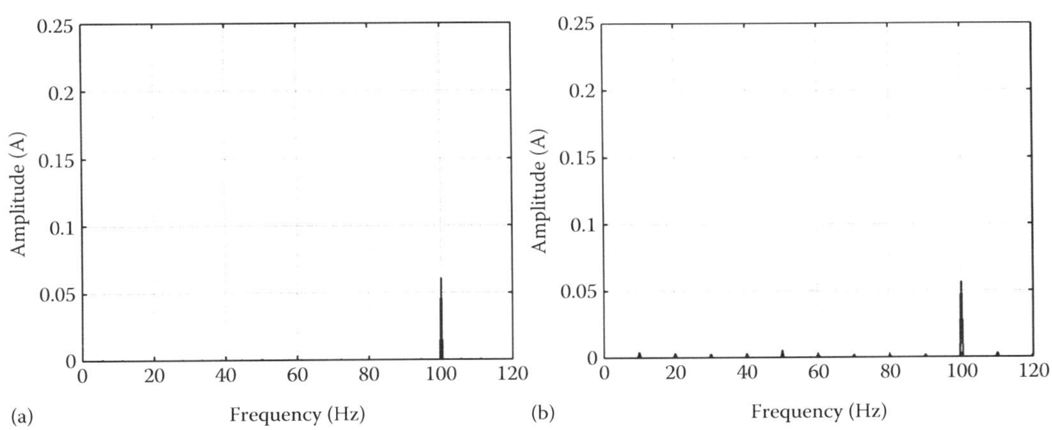

FIGURE 12.19
EPVA signatures of the on-load (a) and the no-load (b) exciting currents for the case of a *YNyn*0 connection and healthy operating conditions (experimental results). (From Oliveira, L.M.R., Cardoso, A.J.M., and Cruz, S.M.A., *Electr. Power. Syst. Res.* 81, 1206, 2011.)

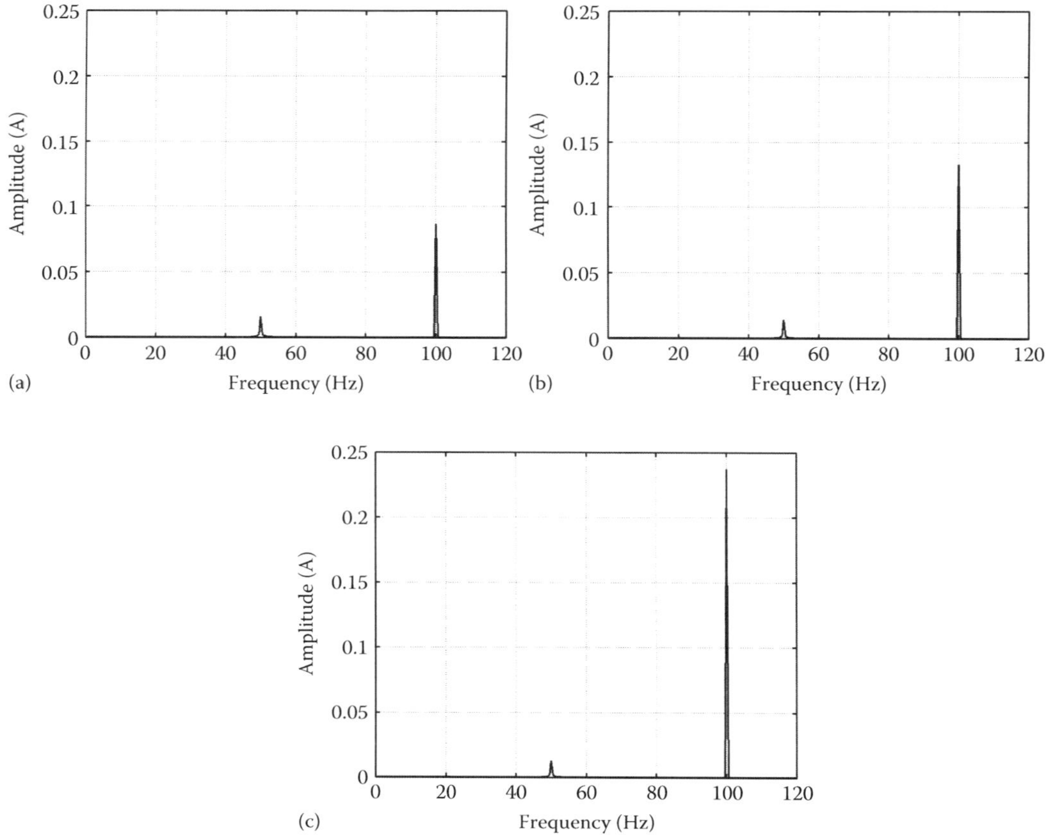

FIGURE 12.20
EPVA signature for the case of two shorted turns in the primary winding of phase *R*, with the following values of the current in the faulty turns: (a) $I_b = I_{1n}$, (b) $I_b = 2 \times I_{1n}$, and (c) $I_b = 4 \times I_{1n}$ (simulated results). (From Oliveira, L.M.R., Cardoso, A.J.M., and Cruz, S.M.A., *Electr. Power. Syst. Res.* 81, 1206, 2011.)

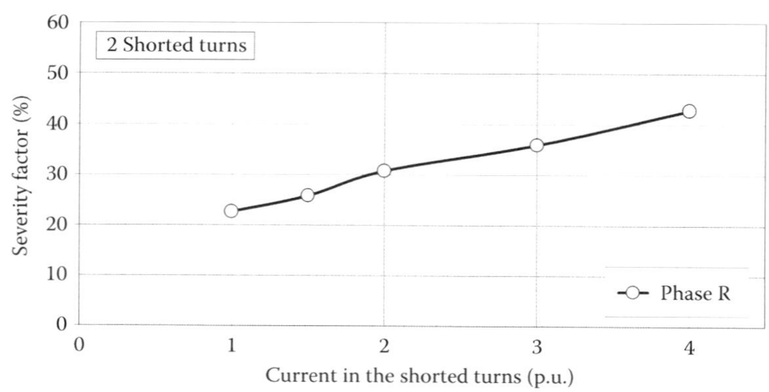

FIGURE 12.21
Evolution of the SF with the current in the two shorted turns (simulated results). (From Oliveira, L.M.R., Cardoso, A.J.M., and Cruz, S.M.A., *Electr. Power. Syst. Res.* 81, 1206, 2011.)

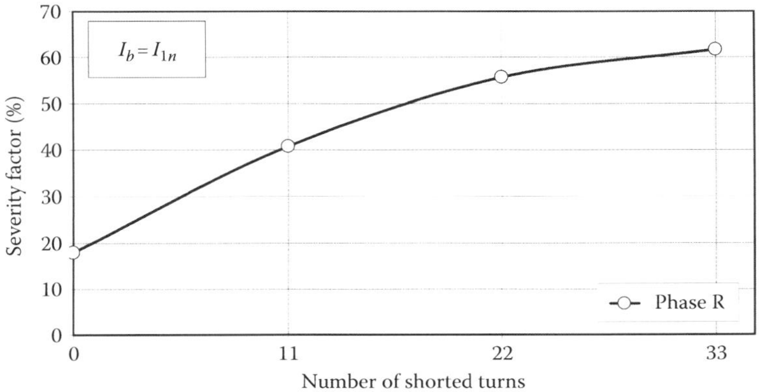

FIGURE 12.22
Evolution of the SF with the number of shorted turns (experimental results). (From Oliveira, L.M.R., Cardoso, A.J.M., and Cruz, S.M.A., *Electr. Power. Syst. Res.* 81, 1206, 2011.)

References

1. Bengtsson C (1996) Status and trends in transformer monitoring. *IEEE Trans Power Deliv* 11:1379–1384.
2. Goulden Reports (2003) The world market for transformers 2003.
3. Global Industry Analysts (2008) Transformers (electricity), a global strategic business report.
4. CIGRE/SC 12/WG 05 (1983) An international survey on failures in large power transformers in service. *Electra* 88:21–48.
5. IEEE Standard 493-1990 (1990) *IEEE Recommended Practice for the Design of Reliable Industrial and Commercial Power Systems*. IEEE, New York.
6. Cardoso AJM, Oliveira LMR (1999) Condition monitoring and diagnostics of power transformers. *Int J COMADEM* 2:5–11.
7. Cardoso AJM (1991) *Fault Diagnosis in Three-Phase Induction Motors* (in Portuguese). Coimbra Editora, Coimbra, Portugal.
8. Inoue Y, Suganuma K, Kamba M et al. (1990) Development of oil-dissolved gas detector for diagnosis of transformers. *IEEE Trans Power Deliv* 5:226–232.
9. Zhang Y, Ding X, Liu Y et al. (1996) An artificial neural network approach to transformer fault diagnosis. *IEEE Trans Power Deliv* 11:1836–1841.
10. Huang YC, Yang HT, Huang CL (1997) Developing a new transformer fault diagnosis system through evolutionary fuzzy logic. *IEEE Trans Power Deliv* 12:761–767.
11. Lin CE, Ling JM, Huang CL (1993) An expert system for transformer fault diagnosis using dissolved gas analysis. *IEEE Trans Power Deliv* 8:231–238.
12. Tomsovic K, Tapper M, Ingvarsson T (1993) A fuzzy information approach to integrating different transformer diagnostic methods. *IEEE Trans Power Deliv* 8:1638–1646.
13. Kawaguchi Y, Yanabu S (1969) Partial-discharge measurement on high-voltage power transformers. *IEEE Trans Power Appar Syst* 88:1187–1194.
14. Fuhr J, Haessig M, Boss P et al. (1993) Detection and location of internal defects in the insulation of power transformers. *IEEE Trans Electr Insul* 28:1057–1067.
15. CIGRE/SC 12/WG 09 (1990) Direct measurements of the hot-spot temperature of transformers. *Electra* 129:47–51.
16. Mechefske CK (1985) Correlating power transformer tank vibration characteristics to winding looseness. *Insight* 37:599–604.

17. Sanz-Bobi MA, García-Cerrada A, Palacios R et al. (1997) Experiences learned from the on-line internal monitoring of the behaviour of a transformer. *IEEE International Electric Machines and Drives Conference*, Milwaukee, WI, pp. TC3/11.1–TC3/11.3.
18. Kulikowski J, Lech W, Rachwalski J et al. (1968) Expérience acquise dans les essais de court-circuit des transformateur. CIGRE rapport 12–13.
19. Dick EP, Erven CC (1978) Transformer diagnostic by frequency response analysis. *IEEE Trans Power Appar Syst* 97:2144–2153.
20. Rogers EJ, Gillies DA, Humbard LE (1972) Instrumentation techniques for low voltage impulse testing of power transformers. *IEEE Trans Power Appar Syst* 101:1281–1293.
21. Bak-Jensen J, Bak-Jensen B, Mikkelsen SD (1995) Detection of faults and ageing phenomena in transformers by transfer functions. *IEEE Trans Power Deliv* 10:308–314.
22. Lachman MF (1994) Field measurements of transformer single-phase exciting current as diagnostic tool, and influence of load tap changers. *IEEE Trans Power Deliv* 9:1466–1475.
23. Rickley AL, Clark RE, Povey EH (1981) Field measurements of transformer excitation current as diagnostic tool. *IEEE Trans Power Appar Syst* 100:1985–1988.
24. IEEE Standard 37.91–2000 (2000) *IEEE Guide for Protective Relay Applications to Power Transformers*. IEEE, New York.
25. Barkan P, Damsky BL, Ettlinger LF et al. (1976) Overpressure phenomena in distribution transformers with low impedance faults, experiment and theory. *IEEE Trans Power Appar Syst* 95:37–48.
26. Stigant SA, Franklin AC (1973) *The J&P Transformer Book*, 10th edn. Newnes-Butterworths, London, U.K.
27. Lunsford JM, Tobin TJ (1997) Detection of and protection for internal low-current winding faults in overhead distribution transformers. *IEEE Trans Power Deliv* 12:1241–1249.
28. Abed NY, Mohammed OA (2007) Modeling and characterization of transformers internal faults using finite element and discrete wavelet transforms. *IEEE Trans Magn* 43:1425–1428.
29. Gómez-Morante M, Nicoletti DW (1999) A wavelet-based differential transformer protection. *IEEE Trans Power Deliv* 14:1351–1359.
30. Wang H (2001) Models for short circuit and incipient internal faults in single-phase distribution transformers. PhD thesis, Texas A&M University, College Station, TX.
31. Oliveira LMR, Cardoso AJM, Cruz, SMA (2011) Power transformers winding fault diagnosis by the on-load exciting current Extended Park's Vector Approach. *Electr Power Syst Res* 81:1206–1214.
32. Oliveira LMR, Cardoso AJM (2010) A permeance-based transformer model and its application to winding interturn arcing fault studies. *IEEE Trans Power Deliv* 25:1589–1598.
33. Oliveira LMR, Cardoso AJM, Cruz SMA (2002) Transformers on-load exciting current Park's vector approach as a tool for winding faults diagnostics. *Conference Record ICEM*, Brugge, Belgium.
34. Oliveira LMR, Cardoso AJM (2008) Intermittent turn-to-turn winding faults diagnosis in power transformers by the on-load exciting current Park's vector approach. *Proceedings of ICEM*, Vilamoura, Portugal.
35. Oliveira LMR, Cardoso AJM (2004) Incipient turn-to-turn winding fault diagnosis of power transformers by the on-load exciting current Extended Park's vector approach. *Proceedings of ARWtr04*, Vigo, Spain, pp. 134–139.
36. Cruz SMA, Cardoso AJM (2001) Stator winding fault diagnosis in three-phase synchronous and asynchronous motors, by the Extended Park's vector approach. *IEEE Trans Ind Appl* 37:1227–1233.

13

Transformer Reliability: A Key Issue for Both Manufacturers and Users

Adolf J. Kachler

CONTENTS

13.1 Introduction

Growing economic pressure and liberation of the electricity market are forcing utilities to seek new ways of asset management and cost/risk assessment. Not only the initial cost but the total costs—that is, the operating costs (loss evaluation, maintenance and outage costs)—are becoming a top priority in asset management [1–3]. This all leads to *reliability* considerations.

Mr. Finance is demanding, on the one side, strong cost reduction measures, and at the same time, he wants to move from well-proven "time-based" to "condition-based" maintenance. Replacement planning is partly cut out, and the engineer is forced into *reliability considerations* and into *economic* risk assessment.

Reliability/condition assessment needs more information than in the past in order to be successful.

Growing age and growing loading of the existing transformer population (30%–40% are older than 25–30 years, and loading has been considerably increased—e.g., in the United States, from an average of 60% to >80% in the past 15 years) stipulate the need for expert diagnosis.

Transformer reliability depends on a large number of factors and responsibilities both for manufacturers and users.

13.2 Definition of Reliability

To date, there is no unique definition of *reliability* for power transformers available.

Hence, the practice to assess reliability in service differs a lot worldwide. China, for instance, is leaning on failure rates and outage rates and calculates a percentage of *availability* due to in-operation ability.

The only standard [4], ANSI C57.117 has made an attempt to define failures (as forced outages of more than 1 day and which require in-tank/internal measures). Based on this definition, in-service failure statistic and reliability values can be derived. In [2], we have elaborated this aspect, and the essence is given in Figures 13.1 through 13.4.

As a start, this background, based on forced outages, could very well serve to generate more transparency and comparability in reliability assessment.

- For pre-qualifications
- To assess the competence and expertise of a manufacturer we need
- Design reviews
- Production quality reviews
- Process competence and experience analysis
- Internal test floor failure statistics
- The delivery reliability
- The in-service reliability statistics
- Finally for an utility also the Support functions for off-line and on-line diagnosis or monitoring of a manufacturer are of importance

FIGURE 13.1
Reliability evaluations.

Based on ANSI C57. 117 (1986), we propose the following:

- Failure: "Termination of a transformer to perform its specified functions"
- Failure with forced outage: "Failure of a transformer that requires its immediate removal from the system"
- Failure with scheduled outage: "Failure for which a transformer must be deliberately taken out of service at a selected time"
- Defect/non-conformity: "Inspection or partial lack of performance that can be corrected without taking the transformer out of service"
- Failure rate/reliability parameters
  - Failure rate (FR): The ratio of the number of failures with forced outage of a given population over a given period of time to the number of accumulated service years for all transformers in that period of time
  - Reliability parameter: Mean time between failures (MTBF) = 1/FR (years)

FIGURE 13.2
Failure definition.

Floating time period: 10 years

- N: Number of units in service in that 10 year time period
- SY: Number of service years accumulated with [N] units in service
- nF: Number of failures or number of units from N units in service
- Failure rate FR
  FR = (nF/SY) 100[%]
- Reliability indicator "mean time between failures"
  MTBF = (1/FR) [years]

FIGURE 13.3
In-service failure statistics or reliability parameters.

	PTD T w.w	BUT	KPT	FS	SAT	STC *	PD	TUSA	PN + T
N	6957	217	424	582	546	391	601	937	3259
SY	31612	712	1793	2658	3148	1185	2660	4790	14666
n_F	80	4	2	0	8	1	2	23	40
FRe (%)	0.25	0.56	0.11	0	0.25	0.08	0.08	0.48	0.27
MTBF	395	178	897	∞	394	1185	1330	208	367

N = No of delivered units
SY = No of Service years
n_F = No of units failed
FRe = Failure rate = $(n_F/SY)100\%$
MTBF = Mean time between failure = (1/FR) years

FRe ≤ 0.5%	excellent
0.5 < FRe ≤ 1.10%	good
1.0 < FRe ≤ 1.5%	satisfactory
1.5 < FRe ≤ 2.0%	acceptable
FRe> 2.0%	not acceptable

* 1995–2003

FIGURE 13.4
Example: Siemens power transformer in-service failure statistics.

13.3 Main Aspects of Reliability

13.3.1 Part 1: Manufacturer's Responsibilities

The manufacturer's responsibilities can be grouped into three major aspects:

- Design (engineering, competence, tools, progress, and related aspects)
- Production quality (tools, machines, processes, materials, competence)
- Transport/erection/commissioning quality

All papers [1,3,5–9] elaborate extensively on the progress of transformer engineering tools, production processes, and materials.

13.3.1.1 Design Impact Reliability and Related Issues

The state of the art is a result of long years of experience and proven evidence by full-scale and model testing. For instance, Siemens has more than 100 years of power transformer

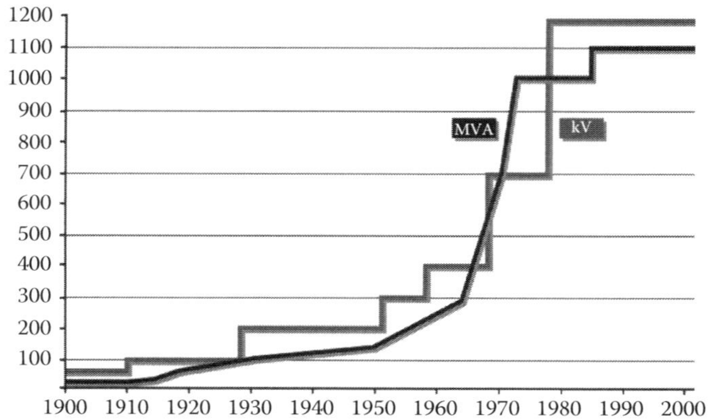

FIGURE 13.5
Siemens experience curves of kV/MVA development.

design and production experience and has participated in all steps of development up to 1200 kV system voltage and 1100 MVA rating. Figure 13.5 shows this background.

The integrity of the design means fulfillment of specification and all operating requirements in respect to the following:

- Dielectric integrity (electric field, stress calculation)
- Thermal integrity (magnetic field, stray field, loss calculation, cooling effectivity, hot spots within guarantees)
- Dynamic integrity under external short circuits, according to system conditions

For electric field and stress calculation, 2D and 3D calculation programs are available. For transient voltage distribution calculation (LI, SI, part winding resonance), extremely effective programs have been developed [6–8]. Examples are given in Figures 13.6 and 13.7.

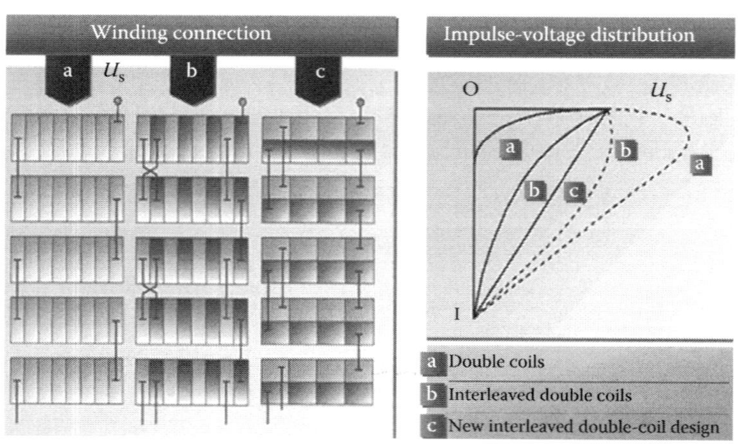

FIGURE 13.6
Impulse voltage distribution for various kinds of windings.

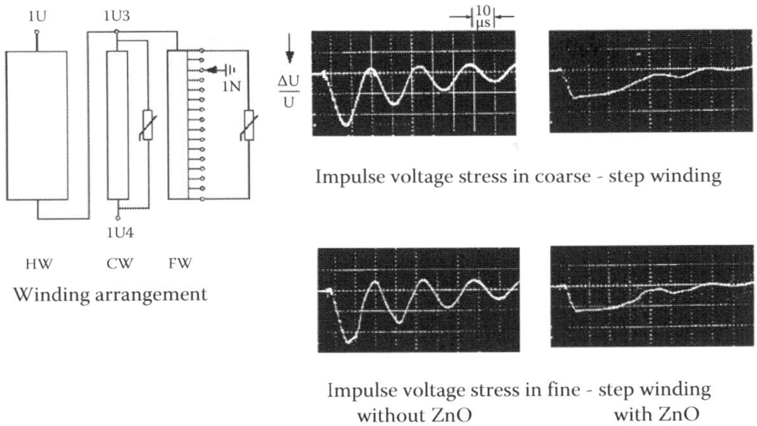

FIGURE 13.7
ZnO protection for regulating windings.

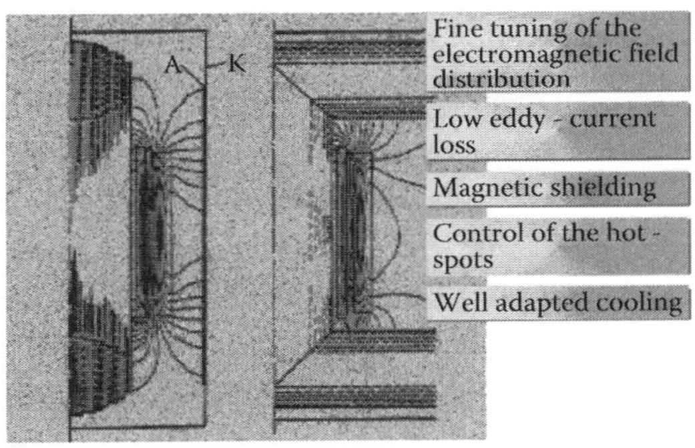

FIGURE 13.8
Electromagnetic field calculation.

For thermal integrity, the calculation of no-load-loss/load-loss new core and conductor materials are employed. Magnetic field and stray field calculation allows determining accurately the losses and temperatures [6–8] (Figure 13.8).

Much progress has been achieved to reduce no-load losses, load losses, and noise in the last two decades. This produces higher efficiency, higher service reliability, and more economic solutions (Figure 13.9).

Dynamic integrity is based on both well-proven design and production rules (derived from model and full-size testing and qualified production processes).

In this context, adequate winding production and drying/pressing processes are of vital importance. A design review must always include production quality aspects. Most important in design reviews is to critically check the customer specification and any special requirements.

13.3.1.2 Production Quality and Related Issues

The best design is not valid if the production cannot maintain the elementary tolerances for all steps of production. This requires a close communication between engineering and

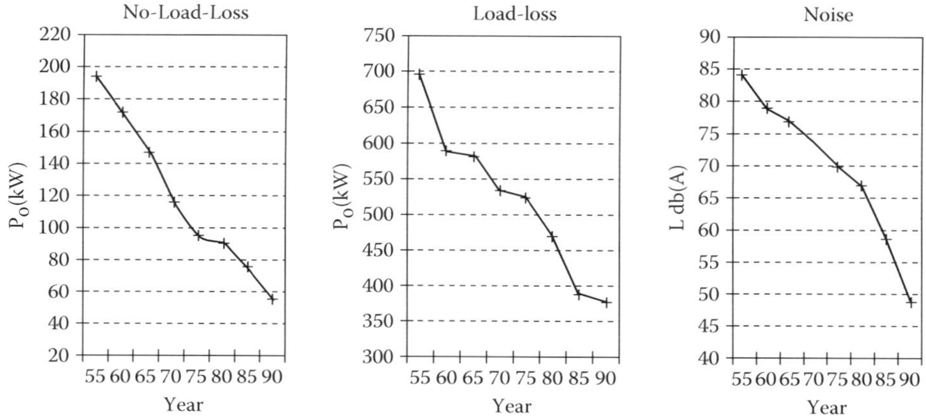

FIGURE 13.9
Progress in no-load loss, load loss, and noise with time according to Ref. [3].

production. Furthermore, all machines and tools must be kept under consequent maintenance, and the good/bad criteria (tolerances) must be clearly defined and observed by the production employees.

In addition to that, an efficient in-process inspection by qualified workers must provide objective evidence of all critical parameters.

In winding production, the tolerances for length, width, and inner and outer diameter, as well as winding height by means of axial and radial compression, as well as in-winding length alignment, are of high importance. Finally, strict rules for cleanliness throughout the whole production must be observed.

The dry and treat process (vapor phase), the pressing process, and the oil filling and impregnation process are of importance for high reliability.

Here are some examples from Siemens (Figures 13.10 and 13.11):

Improved overlapping results in lower losses and lower noise.

In the winding production, wherever possible, axial and radial compression modalities produce extremely compact windings. All of our winding machines have conductor and drum brakes.

Furthermore, the windings are vapor phase dried and oil stabilized under constant pressure. This secures simultaneously both drying and compression shrinkage.

With in-winding length adjustment, we obtain very accurate winding lengths which enables effective common pressing.

All these measures (design and production quality) result in an excellent in-service failure or *reliability* statistic (see Figure 13.4).

Final testing (routine, special, or type) must also be performed under rated conditions according to specification and standards.

Powerful diagnostic controls assure that no faulty transformer leaves the factory.

13.3.1.3 Transport/Erection/Commissioning Quality and Maintenance Instructions

The best design and production quality can be destroyed by means of inadequate transport/erection and commissioning and maintenance processes.

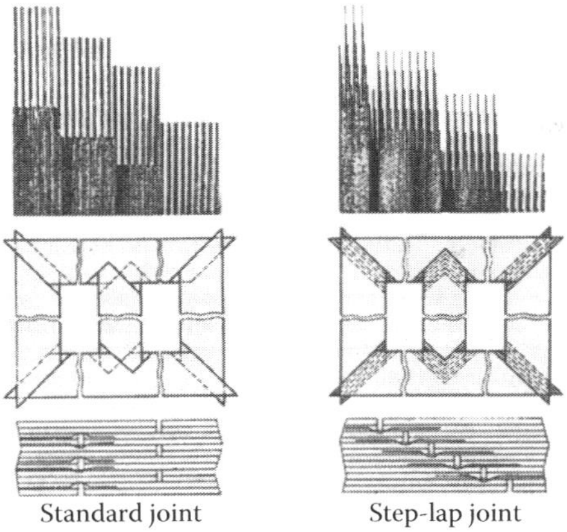

FIGURE 13.10
Core productions: advantages of step-lap versus normal-step cores.

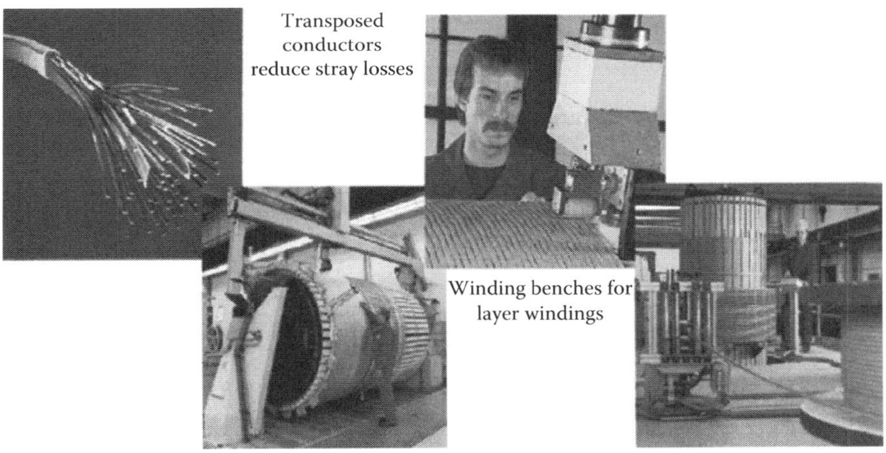

FIGURE 13.11
Example: winding production with axial and radial compression.

The following points are important:

The transported transformer must be equipped with 3D, calibrated shock recorders. Read out whenever the transport means are changed (car–railway–boat and vice versa). Transport damages today are more frequent than in-service failures due to inexperience, inadequate transport means, high speed, and poor rigging. Some examples are shown in Figures 13.12 and 13.13.

The customer is well advised to only employ qualified site erection and commissioning people who have profound knowledge of transformer and involved materials and take adequate records of all indicators (oil level, over pressure, dew point, etc.). Also, only qualified process equipment is mandatory (Figures 13.14 and 13.15).

The instructions in the handbook must duly be followed, and clear records must be maintained for future reference, fingerprints, and trend analysis (Figure 13.16).

FIGURE 13.12
240 MVA/220 kV—transport damage due to excess speed.

FIGURE 13.13
Development of a floating transformer required.

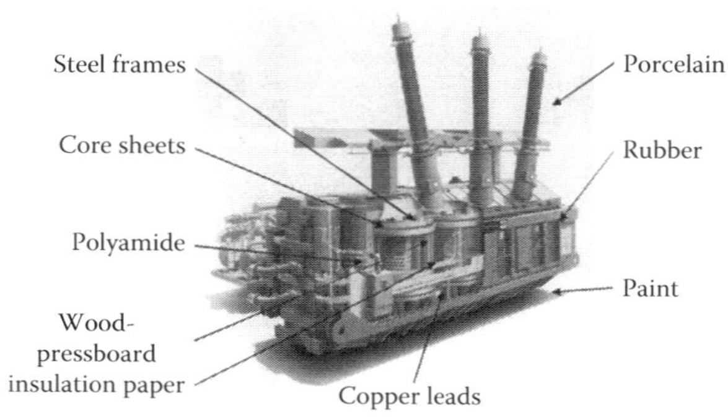

FIGURE 13.14
Overview of materials involved (850 MVA generator transformer).

102.31 Checklist outside inspection

Reference:__________________ Type: __________________
Serial no: __________________ MVA: __________________
Installation project no.:____________ Date: __________________

1. Transport:
   - ☐ With oil above the cover
   - ☐ With oil .... mm below the cover
   - ☐ Without oil
   - ☐ Filling gas is nitrogen
   - ☐ Filling gas is dry air
2. Gas top-up device:
   - ☐ Not provided
   - ☐ Internal pressure within the gas bottles: .... bar
3. Gas gauge pressure in the tank:
   - ☐ .... mbar
   - ☐ None
4. Moisture replica:
   - ☐ Not provided

   Moisture indicator:
   - ☐ Moisture below 8%
   - ☐ Moisture above 8%
5. Control cabinet:
   - ☐ Not available
   - ☐ In order
   - ☐ Damaged
9. Tapchanger head:
   - ☐ Not available
   - ☐ In order
   - ☐ Damaged
10. Wheels / -sets / fastening devices:
    - ☐ Not available
    - ☐ In order
    - ☐ Damaged
11. Transformer-mounted accessories:
    - ☐ Not available
    - ☐ In order
    - ☐ Damaged
12. Impact recorder:
    - ☐ Not provided
    - ☐ Provided but has so far not recorded severe impacts
    - ☐ Transmitted for evaluation to: ....
13. Tank damage:
    - ☐ Not damaged
    - ☐ Damaged (notified, see Annex)
14. Paint finish

FIGURE 13.15
Example for fundamental checks after on-site arrival.

Outside:	Visual check
	Register shock recorder
	Gas bottles still working
	Leakage test
Inside:	Dew point measurement
	Visual inspection

Fingerprint: Frequency response measurement
Ratio, Io—Distribution, DC – Resistance, Insulation resistance measurement, PDC—Analysis, etc.

FIGURE 13.16
Example: commissioning checklist.

13.3.2 Part 2: Customer Responsibilities

13.3.2.1 Adequate Specification

Designs are based on a combination of national/international standards and customer specifications and site requirements [1]:

> The customers' Specification define how the transformers are used, the loading regimes and the system conditions to be envisaged. In some cases the specification

fails to fully cover the operating condition, in terms of loading, switching operations, system transients, insulation coordination and environmental issues. In such cases a transformer designed to meet the specification may be inappropriate for the actual conditions.

We have diagnosed a number of failures of different manufacturers worldwide, where the actual system conditions were more severe than the specification—they were simply unknown and unexperienced. The transient conditions (of dielectric, thermal, and dynamic nature) severely affect design and/or reliability [1,3,6–9].

13.3.2.2 Consequent Maintenance and Service

Transformer life management, condition assessment, and reliability are highly dependent on the mode of operation, frequency of overloading, and transient condition in respect to dielectric, thermal, and dynamic stresses [10–12] (Figure 13.17).

Reliability and transformer life are severely influenced by consequent maintenance according to transformer manual and regular service measures as well as by condition assessment—starting from fingerprinting to regular trend analysis by means of off-line diagnostics and online monitoring.

In [11–13], we have provided for our clients the essentials and guidelines for laboratory diagnostics to quantitatively determine loss of life for condition assessment (Figures 13.18 and 13.19).

Also, mechanical controls are of high importance, for example, tap changer functionality. Torque is frequently an important indicator of contact wear or looseness.

13.3.2.2.1 Online Monitoring

The whole complex of diagnostic and trend analysis will be strongly supported by online monitoring—not in the sense of a self-supporting expert system but as an effective early warning system. The strategies to employ online monitoring vary greatly worldwide, but there is a clear tendency to employ such devices at least for strategically important transformers. Today, it is possible to survey, with one master CPU, up to 20 transformers by means of slaves and PROFIBUS interlinks [14].

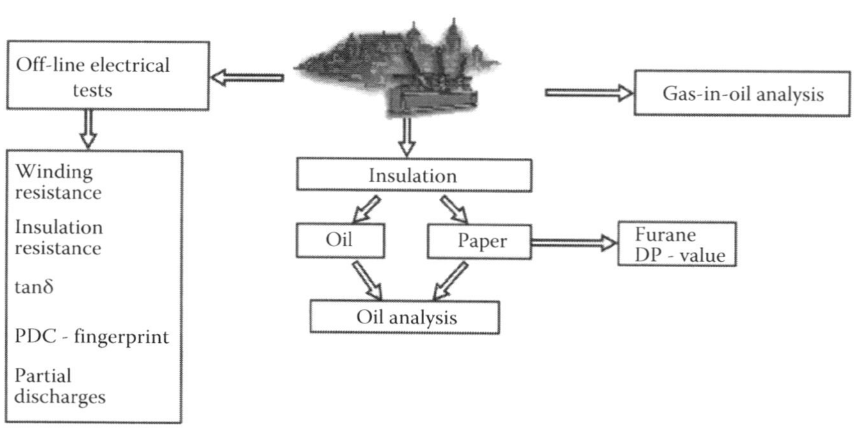

FIGURE 13.17
The most powerful diagnostics to determine the aging status.

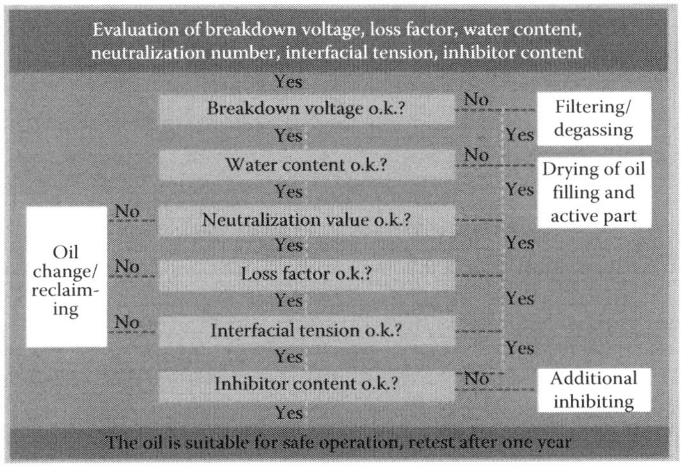

FIGURE 13.18
Example of evaluation of oil analysis.

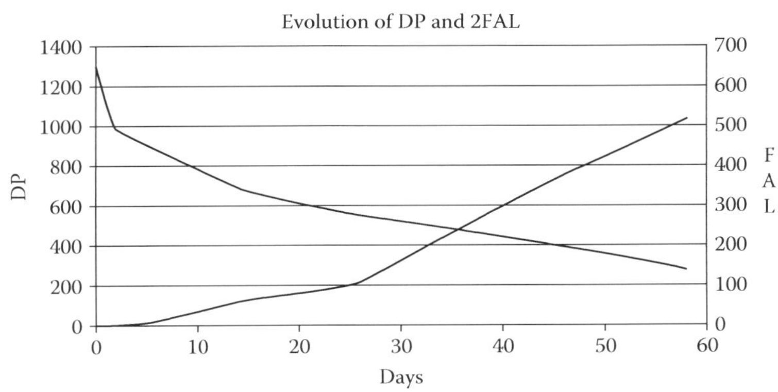

FIGURE 13.19
Principal correlation between *furane* compounds and degree of polymerization as a first approximation.

The costs for such systems are negligible compared to outage costs ranging from US$0.5–1 million/day.

13.3.2.3 Consequent Life Extension/Maintenance Measure

Transformer life management and reliability can be greatly improved by means of life extension techniques. Basis, however, is the application of all the maintenance and service measures described in this section and in Section 13.3.2.2.

The main life extension measures are given in Figure 13.20.

In addition,

- Core and coil drying
- Online monitoring of temperatures, fault gases, moisture, etc.
- Regular sampling controls are suggested.

- A regular trend analysis over gas-in-oil analyses and oil analysis
- Reducing then moisture content in the insulating system through drying
- Degassing
- Oil reclamation/oil change in case of high acidity and/or low interfacial tension. Use of inhibited oils
- Online monitoring of fault gases, especially hydrogen
- Monitoring of moisture
- Monitoring of the operating temperature
- The purpose of maintenance and life prolongation should be
- Recovery of the dielectric properties of the oil/paper insulation to the largest possible extend
- Reduction of the future aging rate

FIGURE 13.20
Required actions against aging and life extension measures.

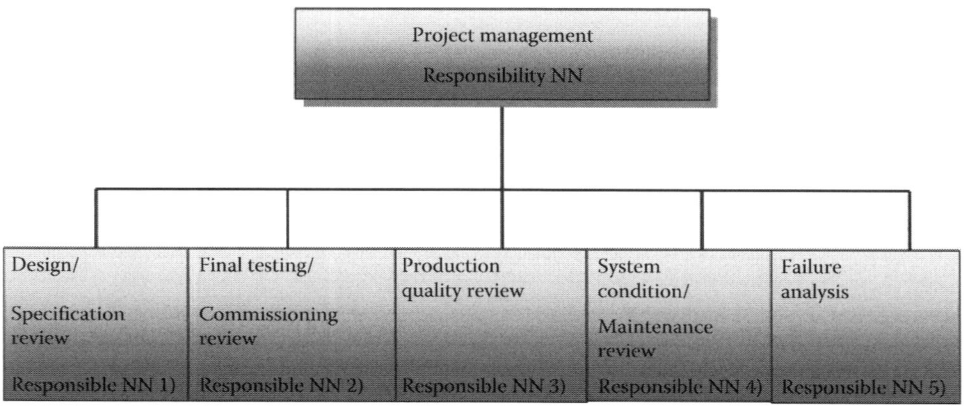

Based on the unbiased findings of the processes 1).....5), the most likely root cause is detected in >95% of the cases

This is a precondition to improve designs, production quality, system condition.

FIGURE 13.21
Joint root cause analysis of field failures.

13.3.2.4 Joint Root Cause Failure Analysis in Case of Failure/Defects

This task is a must for both users and manufacturers. It must account for a critical analysis of the following:

- Design features (design review versus specification test result control)
- Production quality review
- System condition review
- Final failure analysis (see Figure 13.21)

13.4 Conclusions

In Section 13.3.1, we have debated the *prime responsibilities of manufacturers* for assurance of high reliability of power transformers:

- Assure high competence in design (experiences, man power, tools, calculation, and design programs).
- Perform effective design review versus specification/standards and special site requirements; learn from failures (test floor and in service).
- Assure top competence in production quality (tools, machines, maintenance, tolerances, processes, in-process inspection).

Material quality: wherever possible, use world supply management/supplier development for materials (steel, copper, insulation, oil, etc.). Only qualified and by Materials laboratory released materials are allowed.

In Section 13.3.2, we presented the essential customer responsibilities for high reliability:

- Adequate specification in respect to standards, loading, and transient conditions (dielectric, thermal, dynamic)
- Consequent maintenance, service and monitoring and competent diagnostics, condition assessment in the form of fingerprinting, and trend analysis at regular intervals (DGA, all oil values, mechanical control, and moisture and furane control)
- Online monitoring as early warning system
- Application of consequent life extension measures: degassing/oil reclaiming/oil changing and core and coil drying
- Wherever possible: online monitoring of temperature/fault gases and moisture

Consequent root cause analysis of failures/defects jointly with the manufacturers.

References

1. Allan DJ (2001) *Why Transformer Fail*. EuroTechCon, U.K., pp. 99–112.
2. Kachler J (2001) Unique transformer failure statistics an important contribution to economics of power transformer management. *CIGRE*, Dublin, Ireland, Colloquium SC12, June 18–20.
3. Baehr R (2001) Transformer technology, state of the art and trends of future development. *Electra*, 198 (13), 13–19.
4. ANSI Standard C 57.117, IEEE (1986) Guide for reporting failure data for power transformers and shunt reactors on electric utility power systems.
5. Moore HR (2001) *The Impact of Design and Application on Costs, Performance and Reliability of Power Transformers*. EuroTechCon, U.K., pp. 169–180.
6. Kirchenmayer E, Knorr W (2001) Power transformer state of the art and new developments. *3rd Beijing International Conference on Power Transmission Technology*, Beijing, China, November 26–30.
7. Eckholz K, Heinzig P et al. (2001) Design features of very large generator step, up transformers. *3rd Beijing International Conference on Power Transmission Technology*, Beijing, China, November, 26–30.
8. Eckholz K (2003) Transformers special design. *Doble Conference*, Rome, Italy, October 26–28.
9. Kachler J (1997) Diagnostic and monitoring technology for large power transformers (*Fingerprinting, Trend Analysis from Factory to On-Site Testing*). *CIGRE*, Sydney, Australia, Colloquium SC12, October.

10. Sokolov V (1999) What research and technique are required to improve life management of transformers. *Transformer'99*, Kolobrzeg, Poland, April 27–29.
11. Kachler AJ (2000) On site diagnosis of power and special transformers. *ISEI 2000*, Anaheim, CA, April 2–5.
12. Höhlein I (2003) Transformers and service from Siemens. *Siemens Service Symposium*, Nuremberg, Germany, July 9–10.
13. Höhlein I, Kachler AJ et al. Transformer life management a customers' guideline for ageing analysis and laboratory diagnostics. Service Guidelines for Transformers in Service, Siemens Brochure: E50001-U410-A34 Order Number.
14. Kachler AJ, Kutzner R (2000) Transformer life management, SISTRAM+, The Siemens power transformer on, Line monitoring system, *Post CIGRE Symposium*, Siemens, Berlin, September 2–5.
15. CIGRE (2003) Guide for life management techniques for power transformers. Brochure No. 227.

14

Economics in Transformer Management: Focus on Life Cycle Cost, Design Review, and the Use of Simple Bayesian Decision Methods to Manage Risk

Kjetil Ryen

CONTENTS

14.1 Introduction

General strategies for dealing with risk include the following [1]:

- Investing for flexibility so that the changes can be made easily and inexpensively
- Investing in projects that are robust, that is, those that can perform across a variety of future possibilities
- Hedging against uncertainty
- Ignoring risk

The increased focus on profit in the short term often means deferred capital expenditure and reduced operational expenditure including maintenance and repair costs, without considering the inherent risk in doing so.

This is especially valid for power transformers with the high initial capex and low opex. This may give a low focus on costs generated late in the transformer life and the life expectancy itself.

This keynote paper on economics in transformer management aims to focus on two major decisions in the life of a transformer:

- Purchase, securing life cycle cost, and meeting performance requirements
- Decisions on suspect/defective transformers—whether to scrap or repair.

In the field of utilities and power generation, power transformers are decisive to the performance, reliability and productivity of the system. These factors are important because, in part, they determine the financial performance and economic sustainability of this business. Specifically, the technical complexities, high capital costs, and long life expectancy of transformers pose unique decision-making challenges to asset managers.

Another factor that increases the inherent business risk is the rulings of business regulators. For example, in 2001 the Norwegian power business regulator Norwegian Water Resources and Energy Directorate (NVE) introduced penalties for Energy Not Serviced (ENS) toward the grid operators. This system was introduced to give incentives to keep the reinvestments at a level that secures the long-term reliability of the system.

Hence, the need for a risk-based attitude to justify reinvestments is identified.

This chapter gives the background information for a keynote lecture on economics for which I have referred to many sources. I am grateful to Professor Bent Natvig for allowing me to use some of his accounts on risk [2,3] some years ago. The example of using simple Bayesian methods in a decision process was originally for a physician versus patient situation [3]. I identified the usefulness of Professor Natvig's approach and adapted it to a situation where a transformer engineer has to decide what to do with a possibly faulty transformer. I have also used the risk introduction [2] and the example given in the following text and in the CIGRÉ WG A2.20 guide on "Economics of transformer management" [4].

14.2 Risk and Risk Aversion

The focus of this chapter is to demonstrate that risk awareness, or risk aversion, has a clear mathematical basis that makes it a sound background for understanding how to manage a risk profile in a company.

Viktoria Neimane writes in her doctoral dissertation [1]:

> Uncertainty and risk management is becoming an essential an integrated part of the planning process in the electric power industry. For the past decades the uncertainties in load growth, capital cost and regulatory standards have challenged the planners. Uncertainties impose risk and explicit risk management is being developed.

General strategies for dealing with risk include the following:

- Investing for flexibility so that changes can be made easily and inexpensively
- Investing in projects that are robust, that is, those that can perform across a variety of future possibilities
- Hedging against uncertainty
- Ignoring risk

This approach is now more common in the planning process, but it may be claimed that this is not so evident in the specification and purchasing process of industry and power transformers. The increased focus on profit in the short term often means deferred capital (capex) expenditure and reduced operational expenditure (opex) including maintenance cost, without considering the inherent risk in doing so.

Let us assume that we have a utility where the management wants to reduce its risk profile. Operating high voltage equipment induces an appurtenant risk which may lead to a loss of utility by an unwanted event. This loss of utility may be property damage, third-party liability, value of energy not served, or increased unplanned maintenance such as repair/reinvestments after failures.

The issue here is not how the grid owner should spend money to reduce risk profile, but alternatives may be increased testing, monitoring, maintenance, reinvesting, or insuring. The first four alternatives may reduce the failure rate of the equipment amended, but insurance only reduces the monetary risk and not the inherent technical and business risk in operating the utility.

Here in Figure 14.1 we have assumed that it is possible to give a simple and linear numerical expression for the loss of welfare, or loss of utility, that we experience when losing the amount of x (in some currency). We name this loss of utility, or the loss function $L(x)$.

It may seem reasonable that a loss of, say, 2000 is considered twice as serious as a loss of 1000.

This is not necessarily the case. It may be as rational to evaluate the loss of the first 1000 to be less severe than the next 1000, which is additional. The better off we are, the easier it is to dispense with a small amount (refer to Daniel Bernoulli's statement). On the other hand, if we have lost most of our fortune, the loss of an additional 1000 may have severe personal consequences like bankruptcy and selling under execution, loss of credit and credibility, and, hence, a severely reduced standard of living and loss of social position and prestige.

This may give the background for asserting that the loss function should rise steeper the higher amount lost; refer to Figure 14.2. A marginal loss of, say, 1000 is now evaluated higher based on how high the previous loss was. This loss function is in accordance with the factual conduct of insurance in the market, and not the linear function of Figure 14.1.

Assume we have a risk of the simplest kind. There is a probability p for an event to occur including a loss of utility, for instance property damage, during the year. The loss

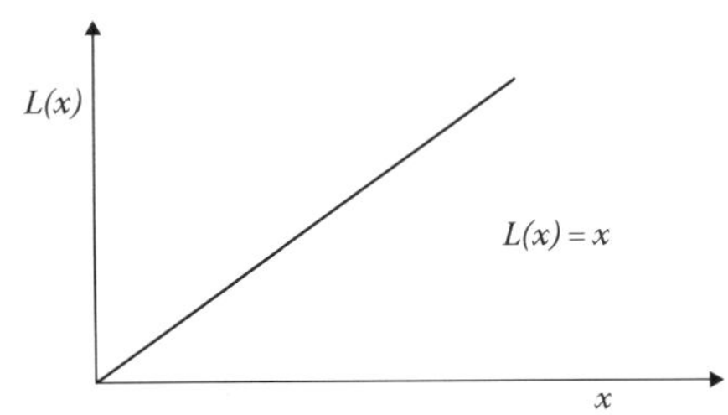

FIGURE 14.1
Linear loss function $L(x)$.

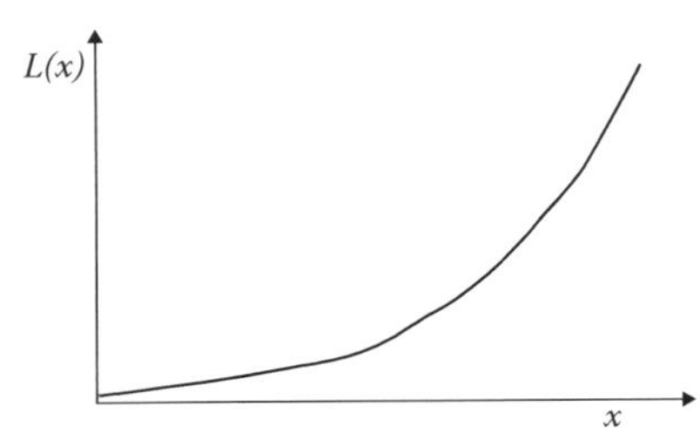

FIGURE 14.2
Curved loss function $L(x)$.

of utility is a fixed amount, z. The probability p may be interpreted as a (idealized) relative frequency. Assume for example, $p=0.01$. If a ("infinite") number of risks of the same kind are exposed in 1 year, 1% of these will cause loss of utility. The average annual loss of utility per risk is then

$$pz \tag{14.1}$$

The common interpretation of this average is expressed as "expected" annual loss of utility for the one risk we are considering. Correspondingly we find our expected annual loss of utility is

$$pL(z) \tag{14.2}$$

Let us now assume that we get an insurance company to offer a cover for full reimbursement of the risk for an annual premium of π, and further assume the insurance company covers a huge number of similar risks under the same conditions in the insurance. To not lose money on this way of providing insurance (with our conditions and assumptions), the annual premiums must be equal to or higher than the annual sustained claims:

$$pz \leq \pi \tag{14.3}$$

By insuring we experience a loss of utility $L(\pi)$. Insuring is beneficial for us if

$$L(\pi) < pL(z) \tag{14.4}$$

The L(loss) function is obviously a growing function and $pz \leq \pi$; hence,

$$L(pz) \leq L(\pi) \tag{14.5}$$

And combining (14.4) and (14.5) gives

$$L(pz) < pL(z) \tag{14.6}$$

for all $z>0$ and all p between 0 and 1. Dividing with pz on both sides, and putting $y=pz$ gives

$$\frac{L(y)}{y} < \frac{L(z)}{z}, \quad \text{for all } y \text{ and } z, \quad 0 < y < z \tag{14.7}$$

Figure 14.3 gives a geometrical representation of this relation. From (14.7) we see that this just implies the form of the curve in Figure 14.2. Now we can infer that if we want insurance covering the risk despite the fact the annual premium is higher than the annual expected loss of utility, our interpretation of this loss of fortune is really represented by an upwards curving loss function as in Figure 14.2, and not by the linear curve in Figure 14.1.

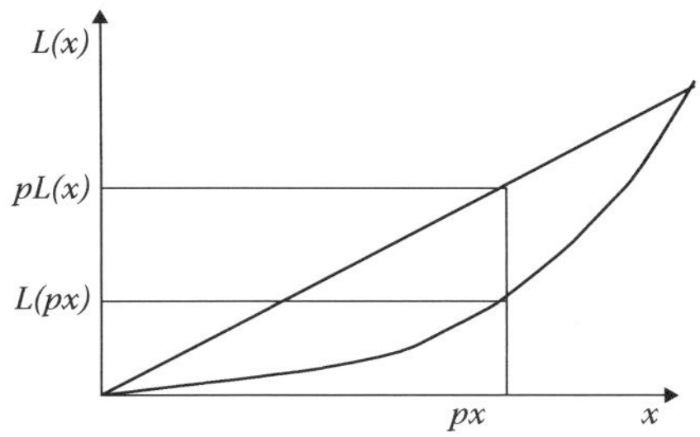

FIGURE 14.3
Risk evaluation at insuring.

If we have two alternatives: risk R has the probability p for loss and the amount lost is z, and risk S has the probability q for loss and the amount lost is y (which is not the same y as mentioned earlier).

- Risk R has probability p for loss z.
- Risk S has probability q for loss y.

If we put

$$L(px) < pL(x) \tag{14.8}$$

and we assume the probability q is greater than p, but at the same time the loss y is much lower than z, in such a way that the expected loss per year is identical in both cases:

$$pz < qy, \quad \text{for } p < q,\ z > y \tag{14.9}$$

The expected loss of utility for S, when using $0 \le (p/q) \le 1$ and (14.8), is

$$S = qL(y) = qL\left(\frac{p}{q}z\right) \le q\frac{p}{q}L(z) = pL(z) \tag{14.10}$$

This is the *expected loss of utility* for R.

We could also have multiplied (14.7) with qy on the lhs and with pz on the rhs directly as (14.7) applies when $y < z$.

We may conclude that we prefer risk S as compared to risk R, in spite of the fact that both cases have the same annual expected loss of utility.

In practice in the insurance business, the annual premium is always fixed far higher than the insurance compensation paid because the insurance company must cover administration costs and a rate of return to its shareholders. The equality sign as stated in (14.3) ($pz \le \pi$) is in fact never there.

But, this is no critique of sound practice based on long experience in the insurance business. Here, it is important to remember that we have considered a theoretical approach with an infinite number of equal insurance cases. In the real world the premium will be decided upon by the insurance company based on the specific risk profile of the policy holder. Hence, the theoretical considerations mentioned earlier do not say anything about a specific insurance case.

Based on the aforementioned theory, why do we consider insuring? If we are risk averse, we prefer a certain annual premium to an outcome with very low probability, but with a heavy cost consequence, in spite of the fact the premium "in average" is higher than the loss by carrying the risk.

The prudent transformer engineer may advise the management to look upon all maintenance, repair, and reinvestments as a methodical system to manage risk. This highlights methods for condition monitoring, test methods, and R&D activities trying to establish a better knowledge of how all our transformer maintenance activities affects the failure rates.

I do not state that the risks associated with a premium cost are *always* higher than the risk of a failure cost. This is true if the risk only involves one piece of equipment, but the cumulative effect of failure probabilities will normally make $p*L(z)$ larger, that is, the probability for a bushing failure + short circuit + fire + collateral damage or the maximum possible loss that involves a piece of equipment.

Insurance premiums are normally based on a location value percentage using expected equipment costs and probability of an insurance related occurrence. The insured risk assumption levels defined by the purchased insurance normally relate to deductibles and/or exclusions, which may be equipment specific. The deductible/exclusion is a tool used to control premium levels and risk assumption by the insured and the insurer. A higher deductible may mean a lower premium but also a higher insured risk assumption level.

With regard to possible future research in the field of economics, there is a need for methods to find the risk aversion profile *of a utility*. This means finding the risk tolerance threshold (risk compensating factors) of the public, management, investors, and regulators.

14.3 Risk Reduction by LCC Focus and Design Review

Risk and uncertainties is an important part of the life cycle cost (LCC) method. This chapter will not be an introduction on how to perform an LCC analysis, but only point at a few important facts in the context of risk mitigation versus transformer life. The very long time perspective is an important fact, as also the difficulties in obtaining relevant and reliable data. The premises and assumptions that are important for assessing transformer life are notoriously difficult to decide upon with the long lifetime of this component in the grid.

Hence, it is of utmost importance to know the design and where the parameter variations are, because these possible variations may be of importance to hedge against variations in the planning assumptions. Here performing a design review is an important hedging technique against the unknown and should be a normal part of risk mitigation when purchasing transformers.

The CIGRÉ WG 12.22 writes in Ref. [5] as follows:

> A design review is a planned exercise to ensure there is a common understanding of the applicable standards and specification requirements, and to provide an opportunity to scrutinise the design to ensure the requirements will be met, using the manufacturer's proven materials and methodology.

This document is an outline, highlighting the various design features and technical requirements which will be reviewed to ensure compliance with the contract. It does not include design limits or parameters. It is the responsibility of the purchaser to ensure that he or she has sufficient expertise to understand and evaluate the design. This document does not supplant the responsibility for the adequacy of the design, or the design limits, which remains with the manufacturer. Deficiencies, which are identified, shall be corrected. However, any changes that are a "betterment" to the design shall be subject to commercial resolution between the purchaser and the manufacturer.

Because the review will include information that is proprietary in nature, it is considered essential that the discussions and information exchanged during the design review process be kept confidential. In my opinion, the purchaser may ask for the documentation of any calculations, design principles, drawings or internal documents relevant to the transformer purchased, to be reviewed before, during or after the (initial) design review. The purchaser should, if requested, sign a confidentiality agreement concerning these documents. The basis for showing the customer the drawings and the calculations is of course the necessary trust between the parties.

The most important decision in the lifetime of the transformer is deciding the design. Hence, it is a bit surprising that it is still not common to perform a thorough design review when purchasing medium and large transformers. For large transformers it is necessary to perform a design review after short listing but before negotiations. This may be done either by prequalifying a necessary number of potential manufacturers before tender documents for the specific purchase are available, or after the bids are received. After the contract is signed, a formal design review is necessary both for medium and large transformers, and even for small transformers, for example, auxiliary transformers in power stations or distribution transformers bought in bulk.

It is surprising that only a few transformer purchasers in the Nordic countries are performing design reviews. In this context of risk management the use of a design review is one of the most important mitigation techniques to secure the compliance between the internal planning process, the technical specifications, and the final product. Figure 14.4 shows the basic perspective on the LCC method.

Here, it is important to recognize the rather narrow window where it is possible to affect the design from the customer side. It is far too common to travel to the manufacturer only at the factory acceptance tests (FAT), or even only at the site acceptance tests (SAT). The unfortunately true horror story is the customer seeing the transformer for the first time at

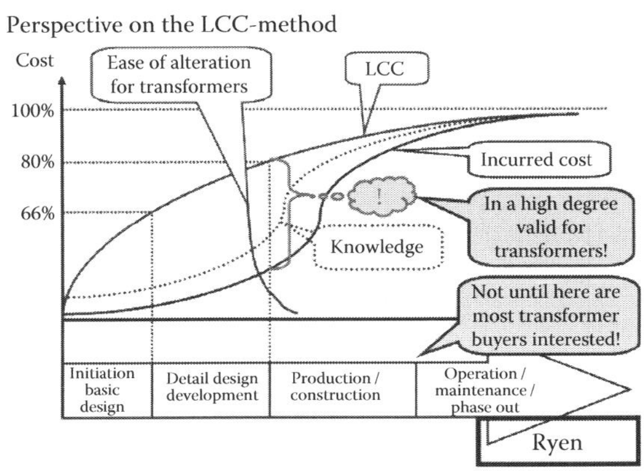

FIGURE 14.4
Short design alteration period for transformers.

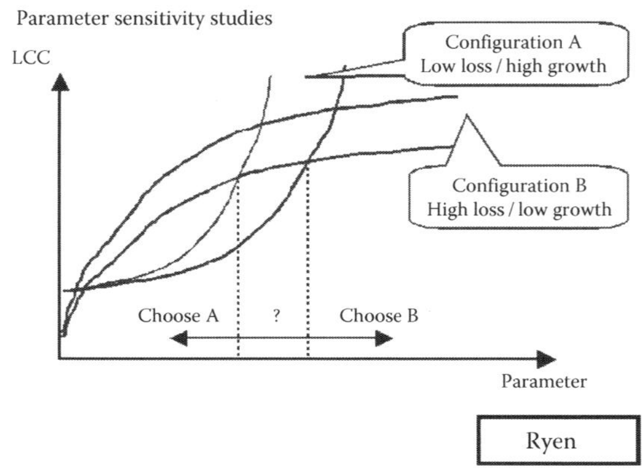

FIGURE 14.5
Parameter sensitivity study may aid choice decision.

the FAT exclaiming, "But, it doesn't have a tap changer!" This is of course no win–win situation, but a real disaster both for the customer and the manufacturer. Even a very simple design review before construction started would have avoided this malady.

In general, a sensitivity study should be performed, but it may be difficult because of the high number of parameters and the fact that a change of a single parameter may affect others. In Figure 14.5 I have tried to illustrate a sensitivity study of a choice between two transformers:

1. *Configuration A*, a transformer with low load losses, high initial capex, and the assumption of a high load growth.
2. *Configuration B*, a transformer with high load losses, low initial capex, and a low load growth.

A sensitivity study may be based on observed data/assumptions or theoretical calculations and simulations. The amount of data may prohibit an extensive study. Many of the parameters changed may also influence other parameters, and a multiparameter study may require a Monte Carlo study, which may not be justifiable.

Some of the uncertainties may be as follows:

- *Uncertainty linked to perception of reality*: A manufacturer told me that one customer demanded a GSU transformer to have a designed lifetime of 13 years, probably the same period as used in the net present value (NPV) analysis. The manufacturer stated they did not know how to make such a transformer. Separating the engineers totally from the purchasing process after they write the technical specifications is not a viable way to secure a minimum LCC.
- *Uncertainty in input data*: Load growth, hot-spot temperature, paper degradation velocity, future voltage change, etc.
- *Uncertainty in the modeling process:* Real operation of the grid from the dispatch centre (real risk aversion conduct), grid parameters, maintenance policies, real ambient temperatures at location, real overload time constants, hot-spot temperature overshoot factor, etc.

- *Uncertainties in the result of the analysis*: Real load flow versus calculated/planned.
- *Uncertainties in the usage of the results*: The capability of the planners to evaluate the result of an analysis may influence what is used as basis for a decision, and what is given to the manufacturer in the specification. Prejudice and old habits in the organization may hamper the change to new practices using changed and better parameters.

It is important to remember that many LCC analyses focus only on the costs. Using this approach in an NPV calculation will undervalue future maintenance and repair costs. In the long term, the increased probability for serious or catastrophic failures for a transformer can have a major impact on the appurtenant risk of having many older transformers in operation.

Especially with the risk of a $N-2$, or worse, situation is often underrated. In catastrophic failures involving transformer fires, fire brigades often demand that the station be dead before fire fighting actions commence, and these situations are often neglected in power system planning scenarios due to the very low probability and the inherent workload preparing for these situations. A sound design, a sound maintenance, monitoring, and reinvestment policy have a positive aspect of deferred outage and repair cost which should be part of an LCC analysis.

Considering the aging of transformer population in many countries with the possibly increasing probability of catastrophic failures, it is no longer possible to continue using the historical very low failure rate (without catastrophic failures) of power transformers, and prognosticate the same failure rate into the future.

Future research should focus on the connection between maintenance policies, physical degradation models (taking into account paper/metal/paper/humidity/oxygen), and their consequences for the development of the failure rate.

14.4 Simple Bayesian Decision Methods

The goal of this chapter is to give a relevant example of how to use simple Bayesian methods and to underline that it is possible to manage uncertainties without getting lost in the process. This chapter is based on a lecture given to Norwegian physicians in 1997 [3]. When I read this lecture in 2000, I recognized this as a relevant example, after making some necessary modifications, for the transformer engineer facing a difficult decision regarding a possible faulty transformer. This modified example is also used in the CIGRÉ guide 248: "Economics of transformer management" made by WG A2.20 where I was editor for risk management [4]. Here, a more fundamental part is added to connect the example to theory.

The power transformer engineer is really in more or less the same situation as a veterinarian: It is not possible to ask the patient relevant questions and a consultation of a patient assumed ill will always includes uncertainties. The latter is also a fact for physicians dealing with humans. Most of us know illness may exist even with normal consultations and test results (i.e., the test results are "normal," but later the patient returned ill).

Less known is probably the fact that consultation and test results may be positive, without the patient being ill (i.e., the tests "proved illness," but the patient was healthy). In the latter case we talk about "false positive." Uncertainties and the possibility of making a fault is a relevant normal situation in many areas and shows the scientific methods cannot prove anything, only state an argument with a certain probability of making a fault (Lat. *probare*, to prove).

So, when a veterinarian, physician or power engineer gives her opinion on a case by trying to identify a disease/condition from its signs and symptoms, that is, by stating a diagnosis, this is not an absolute truth; it is only a specific probability for the condition, disease or illness to exist.

The other way around, the diagnosis "healthy" is only made probable with a various degree of certainty. For the physicians the probability of being right is fortunately rather high, especially with important and serious diseases where many different tests are taken. But it is important to remember that with some conditions the physicians are able to give a correct diagnosis in less than 50% of the cases.

The situation for the power transformer engineer is unfortunately not much better; many of us have performed a post mortem on scrapped transformers only to find some of the transformers in a disappointingly "healthy" condition.

The probability theoretical hand tool being a glove to the hand regarding these questions is the *Bayes* formula. There exists a general agreement on whatever people think of the interpretation of, or basis for, Bayesian statistics.

The conditional probability for an event B, given A, is $P(B|A)$. If A is "component 1 is not functioning," and B is "component 2 is not functioning," the conditional probability $P(B|A)$ then states that the probability component 2 is not functioning, *given* we know component 1 is not functioning:

$$P(B|A) = \frac{P(B \cap A)}{P(A)} \tag{14.11}$$

$$P(B \cap A) = P(B|A) \cdot P(A) = P(A|B) \cdot P(B) \tag{14.12}$$

Solved for $P(B|A)$:

$$P(B|A) = \frac{P(A|B) \cdot P(B)}{P(A)} \quad \text{(Bayes formula)} \tag{14.13}$$

Let us illustrate the use of Bayesian methods considering an assumed, but relevant, example for a power transformer engineer or manager of a transformer fleet.

Let us suppose we are testing the "patient," a specific power transformer, assuming the winding insulation paper is in a very bad condition having low DP values.

We are defining the following events:

- $S = \{$The transformer is in a serious bad condition$\}$
- $L = \{$The transformer is in a little/slightly bad condition$\}$
- $N = \{$The transformer is in a normal/good condition$\}$
- $+ = \{$The test gives a positive result$\}$
- $- = \{$The test gives a negative result$\}$

From international, national, and/or our own transformer statistics (e.g., from inspecting scrapped or failed transformers being repaired) we know/believe that 2% of the transformer population is in a serious bad paper condition, 10% is in a somewhat better condition, but still slightly bad, and 88% of the transformer population is in a good paper condition.

From this statistics we may assume, if we do not utilize more subjective supplementary information, that these frequencies may represent the probabilities of the conditions:

- $P(S) = 0.02$
- $P(L) = 0.10$
- $P(N) = 0.88$

Also assume that we have, on the basis of a rather extensive database of experience, the following knowledge of the qualities of the testing method:

1. Considering one sample, the test in 90% of the cases gives a positive result when taken from a transformer in a seriously bad condition (i.e., 10% false negative).
2. The test gives a positive result in 60% of the cases when applied on a transformer in a slightly bad condition.
3. Finally, the test gives a completely wrong answer in 10% of the cases when administered on a healthy transformer (i.e., 10% false positive).

In our transformer case we may, for instance, take paper samples near the lid through manholes and testing for DP values. We take a paper sample near the lid where the highest average temperatures are present. Some transformer owners place a wire mesh basket with paper samples and presspan under a manhole for easy access through the lid for taking samples over the lifetime of the transformer.
From this information we can further assume that

- $P(+|S) = 0.90$ (i.e., the probability for a positive test result given the transformer is in a serious bad shape)
- $P(+|L) = 0.60$
- $P(+|N) = 0.10$

Now we want to calculate the probability for the transformer, really being in a seriously bad shape, given the test result is positive, that is, we want to find $P(S|+)$:

$$P(S|+) = \frac{P(+|S)\cdot P(S)}{P(+|S)\cdot P(S) + P(+|L)\cdot P(L) + P(+|N)\cdot P(N)} \tag{14.14}$$

If we now insert the numerical values for the probabilities on the rhs of the sign of equation, we get

$$P(S|+) = \frac{0.90\cdot 0.02}{0.90\cdot 0.02 + 0.60\cdot 0.10 + 0.10\cdot 0.88} = 0.11 \tag{14.15}$$

Hence, the probability for the transformer to be in a seriously bad condition, given that the result of the test was positive, is as low as 0.11. This is problematically low if this result is to be the foundation for a multimillion Euro reinvestment. Some will assert the test really is unfit for use.

The situation may be alleviated by taking another, independent, test, to gain more information. In our case, we take another paper sample from the transformer. This corresponds to the A and B samples in doping tests. Both of these tests must be positive to give a negative judgment.

Now we of course want to calculate the probability for our "patient," the transformer, really being in a seriously bad shape, given both test results are positive.

Let us introduce the following notation:

- +1 = {First test is positive.}
- +2 = {Second test is positive.}

Hence, we hunt: $P(S \mid +1 \cap +2)$, that is, the probability for the transformer, really being in a seriously bad shape, given both test results are positive.

Now we look at the situation after the first test is performed and has given a positive result. Instead of using the starting point $P(S)$, $P(L)$, and $P(N)$, based on our general transformer statistics, we now take the starting point with the *updated* probabilities $P(S \mid +1)$, $P(L \mid +1)$, and $P(N \mid +1)$ based on the information, the known fact, the result of the first test was positive.

By utilizing *Bayes* formula we earlier calculated that $P(S \mid +1) + P(S \mid +) = 0.11$.

Correspondingly we get

$$P(L \mid +) = \frac{0.60 \cdot 0.10}{0.90 \cdot 0.02 + 0.60 \cdot 0.10 + 0.10 \cdot 0.88} = 0.36 \tag{14.16}$$

$$P(N \mid +) = \frac{0.10 \cdot 0.88}{0.90 \cdot 0.02 + 0.60 \cdot 0.10 + 0.10 \cdot 0.88} = 0.53 \tag{14.17}$$

The independence of the tests may be interpreted like this: If we want to calculate the probability for the second test result being positive, given the transformer is in a serious bad shape, the result of the second test is not dependent on the result of the first test.

Hence, we have

$$P(+2 \mid S \cap +1) = P(+2 \mid S) = P(+ \mid S) = 0.90 \tag{14.18}$$

$$P(+2 \mid L \cap +1) = P(+2 \mid L) = P(+ \mid L) = 0.60 \tag{14.19}$$

$$P(+2 \mid N \cap +1) = P(+2 \mid N) = P(+ \mid N) = 0.10 \tag{14.20}$$

which are exactly the same probabilities as we used in the calculation of $P(S \mid +)$ earlier.

With this we replace $P(S)$, $P(L)$, and $P(N)$ with $P(S \mid +1)$, $P(L \mid +1)$, and $P(N \mid +1)$; and by again using *Bayes* formula, we get

$$P(S \mid +1 \cap +2) = \frac{0.90 \cdot 0.11}{0.90 \cdot 0.11 + 0.60 \cdot 0.36 + 0.10 \cdot 0.53} = 0.27 \tag{14.21}$$

This is much better than 0.11, but still rather low.

This method has been quite uncontroversial for many years because all the information is based on "hard facts." The backwards style emerges when the starting point is subjective supplementary information about the specific transformer instead of the general transformer statistics. It is difficult to understand this.

Let us pursue our dubious power transformer also after both tests of the paper samples have given positive results, that is, a low DP value.

Let us also assume the power engineer is told by management they need an advice for a decision based on the updated probabilities.

We have $P(S \mid +1 \cap +2) = 0.27$. Correspondingly, we find

$$P(L \mid +1 \cap +2) = \frac{0.60 \cdot 0.36}{0.90 \cdot 0.11 + 0.60 \cdot 0.36 + 0.10 \cdot 0.53} = 0.59 \tag{14.22}$$

$$P(N \mid +1 \cap +2) = \frac{0.10 \cdot 0.53}{0.90 \cdot 0.11 + 0.60 \cdot 0.36 + 0.10 \cdot 0.53} = 0.14 \tag{14.23}$$

Based on these two tests the transformer most probably is only in a slightly (not too) bad condition, which should not initiate costly investigations, for instance by moving the transformer to factory and opening it for a thorough inspection at this instance.

On the other hand, there is not an insignificant probability for the transformer to really be in a seriously bad condition and should maybe be tended to immediately. Hence, the power engineer/management is faced with a decision with great uncertainties: whether to move the transformer to the factory or not.

Let us again introduce the following notation for the two alternatives (decisions D):

- $D1 =$ {The transformer is moved immediately to the factory for inspection and possibly repair.}
- $D2 =$ {Wait some time and see.}

To come to a decision in this question, it is not possible to avoid a *partly subjective* estimation of the loss of utility by choosing $D1$ or $D2$ related to the factual condition of the transformer, S, L or N. If the engineer advises management to move the transformer to the factory and it really is in a healthy (enough) condition, this action represents wasted money entirely, for the factory which could have repaired or manufactured another, needier, transformer and for the utility which should have used the money on another transformer. The transformer itself is certainly not better off transported to and from the factory without any need, with increased probability for being damaged in the process.

Let us assume that the power engineer assessing the transformer has tabulated the loss of utility (in some currency).

We see from Table 14.1 the loss of utility is 0 if the transformer is in a seriously bad condition and the engineer rightly advises the management to "hospitalize" the transformer at once. If the engineer advises wrongly to wait and see in this situation, the most serious fault occurs when the transformer fails with the loss of utility being 200,000, due to the cost of energy not served and the extra cost for unplanned moving and repair.

TABLE 14.1

Estimated Loss of Utility for the Two Alternative Decisions

		Transformer Condition		
		S	*L*	*N*
Decision	*D*1	0	50,000	50,000
	*D*2	200,000	25,000	0

If the transformer is in a slightly bad condition, the loss of utility is the cost for transportation and inspection, 50,000, which is the same in both cases $(D1-L)$ and $(D1-N)$. If it is necessary to dry and reclamp the windings in case $(D1-L)$, the cost for doing this is not a loss of utility, and may even lead to a reduction in the loss of utility due to the increased life time expectancy, and the reduced failure rate, of the refurbished transformer.

If the transformer is in a good condition and the engineer advises to wait and see, again the loss of utility is 0. If, in this good condition, the transformer is moved to the factory, the second worst decision is taken with a loss of utility of 50,000. The loss of utility may even be higher as the moved, inspected, and unrepaired transformer may be damaged during transportation and handling and thus may have a higher probability for a future failure than before moving.

Here, we use the common definition of risk as the *expected loss of utility*. This means that it is necessary to weigh together the loss of utility for the different thinkable conditions that may arise, where weights are the best estimated probabilities for these conditions.

Hence, the appurtenant risk with the decision to "hospitalize" the transformer at once, *RD*1, is

$$\begin{aligned} RD1 &= 0\cdot P(S\,|\,{+1}\cap{+2}) + 50{,}000\cdot P(L\,|\,{+1}\cap{+2}) + 50{,}000\cdot P(N\,|\,{+1}\cap{+2}) \\ &= 0\cdot 0.27 + 50{,}000\cdot 0.59 + 50{,}000\cdot 0.14 = 36{,}500 \end{aligned} \tag{14.24}$$

Correspondingly, the appurtenant risk for "wait and see," *RD*2, is

$$RD2 = 200{,}000\cdot 0.27 + 25{,}000\cdot 0.59 + 0\cdot 0.14 = 68{,}750 \tag{14.25}$$

Consequently we see the expected loss of utility, or risk, for a "wait and see" decision is, in this example, much higher than the risk for "hospitalizing" the transformer at once.

It is quite obvious that if this was the only available information on the condition of the transformer, the engineer should advise the management to decide upon the first decision (*RD*1), regardless that the best estimated probability for the transformer being in a seriously bad condition is as low as 0.27. A brief sensitivity analysis shows that this decision is robust for the relatively gross changes in the costs chosen.

Assume we need to choose between k possible decisions $B_1, B_2, \ldots, B_k$. The decision maker must subjectively estimate the appurtenant loss of utility for choosing the decisions $B_1, \ldots, B_k$ in relation to the true value of the parameter Θ. This loss of utility is given by the *lossfunction* $L(\Theta, Bi)$, which gives the loss of utility for coming to the decision B_i when the parameter takes the value Θ. The loss function is consequently a function of two variables.

The risk, or, *à posteriori expected loss of utility*, by decision B_i is given by

$$\sum_{j=1}^{\infty} L\left(\Theta_j, B_j\right)\pi\left(\Theta_j|D\right) \tag{14.26}$$

if Θ can take values in $\Theta = \{\Theta_1, \Theta_2, \ldots\}$.

If Θ is part of a set, Θ, of real numbers, the summation must be replaced by integration

$$\int_{\Theta} L\left(\Theta_j, B_j\right)\pi\left(\Theta_j|D\right)d\Theta \tag{14.27}$$

The appurtenant risk by taking the decision B_i is consequently a weighted average of the expected loss of utility, where the weights are the à posteriori probability density (point probability):

$$\pi(\Theta\,|\,D)$$

The á posteriori Bayes decision is then the decision from among B_i, $i=1, \ldots, k$ which has the lowest appurtenant risk.

In general, the decision maker must subjectively evaluate the loss function. In this subject, the number of people resisting using subjectively estimation methods also in deciding on the probabilities is remarkably high.

Future research should develop better Bayesian methods used in PC applications used *interactively* during the decision process by a set of skilled engineers discussing the subjectively added information available on the specific transformer family (design), or a single transformer evaluated making better probability distribution for the different failure modes. Applications combining this with risk evaluation methods would be useful.

14.5 Which Probability Rate to Use

When considering risk at the last major lifetime decision in a transformer's life, that is, when to scrap, we must have some ideas of what failure rate to use in our considerations. Let us first look at the dominant failure modes and their failure rates.

From Ref. [6]

> Model experiments show that an ac pre-stressed insulation (3–4 kV/mm) gets a 25–30% reduction in the switching impulse withstand level (SIWL), when wet fibres (2–3% water in cellulose) are present. Electrostatic forces dragging particles into highly stressed oil volumes explain this.

From Table V in Ref. [7] ("The Bossi Report") we find that substation transformers with OLTC for voltages 60+ to <100 kV (winding highest voltage) have a failure rate 2.1%, of this 1.5% is with forced trip. For 100+ to <300 kV the failure rate is 2.3%, of this 1.6% is with forced trip.

From the same report, 70% of all transformer failures cause a forced trip. It is worth making note of Figure 13 in Ref. [7], showing all the failures with forced trips—dielectric failures caused repair times, mainly those more than 30 days.

Of all failures with dielectric cause on transformers, OLTC failures constitute 31%. For winding failures the cumulative probability for outage duration (refer to Figure 16 in Ref. [7]) shows that 80% of all failures have an outage duration more <300 days, while 40% of the failures have an outage duration <100 days, and 20% have outage time <50 days or less.

Considering all the failures, it may be noted, with reference to the first component involved, that about 41% of the failures were due to the OLTC and about 19% due to the windings; the failure origins were 53% mechanical and 31% dielectric.

From an American investigation performed by Hartford Steam Boiler Inspection and Insurance Co. (HSB), we see less than 21% of all failures had a dielectric cause, and 9.3% of all failures originate from moisture [8]. Notice the rather high explosion rate of 2.4% (Table 14.2).

Moisture here refers to both external leakages and confirmed high water content in the oil, so this frequency cannot be used for calculating the number of transformers with high moisture in the oil.

The major impact of "Insulation Issues" is confirmed in Ref. [9]. The Doble and ZTZ service (A Ukrainian-based institution) return of experience demonstrates the following facts [7]:

- 70% of the failures occur in units older than 30 years.
- 50% of the failures originate from OLTCs and bushings.
- 15%–20% are due to reduction in impulse withstand voltage due to water and particles in the insulation.
- 3%–5% are due to excessive aging (e.g., poor cooling).
- 10%–15% are due to mechanical weakness/winding distortion.

It is of interest to look at the experience of the American insurance company (HSB) regarding the extent of failures with dielectric causes *"Deterioration of Insulation"* [8]. Insulation

TABLE 14.2

Percent of Occurrences within Defined Categories from HSB Investigation

Type of Failure	Percent of Occurrences
Electrical disturbances	25.4
Insulation issues	21.0
Lightning	13.7
Overload	10.1
Moisture	9.3
Loose or high resistance connection	6.5
Other	5.6
Poor workmanship	4.0
Explosion	2.4
Maintenance issues	1.6
Sabotage	0.4

deterioration was the second leading cause of failure over the last 20 years. This category includes only those failures where there was no evidence of a line surge. There are actually four factors that are responsible for insulation deterioration: pyrolysis (heat), oxidation, acidity, and moisture. But moisture in the oil is reported separately. The average age of the transformers that failed due to insulation deterioration was 17.8 years—a far cry from the expected life of 35–40 years.

In Australia the failure rate for "costly failures" is 0.4% per transformer year. It is assumed these are mainly windings damage [10].

To consider the risk level for a transformer, the use of monitoring, measuring parameters at interval, and evaluating the condition based on an interpretation of the measured results is a viable method. The use of a failure tree is also convenient. The idea of using a probability tree is taken from a CIGRÉ transformer session paper in 2002 [11]. I have developed the probability tree further to include the probable reduction in repair cost due to early detection of faulty conditions through monitoring. This developed probability tree is taken from a study of using continuous degassing of transformer oil.

The reduced risk for bubbling with a good maintenance, and the ability to increase loading should also be taken into account [12].

The inherent problem with failure statistics is relevant for the transformer population: age, area, voltage classes, weather, design, maintenance, etc.

A major challenge is to estimate the development of the failure rate during the life time. Currently there is no indication for a *general* increase in the failure rate for power transformers. Hence, it is difficult to claim that the bathtub curve is very evident for power transformers. The left hand part of the bathtub curve is probably part of the FAT and the right hand part of the curve is simply not there (Europe), with the exception of catastrophic failures. The inherent failure rate for stochastic failures is more or less constant through the lifetime. These stochastic failures are maybe the main failure mode for power transformers.

In CIGRÉ WG A2.20 [4] this subject was discussed and Table 14.3 shows our viewpoint on a possible model for the development of the failure rate for different transformers versus age.

The next major question is when to get worried. In the CIGRÉ working group we considered a failure rate >2% for unwanted, given a "normal" risk level (or expected loss of utility). CIGRÉ has started work on getting updated failure statistics for all the important components in the substations.

A convenient way of presenting the distribution of failure rate on different failure modes is the probability tree. The idea with the probability tree is the necessity to modify the known failure statistics with subjective knowledge and beliefs about the future. A relevant discussion of which failure frequency to use is possible using competence in our own organization: planning dept., dispatch centre, maintenance dept./contractors, using

TABLE 14.3

Failure Rate over Lifetime

Transformer Age (Years)	Substation Units (%)	Generator Units (%)
≤15	0.5	0.8
16–24	1.0	1.5
25–34	1.5	2.0
35–50	2.0	2.5
>50	3.0	3.5

reported data from IEEE, IEE, CIGRÉ and data from own IT-systems, like an Enterprise Resource Planning system (ERP). Such discussion may also reveal the internal myths and misunderstandings, and reveal data and knowledge not known by all the relevant departments in the company. Such discussions are important in the context of developing the *learning organization* (Figure 14.6).

A relevant example is the previous probability tree, based on the work presented in Ref. [11].

Point 1

From Ref. [7], an average failure rate of 2% p.a. is chosen when no better data or subjective information is known. An underreporting in the failure statistics is often seen, due

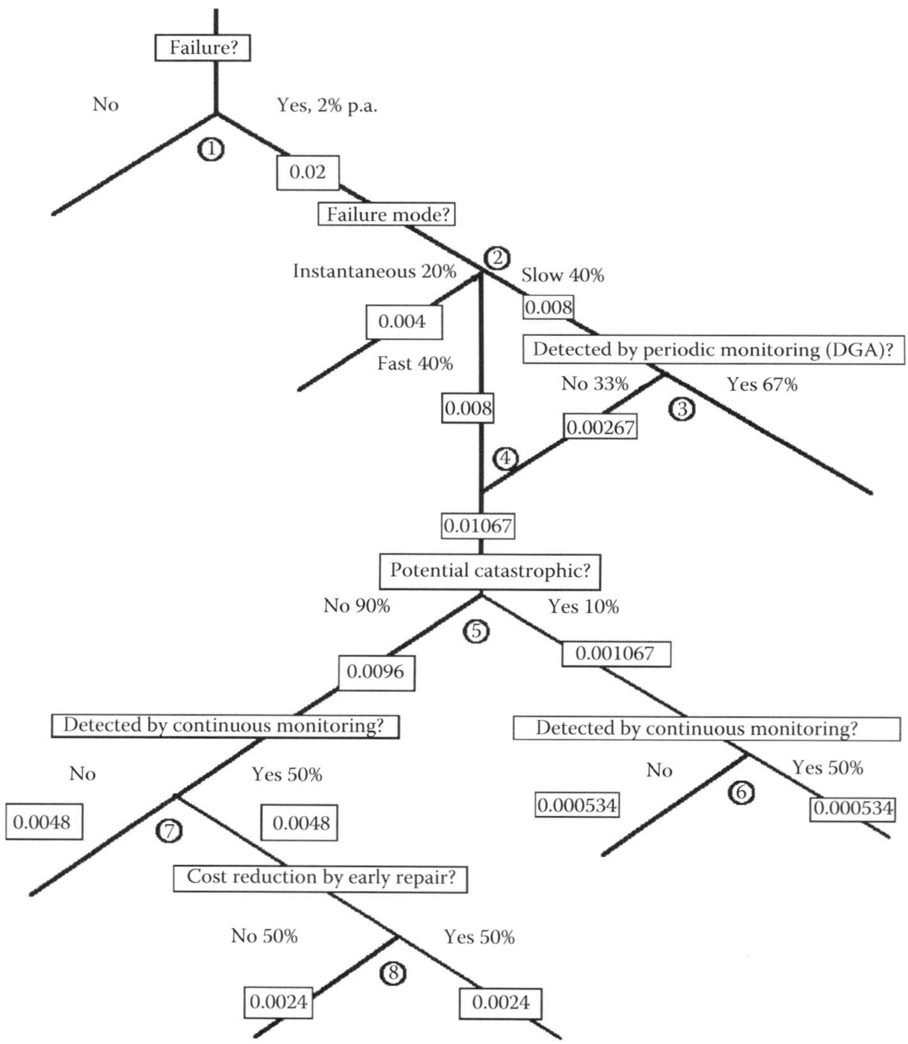

FIGURE 14.6
Probability tree for monitoring function.

to the problem of getting reported faults detected during maintenance, disconnections done manually by the dispatch centre, no reporting of failures giving no trip, too high focus on reporting demands by regulator, etc. Canada has experienced higher failure rates than 2% [6].

Point 2

I distinguish between the three failure modes: instantaneous failures, fast (developing) failures, and slow failures.

Instantaneous failures

These faults have principally two conditions:

- The inherent failure rate constant during the whole lifetime of the transformer. One obvious reason for this constant failure rate is that many of the parameters are chosen for a specific stress level. For instance, the insulation system is designed for maximum normal field strength and a specific BIL level, based on the customer specification and/or international standards. Stress higher than this level increases the failure probability. Aging reduces the withstand level: aging of oil, humidity in oil and paper, paper degradation, etc.
- The further nexus of cause and effect for the instantaneous faults are external stress factors like lightning giving stress levels higher than design, or actual withstand level, causing failures like internal flash over. No surge arrestors, or arrestors failing at first strike and the transformer failing at next (re)strike, may cause serious consequences. Moisture in the insulation paper reduces the withstand level significantly. High moisture level in the oil and insulation (here typical 2%–3% water in the insulation paper) may give a reduction 25%–30% of the withstand level with long wet fibers in a harmful orientation in the field and location in the insulation system [6,13].

Ref. [9] shows that 15%–20% of all failures are caused by reduced insulation withstand level. The American Insurer HSB says 21% of all failures have dielectric causes. I have chosen 20% for instantaneous failures, including the inherent failure rate for "unknown cause of dielectric failure" when investigating a failed unit.

Fast faults

Principally these faults occur between each periodic check of condition. With the now normal use of oil parameter screening, these faults occur between oil samplings for oil and DGA. The probability for such a fault to develop into a failure causing a disconnection is rather high.

Slow faults

These failures occur in the periods between condition monitoring which may be inspection, revision or an oil analysis/DGA.

It is difficult to find a well-documented basis for dividing up the failure rate. Here I have chosen to split the failure rate into fast and slow faults:

$$\frac{(2-2\cdot 0.2)}{2(\%)} = 0.8(\%) \tag{14.28}$$

Instantaneous faults	20%:	0.4%
Fast faults	40%:	0.8%
Slow faults	40%:	0.8%

Point 3

It is normally the slow developing faults which are discovered by regular gas analysis of oil samples by DGA. It may be reasonable to assume most of the slow faults are discovered by the regular (annually–biannually) DGA. In default of other data, I choose the same as the National Grid/British Energy in United Kingdom in Ref. [11].

Point 4

This is a summation point for the fast and slow faults detected by regular DGA

Point 5

Here is a distinction between faults with a catastrophic outcome, and the rest. The division for the failure rates with 10% catastrophic (and the rest 90%) is based on Ref. [7] and own evaluation.

Point 6

The devices for continuous fault gas monitoring (like Hydran, Kelman) give an alarm for increased rate of gassing, and an additional oil sample for a DGA may be taken.

Point 7

It is assumed that the monitoring function detects 50% of the non-catastrophic failures [11].

Point 8

Further it is assumed that the early detection of the non-catastrophic failures brings cost-reductions for half of the failures. These costs are both the direct cost of repair and the regulatory costs, for example, cost of Energy Not Serviced (ENS), or goodwill (bad-will) costs for other business sectors of the concern.

The same remarks on future research and developments may be given here: Better Bayesian methods used in PC applications used *interactively* during the decision process by a skilled set of engineers discussing the subjectively added information available on the specific transformer family (design), or a single transformer evaluated making better probability distribution for the different failure modes. Applications combining this with risk evaluation methods would be useful.

References

1. Neimane V (2001) On development planning of electricity distribution networks, Doctoral dissertation, Kungliga Tekniske Högskolan, Royal Institute of Technology, Stockholm, Sweden. http://media.lib.kth.se/dissengrefhit.asp?dissnr=3253.

2. Natvig B (1997) *Sannsynli ghetsvurderinger i atomalderen (Probability Evaluations in the Atomic Age)* (ISBN 82-00-03325-2), pp. 30–35, Universitetsforlaget AS, Oslo, Norway (in Norwegian).
3. Natvig B (1997) *Hvordan tenkte Thomas Bayes?* (in Norwegian) *(How Did Thomas Bayes Think?)* (ISBN 82-553-1102-5), pp. 4–9, Statistical Memoires, University of Oslo, Oslo, Norway.
4. Economics in transformer management (2004) CIGRÉ 04, Brochure No. 248.
5. Guidelines for conducting design review for transformers 100 MVA and 123 kV and above (2002) CIGRÉ 02, Brochure No. 204.
6. Lundgaard L, Linhjell D, Hansen W, Anker MU (2001) *Ageing and Restoration of Transformer Windings*, SINTEF Energy Research (TR A5540/EBL-K 43-2001), Trondheim, Norway.
7. Bossi A (1983) An international survey on failures in large power transformers in service–Final report of CIGRÉ Working Group 12.05, *Electra*, No. 88, pp. 22–48.
8. Bartley WH (2001) Failure history of transformers—Theoretical projections for random failures. *Proceedings of TechCon*, Mesa, AZ.
9. Sokolov V, Berler Z, Rashkes V (1999) Effective methods of assessment of insulation system conditions in power transformers: A view based on practical experience. *Electrical Insulation Conference and Electrical Manufacturing & Coil Winding Conference, Proceedings*, Cincinnati, OH, ISBN: 0-7803-5757-4.
10. Austin P (June 2001) Transformer economic issues in Australia & New Zealand. *CIGRÉ*, Merida, Mexico, Colloquium SC 12.
11. Breckenridge T, Harrison TH, Lapworth JA, Mackenzie E, White S (2002) The impact of economic and reliability considerations on decision regarding the life management of power transformers, CIGRÉ 02, Report 12-115.
12. McNutt WJ (1998) A new tool to guide loading decisions for power transformers. *Doble Conference*, USA.
13. Particles in oil (2000) CIGRÉ 00, Brochure No. 157.

Part II

Instrument Transformers

Elzbieta Lesniewska

Introduction

Instrument transformers are parts of electric power systems. Their aim is to transfer operating values of currents and voltages for measuring or protective applications. Nevertheless, while the principle of operation of inductive instrument transformers is the same as for the power transformer, the requirements put to them concern the transformation of currents and voltages with high accuracy.

The beginnings of the formation of electric power systems were the beginnings also of instrument transformers. The first primitive instrument transformers appeared over 100 years ago. In 1899, the first current transformer was patented by Siemens (Electrical Measuring Instrument). At the beginning of the measuring of electric energy, instrument transformers were not treated seriously. The problems of magnetic circuits, insulations, accuracy, and transient states did not exist. The first papers on instrument transformers appeared around 1906–1909. The increase of power and rated voltages of transmission lines caused an increase in the importance of instrument transformers as measurement devices and the stepping up of the requirements of their accuracy. Although instrument transformers were applied earlier, only at the end of 1950s was an interest in electromagnetic processes taken in them. The development of construction designs has depended on progress in materials and technology. At the end of the 1950s publications concerning electric and magnetic values versus time and transient states first appeared.

A very important moment for the development of instrument transformers was the introduction of cold-milled silicon steel, which resulted in the improvement of instrument transformer constructions and, consequently, a high accuracy at decreased ampere-turns. In the case of very high accuracy, instrument transformers permalloy or mumetal can now be applied.

New constructions of instrument transformers have appeared for which the principles of operation are different. We now have electronic devices that use very small signals from Rogowski's coil or from resistive or capacitive voltage dividers. We also have optical instrument transformers using Faraday's law or Pockels' effect and fiber-optics to transmit signals and microwave devices using "Bluetooth" technology.

Inductive instrument transformers are, however, still the biggest group of instrument transformers, and the problems concerned with their design are different than for power transformers.

Engineers have constantly worked to improve their constructions and use better and better methods to design them. For more than a decade producers have been interested in combined constructions that join, in one case, current and voltage transformers working as two independent devices. The combined instrument transformer brings new, difficult challenges for designers.

In the following chapters, a new approach toward construction design is presented. The authors present the results of the use of numerical methods in the design process. These methods can be used to optimize constructions.

Field methods help in computing magnetic and electric field distribution and in designing magnetic circuits as well as insulation systems and operating and transient state parameters of instrument transformers. The authors also cover the elimination of mutual interactions between devices during the design process and protection against transmission disturbances. Capacitive couplings make the transfer of HF disturbances possible. The problem is essential from the perspective of the correct operation of measuring and protecting equipment installed in the secondary circuits of current and voltage transformers. Thus, research concerning the transfer of conducted disturbances in the frequency range of 100 kHz–30 MHz or even 50 MHz, covering both simulation analysis and experimental research on real-life models of instrument transformers, becomes increasingly important. In order to gain an insight into the problem, HF disturbance transfers have been examined both for harmonic and pulse signals.

15

Applications of Field Analysis during Design Process of Instrument Transformers

Elzbieta Lesniewska

CONTENTS

15.1 Introduction

Instrument transformers are very important parts of electric power systems. They are indispensable for the proper functioning of the system, because they are elements of the measurement systems as well as the protection systems. The requirements put to them concern the transformation of currents and voltages, with high accuracy, and they are different for measuring and protective instrument transformers. The measurement characteristics of current, voltage errors, and phase displacement decide about the accuracy class of the instrument transformers. Consequently, the cores and windings must be constructed to fit that class.

Nevertheless, while the principle of operation is the same as for the power transformer, the problems concerned with the design are different.

The aim of constructor is to obtain the best parameters at assumed dimensions, for example, obtaining the highest rated power at a given accuracy class as well as shaping an insulation system to obtain the proper electric strength of the system.

The nonlinear magnetic characteristic of the magnetic circuit and eddy-current losses in the core should be taken into account. The instrument transformers can be constructed as combined instrument transformers, as they consist of current and voltage transformers in the same case. Prevention of possible couplings between circuits should also be taken into account.

HV and medium voltage devices also create many problems concerning the electric strength of the insulation system of the instrument transformer, especially for the combined instrument transformer.

All these phenomena have influence on the accuracy of the instrument transformer.

Application of electromagnetic field simulations [18,19] in the design process of instrument transformers enables the solution of some design problems, which cannot be solved using traditional methods.

This new approach was presented and discussed during this workshop. The presented methods are widely illustrated with practical examples that have been verified by experimental research conducted by the author and by the experience of the manufacturers of instrument transformers with whom the author has cooperated for a number of years.

15.2 Solution of Constructional Problems

Application of 2D and 3D field methods during design of instrument transformers creates wide possibilities for constructors:

15.2.1 Computations of Electromagnetic and Magnetic Field Distributions

Give the possibility of determining the equivalent circuit parameters and shaping those of them which have influence on the measurement properties of the instrument transformers. They are the leakage reactance of the secondary windings in the case of current transformers and the primary and secondary windings in the case of voltage transformers. Effect of limited reactance can be obtained by shaping the cores and windings or by using magnetic or electromagnetic shields. The shape and usefulness of the shield can also be estimated using field analysis.

15.2.1.1 Determining the Equivalent Circuit Parameters [1]

The magnetic harmonic field may be described, with the use of the magnetic vector potential **A** ($\mathbf{B} = \nabla \times \mathbf{A}$.), by Poisson's equation:

$$\nabla^2 \underline{\mathbf{A}} - \mu \nabla \left(\frac{1}{\mu} \right) \times \nabla \times \underline{\mathbf{A}} = -\mu \underline{\mathbf{J}}_w \tag{15.1}$$

with Neumann's and Dirichlet's boundary conditions.

Only the 3D analysis allows the correct estimation of the leakage field distribution for windings wound around cores. Applying the numerical finite element method in 3D systems makes it possible to solve this equation. The professional software OPERA3D uses instead of this partial differential equation to calculate magnetic field distribution another partial differential equation for the reduced scalar potential ϕ:

$$\nabla \cdot \mu \nabla \phi - \nabla \cdot \mu \left(\int_{\Omega_J} \frac{\mathbf{J} \times \mathbf{R}}{|\mathbf{R}|^3} \right) d\Omega_J = 0 \tag{15.2}$$

Total leakage reactance of the voltage transformer in terms of its secondary side can be determined from the energy of leakage magnetic flux using magnetic field distribution of the voltage transformer:

$$X_r = \omega L_z = \omega \frac{2w_m}{I_{2m}^2} \tag{15.3}$$

where

$$w_m = \frac{1}{2} \int_V \left(\int_0^B \mathbf{H} \cdot \mathbf{B} \right) dv = \frac{1}{2} \int_V \mathbf{H} \cdot \mathbf{B} dv \tag{15.4}$$

Using field methods makes it possible to determine instrument transformer parameters, especially the leakage reactance of the secondary and primary windings, independently from the shape or localization of the core and the windings. The analytical formulas are true only for cylindrical coaxial windings located on the same column of the core [2]. These formulas have been derived with the assumption that the distribution of the flux in the gap between the windings is uniform and that the flux is divided equally between the primary and the secondary windings, which is not true in many cases.

Consider this example (HV voltage transformer shown in Figure 15.1) where the value obtained based on field computation equals 0.187 Ω and the value obtained by testing on a real-life model is 0.190 Ω.

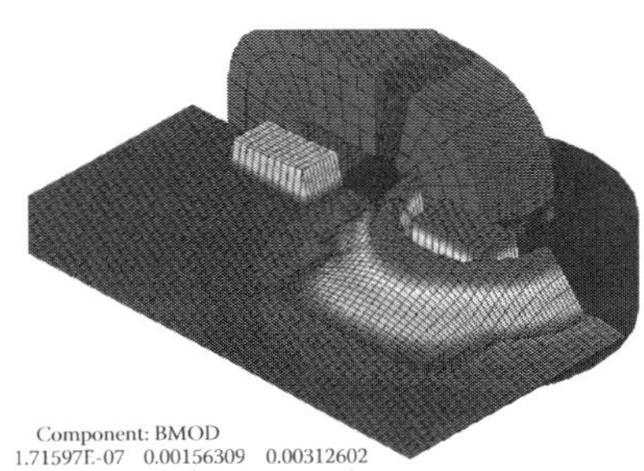

FIGURE 15.1
Magnetic flux density distribution of leakage flux of a voltage transformer 100 VA and ratio $15:\sqrt{3}/0.1:\sqrt{3}$ kV.

15.2.1.2 Choosing an Optimal Design Version [3]

In regions with currents of known and externally forced distribution, the harmonic electromagnetic field may be described, using the magnetic vector potential **A** ($\mathbf{B} = \nabla \times \mathbf{A}$.), by an equation of Helmholtz type:

$$\nabla^2 \underline{\mathbf{A}} - \mu \nabla \left(\frac{1}{\mu} \right) \times \nabla \times \underline{\mathbf{A}} - j\omega\mu\gamma \underline{\mathbf{A}} = -\mu \underline{\mathbf{J}}_w \tag{15.5}$$

with Neumann's and Dirichlet's boundary conditions.

Applying the commercial software OPERA3D based on the numerical finite element method makes it possible to solve this equation and estimate the field distribution in each region.

For 3D problems, the number of elements is limited by the possibilities of software. Therefore, the method of joining solution in different regions ought to be used [4].

In conductive regions without currents, Equation 15.5 is transformed into the homogeneous vector Helmholtz equation:

$$\nabla^2 \underline{\mathbf{A}} - \mu \nabla \left(\frac{1}{\mu} \right) \times \nabla \times \underline{\mathbf{A}} - j\omega\mu\gamma \underline{\mathbf{A}} = 0 \tag{15.6}$$

In linear nonconductive regions with currents, the total field intensity **H** is defined using the reduced field intensity $\mathbf{H}_m$, $\mathbf{H}_e$, and $\mathbf{H}_s$ ($\mathbf{H}_e$ is field from induced eddy currents in conductors):

$$\mathbf{H} = \mathbf{H}_m + \mathbf{H}_e + \mathbf{H}_s \tag{15.7}$$

where $\mathbf{H}_m + \mathbf{H}_e$ is computed using reduced magnetic scalar potential φ.

$$\mathbf{H}_m + \mathbf{H}_e = -\nabla \phi \tag{15.8}$$

in linear nonconductive regions without currents, where $\nabla \times \mathbf{H}_m = 0$ and the electromagnetic field may be described by the equation:

$$\nabla^2 \phi = 0 \tag{15.9}$$

and $\mathbf{H}_s$ is described by Biot–Savart's formula.

In a region of free space that does not include source currents, it is generally found to be most efficient to replace the magnetic field by the gradient of a magnetic scalar potential, ψ, so that

$$\mathbf{H} = -\nabla \psi \tag{15.10}$$

$$\nabla \cdot \mu \nabla \psi = 0 \tag{15.11}$$

The leakage reactance X_{2r} depends on the leakage magnetic flux related to the secondary winding, which is expressed by the definition:

$$X_{2r} = \omega \frac{\hat{\psi}_{2r}}{\sqrt{2}I} \tag{15.12}$$

where $\hat{\psi}_{2r}$ is the maximum of the leakage magnetic flux linked with the secondary winding.

The leakage magnetic flux linked with the secondary winding was calculated as a sum of magnetic flux linked with each turn based on the computed magnetic field distribution.

Analysis of magnetic field distribution gives the possibility of determining the leakage reactance of the second winding, which decides the accuracy class of the current transformer, and of choosing the most profitable arrangement of the cores. As an example was consider two constructional versions of a current transformer (Figure 15.2). In one version,

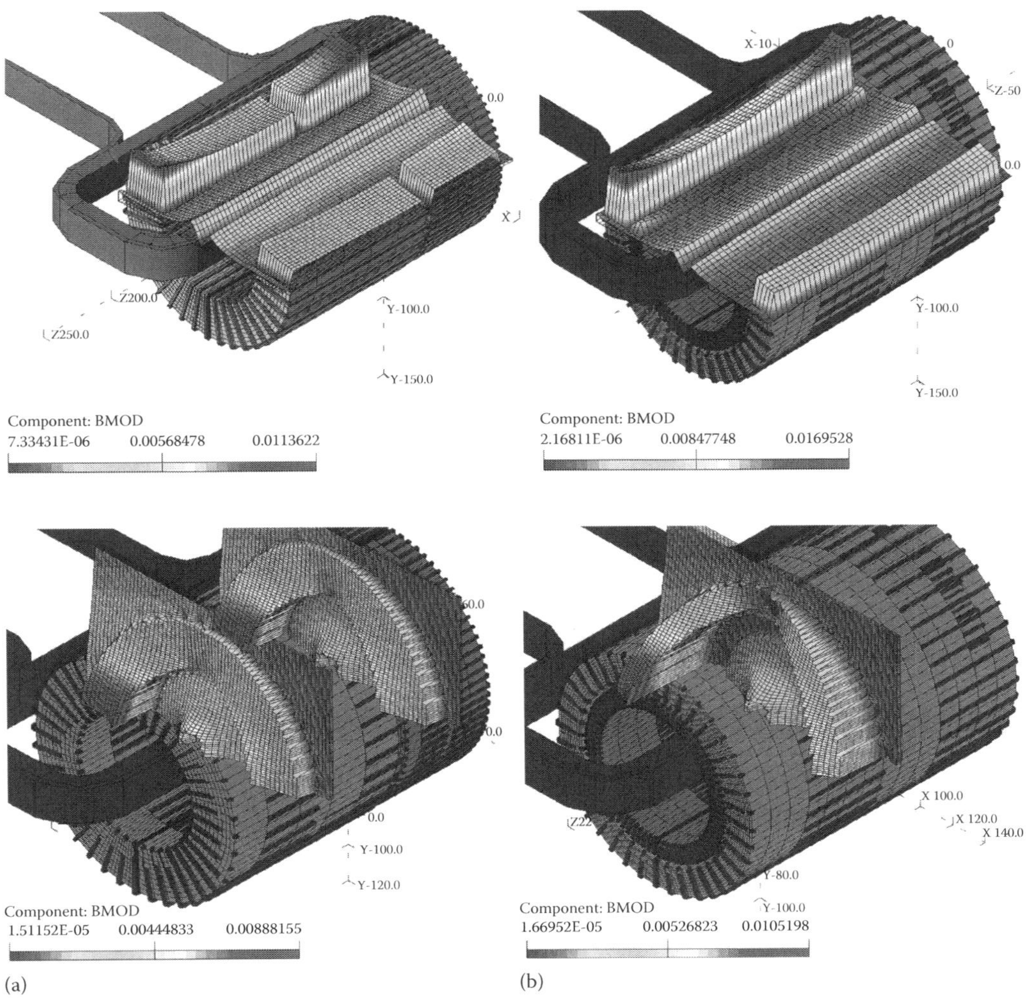

FIGURE 15.2
Distribution of magnetic flux density (T) of leakage flux at the surfaces passing through the centers of cores for two models of a current transformer 50/5/5 A: (a) model 1 and (b) model 2.

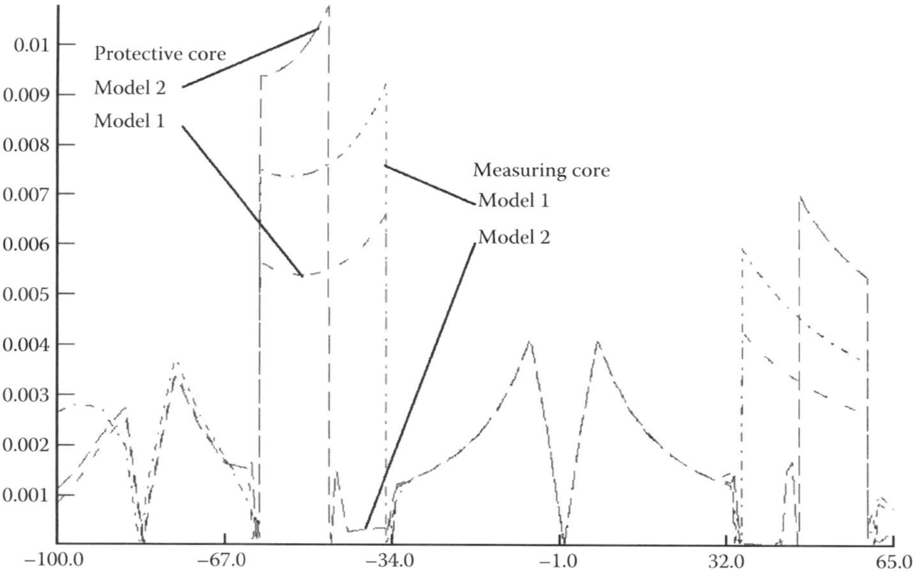

FIGURE 15.3
Distribution of magnetic flux density (T) along the axis passing through the centers of cores for both models.

cores with measuring and protective windings are situated one by one, and in second version, core with measuring winding is inside the protective one. The best version is when the core with wound measuring winding is inside and the core with protective winding is outside, which shields the inner winding from the influence of the magnetic flux of the return conductor (Figure 15.3). Also the instrument security factor K_{FS} is better in this version of the current transformer. The test results are a confirmation of the results of the numerical analysis.

15.2.1.3 Selection of Shields [5]

In regions with currents of known and externally forced distribution, the harmonic electromagnetic field in two dimensions may be described, using the magnetic vector potential $\mathbf{A} = \mathbf{1}_z \underline{A}_z$ ($\mathbf{B} = \nabla \times \mathbf{A}$), by an equation of Helmholtz type (15.4) (Figures 15.4 and 15.5):

$$\nabla^2 \underline{A}_z - \mu \nabla \left(\frac{1}{\mu} \right) \times \nabla \times \underline{A}_z - j\omega\mu\gamma \underline{A}_z = -\mu \underline{J}_{zw} \tag{15.13}$$

Application of 2D field methods in the considered example, where there is a problem with elimination of the influence of the return conductor (HV current transformer 2kA/1A with primary winding type U), permitted the selection of optimal shielding. The choice of the way of shielding the measuring secondary winding of a current transformer under the influence of field of return conductor in order to obtain required accuracy class without increasing power losses can be based on electromagnetic field distribution. Some variations of shield design, electromagnetic and magnetic shields, different materials and shapes of these shields were considered. Finally, an open electromagnetic shield, made

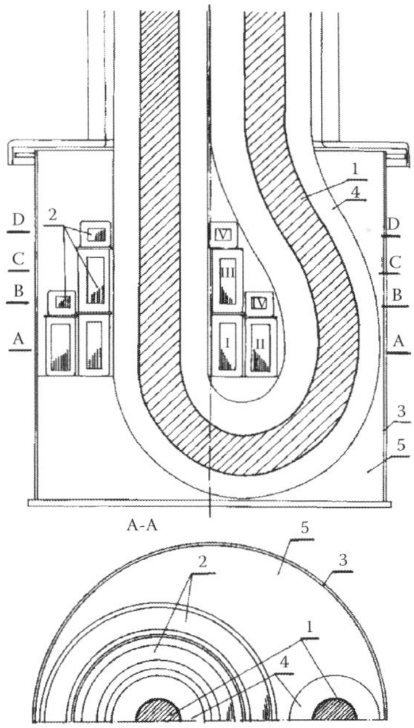

FIGURE 15.4
HV current transformer 2 kA/1 A: 1, primary winding type U; 2, cores with secondary windings; 3, housing; 4, paper–oil insulation; and 5, oil.

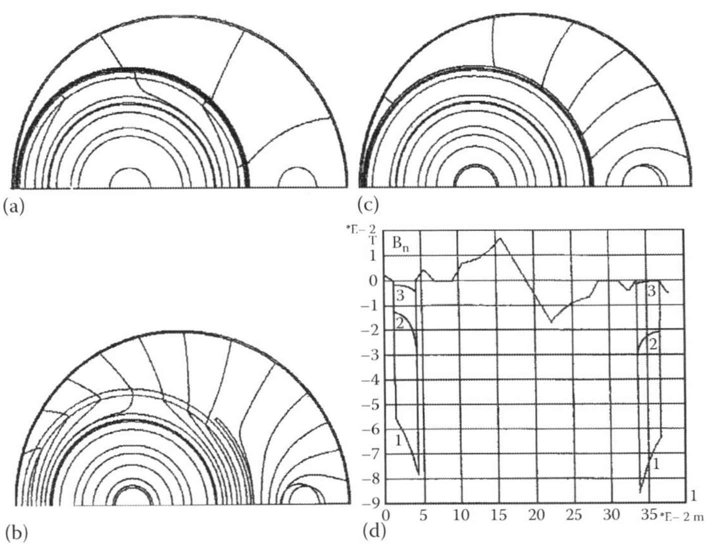

FIGURE 15.5
Distribution of magnetic vector potential in A–A cross section of HV current transformer 2 kA/1 A with primary winding type U with the application of (a) the closed magnetic shield ($A_{max} = 1.18 \times 10^{-2}$ Wb/m, $A_{min} = -2.43 \times 10^{-3}$ Wb/m), (b) the open copper electromagnetic shield ($A_{max} = 1.75 \times 10^{-3}$ Wb/m, $A_{min} = -1.39 \times 10^{-3}$ Wb/m), (c) the closed steel electromagnetic shield ($A_{max} = 2.05 \times 10^{-3}$ Wb/m, $A_{min} = -2.66 \times 10^{-3}$ Wb/m), and (d) distribution of the normal component of magnetic flux density along symmetry axis inside the housing of an HV current transformer with the application of the following: 1, the closed magnetic shield; 2, the open copper electromagnetic shield; and 3, the closed electromagnetic shield made of constructional steel.

from copper, was applied because electromagnetic steel shield is most effective but caused larger power losses in the housing (489.4 and 112.1 W/m at the open copper electromagnetic shield).

15.2.2 The Characteristics of Current Error and Phase Displacement

The field-and-circuit method, based on computed electromagnetic field distributions, is the most direct method of determining *the characteristics of current error and phase displacement* for numerical models of the designed instrument transformers [6,7]. The demand for accuracy of the designed instrument transformers can be obtained by redesigning them.

Modern commercial software packages for calculating electromagnetic fields using the finite element method permit to couple electric circuit equations and magnetic field equations. Thus, the equivalent circuit of a current transformer and simultaneously the time- and labor-consuming stage of determining the parameters of the equivalent circuit is eliminated using the field method.

The field-and-circuit method based on solution of Helmholtz equation (15.13) for 2D electromagnetic field determines secondary voltage and compares it with circuit equation of the same secondary voltage:

$$u_2 = -\left(\frac{n_c}{S_c}\right)^2 \int_{\Omega_c} \frac{i}{\gamma} dv + \frac{n_c}{S_c} \int_{\Omega_c} \frac{\partial A_z}{\partial t} dv \tag{15.14}$$

$$u_2 = Ri + L\frac{di}{dt} \tag{15.15}$$

where R and L are resistance and inductance of the burden of the secondary winding of the current transformer.

Application of field-and-circuit method gives the possibility of computing the complex values of the secondary current and then determining as well the current error

$$\Delta I_n = \frac{I_2 - I_1''}{I_1''} 100\% \tag{15.16}$$

as the phase displacement between primary and secondary currents.

It also permits the determination of the characteristics of the current error and the phase displacement versus load power of the current transformer. These characteristics indicate the accuracy class of design current transformer.

The calculations are carried out, allowing for geometrical and material data of the system, including the magnetizing characteristics of the core, as well as the load, connected to the secondary winding coil, while assuming the current in the primary winding is forced. In the work, the method based on introducing a hypothetical homogeneous core of conductivity γ equivalent to the real core of same dimensions was taken. The dependence of power losses in the core on the magnetic flux density may be obtained in approximation, taking the dimensions of the core and using the characteristics of the transformer lamination.

The characteristics shown in Figure 15.6 were obtained using different methods. They are close to each other. But the field-and-circuit method is the most direct because it eliminates the equivalent circuit.

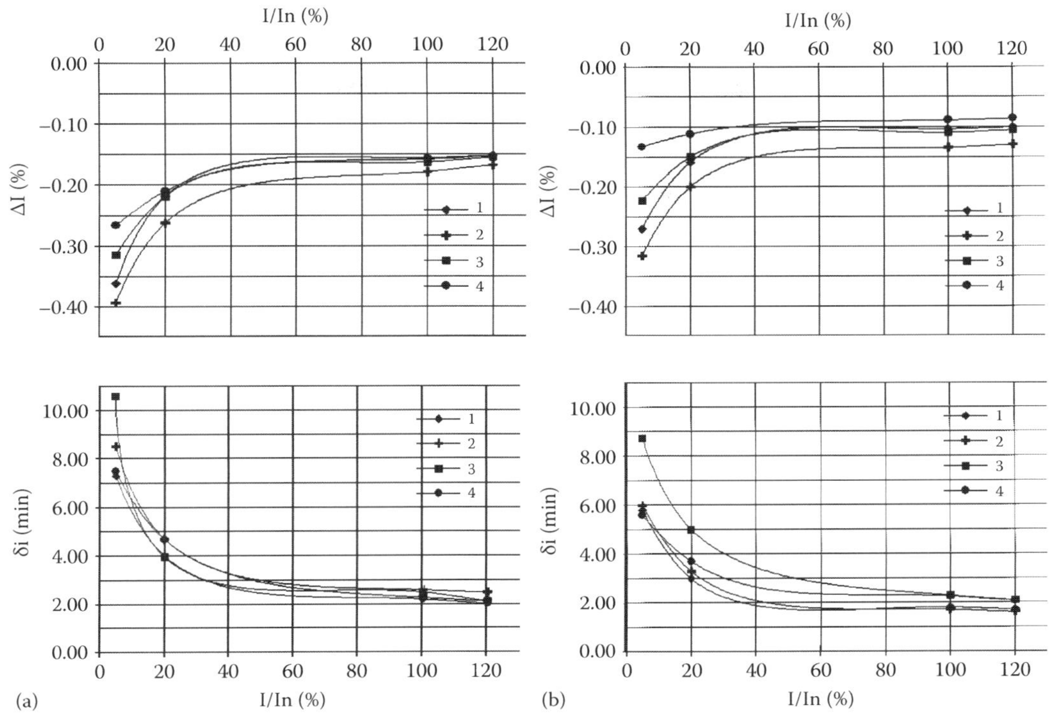

FIGURE 15.6
Comparison of the current error and the phase displacement characteristics of a 5/5 A current transformer for (a) S = Sn = 10 VA, cos φ = 0.8; (b) S = 2.5 VA, cos φ = 0.8; 1, obtained using the field method to determining the parameters of the equivalent circuit of the current transformer and solving a nonlinear circuit with; 2, the analytic method; 3, the measurement; and 4, the characteristics obtained using the field-and-circuit method.

15.2.3 The Composite Error

The field-and-circuit method application also makes it possible to determine *the composite error* and thereby the instrument security factor K_{FS} for the protective current transformers, as well as for measuring current transformers the accuracy limit factor K_{ALF} and shaping them [8]. Determination of the composite error (Figure 15.7)

$$\varepsilon = \frac{100\%}{I_1}\sqrt{\frac{1}{T}\int_0^t \left(i_2 K_n - i_1\right)^2 dt} = \frac{I_0}{I_1}\cdot 100\% \tag{15.17}$$

Applying the field-and-circuit method for different values of forced primary current allows the determination of the complex values of the secondary current and then the computation of the complex error characteristic $\varepsilon = f(I1)$. Both the instrument security factor K_{FS} and the accuracy limit factor K_{ALF} can be determined based on this characteristic (Figure 15.8). These are two important parameters of special concern to engineers designing overcurrent protective and measuring systems.

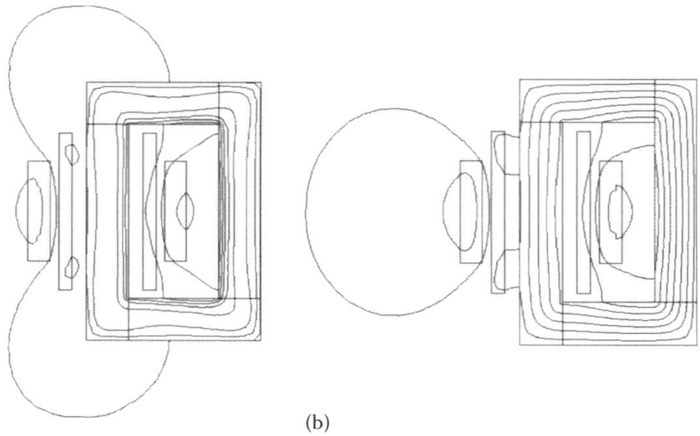

FIGURE 15.7
Distribution of magnetic vector potential **A** in a current transformer 5/5 A (a) at rated state $I_1 = I_{1n}$ (Amax = 2.0×10^{-3} Wb/m, Amin = $-1 \times 7 \times 10^{-3}$ Wb/m) and (b) at overcurrent state with rated burden $I_1 = 157$ A ($\varepsilon = 10$) (Amax = 6.0×10^{-3} Wb/m, Amin = -3.8×10^{-2} Wb/m).

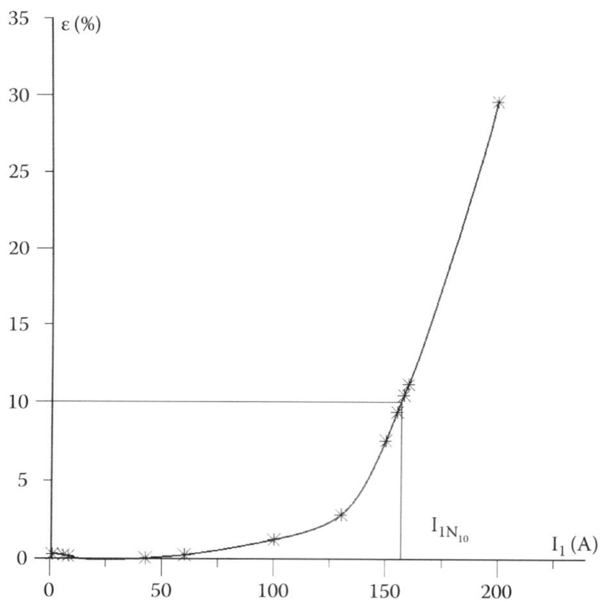

FIGURE 15.8
The composite error versus the primary current of a laboratory model 5/5 A current transformer, at an overcurrent state, obtained on the basis of field-and-circuit calculations—determining the accuracy limit factor K_{ALF} at composite error $\varepsilon = 10\%$ ($I_1 = 157$ A therefore $K_{ALF} = 31.4$).

15.2.4 A Break in the Secondary Circuit of a Current Transformer

At rated current forced in its primary circuit, is the cause of a very high voltage being generated at the secondary terminals. The designer of current transformers should be aware of the magnitude of peak voltages that may be generated at the terminals of the designed current transformer as well as the consequences this would bring. In the course of the

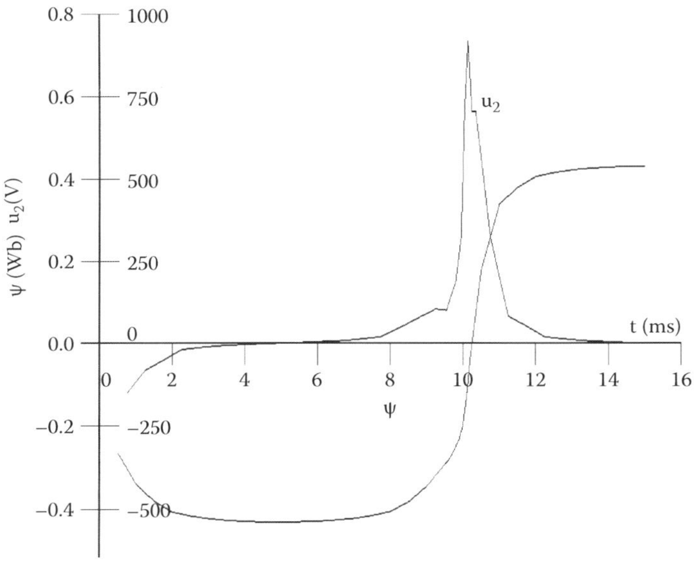

FIGURE 15.9
Flux and secondary voltage versus time, at an open secondary circuit and rated primary current, calculated for a laboratory model 5/5 A current transformer (core losses taken into account).

research, an attempt was made to determine the peak value of voltage, based on the field analysis [8]. It is impossible to solve this problem using the complex method, since the wave shapes of field quantities are considerably deformed (Figure 15.9).
A full set of time-dependent differential equations must be solved instead.

$$\mathrm{rot}\left(\frac{1}{\mu}\mathrm{rot}1_z A_z\right)+\gamma 1_z\left(\frac{\partial A_z}{\partial t}+\mathrm{grad}_z V\right)=1_z J_w \tag{15.18}$$

Space-time 2D analysis allows the computation of magnetic flux coupled with the secondary circuit versus time and the secondary voltage versus time assuming a sinusoidal wave shape of the primary current. In order to solve this problem, the TR module of OPERA commercial software was used.

The calculation results were used for plotting the magnetic flux coupled with the secondary circuit versus time and, subsequently, the secondary voltage versus time, in accordance with the following equation:

$$u_2=\frac{d\Psi}{dt} \tag{15.19}$$

where $\Psi = z\int_S \mathbf{B}\cdot d\mathbf{s}$ is the magnetic flux coupled with the secondary winding.

The maximum value of secondary voltage is also a very important parameter of a current transformer and must be limited by constructors during design. Previously, without field analysis, constructors had empirical and approximate formulas only (Table 15.1).

TABLE 15.1

Peak Values of the Secondary Voltage, Determined on the Basis of a Field Analysis and Calculated Using Approximate Formulas

Based on the Field Analysis	$u_{2s} \approx \frac{1}{\sqrt{1+(\omega L_\mu/R_{Fe})^2}} \mu_{max}\sqrt{2}\frac{S_{Fe}}{l_{Fe}}\omega z_2^2 I_1''$	$u_{2s} \approx \sqrt{2}\frac{S_n}{I_{2n}}\frac{100}{kl}$
V	V	V
918.0	2412.2	1414.2

15.2.5 Transient State

Joining the field-and-circuit method and space-and-time analysis to determine the operation of protective current transformers during *transient state* at short circuit in a power network is advisable [9,10,20]. The requirements put to protective current transformers concern the transformation of currents, with high accuracy especially at transient states. IEC standard [22] obliges designers to determine the instantaneous error current versus time. During a short circuit in a transmission line, the primary current has an exponential component.

$$i_p'' = I_{psc}''\{\cos(\omega t + \beta) - \cos\beta e^{-(t/T_P)}\} \tag{15.20}$$

The secondary current versus time curve may be deformed because of saturation of the core. By current IEC standard definition, the difference between primary current in terms of the secondary windings and secondary current determines instantaneous error current versus time and transformation errors.

Field analysis allows the computation of the secondary current versus time assuming a nonsinusoidal wave shape of the primary current. Until now, to agree with standard requirements, these currents could only be measured during tests on real-life models, because without the application of field-and-circuit method, the computation of the current versus time was very inaccurate using traditional methods.

Using the full set of time-dependent differential Equations 15.18 and 15.14 and circuits equation

$$u_2 = Ri \tag{15.21}$$

(because resistance, R, is the burden of secondary winding) gives primary and secondary currents versus time.

The toroidal protective current transformer 600/1 A at transient state was considered.

Figures 15.10 through 15.12 show the obtained results of the computed secondary current and instantaneous error current versus time at a transient state. The difference between primary current in terms of the secondary windings and secondary current determines instantaneous error current versus time and transformation errors agree with current IEC standard.

Instantaneous error current versus time, required by IEC standard [22], can be computed before building the prototype. Some of the illustrations present these possibilities as practical examples.

As a practical example, a toroidal current transformer class TPZ 2400/1 A [11,12] is considered. The aim of this research was the comparison of tested and computed results

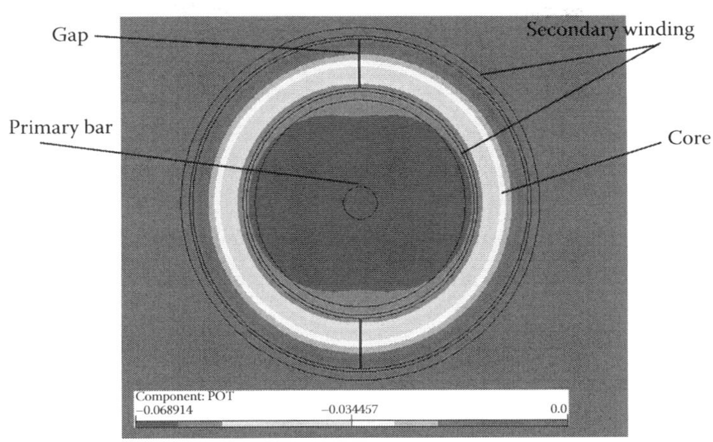

FIGURE 15.10
Distribution of magnetic vector potential **A** of toroidal protective current transformer 600/1 A at transient state for induced primary current $i_1 = 8(\cos 314.16t - e^{-20t})$ kA and time t = 5 ms.

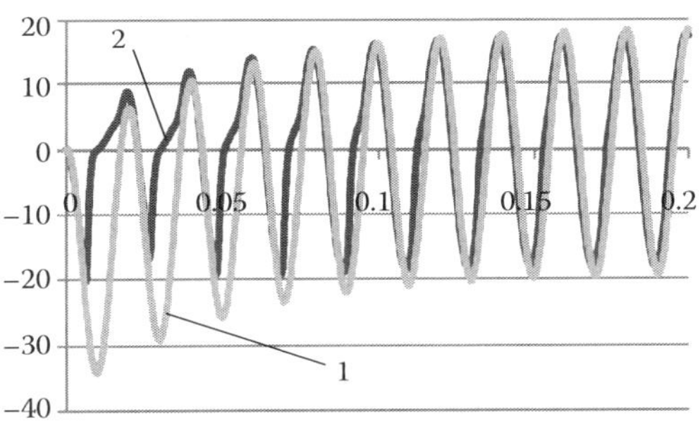

FIGURE 15.11
Primary and secondary currents versus time of toroidal protective current transformer 600/1 A at saturation state: 1, induced primary current $i_1 = 8(\cos 314.16t - e^{-20t})$ kA in terms of secondary winding; 2, secondary current.

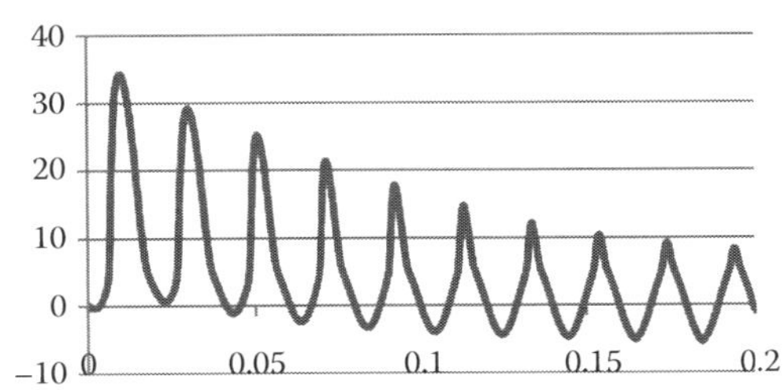

FIGURE 15.12
Instantaneous error current versus time of toroidal protective current transformer 600/1 A at saturation state.

for different loads. The current difference test arrangement was carried out for direct determination of instantaneous error current versus time of a current transformer at transient state (Figure 15.13).

The analysis was performed for different forced primary currents and for different loads. Sometimes it is very difficult to fulfill the standard requirements during laboratory

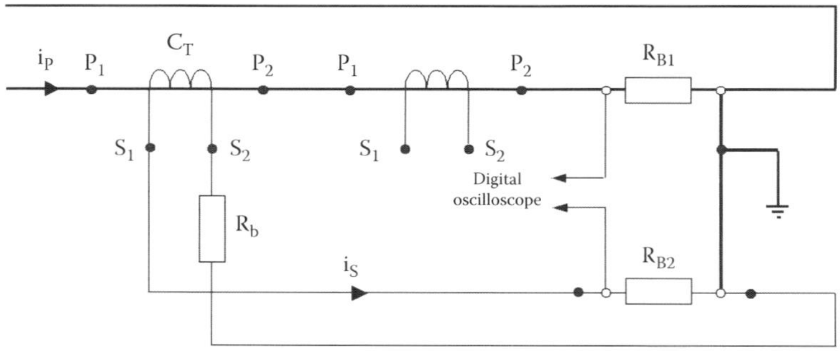

FIGURE 15.13
Differential scheme of connection of the measuring system used for direct determination of instantaneous error current versus time of a current transformer at transient state [Ref.22]. C_T, tested current transformer; R_{B1}, resistor inductive-less type LB1 30 A/60 mV; R_{B2}, shunt inductive-less type LB1 1.5 A/60 mV; and R_b, resistor inductive-less.

testing. In the case of this current transformer, it means achieving the high value of primary current (40800 A because rated symmetrical short-circuit current factor K_{scc} is equal to 17) and specific primary time constant (50 ms) to obtain the proper shape of primary current curve.

In order to carry out the test, a winding of 120 turns was wound on this current transformer instead of the bar with 2400 A, and in this case, the current 20 A gave also 2400 At, and the ratio of the same current transformer was 20/1 A. In the laboratory, it was only possible to achieve a primary time constant equal to 22 ms. Figure 15.14 presents the

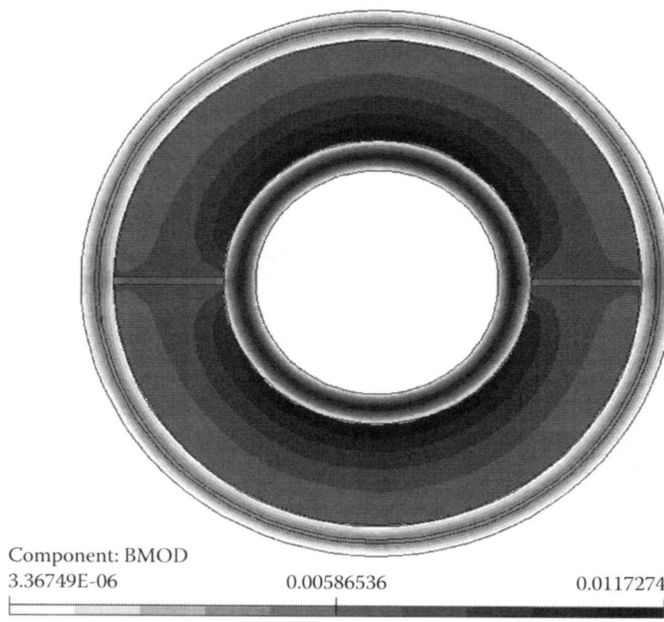

FIGURE 15.14
Distribution of magnetic flux density of toroidal protective current transformer 20/1 A at transient state for load R = 12.2 W, induced primary current $i_p = 57.5\sqrt{2}(\cos 314.16t - e^{-45.5t})\,A$, and time t = 4 ms.

TABLE 15.2

Computed and Measured Peak of Instantaneous Alternating Current Error of the Current Transformer 20/1 A for Different Loads

Rated Primary Short-Circuit Current, I_p (A)	Primary Time Constant, T_p (ms)	Rated Resistive Burden, R_b (Ω)	Peak Instantaneous Alternating Current Error, $\hat{\varepsilon}_{ac}$ (%)	
			Computation	Test
57.5	22	12.2	2.73	2.85
57.5	22	30.8	4.45	4.71
57.5	22	63	8.54	8.69
57.5	22	119	14.9	15.5

field distribution of the protective current transformer 20/1 A during transient state in the considered model.

Table 15.2 gives a comparison of peak instantaneous alternating current errors, required by IEC standard, computed for different load based on curves obtained applying the field-and-circuit method and test (e.g., Figures 15.15 and 15.16). The values obtained by testing on a real model and by computing are almost the same.

The computation using field-and-circuit method permits the determination of the value of peak instantaneous alternating current error at proper conditions, even when it is impossible to obtain them in the laboratory. Figure 15.17 shows instantaneous error current versus time obtained from computation for the shape of the primary current curve, required by IEC standard, and for the same value of primary current but with different specific primary time constant (22 ms).

The next problem especially for class TPX and TPY current transformers is the recognition of instantaneous alternating current error versus time for duty cycle of switching C-0-C-0 where the flux level at the beginning of the second energization remains unchanged.

For the considered current transformer, the same computation using field-and-circuit method and space-and-time analysis for the C-0-C cycle was carried out. Two cases were taken into account. One of interruption of the primary short-circuit current at moments of

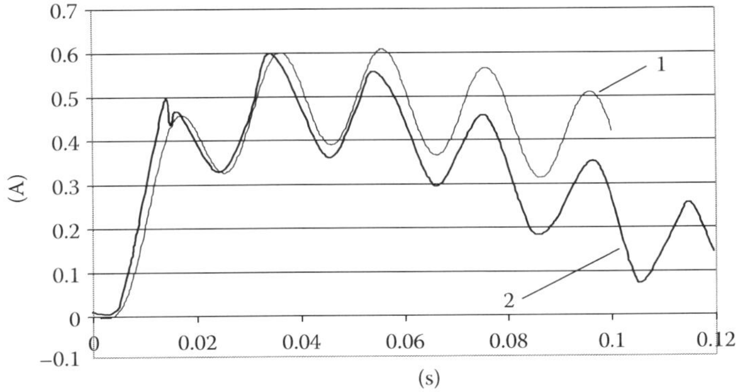

FIGURE 15.15
Instantaneous error current versus time of toroidal protective current transformer 2400/1 A for load R = 12.2 Ω, induced primary current $i_p = 57.5\sqrt{2}(\cos 314.16t - e^{-45.5t})$ A: 1, computation; 2, test.

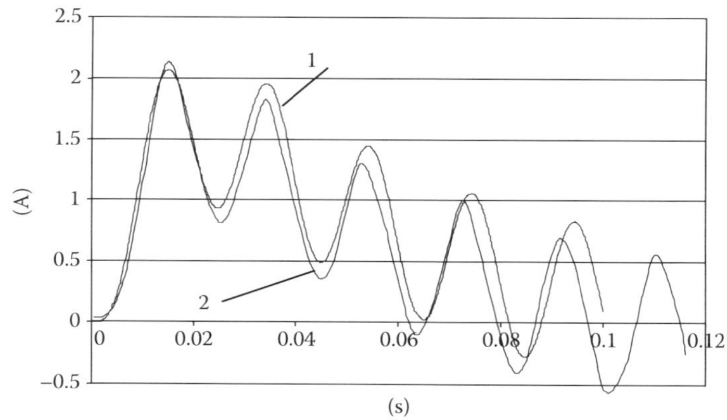

FIGURE 15.16
Instantaneous error current versus time of toroidal protective current transformer 2400/1 A for load R = 119 Ω, induced primary current $i_p = 57.5\sqrt{2}(\cos 314.16t - e^{-45.5t})$ A: 1, computation; 2, —test.

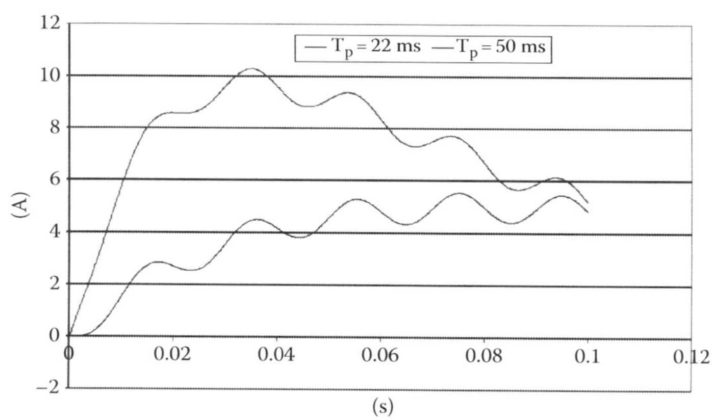

FIGURE 15.17
Computed instantaneous error current versus time of toroidal protective current transformer 20/1 A agree with IEC standard [Ref.22] computed for load R = 10 Ω and induced primary current $i_p = 340\sqrt{2}(\cos 314.16t - e^{-20t})$ A (K_{ssc} = 17, I_p = 20 A, T_p = 50 ms) and for T_p = 22 ms.

peak value of the primary current and second of interruption in time when instantaneous alternating current error achieves a maximum value.

The dead time (during autoreclosing) in both case amounted t_{fr} = 100 ms.

Figure 15.18 shows the results of the computed instantaneous alternating current error versus time for the duty cycle of switching C-0-C-0 in these two cases.

In the second case (Figure 15.8b), the current transformer achieves saturation of the core, and current transformation is incorrect. Determination of peak instantaneous alternating current error is impossible.

The use of this method permits the computation and shaping of the parameters of the designed current transformer in accordance with the IEC standards.

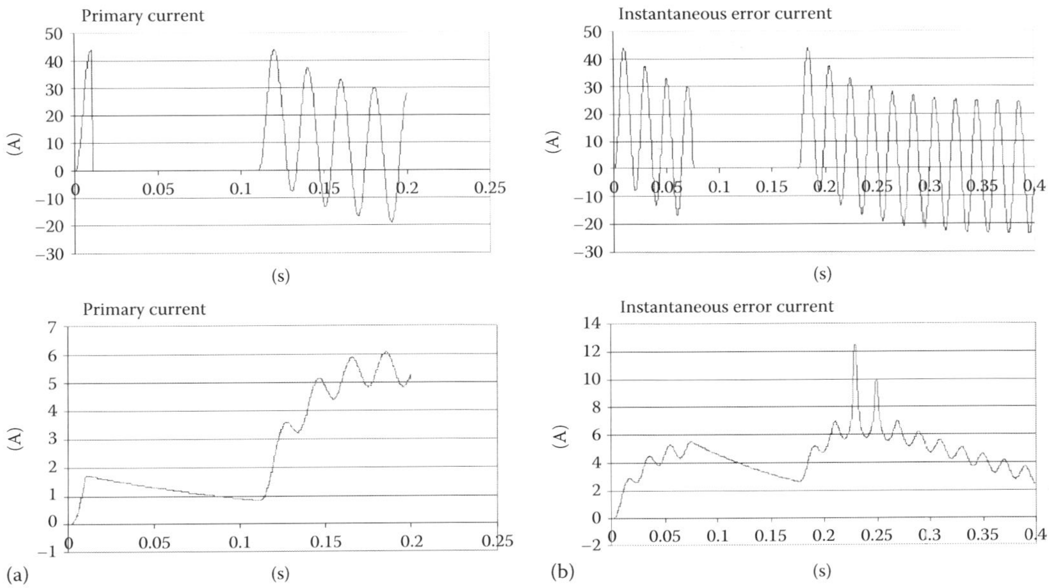

FIGURE 15.18
Primary current during transient state and computed instantaneous error current versus time of toroidal protective current transformer 20/1 A at C-0-C-0 switching: (a) interruption of the primary short-circuit current in time of maximum value of primary current and (b) interruption in time when instantaneous alternating current error achieves maximum value.

15.2.6 Computing the Electric Field Distributions

During design gives, as was previously unattainable, the possibility of shaping the insulation systems. It is of great importance in designing complicated geometrical systems, to allow for shaping the elements of the insulation system, checking the effect of applying specific construction solutions and materials, predicting the effects of joining different kinds of materials, and introducing and shaping electrostatic shields.

A lot of attention was devoted to the methodology of using the calculations of electric field distribution with the aim of choosing the elements that control the electric field and for the assessment of the insulation systems of real-life instrument transformers with respect to the maximum values of electric field intensity as well as the uniformity of the electric field.

The electric field in an insulation system (after introducing the scalar electric potential $\mathbf{E} = -\nabla V$, where $\mathbf{E}$ is the electric field strength) is described by Laplace's equation:

$$\nabla^2 V = 0 \tag{15.22}$$

with Neumann's and Dirichlet's boundary conditions.

Applying the numerical finite element method in 3D systems also makes it possible to solve this equation. Only 3D analysis allows the estimation of the field distribution in the whole system and makes it possible to determine the maximum electric field strength in insulation. The biggest problem of modeling the physical device is to reproduce the real shape of device, which may be sometimes very complicated.

15.2.6.1 Choosing of Design Version of Insulation System of HV Voltage Transformer [1]

In the considered HV voltage transformer, paper–oil insulation was applied. The problem was in shaping the paper–oil insulation, so the electric field distribution was uniform. The main problem was field distribution in the core window and the edge effect near the coil edges. In Figures 15.19 and 15.20, field distribution in the chosen design version and peaks of electric field strength limited by shaping paper insulation may be observed. Table 15.3 gives the maximum values of electric field strength in paper and in transformer oil.

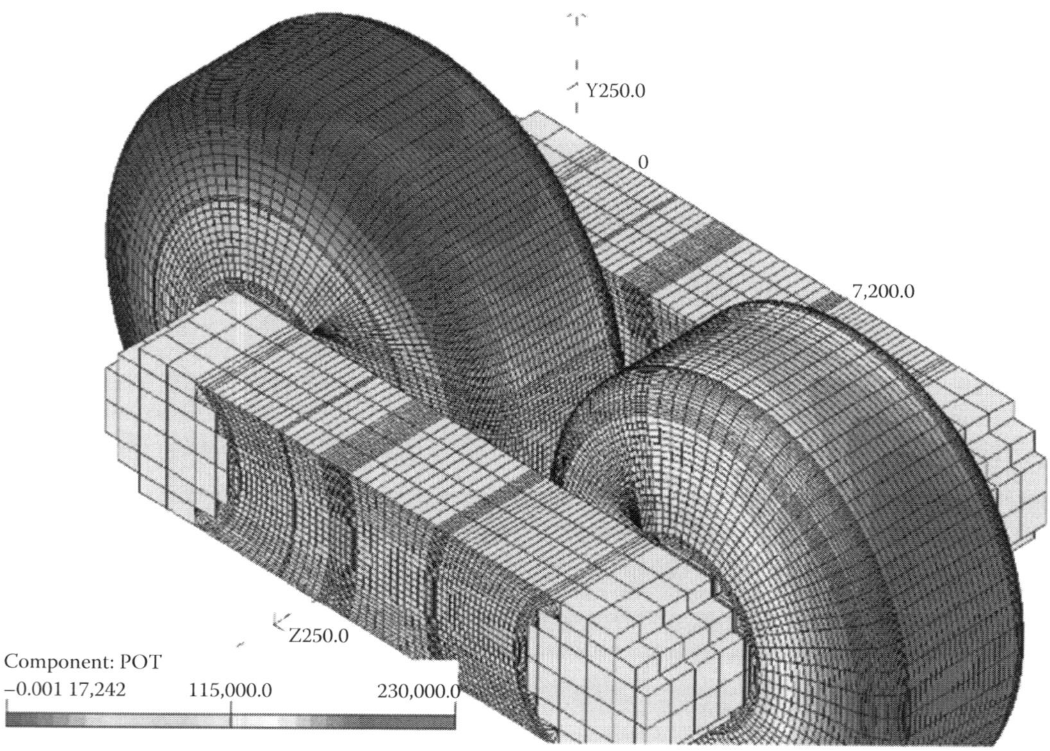

FIGURE 15.19
Distribution of the electric potential (V) in the insulation system of the HV voltage transformer $110:\sqrt{3}\,kV/100:\sqrt{3}\,V$.

TABLE 15.3
Maximum Values of the Electric Field Strength in Paper–Oil and in Oil Insulation in Considered Voltage Transformer

In Paper–Oil Nearby Edge of the Primary Coil	In Paper–Oil Nearby Edge of the Secondary Coil	In Transformer Oil	Area Where Electric Field Strength Exceeded 5 kV/mm
8.89 kV/mm	6.70 kV/mm	3.17 kV/mm	42.5 mm^2

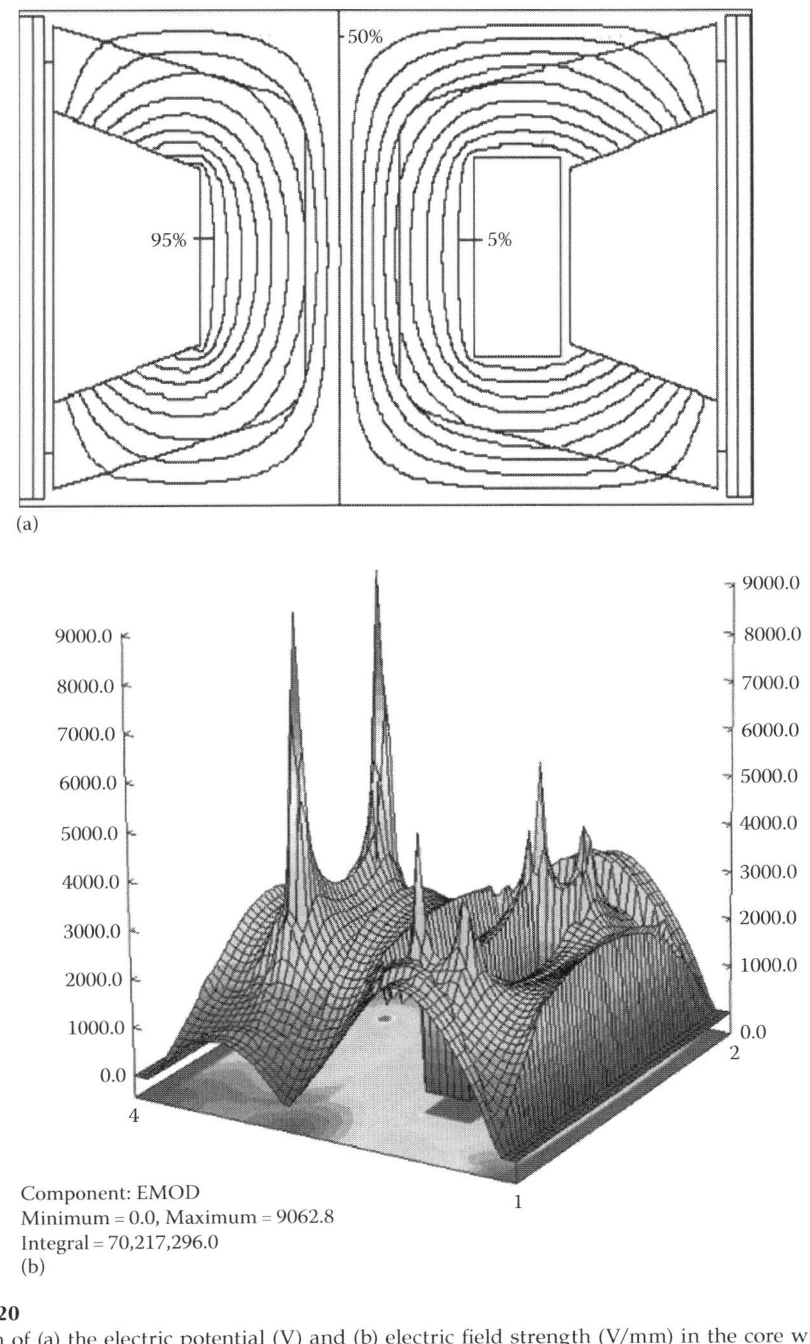

FIGURE 15.20
Distribution of (a) the electric potential (V) and (b) electric field strength (V/mm) in the core window of HV voltage transformer $110:\sqrt{3}$ kV/$100:\sqrt{3}$ V.

15.2.6.2 *Improvement of the Electric Strength of an Insulation System of a Medium Voltage Instrument Transformer [13]*

The subject of analysis is a voltage transformer with two insulated voltage terminals and a transformation ratio of 15 kV/100 V. The aim of redesigning the voltage transformer was the achievement of a greater rated power at the same accuracy class and greater electric strength in the same dimensions. This modernization considered the shape of its core and windings. The main assumption was the same external dimensions of both versions. Greater rated power requires an increase in the cross-sectional area of the core, causing a decrease in the thickness of the insulation layer and, in consequence, lowers the electric strength of the whole insulation system of the voltage transformer (Figure 15.21).

The behavior of voltage transformer was checked under test conditions and under operating conditions. Under test, terms of primary winding are joined together, and the potential 38 kV was forced. The core and secondary winding are grounded. Under operating conditions, terms are joined to two different phases.

The computations show that the electric field concentrates nearby the primary coil and especially in the core window. The level of the electric field strength in the voltage transformer insulation is below 8 kV/mm at the test voltage of 38 kV. But the peak values of the electric field strength may be observed by the edges of both parts of the primary coil in the foils of trivolton and achieved, under test voltage, 16.2 kV/mm. The computation done at the operating state shows the level of electric field strength to be 3 kV/mm and the peak values in trivolton to be 6.4 kV/mm (Figures 15.22 and 15.23).

The electric strength of epoxide resin as well as trivolton is higher than the level of the values of the electric field strength in the insulation of the voltage transformer, and there are no dangers of breakdowns of insulation or partial discharges neither at voltage test nor at rated operating state. Every construction improvement is carried out in order to obtain a better device. For the voltage transformer, it means achieving a greater rated power at the same accuracy class and greater electric strength at the assumption of the same external dimensions.

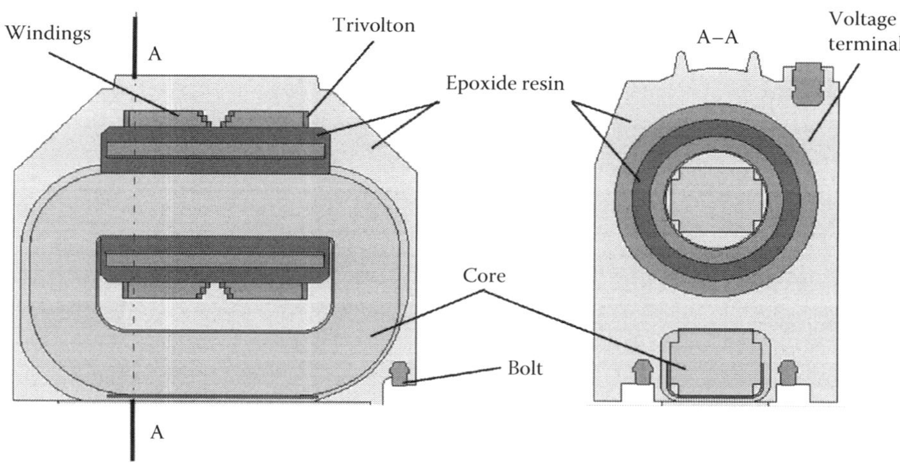

FIGURE 15.21
Constructional variant of the voltage transformer 15 kV/100 V.

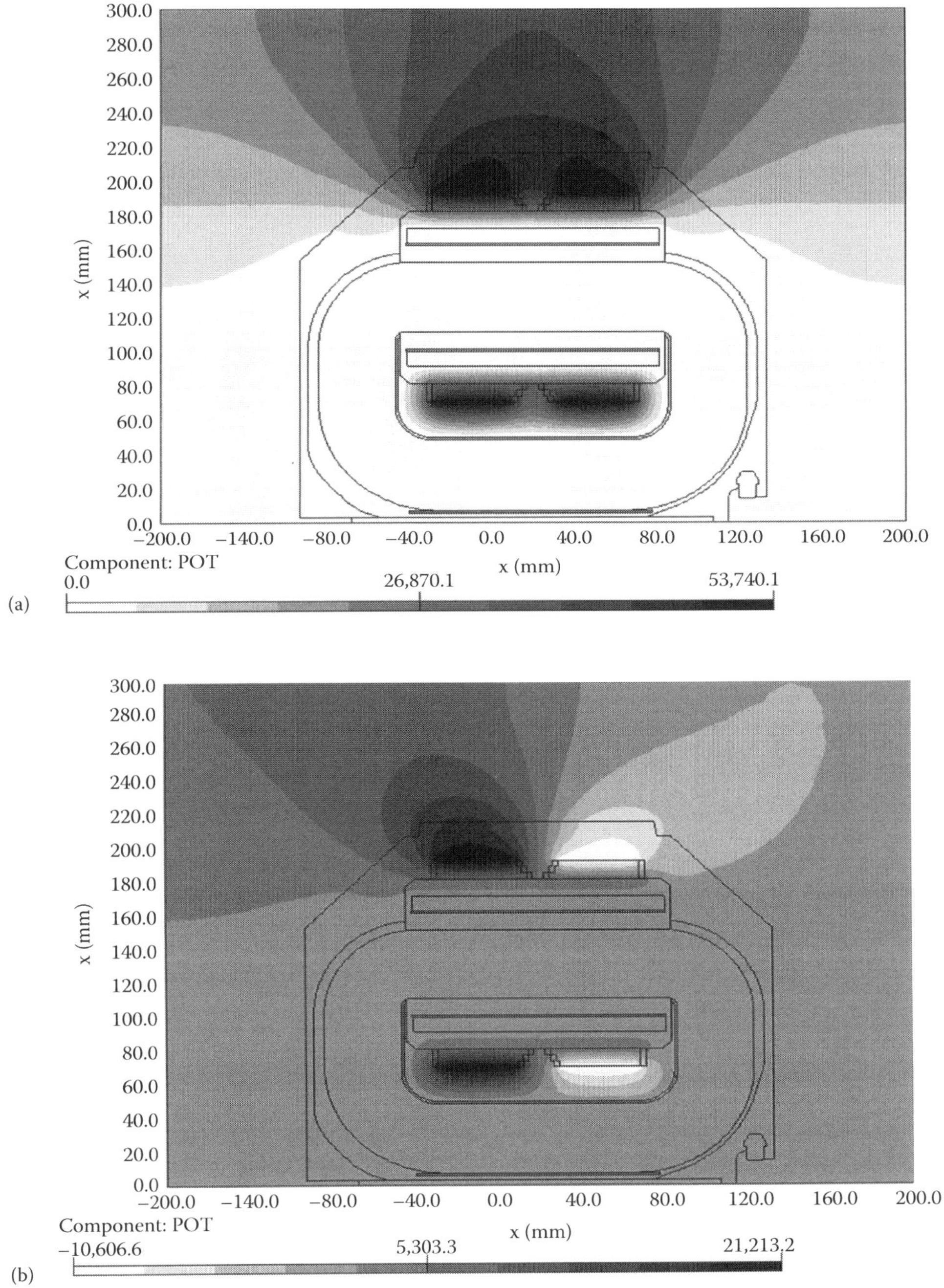

FIGURE 15.22
Distribution of electric potential (V) at the surfaces passing through the centers of the voltage transformer after redesign (a) under test conditions (38 kV-1 min) and (b) under operating conditions.

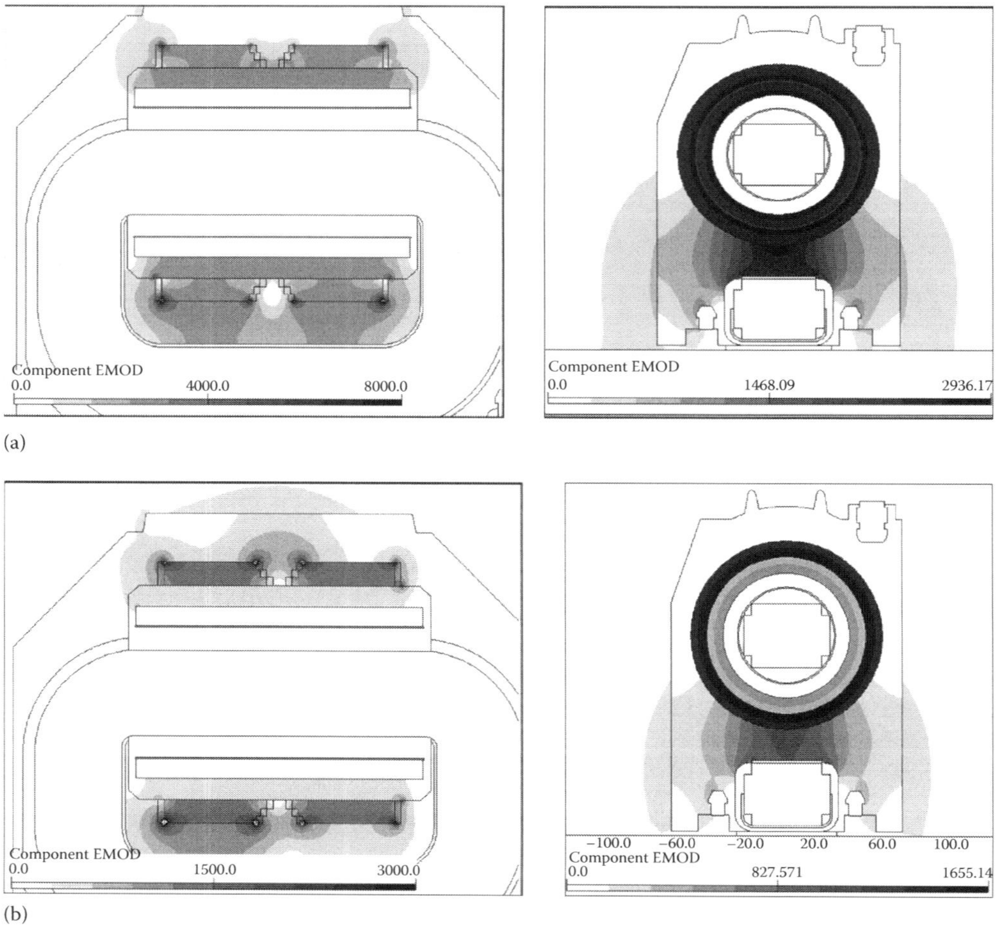

FIGURE 15.23
Distribution of the electric field strength (V/mm) in the area of the voltage transformer after redesign under (a) test conditions (38 kV-1 min) and (b) operating conditions.

15.2.6.3 Selection of Electric Shields [14]

For HV switchgear SF_6 gas is used as insulation. Maintaining identical insulation conditions through the switchgear requires making voltage transformers with SF_6 gas insulation. The subject of analysis is an HV voltage transformer with SF_6 gas insulation and a transformation ratio of $110:\sqrt{3}$ kV/$100:\sqrt{3}$ V (Figure 15.24).

Electric shields' aim is to shape electric field distribution in insulation. Figure 15.25 and Table 15.4 show the results of using grounded shield. It is not possible to eliminate the grounded shield as it locates the electric field far from the sharp core edges and roughness of which could cause a distortion of the electric field; this could result in discharges. In this design, the grounded shield with a maximum radius, bent out in direction of the housing, has been used, and the top of housing has been changed to a trapezoid cone (Figure 15.26). It is compromise between two contradictory trends. It does not cause concentration of potential lines and protects the core area from discharges.

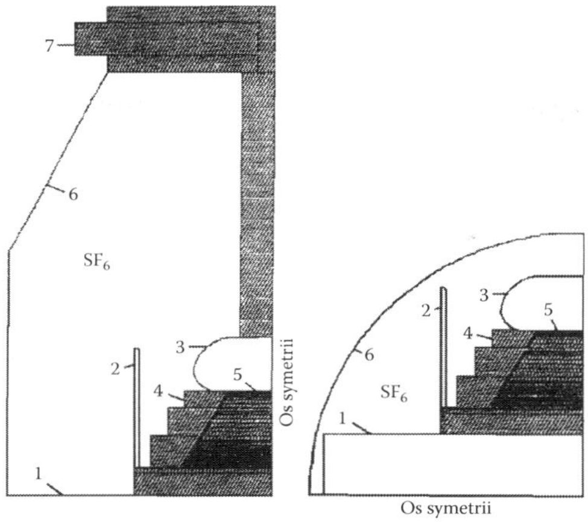

FIGURE 15.24
HV voltage transformer $110:\sqrt{3}\,\text{kV}/100:\sqrt{3}\,\text{V}$: 1, core; 2, grounded shield; 3, HV shield; 4, foil; 5, secondary coil; 6, housing; and 7, cover.

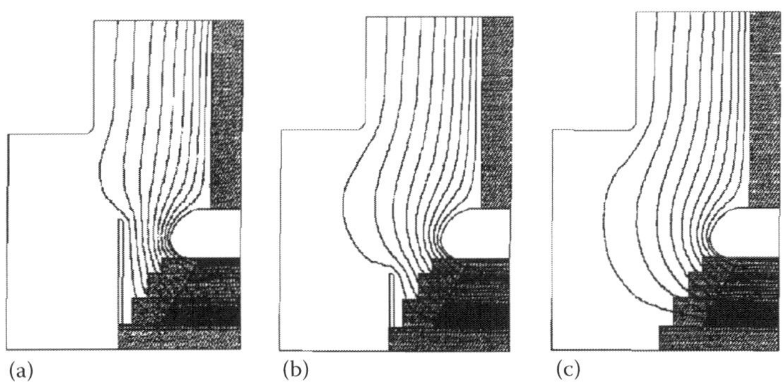

FIGURE 15.25
Distribution of electric potential (V) in insulation of HV voltage transformer when grounded shield has radius (a) 190 mm, (b) 147 mm, and (c) without shield.

TABLE 15.4
Maximum Values of the Electric Field Strength in Insulation of Voltage Transformer

Radius of Grounded Shield (mm)	Area Where Electric Field Strength Exceeded 8 kV/mm (mm²)	Maximum Values of the Electric Field Strength (kV/mm)
190	79.6	15.8
147	23.3	15.2
0	11.4	14.5

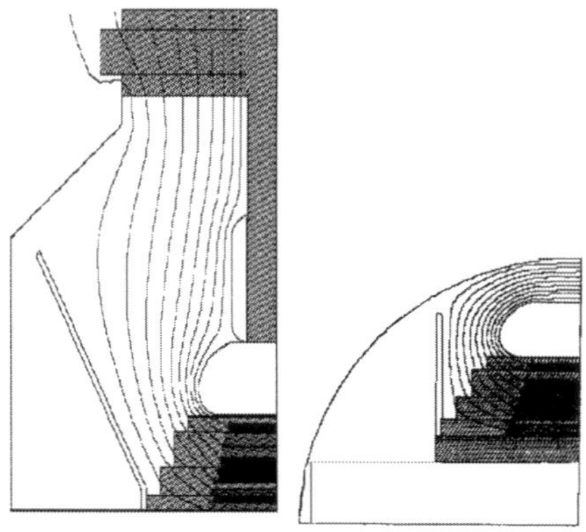

FIGURE 15.26
Distribution of electric potential (V) in insulation of the last design version of HV voltage transformer.

15.2.6.4 Design of an Insulation System with Capacitance Control [15,16]

Sometimes in complicated insulation system, the uniform field distribution can be obtained by using capacitance control. This means using aluminum electrostatic shields between the paper layers.

In the considered combined instrument transformer, the endings of the secondary windings of the current part of the combined instrument transformer at potential zero and the lead conductor of the primary winding of the voltage part at potential 230 kV are both in the ceramic insulator. The main insulation system of the examined instrument transformer is made of paper–oil (Figure 15.27).

Therefore, this construction solution is applied in this case, as shown later. The design process required mutual interaction of two computer programs (Figure 15.28).

It concerns the design of a system of electrostatic control shields for both parts of the insulation and in computing the electric field distribution in all elements of the insulation system of a combined instrument transformer. The potentials of the electrostatic shields (Dirichlet's boundary conditions) needed by the program OPERA3D for computing field distributions are obtainable using computed potentials by the author's software SHIELDS. The distribution of potentials in any paper–oil main insulation and, as a result, the distribution of electric field strength are determined by the capacitance between the electrostatic shields.

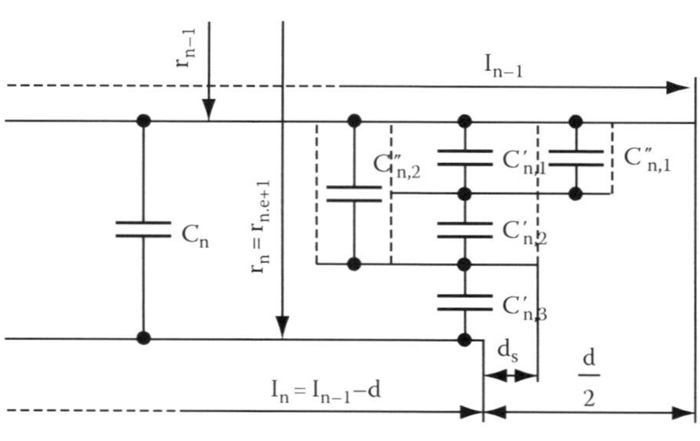

FIGURE 15.27
System of capacitors between the main shields.

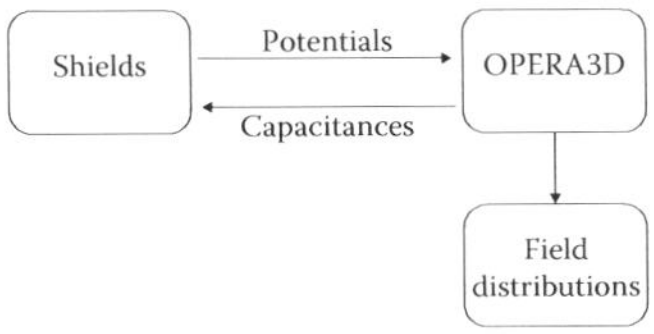

FIGURE 15.28
Mutual interaction of programs.

SHIELDS was used to optimize the number, dimensions, and position of each electrostatic control shield in the insulation of the voltage and current parts of the combined transformer. Criteria of choosing the parameters of the electrostatic shields were the following: not exceeding the permissible electric field strength, uniform distribution of the electric field strength in the axis direction, and the minimum difference of its value in the radius direction [2,16]. Each modification causes a change of capacitance and as a consequence a change of potential distribution. SHIELDS requires the values of capacitance (as there are multiple capacitances, each must be computed and entered) to optimize field distribution in a paper–oil insulation (Figures 15.29 and 15.30).

The capacitance of every complicated shape should be computed based on its electrical energy:

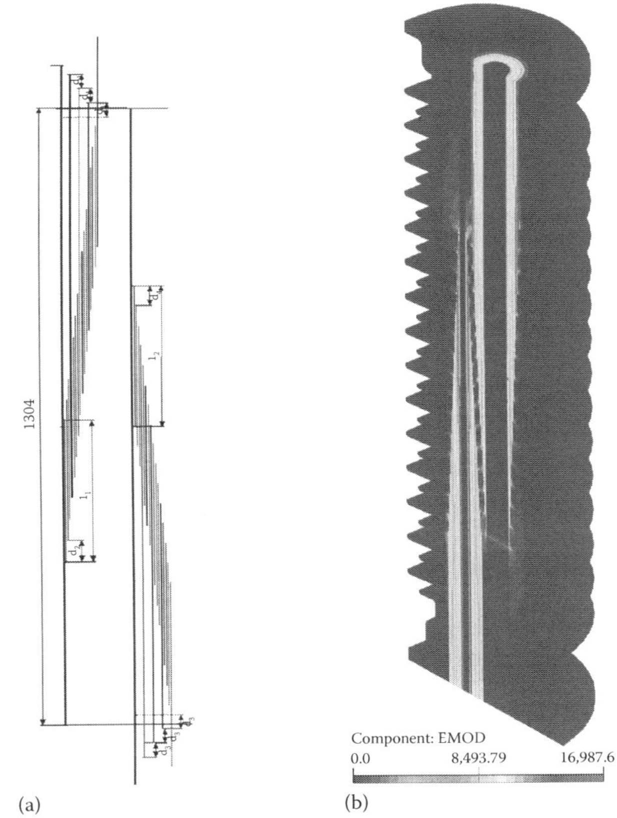

FIGURE 15.29
Combined instrument transformer. (a) The proposed design of electrostatic control shields. (b) Distribution of the electric field strength (V/mm) in the insulation system of the ceramic insulator of the combined instrument transformer.

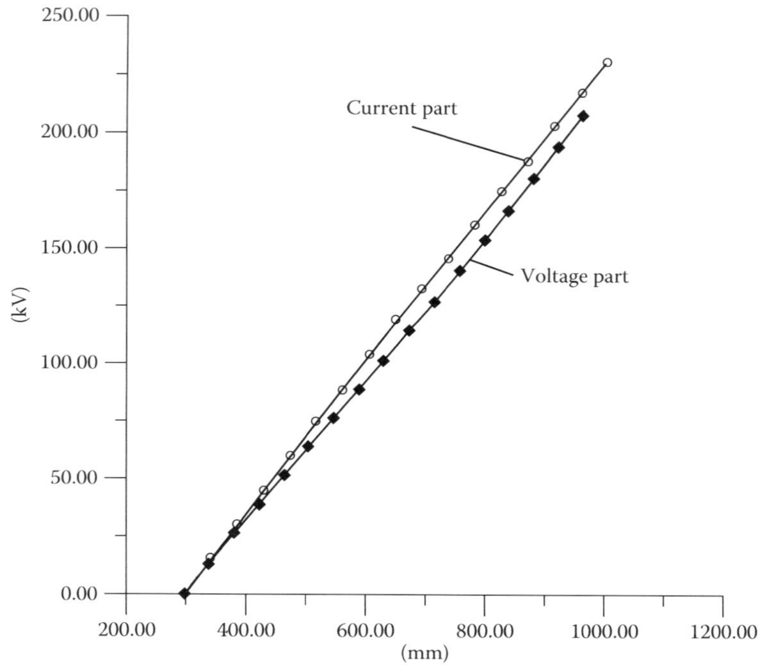

FIGURE 15.30
Combined instrument transformer. Distribution of the electric potential along the height of the ceramic insulator.

$$C = \frac{2W_e}{U^2} \tag{15.23}$$

This electrical energy should be computed based on the field distribution between the electrodes:

$$W_e = \int_v \frac{D \cdot E}{2} dv \tag{15.24}$$

15.2.6.5 Design of Insulation System of the HV Combined Instrument Transformers [15,16]

The combined instrument transformer is composed of two parts: current and voltage transformers in a common housing [21]. Therefore, it has a very complicated insulation system. The design is a process of forming parts of its insulation system. The electric strength of an insulation system during the design process is achieved using an iterative approach. One modification in the design produces other modifications. The design process continues until the permissible electric field strength is not exceeded in any part of the insulation system and the homogeneity of the electric field is maintained. The values of the electric field strength must not exceed 15 kV/mm, and the homogeneity of the electric field must be maintained. These conditions result in the almost complete elimination of the possibility of insulation breakdowns and partial discharges. The biggest

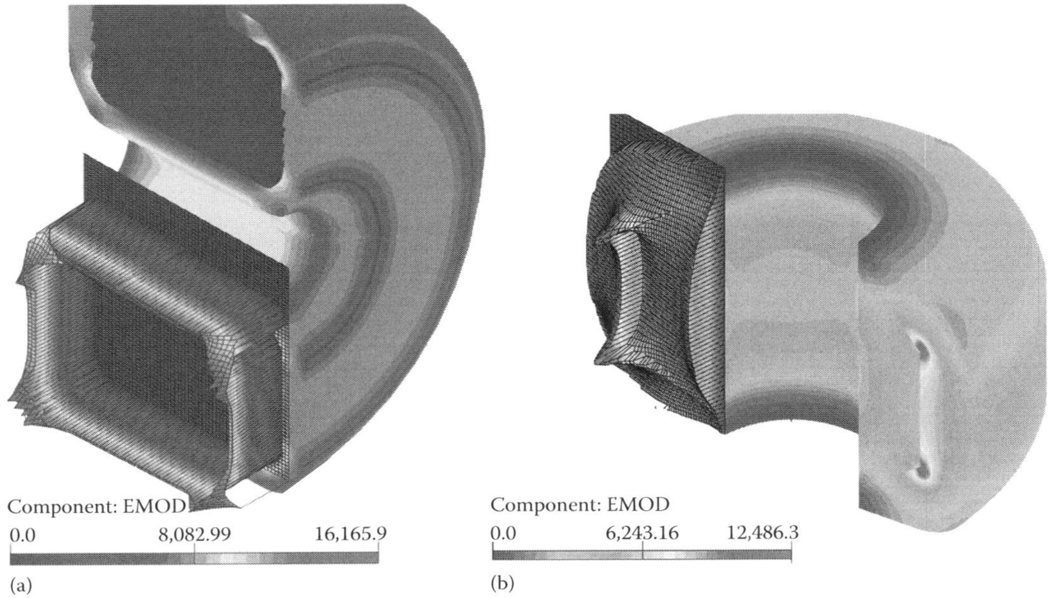

FIGURE 15.31
Distribution of the electric field strength (V/mm) in the insulation system of the combined transformer (a) of the casing with cores (the current part) and (b) of the HV voltage coil (primary winding of the voltage part).

problem of modeling the physical device is the reproduction of the real shape of device, which may sometimes be very complicated. Figure 15.31 presents the obtained field distributions in some parts of the insulation systems of an HV combined instrument transformer. The designed prototype of the considered combined transformer successfully passed all HV tests.

15.2.7 Inner Electromagnetic Compatibility

In the case of combined instrument transformers, the great importance is the problem of *inner electromagnetic compatibility*. It is very important during the design process of the combined transformer to take care of the mutual couplings between the voltage and the current transformers. Because of the low operating frequency, 50 or 60 Hz, it may be assumed that the dominant couplings are through the magnetic field and the electric field. One method of the IEC 60044-3 standard [21], concerning combined transformers, estimates mutual coupling effects by determination of variations of current and voltage errors and phase displacements during a simultaneous operating state.

Application of the field method permits the determination of the coupling parameters (mutual inductance and mutual capacitance) for any geometrically complicated model of a real electric device [17]. The design should eliminate or reduce the mutual interaction to acceptable values so that the current transformer as well as the voltage transformer can operate as independent devices. This means that they both have a negligible variation of errors and phase displacements caused by their interaction.

In the considered combined instrument transformer, the dominant couplings are through the magnetic field and the electric field. And so the dominant influence of the

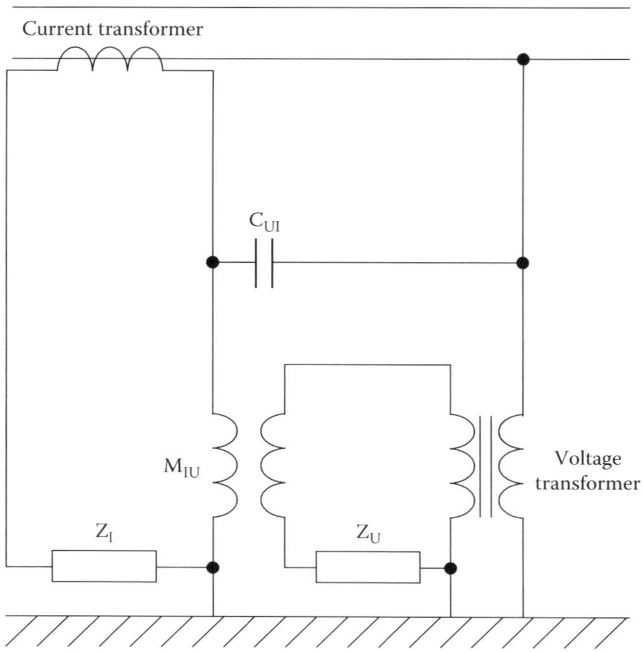

FIGURE 15.32
Circuit model of couplings in a combined transformer where C_{UI} is the mutual capacitance (coupled by the electric field), M_{UI} is the mutual reactance (coupled by the electromagnetic field), and Z_U, Z_I are the burdens of the secondary windings of the voltage and current transformer.

current transformer on the voltage transformer is the influence through the magnetic field, and of the voltage transformer on the current transformer is through the electric field (Figure 15.32).

The mutual reactance between the current and voltage parts of the combined transformer may be determined using formula (15.25) for magnetic flux linked with the secondary winding of the voltage transformer based on the computed magnetic field distribution (using Equation 15.1 or 15.5) during the operation of the current transformer:

$$X_m = \omega M_{UI} = \omega \frac{\hat{\psi}_{IU}}{\sqrt{2}I} \tag{15.25}$$

where $\hat{\psi}_{IU}$ is magnetic flux linked with the secondary winding produced by I current in the secondary or primary winding of the current transformer.

The primary winding of the current transformer, with a current of the order of 3000 A, is placed in the head of the combined transformer. However, the secondary circuits of the voltage transformer, subject to induced current by the current transformer, are placed in the bottom part of the housing. The core of the voltage transformer was positioned, in relation to the primary bar of the current transformer that creates magnetic flux, so that it could not pass through the coils (Figure 15.33b).

In this case, induced additional voltages in the secondary windings change from 41.4 mV (for position shown in Figure 15.33a) to 2.2 mV.

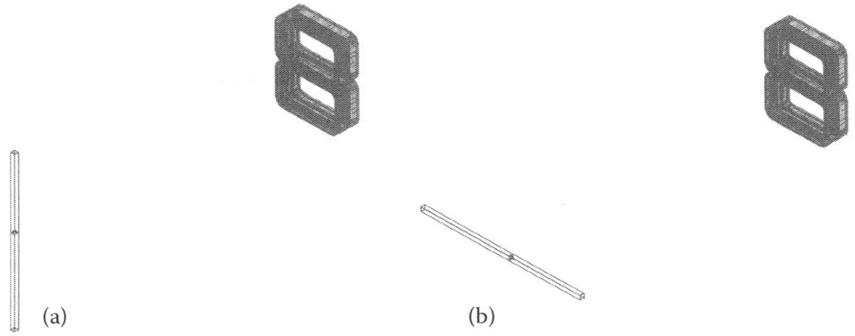

FIGURE 15.33
Possible location of the primary bar with current 3000 A of the current transformer with respect to the voltage transformer core in the bottom part of the housing: (a) perpendicular to the middle column of core and (b) parallel to it.

The parallel position of the bar and the voltage transformer core and the long distance, about 2000 mm, causes that the influence of the primary current is practically negligible.

The wires, according to design, are twined together so that the total ampere-turns are zero. Therefore, they create negligible leakage magnetic field (Figure 15.34).

The mutual capacitance C_{UI} of every complicated shape should be computed based on its electrical energy:

$$C_{UI} = \frac{2W_e}{U^2} \tag{15.26}$$

The mutual capacitance is determined only by this part of the electric field energy, which links both circuits. The electrical energy (as in Equation 15.24) should be computed based on the field distribution between the electrodes (15.22).

In Figure 15.35, coupling by electric field between current and voltage parts of the combined transformer can be observed. Inside the ceramic insulator, the electric field passes into the oil because here in the main paper–oil insulation of both parts shields end. In this area, couplings between both parts through the electric field occur.

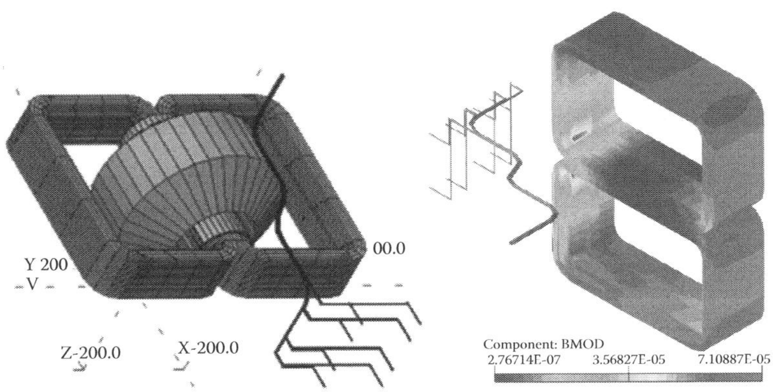

FIGURE 15.34
Distribution of the magnetic flux density (T) in the core of voltage part of the combined instrument transformer caused by secondary currents of the current parts.

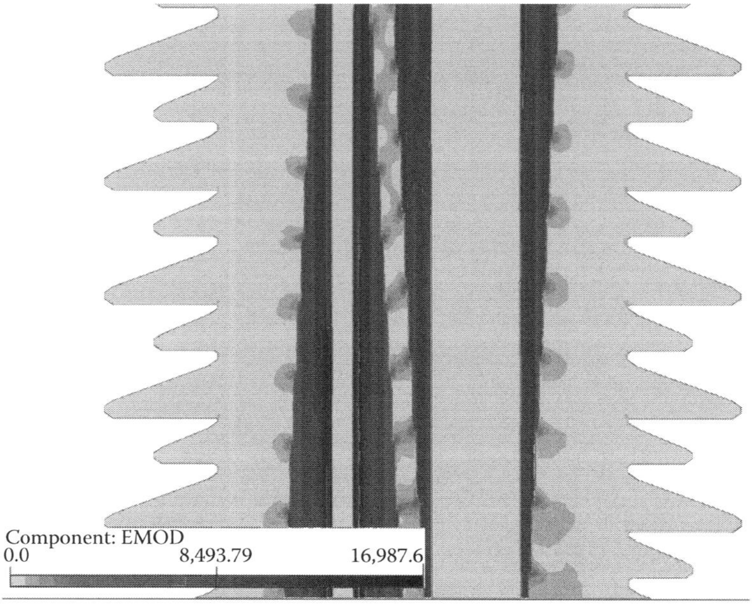

FIGURE 15.35
Distribution of the electric field strength (V/mm) in the area of the ceramic insulator, where the electric shields end (a blow up from Figure 15.29b).

TABLE 15.5
Result of Computations and Measurements of the Variations of Current and Voltage Error for the Final Version of Construction of the Combined Transformer

	The Variation of Current Error, $\pm\varepsilon_i$ (%)	The Variation of Voltage Error, $\pm\varepsilon_v$ (%)
Results of computing	0.033	0.004
Test results	0.034	0.02

In order to avoid this situation, the endings of the secondary windings of the current transformer were placed inside the grounded metal pipe, which also hardens the mechanical construction. The paper–oil insulation with capacitance control is wound directly on the pipe. The pipe conducts the current to ground and protects the secondary windings of current transformer from the influence of the voltage transformer operating nearby.

The mutual inductance M_{IU} and the mutual capacitance are very small for the final construction of the combined instrument transformer after limited mutual influences and equal to 0.12 μH and 1.95 pF, respectively (Table 15.5).

Prevention of possible couplings before making the prototype is simpler and less expensive than the elimination of interference during the operation of individual instrument transformers caused by placing them in the same case.

15.3 Conclusion

Analysis of physical field distribution using accurate numerical methods gives wide possibilities especially for constructors of precision electrical devices, which require very high accuracy. By applying the field method, the constructor is able to predict all steady-state and transient-state parameters of the designed device before making a real-life prototype. In the case of instrument transformers, it is very important to accurately compute the field distribution. Therefore, the reproduction of the shape of the device must be very close to the real shape without any oversimplified assumptions. The agreement of computing and test results depends closely on field distributions in real-life devices. The accuracy class of the designed instrument transformer and electric strength of its insulation system depends on the accuracy of its numerical model.

The tests on the real-life model are very expensive, so it is vitally important to carry out careful experimentation using computer software first to perfect the design prior to building and testing real-life models.

References

1. Lesniewska E (1997) Application of electromagnetic field simulations in the design process of instrument transformer. Zeszyty Naukowe PL, nr 766, Rozprawy Naukowe z.236 (in Polish).
2. Koszmider A, Olak J, Piotrowski Z (1985) *Current Transformers*. WNT, Warszawa, Poland (in Polish).
3. Lesniewska E, Chojnacki J (2002) Influence of the correlated localisation of cores and windings on measurement properties of current transformers. In *Studies in Applied Electromagnetics and Mechanics*, A.Krawczyk and S.Wiak (eds.), Vol. 22, IOS Press, Ohmsha, Tokyo, Japan, pp. 236–241.
4. *OPERA-3D User Guide* (2004) Vector Fields Limited, England, U.K.
5. Lesniewska E, Borowska-Banas I, Komęza K (1995) Influence of constructional parameters on leakage reactance of current transformer secondary winding. *Archives of Electrical Engineering*, XLIV: 245–259.
6. Lesniewska E, Jalmuzny W (1992) The estimation of metrological characteristics of instrument transformers in rated and overcurrent conditions based on the analysis of electromagnetic field. *COMPEL*, 11(1): 209–212.
7. Lesniewska E (1995) The comparison of the methods used for estimation of current transformer metrological properties. *COMPEL*, 14(4): 75–78.
8. Lesniewska E (1998) A field analysis of an overcurrent state and an incidental break in the secondary circuit of a current transformer. *COMPEL*, 17(1/2/3): 267–272.
9. Yatchev I, Milenov I (2002) Computation of transient current transformer error based on coupled field-circuit analysis. *Proceedings of the 10th International IGTE Symposium on Numerical Field Calculation in Electrical Engineering*, Graz, Austria, pp. 300–303.
10. Lesniewska E, Ziemnicki J (2006) Transient state analysis of protective current transformers at different forced primary currents. *Przegląd Elektrotechniczny*, 82 (5): 57–60.
11. Koszmider A, Jalmuzny W, Brodecki A (2005) Estimation of measurement properties of current transformers type TPZ. Report from Research, Lodz, Poland (in Polish).
12. Ziemnicki J (2005) Application of field-and-circuit method in the estimation of measurement properties of current transformers in transient state, PhD thesis, Technical University of Lodz, Lodz, Poland (in Polish).

13. Lesniewska E (2005) *Improvement of the Electric Strength of an Insulation System of a Medium Voltage Instrument Transformer Using Field Analysis Computer Engineering in Applied Electromagnetism*. Springer, West Sussex, England, pp. 143–148.
14. Lesniewska E, Kowalski Z (1991) Designing voltage transformer insulation system with SF6 insulations using CAD. *Archiv für Elektrotechnik*, 74: 427–432.
15. Lesniewska E (2001) Application of 3D field analysis for modelling the electric field distribution in ceramic insulator of HV combined instrument transformer. *Journal of Electrostatics*, 51–52: 610–617.
16. Lesniewska E (2002) The use of 3D electric field analysis and the analytical approach for improvement of a combined instrument transformer insulation system. *IEEE Transactions on Magnetics*, 38(2): 1233–1236.
17. Lesniewska E, Koszmider A (2004) Influence of the interaction of voltage and current parts of a combined instrument transformer on its measurement properties. *IEE Proceedings on Science, Measurement and Technology*, 151(4): 229–234.
18. Turowski J (1993) *Elektrodynamika Techniczna*. WNT, Warszawa, Poland (in Polish).
19. Turowski J, Sikora R, Pawluk K, Zakrzewski K (1990) Analiza i synteza pol elektromagnetycznych. Wydawnictwo Polskiej Akademii Nauk, Wroclaw, Warszawa, Krakow, Gdansk, Lodz (in Polish).
20. Biddlecombe CS, Simkin J, Jay AP, Sykulski JK, Lepaul S (1998) Transient electromagnetic analysis coupled to electric circuit and motion. *IEEE Transactions on Magnetics*, 34(5): 3182–3185.
21. INTERNATIONAL STANDARD IEC 60044-3 Ed2.02002, Instrument transformers. Part 3: Combined transformer.
22. INTERNATIONAL STANDARD IEC 60044-6, Ed1.0 modified, 1992 Instrument transformers. Part 6: Requirements for protective current transformers for transient performance.

16

CAD System-Boundary Integral Equation Method for 3D Electric Field Analysis of Voltage Transformers

Ivan Yatchev and Radoslav Miltchev

CONTENTS

16.1 Introduction

Boundary integral equation methods (BIEMs) are widely used for computation of three-dimensional (3D) electric fields [1–11]. They offer abilities to solve problems in unbounded regions and an easier method of building surface mesh than volume mesh, as needed when finite element method is employed for solving 3D problems.

In the present chapter, BIEM in its indirect variant is employed for the analysis of the 3D electric field of a voltage transformer. A general-purpose CAD system has been used at pre- and post-processing levels. This numerical analysis has been incorporated into a CAD system of instrument transformers.

16.2 CAD of Instrument Transformers

Electric field analysis is an important element in the CAD of instrument transformers. Development of a system for CAD of instrument transformers includes design and realization of fundamental tasks and auxiliary tools [12–16] to increase the benefits of standard software equipment used in design process of instrument transformers.

The model of system for computer-aided design of instrument transformers is shown in Figure 16.1. It was created and proposed as a result of generalization of investigations in the area of design, manufacturing, and study of instrument transformers and incorporation of

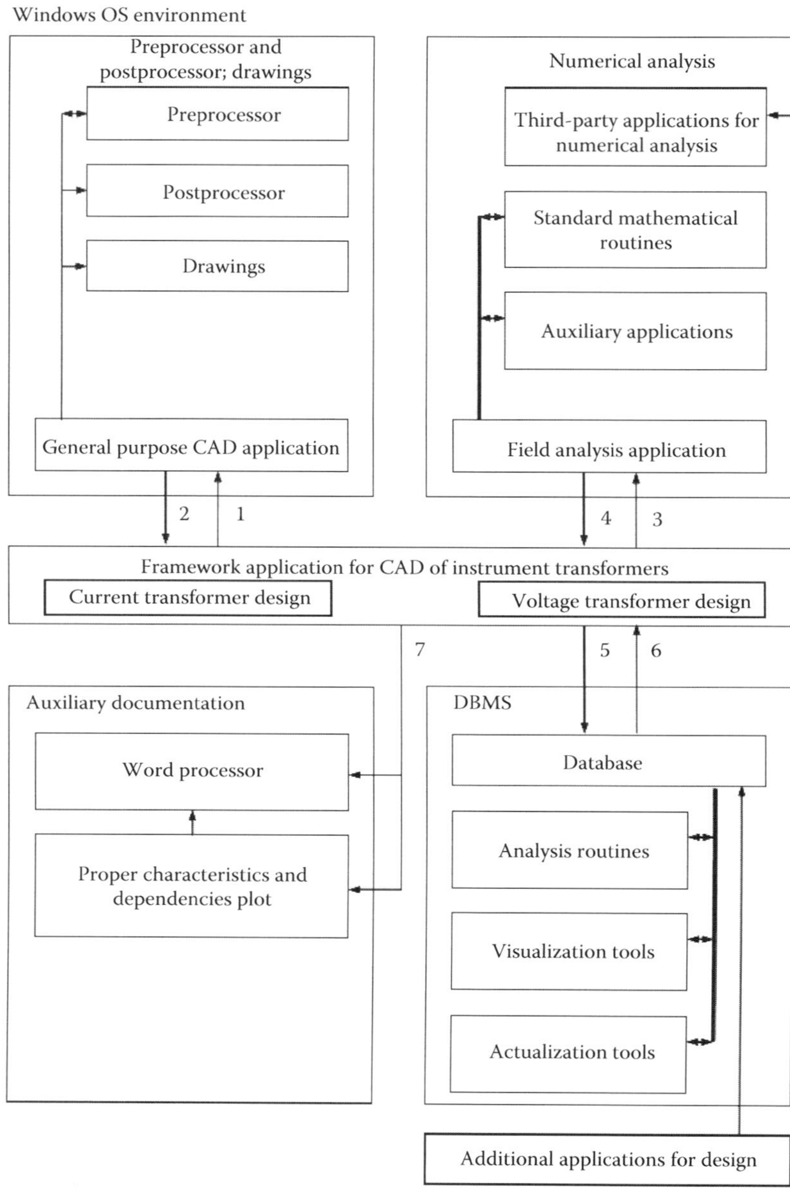

FIGURE 16.1
Model of a system for CAD of instrument transformers.

two-dimensional (2D) and 3D field analysis based on the use of different numerical methods. The present model revises theoretical and practical investigations done in [17–21] and summarizes features of the system for computer-aided design of instrument transformers based on the interaction of self-developed software, commercial software and classical methods of design. The system of such kind reflects the specifics of instrument transformers as electromagnetic devices, interconnected elements and parts, and various physical phenomena that describe the real system behavior.

Realization of the system based on the proposed model is carried out in the Windows® operating system environment. The architecture based on the use of several applications assumes that all of them can transfer their results to the others. The integration between all the described software applications is accomplished by using COM (Component Object Model) technology that gives object-oriented, scaleable, and unified protocol supported by many vendors. The main part of COM is ActiveX Automation technology. This was a built-in feature in many recently developed powerful contemporary Windows-based general-purpose commercial applications. The benefits of using technology include the increasing potential of software and addition of new functionality. As [14] shows, such an approach is in agreement with the new trends in software creation for scientific purposes.

The main part of the developed system for computer-aided design of instrument transformers is framework application. The main features of the framework application are the following:

- The ability to maintain containers with design procedures of different types of instrument transformers (current transformers, voltage transformers) based on classical methods of design
- Governing all necessary input/output information used or created by the containers
- Accomplishing bilateral communication with widely used commercial software, self-developed applications, and third-party applications
- Support of database with materials and physical properties of the materials used in manufacturing
- Management of results obtained from the design of different instrument transformers
- Creation of high-quality technical documentation including drawings, tables, and characteristics

As shown in Figure 16.1, the proposed system is organized by the communication between the framework application and four main modules, which are developed and incorporated in software applications in different areas of interest:

- Module for work in the area of general-purpose CAD system
- Module for field analysis with numerical methods
- Module for work in DBMS environment
- Module for preparation of auxiliary technical documentation

The framework application is controller application. Its procedures give access to the existing object models of different software applications. This includes access to general-purpose CAD system like AutoCAD, office applications like Microsoft Word processor and Microsoft Excel spreadsheet, database management system like Microsoft Access. An investigation shows that framework application can use successfully built-in functions of MATLAB® also. The applications that do not support ActiveX can be used in shell mode.

The proper GUI (Graphical User Interface) of the framework application strongly facilitates the process of computer-aided design of instrument transformers. To increase the abilities of GUI and to obtain better functioning, ergonomics and communication with the user, the mainframe application uses Tabbed Dialog graphical component. Based on this control we can divide user interface into several sections: section for input data, section for auxiliary data, section for B–H characteristics of used magnetic materials by table and diagram, section for output information, and section for specific plots. The framework application creates its user interface dynamically depending on the type and quantity of input information for the design of chosen instrument transformer built-in into the database. Snapshots for the container with design procedure for voltage transformer are shown in Figure 16.2.

Relationships between the framework application and the other modules of the system for CAD of instrument transformers are also shown in Figure 16.1. Information stream denoted by 1 unites the results received from the computer-aided design of instrument transformer. Results are obtained by classical methods; and the stream contains information about the creation of high-quality drawings based on standards for technical documentation, creation of 2D or 3D model for numerical field analysis, and visualization of postprocessor results. Information streams 2 and 3 contain information about field analysis and include the data for mesh generation of the studied model of instrument transformer, boundary conditions, and physical properties of materials. The stream denoted by 3 includes information from stream 2, and stream 4 contains the result information from the field analysis using numerical methods (numerical quantities of desired variables).

The framework application has possibilities to take numerical data input using full DBMS (Database Management System) maintenance including information stream 5 with results from the computer-aided design of instrument transformers stored in proper format for future use and stream 6 with data about standard materials (steels, wires, insulation, etc.) and their properties which are essential parts of design and study process. The results from CAD of instrument transformers that are necessary for the creation of

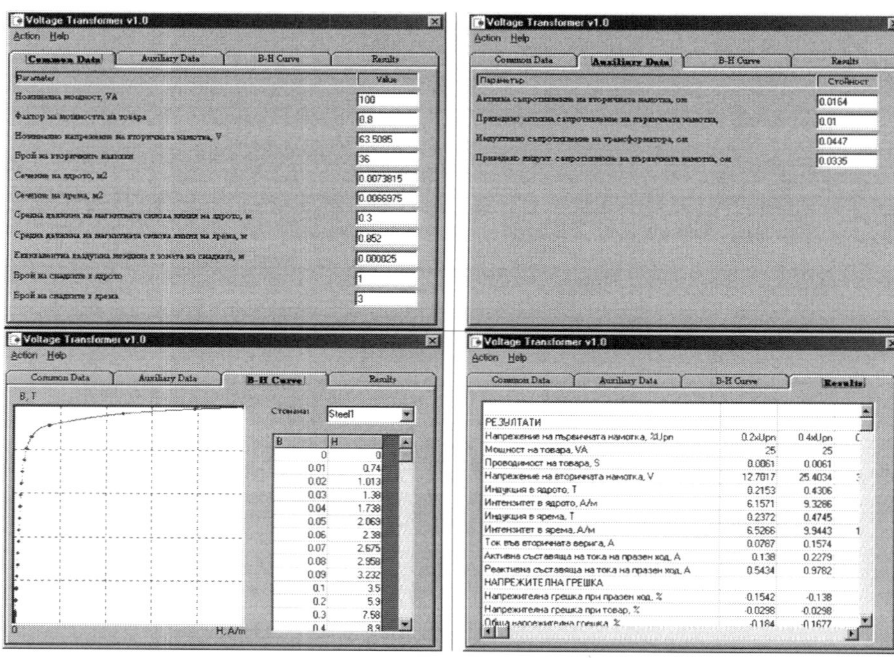

FIGURE 16.2
General view of framework application graphical user interface (in the case of built-in procedure for CAD of voltage transformers).

auxiliary technical documentation like tables, diagrams, and texts are in the base of information stream 7.

All modules are built in such a way to form a distributed system, the parts of which can operate independently, semi-independently or together in the context of computer-aided design of the instrument transformers.

The module for work in the area of a general-purpose CAD system is a very important one as usually design engineers employ it. Its development was based on the solution of several common problems using one standard environment:

- Creation of proper tools that operate in the chosen environment with abilities to build models of studied instrument transformers and generating the relevant meshes for their analysis using numerical methods, that is, to perform all the tasks at the preprocessor level
- Creation of proper tools that operate in the chosen environment with abilities to maintain postprocessor results from numerical field analysis, including representation of dynamic and static results
- General use of environment for requirements to create all the drawings needed for manufacturing of the designed instrument transformer using built-in drawing procedures of the containers and communication channel of mainframe application
- Export models and postprocessor results in VRML (Virtual Reality Modeling Language) format

To solve these common problems, a set of user tools were created in the AutoCAD® 2000 environment and drawing procedures for designed instrument transformer were developed. The procedures for drawing preparation use a communication channel between the framework application and the general-purpose CAD system. The tools created are available due to the object model of CAD system. They are grouped into two general toolbars "Preprocessor" and "Postprocessor" and include the following functions: managing project parameters; applying boundary conditions, material properties, and area and volume attributes; mesh generation (supporting 2D and 3D structured and unstructured meshes); exporting model and postprocessor result in VRML shape; and postprocessor results using isolines, isosurfaces, color maps, and animations.

The framework application for CAD of the instrument transformers uses developed full-feature DBMS for storage, management and analysis of input data and results from design as a separate module. Design and development of this module was based on the research of information streams as a part of computer-aided design of instrument transformers including: search for unique information streams; simultaneous correspondence between table structures in database and information stream; simplify information that must be included into one field of the table; ensure conditions to connect tables. Considering all special features investigated earlier and connected with relational databases, three common types of tables are developed:

- Tables to summarize data for used materials and standard products necessary for CAD of current and voltage transformers and lookup tables to manage B–H curves for magnetic material, physical constants, standard conductors, and electromagnetic steels.

- Tables with description of input and output information for CAD of instrument transformers.
- Lookup tables with data from industrial standard for the design and use of instrument transformers (different types of standard errors, rated currents, rated voltages, etc.).

The snapshots shown in Figure 16.3 represent a part of the interfaces developed to automate the work in the DBMS module. From top to bottom they are: common screen with lookup tables, steel materials and sections with the basic characteristics and physical properties of used materials.

The module for field analysis with numerical methods in the case of voltage transformers includes procedures for BIEM in three dimensions for electric field analysis and finite element method in two dimensions for magnetic field analysis. The magnetic field analysis is used for obtaining the leakage inductance of studied transformers using the

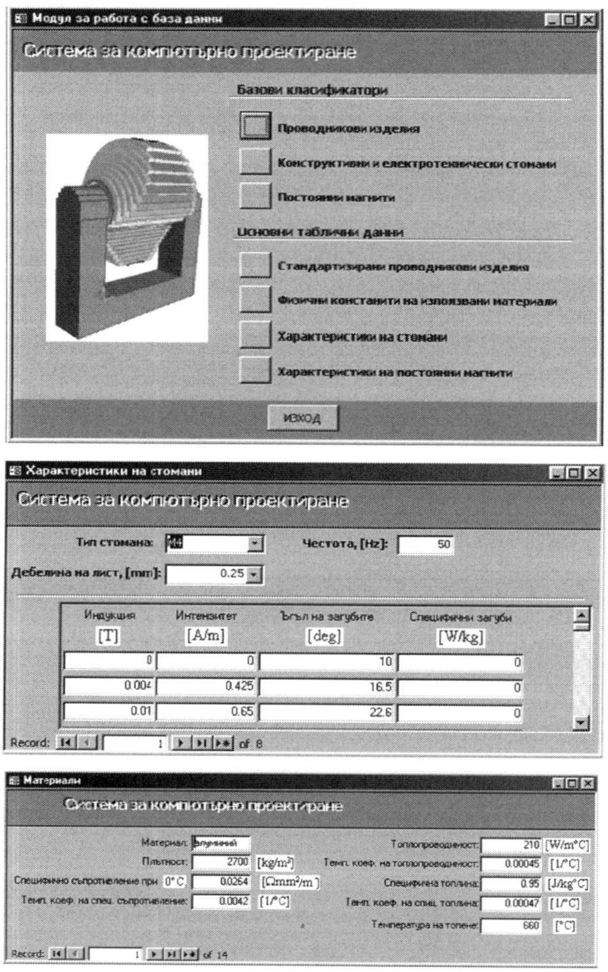

FIGURE 16.3
General view of several interfaces representing DBMS module environment.

program FEMM [22]. For the electric field analysis, own computer program is used, based mainly on the mathematical model and numerical realization [23,24], presented in the next sections.

16.3 Mathematical Model

The idea of an indirect BIEM is the first to introduce surface charges of unknown density over all boundaries—electrodes and dielectric boundaries. Then the surface charge distribution over all the boundaries is obtained after solving a system of integral equations. The dielectrics are removed and all needed results are obtained with the help of integrals over the boundaries in free space.

Thus, the mathematical model of the electric field consists of Fredholm integral equations of the first and second kind [1,2]. They are obtained after defining the surface charge density for all boundaries and forming the system of integral equations for this charge density.

The surface charge density σ for electrodes is defined as

$$\sigma = \varepsilon_0 E_n \tag{16.1}$$

where E_n is the normal component of the field intensity on the electrode boundary.

The surface charge density in this case directly corresponds to the electric field intensity, which at the boundary of an electrode has only a normal component. For a dielectric boundary the surface charge density is defined as

$$\sigma = 2\varepsilon_0 \lambda E_n \tag{16.2}$$

where $\lambda = \varepsilon^{(i)} - \varepsilon^{(e)}/\varepsilon^{(i)} + \varepsilon^{(e)}$; $\varepsilon^{(i)}$, $\varepsilon^{(e)}$ are the dielectric permittivities of the inner and outer media. The normal vector is directed to the medium of $\varepsilon^{(e)}$.

Consider a general system containing n boundaries, of which m are boundaries of electrodes of known potential U_k, $k = 1, \ldots, m$, p are dielectric boundaries and s are conducting parts of unknown potential but of known total charge q_k, $k = m + p + 1, \ldots, n$.

Assuming surface charge density distribution as a simple layer potential, the following boundary integral equations are obtained:

- *For electrode boundaries of known potentials,* Fredholm integral equation of the first kind

$$\frac{1}{4\pi\varepsilon_0} \sum_{j=1}^{n} \int_{S_j} \sigma(M_j) \frac{1}{r_{Q_k M_j}} dS_j = U_k; \quad k = 1, \ldots, m \tag{16.3}$$

- *For dielectric boundaries,* Fredholm integral equation of the second kind

$$\sigma(Q_k) - \frac{\lambda_k}{2\pi} \sum_{j=1}^{n} \int_{S_j} \sigma(M_j) \frac{\cos(\mathbf{r}_{Q_k M_j}, \mathbf{n}_{Q_k})}{r_{Q_k M_j}^2} dS_j = 0; \quad k = m+1, \ldots, m+p \tag{16.4}$$

- *For boundaries of conducting bodies of unknown potential,* Fredholm integral equation of the first kind and the total charge equation for each conducting part

$$\frac{1}{4\pi\varepsilon_0}\sum_{j=1}^{n}\int_{S_j}\sigma(M_j)\frac{1}{r_{Q_kM_j}}dS_j - U_k = 0; \quad k = m+p+1,\ldots,m+p+s \tag{16.5}$$

$$\int_{S_k}\sigma(M_k)dS_k = q_k; \quad k = m+p+1,\ldots,m+p+s \tag{16.6}$$

In (16.3 through 16.6), the following notations are used:
k represents the number of the boundary
U_k represents the potential of the k-th electrode or k-th conducting part
S_j represents the boundary j
M_j represents the influence point
Q_k represents the observation point
$\mathbf{r}_{QkMj}$ represents the radius vector from point Q_k to point M_j
$\mathbf{n}_{Qk}$ represents the outward unit normal vector to the boundary S_j
$\lambda_k = \varepsilon_k^{(i)} - \varepsilon_k^{(e)}/\varepsilon_k^{(i)} + \varepsilon_k^{(e)}$

16.4 Numerical Approach Employed

The system of integral Equations 16.3 through 16.6 is solved by the method of mechanical quadratures and constant boundary elements.

The discrete analogues of (16.3 through 16.6) are obtained by dividing all the boundaries into boundary elements and representing the integrals over each boundary as a sum of integrals over its boundary elements:

$$\frac{1}{4\pi\varepsilon_0}\sum_{j=1}^{n}\sum_{b=1}^{N_j}\int_{S_{jb}}\sigma(M_{jb})\frac{1}{r_{Q_{ki}M_{jb}}}dS_{jb} = U_k; \quad i = 1,\ldots,m_{be} \tag{16.7}$$

$$\sigma(Q_{ki}) - \frac{\lambda_k}{2\pi}\sum_{j=1}^{n}\sum_{b=1}^{N_j}\int_{S_{jb}}\sigma(M_{jb})\frac{\cos(\mathbf{r}_{Q_{ki}M_{jb}},\mathbf{n}_{Q_k})}{r^2_{Q_{ki}M_{jb}}}dS_{jb} = 0; \quad i = m_{be}+1,\ldots,m_{be}+p_{be} \tag{16.8}$$

$$\frac{1}{4\pi\varepsilon_0}\sum_{j=1}^{n}\sum_{b=1}^{N_j}\int_{S_{jb}}\sigma(M_{jb})\frac{1}{r_{Q_{ki}M_{jb}}}dS_{jb} - U_k = 0; \quad i = m_{be}+p_{be}+1,\ldots,N_{be} \tag{16.9}$$

$$\sum_{b=1}^{N_j}\int_{S_{kb}}\sigma(M_{kb})dS_{kb} = q_k; \quad k = m+p+1,\ldots,m+p+s \tag{16.10}$$

where

N_j is the number of boundary elements on the boundary j

S_{jb} is the b-th boundary element of the j-th boundary

m_{be}, p_{be} are the total number of boundary elements on the boundaries of electrodes, and dielectrics, respectively

N_{be} is the total number of boundary elements

Equations 16.7 through 16.10 constitute a system of linear equations with respect to the values of the surface charge density at the nodes of the boundary elements and for the unknown potentials of conducting parts. For overcoming the singularities appearing in the integral equations of the first kind—in the linear equation system they cause singularities in the diagonal coefficients of the matrix—analytical expressions for rectangular elements are used. For a rectangular boundary element of sides a and b, the integral used for diagonal coefficients is

$$\int_{S_{ii}} \frac{1}{r_{QM}} dS_M = 2a \ln \frac{\sqrt{a^2+b^2}+b}{\sqrt{a^2+b^2}-b} + 2b \ln \frac{\sqrt{a^2+b^2}+a}{\sqrt{a^2+b^2}-a} \tag{16.11}$$

For all other integrals numerical integration using Gaussian quadratures is used.

After solving the system (16.7) through (16.10) and obtaining the surface charge density σ, the potential U and the electric field intensity **E** at an arbitrary point P are obtained by the well-known expressions

$$U(P) = \frac{1}{4\pi\varepsilon_0} \sum_{j=1}^{n} \int_{S_j} \sigma(M_j) \frac{1}{r_{PM_j}} dS_j \tag{16.12}$$

$$\mathbf{E}(P) = \frac{1}{4\pi\varepsilon_0} \sum_{j=1}^{n} \int_{S_j} \sigma(M_j) \frac{\mathbf{r}_{PM_j}}{r_{PM_j}^3} dS_j \tag{16.13}$$

16.5 Approach for Building the 3D Model

Creation of a proper 3D model for numerical computation is one of the most important tasks of the study and design. The geometry should be described in a way suitable for the numerical technique. The BIEM for solving 3D problems requires a specific geometry description. Approach and tools were developed for describing geometry, applying boundary conditions, physical properties and loads in 3D problems solved both by the Finite Element Method and by the BIEM. The essence of the approach is that pre- and postprocessor levels are carried out in the environment of general-purpose CAD program with the help of additional software for coupling with the processor. For building the model of the studied voltage transformer, a mixed approach with solid and surface geometrical objects was employed. The approach is illustrated in Figure 16.4. Solids are transformed automatically to surface objects taking into account the direction of the normal vector.

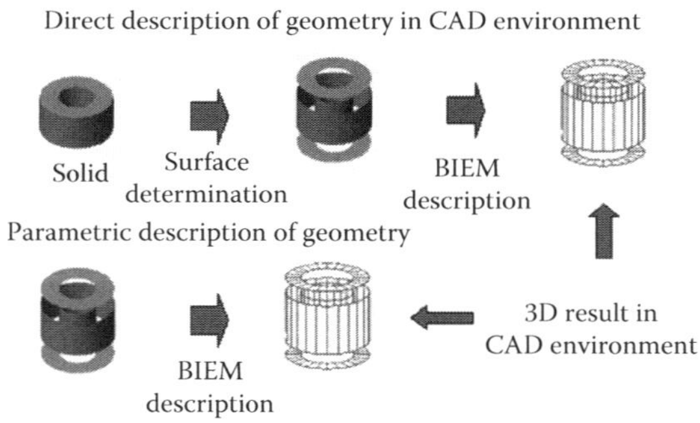

FIGURE 16.4
Approaches employed for building and verification of the 3D model.

The AutoCAD system has been employed. Additional software tools have been developed for visualization at pre- and postprocessor levels, making use of ActiveX Automation technology in the AutoCAD environment.

For verification of the developed tools, the procedures for parametric description of geometry of the model and automatic building of it in the AutoCAD environment have been created. These procedures are the base for mesh generation and creation of the appropriate postprocessor. For defining the geometry, the parametric description of all boundaries is employed. Each surface is described as a function of two parameters—u and v. After generating the mesh, each boundary element is defined by lower and upper values of u and v. Thus, each boundary element is a part of a surface, defined by four points. The following basic types of elements have been implemented so far:

- Rectangle (parameters u and v are the co-ordinates along the two sides)
- Part of a cylinder (u is the angle, v is the height co-ordinate)
- Part of a disk (u is the radius, v is angle)
- Part of a torus (u and v are angles)

The system is open and any other surface type described parametrically can be added. Regular and cosine distribution of the boundary elements is supported.

The quality of generated mesh affects directly the quality of the postprocessor results. For better presentation of the results, additional mesh refinement is carried out at postprocessor level, followed by approximation of the surface charge density over the refined mesh. This improves the visualization of the results, especially of the color maps. An example of the mesh refinement in azimuthal direction is shown in Figure 16.5.

Employment of general-purpose CAD system for solving problems of the postprocessor enables receiving a lot of different viewpoints in 3D space, real time visualization, and manipulation. It is possible to obtain images representing the geometry of the model and the mesh, contours, equipotentials, and areas in 2D and 3D space. For the post-processor level, a tool for visualization of the surface charge density has also been developed.

Another advantage of using general purpose CAD system is the fact that many of the design engineers use this system in their everyday work. Having additional tools at their disposal, it becomes easy for them to deal with field analysis as a real opportunity.

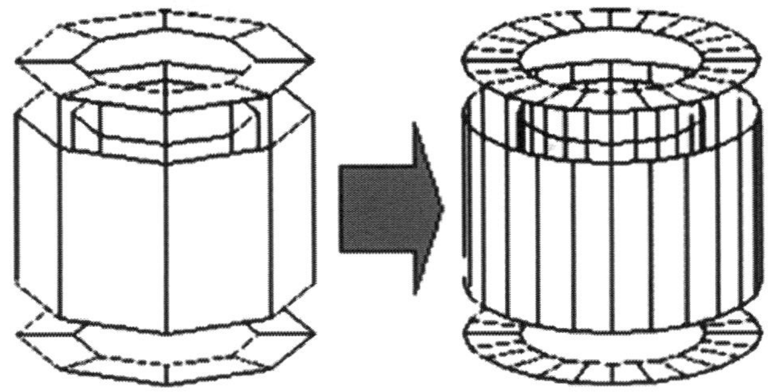

FIGURE 16.5
Improvement of postprocessor results based on parametrical output.

16.6 Electric Field of a Voltage Transformer

As an example of the presented approach, a 126 kV voltage transformer has been modeled and its electric field has been analyzed. An idea about the geometry of the voltage transformer can be obtained from Figure 16.6 where part of the high-voltage winding has been removed. The orientation of the co-ordinate system is also shown. The cross section of the high-voltage winding is stepwise, covered by solid insulation.

Two principal constructions have been studied—without additional HV shields (referred to as "basic VT") and construction with additional two HV shields located over the edges of the uppermost layer of the HV winding (referred to as "VT with shields"). The shields are in the shape of torus. Different small radii of the shields have been considered.

The geometry of the transformer is such that it cannot be modeled as 2D and that is why 3D analysis is needed. For building the geometric model a total of about 130 macro-elements are used, resulting in about 5000 boundary elements after the mesh generation. Cosine discretization was used in the x-direction of the winding elements.

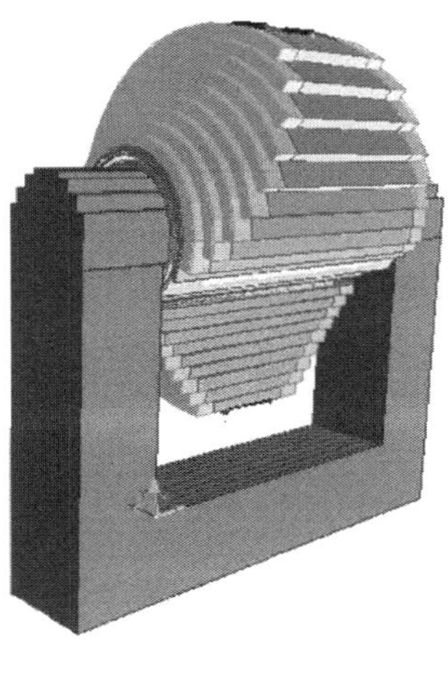

FIGURE 16.6
Geometry of the voltage transformer (VT).

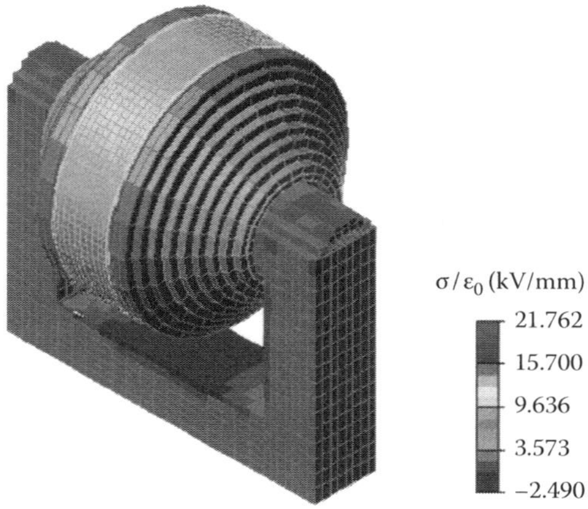

FIGURE 16.7
Surface charge density for the basic VT.

Figure 16.7 shows the color map of the obtained charge density on the boundary surfaces for the basic VT. The approach with mesh refinement in the azimuthal direction and approximation of the surface charge density has been employed. On electrode boundaries, that is, on the windings and the core, the charge density is proportional to the field intensity. The most electrically loaded area is on the end part of the high-voltage winding on its lower part (i.e., close to the yoke). It is practical to present the results for the charge density divided by ε_0, thus having the dimension of the electric field intensity.

In Figures 16.8 and 16.9, maps of equipotential lines are shown for the *XZ* and *YZ* planes. Both *XZ* and *YZ* planes are chosen to coincide with the planes of symmetry of the

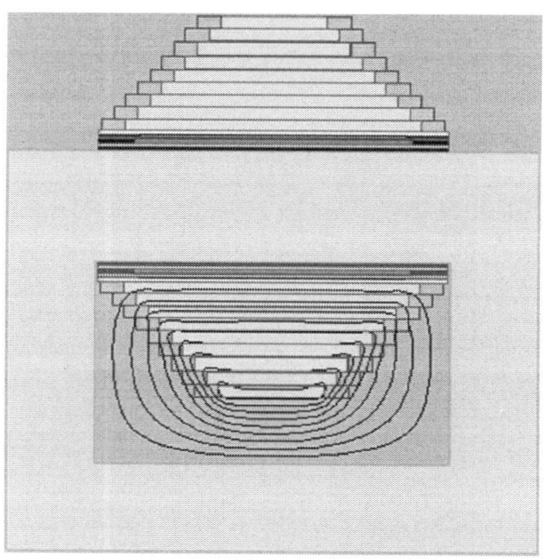

FIGURE 16.8
Map of equipotential lines at *XZ* plane for the basic VT.

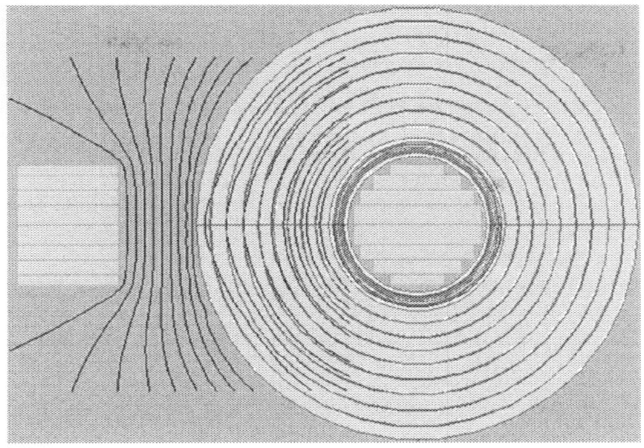

FIGURE 16.9
Map of equipotential lines at *YZ* plane for the basic VT.

transformer. The difference between the potential values of two adjacent equipotential lines is 10% of the applied voltage.

Similar results are obtained in the case of VT with shields. The shields are placed over the high-voltage winding and connected to the maximal potential. Several small radii of the torus were employed—from 3 to 7 mm, denoted, respectively, by *r*3 to *r*7. In Figure 16.10, color map of the surface charge density for the radius *r*3 is shown. The presence of shields leads to decreasing about two times the maximal value of the surface charge density (corresponding to the field intensity), which is now moved from the outer HV layer to the shield. The map of equipotential lines for the same case is given in Figure 16.11 for the *XZ* plane and in Figure 16.12 for the *YZ* plane.

It can be seen that, while the maps in *XZ* planes are different due to the shields, those in *YZ* plane are almost the same. The maximal field intensity is obtained in the region of the shields.

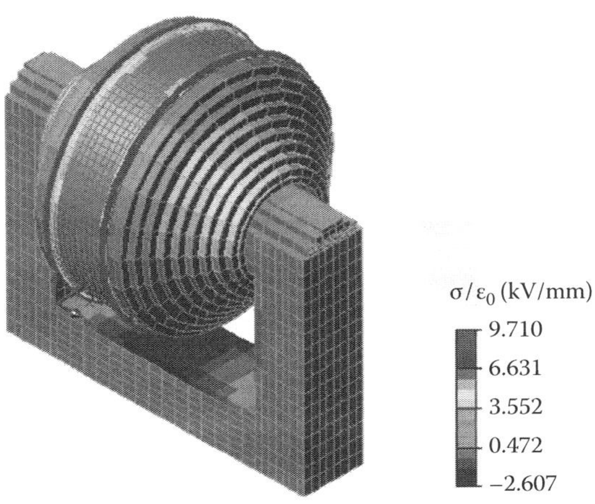

FIGURE 16.10
Surface charge density for the VT with shields (*r*3).

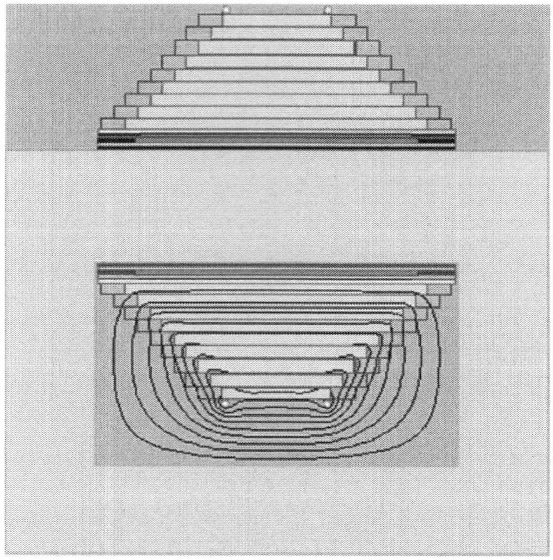

FIGURE 16.11
Map of equipotential lines at XZ plane for VT with shields ($r3$).

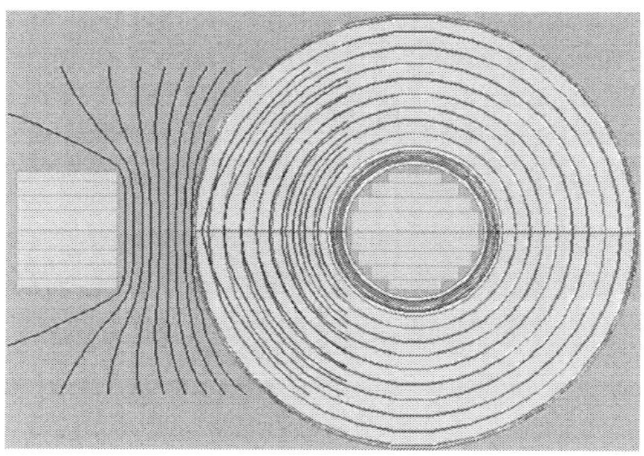

FIGURE 16.12
Map of equipotential lines at YZ plane for VT with shields ($r3$).

For visualizing the electric field intensity on the electrodes only, the dielectrics can be hidden and absolute value of the charge density can be taken. The normal electric field intensity for the case $r7$ is shown in Figure 16.13. For this case the maximal intensity is about 35% less than for the case $r3$.

The dependence of the maximal value of the field intensity on the small radius of the shield is shown in Figure 16.14.

Additional possibility of output result is the distribution of the field intensity along given line segment. As an example of such a segment the one connecting the end of the HV winding and the core (parallel to z-axis) was chosen and the electric field intensity was calculated at its 12 points. The results for different shield cases are shown in Figure 16.15. The points that are inside the shield are removed. It is seen that the field intensity along the line increases with the shield radius, while at the same time the maximal intensity gets lower.

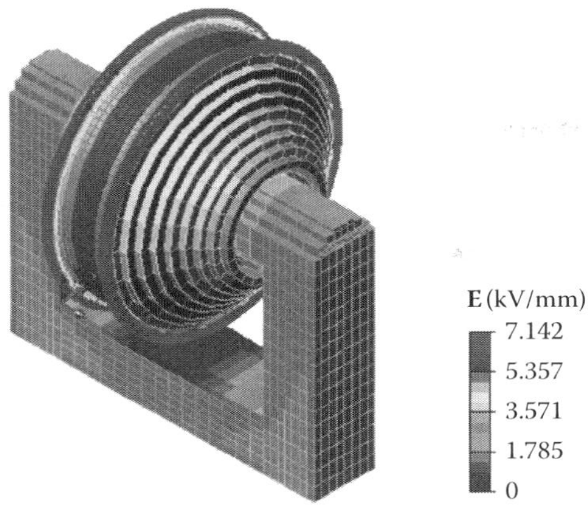

FIGURE 16.13
Electric field intensity for the VT with shields (*r*7).

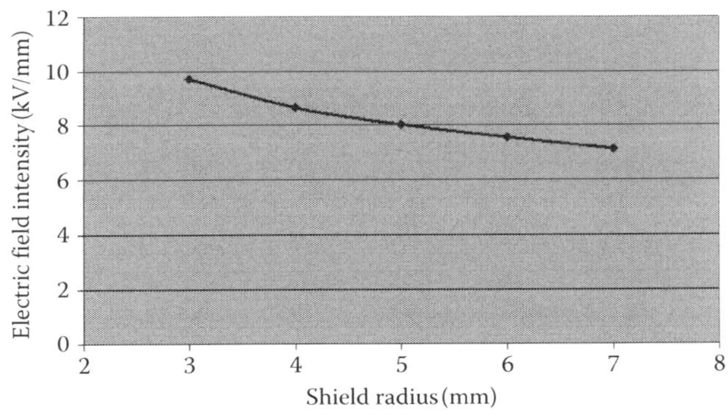

FIGURE 16.14
Maximal electric field intensity as a function of the shield radius.

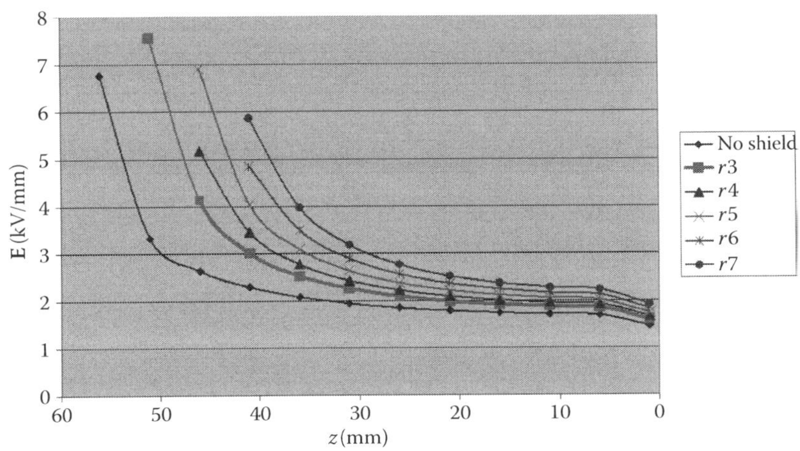

FIGURE 16.15
Electric field intensity along a line segment for different shields cases.

16.7 Conclusions

An approach for 3D electric field analysis has been presented for a typical insulating system. The approach employs indirect BIEM as a mathematical basis, parametric description of the boundaries and general purpose CAD system for visualization at pre- and post-processor levels. This approach has been incorporated into a CAD system of instrument transformers.

As an illustration of the presented approach, the 3D electric field of a 126 kV voltage transformer has been analyzed. Two constructions are considered—without and with additional torus shields. The construction with additional shields features more than two times lower value of the maximal field intensity and it can be subject to further improvement by varying the position and size of the shields.

The approach can be of help to design engineers making it easy for them to deal with field analysis in their work.

References

1. Tozoni O, Mayergoyz I (1974) *Calculation of Three-Dimensional Electromagnetic Fields*. Tehnika, Kiev (in Russian).
2. Kolechitsky E (1983) *Calculation of Electric Fields in High Voltage Equipment*. Energoatomizdat, Moscow, Russia (in Russian).
3. McWhirter JH, Duffin RJ, Brehm PJ, Oravec JJ (1979) Computational methods for solving static field and eddy current problems via Fredholm integral equations. *IEEE Trans Magn*, MAG-15:1075–1084.
4. Andjelic Z et al. (1992) *Integral Methods for the Calculation of Electric Fields—For Application in High Voltage Engineering*. Scientific Series of the International Bureau, Forschungszentrum Jülich, Jülich, Germany, Vol. 10.
5. Misaki T, Tsuboi H, Itaka K, Hara T (1982) Computation of three-dimensional electric field problems by a surface charge method and its application to optimum insulator design. *IEEE Trans Power Appar Syst*, PAS-101:627–634.
6. Nabors K, White J (1991) Fastcap: A multipole accelerated 3-d capacitance extraction program. *IEEE Trans Comput Aided Des*, 10:1447–1459.
7. Gutfleisch F, Singer H, Förger K, Comollon JA (1994) Calculation of high-voltage fields by means of the boundary element method. *IEEE Trans Power Delivery*, 9(2): 743–749.
8. Maeder G, Uhlmann H (1995) Computation of capacitance coefficients of three-dimensional integrated structures with the boundary element method. In *Advanced Computational Electromagnetics*, Honma T. (Ed.), Elsevier Science, Amsterdam, the Netherlands, pp. 13–23.
9. Yatchev I (1996) Axisymmetrical electric field computation using boundary integral equation method. *J Electrost*, 36:277–284.
10. Lesniewska E (2002) The use of 3D electric field analysis and the analytical approach for improvement of a combined instrument transformer insulation system. *IEEE Trans Magn*, 38(2):1233–1236.
11. Que W, Sebo SA (2003) Electric field and voltage distribution along non-ceramic insulators. *Proceedings of 2003 World Conference and Exhibition on Insulators, Arresters and Bushings*, Marbella, Spain, pp. 361–375.
12. Conrad A (1985) Electromagnetic devices and the application of computational techniques in their design. *IEEE Trans Magn*, 21:2382–2387.

13. Sabonnadiere J-C, Coulomb J-L (1988) *Finite Element Methods in CAD*. North Oxford Academic, London, U.K.
14. Kornbluh K (1999) Math and science software. *IEEE Spectr*, 36:88–91.
15. Mitroznik M, Prather D (1997) How to choose electromagnetic software. *IEEE Spectr*, 34:53–58.
16. Moham N, Undeland T, Robins W (1995) *Power Electronics, Converters, Applications and Design*, 2nd edn., Wiley, New York.
17. Miltchev R (1999) An approach of combining general-purpose CAD systems with finite element method software. *Proceedings of SIELA'99*, Plovdiv, Bulgaria, Vol. 2, pp. 138–143.
18. Miltchev R, Yatchev I (2000) Electrical apparatus CAD using controlled interaction between conventional software. *Proceedings of ICATE'2000*, Craiova, Romania, pp. 149–152.
19. Miltchev R, Yatchev I, Ritchie E (2000) Approach and tool for computer animation of fields in electrical apparatus. *Proceedings of JBMSAEM'2K*, Ohrid, Macedonia, pp. 149–155.
20. Miltchev R, Yatchev I, Terziisky A (2002) Computer aided design of current transformer. *Proceedings of ICATE 2002*, Craiova, Romania, Vol. 1, pp. 89–93.
21. Miltchev R, Yatchev I, Terziisky A (2002) Computer aided design of voltage transformer. *Proceedings of ELMA 2002*, Sofia, Bulgaria, Vol. 2, pp. 44–51.
22. Meeker D (2003) *Finite Element Method Magnetics Version 3.3—User's Manual*, Foster-Miller Inc., Waltham, MA.
23. Yatchev I, Miltchev R, Terziisky A (2001) 3D electric field analysis of a voltage transformer using boundary integral equation method. Invited paper, *Proceedings of Fifth International Conference on Applied Electromagnetic, PES 2001*, Nis, Serbia, pp. 205–208.
24. Yatchev I, Miltchev R, Terziisky A (2002) 3D electric field analysis of a voltage transformer using boundary integral equation method and general-purpose CAD system. *Facta Universitatis (Nis)*, 15(2):227–236.

17

Instrument Transformers' Insulation Behavior to High-Voltage and High-Frequency Transients

A. Ibero

CONTENTS

17.1 Introduction

There are three different failure modes that exist, those that can affect dielectric insulation of this type when exposed to this class of transient.

17.2 Possible Failure Modes

The initial effect of a high-voltage (HV), high-frequency (HF) transient on insulation is to produce a mechanical shock wave in the insulation. This mechanical shock wave will produce a vibration that compresses and expands the insulating material. This shock wave could cause immediate fracture of the insulation at its weakest point, which will normally coincide with the maximum electrical gradient, where the mechanical gradient is also highest [1].

To prevent this failure, the dimensions of the insulation must be calculated so that the electrical gradients are less than the value that will cause electrical perforation or mechanical breakage (Figure 17.1).

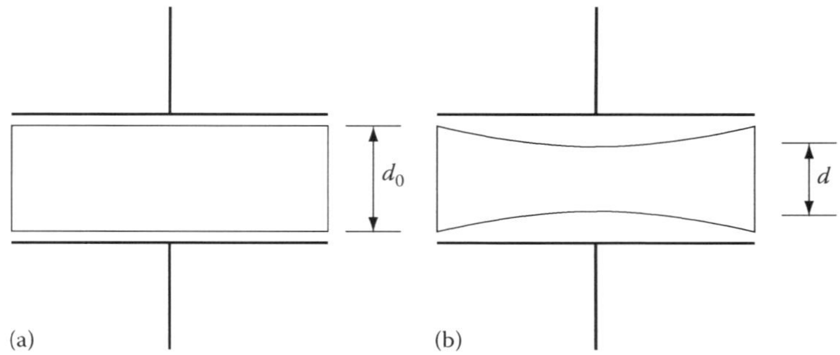

FIGURE 17.1
Representation with no electric field and with an applied electric field.

The pressure produced by the electrical field is

$$P_C = \frac{1}{2}DE = \frac{1}{2}\varepsilon_0\varepsilon_r\frac{V^2}{d^2} \tag{17.1}$$

Hooke's Law for high compression fatigue gives

$$P_C = Y\ln\left(\frac{d_0}{d}\right) \tag{17.2}$$

At the equilibrium point, making the two pressures equal, we have

$$V^2 = \frac{2Y}{\varepsilon_0\varepsilon_r}d^2\ln\frac{d_0}{d} \tag{17.3}$$

This equilibrium becomes unstable when $d < 0.6d_0$.

Fortunately, this behavior can be avoided by making the insulation thick enough and sufficiently compact. However, instantaneous mechanical fracture or perforation is not the only possible failure.

There is another failure mode derived from this one that is caused by the mechanical compression and expansion of the insulation.

A second failure mode, which is a consequence of the first one, is partial discharge. These partial discharges are caused by inadequate impregnation of the insulation. The mechanical shock waves produce an internal cavitation phenomenon in the oil, releasing any gases that have remained occluded in it. For this reason, it is essential to impregnate the paper with oil that has been thoroughly degassed, and the insulation must be very compact.

The third failure mode, which bears no relationship to the mechanical shock waves nor to the partial discharges produced inside the insulation by these, involves partial discharges caused by high resistivity of the Faraday screening of the insulation. In fact, as we shall see later, the values of current (kA) and frequency (MHz) involved make the design of the Faraday screening a complex task since very flexible electrodes must be achieved, with a low transversal resistance to allow these high currents to flow, while at the same time, they must not present difficulties when it comes to their use in the manufacture of the equipment.

17.3 Equivalent Circuit of the Insulation of an Inverted-Type Current Transformer

Figure 17.2 shows a cross section of the insulation of a current transformer (CT), with the different elements that it consists of.

The analysis of the transient response of a CT exposed to HF transients is based on the study of the equivalent circuit shown in Figure 17.3 and its response to such transients.

It should be noted that this circuit can be used when the following condition is satisfied: the applied voltage transients must have rise or fall times or equivalent oscillation frequencies such that this circuit is valid in the space-time domain. In other words, the propagation times of the electromagnetic phenomena through the length and breadth of the insulation must be much shorter than the rise or fall times of the transient voltage waveforms applied to the transformer.

This is important in 420 kV very-high-voltage equipment and becomes really critical at 525 and 765 kV.

In order to study this matter, we shall first define some of the concepts involved:

When we mention HV, HF transients, we are really talking about overvoltage phenomena between the line and ground due to lightning and/or circuit-breaker operations.

In reality, the most damaging phenomenon for the insulation is a lightning-type overvoltage spike with an abrupt leading or trailing edge caused, for example, by starting external spark gaps.

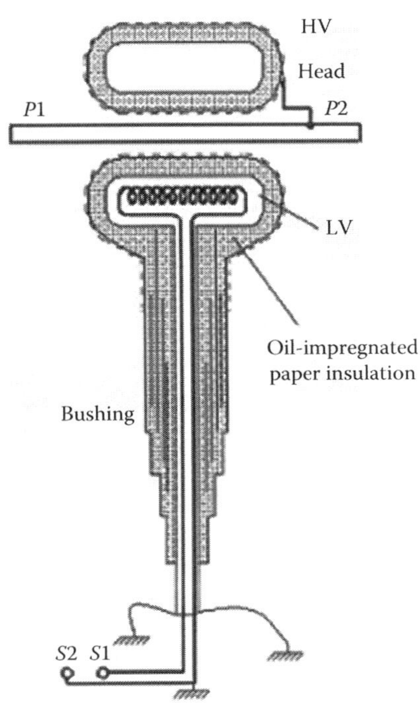

FIGURE 17.2
Cross section of the insulation of a CT.

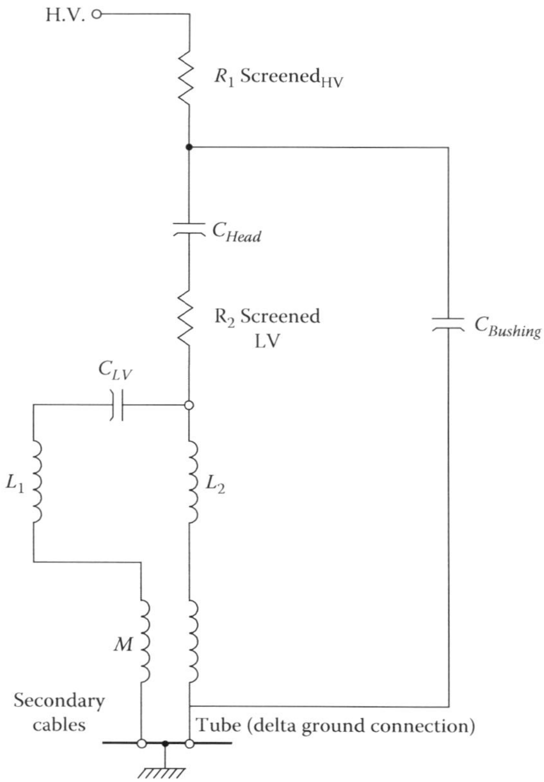

FIGURE 17.3
Equivalent circuit of the insulation of a CT.

This type of overvoltage provokes a *stepped wave* that propagates along the line until it reaches the instrument transformer, inducing HF overvoltage and overcurrent surges in its insulation, which must be designed with adequate dimensions to withstand these.

In practice, and given the dimensions of the lines, the connections between circuit breakers and CTs, the physical dimensions of the CT itself and its pedestal, etc., the stepped wave is transformed into a spike with a minimum fall time of 0.3 μs (Figure 17.4).

Under these conditions, the CT is subjected to this voltage between the HV line and ground. The following propagation paths are available through the insulation:

The wave enters the dielectric through the HV head as conduction current and voltage. It passes through the HV Faraday screening and is converted into a displacement current and electric field to cross the dielectric material and reach the low-voltage (LV) electrode (Figure 17.5).

If we take into account that we are talking about thicknesses of 70–100 mm of dielectric and that this dielectric is oil-impregnated paper with a propagation constant of

$$\frac{1}{\mu_r \varepsilon_r} = \frac{1}{\sqrt{2.9}} \tag{17.4}$$

and therefore, a velocity of

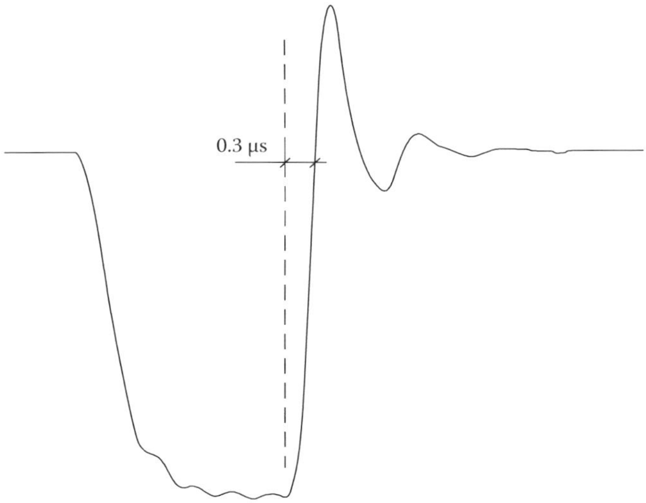

FIGURE 17.4
Stepped wave.

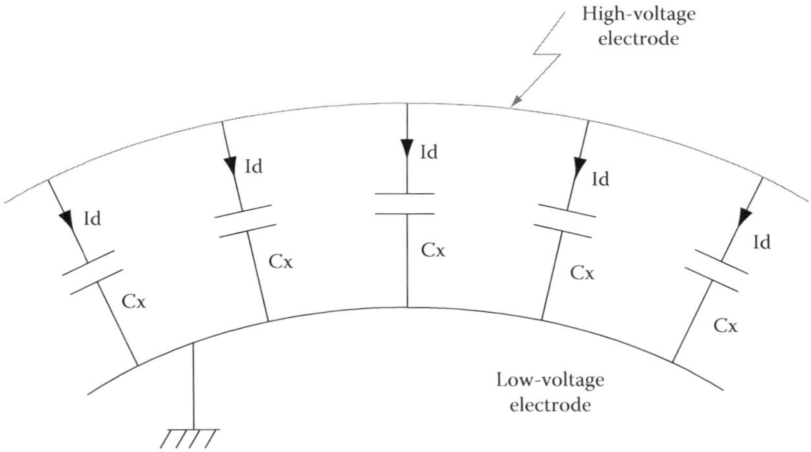

FIGURE 17.5
Representation displacement current and electric field to cross the dielectric material.

$$V = \frac{C}{\sqrt{\mu_r \varepsilon_r}} = \frac{3\times10^8 \text{ m/s}}{\sqrt{2.9}} = 1.76\times10^8 \text{ m/s} \tag{17.5}$$

To cover 100 mm = 10 × 10^{-2} m, the time required is

$$t = \frac{10\times10^{-2} \text{ m}}{1.76\times10^8 \text{ m/s}} = \frac{10}{1.76}\times10^{-10} \text{ s} = 5.68\times10^{-4} \text{ μs} \tag{17.6}$$

Therefore, tpropagation << tperturbation since 5.68 × 10^{-4} μs << 0.3 μs. The other propagation path is via the length, breadth, and thickness of the insulating bushing. The

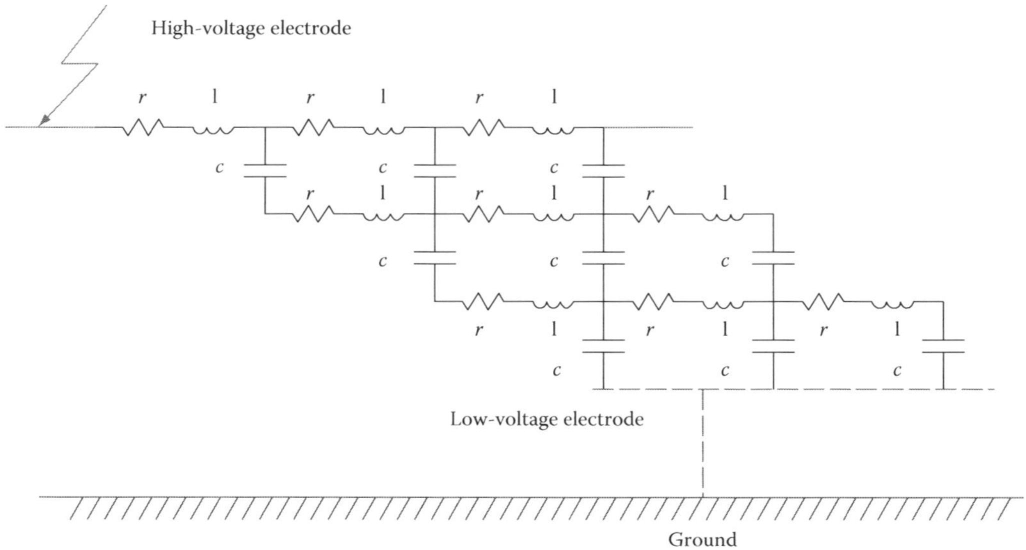

FIGURE 17.6
Equivalent circuit of the insulating bushing of a CT.

insulating bushing is really a bushing that allows the insulation to be reduced in size and compacted in order to make an element with the smallest possible physical dimensions (Figure 17.6).

We have already seen that the thickness of the insulation is such that the propagation time through it is much shorter that the duration of the perturbation. However, an analysis is required of the length and width of the insulating bushing to ensure that it behaves as a short transmission line from the point of view of the HV, HF transients that are applied.

If it behaves as a long transmission line, undesirable delay phenomena may occur in the voltage propagation and distribution that may cause failures in the insulation.

In the case of the cone of an insulating bushing, we are dealing with a length of up to about 3 m at most. The propagation velocity through oil-impregnated paper is $v = 1.76 \times 10^8$ m/s.

Therefore, the propagation time or delay along the length of the bushing will be

$$t_r = \frac{3\text{ m}}{1.76 \times 10^8\text{ m/s}} = 1.7 \times 10^{-8}\text{ s} = 1.7 \times 10^{-2}\ \mu\text{s} \qquad (17.7)$$

The duration of the perturbation is 0.3 μs.

A reasonable criterion that can be used to determine whether the bushing is a long or a short transmission line with respect to the applied voltage is $0.1 \times \text{tperturbation} \geq \text{tdelay}$ [2].

In the present case, $0.1 \times 0.3\ \mu\text{s} = 0.03 \geq 0.017\ \mu\text{s}$.

Therefore, this case can be regarded as a short transmission line. In addition to being physically short, for this to be valid, it is also necessary that the electrical resistance of the conductors in the "transmission line" be practically zero (Figure 17.7). This implies that the screens used to make the bushing must be metallic; otherwise, the delay time will

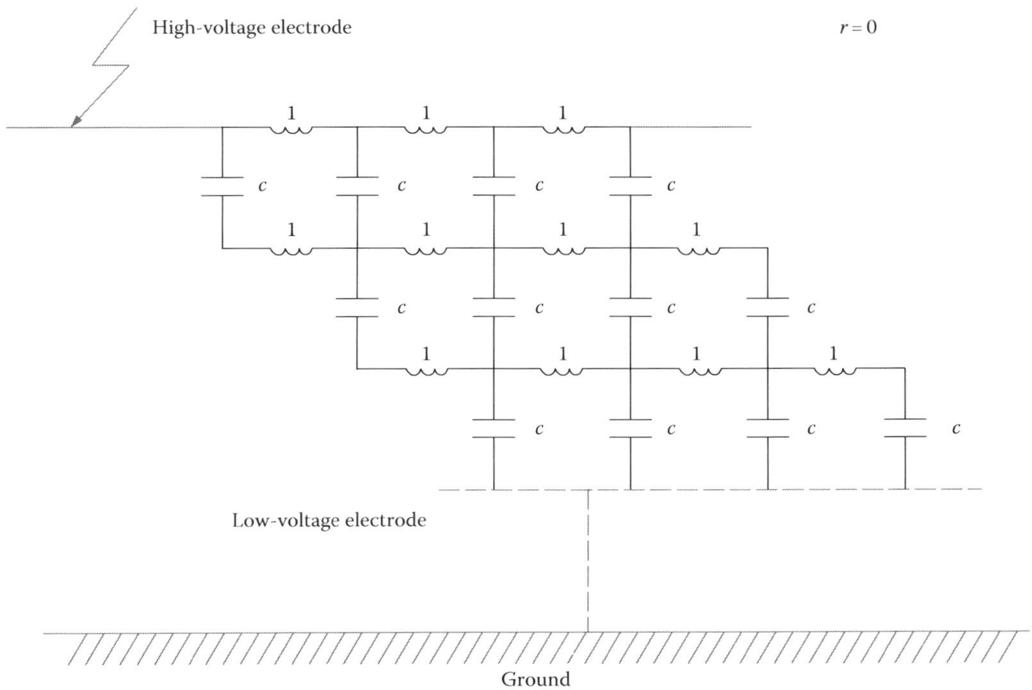

FIGURE 17.7
Equivalent circuit of the insulating bushing of a CT regarded as a short transmission line.

increase to inadmissible values. This would no longer be a short transmission line, leading to problems with reflected waves along the line and consequently undesirable voltage distributions in the bushing.

If the previous conditions are satisfied, the circuit shown in Figure 17.2 can be taken as valid, with the discrete values of capacitance, resistance, and self-induction shown in the figure representing the electrical behavior of the insulation of the CT.

The terms "long transmission line" and "short transmission line" are used here to distinguish between cases in which an equivalent circuit based on discrete components can be used and cases in which it is necessary to use a circuit with distributed constants, using the wave propagation equations of D'Alambert or Heaviside.

17.4 Description of the Parameters of the Equivalent Circuit of an Inverted-Type Current Transformer

The values of $R_{1\,HV}$ and $R_{2\,LV}$ correspond to the transversal values of electrical resistance of the HV and LV electrodes. These resistances are on the order of a few hundredths of an ohm.

The values of C_{HEAD} and $C_{BUSHING}$ correspond to the values of capacitance between the two electrodes, HV and LV, and the capacitance of the bushing, respectively. These values are on the order of a few hundred picofarads for the head and 100 or 200 pF in the bushing.

The circuit formed by C_{LV}, L_2, L_1, and M represents the coupling between the LV winding and the LV electrode. What is required in this case is good coupling here to avoid undesirable voltages appearing between the LV electrode and the secondary winding.

The HF current (a few kiloamps) that flows between the HV and LV electrodes in the head will flow to ground through the LV tube. This current will be coupled to the output wires of the secondary winding and will generate an electromotive force opposed to the magnetic field that induces it.

In this case, the requirement is that the coupling coefficient $K = M/\sqrt{L_1L_2}$ be as close as possible to 1 so that the voltage that appears across C_{LV} is almost zero.

The values of L_1, L_2, and M are on the order of microhenries. The value of C_{LV} is a few thousand picofarads.

17.5 Equivalent Circuit of the Current Transformer Assembly: Chopped Lightning-Type Impulse Wave Test Simulator

The following section describes the laboratory procedure used to test the transient response by applying lightning chopped impulse waves generated using spark gaps (Figure 17.8).

A system to measure the voltages and currents in the insulation is included. The recorded waveforms are shown in Figure 17.9.

It can be seen that the applied voltage has a peak value of 1270 kV peak and a current through the insulation of 5.070 kA peak at a frequency of about 1600 kHz.

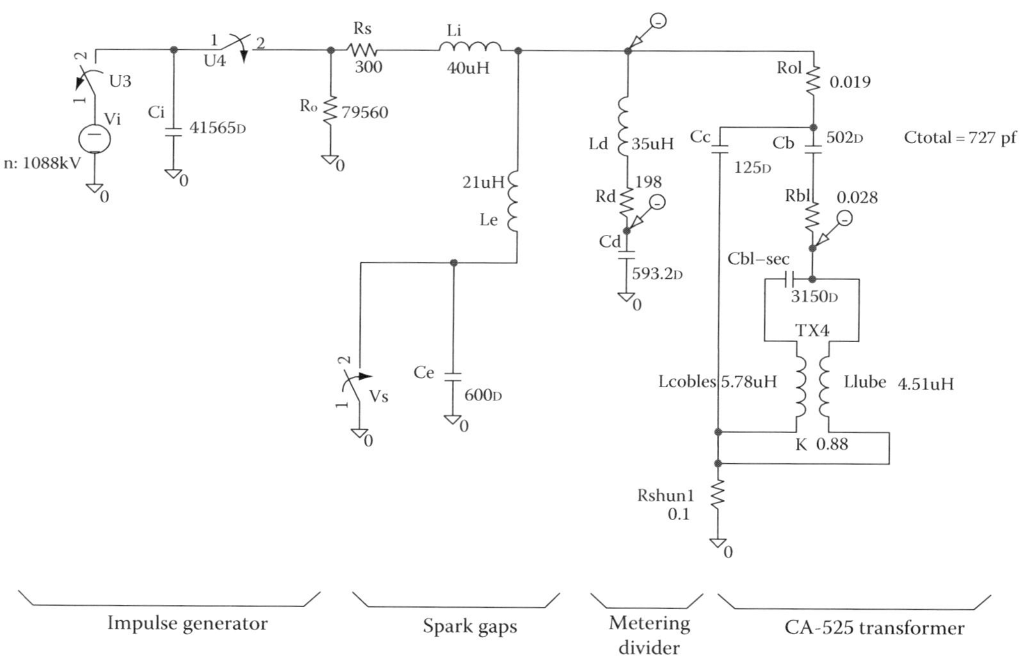

FIGURE 17.8
Equivalent circuit of the CT and the chopped lightning impulse installation.

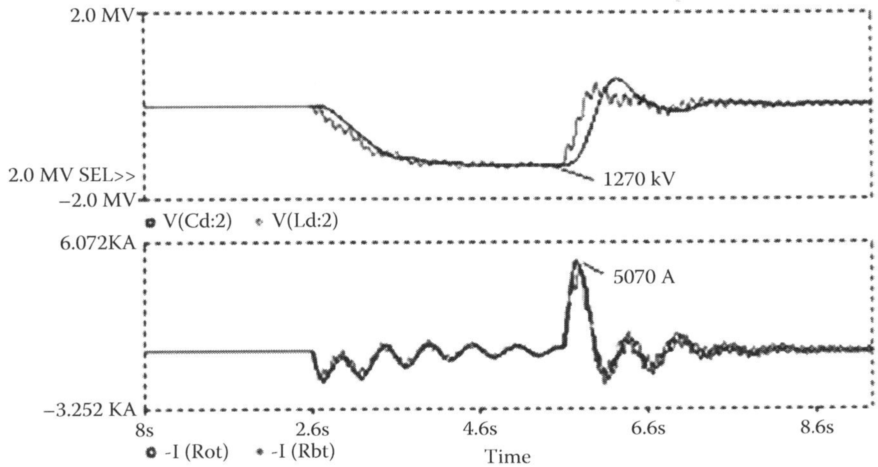

FIGURE 17.9
Waveform in the chopped lightning impulse test.

This test was repeated 600 times on the unit, and the insulation was tested by dissolved gas analysis (DGA) of the oil (comparing a sample taken before the test with a sample taken 3 days after the test) and by applying the routine tests (partial discharge tests) after this test.

17.6 Results Obtained

Table 17.1 shows examples of the results of DGA before and after these tests. It can be seen that the variations in the results of the DGA are due only to the intrinsic precision of the method used to analyze the gases and that there is no real increase in the amount of gas.

TABLE 17.1
Examples of the Results of DGA before and after These Tests

Before		After	
REPORT	TM-87/99	Report	TM-88/99
DATA SAMPLING	08/06/99	Sampling Data	11/06/99
SAMPLE POINT	FC	Sampling Point	FC
H_2	5	H_2	5
O_2	17.615	O_2	14.900
N_2	45.700	N_2	40.700
CO	0	CO	0
CO_2	82	CO_2	90
CH_4	0	CH_4	0
C_2H_4	0	C_2H_4	0
C_2H_2	0	C_2H_2	0

Applicable standards allow increases in gas concentrations of a few ppm. Significant increases in gas content, for example, acetylene, may indicate problems associated with the resistance of the Faraday screening, the shape of the electrodes, or the materials used, and although this may indicate a poor design as regards the results in this test, this does not mean that problems will appear during service.

17.7 Conclusions

This test allows definitive testing of the resistance of the insulation to this type of transient.

Problems may be expected when there are changes in the gas concentrations that cannot be explained by other phenomena, or the test pieces do not pass the partial discharge tests after being subjected to this test.

References

1. Lucas JR (2001). High voltage engineering.
2. Schwab AJ (1988). Field theory concepts.
3. Electroténica Arteche Hnos SA, Internal technical notes for design.

18

Transfer of HF Disturbances through Instrument Transformers

Wieslaw Jalmuzny and Andrzej Koszmider

CONTENTS

18.1 Introduction

In its practical aspect, the most common causes of electromagnetic high-frequency (HF) disturbances in the primary circuits of instrument transformers are the following: lightning strokes, bursts of very short transients generated by the circuit switching operations, transient HF disturbances, as well as the continuous disturbances in the form of waves of wide frequency band. Instrument transformers (also electronic instrument transformers with inductive parts) [1,2], by coupling the primary circuits with secondary circuits, make the propagation of electromagnetic disturbances generated on their primary circuits possible. For the electronic devices and information technology equipment installed on the secondary circuits, instrument transformer is the source of disturbances. The values of the disturbances penetrating secondary circuits depend on the perturbations occurring in the primary circuits as well as on the degree of their damping by the instrument transformer. Favorable or adverse conditions for the transfer of disturbances to secondary circuits occur, depending on the design features of the instrument transformers.

The devices and equipment operating in secondary circuits, according to the relevant standards, have to demonstrate a certain required level of electromagnetic immunity, up to which the value of disturbance transferred by instrument transformers from the primary circuits should be limited in a deliberate way.

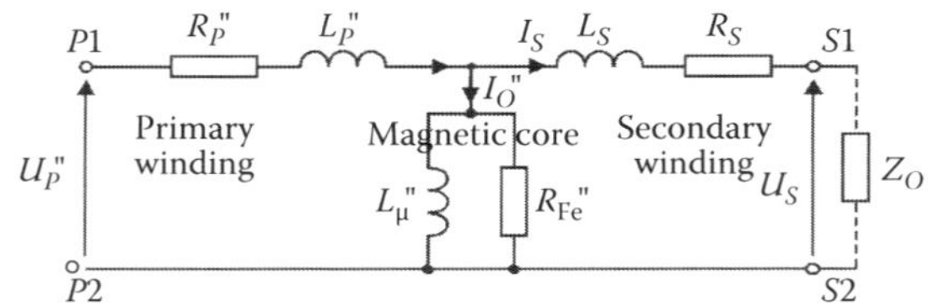

FIGURE 18.1
Simplified equivalent circuit of one-core CT for the power frequency; the values of the primary winding referred to the secondary.

For the power frequency range, the instrument transformer windings may be treated as two self-inductances coupled with the mutual inductance. The equivalent circuit for power frequency takes into account resistances of both windings (R_P, R_S), their self-inductances (L_P, L_S), inductance L_μ corresponding to flux coupled with the both windings, as well as losses in the iron of the core represented by the resistance R_{Fe} (as a result of the hysteresis and eddy currents). Figure 18.1 shows equivalent circuit for a current transformer (CT) for the power frequency (50 or 60 Hz).

The model, corresponding to the network presented in Figure 18.1, cannot be used to analyze the transfer of HF disturbances because it does not include parasitic capacitances unimportant for power frequency. On the other hand, the large values of exciting inductance L_μ cause the currents in the exciting branch to be ignored at the frequency range of a few hundred kilohertz.

According to the EMC standard requirements [3], conducted disturbances should be limited and so measured within the range of frequency from 150 kHz to 30 MHz. The surge rise times correspond to approximately the same frequency range. Thus, in reality, the circuit shown in Figure 18.1 is far from sufficient for the transfer analysis of that type of disturbances. The instrument transformer for electromagnetic disturbances of frequencies of up to 30 MHz has an extremely complicated electromagnetic structure in which, additionally, occur capacity couplings, eddy currents in structural elements (e.g., the core, the shields, the casing), as well as voltage drops on inductances of conductors and on the terminals.

At high frequencies, all those phenomena have an influence on the transfer of disturbances to the secondary circuits. Electromagnetic couplings of an instrument transformer can also lead to the propagation of disturbances between structural elements (windings, shields, core, casing). The layout of parasitic capacitances playing the most important role in the transfer of disturbances is presented in Figure 18.2.

The disturbance signals in the primary circuits can appear as either common (CM) or differential mode (DM) currents. In the secondary circuits of instrument transformers, as of one of the terminals being always grounded (earthed), CM signals are equal to the DM ones. According to what has been discussed earlier, we can study the transfer of primary CM disturbance signals or the transfer of primary DM disturbance signals, generating the DM signals in the secondary circuit. The results of the experimental testing [4–6] prove, though, that the transfer of CM signals is far more significant; for example, in the conditions for which the CM transfer factor was approximately 30%, DM transfer coefficient was smaller than 1%. What follows is that DM signals are dumped with very positive results by the instrument transformers, and when considering the device-related threats of disturbances in secondary circuits, only the transfer of CM signals can be investigated.

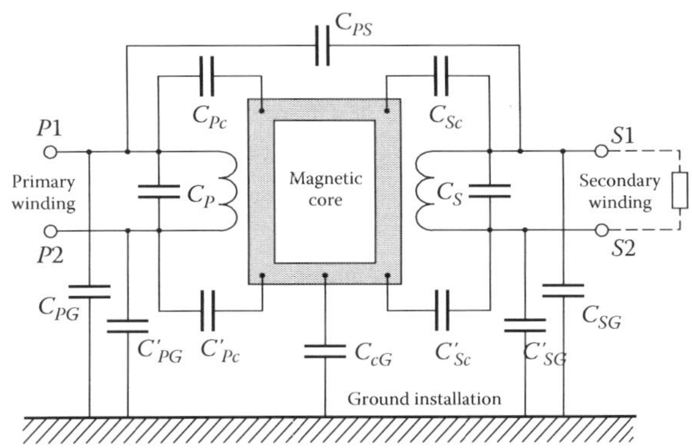

FIGURE 18.2
Simplified equivalent circuit of a CT illustrating the capacitive couplings for the disturbing signals. C_{PS}, capacitance between primary winding and secondary winding; C_P, C_S, capacitances between terminals of the windings; C_{Pc}, C'_{Pc}, capacitances between primary winding and the core; C_{Sc}, C'_{Sc}, capacitances between secondary winding and the core; C_{PG}, C'_{PG}, capacitances to earth of the primary winding; C_{SG}, C'_{SG}, capacitances to earth of the secondary winding; C_{cG}, the capacitance to earth of the core.

18.2 Equivalent Circuit of the Instrument Transformer in the Research of the Transfer of HF Disturbances

Apart from parasitic capacitances, the inductances and resistances of conductors which constitute the leads can be of considerable importance, similarly to parallel resistances representing the losses in insulating elements of the instrument transformer. For the electric insulation materials used in the majority of instrument transformers, the dielectric loss factor is very small (e.g., for the epoxy resin = 0.03, for the electrical insulating paper = 0.002/0.003), as a result of which the conductance constitutes a small part of the susceptance and, consequently, the series admittance of insulating elements.

The performed analysis indicates that the model which excludes the effect of the conductance of isolation is sufficiently accurate. Parasitic capacitances will play a vital part in this kind of model. This conclusion means that the propagation of disturbances is primarily due to the energy transfer out of the electric circuits of the instrument transformer.

Between the terminals of both primary and secondary windings, there are also series inductances, which, for certain frequencies, may cause the phenomenon of resonance as well as ferroresonance to occur. The equivalent circuit of the instrument transformer winding may be defined with three basic parameters: the inductance of winding L, the parasitic capacitance C, as well as the resistance of winding R (Figure 18.3).

An example of the impedance characteristic between the secondary winding terminals of the CT is shown in Figure 18.4.

Above the range of 30 MHz, there appear the phenomena which make the instrument transformer equivalent circuit impossible to formulate within this frequency range. Within this range, despite the electric parameters being treated as distributed parameters, there appear the wave phenomena difficult to analyze (e.g., reflections) as well as the couplings between different elements of circuits and frequently between the casing elements.

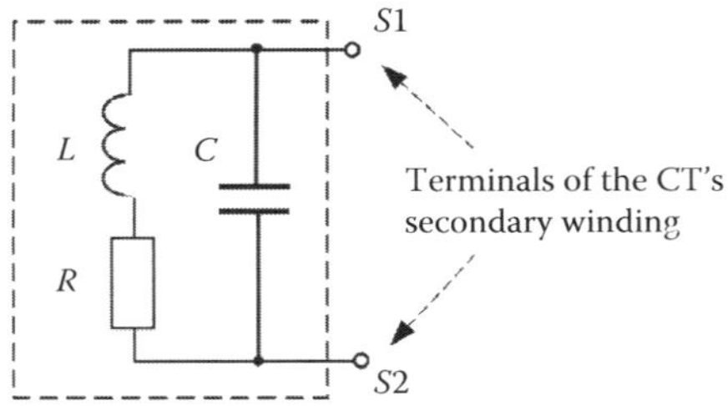

FIGURE 18.3
Equivalent network seen from the secondary terminals of a CT.

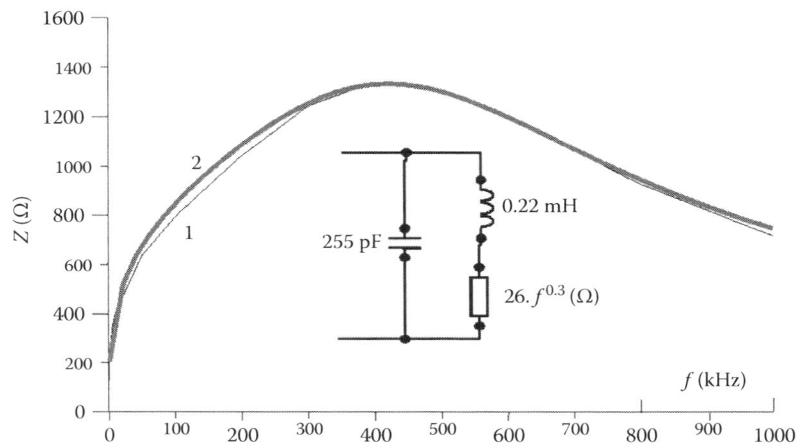

FIGURE 18.4
Characteristic of the impedance between the secondary terminals of a CT (1, measurement; 2, calculation).

Formulating an equivalent network as well as the definition of its elements is important from the quality perspective, which allows understanding the role of individual elements of the circuit. In order to develop the equivalent circuit of an instrument transformer which would work as expected, it should have the following features:

- Uncomplicated structure (appropriate simplifications)
- Possibility of quick calculation of electric parameters of all elements in the network
- Possibility of performing the analysis of the circuit in the different configurations of appearing electromagnetic disturbances
- Model accuracy, sufficient for the sake of the analysis

The equivalent circuit presented in Figure 18.5 was recognized as applicable for testing and determining the degree of transfer disturbances to secondary circuits for the frequency range (150 kHz–30 MHz) of conducted disturbances. It does not cover all the phenomena related to the transfer of disturbances on purpose. However, it considers those phenomena, which have a decisive influence on disturbance penetration from primary to secondary circuits of instrument transformers.

In the assumed model of the instrument transformer used to elaborate the equivalent circuit, the following simplifications were applied:

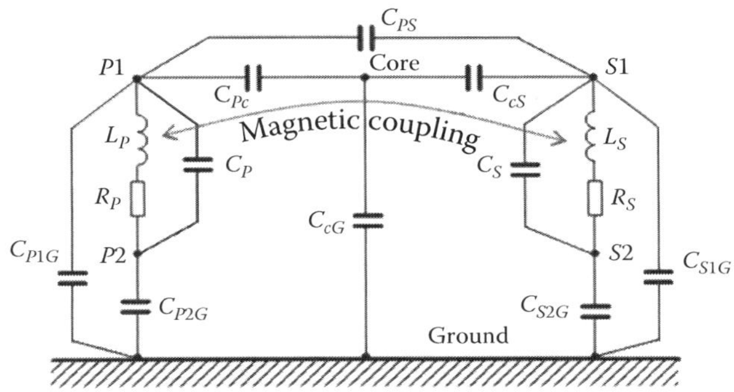

FIGURE 18.5
Equivalent circuit of the CT (within the HF range → the magnetic coupling is negligible).

- Magnetic coupling between primary and secondary windings was omitted.
- Existence of merely electric couplings between the elements of the instrument transformer model was accepted.
- The number of all elements was limited to visible units only, that is, to those to which measuring conductors can be connected without any interference in the internal structure of the instrument transformer.
- The effect of conductance in insulating elements was neglected.
- Constant value of internal resistance of windings was assumed.
- Wave phenomena were ignored.

As a result, an equivalent circuit was assumed, which, however, due to different instrument transformers constructions, can be modified according to the simplifications discussed earlier. For example, for an additional core with the second secondary winding, it introduces an additional branch in the circuit, as well as the additional capacitances between the introduced element and the rest of existing network.

Parasitic capacitances can be determined according to the method developed in Ref. [7]. It is about measuring the resultant capacitances between the CT or VT terminals and the earth (ground) in all the possible arrangements of their direct electrical connections. The solution of the received system of equations gives, as a result, the sought for capacitances values. The remaining parameters related to the measuring circuit (conductors) are possible to measure by means of the bridge method. Figure 18.6 shows the equivalent circuit of a CT with two cores and two secondary windings.

18.3 Simulation Studies of Transfer of Disturbances

Despite the limitations concerning the verity of the mathematical model represented by the equivalent circuit, the results of simulation will be shown later, demonstrating, thus, to which degree the circuit might prove applicable.

The *CircuitMaker* software was used for the computer simulation. A mathematical model for the computer simulation was produced according to the equivalent circuit of the CT.

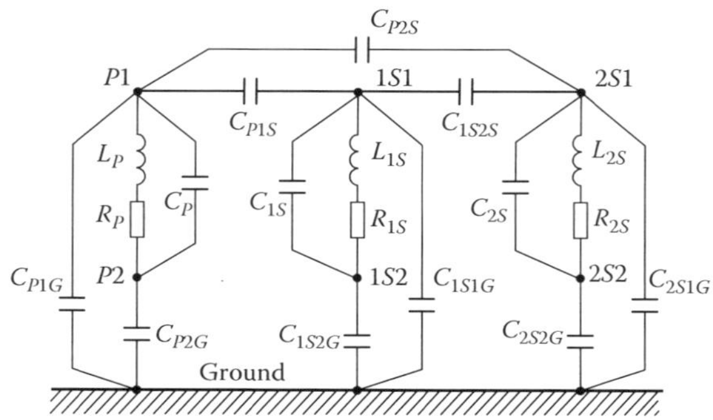

FIGURE 18.6
Equivalent circuit of a CT (two cores and secondary winding: 1S1–1S2 and 2S1–2S2 model). C_{P1S}, capacitance between the primary (P1–P2) and secondary (1S1–1S2) windings; C_{P2S}, capacitance between the primary (P1–P2) and secondary (2S1–2S2) windings; C_{1S2S}, capacitance between two secondary windings; C_P, the equivalent capacitance of primary winding; L_P, the equivalent inductance of the primary winding; R_P, the equivalent resistance of primary winding; C_{1S}, the self-capacitance of secondary winding 1S1–1S2; L_{1S}, the equivalent inductance of secondary winding 1S1–1S2; R_{1S}, the equivalent resistance of secondary winding 1S1–1S2; C_{2S}, the self-capacity of secondary winding 2S1–2S2; L_{2S}, the equivalent inductance of secondary winding 2S1–2S2; R_{2S}, the effective resistance of secondary winding 2S1–2S2; C_{P1G}, C_{P2G}, the capacitances to earth of primary winding; C_{1S1G}, C_{1S2G} the capacitances to earth, of secondary winding 1S1–1S2; C_{2S1G}, C_{2S2G}-the capacitances to earth of secondary winding 2S1–2S2.

The model covers capacitances, inductances, and resistances of connecting conductors, the parameters of measuring devices and resistance of the instrument transformer burden R_o. The simplified network illustrating the studies options was presented in Figure 18.7.

Due to the software limitations (program does not accept the nonlinearity of elements), the fixed average values of parameters were used for specified frequency ranges in the winding circuit of the tested instrument transformer. A simulation analysis of the instrument transformer was carried out for the different connection configurations at sinusoidal disturbing signals.

The effect of secondary windings and core grounding was examined on the disturbance transfer from the primary to the CT's secondary side.

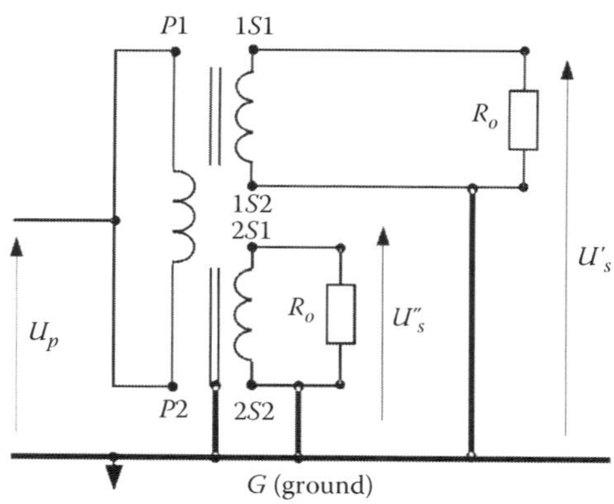

FIGURE 18.7
Diagram of instrument transformer connections (the bold line, the grounding conductors connected or disconnected during the tests).

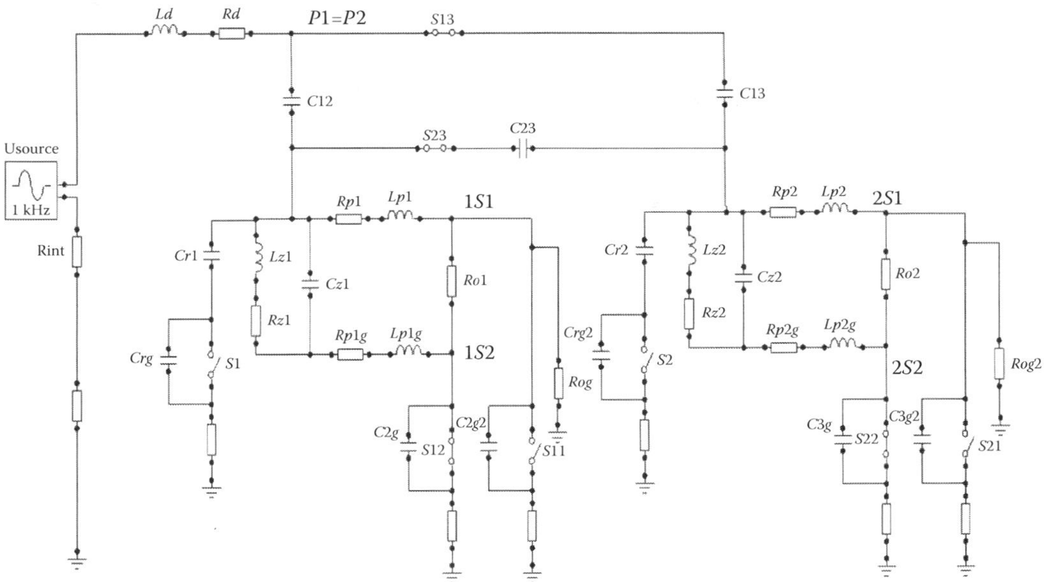

FIGURE 18.8
Network of instrument transformer connections subjected to the simulation of sinusoidal signals (the general two-core model).

The modification of that model into the one-core model only involves opening the switches *S*12 and *S*23 of the model from Figure 18.8. The core earthing in the analyzed instrument transformer means closing the switch *S*1 or *S*2. Grounding of the secondary winding terminal eliminates the capacity to earth of the winding (e.g., *C*2*g* by closing switch *S*12).

The simulation analysis was done within the frequency range of the conducted signals, that is, from 100 kHz to 30 MHz. The sine-shaped disturbance signal of the fixed value $U_p = 250$ mV within the frequency range of the conducted signals was applied to the primary winding terminals. For this disturbance, the perturbation signals which appeared on the secondary winding terminals were calculated. One-core and two-core models were studied in various core earth and burden configurations. Figures 18.9 through 18.11 show the simulation results.

The simulation analysis of the instrument transformer was also carried out for different connection configurations at the pulse disturbance signals. During the simulation, pulses of the rise time of several microseconds and nanoseconds [6,8] were modeled. The pulses of a longer rise time were obtained by connection of an additional capacitance on the generator terminals. The effect of secondary windings on earth was examined, as well as the influence of the core earth on disturbance transfer from the primary to the secondary CT's side.

The pulse disturbance of the maximum value $U_p = 1$ kV was applied to the primary winding terminals. For this disturbance, the perturbation signals appearing on the secondary winding terminals were calculated. One-core and two-core models were tested at different burden and core grounding configurations. The simulation results for microsecond pulses in the one-core model are presented in Figures 18.12 and 18.13.

The simulation results for nanosecond pulses with the rise time 10, 20, and 40 ns (Figure 18.14) in the event of one-core model are shown in Figures 18.15 and 18.16.

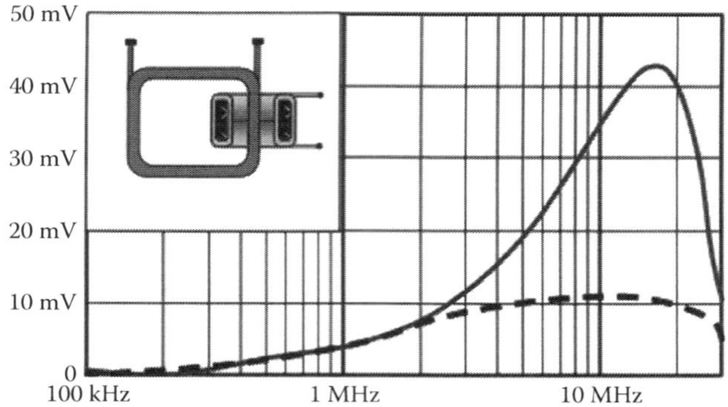

FIGURE 18.9
Frequency characteristic of the signal U_s' on the terminal 1S1 at the earthed terminal 1S2 and the burden $R_o = 50$ Ω (solid line, the non-earthed core; broken line, the earthed core).

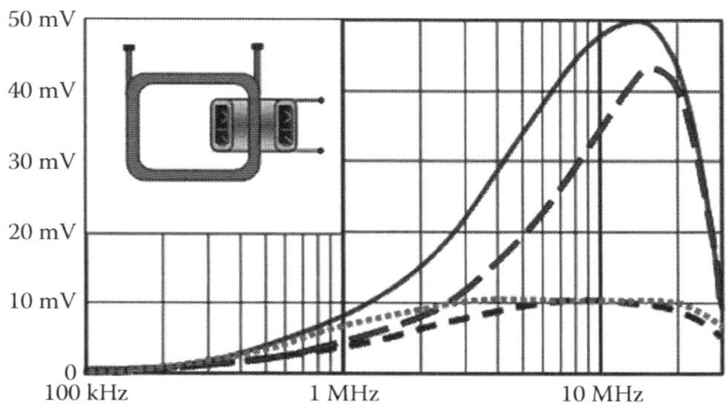

FIGURE 18.10
Frequency characteristic of signal U_s' at the earthed terminal 1S2 (the solid line [$R_o = 1$ MΩ] and the broken line [$R_o = 50$ Ω], the non-earthed core; the dotted line [$R_o = 1$ MΩ] and the short broken line ($R_o = 50$ Ω], the earthed core).

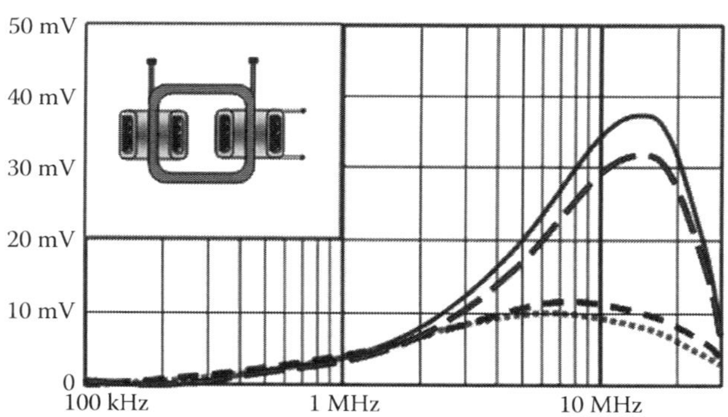

FIGURE 18.11
Frequency characteristic of the signal U_s' on the terminal 1S1 at the earthed terminal 1S2 and the burden $R_o = 1$ MΩ (solid line, the non-earthed cores *c*1 and *c*2; broken line, the earthed core *c*2; dotted line, the earthed cores *c*2 and *c*1; short dotted line, the earthed core *c*1).

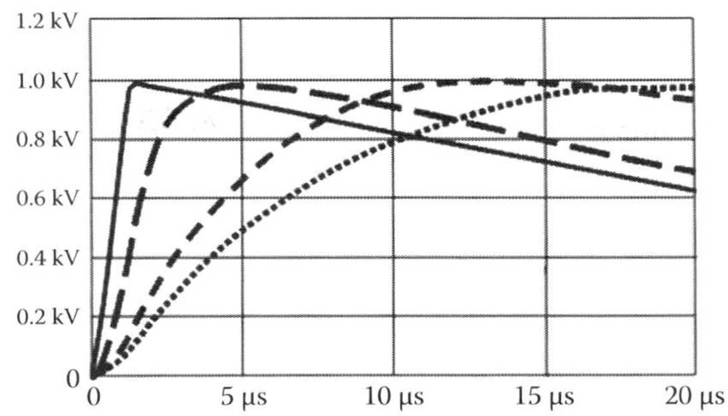

FIGURE 18.12
Microsecond disturbing signals U_p applied to the primary terminals at different rise time of the test pulses.

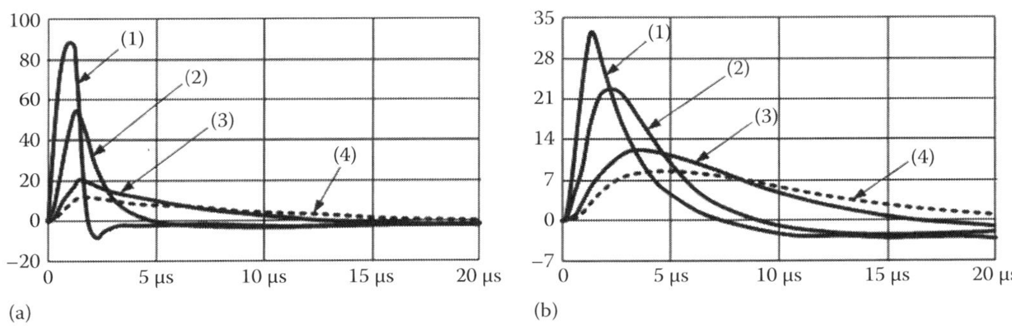

FIGURE 18.13
Disturbance signals U_s' (V) on the terminal 1S1 at the earthed terminal 1S2 for different rates of rise of the disturbance test pulse U_p; (a) the non-earthed core, (b) the earthed core; (1) for $T1 = 1$ μs, (2) for $T1 = 5$ μs, (3) for $T1 = 10$ μs, and (4) for $T1 = 20$ μs.

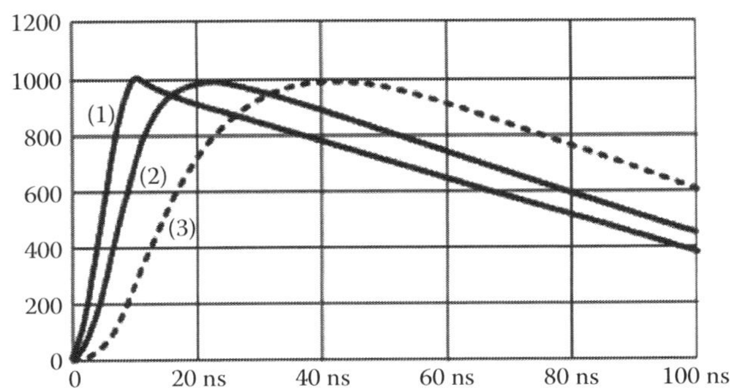

FIGURE 18.14
Nanosecond disturbance signals U_p (V) applied to the primary terminals at different rates of rise of the test pulses; (1) for $T1 = 10$ ns, (2) for $T1 = 20$ ns, and (3) for $T1 = 40$ ns.

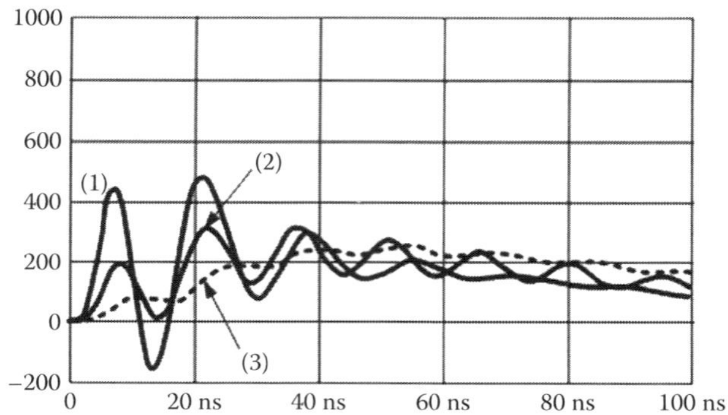

FIGURE 18.15
Disturbance signal U_s' (V) on the terminal 1*S*1 at the earthed terminal 1*S*2 for different rates of rise of the disturbing pulse; (1) for $T1 = 10$ ns, (2) for $T1 = 20$ ns, (3) for $T1 = 40$ ns.

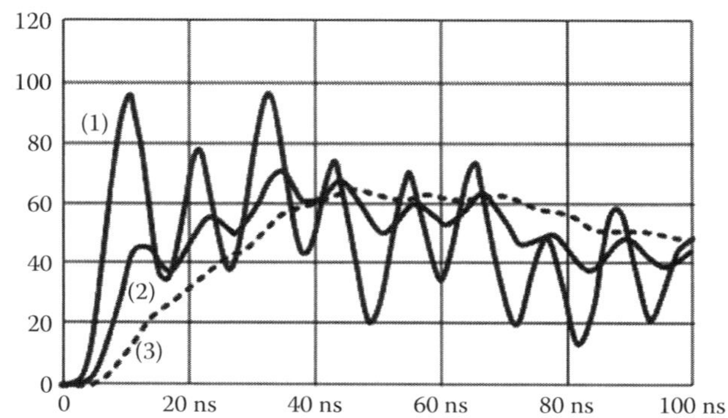

FIGURE 18.16
Disturbance signal U_s' (V) on the terminal 1*S*1 at the earthed terminal 1*S*2 and earthed core *c*1 for different rates of rise of the disturbing pulse; (1) for $T1 = 10$ ns, (2) for $T1 = 20$ ns, (3) for $T1 = 40$ ns.

The simulation studies for pulse signals have been divided into two parts, depending on the kind of pulses, namely, the testing of microsecond pulse transfer (of the rise time 1.2 μs and more) and of nanosecond pulse transfer (of rise time of 5 ns and more). The rate of rise of the disturbance pulse influences the values of disturbance signal in the secondary circuit U_s. For example, for the *microsecond* pulse, the maximum transfer factor (the $KT = U_s/U_p = 90/1000$) is about 9%, and for the remaining ones, of lower rise times—5%, 2%, and 1%, respectively. The grounding of the core also reduces the disturbance signal U_s. With the earthed core, the transfer rate achieves the values of 3%, 2%, 1%, and 0.7%, respectively.

What results from the studies of the *nanosecond* pulse transfer is that, the faster the disturbance, the greater the disturbance signal will appear on the secondary side of the instrument transformer. Moreover, while analyzing the transfer factor, it can be observed that, for so fast pulses, the transfer factor values may even reach 50%.

What seems interesting is the analysis of the secondary voltage shape caused by pulse disturbances.

In the case of pulses of lower rate of rise, which is in the case of the microsecond pulses, the secondary waves acquire the characteristic shape of the differential circuit response. Such a shape has typically a biexponential wave, disappearing to the zero value and then, before the sign changes and after exceeding the maximum negative value, slowly disappearing. The response for the faster pulses, that is, the nanosecond ones, changes completely its nature. The oscillate nature of voltage in secondary circuits is the predominant feature of secondary waves in this case.

18.4 Experimental Studies of Disturbance Transfer

18.4.1 Testing of Transfer of Harmonic Disturbances

The physical model of an MV CT not yet resin insulated was used for the studies in order to assure the accessibility of all the elements of the instrument transformer, enabling, thus, the measurement of its parameters as well as the introduction of corresponding modifications. The diagram of the CT construction is shown in Figure 18.17. The testing was done for the model with one or two magnetic cores.

The measurements were carried out by means of *EMC Analyzer* by Hewlett Packard. The analyzer has a built-in *tracking generator*, enabling the generation of fixed value signals of wide frequency range. The impedance of entry and exit channels is 50 Ω. The measurements were arranged as in Figure 18.18.

For the connections, two concentric cables BNC (50 Ω), each 20 cm long, were used.

Figure 18.19 shows the typical frequency characteristic of the transfer factors within the range of 1–30 MHz. Below 1 MHz, the values of coefficients are very small, and in fact, the range of above 5 MHz is worth our attention.

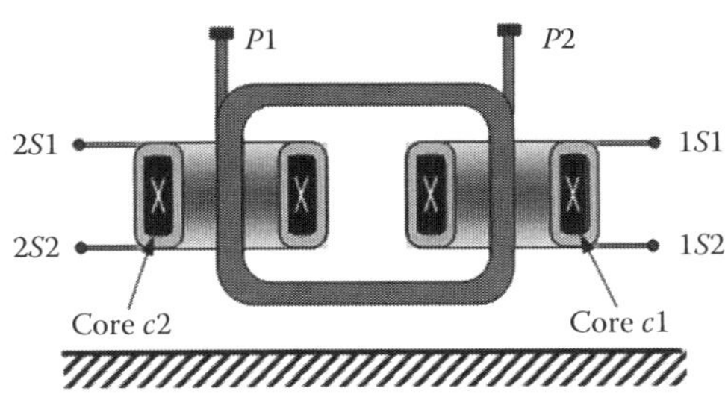

FIGURE 18.17
Diagram of tested CT.

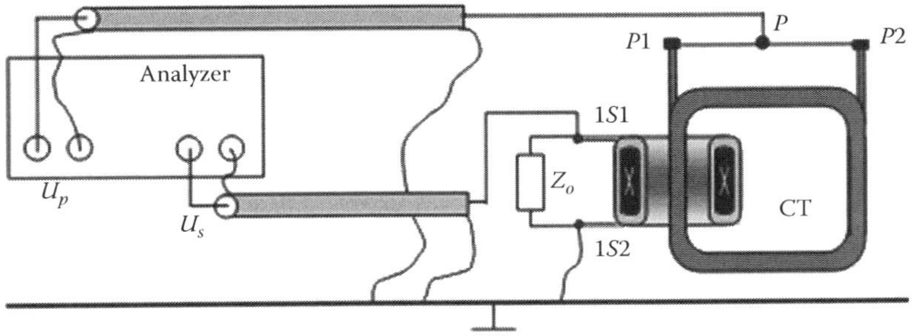

FIGURE 18.18
Schematic diagram of the measuring arrangement.

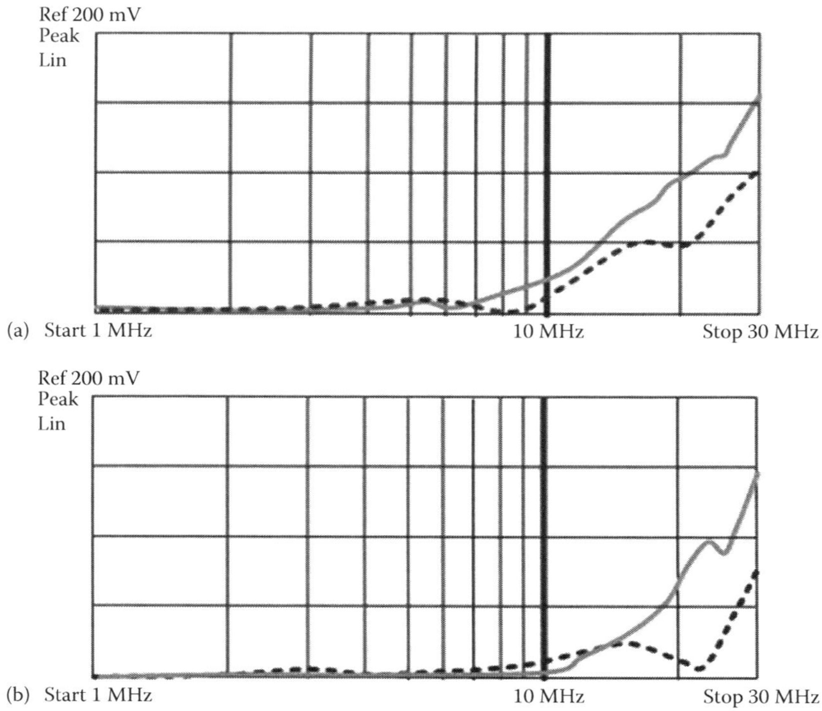

FIGURE 18.19
Signal U'_s measured between terminals 1*S*1 and 1*S*2 (a) and signal U''_s measured between terminals 2*S*1 and 2*S*2 (b) where 1*S*2 and 2*S*2 were earthed (broken lines with earthed core *R*2).

Grounding of the core reduces, to a large extent, the transfer of HF signals to those windings which are located on the earthed core. In this case, the transfer of signals is small. However, this relationship is not linear, as within a certain range, frequency transfer reduction is greater and within others, smaller, as shown in Figure 18.19.

Following the studies of CTs, which were done for harmonic disturbances, the following conclusions can be drawn:

- The effect of the CT burden change on the disturbance transfer within a wide frequency range is small.
- The study of the disturbance transfer with the resistance in secondary circuit $R_o = 50\ \Omega$ gives representative results.
- From the point of view of disturbance transfer, the recommendable design features are those with a large distance between the primary and secondary windings.
- The earth of the core reduces the transfer of disturbances of above 5 MHz; thus, every core would have to be earthed.

18.4.2 Testing on Pulse Signal Transfer

Measurements were carried out by means of the digital oscilloscope by *TEKTRONIX* of the frequency range of up to 1 GHz and the generator of microsecond pulse waves (surge)

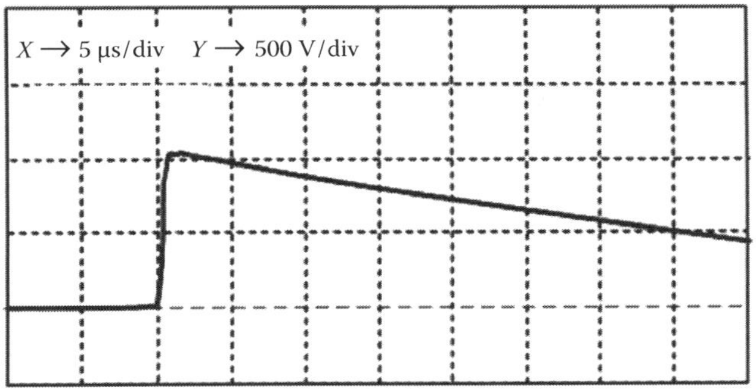

FIGURE 18.20
Disturbance signal introduced between the primary winding and ground U_p (P–G).

(Figure 18.20) and the fast nanosecond transients (burst). Microsecond signals were measured at the sampling frequency of 25 MS/s, while the nanosecond signals were measured at the sampling frequency of 5 GS/s.

Waveforms of signal U_s measured between the terminals 1S1 and 1S2 at the earthed terminal 1S2 for one-core CT are shown in Figure 18.21. Figure 18.22 presents the effect of the core number on the transfer factor. In this case, a one-core instrument transformer was compared with a two-core instrument transformer.

Based on the measurements, the following conclusion could be drawn: the instrument transformers with a greater number of cores have a greater factor of disturbance transfer.

The testing on the effect of the core earth was executed in such a way that the voltage of disturbances was measured on two windings, where one was on the core being earthed or not and the other on the non-earthed core. The results are shown in Figure 18.23.

Based on the testing results, the following conclusions can be drawn:

- Grounding of the core considerably reduces the transfer of pulse disturbances. The measurements obtained prove that the transfer factor decreased nine times.
- In the case of the two or more core instrument transformers, in order to decrease the transfer of disturbances, it would be recommendable to earth every core.

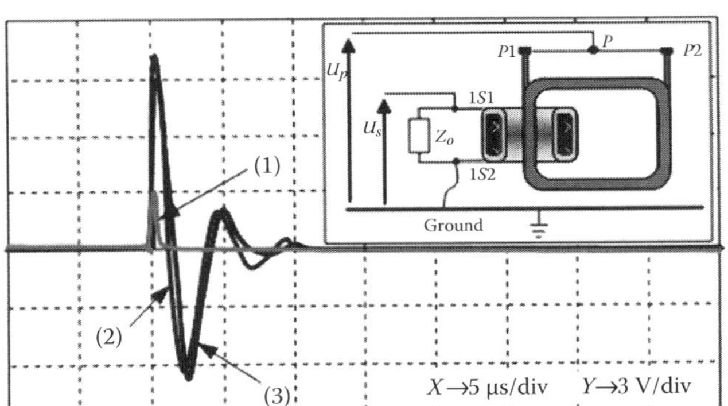

FIGURE 18.21
Signal U_S measured between the terminals 1S1 and 1S2 at the earthed terminal 1S2: 1, $R_o = 50\ \Omega$; 2, $Z_o = 1\ \Omega$ ($\cos\varphi = 0.8$); 3, $R_o = 1\ \mathrm{M\Omega}$.

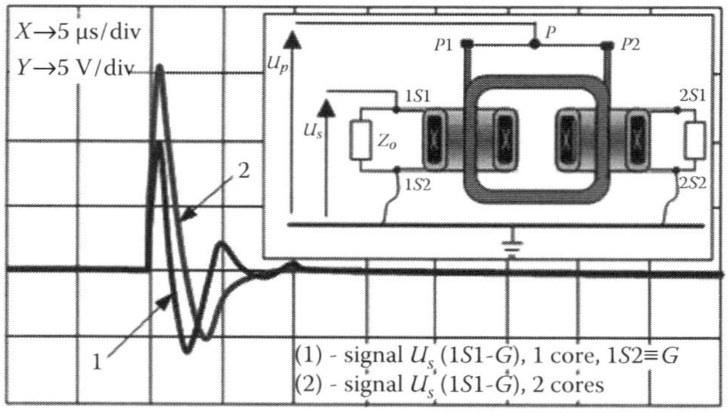

FIGURE 18.22
Signal U_2 measured between the terminals 1S1 and 1S2, where the terminal 1S2 was earthed: 1, one-core model; 2, two-core model.

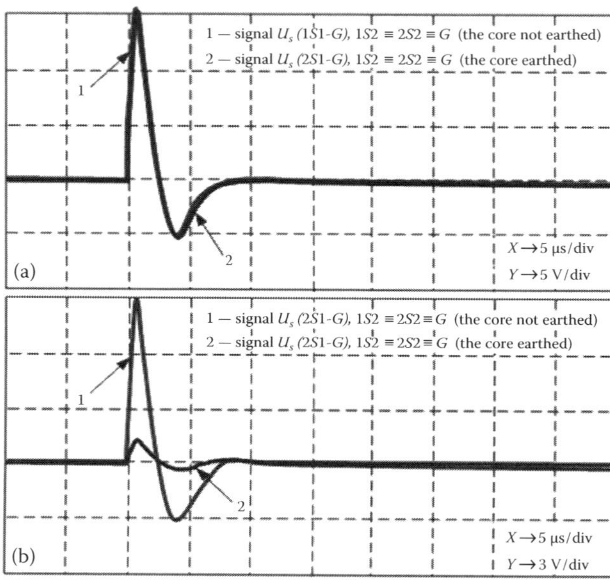

FIGURE 18.23
Signal U_s' measured between the terminals 1S1 and 1S2 on the core with no connection to earth possible (a) and signal U_s' measured between the terminals 2S1 and 2S2 on the core with connection to earth possible (b).

For the two-core model, the effect of core connection to earth on the transfer of nanosecond pulse disturbances on the secondary side of the instrument transformer was studied. The waveforms of disturbance signal U_p and signals U_s' and U_s'' are shown in Figures 18.24 and 18.25.

Based on the pulse disturbance measurements, the following conclusions can be formulated:

- Effect of the burden of value of Z_o on the pulse transfer is negligible.
- Measurements of disturbance transfer at the secondary side loaded with the resistance $R_o = 1\,M\Omega$ can be fully representative due to the largest coefficient of pulse disturbance transfer obtained.

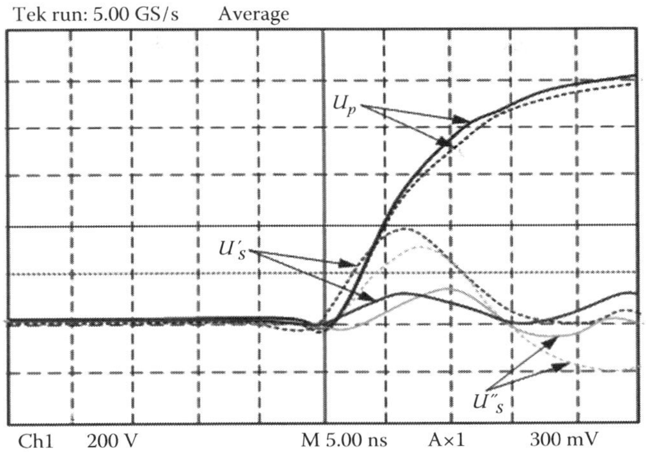

FIGURE 18.24
Waveforms of disturbance signal U_p and signals U_s' and U_s'' for the nanosecond type pulse ($t_r = 5$ns) (solid line represents the earthed core *c*2, broken line represents the non-earthed core *c*2; broken line (U_s') represents the winding on the earthed core, broken line (U_s'') represents the winding on the non-earthed core).

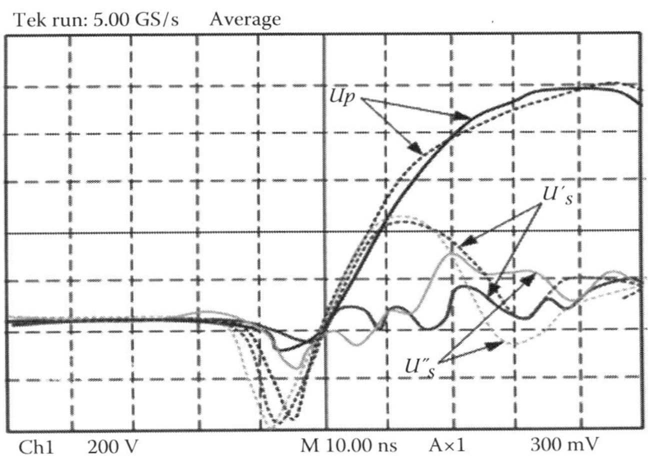

FIGURE 18.25
Waveforms of disturbance signal U_p and the signals U_s' and U_s'' for the nanosecond pulse type ($t_r = 10$ns) (solid line represents the earthed core *R*2; broken line represents the non-earthed core *R*2).

- The design features allowing comfortably large distances between the primary and secondary windings.
- The connection to earth of the core considerably reduces the transfer of disturbances to the winding situated on the earthed core. With nanosecond pulses, the decrease of transfer to the windings on non-earthed cores also takes place.
- The rate of rise of disturbance pulses influences substantially the values of signals transferred on the secondary side. Remarkably, the faster the pulse, the greater the value of the transfer factor is.

- The number of cores (windings) has a considerable influence on the disturbance transfer rate. For the one-core model, at the earthed terminals of secondary windings, the coefficient KT reaches the value of 20%, while for the two-core, it approaches 40%.

18.5 Conclusions

- The study of electromagnetic disturbance transfer by instrument transformers has resulted in fundamental conclusions, concerning the effect of their design features on the transfer. The important conclusion, which refers to the whole study, is that quality-wise findings are independent of the kind of disturbance signal. The study was carried out for harmonic signals as well as two kinds of pulse signals: microsecond and nanosecond signals.
- A general feature influencing the transfer of signals is the "compactness" of the structure. What is meant by compactness is the extent to which the windings fill the magnetic core window. The larger filling factor of the magnetic core window causes the windings to be located closer to each other, and this makes the coupling capacitance larger. Obviously, the coupling capacitances are also dependent on different features such as, for example, form of the primary winding, which, in most general terms, may be the coil or the bar winding one. The practical effect of the propriety described earlier is that in general, the transfer factor of disturbances in two or more core instrument transformers will be larger than in the one-core instrument transformers. The compactness of the construction is the feature welcome from the perspective of operating characteristics of an instrument transformer. Thus, the smaller the spacing between the primary and secondary windings, the smaller the leakage impedances, which, as series elements, have a negative influence on the metrological characteristics of instrument transformers.
- The connection to earth of the magnetic core fundamentally reduces the transfer of disturbances to the earthed core. The reduction of disturbances caused by the magnetic core grounding reaches 60%–70%. This reduction, however, refers to microsecond pulses as well as the frequencies below 5 MHz of only the secondary winding wound on the earthed core. The impact on the transfer to the windings wound on the non-earthed core is noticeable for the highest frequencies (above 5 MHz) or for nanosecond pulses. The results are the same for all the study conditions. Apparently, from the point of view of the effect of the design features of instrument transformers on the transfer of electromagnetic disturbances to secondary circuits, it is recommendable that the cores are earthed.
- The effect of connection of cores to earth is noticeable for the frequencies of 1 MHz, and fundamental impact occurs at above 10 MHz. It is possible to lower the frequency for which the reduction of disturbance signals starts through diminishing the impedance for the disturbance signals. The connection with the ground installation during the studies was made as a connection with a copper conductor. A considerable reduction of the inductance of this connection was possible as it was made of a woven flat conductor. Connection of cores to earth may be applied both

in the case of current and voltage transformers (VTs). The way of operating and the effectiveness are the same.

- The disturbances, which can be transferred from the primary circuit to the secondary circuits of instrument transformers, can be of different nature.
- The research shows that the effect of design features and operating conditions of instrument transformers on the degree of transfer of disturbances is independent of the type of the disturbance. The conclusion can be thus drawn that the coefficient of transfer in given conditions may be approximately the same for harmonic and pulse disturbances. Such reasoning allows reaching the conclusion that complicated testing of transfer of pulse disturbances [8] can be replaced with the considerably easier fundamental study of harmonic disturbance transfer. On the other hand, the studies of the transfer of harmonic disturbances, compared with the studies on the pulse transfer, offer a substantially larger amount of information about EMC proprieties of the instrument transformer within the whole frequency band, interesting from the perspective of the EMC standards.
- The comparison of the results of studies shows, with high accuracy, that the factors of transfer for pulses and for harmonic waves of equivalent frequencies are the same. Therefore, it is recommended that the research of transfer of pulses should be replaced by the studies of transfer of harmonic waves. For the studies to cover fast pulses (5 ns), too, the frequency limits of harmonic waves should be extended to 50 MHz. After determination of the transfer coefficients within the range of 150 kHz–50 MHz with low-voltage signals, the transfer coefficients should be evaluated and compared with the limit value, or else, for the given peak value primary pulses, the peak values of secondary pulses should be calculated.

References

1. IEC 60044-7 (1999) Instrument transformers. Part 7: Electronic voltage transformers.
2. IEC 60044-8 (2002) Instrument transformers. Part 8: Electronic current transformers.
3. IEC 61000-2 Electromagnetic compatibility (EMC). Part 2-2: Environment. Compatibility levels for low-frequency conducted disturbances and signalling in public low-voltage power supply systems.
4. Jalmuzny W (1995) Rapport de stage TEMPRA. Ecole Centrale de Lyon (Centre de Génie Electrique de Lyon) CNRS URA 829), Lyon, France (in French).
5. Koszmider A, Jalmuzny W, Brodecki D (1997) The transfer of interferences trough instrument transformers. *Fourth International Conference on Electrical Power Quality and Utilisation*, pp. 495–505, 23–25 September 1997, Cracow, Poland.
6. Koszmider A, Brodecki D (2004) Determination of EMC properties of instrument transformers for pulse disturbances in their primary circuits. Report from the realization of project of KBN (Committee on Scientific Research—Poland) (in Polish).
7. Brodecki D, Mana H (1997) Determination of parasitic capacitances of the inductive current/voltage transducers. *XII Symposium on Instrumentation Transform*, pp. 25–27, June 1997, Inowlodz, Poland (in Polish).
8. Koszmider A (2003) Characteristics of HV inductive instrument transformers due to requirements of EMC Standard. *ZN PL ELEKTRYKA (Scientific Bulletin Electricity of Technical University of Lodz)*, 100: pp. 171–182 (in Polish).

Part III

High-Frequency Transformers

H. Bülent Ertan

Introduction

The transformer is a simple device and yet plays a very important role in the transmission and distribution of energy, in measuring electrical variables, and in electronic applications. Miniaturization and portability are very important issues in modern applications such as in telecommunication equipment, in computer products, and in biomedical applications. Products in these areas and many others use transformers in their electronic circuits.

Transformers may be bulky, heavy, and costly if they are not properly designed. This is the main obstacle that needs to be overcome in any size or weight reduction effort. One of the well-known methods for size reduction is to increase the frequency at which a transformer is operated.

Since their introduction, voltage and current ratings of power semiconductors have shown continuous improvement along with the frequency of operation. This development has encouraged transformer designers to explore new technologies and new materials to design smaller and better performance transformers, operating at higher frequencies for existing or new applications.

This part of this book addresses the segment of transformer applications, where significant developments have taken place since the 1970s. Naturally, core materials are very important in designing a high-frequency transformer. Conventionally, silicon steel and its alloys are used for manufacturing transformer cores; at higher frequencies, laminations get thinner. However, the operating frequencies are limited to the kHz range with conventional materials. Metallic glass cores, powder metal cores, and ferrite cores allow operating at higher frequencies, reaching the MHz range. At the highest frequency range, coreless transformers are used to solve core loss problems.

Chapters 19 and 20 introduce ferrite and powder metal cores and consider their properties from the point of view of transformer applications.

High-frequency transformer modeling, analysis, and design are complex issues because of parasitic effects. Similarly, the measurement of transformer parameters for the prediction of its performance is an issue that requires experience. Subsequent chapters of the book address these problems from the point of view of magnetic and electric circuits.

Chapters 21 and 22 deal with high-voltage, high-frequency (kHz range), and high power applications. Two examples of high-voltage, high-frequency transformer applications are considered, and the issues mentioned in the previous paragraph dealt with.

Chapter 23 is devoted to the highest frequency range, where coreless transformers are utilized. Finally, Chapter 24 considers planar transformers, which play an important role in miniaturizing electronic circuits.

19

High-Frequency Transformer Materials

Ralph Lucke

CONTENTS

19.1 Introduction

The materials used for high-frequency power transformers are under permanent improvement. Soft magnetic ferrites designed for high operating frequencies and equipped with increasing DC-bias behavior are state of the art for a lot of applications. There are installed capacities for the production of more than 300,000 tons soft ferrites per year. Most of them go into power applications replacing metal sheet transformers and opening new application fields. Metallic powders and amorphous metals have advantageous properties for simple component geometries and become payable even for mass production. Basically the different materials have common application-related basics. Principal differences will be discussed appropriately.

19.2 Application-Related Basics

Power applications of high-frequency products comprise transformers in switch-mode power supplies, power chokes, and electronic lamp ballast devices. There are two important trends for the development: the miniaturization of power transformers and the demand for increasing transmitted power. The latter includes transformers for wind power plants, contactless chargers for electrical cars (safety!), and contactless energy transmission to robotic arms (unlimited twisting in all directions). For these purposes, large core shapes are introduced (>100 mm).

The target of a high-volume-related power density is typical for all the mentioned applications. This can be obtained by the application of a high working frequency f and/or a high excitation ΔB:

$$P_{trans} = C \cdot f \cdot \Delta B \cdot A_e \cdot A_N j \quad (19.1)$$

where
- P_{trans} is the transmissible power (W)
- f is the frequency (s^{-1})
- ΔB is the flux density swing [$T=Vs/m^2$]
- A_e is the effective magnetic cross section (m^2)
- A_N is the winding cross section (m^2)
- j is the current density (A/m^2)
- C is a coefficient characterizing the converter topology [1,2]

The switching frequency and the flux density swing cannot be increased unlimited because of the heat generation by power losses within the ferrite transformer. Equation 19.2 is a modified formulation of the transmissible power more precisely considering the transformer design. The transmissible power is influenced by the converter topology, the material properties, the thermal design, the winding design, and the core geometry.

The so-called performance factor (*PF*) is defined as the product of frequency and flux density swing, necessary for generating exactly 300 kW/m³ core losses at 100°C. Soft ferrites for power applications can be objectively compared this way even between different suppliers. The higher the *PF*, the better is the ferrite material. The *PF* is generally plotted against frequency (see Figure 19.1). That way one can make an optimal material choice for a circuit-dependent frequency. A material with a higher *PF* is suited for a higher transmissible power under the conditions of the same heat dissipation and the same core shape. On the other hand, the core shape can be reduced with the same power capacity introducing a material with higher *PF*.

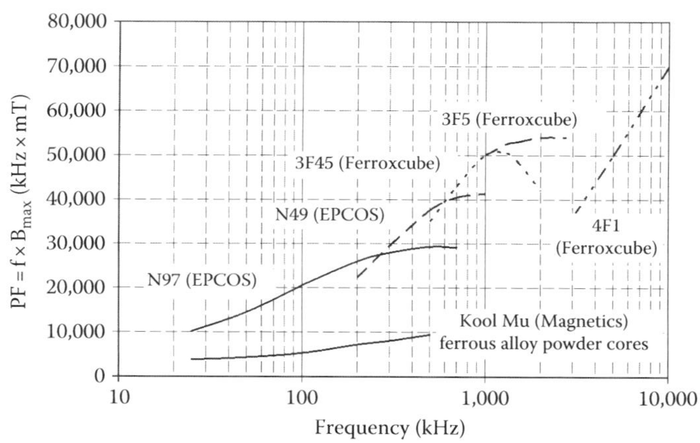

FIGURE 19.1
PF dependent on the frequency for power transformer materials by different manufacturers ($T = 100°C$, $P_C = 300\,kW/m^3$). (From EPCOS AG, *Book Ferrites and Accessories*, edn. 09/2006, 2006; *Ferroxcube: Databook Soft Ferrites and Accessories*, edn. 09/2008; Kool Mu®, *A Magnetic Material for Power Chokes.* Technical Bulletin, Magnetics—A Division of Spang & Company, Pittsburgh, PA, 1999.)

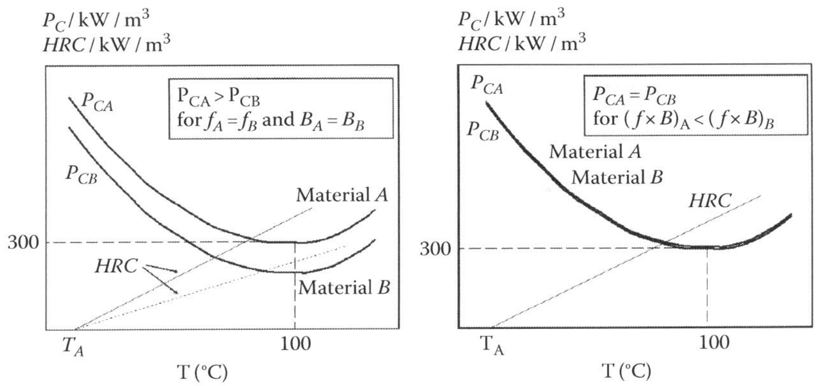

FIGURE 19.2
Temperature dependence of the core losses P_C and the heat removal capacity (HRC).

$$P_{trans} = C \cdot \frac{PF}{\sqrt{P_C}} \cdot \frac{\Delta T}{R_{th}} \cdot \sqrt{\frac{f_{Cu}}{\rho_{Cu}}} \cdot \frac{1}{CLF} \tag{19.2}$$

where
PF is the Performance factor (kHz · mT)
P_C is the core losses (kW/m^3)
ΔT is the temperature rise of the ferrite core (K)
R_{th} is the thermal resistance of the ferrite core (incl. coil) (K/W)
f_{Cu} is the copper filling factor (coil)
ρ_{Cu} is the electrical winding resistance
CLF is the core loss factor

The temperature rise ΔT is limited by the position of the minimum of the ferrite power losses. This material-dependent value is related to the ferrite core operation temperature, necessarily to be lower than the position of the power loss minimum (see Figure 19.2). Some ferrite manufacturers publish maximal ΔT values for power transformer ferrites in data books for designing purposes (see Table 19.1).

Significant efforts have been made by the ferrite manufacturers in order to reduce the core losses P_C since the latter strongly influence the transmissible power of a converter. Other ferrite material properties also have application relevant features that are subject of a separate paragraph.

19.3 Material-Related Demands on Power Transformer Ferrites

The properties of ferrite materials and components containing ferrites are dependent on the following surrounding conditions:

- Composition of the ferrite material
- History of the component
- Operating conditions for the component

TABLE 19.1

Recommended Maximal Temperature Rise ΔT for Power Transformer Ferrites by EPCOS AG

Material	ΔT_{max} (K)	Position of the Power Loss Minimum (°C)[a]
N27	30	80
N41	30	60
N49	20	70
N51	10	50
N72	40	80
N87	50	100
N92	50	100
N95	50	100
N97	50	100

Source: EPCOS AG, *Book Ferrites and Accessories*, edn. 09/2006, 2006.

[a] At typical operating conditions.

Consequently, material properties and core shape geometry cannot be considered separately. The characterization of power transformer ferrites usually includes the following properties measured at toroidal core shapes:

- Initial permeability μ_i
- Core losses P_C dependent on excitation, frequency and temperature
- Amplitude permeability μ_a dependent on excitation and temperature
- Hysteresis loop for determining the saturation flux density B_S and the coercive field strength H_C dependent on the temperature

The publication of these material properties is an application support on the one hand and a tool for the objective comparison between manufacturers on the other hand. The measurement conditions are part of international agreed IEC standards [3].

The initial permeability μ_i is dependent on composition, temperature and material density, as well as on exterior influences coming from mechanical loads. Their value determines the number of turns N necessary for obtaining a certain inductivity L. The value of the effective permeability of gapped cores μ_e can be calculated by means of the initial permeability μ_i and the gap width s.

$$L = \frac{N^2 \cdot \mu_O \cdot \mu_e \cdot A_e}{l_e} \tag{19.3}$$

and

$$\mu_e = \frac{\mu_i}{1 + (s/l_e) \cdot \mu_i} \tag{19.4}$$

for $s \ll l_e$.

where

μ_e is the effective permeability
l_e is the effective magnetic path length (m)
s is the width of air gap (m)

Usual initial permeability for power transformer ferrites ranges between $\mu_i = 800$ and 3000 at ambient temperature (see Figure 19.3). Their basic composition (and consequently their crystal anisotropy and magnetostriction) clearly differs from broadband transformer ferrites resulting in a lower μ_i level (see Figure 19.4).

The inductivity at working temperature is important for the application of the transformer. The working temperature may not exceed the temperature of material power loss minimum (see section 19.2) in order to prevent a thermal runaway. A second important influence on the inductivity is the excitation level. There is a material-dependent characteristic value A_{L1} for each core shape, determining the minimal inductivity at a well-defined excitation (320 mT) and temperature (100°C):

$$A_{L1} = \mu_O \cdot \mu_a \cdot \frac{A_e}{l_e} \tag{19.5}$$

with the magnetic field constant $\mu_O = 4\pi \times 10^{-7}$ H/m.

The amplitude permeability μ_a is a material property necessary for the characterization of temperature-dependent properties at high AC excitation. It is included in ferrite material data sheets and in data books (see Figure 19.5).

The core loss is the most important property of a power transformer ferrite. Its reduction is subject of a long period of efforts by various ferrite manufacturers. The theoretical understanding of the effect of different parameters on the core losses leads to a stepwise reduction of losses at the manufacturing stage.

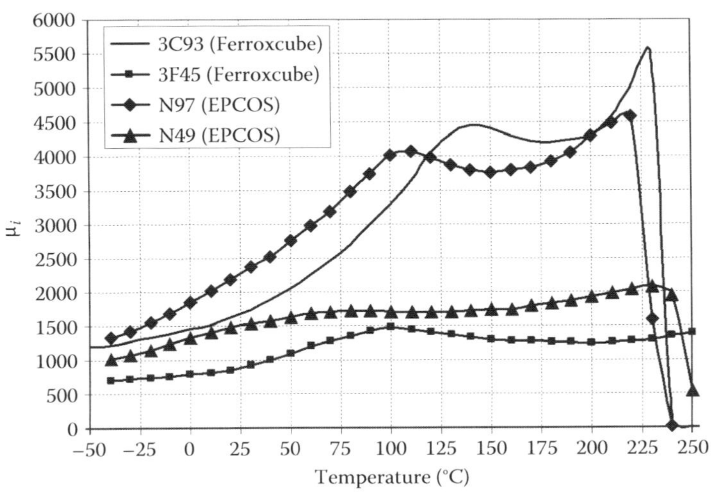

FIGURE 19.3
Typical behavior of the initial permeability of power transformer ferrites dependent on temperature. (From EPCOS AG, *Book Ferrites and Accessories*, edn. 09/2006, 2006; *Ferroxcube: Databook Soft Ferrites and Accessories*, edn. 09/2008.)

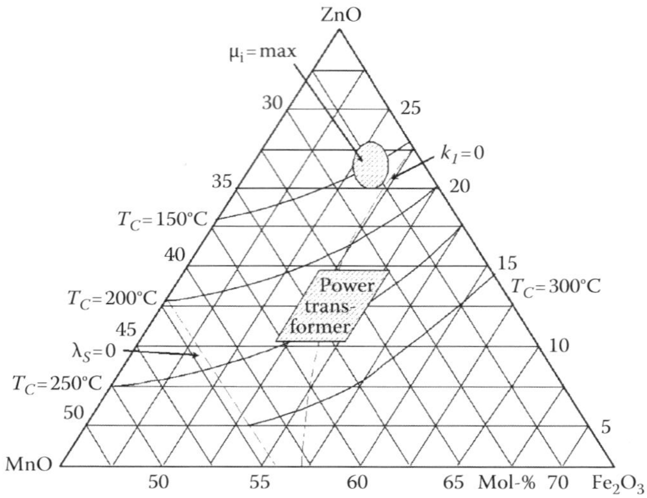

FIGURE 19.4
Three-component diagram for Mn–Zn ferrites and position of the most important power transformer ferrites ($k_1 = 0$ and $\lambda_S = 0$ are the zero lines of the crystal anisotropy and the magnetostriction, respectively).

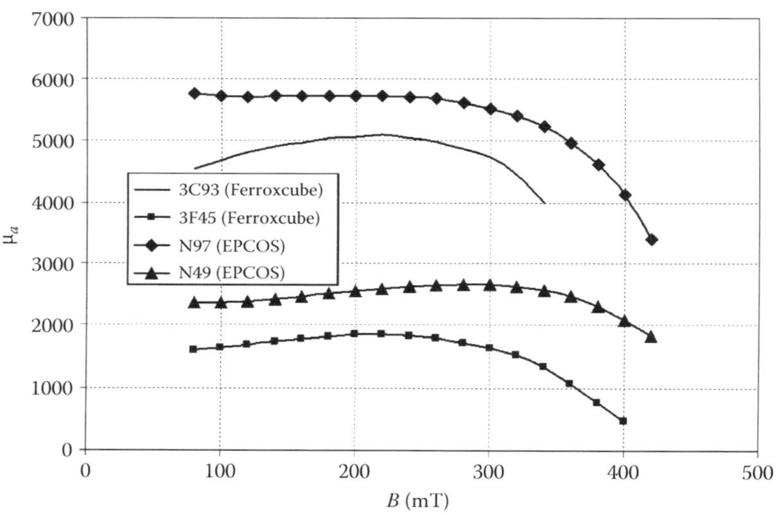

FIGURE 19.5
Amplitude permeability μ_a variation with the excitation B ($T = 100°C$). (From EPCOS AG, *Book Ferrites and Accessories*, edn. 09/2006, 2006; *Ferroxcube: Databook Soft Ferrites and Accessories*, edn. 09/2008.)

The core loss P_C occurs due to the following loss mechanisms (see Figure 19.6):

$$P_C = P_H + P_E + P_R \tag{19.6}$$

where

P_H is the hysteresis losses
P_E is the eddy current losses
P_R is the residual losses

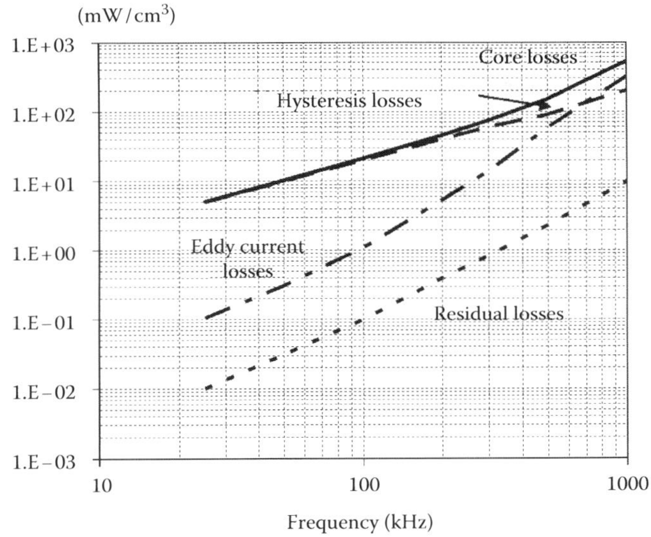

FIGURE 19.6
Schematic image of different power loss shares.

P_H: hysteresis losses arise from energy dissipation by aligning magnetic domains and hindered domain wall movement caused by ferrite material lattice defects. Each magnetic reversal results in a run through a part of the hysteresis loop with the enclosed area as a measure for the hysteresis losses.

$$P_H = f\int HdB \tag{19.7}$$

Hysteresis losses can be minimized by choosing a basic composition close to the zero line of the crystal anisotropy constant and the magnetostriction constant. Moreover, a homogeneous microstructure and a low defect rate are advantageous. All these measures support an unhindered domain alignment and domain wall movement.

P_E: eddy current losses are initiated by the electrical conductivity of the ferrite microstructure. Mn–Zn ferrites belong to the material group of semiconductors since they have low activation energy for electron transfer ($\leq$0.1 eV). Along with the limited DC resistance, the dielectric properties of the grain boundaries have an additional influence on the frequency-dependent drop of the AC resistance. The importance of the specific electrical resistance ρ becomes evident in the following equation:

$$P_E = \frac{C_D \cdot d^2 \cdot \hat{B}^2 \cdot f^2}{\rho} \tag{19.8}$$

where
- C_D is the dimensional constant
- d is the mean eddy current diameter (m)
- $\hat{B}$ is the flux density swing (T)
- f is the frequency (s^{-1})
- ρ is the specific electrical resistance (Ω m)

The reduction of the eddy current losses can be obtained by an increasing specific electrical resistance (ρ ↑) and/or by a more fine grained microstructure (d ↓). In this connection, the addition of oxidic dopants on the order of 10–1000 ppm becomes important. The added metal ions move toward the grain boundaries during sintering because of their ion radii. Amorphous regions with a high electrical resistance are formed there. The higher the area of grain boundaries per unit volume and the thicker the grain boundaries are, the higher is the specific resistance of the ferrite. On the other hand, the specific electrical resistance can also be influenced by the composition of the grain itself. There are certain metal ions compatible to the spinel lattice and making the electron transfer more difficult.

P_R: residual losses are caused by the resonance of the spin rotation within the domain walls [4]. The spin rotation frequency f_S is dependent on the excitation and frequency according to following equation:

$$f_S = \frac{D \cdot B_m \cdot f}{B_S \cdot \delta} \tag{19.9}$$

where
- D is the domain size (m)
- B_m is the magnetic flux density (T)
- f is the frequency (s^{-1})
- B_S is the saturation flux density (T)
- δ is the domain wall thickness (m)

The domain size D is weakly dependent on the frequency. It decreases with increasing frequency f. As a result, there are values for the spin rotation frequency f_S possible above the cutoff frequency according to *Snoek*'s law. This underlines the importance of the residual losses for high working frequencies.

In case of disadvantageous geometry of the ferrite cores, the dimensional resonance f_d can have a contribution to the core losses. Operation at this working condition should be avoided.

$$f_d = \frac{c_o}{2 s_w \sqrt{\varepsilon \cdot \mu}} \tag{19.10}$$

where
- $c_o = 3 \times 10^8$ m/s (light speed in vacuum)
- s_w is the "effective" thickness (m)
- ε is the permittivity
- μ is the permeability

The hysteresis losses are basically responsible for the temperature dependency of the core losses. Their tight connection to the crystal anisotropy and the magnetostriction causes the origin of a power loss minimum at the same position as the secondary permeability maximum. The position of the power loss minimum can be varied by the ferrite composition. The power loss level can be reduced by shifting the minimum of power losses toward lower temperatures (see Figure 19.7). This can be explained by increasing AC resistances toward lower temperatures in case of low working frequencies (e.g., 100 kHz).

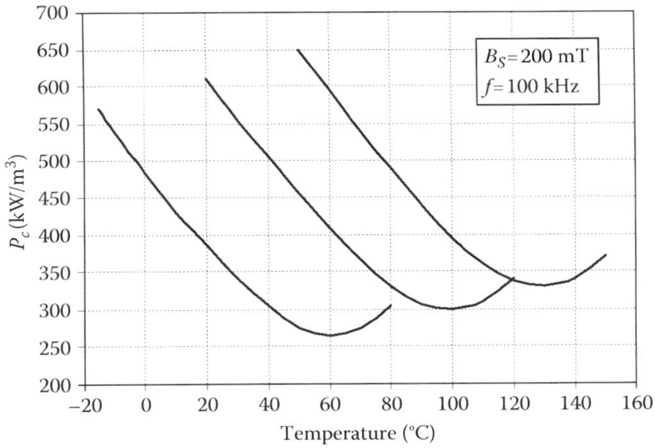

FIGURE 19.7
Core losses of ferrites with modified compositions.

The application of power transformer ferrites at different frequencies results in shifted positions of the power loss minimum. As an example, the temperature of the power loss minimum decreases with increasing working frequency under identical excitation B_m. This is a result of the changing contribution of the different loss mechanisms to the core losses (see Figure 19.8). Compared to low frequencies (e.g., 100 kHz), the temperature dependence of the AC resistance is very low at high frequencies (e.g., 1 MHz). On the contrary, the resonance losses (which are only relevant at high frequencies) are increasing with the temperature, resulting in an overcompensation of the low hysteresis losses.

The saturation magnetization B_S is the condition of the ferrite when all magnetic dipoles are aligned parallel to a sufficient strong magnetic field H_S. The saturation magnetization

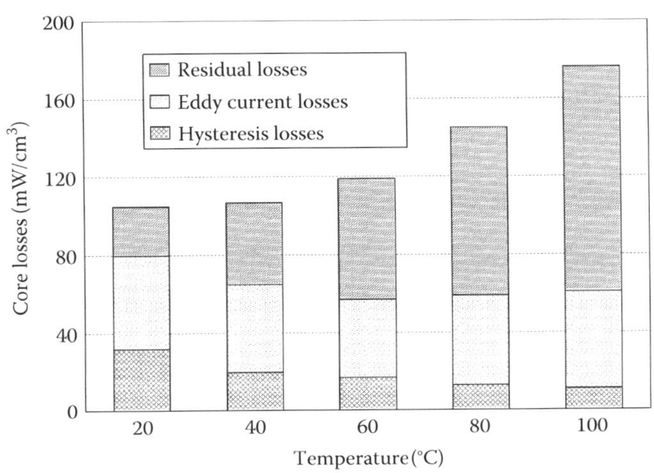

FIGURE 19.8
Power loss contributions to total loss of a Mn–Zn ferrite at different temperatures (f=1 MHz). (From Suh, J.J. et al., *IEEE Trans. Magn.* 36(5), 3402, 2000.)

is a measure for the maximum excitation of a ferrite typical for power transformer applications. It is also an indicator for the behavior under DC-bias magnetization as well as for the amplitude permeability μ_a.

B_S is dependent on the magnetic moments per unit volume. In practice there exists an optimal ion distribution in a suited crystal lattice (e.g., spinel lattice) contributing to the resulting number of magnetic moments. A reduction of the theoretical calculated saturation magnetizations occurs by microstructural influences like lattice distortions, pores, and grain boundaries. Since the ferromagnetic properties of ferrites disappear above the *Curie* temperature T_C, it is useful for high excitations to keep a rather high distance between the working temperature and T_C (see Figure 19.9).

The properties of Mn–Zn ferrites can be systematized by targeting one important feature and varying at least one relevant influence. The resulting "Material Design Tool" is suited for the prediction of material properties as initial permeability, core losses or saturation flux density by modification of composition, sintering conditions, and measuring conditions [10]. There is a considerable acceleration possible for the development times of customized power transformer ferrites. Basic trends can be transferred to other groups of ferrite materials. The principle of Material Design Tools leads to comprehensive results by rather low efforts (see Figure 19.10).

The obtained results can be transferred without any problems into the production. Logistic advantages in the production sequence can be realized, finally resulting in a more stable and improved quality.

Apart from the magnetic properties, ferrite cores have to meet also mechanical demands. Thermomechanical stresses play an important part. Power transformer ferrites use to warm up during the operation according to dissipating power losses (see section 19.2). Different thermal expansion coefficients, caused by the assembling with coil formers, by

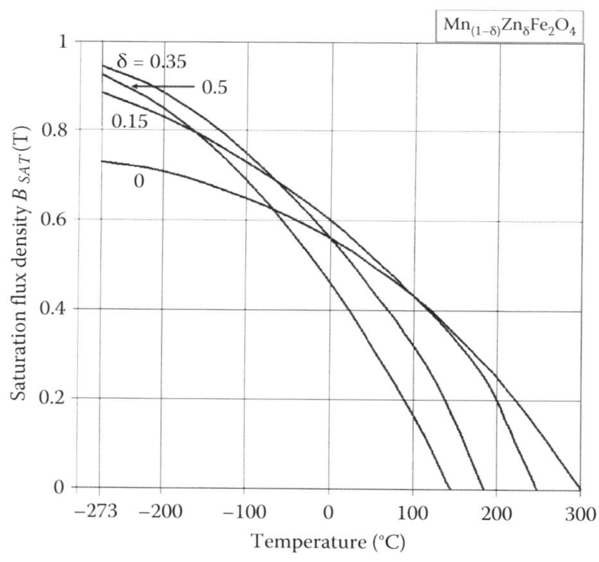

FIGURE 19.9
Temperature dependence of the saturation flux density B_{Sat} of Mn–Zn ferrites. (From Snelling, E.C., *Soft Ferrites—Properties and Applications*, 2nd edn., Butterworths, London, U.K., p. 69, 1988.)

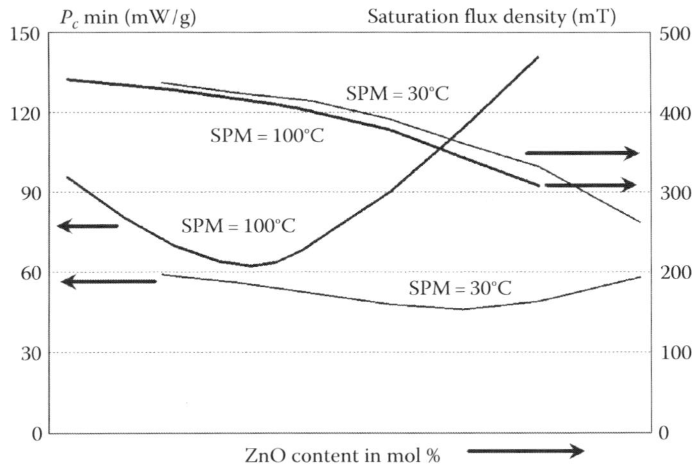

FIGURE 19.10
Properties of power transformer ferrites in dependence on the ZnO content applying a Material Design Tool ($P_{C\,min}$ stands for minimal achievable core losses; SPM is the temperature of the loss minimum).

gluing and coating with polymers, or by clipping, result in thermomechanical stresses that may not cause any irreversible modification of the ferrite core. These stresses do directly interfere with the crystal anisotropy and magnetostriction. Consequently, magnetic properties as the permeability or hysteresis losses can be deteriorated.

Compared to these long-term stresses, there are short-term stresses caused by grinding of gapped cores or by reflow solder. They can result in mechanical failure of ferrite cores.

There are methods of mechanical characterization well known from structural ceramics. The following measuring methods are applied to ferrites:

Short-term mechanical properties

- Fracture strength σ (statistical approach according to Weibull) (see Figure 19.11)
- Fracture toughness K_{IC}
- Static fatigue
- Measurement of the *R*-curve behavior (crack strengthening)
- Measurement of the subcritical crack propagation (*vK*-curve)
- Cyclic fatigue
- Cycles to failure
- Life time (see Figure 19.12)

The measurement results include detailed information about the mechanical properties under certain stresses and allow conclusions for a practical relevant material design. It is an instrument for the construction of real multifunctional materials. As valid for each multiparameter optimization, the technological processing window becomes smaller, resulting in advanced technology.

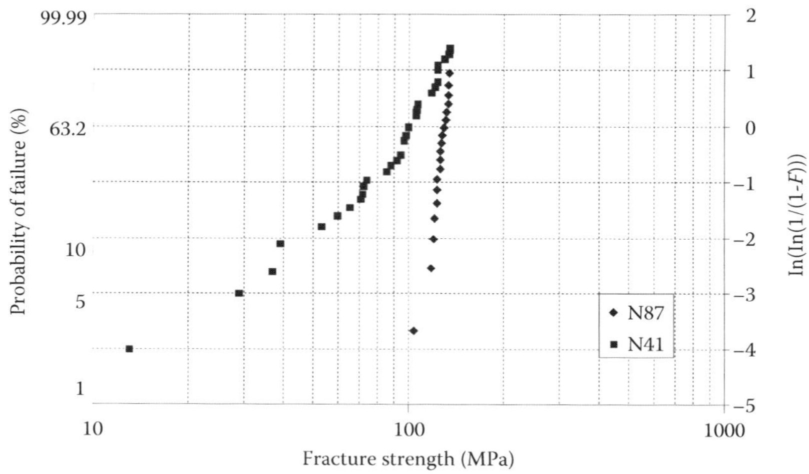

FIGURE 19.11
Weibull distribution of the two power transformer ferrites N41 and N87 (by EPCOS). (From Lucke, R. and Hahn, I., Mechanical reliability of Mn–Zn ferrites, *Proceedings of the Eighth International Conference on Ferrites*, Kyoto, Japan, p. 1117, 2000.) The probability of failure is drawn logarithmically per definition in order to get straight lines for the dependency.

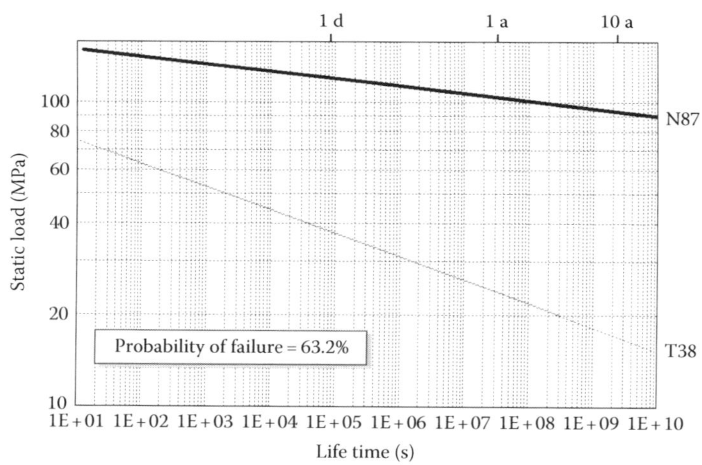

FIGURE 19.12
Mechanical lifetime of N87 and T38 (by EPCOS). (From Lucke, R. and Hahn, I., Mechanical reliability of Mn–Zn ferrites. *Proceedings of the Eighth International Conference on Ferrites*, Kyoto, Japan, p. 1117, 2000.)

19.4 Core-Shape-Related Peculiarities

The manufacturing of ferrite cores requires considerable technological measures in order to transfer typical material properties to extraordinary component geometries. The following discrepancies become obviously between toroidal core shapes and real customized core shapes:

- The 3D distribution of the magnetic flux is rather inhomogeneous at real core shapes compared to toroids. Consequently, there are areas with above-averaged excitation and finally modified characteristics (permeability, power losses).
- The manufacturing of real ferrite core shapes is bound up with processing conditions. The latter can influence 3D gradients of the material properties over the ferrite core volume.
- The introduction of gapped core shapes will reduce the dependence of the component characteristic on the working and surrounding conditions as well as on varying material properties.

A realistic adjustment of the specifications to the actual values is therefore necessary for the compilation of core-shape-related properties.

The trend to rather flat core shapes of electronic circuits forces the application of extremely flat ferrite core shapes. Associated with that there are changed winding arrangements as well as a number of advantages for the circuit design:

- Excellent reproducibility of electrical parameters
- Superior thermal behavior (heat removal)
- High level of integration

The core-shape-related working conditions to be considered for the transformer design are subject of recommendations in data books or design tools. They are not discussed in this publication.

19.5 Newest Results of Ferrite Material Development

Improvements of the material properties for power transformer ferrites mentioned in section 19.3 are subject of worldwide, intensive development activities. The emphasis is placed on the following subjects (Figure 19.13):

- Improvement of the material performance for the medium frequency range (f = 100–500 kHz, minimized core losses in a wide temperature interval, maximized saturation flux density) (see Figure 19.14)
- Minimized core losses for high-frequency applications (f = 500–3000 kHz, optimized saturation flux density)
- Optimization of the performance under extreme DC-bias magnetization (f = 100–300 kHz, maximized saturation flux density)

The consequent application of the knowledge about loss mechanisms, saturation, and permeability allows tailor-made solutions for the customer. The manufacturing demands require reproducible processes at highest technological level (Figure 19.15).

The following ferrite materials by different manufacturers represent the state of the art at the market.

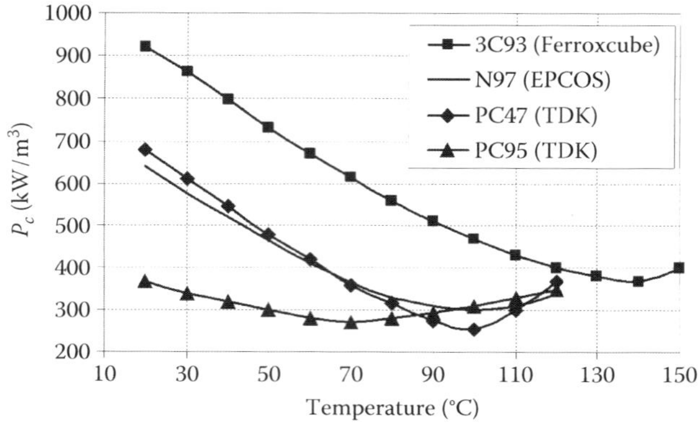

FIGURE 19.13
Minimized core losses in a wide temperature range (100 kHz/200 mT). (From EPCOS AG, *Book Ferrites and Accessories*, edn. 09/2006, 2006; *Ferroxcube: Databook Soft Ferrites and Accessories*, edn. 09/2008; Ferrites for switching power supplies, TDK Product Catalogue, 2007.)

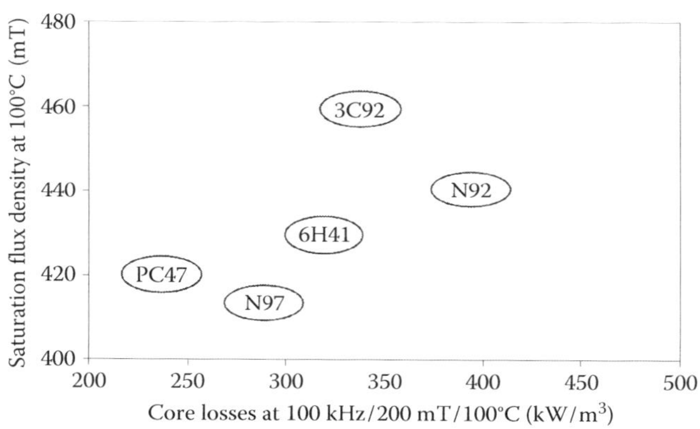

FIGURE 19.14
Optimization of the performance (low losses and maximal saturation flux density) PC47 (TDK) (From Ferrites for switching power supplies, TDK Product Catalogue, 2007.), 6H41 (FDK) (From Ferrite cores for transformers & choke coils, FDK Products, 2008.), N97 + N92 (EPCOS) (From EPCOS AG, *Book Ferrites and Accessories*, edn. 09/2006, 2006.), and 3C92 (Ferroxcube). (From *Ferroxcube: Databook Soft Ferrites and Accessories*, edn. 09/2008.)

19.6 Potentials and Limits of Conventional Ferrite Materials

The material improvements within reach are getting smaller, and future progresses will be made rather in quality then in quantity. For some products, the physical limits of the system Mn–Zn ferrite are close. Nevertheless, big efforts are done at a high technological level in order to use the existing experiences for that material system.

The trend to higher working frequencies is a real challenge for Mn–Zn ferrites because of their low specific electrical resistance at interesting compositions (Fe excess). Working frequencies of about 3 MHz are considered as useful limits (see Figure 19.16). The necessary ferrite microstructure has to be very fine grained and equipped with well-isolating grain boundaries.

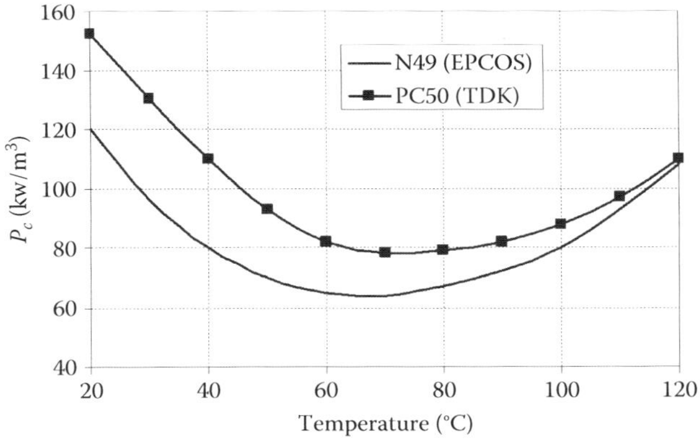

FIGURE 19.15
Minimization of the core losses for high-frequency applications (500 kHz/50 mT). (From EPCOS AG, *Book Ferrites and Accessories*, edn. 09/2006, 2006; Ferrites for switching power supplies. TDK product Catalogue, 2007.)

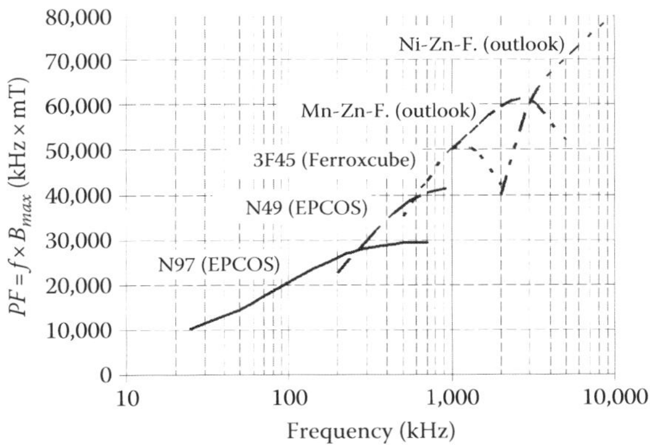

FIGURE 19.16
Power transformer ferrites with the trend to higher working frequencies (PF versus frequency, $T = 100°C$, $P_C = 300\,kW/m^3$).

The frequently necessary robust behavior toward DC-bias magnetization requires a high material density as well as a microstructure with a low defect rate [14]. This compromise can be found by exact control of all technological steps for the manufacturing.

Considering higher working frequencies (>3 MHz), there are alternative ferrite materials. Materials with clearly higher specific electrical resistance, as Ni–Zn ferrites, seem to be better qualified for power transformers with high working frequencies. Apart from the eddy current losses, the loss mechanisms left have to be reduced.

19.7 Comparison of Ferrites with Other Soft Magnetic Materials

The main difference between ferrites and most other magnetic materials is due to their material nature. Ferrites consist of oxides, causing a dilution effect of the magnetic moments by oxygen ions in the crystal lattice. Pure metals, like iron, nickel and cobalt,

TABLE 19.2

Properties of Soft Magnetic Materials

Material	Initial Permeability μ_i	Saturation Flux Density B_S (T)	*PF* at Operating Frequency (kHz × mT)	Curie Temperature T_C (°C)	Specific Resistivity (Ω cm)	Typical Operating Frequency
Fe	250	2.2	—	770	10×10^{-6}	50–1,000 Hz
Si–Fe (unoriented)	400	2.0	—	740	50×10^{-6}	50–1,000 Hz
Si–Fe (oriented)	1,500	2.0	—	740	50×10^{-6}	50–1,000 Hz
50–50 Ni Fe (grain oriented)	2,000	1.6	—	360	40×10^{-6}	50–1,000 Hz
79 Permalloy®	12,000–100,000	0.8–1.1	—	450	55×10^{-6}	1–75 kHz
Amorphous alloys	3,000–20,000	0.5–1.6	—	200–380	140×10^{-6}	to 250 kHz
Permalloy® powder	14–550	0.3	3,000–12,000	450	1	10 kHz–1 MHz
High flux® powder	14–160	1.5	—	360	—	10 kHz–1 MHz
Kool Mu® powder	26–125	1.0	3,000–10,000	740	—	to 1 MHz
Iron powder	5–80	1.0	—	770	10^4	100 kHz–100 MHz
Mn–Zn ferrites	750–20,000	0.3–0.55	6,000–55,000	100–300	10–100	10 kHz–3 MHz
Ni–Zn ferrites	10–2,300	0.3–0.5	20,000–70,000	150–450	10^6	1–100 MHz

Source: Kool Mu®. *A Magnetic Material for Power Chokes.* Technical Bulletin, Magnetics—A Division of Spang & Company, Pittsburgh, PA, 1999, and own measurements.

and some rare earth elements at low tempreatures show a strong ferromagnetism derived from unpaired electron spins causing highest saturation magnetizations. The oxygen ions serve a useful purpose, however, since they strongly increase the resistivity. This makes the ferrites interesting especially for higher-frequency applications.

The Table 19.2 summarizes properties of soft magnetic materials interesting for power applications.

Especially for power chokes, there is an advantage of the metal-based materials due to their high saturation flux density. If one succeeds in a reproducible resistance increase of the grain boundary phase under manufacturing conditions, there will be applications up to 1 MHz under high excitations. On the other hand, the prices of the metal-based materials are still clearly above the ferrite core prices. Finally, application-related requirements like size, costs, layout, and performance decide about the choice of component material.

References

1. Roespel G (1978) Effect of the magnetic material on the shape and dimensions of transformers and chokes in switch-mode power supplies. *J. Magn. Magn. Mater.* (9), 145–149.

2. Esguerra M (1999) New generation of application-tailored soft ferrites. *Proceedings of High frequency Magnetic Materials Conference*, Santa Clara, CA, pp. 8–24.
3. IEC-Standard 62044 Cores made of soft magnetic materials—Measuring methods. Part 3: Magnetic properties at high excitation level.
4. Yamada S, Otsuki E (1997) Analysis of eddy current loss in Mn–Zn ferrites for power supplies. *J. Appl. Phys.* 81(8), 4791–4793.
5. EPCOS AG (2006) *Book Ferrites and Accessories*, edn. 09/2006.
6. *Ferroxcube: Databook Soft Ferrites and Accessories*, edn. 09/2008.
7. Suh JJ, Song BM, Han YH (2000) Temperature dependence of power loss of Mn–Zn ferrites at high frequency. *IEEE Trans. Magn.* 36(5), 3402–3404.
8. Snelling EC (1988) *Soft Ferrites—Properties and Applications*, 2nd edn. Butterworths, London, U.K., p. 69.
9. Lucke R, Hahn I (2000) Mechanical reliability of Mn–Zn ferrites. *Proceedings of the Eighth International Conference on Ferrites*, Kyoto, Japan, p. 1117.
10. Lucke R, Ahne S, Wrba J (2002) The present and future of ferrite materials for power applications. *Proceedings of PCIM Conference*, Shanghai, China, pp. 19–123.
11. Kool Mu® (1999) *A Magnetic Material for Power Chokes*. Technical Bulletin, Magnetics—A Division of Spang & Company, Pittsburgh, PA.
12. Ferrites for switching power supplies. TDK Product Catalogue, 2007.
13. Ferrite cores for transformers & choke coils. FDK Products, 2008.
14. Lucke R, Esguerra M, Wrba J (2004) Processing requirements for high end Mn–Zn ferrites. *Proceedings of Ninth International Conference on Ferrites*, San Francisco, CA, pp. 245–250.

20

Powder Core Materials

Arcan F. Dericioglu

CONTENTS

20.1 Core Materials in General

Transformers are used in the conversion process in power electronics which generally constitute the heaviest and bulkiest component in the conversion circuit [1]. Specifically, the high-frequency transformer of a switched-mode power supply (SMPS) constitutes about 25% of its overall volume and 30% of its overall weight [2]. However, with the advancing technology, there is an increasing demand for miniaturized high-performance electronic devices such as portable computers, cellular phones, etc. [3], utilizing such components. In addition to this, increasing transmitted power requirements for high-efficiency power generators as well as contact-less applications are other important issues in transformer development. As a result, besides the efforts on optimization the design of the transformers, the development of highly efficient low loss core materials became crucial.

As introduced by R. Lucke in Chapter 19, transmitted power, P_{trans}, is directly proportional to *performance factor*, *PF* (kHz · mT), while it is inversely proportional to *core losses*, P_c (kW/m^3 or mW/cm^3), and *core loss factor*, *CLF*:

$$P_{trans} \propto PF, \frac{1}{P_c}, \frac{1}{CLF} \tag{20.1}$$

Among these parameters, *PF* and P_c are controlled by the intrinsic properties of the core material, while *CLF* is core geometry dependent. This follows that core materials with specific and desirable property sets should be developed in order to maximize PFs while suppressing core losses. Consequently, advances in material properties can help reducing core shape complexity leading to lessened processing constraints.

Energy losses which were previously termed as core losses are inevitable in transformers and chokes. These energy losses cause heat to be generated in the structure and hence lead to thermal problems. The sources of the losses in transformers, chokes, or inductors can be classified as follows.

Hysteresis loss results from the sweeping of magnetic flux from positive to negative and is quantified by the area enclosed by the magnetic flux versus magnetization force loop. Hysteresis loss is a consequence of the materials intrinsic properties related to the energy used to align and realign the magnetic domains.

Varying electric or magnetic field that passes through a conducting material induces eddy currents opposing this variation. As a result of this opposing currents and the electrical resistance of the conducting material, heat losses termed as eddy current losses appear in the system.

Copper or winding loss is caused by the ohmic resistance of the coil. It is dependent on the wire size, switching frequency, etc. [4].

From the application parameters point of view, the type of a transformer core depends on operational factors such as applied voltage, current, and frequency. In addition to these, size limitations and construction costs are further factors to be considered. Conventionally applied core materials are air, soft iron, and steel. Selection of these conventional core materials depends on the particular application and frequency range under consideration. Each of these media might be suitable for particular applications while unsuitable for others. Generally, air-core transformers are used when the voltage source has a frequency above 20 kHz, while iron-core transformers are usually used when the source frequency is below 20 kHz. Transformer cores composed of laminated steel sheets isolated with a nonconducting material provide highest efficiency among conventional core materials primarily due to better dissipation of heat generated during the process.

As formerly mentioned, rapidly increasing working frequencies in various power electronics applications such as power conversion banks, uninterruptible power supplies, magnetic filters, etc., initiated a growing interest in the use of special soft ferrite cores for transformers and reactances [5]. Leaving the details about ferrite cores in Chapter 19, it should here be mentioned that although ferrites are popular core materials due to their attractive magnetic properties at high frequencies, they are structure-sensitive materials, the properties of which critically depend on the manufacturing process [6]. Generally, the ferrites are commercially produced by applying conventional ceramic processing methods, and it is difficult to achieve high purity and homogeneous powders on microscopic scale [7], which constitutes a considerable drawback at the stage of their application.

Operational efficiency and ease of manufacturing of core materials have long been the interest of both academic research and industry. As a result of this, powder metallurgical processes have been applied in order to simplify the process of producing magnetic cores (powder cores) to improve the efficiency percentage of the core materials and to facilitate making the cores in any shape [8]. Compared with ferrite cores, powder cores have the advantage of higher saturation induction making them suitable for large current applications [3]. In addition to this, powder core materials are effective in suppressing the eddy current losses due to the insulation of metallic powders by a nonconductive layer, as schematically illustrated in Figure 20.1. Selection of a ferrite or powder component as a core material depends on the frequency band of application (wideband or narrowband) and the signal power which will be handled. For a given core size, ferrite material will saturate at a much lower magnetic flux density than the powder core. Furthermore, temperature coefficient of ferrite cores is greater than that of the powder cores because of

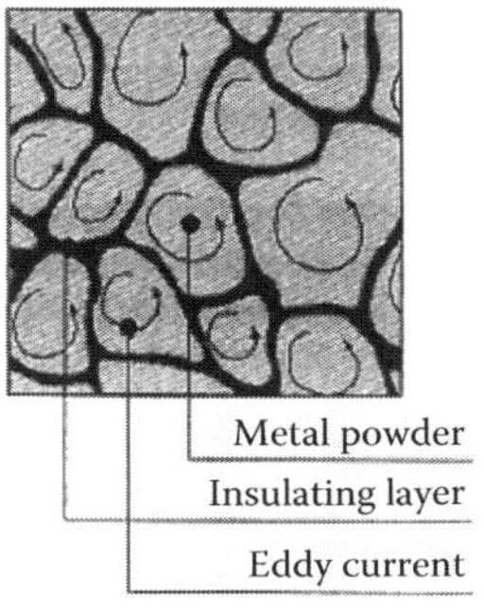

FIGURE 20.1
Eddy current confinement in powder cores. (From Online, Terminology and definitions of terms 1, http://www.cwsbytemark.com/CatalogSheets/MPP%20PDF%files/53.pdf, 2006.)

their relatively higher permeability. Consequently, powder cores are more suitable for narrowband applications where expected inductance values do not vary with temperature changes, which is not so critical in wideband applications.

20.2 Powder Cores

Powder cores are components obtained by compressing metallic powders which are electrically insulated from each other by an insulation layer at the particulate surface providing a distributed air gap in the final material. The resulting density of the compact leads to a variation in permeability values to be obtained. Small air gaps distributed evenly throughout the cores increase the amount of DC that can be passed through the winding before core saturation takes place. Powder cores that are not pressed with an organic binder do not exhibit any thermal aging effects.

A typical processing route used for the production of powder cores is illustrated in Figure 20.2. Firstly, powders are prepared by atomization from a molten alloy with desired chemical composition. Atomization is a powerful powder fabrication technique in which

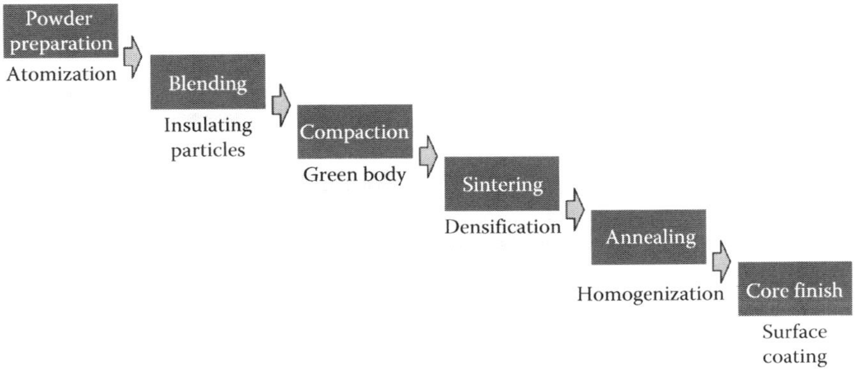

FIGURE 20.2
Typical processing route used for the production of powder cores.

FIGURE 20.3
Scanning electron micrograph of a spray-dried Fe–Si–Al (Sendust) powder containing 0.5 wt.% polyvinyl alcohol and 0.15 wt.% glycerin. (From Kishimoto, Y. et al., *J. Mater. Sci.*, 38, 3479, 2003.)

the powder chemistry and shape characteristics can fully be controlled. Figure 20.3 presents a typical spray-dried powder. Alternatively, powder material can also be obtained through grinding high permeability metallic materials. If necessary, powders can be classified by sieving before the next step in order to obtain a desired particle size distribution. Following powder preparation, the particles are mixed with isolation means and binders to obtain a homogeneous distribution. Mixed powders are compacted in a mold applying an external pressure to deform them which eventually leads to shape and dimensional control in the final component. Binders are used to improve the strength and hence handling of the compacted bodies (green bodies) where necessary.

As the next step, green bodies are sintered at an elevated temperature specific for the alloy under consideration, which is usually below its melting point. This process creates bonding between the particles by atomic transport events that leads to the ultimate densification in the final product. An annealing step might succeed sintering in order to homogenize the microstructure. Figure 20.4 shows the optical micrograph of a powder core sintered at 1523 K in hydrogen atmosphere. This micrograph reveals that a relatively high-density structure almost free from residual voids can be obtained after the sintering process. Finally, during the finishing step, core surfaces are covered with an organic coating which provides a tough chemical resistance and high dielectric protection for the cores.

Application of powder metallurgical methods to produce core materials has various advantages from both the processing and final material properties point of

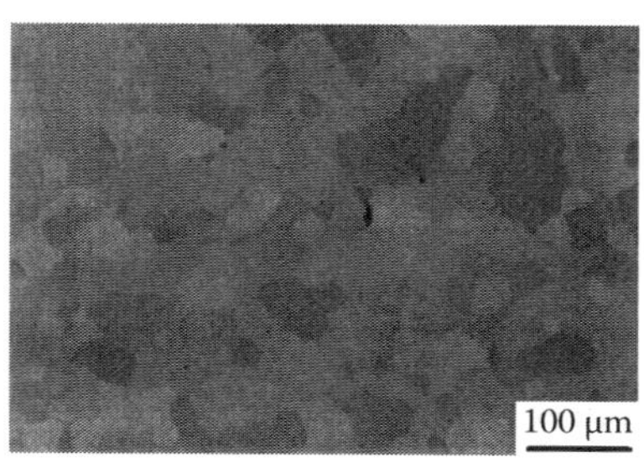

FIGURE 20.4
Optical micrograph of a Sendust core sintered at 1523 K in hydrogen atmosphere. Average grain size is approximately 100 μm where no extensive residual voids are observed. (From Kishimoto, Y. et al., *J. Mater. Sci.*, 38, 3479, 2003.)

views [10]. Applying powder metallurgy, it is possible to produce close-to-size components reducing or eliminating secondary operations as well as process scrap resulting in cost efficiency. With the use of homogeneous prealloyed powders with controlled size distributions, chemical composition fluctuations can be suppressed in the final component to a great extent leading to uniform material properties. Consequently, near net shape powder cores with predictable magnetic properties can be produced with a cost-efficient method.

Powder core materials which are mainly used in low loss inductors for SMPSs, switching regulators, noise filters, pulse and flyback transformers, and many other applications can be divided into four main groups according to their compositions:

1. "Iron powder" cores (Fe)
2. Molybdenum permalloy powder ("MPP") cores (Ni–Mo–Fe)
3. "High flux" cores (Ni–Fe)
4. "Sendust" cores which are also known as "Kool Mμ™" cores (Fe–Si–Al)

Basically, iron powder cores are prepared from extremely small iron particles with purity higher than 99%. These types of cores are available in two classes: carbonyl iron and hydrogen-reduced iron. The former reveals lower core losses and exhibits high-quality factor (Q factor) for RF applications while it is relatively more costly. Iron powder cores are available in permeabilities from 1 to 100 [12]. However, in general, iron powder cores are the most cost-effective of all the powder cores. MPP-type powder cores are molybdenum permalloys of composition 81% nickel, 2% molybdenum, and 17% iron by weight [13]. MPP cores are available in initial permeabilities (μ_i) of 26, 60, 125, 160, 173, 200, and 550 [12], and their normal effective permeability range changes between 14 and 350 [13]. Compared to other materials, MPP cores reveal the highest manufacturing cost. High flux cores are composed of 50 wt.% nickel and 50 wt.% iron alloy powder where the base material is similar to the regular nickel iron lamination in tape wound cores [12]. They are produced with permeabilities ranging from 14 to 160 [13]. Sendust powder cores which are also known as Kool Mμ are produced from a ferrous alloy powder. The base material is approximately 85% iron, 6% aluminum, and 9% silicon by weight where the specific alloy composition is shown in Fe–Al–Si ternary phase diagram (Figure 20.5). Sendust cores are available in initial permeabilities of 60–125. Generally, Sendust cores cost less than MPPs or high fluxes, but are slightly more expensive than iron powder cores.

Earlier-described powder core types are widely applied in numerous applications, and all of these types are commercially available with well-defined performance parameters. As a result, the choice of one type of powder core material over another often depends on the following:

1. DC bias current through the inductor.
2. Ambient operating temperature and acceptable temperature rise. Ambient temperature of over 373 K is now quite common.
3. Size constraint and mounting methods (through hole or surface mount).
4. Cost: iron powder being the cheapest and MPP the most expensive.
5. Electrical stability of the core with temperature changes.
6. Availability of the core material [15].

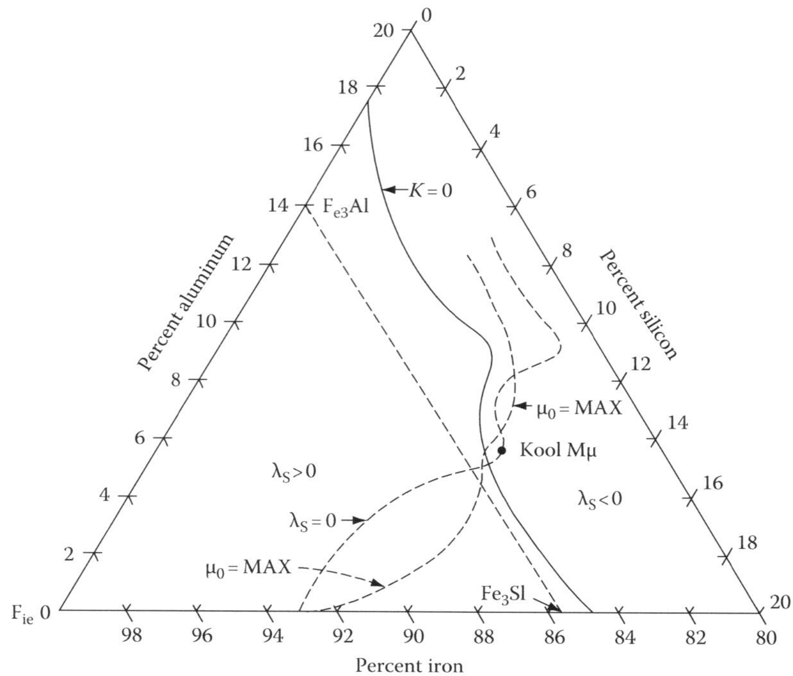

FIGURE 20.5
Ternary phase diagram of Fe–Si–Al alloy system showing the composition Kool Mμ™ (Sendust) powder core material. (From Kool Mμ®, A magnetic material for power chokes, Technical Bulletin No. KMC-S1, Magnetics—Division of Spang & Company, Pittsburgh, PA, p. 2, 1999.)

20.3 Performances of Powder Core Materials

Core losses, the causes of which were mentioned in a preceding section, are one of the most important criteria in the selection of a suitable core material for a specific application. Due to the varying effects of different loss sources under different excitation conditions and operating frequencies, core losses should be investigated as a function of these parameters, and different powder core types should be evaluated accordingly. Figure 20.6 demonstrates the core loss of available powder core materials as a function of AC flux density at 100 kHz for constant permeability for all materials. As it is clear from the figure, MPP reveals the lowest core loss compared to other materials where iron powder core shows the highest. The same tendency is also observed when the core losses of the powder cores mentioned earlier are compared as a function of applied frequency for constant core permeability at a given AC flux density. Figure 20.7 reveals that core losses of MPP and iron powder cores form the lower and upper bounds for the losses among these materials, respectively, while Sendust and high flux cores show comparable loss values with the latter being slightly lower.

Having knowledge about the loss characteristic of a core material is one of the most critical issues in core material selection in order to design the optimum system for a certain purpose. As powder core materials are available from various manufacturers with varying

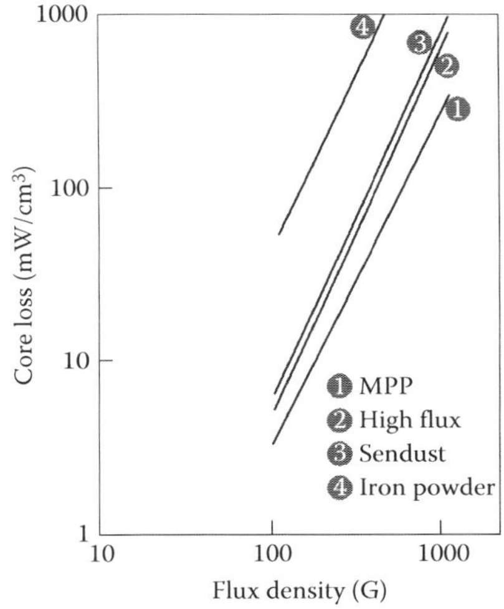

FIGURE 20.6
Core loss of commercially available powder core materials as a function of AC flux density at a frequency of 100 kHz. (From Online, *MPP, High Flux, Sendust Basic Material Characteristics (Magnetic Powder Cores 06–07)*, http://www.cwsbytemark.com/CatalogSheets/MPP%20PDF%20files/6.pdf, 2006.)

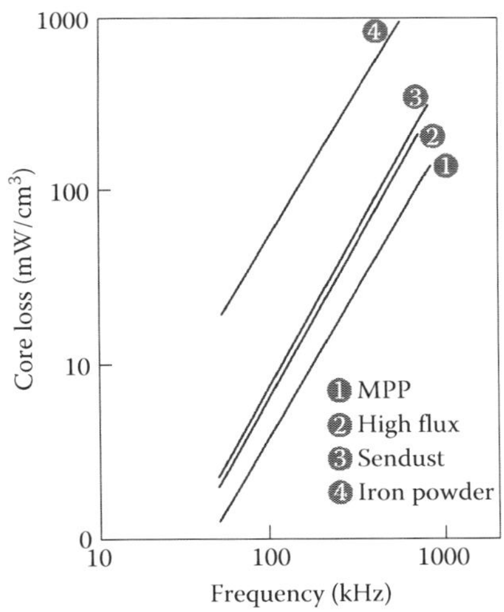

FIGURE 20.7
Core loss of commercially available powder core materials as a function of frequency at an AC flux density of 100 G. (From Online, *MPP, High Flux, Sendust Basic Material Characteristics (Magnetic Powder Cores 06–07)*, http://www.cwsbytemark.com/CatalogSheets/MPP%20PDF%20files/6.pdf, 2006.)

permeability, geometry, and dimensions, a rigorous comparison involves a detailed data search and analysis from many sources—mostly from manufacturer data. In this regard, relatively updated core loss data curves were gathered from the manufacturers' data sheets organizing them with the same units [1]. Gathered data were digitized and approximated in the form of

$$\omega = kf^m B^n \tag{20.2}$$

where
ω is the calculated core loss density (W/kg)
f is the frequency (Hz)
B is the flux density (T)

Figure 20.8 presents one such example where core losses of MPP, high flux, and Kool Mμ cores with 125μ permeability are summarized as a function of AC flux density at different applied frequencies for each core material. These frequencies are the maximum operating values of each specific core material with the given permeability and geometry. The data of Figure 20.8 are similar to those given in Figures 20.6 and 20.7 in terms of the order of the core loss values. However, it should be mentioned that core loss curves of Figure 20.8 were obtained from reported specific relations in the form of Equation 20.2 which makes the comparison possible either as a function of frequency or flux density while keeping the other parameters constant.

Thermal behavior of powder core materials is another important issue in core material selection and design. Table 20.1 shows the thermal properties of MPP, high flux, and Kool

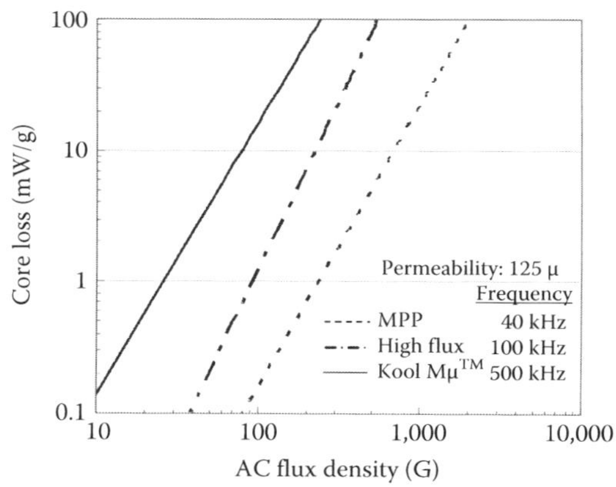

FIGURE 20.8
Core loss of MPP, high flux, and Kool Mμ™ powder core materials with identical permeabilities as a function of AC flux density for varying applied frequency values.

TABLE 20.1
Thermal Properties and Densities of MPP, High Flux, and Kool Mμ™ Powder Cores

Core Types	Curie Temperature (°C)	Density (g/cm³)	Coefficient of Thermal Expansion (°C)	Thermal Conductivity (W/(cm K))
MPP	460	8.7	12.9×10^{-6}	0.8
High flux	500	8.2	5.8×10^{-6}	0.8
Sendust (Kool Mμ)	500	7.0	10.8×10^{-6}	0.8

Source: MAGNETICS®, *Powder Cores Design Manual and Catalog*, pp. 3-1, 3-3, 3-4, 3-14, 3-15, 3-16, 3-19, 3-20, 2002.

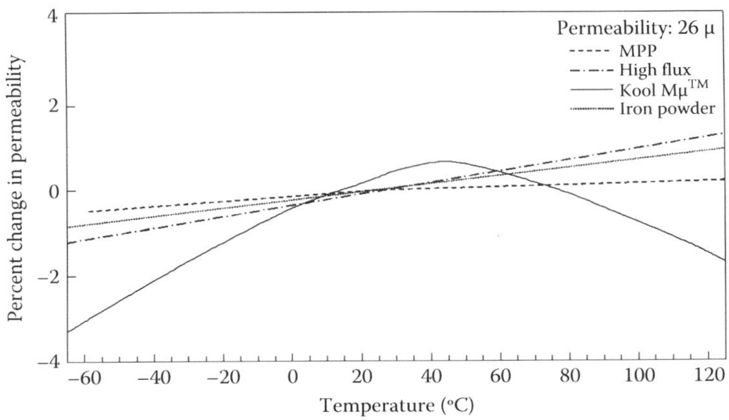

FIGURE 20.9
Change in permeability of available powder core materials (permeability: 26 μ) with temperature. (From MAGNETICS®, Powder Cores Design Manual and Catalog, pp. 3-1, 3-3, 3-4, 3-14, 3-15, 3-16, 3-19, 3-20, 2002; Nicol, D.J., *An Objective Comparison of Powder Core Materials for Inductive Components with Selected Design Examples*, Home Application Notes, Micrometals, Inc., Anaheim, CA.)

Mμ core materials along with their densities. All of these core materials reveal comparable Curie temperatures around 500°C, while thermal expansion coefficient of high flux is relatively lower compared to those of the others which provides better dimensional stability under operational temperature fluctuations.

In this context, Figure 20.9 depicts the temperature stability of available powder core materials with a permeability of 26 μ. Here, temperature stability is quantified as the percent change in permeability (inductance change) of a core material with varying temperature. According to Figure 20.9, it is clear that MPP powder core reveals the highest temperature stability exhibiting less than ±1% permeability change over a temperature range of about 190°C. Iron powder and high flux core materials also reveal considerable temperature stability where the change in their permeability is within ±2% over the same temperature range. Kool Mμ reveals the least temperature stability in this range with a change in permeability reaching to −2% to −4% at 125°C and −65°C, respectively. Here, it should be noted that despite its relatively lower temperature stability, Kool Mμ has no thermal aging concerns associated with iron powder cores, as no organic binders are used during its manufacture [18]. A similar temperature stability behavior is observed for powder cores with higher permeability values (Figure 20.10). As it is shown in Figure 20.10, MPP core with 125 μ permeability reveals almost no permeability change between −65°C and 125°C, while high flux and Kool Mμ cores exhibit permeability changes between about −2 to 2 and −14 to −4 in the same temperature range, respectively.

Saturation characteristic of powder core materials is unique compared to other core materials in that small air gaps (pores) evenly distributed throughout the material increase the amount of DC that can be passed through the winding before the core saturates. Figure 20.11 presents normal magnetization curves of MPP, high flux, and Kool Mμ powder cores with a permeability of 125 μ. Saturation flux density of the powder cores increases gradually with magnetizing force where it reaches to about 14,000 G at 1,000 Oe for high flux material, which is the highest level among these materials. The saturation level is about 10,000 and 8,000 G for Kool Mμ and MPP powder cores at the same magnetizing force, respectively.

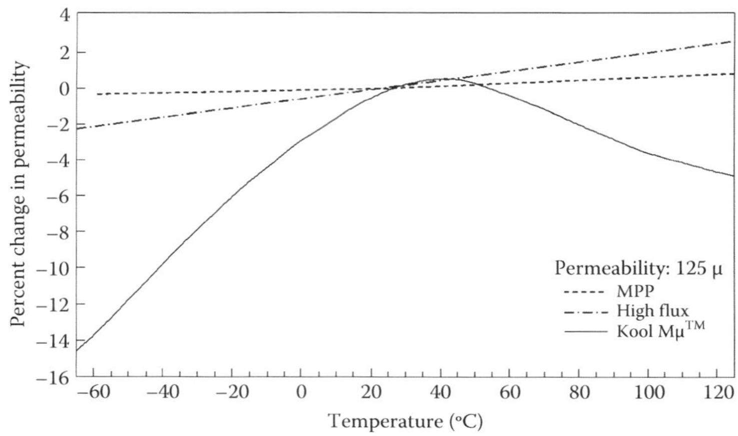

FIGURE 20.10
Change in permeability of available powder core materials with temperature. (From MAGNETICS®, Powder Cores Design Manual and Catalog, pp. 3-1, 3-3, 3-4, 3-14, 3-15, 3-16, 3-19, 3-20, 2002.)

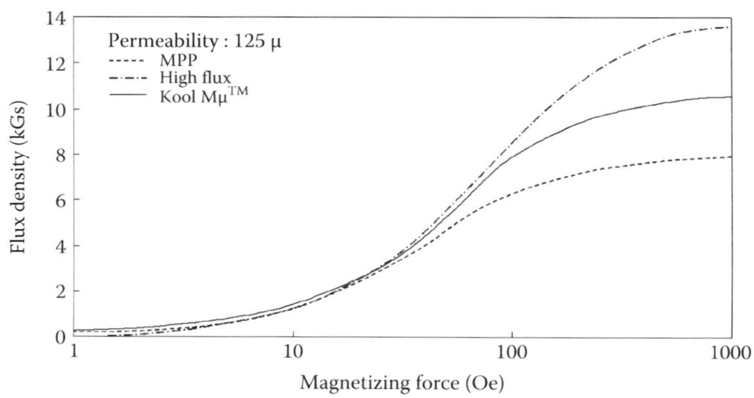

FIGURE 20.11
Normal magnetization curves for MPP, high flux, and Kool Mμ™ powder cores. (From MAGNETICS®, *Powder Cores Design Manual and Catalog*, pp. 3-1, 3-3, 3-4, 3-14, 3-15, 3-16, 3-19, 3-20, 2002.)

The gradual or "soft" saturation characteristic of powder core materials can also be observed through the change of their initial permeability under DC bias condition (Figure 20.12). It is clear from this figure that while the inductance of powder core materials changes gradually with DC magnetizing force, gapped ferrite core reveals a small change in the initial permeability up to a certain magnetizing force above which it sharply drops to considerably lower permeability levels. This behavior can be attributed to the presence of a single discrete gap in the ferrite material that maintains a constant permeability until the core abruptly saturates where inductance plunges [20]. In the case of powder core materials, however, evenly distributed fine gaps in the structure lead to the "soft" saturation characteristics where individual particles of the core do not saturate simultaneously.

Under AC excitation, powder core materials exhibit considerable variations in their inductance values. As it is depicted in Figure 20.13, MPP powder core material with 125 μ

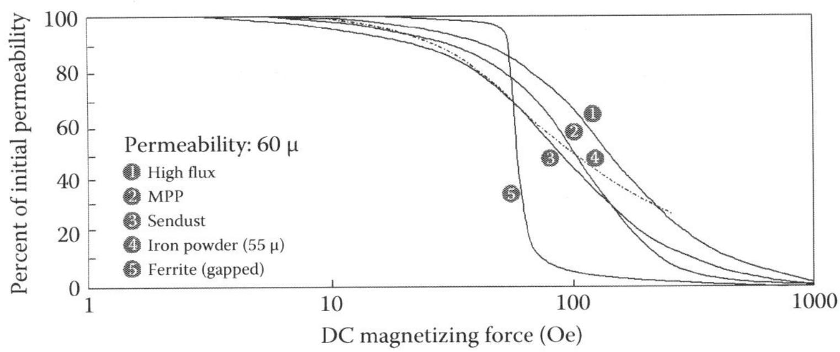

FIGURE 20.12
Permeability versus DC bias behavior of high flux, MPP, Sendust, and iron powder cores compared with that of gapped ferrite. (From Kool Mμ®, A magnetic material for power chokes, Technical Bulletin No. KMC-S1, Magnetics—Division of Spang & Company, Pittsburgh, PA, p. 2, 1999; Online, *MPP, High Flux, Sendust Basic Material Characteristics (Magnetic Powder Cores 06–07)*, http://www.cwsbytemark.com/CatalogSheets/MPP%20PDF%20files/6.pdf, 2006.)

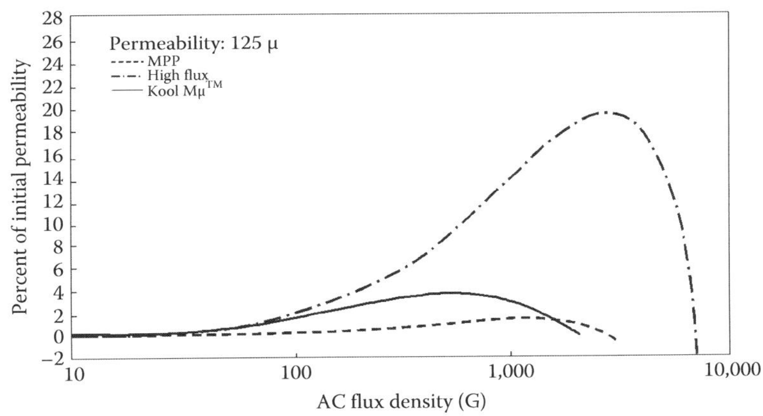

FIGURE 20.13
Permeability of MPP, high flux, and Kool Mμ™ powder cores as a function of AC flux density. (From Online, *Permeability versus AC Flux Density (Magnetic Powder Cores 16–17)*, http://www.cwsbytemark.com/CatalogSheets/MPP%20PDF%20files/16.pdf, 2006.)

permeability is very stable with an inductance change of less than 2% above 2000 G of AC flux density. For the same permeability, Kool Mμ cores reveal less stability with an inductance change of about 4% at around 500 G of AC flux density. High flux powder core exhibits the lowest AC excitation stability among other powder core materials where its inductance change reaches to a peak of about 20% at around 3000 G of flux density after which it sharply drops to lower levels. The change of inductance of the earlier-given powder core materials with identical permeability values is given in Figure 20.14 as a function of frequency. As it is clear from this figure, all of the powder core materials exhibit a gradual decrease in their inductance with frequency. While Kool Mμ reveals the highest stability with 20% inductance decrease up to 10 MHz, MPP and high flux core materials exhibit an inductance change of around 20% at 1 MHz range after which the decrease in inductance accelerates.

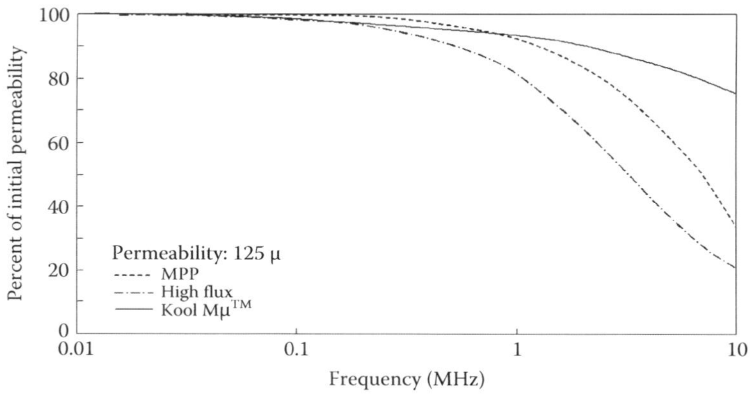

FIGURE 20.14
Change of permeability of MPP, high flux, and Kool Mμ™ powder cores with frequency. (From MAGNETICS®, *Powder Cores Design Manual and Catalog*, pp. 3-1, 3-3, 3-4, 3-14, 3-15, 3-16, 3-19, 3-20, 2002.)

20.4 Concluding Remarks

Basic material characteristics of available powder core materials, the performances of which were discussed in some detail in the preceding section, are summarized in Table 20.2 in comparison with those of the gapped amorphous and ferrite core materials. According to the presented data, MPP cores have the lowest overall core loss and best temperature stability. Specifically, MPP offers high resistivity, low hysteresis and eddy current losses, and very good inductance stability under DC bias and AC conditions. Although MPP cores are the costliest among all powder core materials, their highest quality in terms of core loss and stability makes them desirable for various applications such as high Q filters, loading coils, resonant circuits, RFI filters for frequencies below 300 kHz, transformers, chokes, differential mode filters, and DC-biased output filters.

TABLE 20.2

Basic Material Characteristics of Available Powder Core Materials along with Those of Gapped Amorphous and Ferrite Cores

Core Materials	Core Loss	Permeability versus DC Bias	Relative Cost	Frequency Range	Saturation Flux Density (G)	Temperature Stability
MPP	Lowest	Better	High	1 MHz	7,000	Best
High flux	Low	Best	Medium	1 MHz	15,000	Better
Sendust (Kool Mμ)®	Medium	Good	Low	2 MHZ	10,000	Good
Iron powder	Highest	Poor	Lowest	100 kHz	10,000	Poor
Amorphous (gapped)	Medium	Better	Highest	300 kHz	6,500	Poor
Ferrite (gapped)	Lowest	Poor	Low	1 MHz	4,500	Poor

Source: Online, *MPP, High Flux, Sendust Basic Material Characteristics (Magnetic Powder Cores 06–07)*, http://www.cwsbytemark.com/CatalogSheets/MPP%20PDF%20files/6.pdf, 2006.

High flux powder cores are superior to MPP cores in terms of their DC bias behavior and saturation flux density level. Because of these properties along with their higher energy storage capability, high flux cores are most commonly used in applications such as in-line noise filters where the inductor must support large AC voltages without saturation, switching regulator inductors to handle large amount of DC bias current and pulse transformers as well as flyback transformers as its residual flux density is near to 0 G. Sendust or Kool Mμ cores, on the other hand, exhibit intermediate core loss levels and temperature stability. Additionally, these types of cores also exhibit very low magnetostriction coefficient, and therefore, they are suitable for applications requiring low audible noise. It should also be mentioned that at high frequencies, Sendust cores reveal up to 80% reduction in core loss over iron powder cores lowering the temperature rise in the component. Despite their inferior operational parameters, iron powder cores constitute a design alternative to other powder core materials because of their significant cost-effectiveness. As a result of this, their highest core loss among all powder materials can be compensated by using larger sized cores. Consequently, in applications where space and higher temperature rise are insignificant compared to savings in costs, iron powder cores offer the best solution.

References

1. McLyman T, Colonel Wm (1997) *Magnetic Core Selection for Transformers and Inductors*, 2nd edn., CRC Press, New York.
2. Petkov R (1996) Optimum design of a high-power, high-frequency transformer. *IEEE Trans Power Electron*, 11(1), 33–42.
3. Yanagimoto K, Majima K, Sunada S, Aikawa Y (2004) Effect of Si and Al content on core loss in Fe-Si-Al powder cores. *IEEE Trans Magn*, 40(3), 1691–1694.
4. Online (2012) How transformers, chokes and inductors work, and properties of magnetics. http://www.coilws.com/index. php? main_page=page&id=104
5. Angeli M, Cardelli E, Torre Della E (2000) Modelling of magnetic cores for powder electronics applications. *Physica B*, 275, 154–158.
6. Snelling EC (1998) *Soft Ferrites—Properties and Applications*, 2nd edn., Butterworths, London, U.K.
7. Mangalaraja RV, Ananthakumar S, Manohar P, Gnanam FD (2003) Initial permeability studies of Ni-Zn ferrites prepared by flash combustion technique. *Mater Sci Eng A Struct*, 355, 320–324.
8. Fukui K, Watanabe I, Morita M (1972) Compressed iron powder core for electric motors. *IEEE Trans Magn*, 8(3), 682–684.
9. Online (2012) Terminology and definitions of terms 1. http://www.cwsbytemark.com/CatalogSheets/MPP%20PDF%20files/53.pdf
10. German RM (1989) *Powder Metallurgy Science*, Metal Powder Industries Federation, Princeton, NJ.
11. Kishimoto Y, Yamashita O, Makita K (2003) Magnetic properties of sintered Sendust alloys using powders granulated by spray drying method. *J Mater Sci*, 38, 3479–3484.
12. Online (2012) *MPP, Sendust, Kool Mu, High Flux and Iron Powder Core Properties and Selection Guide*. http://www.coilws.com/index. php? main_page=page&id=49
13. The Arnold Engineering Co. (2003) Soft Magnetics Application Guide, p. 30. 16 February 2003 Rev. A.
14. Kool Mμ® (1999) A magnetic material for power chokes. Technical Bulletin No. KMC-S1, Magnetics—Division of Spang & Company, Pittsburgh, PA, p. 2.
15. Online (2012) Specification & properties. http://www.cwsbytemark.com/mfg/sendust.php

16. Online (2012) *MPP, High Flux, Sendust Basic Material Characteristics (Magnetic Powder Cores 06–07)*. http://www.cwsbytemark.com/CatalogSheets/MPP%20PDF%20files/6.pdf
17. MAGNETICS® (2002) *Powder Cores Design Manual and Catalog*, pp. 3-1, 3-3, 3-4, 3-14, 3-15, 3-16, 3-19, 3-20.
18. Kool Mμ E Cores Technical Bulletin "Magnetics Kool Mμ E-Cores" (2005) Bulletin No. KMC-EI Rev 1, Magnetics–Division of Spang & Company, Pittsburg, PA.
19. Nicol DJ *An Objective Comparison of Powder Core Materials for Inductive Components with Selected Design Examples*. Home Application Notes, Micrometals, Inc., Anaheim, CA.
20. Teng N (2001) Fe-Si-Al powder E-cores compete with gapped ferrites. Magnetics—Division of Spang & Company, Pittsburgh, PA.
21. Online (2012) *Permeability versus AC Flux Density (Magnetic Powder Cores 16–17)*. http://www.cwsbytemark.com/CatalogSheets/MPP%20PDF%20files/16.pdf

21

Measurement and Prediction of Parameters of High-Frequency Transformers

H. Bülent Ertan, Erdal Bizkevelci, and Levent B. Yalçıner

CONTENTS

21.1 Introduction

In modern power electronics, the frequencies employed at the power stages of converters are much higher than those utilized some 15–20 years ago. Modern power switches capable of operating at higher frequencies facilitate this trend. The designers try to reduce the size of the power electronic equipment and also to reduce acoustic noise emitted from such devices, taking this opportunity to design for higher frequencies. Of course attention needs to be paid to the switching losses as the frequencies increase. To address this issue, zero voltage and/or zero current switching power stage topologies are employed. Switch mode power supply frequencies may go up to the MHz range now. However, industrial drives of moderate size employ modulation frequencies on the order of 15 kHz or more if necessary, and at higher power levels, frequencies around 5 kHz are possible.

Transformers are an essential part of quite a number of power devices. They prove useful to isolate the mains from the power stage, to raise the voltage level, or to match impedance just to count some of the reasons for using them. Transformers are expensive and large, and so higher frequencies will significantly reduce their size. However, other problems are introduced: The stray capacitance and leakage inductance of transformers become very important at the frequency levels mentioned earlier, as well as the skin effect and core losses. EMI from such devices is also an important issue. According to literature reviews, much research addressing these issues may be found. This contribution shall concentrate only on the modeling and parameter prediction of certain class of high-frequency power transformers as well as the measurement of their parameters.

An accurate transformer model is essential for design purposes. It then becomes possible to carry out simulations using various simulation tools in order to find out whether any undesirable effects are likely to be observed during operation. The modeling issue has its roots in power transformers. From the 1960s and 1970s, several books are available summarizing the state of the art on transformer design and modeling [1–3]. A more recent book by Flanagan [4] deals with all the practical issues of transformer design. However, interest on this issue remains alive as the frequencies go higher and computers are better used for analysis and design purposes [5–13]. Listing all the research shall not be attempted here. However, some papers addressing important concerns shall be mentioned. For example, Biernacki [5] in a recent paper presents a model for transformers operating in the 10 MHz range. Of course at such frequencies, reflection from the transformer becomes an important issue. The model uses a reflection coefficient technique with the scattering matrix. The scattering matrix relates the voltage wave incident on the ports to those reflected by the ports. This approach presented in this paper is useful as the reflection and transmission coefficients can be accurately and directly measured, as emphasized in the paper.

Reference [6] presents a study on modeling and parameter calculation of lower frequency (tens of kHz) transformers. This reference presents a model, which incorporates the magnetic circuit and electric circuit of a given transformer interacting with each other. All the parameters needed for modeling are taken from the data sheets of the transformer core and winding geometry. The hysteresis effect of the core is considered as well as the skin effect and proximity effect in the model. It must be noted that this chapter reports very good results with this model, although the B–H curve modeling accuracy is a problem and is likely to cause errors when simulating nonsinusoidal excitation of transformers.

As discussed earlier, the frequency range employed in the power stage of modern power electronics covers a range from a few kHz to MHz. Different models need to be used to predict the transformer performance in different frequency ranges [7,10,11]. This chapter will concentrate on the kHz range of operation of high-frequency power transformers.

21.2 Equivalent Circuit of a High-Voltage/High-Frequency Transformer

The general, lumped parameter, equivalent circuit of a kHz range high-frequency transformer is shown in Figure 21.1. In this circuit, L_c represents the magnetizing inductance, and R_c core losses, L_p, L_s leakage inductance, and R_p, R_s winding resistance of primary and secondary, respectively. In addition to these parameters, stray capacitance of the windings is taken into account by the capacitive elements in the equivalent circuit [1]. The capacitances represented with the lumped elements are the following:

- Between primary winding layers, C_p
- Between secondary winding layers, C_s
- Between primary and secondary winding, C_{ps}
- Primary winding to ground, C_{pg}
- Secondary winding to ground, C_{sg}
- Secondary terminals to ground, C_{tg}

In the analysis of the circuit, for most practical purposes, the magnetizing inductance L_c and core loss branch R_c can be neglected without a significant error. This issue is further discussed in Section 21.10.

To simplify the circuit, the equivalent circuit can be referred to the primary side. The referred parameters can be determined in terms of actual values by writing the node equations. The referred capacitances are denoted as C'_p, C'_s, and C'_{ps} in the equivalent circuit. The referred secondary winding resistance and leakage inductance can be added with the corresponding parameters for the primary winding and are denoted as R_{tot} and L_{tot}. If the primary number of turns is N_p, and the secondary number of turns is N_s, then the parameters in the primary referred equivalent circuit of Figure 21.2 can be shown to be

$$C'_p = C_p + C_{ps}\left(1 - \frac{N_s}{N_p}\right) \tag{21.1a}$$

$$C'_s = \left(\frac{N_s}{N_p}\right)^2 C_s + C_{ps}\left(\frac{N_s}{N_p}\right)\left(\frac{N_s}{N_p} - 1\right) \tag{21.1b}$$

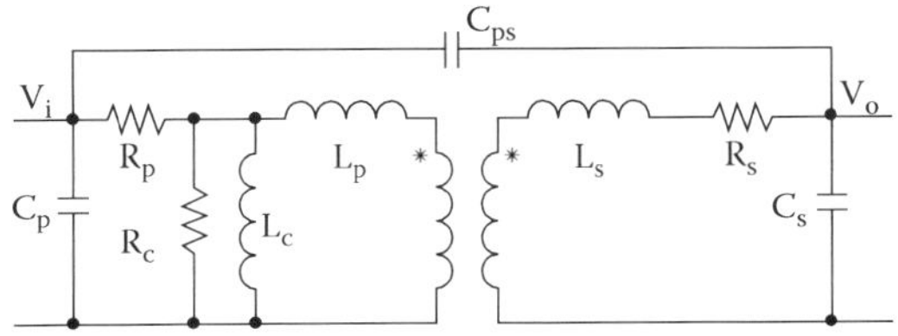

FIGURE 21.1
Equivalent circuit of a high-frequency transformer.

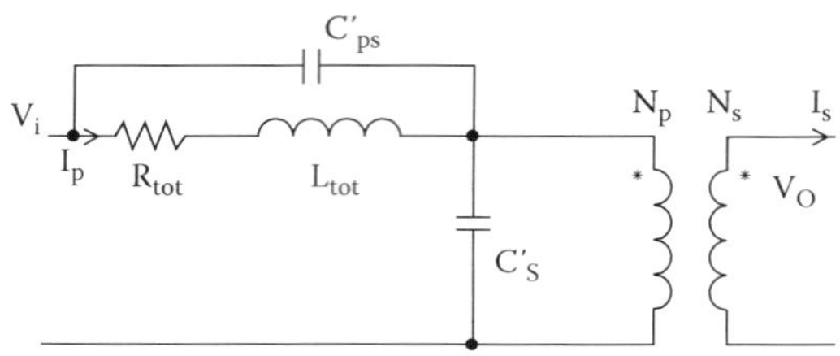

FIGURE 21.2
Primary referred and simplified equivalent circuit.

$$C'_{ps} = \left(\frac{N_s}{N_p}\right) C_{ps} \tag{21.1c}$$

$$R_{tot} = R_p + \left(\frac{N_p}{N_s}\right)^2 R_s, \quad L_{tot} = L_p + \left(\frac{N_p}{N_s}\right)^2 L_s \tag{21.1d}$$

Further simplification of the primary reflected circuit may be possible depending on the particular problem. For example, the turns ratio is high (200–600) in a high-voltage x-ray generator transformer. This fact can be used for the simplification of the expressions for the referred parameters. Furthermore, the referred secondary winding capacitance is usually higher than any other capacitance because of the large surface area and high turns ratio. Therefore, the primary winding capacitance can be neglected with respect to the secondary and the primary–secondary winding capacitance. The reflected and simplified equivalent circuit is a series RLC circuit. Although the reflected primary–secondary capacitance is shown in Figure 21.2, it usually has a negligible effect on the performance of the circuit [4,14].

From the equivalent circuit in Figure 21.2 (neglecting C'_{ps} and C'_s), the output voltage of the transformer can be written as

$$V_o = \frac{V_i - j2\pi \cdot f_s \cdot L_{tot} \cdot (N_s / N_p) I_s}{(1 - (2\pi f_s)^2 L_{tot} \cdot C_{tot})} \left(\frac{N_s}{N_p}\right) \tag{21.2a}$$

where
- V_o represents output AC voltage
- V_i represents input AC voltage
- f_s represents switching frequency
- C_{tot} represents C'_s
- I_s represents output current

In order to decrease the number of turns and turns ratio, which is vital for reducing the size, and thus the cost of the system, natural resonance gain of the series RLC circuit can be used. From Equation 21.2a, it is seen that the output voltage is dependent on resonance gain and load current. Thus, it is important to express the resonant gain in terms of full load current. There is a compromise between primary current I_p and resonant gain

(Equation 21.2b). If resonant gain is increased, input current also increases, which affects the cost of the switching circuitry:

$$I_p = \left(\frac{N_s}{N_p}\right) I_s + j2 \cdot \pi \cdot C_{tot} \cdot \left(\frac{N_p}{N_s}\right) V_{s\_ac} \tag{21.2b}$$

21.3 Determination of the Parameters

Once the equivalent circuit is established, the next issue is to obtain the parameters accurately. This can be done in two ways. If the transformer is available, the parameters can be measured. This issue is discussed in Section 21.7. If, however, a transformer is being assessed at the design stage, it is essential to have some means of prediction of its parameters.

Prediction of the parameters of high-frequency transformers from the material properties and physical dimensions has always been of interest. Much research work is reported in the literature on this issue [2,6,7,14–18]. Early work on transformer parameter prediction naturally relies on approximation of the magnetic field of the device to predict the inductances. Similarly for prediction of the stray capacitances, it is essential to make certain approximations in order to obtain the electrostatic field distribution. At present, most researchers have access to numerical field solution software, which can be used to obtain an accurate representation of the transformer magnetic and electric field. Therefore, it is possible to predict the equivalent circuit parameters with good accuracy. The degree of accuracy depends on how well the material properties are known. Sometimes the physical structure may be complicated, and a 3D solution may be essential. To reduce the computational effort, 2D numerical solution models may be used. This may affect the parameter prediction accuracy.

The parameter prediction issue is the subject of Sections 21.5 and 21.12. Both the analytical methods and the numerical techniques that may be used for parameter prediction are discussed. These approaches are applied to a prototype transformer, and their results are compared. Therefore, an assessment of the degree of accuracy that can be achieved can be made.

Analytical methods are naturally far less time consuming regarding computation time. Their weakness lies in the assumptions involved for approximating the actual field distribution. Numerical field solution results may be used to improve the approximations made, and the accuracy of the analytical methods can be improved for a given type of transformer structure. Analytical methods give great flexibility to the designer as they allow investigation of a large number of alternatives in a short time frame. Using this approach, the designer can assess the effect of each physical dimension will have on the value of the device parameters. So the investigation can easily concentrate on certain dimensions to reach a desired parameter range. Alternatively the design problem can be modeled as a mathematical design optimization problem [17].

In this chapter, the issues mentioned earlier (parameter measurement, analytical parameter prediction, numerical parameter prediction) shall be discussed and illustrated on a prototype high-frequency power transformer.

21.4 Test Transformer

In modern radiology units, the topology of typical power supply is shown in Figure 21.3. The transformer primary in this circuit is supplied by a high-frequency (about 20 kHz or higher) voltage. In this manner, the size of the transformer is reduced. On the DC bus, the voltage level is usually at about the rectified mains voltage level. The secondary voltage is required to reach a level between 40 and 150 kV after rectification. It is often found advantageous to use two transformers supplied from a common DC bus with their secondaries in series to obtain the desired HV level.

The design of the HV transformer constitutes one of the major issues in developing an x-ray power supply. This unit should be small in size, yet HV insulation should be achieved. The excitation frequency should not cause a resonance condition to arise in the transformer. Otherwise, the primary currents may reach unwanted levels. However, the resonance characteristic may be used to an advantage, as the gain of the device may reduce the turns ratio, helping the transformer size to be further reduced.

The transformer under consideration in this study is a 6 kW, 300 V–66 kV transformer designed to operate at about 20 kHz. Figure 21.4 illustrates the general structure of the transformer. The core is manufactured from 0.08 mm %55 NiFe sheet steel. The wire sizes

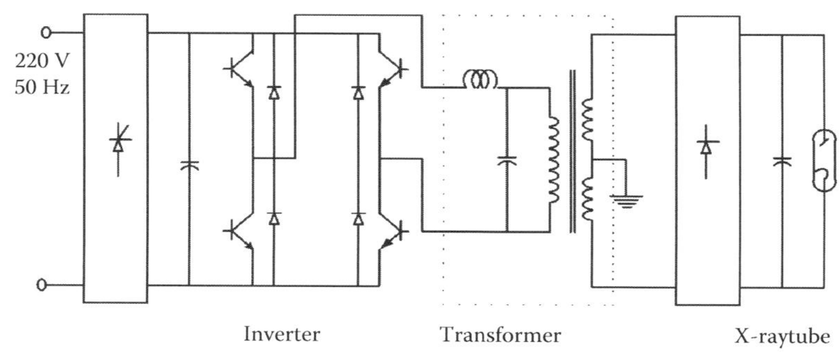

FIGURE 21.3
Typical x-ray power supply.

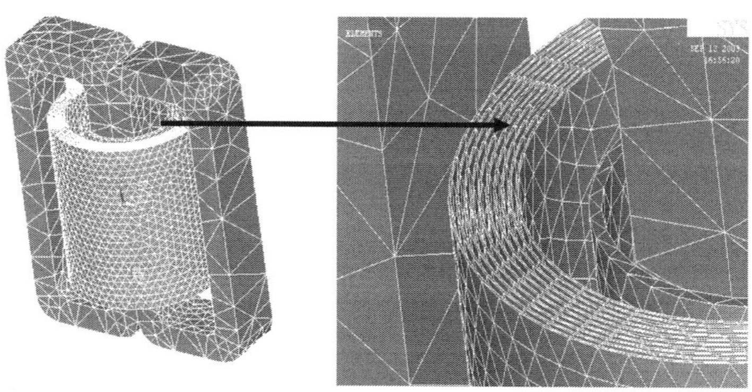

FIGURE 21.4
3D view of transformer finite element model.

are chosen to have a smaller diameter than twice the skin depth. The primary is single layer, while the secondary is formed of 10 layers. In between the primary and secondary winding, a grounded shield is placed. Two of these transformers are utilized in the application here with primaries connected in parallel and the secondary windings connected in series. Hence, 132 kV rms voltage can be obtained at the secondary side. The rest of the power circuit is as given in Figure 21.3.

The transformer described earlier shall be used to illustrate the methods used to measure the transformer parameters and to calculate the transformer performance.

First the calculation of parameters of high-frequency transformers shall be considered. In dealing with this problem, the historic development path will be followed, and the analytical approach used for parameter calculation will be discussed [1,2]. This approach is also useful for gaining an understanding of the problem undertaken.

Consider the equivalent circuit given in Figure 21.1. As will be illustrated in the following sections around the operating point, the shunt elements of the equivalent circuit are not very important as far as the prediction of performance of the device is considered. Therefore, the following section is devoted only to the calculation of leakage inductance and stray capacitances.

21.5 Equations for Parasitic Parameters

For being able to design a transformer with the desired parameters, it is essential to express the parameters of the equivalent circuit in terms of dimensions of the transformer.

21.5.1 Leakage Inductance

The leakage flux in the transformer stores magnetic energy along its path in air (the energy stored in the core is negligibly small due to the high permeability). If the stored energy is calculated from the transformer's geometry, the total energy due to leakage flux can be found [1].

For the transformer in this application, a shell-type core, with primary and secondary windings placed concentrically around the center legs, is chosen (Figure 21.4). In the energy calculations, assumptions have been made which are listed as follows:

- Primary ampere-turns is equal to secondary ampere-turns.
- The leakage flux lines are parallel to the axial symmetry line in the region where coils are placed.
- The density of leakage flux lines decrease linearly with the distance from the MMF source (Figure 21.5).
- The magnetic core has infinite permeability.

The magnetomotive force along a leakage path is integrated on a closed path (Figure 21.5) defined by the $d\vec{l}$ operator. It is assumed that the path length is not influenced by the finite wire thickness of both primary and secondary windings. Thus, a simple scalar equation may be written to define any H-field along a given radius (for the region where flux lines are parallel to the axial symmetry axis) where H_r is

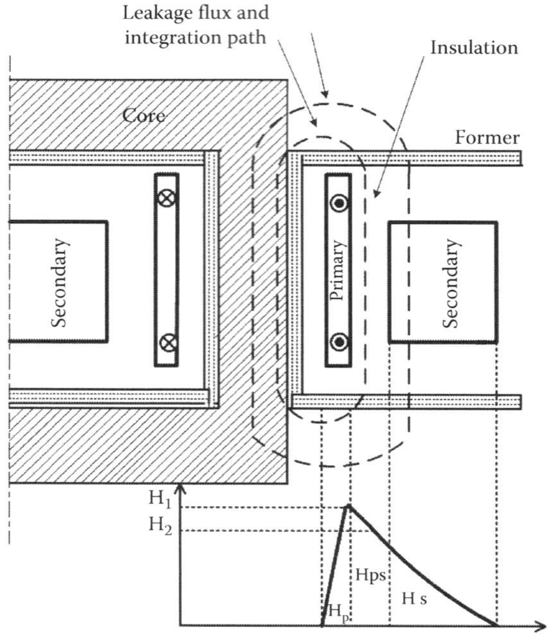

FIGURE 21.5
H-field variation versus transformer's geometry. (H_p indicates the field intensity variation in the primary winding region. H_{ps} field intensity variation between primary and secondary and H_s field intensity variation in the secondary winding region).

$$H_r = \frac{mmf_r}{height\_winding} = \frac{NI}{width\_winding \times height\_winding} radius \tag{21.3}$$

where mmf_r is the mmf acting at radius r and "radius" is the radius at r.

The H-field variation along the radius is plotted in Figure 21.5. The region contributing to leakage flux is divided into a number of shells (concentric cylinders of width dx along the radius). The differential energy of one shell can be calculated as

$$dW_r = \frac{\mu_0}{2} H_r^2 \times d(shell\_volume) \tag{21.4}$$

where μ_0 is the absolute permeability. Note that each shell has its own differential volume operator and H-field variation. The total energy is simply the sum of the contributions of all the shells. Therefore, the total leakage inductance is

$$L_{tot} = \frac{2}{I^2}(W_1 + W_2 + \cdots + W_n) \tag{21.5}$$

The calculation given earlier is naturally approximate as it relies on simplifying assumptions about the distribution of the field. A more accurate approximation can be made by a better approximation of the H-field variation. Alternatively, a correction factor may be introduced by using a comparison of calculated and experimental results of earlier designs. Details of the calculations are given in [14,17].

21.5.2 Winding Capacitance

The high-voltage insulation thickness between the secondary winding layers gives rise to parasitic capacitance. In the same manner, significant capacitance occurs between primary and secondary winding layers. There are also winding to ground capacitance, winding to winding capacitance, or terminal capacitance of the primary winding, but these are neglected here. This is because their value is quite small in comparison with other capacitive components and their contribution to the frequency response around the switching frequency of the inverter is negligible.

The capacitance between two layers of a coil may be calculated by modeling the layers as two concentric cylinders with radius "a" (inner) and "b" (outer) of height "h," having a relative dielectric constant ε_r. Then,

$$C_{static} = \frac{2\pi\varepsilon_0\varepsilon_r h\_winding}{\ln(b/a)} \tag{21.6}$$

In each layer, the voltage distribution is such that the voltage increases from one end of the layer to the other, as shown in Figure 21.6. It is important to determine an equivalent capacitance value that is referred to the full winding potential rather than the interwinding potential.

For a layer having V_1 volts at the beginning (x=0) and V_2 volts at the end (x=width of winding) of a layer and the full winding potential is "V" volts, the referred capacitance to the winding-end potential can be calculated as [17]

$$C = C_{static}\left(\frac{V_1^2 + V_1V_2 + V_2^2}{3V^2}\right) \tag{21.7}$$

The previous equation is used to refer all layer-to-layer capacitance to full winding voltage.

The secondary coil height is lower here than the primary coil in the prototype due to a higher voltage level of this coil and the necessity to provide enough insulation. This

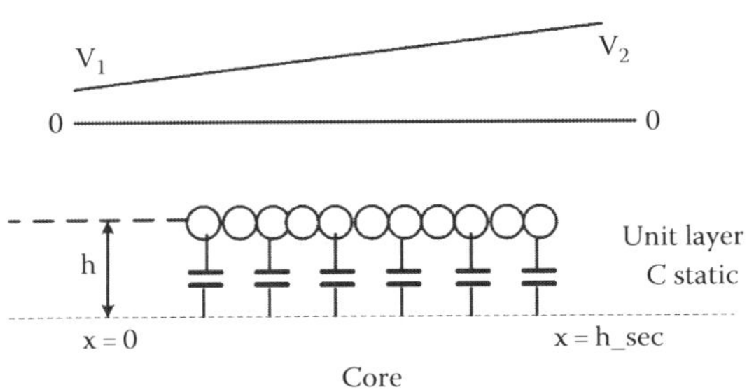

FIGURE 21.6
Voltage distribution across a winding layer.

height difference causes fringing of the electric field between the secondary coil layers and the primary winding which should be taken into account. To account for the fringing effect, the Carter coefficient [14,19] is used. The secondary winding layer capacitance and primary–secondary capacitance are calculated separately.

21.5.3 Secondary Winding Capacitance

Parasitic capacitance occurs between layers of the secondary coil (Figure 21.7). The capacitance of each layer may be treated as if it is connected in series with other layer capacitances. The fringing effect at the end of the layers is neglected here. Note that each layer is at a different potential difference and the capacitance of each layer must be referred to the winding full voltage.

Thus, the secondary winding capacitance is the sum of capacitance of successive layers, referred to the full winding voltage, which can be calculated as

$$Cs = \sum_{k=0}^{n-1}\left(\frac{1}{C_{k\_static}((3k^2+3k+1)/n^2)}\right)^{-1} \tag{21.8}$$

where

n is the number of secondary layers

C_{k_static} is the static capacitance between k and k+1 layers

21.5.4 Primary–Secondary Winding Capacitance

The capacitance between primary winding and secondary winding layers is parallel to each other (Figure 21.8). The fringing effect in the calculation of this capacitance is taken into account using the Carter coefficient. (The situation is viewed as calculation of the capacitance between two parallel plates, one shorter than the other.) The calculated capacitance should be again referred to the full secondary voltage. The capacitance between the primary- and secondary winding layers can be calculated as

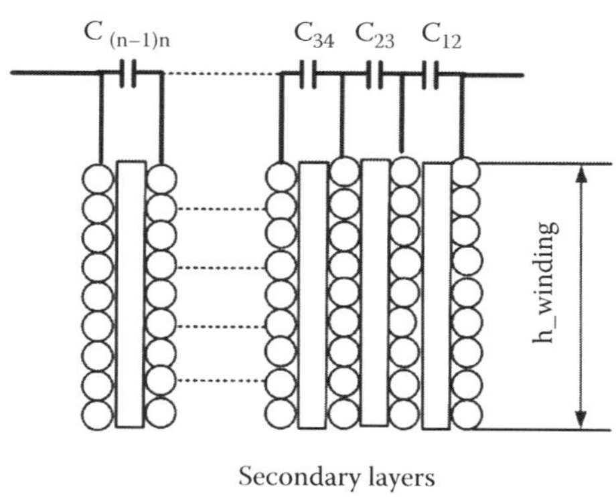

FIGURE 21.7
Secondary coil layer capacitance.

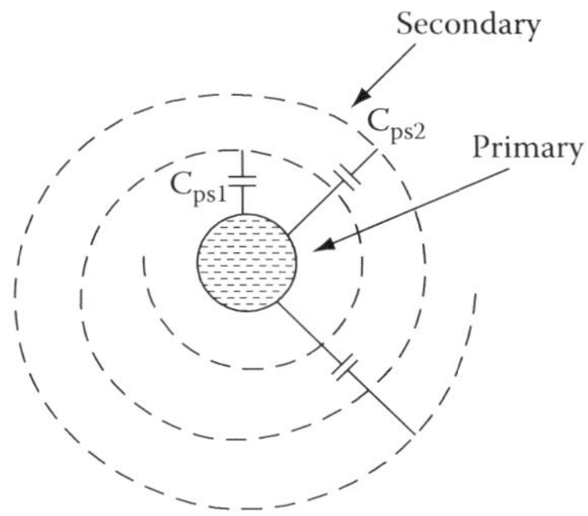

FIGURE 21.8
Primary–secondary winding capacitance.

$$C_{ps} = \sum_{k=0}^{n-1} C_{k\_static} \times \frac{3k^2 + 3k + 1}{3n^2} \times F_{carter}(A,B) \quad (21.9)$$

where
n is the number of secondary layers
C_{k_static} is the static capacitance between primary and k^{th} secondary layer
$F_{carter}(A,B)$ is the Carter coefficient for the kth layer

F_{carter} (A,B) is a function which takes into account the fringing field due to the size difference between primary and secondary. A is the ratio of the difference between primary and secondary winding heights to the distance between the primary and secondary coils. B is the ratio of the difference between primary and secondary winding heights to primary winding height.

21.6 Discussion

There are several assumptions in the determination of various capacitance components from the winding geometry. In the manufacturing process of high-voltage transformers, there may be some unpredictable dimensional errors while winding the coil. These errors affect the accuracy of the capacitance calculations. On the other hand, the approach used in the determination of the parasitic capacitance has certain assumptions, which also affect the accuracy. First of all, the layers are considered as conducting sheets having a linear voltage distribution. Furthermore, the wire to wire or wire to upper cross wire capacitances are neglected. Finally, the Carter coefficient that is used for accounting the fringing field effects is a rough estimation of the real situation. For this reason, some means of correction of the calculations may be needed. One way of decreasing the calculation error may be to use a correction constant, which may be obtained by comparing the calculated and measured results from previously manufactured transformers.

In the previous sections, the calculation of the parameters using analytical equations is discussed. As pointed out earlier, the accuracy of such approach is limited by the validity of the assumptions for the particular problem. An alternative approach to the prediction of the parameters is the use of numerical field solution techniques for this purpose [7,8]. This problem will be addressed in Section 21.12.

21.7 Measurement of Transformer Parameters

Before going any further with other approaches to the calculation of transformer parameters, it is worthwhile to check what accuracy may be expected from the analytical calculation of high-frequency transformer parameters. This necessitates experimental determination of the equivalent circuit parameters of some available devices. However, this is not an easy task.

In the literature, various studies are reported trying to find a solution to this issue [10,12,13,15,16,18,20]. For example, in Ref. [18], several measurement methods are reported; these are applied on a 500 W, 20 kHz transformer, with a ferrite core.

One of the methods discussed in Ref. [18] is based on writing down the voltage and current equations for the open circuit and short circuit conditions both on the primary and secondary of the transformer. In this manner, a total of eight equations are written. Next, measurements are taken at several frequencies under the conditions defined earlier. The resulting equations are simultaneously solved to obtain the parameters. The usual difficulties exist with this approach; for example, it is not possible to perform the tests at rated voltage and current conditions.

The second approach in this reference is based on the application of a square wave pulse to the device and recording the performance. The third approach is via measuring the resonant frequency of the transformer coils with external inductors and capacitors. The authors also summarize the advantages and disadvantages of each method.

In the third method, the leakage inductances of the core are assumed to be negligible. The other methods also have accuracy problems. The authors' findings indicate that some of these measurements lead to considerably different values.

References [10,12,13,15,16,20] present other methods to the parameter measurement problem. However, the approach adopted in these references shall not be discussed here.

The authors of this chapter have previously reported an alternative approach for measuring the stray capacitance and inductance values. In this method [8,17], the equivalent circuit that was used is shown in Figure 21.9. The winding resistances are measured. In that case r_{eq} in Figure 21.9 can be calculated by referring the secondary winding resistance to the primary side. It is assumed that if the wire cross sections are chosen to avoid skin effect, the measured dc resistance values are valid at the operating frequency. In that case an experiment can be performed, by impressing voltages on to the primary, at the desired frequency. The gain and phase of the voltages on the secondary can be measured. Sweeping the frequency of the applied voltage and using the dc value of the resistance, L and C can

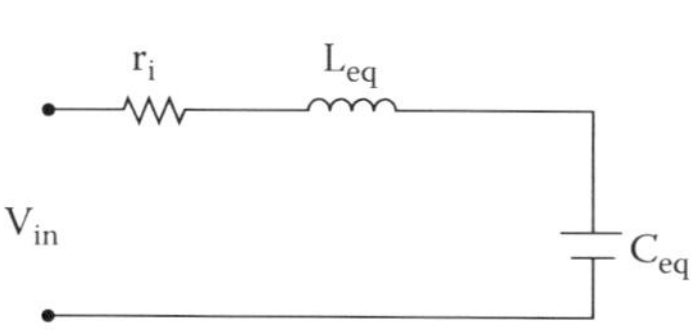

FIGURE 21.9
Simplified equivalent circuit. All parameters are referred to primary.

be found. However, this method is found to be very sensitive to the variations in resistance value and, therefore, found to be not satisfactory.

In the following section of this chapter, the focus is on introducing a different approach to the parameter measurement problem and the verification of its accuracy.

21.8 Method

The approach adopted in this study, for the measurement of the parameters of high-frequency transformers, is based on short circuit and open circuit tests commonly used for transformer parameter identification. In this case, however, the measurements are not done at a single frequency but over 0.5–20 kHz range.

Consider the equivalent circuit of the high-frequency transformer used for tests here, shown in Figure 21.10.

Note that the shield and one end of the secondary are grounded in the normal operation of the transformer. The primary referred equivalent circuit of this transformer is shown in Figure 21.11. The equivalent circuit parameters are as follows:

- C_pw: Lumped, primary interwinding capacitance
- C_pn: Lumped, primary winding to ground capacitance
- C_sw: Lumped, secondary interwinding capacitance

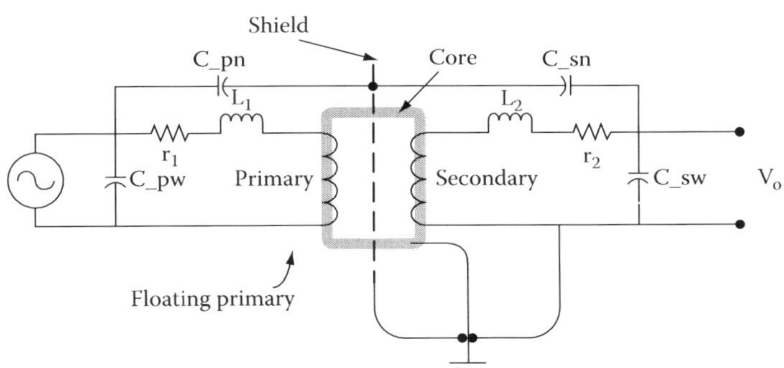

FIGURE 21.10
Equivalent circuit of a transformer.

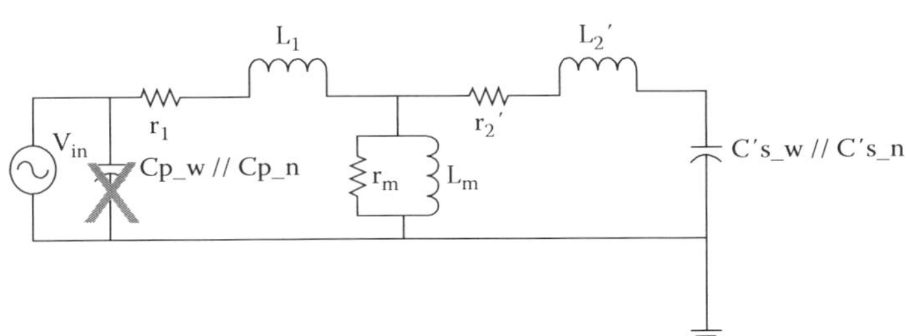

FIGURE 21.11
Primary referred equivalent circuit.

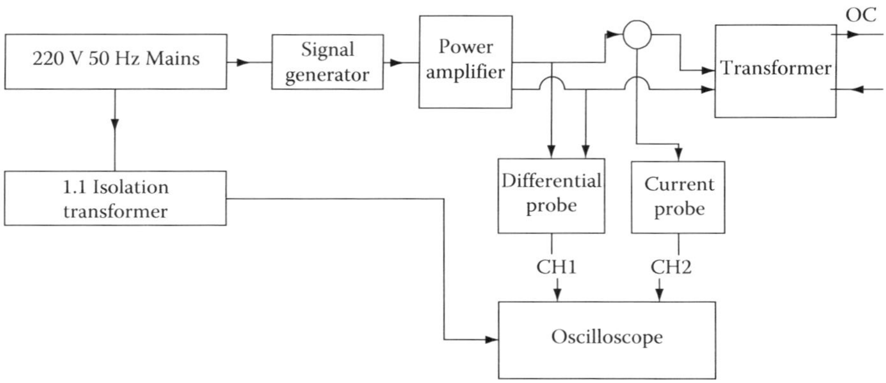

FIGURE 21.12
Open circuit test setup.

- C_sn: Lumped, secondary winding to ground capacitance
- r_1: Primary winding resistance
- r_2: Referred secondary winding resistance
- L_1: Primary leakage inductance
- L_2: Referred secondary leakage inductance
- r_m: Core loss resistance
- L_m: Magnetizing inductance

The "'" sign in Figure 21.11 indicates that the particular parameter is referred to the primary side. The experimental setup used in the tests basically consists of signal generator, a 3 kW power amplifier, a DSO capable of mathematical operations, as shown in Figure 21.12. It must be noted that in order to reduce any unwanted effects during such a test, all leads must be kept very short. Furthermore, if necessary, probe parameters must be taken into account in subsequent calculations.

21.9 Short Circuit Test

With the considerations in the previous section, when the secondary of the transformer is short-circuited, the equivalent circuit becomes that shown in Figure 21.13.

During this test, the sinusoidal supply frequency is stepped with 1 kHz steps. The voltage applied, the current, and its phase angle are measured. The power input to the transformer is computed from the recorded signals. The magnitude of the applied voltage is adjusted to keep the current high but within the amplifier limits, at the particular frequency.

21.9.1 Series Resistance

With the recorded data, the equivalent series resistance ($r_1 + r_2 = R_s$) can be calculated either from the computed power or from phase angle information. Both methods are

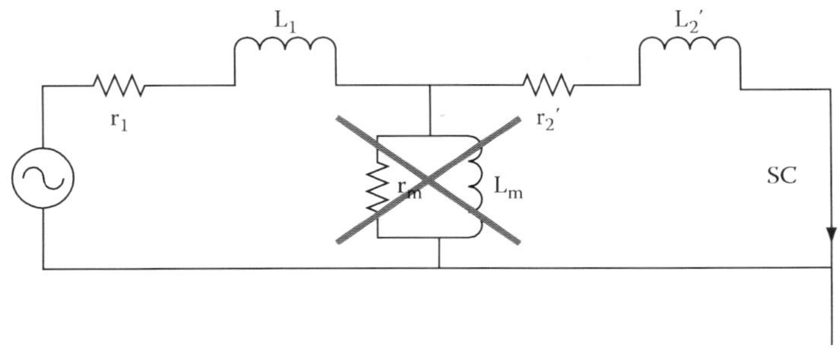

FIGURE 21.13
Short circuit test equivalent circuit.

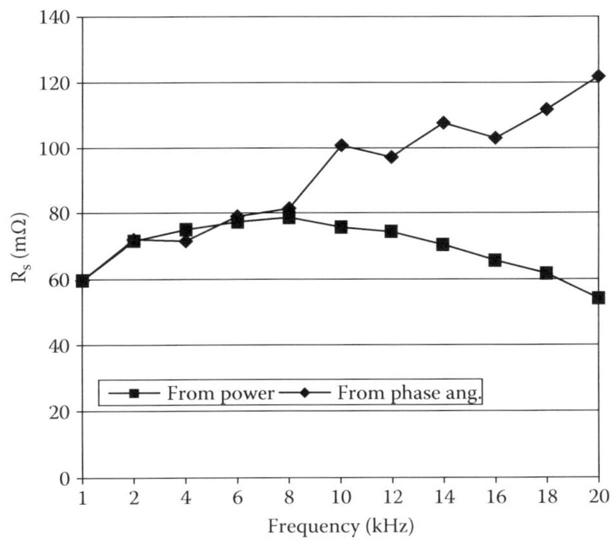

FIGURE 21.14
Measured resistance.

expected to lead to the same results. However, when plotted, the curves shown in Figure 21.14 are obtained.

The dc winding resistances of both coils are also measured, and the referred value of the total resistance is found to be 60.5 mΩ. Figure 21.14 shows that at low frequencies, both methods yield the same result, which matches the measured dc resistance value. As the frequency is increased, both methods return the same value of R_s until 8 kHz. Then, the resistance value found from power measurement considerably deviates. This deviation is certainly because of the smaller sample numbers at increasing frequencies, during measurements.

The calculated resistance value converges to about 60 mΩ at the operating frequency range 18–20 kHz when calculated from phase angle data, indicating that the wire cross sections are correctly chosen and the skin effect is not really important. The increase in the value of the apparent resistance in midfrequencies may be attributed to the distributed stray capacitances.

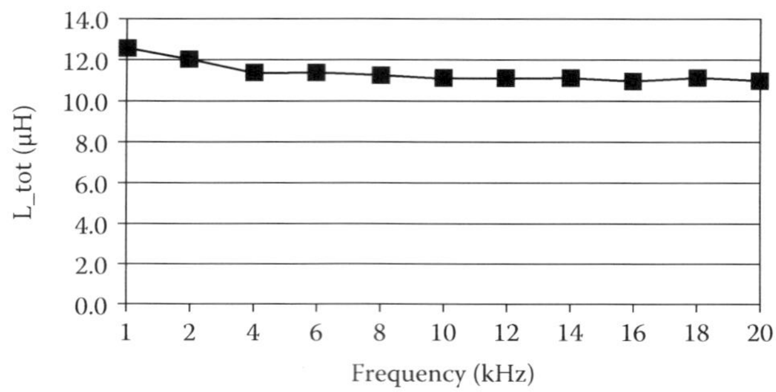

FIGURE 21.15
Total leakage inductance from power measurements.

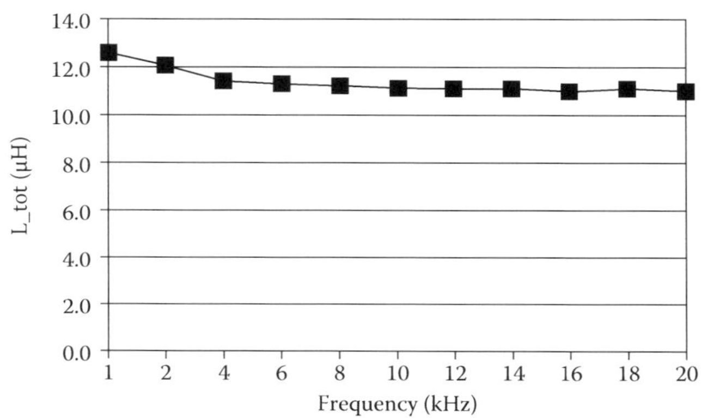

FIGURE 21.16
Total leakage inductance from phase angle measurements.

21.9.2 Series Inductance

From the measured variables, it is easy to determine the apparent inductance in the equivalent circuit. Figures 21.15 and 21.16 display the variation of the series inductance value with frequency both from power and from phase angle measurements. Both measurements are seen to show a similar variation. It is, therefore, concluded that the total primary referred series inductance value can be taken as 11 μH.

21.10 Open Circuit Tests

The equivalent circuit for the open circuit test is shown in Figure 21.11. As discussed earlier, the shunt capacitance on the input side is very small and can be neglected without being an error source. On the other hand, the shunt capacitor referred from the secondary side is not negligible and makes identification of the parameters at a single frequency extremely difficult.

To overcome this problem, tests are performed both at low frequencies (50–500 Hz range) and at the high-frequency range (10–20 kHz). These measurement ranges are decided on the

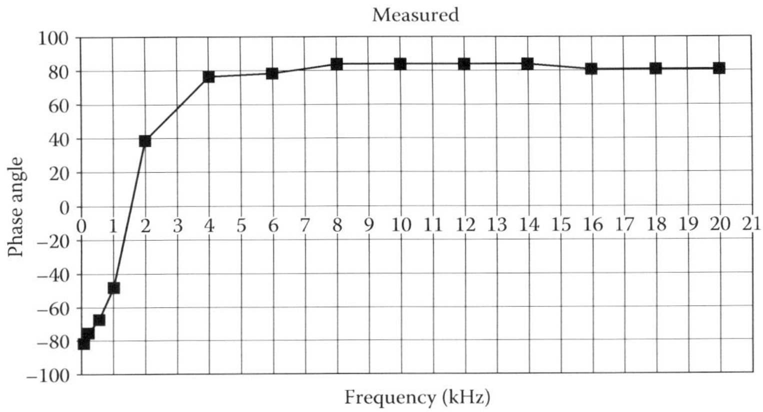

FIGURE 21.17
Variation of the current phase angle during the open circuit test.

basis of measurement of input current phase angle against frequency given in Figure 21.17. This figure shows that at the low-frequency end, the secondary side capacitance has very little effect and the current is lagging the applied voltage by about 90°. Above 4 kHz though, it can be observed that the shunt branch loses its effect and the current has a leading phase angle. Therefore, at the low-frequency end, the secondary shunt capacitance may be treated as an open circuit. For each test frequency, V_{in}, i_{in}, and input power are measured.

In the previous section, it was shown that the value of r_1 can be taken as its dc value. The value of primary leakage inductance on the other hand was not obvious from the measurements. Since this inductance is very small ($L_1 + L_2 = 11\,\mu H$). Its effect on the value of L_m and r_m is small too. To minimize possible errors, however, either a value predicted from analytical expressions or field solutions could be used. Here, a value found from analytical calculation is used in the computations.

21.10.1 Low-Frequency Measurements

Figures 21.18 and 21.19 show the calculated values of r_m and L_m from recorded measurements for the low-frequency range. During measurements, the core flux density is kept

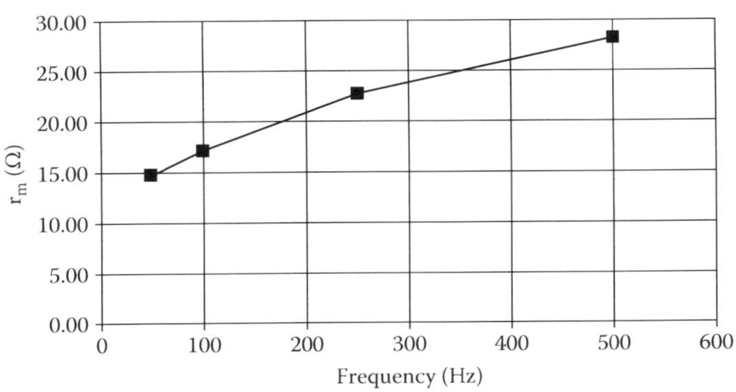

FIGURE 21.18
Measured value of r_m.

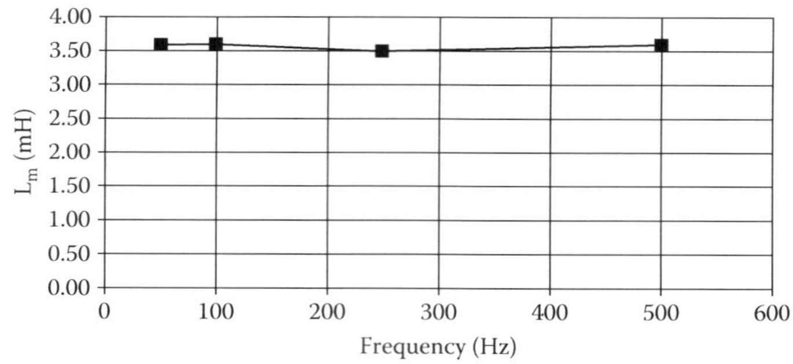

FIGURE 21.19
Measured value of magnetizing inductance.

around 0.1 T. The core loss measured during this test has no meaning as both the flux density and the frequency of the test are well below the actual operational values.

It can be observed from Figure 21.19 that the value of L_m stays pretty well constant at about 3.5 mH. It must be noted that although r_m could not be measured, the shunt branch has very little effect on the performance at the operating frequency (≈20 kHz) of the transformer.

21.10.2 High-Frequency Measurements

At the 10 kHz–20 kHz range, the equivalent circuit of the transformer is approximated, as shown in Figure 21.11. V_{in}, i_{in}, and input power are measured. The values of the series elements are known from the short circuit test results of Section 21.9. Therefore, the secondary side capacitance can be easily calculated. This value is found to increase from 3.49 μF at 10 kHz to 3.68 μF at 20 kHz. Just to make sure that eliminating the shunt branch does not affect the calculations, secondary capacitance is also calculated by substituting the shunt element values found from the low-frequency tests. The findings indicate that the effect of the shunt branch is on the order of 0.1% and is negligible.

21.11 Equivalent Circuit Verification

From the measurements given earlier, the equivalent circuit of the test transformer (referred to the primary side) is shown in Figure 21.20. Note that the shunt branch is not shown in this figure as it is shown to be ineffective at high frequencies.

To verify that this equivalent circuit can be used for the prediction of the transformer performance, gain versus frequency measurements are also made up to 40 kHz and are plotted in Figure 21.21. The same curve is predicted with the measured parameters using the equivalent circuit. The result found is displayed in Figure 21.21 along with the measurements.

It must be noted that the probe capacitances may be important and they must be taken into account in the predictions. The figure indicates that the predicted resonance frequency is within 4% of the actual value. This certainly is acceptably accurate.

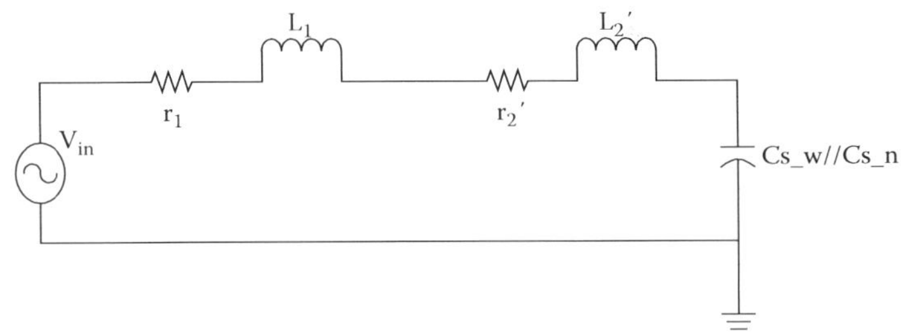

FIGURE 21.20
Equivalent circuit of the test transformer.

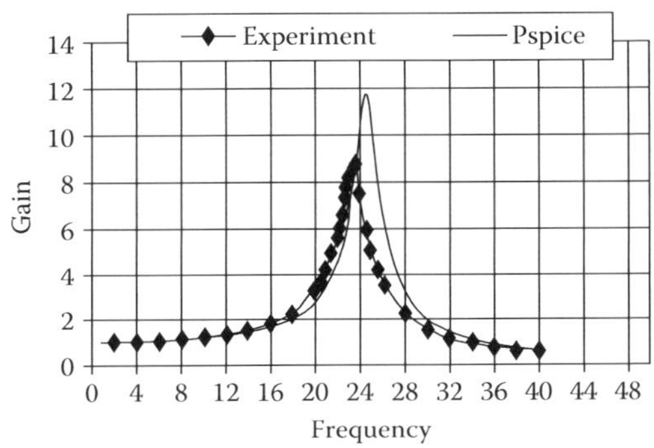

FIGURE 21.21
Gain versus frequency curve of the test transformer.

21.11.1 Comparison with Analytical Calculation Results

The parameters found from measurements are compared with the parameters of the test transformer, analytically calculated using the equations obtained in Section 21.5. The findings may be seen in Table 21.1. It can be observed from the table that the predicted value of total leakage inductance L_s is about 15% less than the measured value and the total secondary winding capacitance is less than that measured by about 11%. This would certainly introduce an additional error in predicting both the resonance frequency and performance of the transformer. As mentioned earlier, an improvement in the predictions from the analytical equations may be possible by developing the analytical equations with better assumptions. However, the accuracy that can be achieved in predicting the transformer parameters is limited to the accuracy of the field solution. The next section is devoted to developing an approach for this purpose.

TABLE 21.1
Results of Capacitance Calculations

Analysis Method	L_s (μH)	L_s Meas (μH)	C_{sw} (μF)	C_{sn} (μF)	C_{total} (μF)	C_{total} meas. (μF)
Field solution	10.08	11	1.04	2.4	3.44	3.68
Analytical calculation	9.37		0.91	2.41	3.32	

C_{total} is the parallel equivalent of the secondary capacitances.

21.12 Finite Element Computations

21.12.1 Determination of the Stray Capacitance

For the purpose of determining the stray capacitances of the transformer, electrostatic solutions are needed. Since a grounded shield exists between the primary and secondary, these two coils may be separately studied. As indicated later, this provides speed in obtaining a solution. Furthermore, the shield is much closer to the coils (compared with the core), and therefore, in the solutions, the core is not included in the model.

21.12.2 Primary Winding Capacitance

The primary winding is a single layer of 19 turns. Therefore, a 3D model is needed to obtain a reasonably accurate answer. Figure 21.22 displays the FE model used for the solution. Each turn is divided into 40 smaller pieces and assigned with a constant voltage. The bottom of the coil is at "0" potential and the other end of the coil at 300 V. The coil model is placed in a cube of 200 mm side length. Considering that the radius of the coil is about 20 mm and the coil height is 70 mm, the boundaries are placed far enough apart. The outermost elements are defined as infinite elements and external nodes are assigned an infinite flag. The electrostatic problem is solved using ANSYS. Using the energy method, the primary winding to shield capacitance C_{pw} is found to be 33 pF.

FIGURE 21.22
Primary coil model.

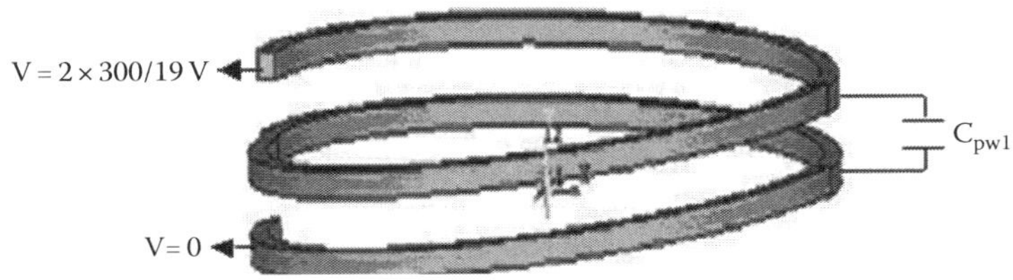

FIGURE 21.23
The model for computation of primary coil interwinding capacitance.

To determine the interwinding capacitance, two turns of the primary winding are considered (Figure 21.23). One end of this coil is assigned "0" potential. Each turn is divided into 40 smaller sections, and each section is assigned a gradually increasing voltage of 300/(19 × 40) V. From the field solution, the energy stored in the field and hence the interwinding capacitance for one turn C_{pw1} is found. Then, the total interwinding capacitance is

$$C_{pw} = \frac{C_{pw1}}{18} \tag{21.10}$$

C_{pw} was found to be 0.4 pF for the test transformer. When compared with the referred values of the secondary capacitance, these values can be observed to be very small. Therefore, the equivalent circuit of the transformer can be simplified as proposed in the previous sections (Figure 21.11).

21.12.3 Secondary Winding Capacitance

The secondary coil of the transformer has 2925 turns and is composed of 10 layers and has axial symmetry. The inner radius of the secondary coil is 27.5 mm, and the outer radius is 35 mm.

Again a 3D model of the problem is formed, and the electrostatic problem is solved using ANSYS. In this case, to shorten the solution time, the layers are modeled as conductive sheets with linearly increasing potential applied along the length of the coil. The sheets are connected in series as in the actual transformer. The shield is also included in the model. Appropriate permittivity values are assigned to the insulating materials. The coil is assumed to be within a sphere of radius 170 mm, and "infinite" boundary elements are chosen. The resulting model has 6569 elements, and the solution takes a few minutes on a modern PC. The computed capacitances are tabulated in Table 21.1. Figure 21.24 displays the electrostatic field distribution determined from the field solution.

When compared with the analytical calculations, it can be observed that the coil to ground capacitances found from field solution match each other well. On the other hand, the interwinding capacitance from the analytical calculation is about 10% less than the numerically computed value.

All of the results have been summarized in Table 21.1. It can be observed that from the analytical solution, the total referred secondary capacitance value is about 10% less than the measured value, while the value found from the field solution is 6.5% less than the measured value of this capacitance.

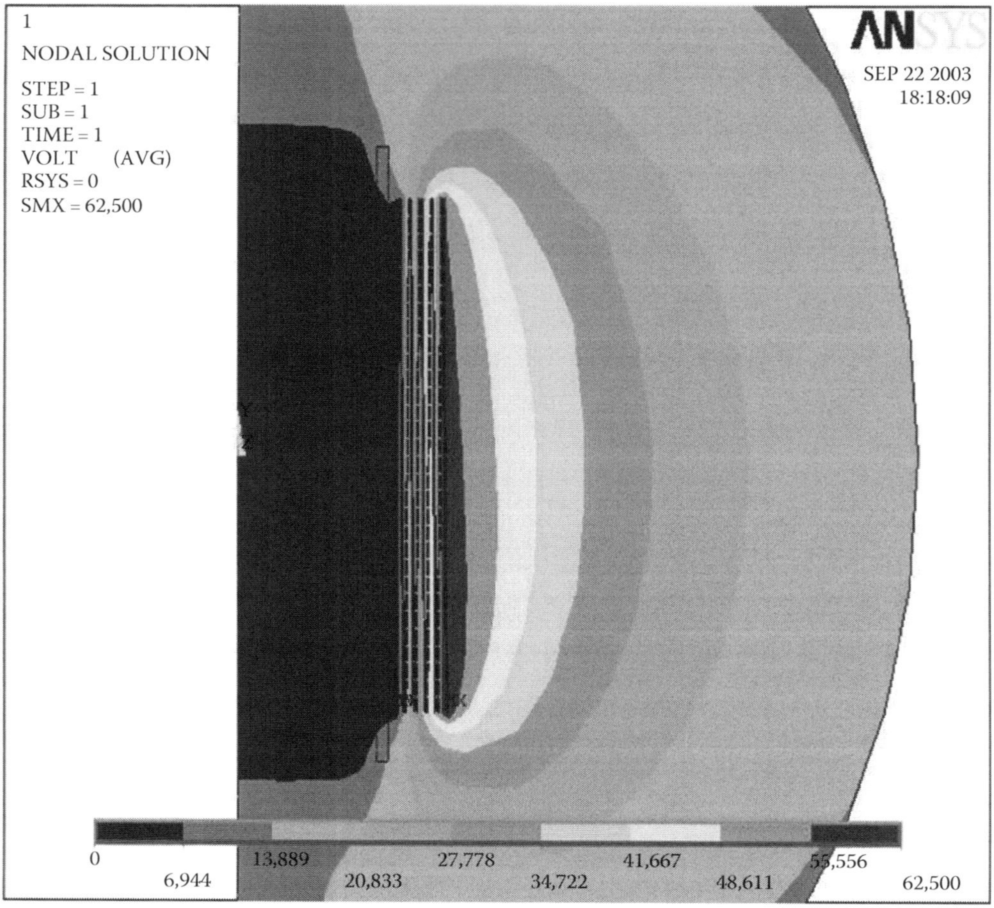

FIGURE 21.24
Secondary winding electrostatic field distribution.

21.12.4 Computation of Leakage Inductance

In this case, the transformer is modeled along with its core in 3D (Figure 21.4). The primary is represented as a single copper sheet, while 10 concentric copper sheets represent the secondary. To determine the inductances, a static solution is thought to be appropriate. Equal and opposite mmf's are applied to the two coils. The transformer is placed in a cube, and the boundaries are assigned Dirichlet conditions. The magnetization characteristic as given by the manufacturer is used.

The model thus created had 189,750 elements, and the solution time was several hours. Once the solution is obtained, the primary referred leakage inductance is found by calculating the total energy stored in the model (W_{tot}).

$$L_{tot} = 2\frac{W_{tot}}{I_{irated}^2} \tag{21.11}$$

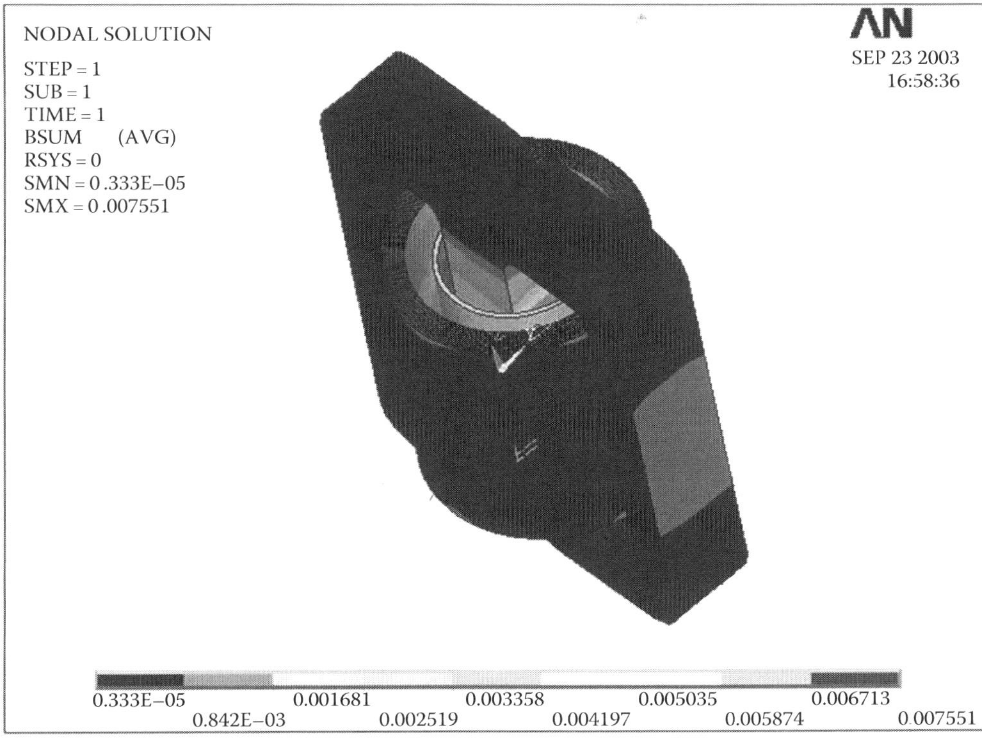

FIGURE 21.25
Flux density distribution from field solution to determine leakage flux.

Figure 21.25 displays the flux density distribution obtained from the field solution. The value of total referred leakage inductance found from the field solution is given in Table 21.1 along with the result of the analytically found value.

It can be observed that the value of inductance found from the analytical calculation is about 15% less than the measured value, while the computed value from field solution is smaller by about 8%.

21.13 Conclusions

This study was initiated to develop a reliable method for predicting the performance of a high-frequency transformer at the design stage. It is shown that if the parameters of the simple usual equivalent circuit can be determined accurately, the performance of the transformer can be well predicted. To prove this point, an easy to apply method for parameter identification has been presented.

When the parameters of the equivalent circuit are predicted from analytical expressions, differences up to 15% are observed. The accuracy of the calculations can be improved if the assumptions used in their derivation can be made with a better understanding of the actual field distribution. For this purpose, the parameters are also predicted using numerical field solution techniques. It is found that the prediction accuracy of the parameters is improved. Therefore, it may be argued that the field solution results can be used to improve the analytical expressions by making better assumptions in deriving the analytical expressions. However, it is clear that the prediction accuracy is limited by the accuracy with which the details of the geometry of the transformer are known.

The ultimate purpose is to design better and smaller transformers. Analytical expressions leading to accurate results can be employed to mathematically search for an optimum design to satisfy the specifications, as suggested in Ref. [17].

References

1. MIT Members (1963) *Magnetic Circuits and Transformer*, 2nd edn. MIT Press, Cambridge, MA.
2. McLyman WT (1978) *Transformer and Inductor Design Handbook*. Dekker, New York.
3. Grosner A (1966) *Transformers for Electronic Circuits*. McGraw-Hill, New York.
4. Flanagan WM (1992) *Handbook of Transformer Design and Applications*. McGraw-Hill, New York.
5. Biernacki J, Czarkowski D (2001) High frequency transformer modeling. *The 2001 IEEE International Symposium on Circuits and Systems, ISCAS 2001*, Sydney, Australia, May 6–9, Vol. 2, pp. 676–679.
6. Dallago E, Sassone G, Venchi G (July 1997) High-frequency power transformer model for circuit simulation. *IEEE Transactions on Power Electronics*, 12(4), pp. 664–670.
7. Fernandez C, Prieto R, Garcia O et al. (2002) Modeling core-less high frequency transformers using finite element analysis. *IEEE PESC'02*, Cairns, Australia, Vol. 3, pp. 1260–1265.
8. Ertan HB, Leblebicioğlu K, Yalçiner B (2004) Measurement and calculation of leakage inductance and stray capacitance of high frequency transformers. *Electromotion*, 10(3), 120–125.
9. Dallago E, Giuseppe V (1999) Analytical and experimental approach to high-frequency transformer simulation. *IEEE Transactions on Power Electronics*, 14(3), 415–421.
10. Zhu JG, Hui SYR, Ramsden VS (1996) A generalized dynamic circuit model of magnetic cores for low and high-frequency applications, part I: Theoretical calculation of the equivalent core loss resistance. *IEEE Transactions on Power Electronics*, 11(2), 246–250.
11. Zhu JG, Hui SYR, Ramsden VS (1996) A generalized dynamic circuit model of magnetic cores for low and high-frequency applications, part II: Circuit model formulation and implementation. *IEEE Transactions on Power Electronics*, 11(2), 246–250.
12. Zhu JG, Ramsden VS, Hui SYR (1999) Measurement and modeling of stray capacitances in high frequency transformers. *IEEE PESC'99*, Charleston, SA, June 27–July 1, pp. 763–768.
13. Zhu JG, Hui SYR, Ramsden VS (2000) Comparison of experimental techniques for determination of stray capacitances in high frequency transformers. *IEEE PESC'00*, Galway, Ireland, pp. 1645–1650.
14. Demirel O (1999) The design, test and optimisation of high frequency resonance transformer for basic radiological unit. MS thesis, Hacettepe University, Ankara, Turkey.
15. Baccigalupi A, Daponte P, Grimaldi D (1993) On circuit theory approach to evaluate stray capacitance of two coupled inductors. *Instrumentation and Measurement Technology Conference, IEEE IMTC/93*, Irvine, CA, May 18–20, pp. 549–553.
16. Blache F, Keradec JP, Cogitore B (1994) Stray capacitances of two winding transformers: Equivalent circuit, measurements, calculation and lowering. *IEEE Industry Applications Society Annual Meeting*, Denver, Colorado, October 2–6, Vol. 2, pp. 1211–1217.
17. Ertan HB, Leblebicioğlu K, Demirel O (2000) An approach to design and optimization of HV high frequency resonant power transformers. *International Conference on Electrical Machines, ICEM 2000*, Espoo, Finland, August 28–30, pp. 183–187.

18. Lu HY, Zhu JG, Hui SYR (2003) Experimental determination of stray capacitances in high frequency transformers. *IEEE Transactions on Power Electronics*, 18(5), 1105–1112.
19. Say MG (1958) *The Performance and Design of Alternating Current Machines,* 3rd edn., Pitman paperbacks, London, U.K.
20. Keyhani A, Chua SW, Sebo SA (1991) Maximum likelihood estimation of transformer high frequency parameters from test data. *IEEE Transactions on Power Delivery*, 6(3), 858–864.

22

*High-Voltage, High-Frequency Transformer Design**

Paul Lefley and Philip Devine

CONTENTS

22.1 Introduction

Conversion of AC mains voltages to a high-voltage (>10 kV), high-power (>10 kW) DC level is a key area of technology which is a requirement in many industrial processes, for example, in particulate emission control using electrostatic precipitators.

In the majority of cases, this high-voltage DC requirement is achieved with a single-phase AC regulator, coupled to a large oil filled 50/60 Hz transformer. Power and voltage control is accomplished by variation of the firing angle of the thyristor set. There has been little progress from this topology over the years due, in part, to the lack of availability of high-frequency, high-voltage transformers of sufficient power ratings. This has made the conventional 50/60 Hz design the more attractive solution. This topology while robust and simple has severe drawbacks as far as operation is concerned:

- Low-quality input currents and low power factor
- Sluggish operating characteristics
- Low power supply efficiency
- Large size, weight, and civil engineering costs associated with the oil-insulated transformer

* This chapter has been reproduced from the original article published in the *IEEE Transactions on Power Delivery* in 2001.

A high-frequency switched mode–based power supply has been developed at the University of Leicester [15] with a maximum output voltage of 50 kV at 1 A continuous rating for electrostatic precipitation. There are several improvements that may be expected from the adoption of high-frequency switched mode operation:

1. High-frequency switching operation will allow much more precise control over the operating parameters such as output voltage level, current level, voltage rise times, and response to variations in load demand.
2. High-frequency switching will allow a significant reduction in the size and weight of the high-voltage transformer. This reduction in size and weight leads to a compact design, which minimizes the installation and maintenance costs.
3. High-frequency switching will allow the reactance of the transformer core to be much lower, and hence, the efficiency of the power supply can be improved.
4. The ability to modulate the output voltage. In some applications, the ability to pulse the DC output voltage of the converter from one level to another at a specific and programmable magnitude, time duration, and period has substantial benefits, for example, in electrostatic precipitators, this method may improve dust/gas particle charging and collection.

The transformer must be designed to be driven from a voltage-sourced H-bridge inverter; see Figure 22.1. There is sufficient flexibility with this type of inverter for using a range of voltage control strategies, for example, pulse width or phase shift control. Furthermore, the load may be made resonant if required. This flexibility in inverter control and operation is important to avoid undesirable affects such as ringing. Such effects may occur because of the nature of the transformer, exhibiting parasitic capacitance and inductance, which affects the voltage and current waveforms in the inverter.

A thorough and careful design of a high-voltage, high-frequency transformer is required to ensure that the electrical and magnetic loadings are optimized, that the electrostatic and thermal stresses are acceptable for the voltage and power requirements, and that the parasitic parameters are minimized.

The design of a high-frequency, high-voltage, high-power transformer differs widely from the standard transformer design methodologies [1]. Several related issues must be analyzed:

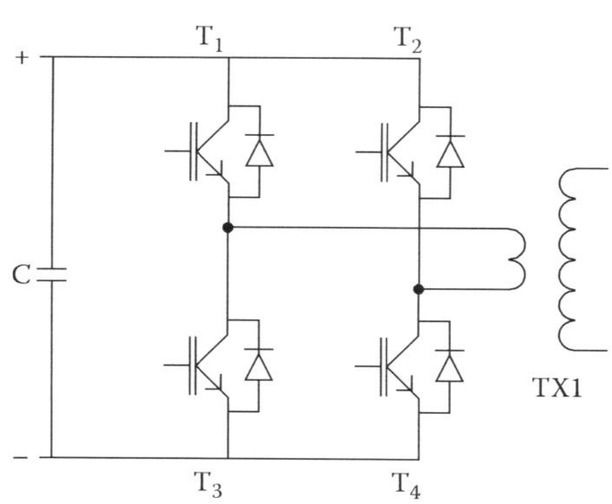

FIGURE 22.1
Voltage-fed H-bridge output stage for driving a high-frequency transformer. (Courtesy of the IEEE, New York.)

1. Insulation requirements
2. Parasitic elements
3. Core loss and heat dissipation
4. Corona effects
5. Rectification

High-voltage transformers generally have a large turns ratio, typically 600:1 to 900:1.

Sufficient insulation thickness between the primary and secondary windings is required to avoid electrical breakdown. Therefore, the electromagnetic coupling of the primary and secondary winding will not be as tight as in conventional low-voltage transformers [2]. This results in a parasitic leakage inductance referred to the primary side that can affect the maximum power throughput of the transformer. Hence, there is a trade-off between insulation distance and leakage inductance.

Furthermore, the high number of turns that is required for the secondary winding causes a high distributed capacitance. When referred to the primary, this capacitance value is multiplied by the square of the turns ratio and therefore is not negligible [3,4]. This parasitic capacitance induces an ineffective current through the secondary winding which results in a loss of transformer efficiency.

Corona discharge can seriously affect the operation and life expectancy of a high-voltage transformer. Any sharp corner or protrusions may lead to an enhanced electric field and corona in this vicinity. A corona will create highly reactive molecules, which may degrade the insulation and lead to electrical breakdown.

The transformer was designed to drive a full-wave rectifier and to have a modular design so that different maximum voltages could be specified. The transformer was therefore designed with two secondary bobbins each driving a separate diode bridge, as shown in Figure 22.2. In order to ensure equal voltage sharing, a capacitive divider was also used.

Bearing these factors in mind, a high-voltage, high-frequency transformer for an electrostatic precipitator power supply was designed with the following electrical specifications:

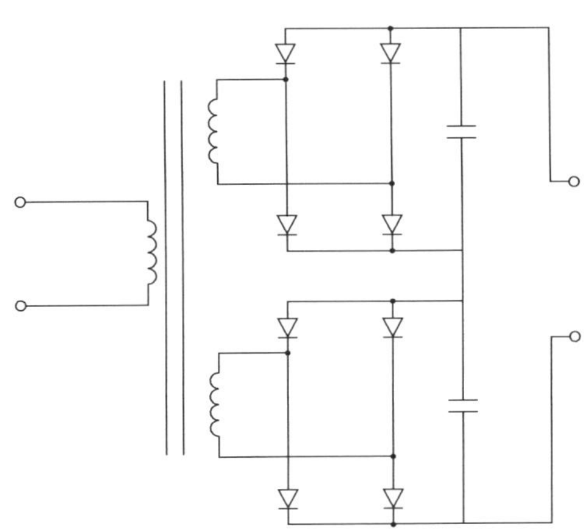

FIGURE 22.2
Transformer with two secondary bobbins driving separate diode bridges. (Courtesy of the IEEE, New York.)

- $V_{primary}$ = 587 V quasi-square wave (rectified output voltage from a three-phase 415 V supply)
- $V_{secondary}$ = 50 kV (nominal output voltage from the proposed power supply)
- Power rating = 50 kVA (required rating)
- Switching frequency = 20 kHz (power electronic specification for low loss and inaudible operation)

22.2 Transformer Design

The transformer design required consideration of the insulation materials, the magnetic material, and the management of electrical, magnetic, and thermal stresses.

22.2.1 Magnetic Design

Because of the high-frequency requirement, it was decided to use a ferrite core, and to ensure low losses, the flux density was limited to 0.25 T. In order to accommodate two bobbins (one for the lower-voltage and one for the higher-voltage winding), a rectangular core consisting of two large "C"-shaped segments was used with a 50 mm × 50 mm (2″ × 2″) cross section and an overall size of 450 mm high × 500 mm wide. This was made from 100 mm × 25 mm × 25 mm (4″ × 1″ × 1″) ferrite sections, since they were the largest available; these were ground to give a good fit between the sections and assembled in a stretcher bond format (similar to house brick construction) for added strength.

A method described by McLyman [5] was used for the magnetic calculations. This method was implemented using MathCAD, the output of which is in Appendix 22.A. This implementation was very useful as it facilitated a design process in which the consequences of a parameter change were immediately reflected on the computer screen and hence enables the designer to focus in on an optimum solution.

The important parameters are the following:

- Two primary windings in parallel, each containing 13 turns of AWG7 wire giving a primary copper loss of 4.6 W
- Two secondary windings in series, each containing 700 windings of AWG21 wire
- Total losses (copper and "iron") of 510 W

22.2.2 Insulation Materials

In addition to the solid insulation required for the bobbins, the insulating encapsulation needed to be defined.

It is clear that the second bobbin needed to be large enough to prevent surface tracking from the high-voltage winding to the transformer core or the primary winding. The thickness of the bobbin also had to be great enough to prevent breakdown through the insulation. Tracking always occurs when the surface electric field exceeds that of either the solid or the encapsulation surrounding it. It may occur at lower fields because of physical and chemical inhomogeneities on the surface and because of geometric field enhancements. Although reliable figures appear to be hard to find, a distance of at least 1 kV/mm is commonly allowed for well-controlled surface conditions. (For example, Rowland and

Nichols [6] found that voltages of at least 15 kV were required to sustain dry band arcing over a 10 mm gap on an "arc resistant thermoplastic compound.")

The voltage waveform to be experienced by the higher-voltage bobbin was rather unusual in this application since it was of high frequency (20 kHz) and had a negative DC offset of half the peak-to-peak AC voltage. The negative DC offset may give rise to space charge injection [7] and consequent field distortion within the insulation. The maximum ("Poissonian") field within the material may therefore be greater than the average ("Laplacian") applied field. The high frequencies may also lead to accelerated aging processes, due, for example, to partial discharges. A polymeric insulation was to be used for ease of machining. Because of the large bobbin size, it was decided not to use a very expensive material such as PTFE. Nylon 6,6 was chosen as a compromise between ease of manufacture, cost, and insulation properties. This transformer was being used in a noncritical application. A more critical search for insulation materials may be appropriate in other applications. The breakdown strength of such materials under these conditions is not readily available, but work by Dissado, Montanari, Crine, Lewis, and others (e.g., [8–10]) suggests that they should be able to withstand such conditions indefinitely at fields up to approximately 10 kV/mm. Because of the unknowns, especially to do with space charge effects, it was decided to limit the electric field in the bobbin to 5 kV/mm. Nevertheless, this is still high; typically 0.5 kV/mm AC or 1.0 kV/mm DC is the maximum used in high-voltage supplies for safety critical applications [11].

Given these considerations for the higher-voltage bobbin, it was necessary to consider the choice of materials for the transformer encapsulation. A solid encapsulation is likely to have many drawbacks, especially during the design phase of a prototype. It is likely to be inferior in transporting heat from the transformer, it would be impossible to make modifications to the transformer once the encapsulation was in place, and if the encapsulation suffered electrical breakdown, it may be necessary to completely rebuild the transformer. A fluid encapsulation such as transformer oil or an insulating gas was therefore considered. Transformer oil is a good medium for heat transport. Under ideal natural convection conditions, it has a heat transfer coefficient of approximately 95 W/m^2 K; this is equivalent to forced air cooling with a flow velocity of approximately 25 m/s [12]. For prototyping, oil causes severe disassembly problems, as the oil needs to be removed from the surfaces of all components. Impurities in oil tend to accumulate at points of high field divergence, that is, at the most critical points, due to dielectrophoresis [13] leading to localized discharging. It is therefore usual to continuously pump and filter the oil; this is inconvenient for a small transformer.

For these reasons, it was therefore decided to use a gaseous encapsulation. Although it may be possible to use air at normal atmospheric pressure (this is the case in many high-voltage laboratory supplies), this would make the transformer rather large as the breakdown strength of air is only 2–3 kV/mm even under ideal parallel plate conditions. The air would also need to be clean and dry and, therefore, maintained in a sealed container. Sulfur hexafluoride, SF_6, has a much better dielectric strength (see Figure 22.3). It can be seen that, even at a pressure of 1 bar, the breakdown strength of SF_6 exceeds the maximum design strength for the bobbin (5 kV/mm). At 2 bar, the breakdown strength is comparable with nylon 6,6, and so tracking resistance is unlikely to be improved above this pressure. It was therefore decided to use SF_6 at a minimum of 2 bar as the encapsulation material for the transformer.

22.2.3 Thermal Analysis

The McLyman technique was also used for checking the thermal design of the transformer. A maximum temperature rise of 50°C was specified. The MathCAD output, shown

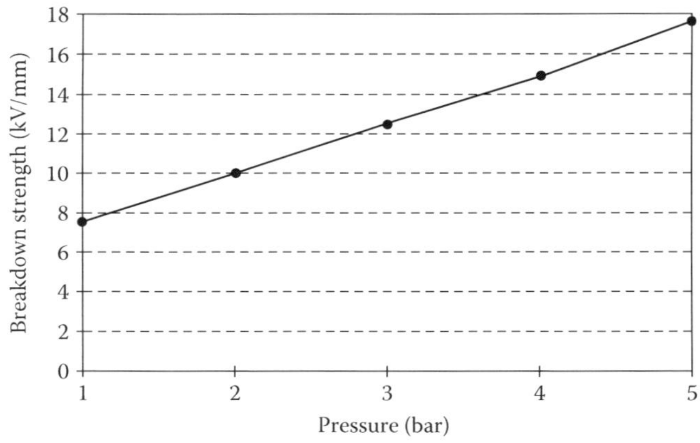

FIGURE 22.3
50 Hz breakdown strength in a homogeneous field for a 20 mm interelectrode gap as a function of absolute gas pressure. (Courtesy of the IEEE, New York.)

in Appendix 22.A, shows such thermal calculations and is based on estimating the heat loss by thermal radiation plus the heat loss by convection. The method indicates a minimum required core volume of 957 cm^3 and a minimum surface area of 533 cm^2, which is considerably less than the core volume of 4250 cm^3 and the surface area of approximately 1500 cm^2 that was actually used (although only about 50% of this area was exposed). A temperature rise of less than 50°C was therefore to be expected. The oversize core was a consequence of the secondary winding layout and the extra insulation thickness in the high-voltage bobbins in order to avoid an electrical breakdown to the core.

It is quite clear from these calculations that the McLyman method can be used in an iterative way to determine the most favorable core volume and surface area in order to satisfy the required magnetic and electrical loadings. Appendix 22.B shows a flow chart scheme of an iterative design process using the McLyman method. However, when designing a high-voltage transformer, it does not take into account the extra dimensions of core that may be required for the purposes of electrical insulation at very high voltages, but serves as a guide to determine the minimum core requirements.

22.2.4 Design of High-Voltage Bobbins and Electrostatic Analysis

Figure 22.4 shows a drawing of the higher-voltage (50 kV) bobbin on which the high-voltage secondary winding was wound. The primary winding for each bobbin was wound on a welded polypropylene former and placed inside a metal sleeve, and the complete assembly is then inserted into the outer bobbin. The metal sleeve (with antishorting slot) provides an inner ground plane surface to maintain a uniform (radial) electrostatic field inside the high-voltage secondary bobbins. This arrangement can be partially seen in Figure 22.9, which also shows the nylon support frame for the bobbins and the rectifier. Figure 22.5 shows a photograph of both the 25 and 50 kV outer bobbins. The bobbins contain two unusual features.

Firstly, in order to obviate the divergent electric stress that would otherwise occur at the ends of the windings, stress relief rings are used. These rings, which are split to prevent shorted turns around the transformer core, are connected to the ends of each secondary winding; indeed they form the connections. The usual way to relieve stress is to alter the

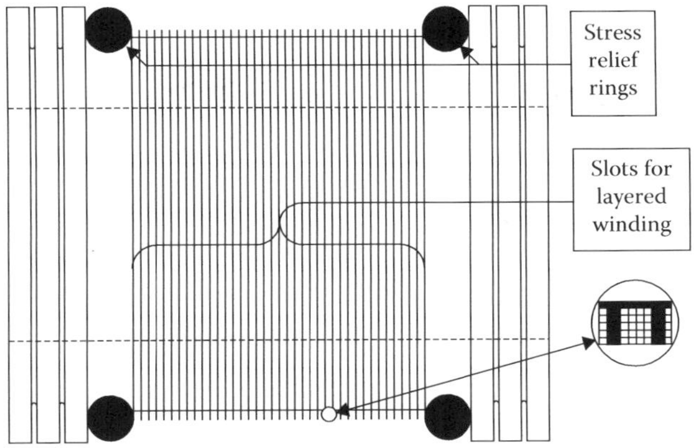

FIGURE 22.4
Side view of the 50 kV bobbin showing stress relief rings and slots for layered windings. (Courtesy of the IEEE, New York.)

FIGURE 22.5
Photograph of the 25 and 50 kV high-voltage bobbins. (Courtesy of the IEEE, New York.)

winding spacing at the ends, but this can be very wasteful of space. The stress relief rings decrease the divergent field (≈V/r) due to the high voltage, V, by increasing the effective radius of the wire, r. In this way, the maximum field on the inside of the stress relief ring is limited to 5 kV/mm.

The other unusual feature is a slotted bobbin that allows layered windings. In this way, it was possible to increase the number of turns considerably without placing large demands on the insulation of the winding wire; indeed this was only rated at 600 V. Furthermore, the parasitic capacitance between the ends of the windings is reduced. A single cylindrical winding along the surface of the bobbin, which only had 300 turns, was found to have a parasitic capacitance of 130 nF at 1 kHz (measured using a dielectric spectrometer). However, the arrangement shown here, which had more windings (700 turns), had a reduced capacitance of 64 nF. Such an arrangement does give rise to locally high divergent fields at the corners at the bottom of the slots, and an electrostatic analysis was made to ensure that a field of 5 kV/mm was not exceeded. The results of this analysis for the 50 kV rated bobbin is shown in Figure 22.6a–c. It is interesting to observe the beneficial effect of the stress relief rings at the ends of the bobbins, Figure 22.6a and b. Figure 22.6c shows the electric field intensity in the 50 kV bobbin without a stress relief ring present. The dashed circle shows an area where the diverging E field is in the region of 5 kV/mm, which could eventually lead to a breakdown in the insulation at the end of the winding, whereas at the same location in Figure 22.6b, there is a nondiverging field in the region of 2 kV/mm.

22.3 Test Results

The transformer was tested as part of the high-frequency, high-voltage power supply described earlier. Initially it was tested at low power but at 60 kV using a large resistive load to prove the integrity of the insulation. It was installed in National Power's 2035 MW coal-fired "Didcot A" power station at Didcot (near Oxford), United Kingdom, where it replaced a conventional 50 Hz transformer rectifier and supplied an electrostatic precipitator field. Precipitators are difficult loads since they are prone to arcs and sparks, during which time they act as virtual short circuits, and the current they draw is very unsteady due to the corona discharge and continuously changing coal dust burden. The power electronics contained protection circuits to ensure that the supply voltage was rapidly reduced during arcing and sparking. The load was therefore a useful test to ensure that the unit was sufficiently powerful and robust, but it was difficult to obtain accurate results because of the unsteady current drawn.

Figure 22.7 shows the measured current–voltage characteristics of the electrostatic precipitator load; the line is an exponential curve drawn to assist the eye. The current increased rapidly above a corona inception voltage of approximately 30 kV. The circled groups of measurements at approximately 0.2 A (~38 kV) and 0.4 A (~43 kV) were almost certainly made as the supply was recovering from a spark in the precipitator. The voltage is increasing at this point, but little corona current is being produced. By extrapolation of the exponential, it may be estimated that the transient arcing current was several amps at the higher voltages. The dotted exponential curve shows the average "steady-state" characteristics of the precipitator.

Conventional transformer–rectifier sets have ripples that are typically 50% of the mean voltage [14]. Arcing occurs if the peak of the voltage ripple exceeds a critical value, and so for much of the cycle, when the voltage is much less than this critical value, little corona current is produced. Typically the mean current is much less than half that produced at the peak. Furthermore, the ripple produces a considerable capacitive displacement current which increases the size of the secondary transformer windings. In this high-frequency supply, very little ripple was observed. At 30 kV, the ripple was 85 ± 10 V; at 44 kV, it was 230 ± 20 V; that is, it was much less than 1% under all conditions. It is therefore possible for the supply to produce more than twice as much current without arcing occurring.

Figure 22.8 shows the primary current and voltage waveforms for an output of 38 kV. As the primary voltage is provided by a high-frequency, phase-controlled H-bridge inverter, there is a dead time between both positive and negative half cycles when the duty cycle is less than 100%. By considering the voltage waveform in Figure 22.8, it can be seen that the period is approximately 50 μs corresponding to a switching frequency of 20 kHz. The duty

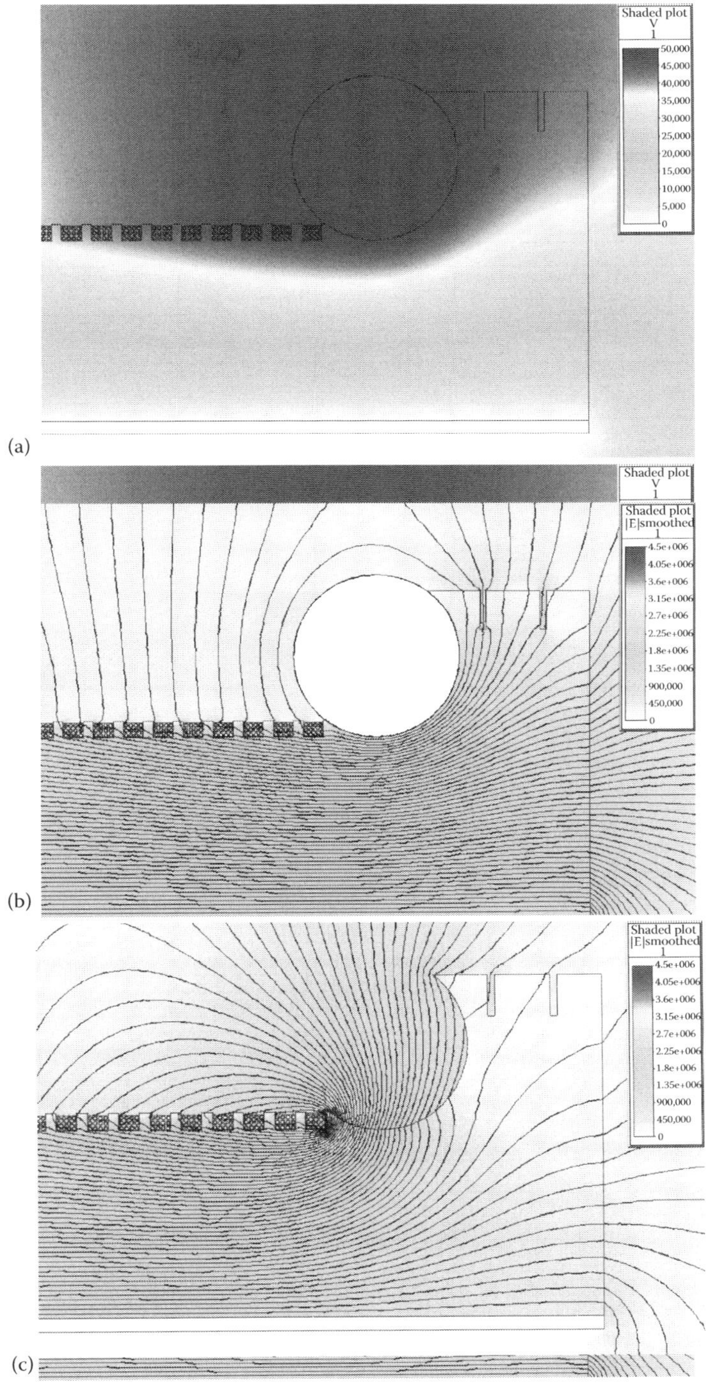

FIGURE 22.6
(a) Electrostatic field analysis around one end of the 50 kV bobbin showing voltage distribution. (b) Electrostatic field analysis of the 50 kV bobbin showing field strength distribution. (c) Electrostatic field analysis of the 50 kV bobbin showing the effect of removing the stress relief ring.

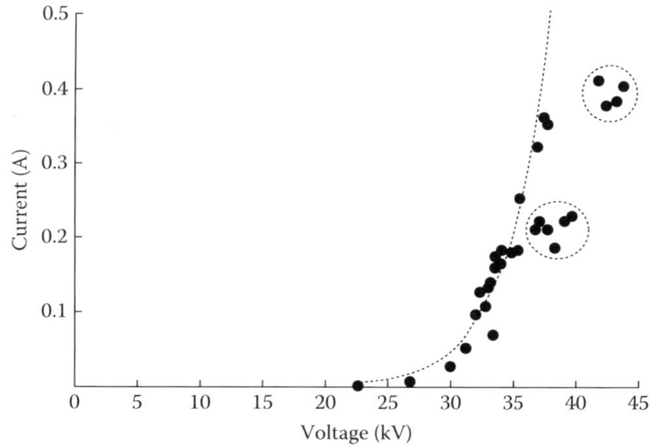

FIGURE 22.7
Measured current–voltage characteristic under electrostatic precipitator load. (Courtesy of the IEEE, New York.)

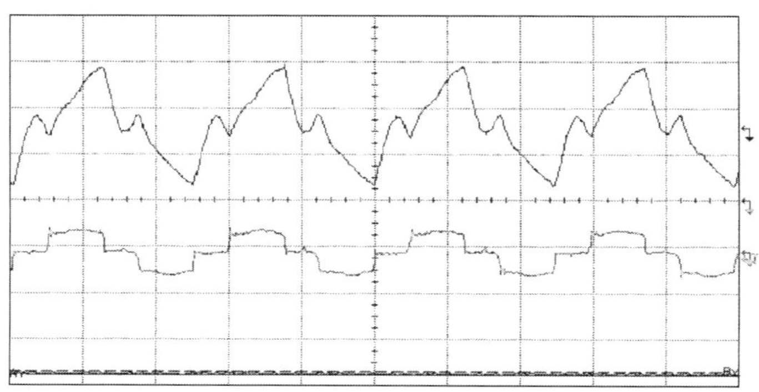

FIGURE 22.8
Primary current (upper trace, 50 A/div) and primary voltage (middle trace, 1 kV/div) for a −38 kV output (bottom trace, 10 kV/div) (timebase 20 ms/div). (Courtesy of the IEEE, New York.)

cycle was set to 40% with the dead times equal to 10 μs and the positive and negative on times being 15 μs. The primary voltage was ±500 V. During the positive (negative) on times, the current increased (decreased) reasonably linearly, due to the magnetizing inductance, to a value of approximately +(−)64 A. During the dead times, all four H-bridge transistors are off, but there is a current path through their flywheel protection diodes. The current during this period is otherwise uncontrolled, and the small current oscillation during the dead time probably corresponds to a resonance of the magnetizing inductance with the parasitic capacitance.

It was difficult to estimate the efficiency of the transformer accurately because of the measurement uncertainty in the continuously fluctuating secondary current and unfortunately the experimental arrangement precluded measurement of the instantaneous secondary current. All our estimates lead us to believe that the efficiency of the transformer with rectifier was better than 97%. The total rectifier voltage drop was 200 V which, at 0.4 A, would give an 80 W power loss and contribute ~0.5% to the overall power lost. The best indication of power loss was the temperature rise of the transformer core, which was

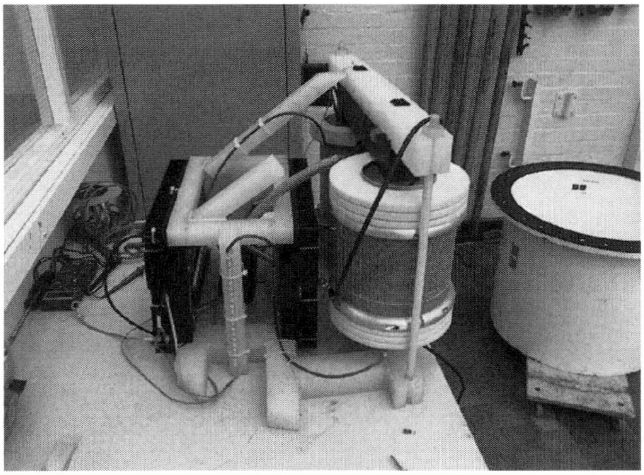

FIGURE 22.9
Complete assembly of the high-voltage, high-frequency transformer ready for pressurized containment. (Courtesy of the IEEE, New York.)

measured using a thermistor placed near the end of one of the bobbins. The temperature rose from 16°C ± 2°C to 33°C, a rise of 17°C ± 2°C, when the output was 45 kV, 0.4 A, a power of 18 kW. The design calculations, Appendix A, suggest that with a loss of 510 W, a temperature rise of 50°C would be found for a core surface area of 533 cm^2. The actual exposed surface area was approximately 800 cm^2. An estimate of the actual power lost is therefore

$$\frac{17°\text{C}}{50°\text{C}} \times \frac{800\,\text{cm}^2}{533\,\text{cm}^2} \times 510\,\text{W} = 260\,\text{W} \tag{22.1}$$

For an output power of 18 kW, this corresponds to an efficiency of 98.5%. The overall efficiency of the power supply was estimated to be >95% which compares very favorably with conventional 50 Hz transformer–rectifier sets with a typical efficiency of around 60%.

Figure 22.9 shows the final transformer assembly with bridge rectifiers and smoothing capacitors before being lowered into the container and pressurized with SF_6.

22.4 Conclusion

This chapter presented a design layout for a high-frequency, high-voltage transformer for use in a high-voltage switched mode power supply. The electrical specifications were determined by the output requirements of the power supply as well as the operating constraints of the power electronics that drives the transformer, for example, 20 kHz switching frequency. From these basic specifications, the McLyman technique was used to determine the minimum core volume and surface area for a given switching frequency, efficiency, and temperature rise. A balance between the electrical and magnetic loadings is addressed in the McLyman technique as this provides the most effective utilization of the copper and iron materials. When designing a high-voltage transformer, the McLyman minima for the core volume and surface area should be regarded as a guide, as the physical constraints on the transformer design are largely dictated by the high-voltage requirements. The high-voltage design should take account of the dielectric breakdown of the secondary bobbin, which for nylon 6,6 should not be any greater than 5 kV/mm. The secondary winding should also be distributed over the length of the bobbin as this minimizes the possibility of an intrawinding breakdown. Furthermore, the use of stress relief rings at very high voltages prevents highly divergent E fields and thus reduces the risk of a breakdown at the ends of the secondary winding or within the insulating bobbin.

The prototype transformer that was designed using the techniques described in this chapter for an electrostatic precipitator power supply worked very well. It was found to be much more efficient and smaller than a conventional 50/60 Hz transformer, and being driven from a power electronic inverter has other benefits, for example, real-time control of the output voltage with overcurrent protection.

Appendix 22.A: Magnetic and Thermal Design Using the McLyman Design Method

("K" values are coefficients defined by McLyman)

$V_p := 587$ V—primary square wave voltage

$V_o := 50{,}000$ V—secondary voltage

$P_o := 25{,}000$ VA—output power

$f := 20{,}000$ Hz—operating frequency

$\eta := 0.98$ efficiency—specified efficiency

$\Delta T := 50°C$—allowable temperature rise

$B_m := 0.25$ Tesla—maximum core flux density

1. Calculation of Power Handling Capability, P_t
   $P_{in} := P_o$ VA—approximate power input
   $P_t := 2P_{in}$ $P_t = 5 \times 10^4$ VA—apparent power handling capability
2. Calculation of Area Product, A_p
   $k := 4$—coefficient indicating square wave input
   $K_u := 0.01$ (window utilization factor) $K_j := 323$ (current density coefficient)

$$A_p := \left(\frac{P_t \cdot 10^4}{k \cdot B_m \cdot f \cdot K_u \cdot K_j} \right)^{1.16} \qquad A_p = 3.243 \cdot 10^4 \text{ cm}^4$$

3. Evaluation of Core Geometry
   3.1 Volume of Transformer Core
   $K_v := 17.9$ (volume coefficient for C-core)

$$\text{Volume} := K_v \cdot A_p^{0.75} \qquad \text{volume} = 4.326 \cdot 10^4 \text{ cm}^3$$

3.2 Transformer Core Surface Area, A_t

$K_s := 39.2$ (surface area coefficient for C-core)

$A_t := K_s \cdot A_p^{0.5}$ $\quad A_t = 7.059 \cdot 10^3\ \text{cm}^2$

3.3 Transformer Current Density, J

$K_j := 468$—current density coefficient for C-core at temperature rise of 50°C

$J := K_j \cdot A_p^{-0.125}$ $\quad J = 127.755\ \text{A/cm}^2$

4. Total Estimated Transformer Losses, P_Σ

$$P_\Sigma := \frac{P_o}{\eta} - P_o \quad P_\Sigma = 510.204\ \text{W}$$

5. Maximum Efficiency When P_e (Core Loss) $= P_{cu}$ (Copper Loss)

Best case efficiency when $P_e = P_{cu}$

$$P_{cu} := \frac{P_\Sigma}{2} \quad P_{cu} = 255.102 \quad P_e := P_{cu} \quad P_e = 255.102\ \text{W}$$

6. Calculation of Core Loss

For Fernk type |100/57/44 (MMG-Neosid) at 0.25 T and 20 kHz:

$P_{Fe25} := 0.2\ \text{W/cm}^3$ at 25°C $\quad P_{Fe100} := 0.3\ \text{W/cm}^3$ at 100°C

So at T = ambient + 50°C, that is, T = 75°C

$$P_{Fe} := P_{Fe25} + \frac{75-25}{100-25} \cdot (P_{Fe100} - P_{Fe25}) \quad P_{Fe} = 0.267\ \text{W/cm}^3$$

We need an *effective* volume of core of greater than $P_e/P_{Fe} = 956.633\ \text{cm}^3$.

The worst case efficiency is for $P_e \gg P_{cu}$, so we need an effective volume of core:

$$\frac{P_\Sigma}{P_{Fe}} = 1.913 \cdot 10^3\ \text{cm}^3$$

7. Calculation of Primary Turns, n_p

Cross-sectional area of MMG 100/57/44 core

$A_c := 4{\cdot}2.464 \cdot 2.4646$ $\quad A_c = 24.291\ \text{cm}^2$

$$n_p := \frac{v_p \cdot 10^4}{4 \cdot B_m \cdot A_c \cdot f} \quad n_p = 12.083$$

To ensure flux density is not exceeded, use $n_p := 13$ turns.

8. Transformer Current Density

$J = 127.755\ \text{A/cm}^2$

9. Primary Current, I_p

$$I_o := \frac{P_o}{V_o} \quad I_o = 0.5 \quad I_p := \frac{V_o \cdot I_o}{V_p} \quad I_p = 42.589\ \text{A}$$

10. Calculate Base Wire Size for Primary, A_W

$$A_w := \frac{I_p}{J} \quad A_w = 0.333\,\text{cm}^2$$

11. American Wire Gauge
Since there are two parallel primary windings, area of each, $A = (A_w/2)\ A = 0.167\,\text{cm}^2$. This is approximately AWG5. It is possible to use AWG7 (area of $0.105\,\text{cm}^2$) if primary losses can be neglected. For 2 in parallel, area $= 0.21\,\text{cm}^2$.
12. Calculation of Primary Resistance
Mean length per turn: $MLT_p := 19.2\,\text{cm}$
Resistivity of copper: $\rho := 1.71881 \cdot 10^{-6}\ \Omega\ \text{cm}$
Resistance per cm: $R_p := (\rho\ /\ 0.21)\ R_p = 8.185 \cdot 10^{-6}\Omega/\text{cm}$
Temperature coefficient of resistance at 75°C: $\xi := 1.24$
$R_p := MLT_p \cdot n_p \cdot R_p \cdot \xi \quad R_p = 2.533 \cdot 10^{-3} \quad R_p \cdot 1000 = 2.533\,\text{mW}$
13. Primary Copper Loss
$I_p^{\,2} \cdot R_p = 4.595\,\text{W}$
14. Calculation of Secondary Turns

$$n_s := \frac{n_p}{V_p} \cdot V_o \quad n_s = 1.107 \cdot 10^3$$

To allow the use of a nominal 80% duty cycle on the primary $n_s := (n_s/0.8)\ n_s = 1.384 \cdot 10^3$
And so $n_s := 1400$ was chosen.
15. Calculate Base Wire Size for Secondary

$$A_w := \frac{I_o}{J} \quad A_w = 3.914 \cdot 10^{-3}\,\text{cm}^2$$

therefore, use AWG21
16. Calculation of Secondary Resistance
Mean length per turn: $MLT_s := 2 \cdot \pi \cdot 15.5\,\text{cm}$
Resistance per cm (interpolating by 1.26513): $R_s := 418.9 \cdot 10^{-6}\ \Omega/\text{cm}$ from AWG table
$R_s := MLT_s \cdot n_s \cdot R_s \cdot \xi \quad R_s = 70.823\,\text{W}$
17. Calculation of Temperature Rise
Assume that $P_\Sigma = P_{cu} + P_{Fe}$.
Assume that thermal energy is distributed throughout the core and winding assembly.
Heat transfer by thermal radiation:
$k_r := 5.7 \cdot 10^{-12}$—Stefan–Boltzmann constant, W/cm² K⁴
$\varepsilon := 0.95$—emissivity
$T_2 := 75 + 273$—hot body temperature, K
$T_1 := 25 + 273$—cold body temperature
$W_r := k_r \cdot \varepsilon\ (T_2^{\,4} - T_1^{\,4}) \quad W_r = 0.0367\,\text{W/cm}^2$ of surface area

Heat transfer by convection:
$k_c := 1.4 \cdot 10^{-3}$
$F := 3.5$—air friction factor: 1 for vertical surface, 1.25 for horz. flat surface; SF_6 is at least 3.5 times better than air at low flow rates
$\theta := 50°C$—temperature rise
$\zeta := 1.25$—depends on surface shape, varies from 1 to 1.25
$P := 2$—relative barometric pressure

$$W_c := k_c \cdot F \cdot \theta^{\zeta} \sqrt{P} \quad W_c = 0.9213\,W/cm^2 \text{ of surface area}$$

Total heat transfer:
$W := W_r + W_c \qquad W = 0.9581\,W/cm^2$
Assuming worst case efficiency, η (i.e., $P_{cu} = \Sigma P$), the minimum surface area required is

$$A_{smin} := \frac{P_{\Sigma}}{W} \quad A_{smin} = 532.538\,cm^2$$

(Actual area is approximately $1500\,cm^2$. Approximately 50% of this area is exposed—so well within spec.)

Appendix 22.B: Iterative Transformer Design Using the McLyman Method

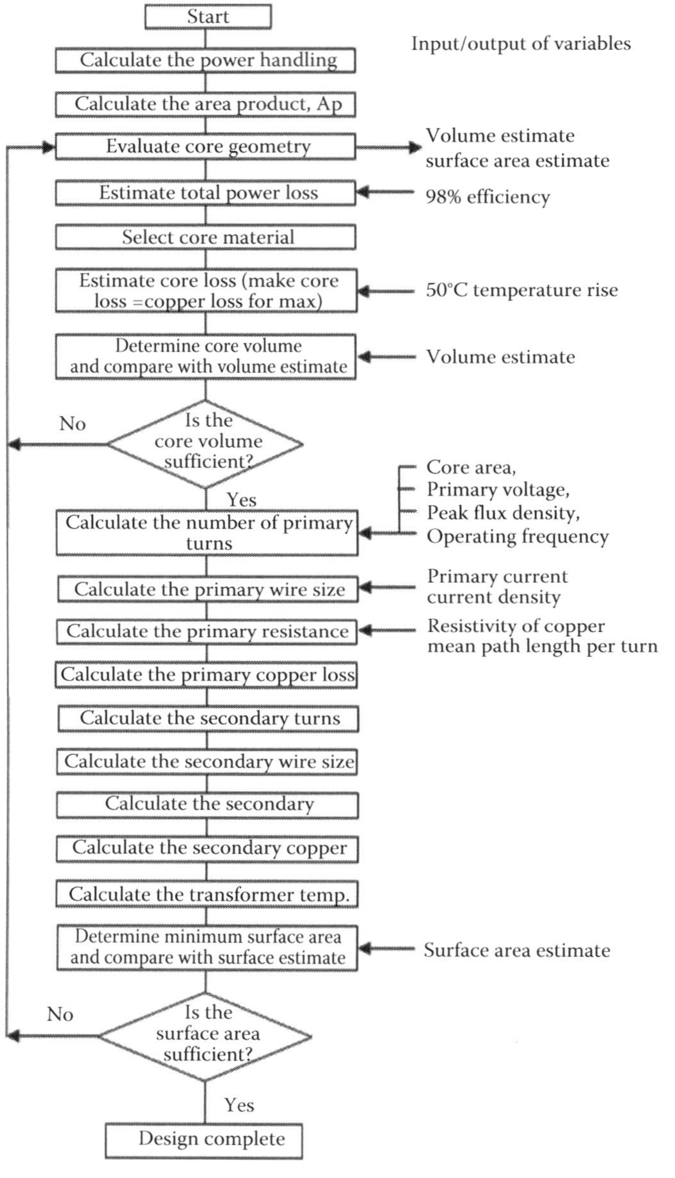

References

1. Devine PJ (1995) The requirements of a high voltage DC power supply. Internal report. Department of Engineering, University of Leicester.

2. Liberati G, Cincinelli L (1992) High voltage, high power switched mode power supply for ion productions in ESP. SAE Technical Paper 929268. doi: 10.4271/929268.
3. Perez MA, Blanco C, Rico M, Linera FF (1995) A new topology for high voltage, high frequency transformers. *IEEE Applied Power Electronics Conference and Exposition,* Vol. 2, Dallas, TX, March 5–9, pp. 554–559.
4. Takano H, Hatakeyama T, Sun JM, Laknath KGD, Nakaoka M, (1996) Feasible characteristic evaluations of resonant inverter linked DC-Dc power converter using high voltage transformer parasitic components. *Sixth International Conference on Power Electronics and Variable Speed Drives IEE,* Conf. Publ. No. 429, Nottingham, U.K., September 23–25, pp. 525–533.
5. McLyman WT (1978) *Transformer and Inductor Design Handbook.* New York: Dekker.
6. Rowland SM, Nichols IV (1996) Effects of dry-band arc current on ageing of self-supporting dielectric cables in high fields. *IEE Proceedings Science, Measurement and Technology,* 143(1), 10–14.
7. Fothergill JC, Dissado LA (eds.) (1998) *Space Charge in Solid Dielectrics.* Leicester, U.K.: Dielectrics Society, pp. 259–272.
8. Dissado LA, Mazzanti G, Montanari GC (1997) The role of trapped space charges in the electrical ageing of insulating materials. *IEEE Transactions on Dielectrics and Electrical Insulation,* 4(5), 496–506.
9. Crine JP (1997) A molecular model to evaluate the impact of ageing on space charges in polymer dielectrics. *IEEE Transactions on Dielectrics and Electrical Insulation,* 4(5), 487–495.
10. Connor P, Jones JP, Llewellyn JP, Lewis TJ (1998) Mechanical origin for electrical ageing and breakdown in polymeric insulation. *Proceedings of the 1998 IEEE Sixth International Conference on Conduction and Breakdown in Solid Dielectrics,* Conf code 49–244, Vasteras, Sweden, pp. 434–438.
11. Williams JW (1995) Navy/industry 'NAVMAT' high voltage guidelines—An overview. *High Voltage Workshop,* Salt Lake City, UT: University of Utah.
12. *Conti-Elektro-Berichte* (July/September 1966), p. 189.
13. Pohl HA (1978) *Dielectrophoresis.* Cambridge, U.K.: Cambridge University Press.
14. Parker KR (ed.) (1997) *Applied Electrostatic Precipitation.* London, U.K.: Chapman and Hall.
15. Fothergill JC, Devine PW, Lefley PW (2001) A novel prototype design for a transformer for high voltage, high frequency, high power use, *IEEE Transactions on Power Delivery,* 16(1), 89–98.

23

Coreless PCB Transformers

Jesús Doval-Gandoy and Moisés Pereira Martínez

CONTENTS

23.1 Introduction and Fundamental Characteristics of Coreless PCB Transformers

Transformers have been designed and widely used over the last century; they are used for electrical isolation, signal coupling, and energy transfer. Normally, a transformer consists of windings wound on magnetic cores.

Magnetic cores are made of ferromagnetic materials, which provide good conducting paths for magnetic flux. They provide a high degree of magnetic coupling and they reduce the leakage inductance. The type of core is selected in the design process depending on the application, the operating frequency, power to be transferred, etc.

In the past, most transformer designs were for low-frequency operations; more recently, the operating frequency of many switched-mode power supplies has been significantly increased to several hundreds of kilohertz or up to a few megahertz. The use of coreless transformers is an alternative to transformers with a core for some applications; the windings can be made of twisted coils [1] or printed planar windings. Identical twisted coils are not easy to manufacture, and some parameters of twisted coil transformers are difficult to control. By using printed planar windings instead of twisted coil, it is possible to manufacture transformers with precise parameters.

The coreless PCB transformer consists of three parts: the primary winding, the dielectric laminate, and the secondary winding. The PCB transformer can be built on the same circuit board with other electronics. It can also be fabricated as a stand-alone device if desired. There is no need to cut a hole on the PCB for accommodating the magnetic cores in coreless PCB transformers.

The use of printed planar windings has been treated in several works during the last decade [2–26]. Most of them study the planar transformers with core. Some of these works study PCB coreless transformers from different points of view. Most of the applications of coreless PCB transformers presented in the literature are for low output power, typically driving MOSFET transistors.

As explained in Ref. [20], the coreless PCB transformers have good features in the high-frequency range from a 500 kHz to a few megahertz, and it is not useful at frequencies smaller than 300–400 kHz. In general, at frequencies smaller than 300 kHz, the magnetizing reactance is too low, then the voltage gain is low and the primary winding current is too large.

Coreless PCB transformers can achieve high power density, and they are suitable for applications in which height requirements are stringent. Coreless PCB transformers do not need space to accommodate the magnetic core and have no core limitations such as core losses and saturation. Their sizes can be smaller than those of core-based transformers. Coreless PCB transformers can be built with high isolation voltage from 10 to 50 kV depending on the PCB materials.

It is possible to obtain PCB transformers without magnetic core with high coupling factor, high voltage gain, and low radiated EMI problems, as presented in some papers [18,19].

High voltage gain can be achieved with coreless PCB transformers by connecting an external capacitor across the secondary windings, the resonant effect of the inductive components of the transformer, and the external capacitor can provide high voltage gain.

Various ways to lay out planar transformers have been presented in the literature. Planar windings of various shapes, see Figure 23.1, have been studied in several papers [23,25,26].

FIGURE 23.1
Planar windings of various shapes.

It has been demonstrated that circular spiral windings provide the greatest inductance among various types of winding configuration, (circular, rectangular, polygonal). Circular spiral windings have less resistance and inter-winding capacitance than square or polygonal spiral windings because circular windings have shorter tracks and less overlap area than the others.

Exact modeling and characterization of coreless PCB transformers is very important for designing electronic circuits with these transformers. This chapter covers aspects related to the characterization of coreless PCB transformers and behavior of the transformer under different load conditions and operating frequency.

Some of the characteristics of coreless PCB transformers treated in this section are

- Equivalent circuit of the transformer and description of the parameters used in the transformer model
- Transfer function of the transformer
- Variation of some of the parameters with frequency
- Effects of loading the transformer with resistive load, capacitive load
- Input impedance and efficiency of coreless PCB transformers
- Description of methods for obtaining self and mutual impedances
- Influence of geometric parameters (conductor thickness, conductor width, outermost radius, etc.) in the electric parameters of the transformer
- Calculation, simulation, and measurement of the coreless transformer parameters

23.2 Modeling of Coreless PCB Transformers

23.2.1 Equivalent Circuit

The coreless PCB transformer is well described by using a high-frequency transformer model as shown in Figure 23.2, where

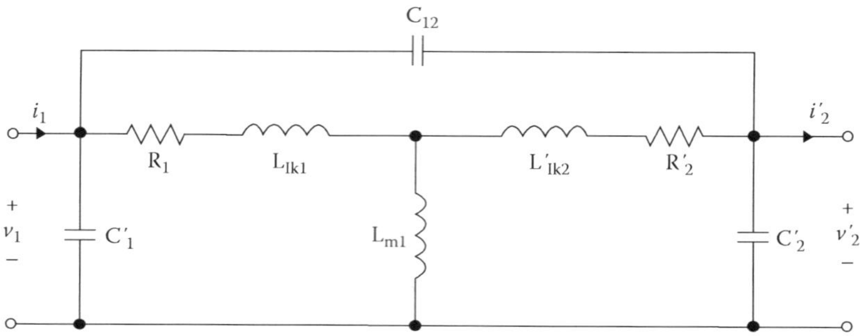

FIGURE 23.2
High-frequency model for coreless PCB transformer.

$$\begin{aligned}
&R_1 : \text{primary winding resistance}\\
&L_{lk1} : \text{primary winding leakage inductance}\\
&C_1 : \text{primary winding capacitance}\\
&R_2 : \text{secondary winding resistance}\\
&L_{lk2} : \text{secondary winding resistance}\\
&C_2 : \text{secondary winding capacitance}\\
&C_{12} : \text{capacitance between primary and secondary windings}\\
&L_{m1} : \text{mutal inductance}\\
&n : \text{turns ratio}
\end{aligned} \tag{23.1}$$

The parameters of the secondary side referred to the primary side are expressed in the following equation [15]:

$$\left.\begin{aligned}
R_2' &= n^2 \cdot R_2\\
L_{lk2}' &= n^2 \cdot L_{lk2}\\
C_1' &= C_1 + \frac{n-1}{n} \cdot C_{12}\\
C_2' &= \frac{C_2}{n^2} + \frac{1-n}{n^2} \cdot C_{12}
\end{aligned}\right\}
\quad
\left.\begin{aligned}
C_{12}' &= \frac{1}{n} \cdot C_{12}\\
V_2' &= n \cdot V_2\\
i_2' &= \frac{i_2}{n}
\end{aligned}\right\} \tag{23.2}$$

Figure 23.3 shows a photograph of six coreless transformers with different geometric parameters. Transformers 1A, 1B, and 1C have the same number of turns, conductor width, and outermost radius, but the thickness of the PCB is 1.66 mm for transformer 1A, 1 mm for transformer 1B, and 0.5 mm for transformer 1C. The outermost diameter of the transformer number 2 is 25 mm. Transformer number 3 has 20 turns in the primary winding and 15 turns in the secondary. Transformer 4 has two secondary windings.

Transformers shown in Figure 23.3 have the primary circular spiral winding on the bottom layer. The primary has 15 or 20 turns, and the conductor width is 0.3 or 0.4 mm. The secondary

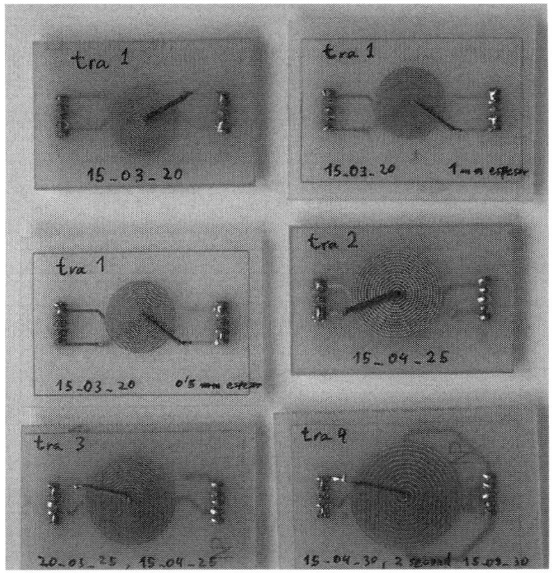

FIGURE 23.3
Photograph of six coreless transformers with different geometric parameters.

circular spiral winding is on the top layer; it has 15 turns and the conductor width is 0.3 or 0.4 mm. For both windings in the six transformers, the outermost radius is 20, 25, or 30 mm, depending on the transformer. The thickness of the PCB used is 0.5, 1, or 1.66 mm. The geometric parameters of transformers presented in Figure 23.3 are listed in Table 23.1.

For the transformers shown in Figure 23.3, the resistance, inductance, and capacitance of both windings; the inter-winding capacitance; and the magnetizing inductance can be measured with an impedance analyzer. The coreless transformers presented earlier were designed to operate at high frequency, from few hundreds of kilohertz to a few megahertz; the model parameters presented in Table 23.2 for transformer 1A were measured at about 2 MHz with an impedance analyzer.

23.2.2 AC Winding Resistance

Due to skin effect, the AC winding resistance depends on the frequency. The winding resistance is given by the following equation presented in Ref. [15], where R_o is the dc

TABLE 23.1
Dimensions of the Transformers under Test

Transformer	N	Conductor Width (mm)	Thickness PCB (mm)	Outermost Radius (mm)
1A	15:15	0.3	1.66	20
1B	15:15	0.3	1	20
1C	15:15	0.3	0.5	20
2	15:15	0.4	1.66	25
3	20:15	0.3: 0.4	1.66	25
4	15:15	0.4: 0.3	1.66	30

TABLE 23.2

Electrical Parameters for the Transformer "1A" at about 2 MHz

C_1 (pF)	C_2 (pF)	C_{12} (pF)	R_1 (pF)	R_2 (pF)	L_{lk1} (pF)	L_{lk2} (pF)	L_{m1} (pF)	L_{m2} (pF)
3.27	3.27	9.127	1.32	1.32	0.63	0.63	1.1085	15:15

resistance of the conductor, f is the operating frequency, and f_a and f_b are critical frequencies of the conductor.

As detailed in Ref. [27], there are three frequency ranges for the resistance of microstrip conductors.

At low frequencies, skin depth is large compared to thickness and current is effectively evenly distributed throughout the conductor volume. A conductor is considered to be electrically thin in this frequency region. The edge singularity is of no consequence, and loss is constant with frequency.

When the resistance per unit length equals the inductive reactance per unit length, the edge singularity begins to form. Loss increases as the edge singularity emerges from the uniform low-frequency current distribution. Once it has completely emerged, except for the effect of dispersion, loss is once more constant with frequency. This transition starts at frequency f_a:

$$R(f) = R_o\left[1+\frac{f}{f_a}+\left(\frac{f}{f_b}\right)^2\right]^{1/4} \tag{23.3}$$

$$f_a = \frac{R}{2\cdot\pi\cdot L}\begin{cases} R = \dfrac{1}{\sigma\cdot w\cdot t} \Rightarrow \begin{cases} R: \text{resist. per unit length} \\ \sigma: \text{bulk conductivity} \\ w: \text{width} \\ t: \text{thickness} \end{cases} \\ L = \dfrac{Z_0}{\upsilon} \Rightarrow \begin{cases} L: \text{induct. per unit length} \\ Z_0: \text{characteristic impedance} \\ \upsilon: \text{velocity of propagation} \end{cases} \end{cases} \tag{23.3a}$$

$$f_b = \frac{4}{\pi\cdot\mu\cdot\sigma\cdot t^2} \Rightarrow \begin{cases} \mu: \text{conductor magnetic permeability} \\ \sigma: \text{bulk conductivity} \\ t: \text{thickness} \end{cases} \tag{23.3b}$$

While this expression for the first transition frequency seems to be intuitively reasonable, it should be carefully noted that it is based purely on empirical observation.

When the conductor is thick compared to skin depth, loss increases because current is increasingly confined to the surface of the conductor. It is in this region in which the classic square root of frequency behavior can occur. The transition frequency is selected to be where the conductor is two skin depths thick as f_b.

23.3 Loading the Coreless PCB Transformers with Resistive Load

This section presents the transfer function and the input impedance for a transformer with resistive load.

23.3.1 Transfer Function

The transfer function of the PCB transformer is evaluated under loaded condition. Figure 23.4 shows the high-frequency model of the transformer referred to the primary side where a resistive load has been added.

From Figure 23.4, it can be obtained the following equations:

$$\frac{v_1 - v_{Lm}}{R_1 + s\cdot L_{lk1}} = \frac{v_{Lm}}{s\cdot L_m} + \frac{v_{Lm} - v_2'}{R_2' + s\cdot L_{lk2}'}$$
$$\frac{v_{Lm} - v_2'}{R_2' + s\cdot L_{lk2}'} + \frac{v_1 - v_2'}{1/(s\cdot C_{12}')} = \frac{v_2'}{1/(s\cdot C_2')} + \frac{v_2'}{n^2\cdot R_L} \tag{23.4}$$

From (23.2) and (23.4), the transfer function of the transformer can be obtained and it is shown in (23.5):

$$\frac{v_2}{v_1} = \frac{1/n\cdot\left\{s\cdot C_{12}'\left[\left(R_2' + s\cdot L_{lk2}'\right)\left(\frac{1}{R_1 + s\cdot L_{lk1}} + \frac{1}{s\cdot L_m}\right) + 1\right] + \frac{1}{R_1 + s\cdot L_{lk1}}\right\}}{\left[\left(R_2' + s\cdot L_{lk2}'\right)\left(\frac{1}{R_1 + s\cdot L_{lk1}} + \frac{1}{s\cdot L_m}\right) + 1\right]} \cdot \frac{1}{\left(\frac{1}{R_2' + s\cdot L_{lk2}'} + s\cdot C_{12}' + s\cdot C_2' + \frac{1}{n^2\cdot R_L}\right) - \frac{1}{R_2' + s\cdot L_{lk2}'}} \tag{23.5}$$

23.3.2 Input Impedance

From Figure 23.4 and from (23.5), the input impedance of the transformer referred to the primary side can be calculated by using the following equation:

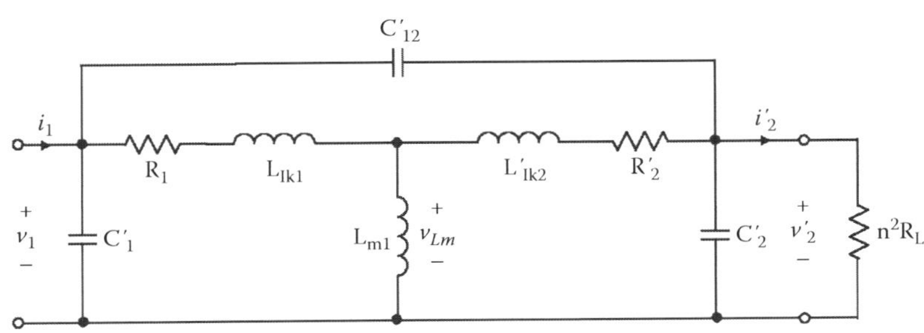

FIGURE 23.4
High-frequency model for coreless PCB transformer with resistive load.

$$Z_{in} = \frac{1}{s \cdot C_1' + s \cdot C_{12}' \cdot (1 - n \cdot (v_2 / v_1)) + ((1 - W)/(R_1 + s \cdot L_{lk1}))} \tag{23.6}$$

where

$$W = \frac{s \cdot C_{12}' + \dfrac{R_2' + s \cdot L_{lk2}'}{R_1 + s \cdot L_{lk1}} \cdot \left(\dfrac{1}{R_2' + s \cdot L_{lk2}'} + s \cdot C_{12}' + s \cdot C_2' + \dfrac{1}{n^2 \cdot R_L} \right)}{\left[(R_2' + s \cdot L_{lk2}') \left(\dfrac{1}{R_1 + s \cdot L_{lk1}} + \dfrac{1}{s \cdot L_m} \right) + 1 \right]} \cdot \frac{1}{\left(\dfrac{1}{R_2' + s \cdot L_{lk2}'} + s \cdot C_{12}' + s \cdot C_2' + \dfrac{1}{n^2 \cdot R_L} \right) - \dfrac{1}{R_2' + s \cdot L_{lk2}'}}$$

23.3.3 Resonant Frequency

The resonant frequency depends on the equivalent inductance and capacitance of the transformer model and is given by the following equation:

$$f_o = \frac{1}{2 \cdot \pi \cdot \sqrt{L_{eq} \cdot C_{eq}}} \quad \begin{cases} C_{eq} = C_{12}' + C_2' \\ L_{eq} = L_{lk2}' + \dfrac{L_{lk1} \cdot L_m}{L_{lk1} + L_m} \end{cases} \tag{23.7}$$

Figure 23.5 presents the voltage gain versus frequency for the transformer "1A." From Figure 23.5, it can be observed that the voltage gain is found to have a wide operating range from few hundred kilohertz up to 10 MHz.

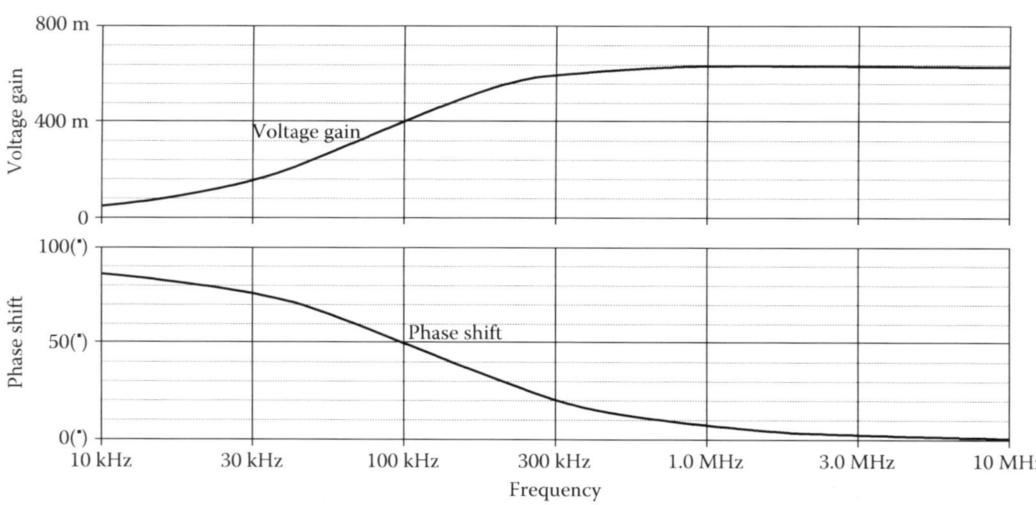

FIGURE 23.5
Voltage gain and phase shift of the transformer with resistive load.

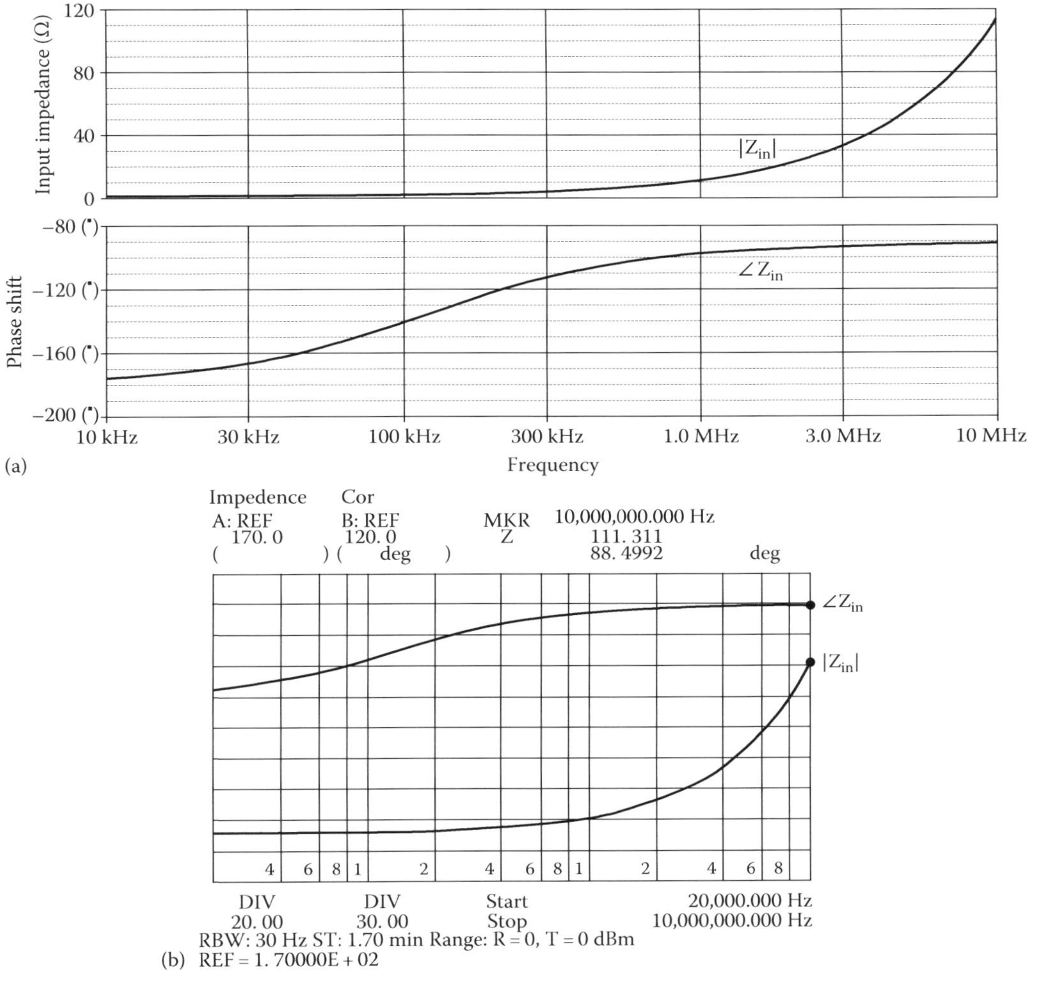

FIGURE 23.6
(a) Input impedance of the transformer with resistive load. (b) Input impedance of the transformer with resistive load versus frequency. Experimental results obtained with an impedance analyzer (x-axis: frequency/y-axis: input impedance).

Figure 23.6a shows the input impedance versus frequency for the transformer with resistive load. Figure 23.6b shows the graphical results presented by a network analyzer when it measures the input impedance of the transformer.

From Figure 23.6a and b, it can be appreciated that the input impedance is predominantly inductive at high frequencies.

23.4 Loading the Transformer with Resistive and Capacitive Load

In this section, the coreless PCB transformer with resistive and capacitive load is going to be analyzed.

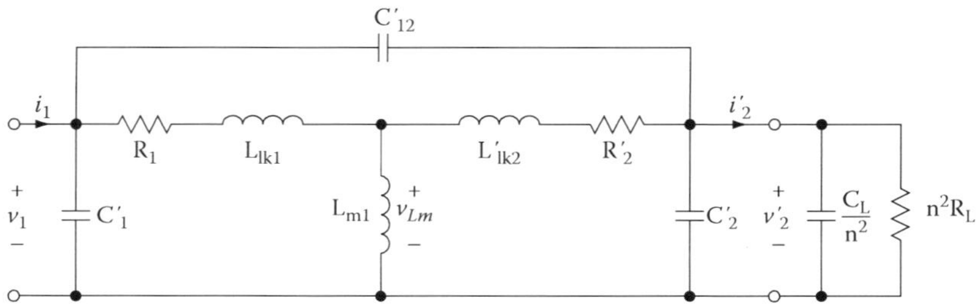

FIGURE 23.7
High-frequency model for coreless PCB transformer with resistive and capacitive load.

Figure 23.7 shows the high-frequency model for the coreless PCB transformer with resistive and capacitive load.

Coreless PCB transformers have been used as electrical isolation in the MOSFET/IGBT gate drive circuits; the gate-source load of the power transistors can be treated as a capacitive and resistive load. The gate capacitance ranges from several hundreds of picofarads to a few nanofarads. In order to simulate the resistive component of the typical gate drive circuit, a parallel resistor is included.

23.4.1 Transfer Function

The transfer function of the PCB transformer is evaluated under loaded condition shown in Figure 23.7. From Figure 23.7, the following equation can be obtained:

$$\frac{v_1 - v_{Lm}}{R_1 + s \cdot L_{lk1}} = \frac{v_{Lm}}{s \cdot L_m} + \frac{v_{Lm} - v_2'}{R_2' + s \cdot L_{lk2}'}$$

$$\frac{v_{Lm} - v_2'}{R_2' + s \cdot L_{lk2}'} + \frac{v_1 - v_2'}{1/(s \cdot C_{12}')} = \frac{v_2'}{1/(s \cdot (C_2' + (C_L / n^2)))} + \frac{v_2'}{n^2 \cdot R_L} \tag{23.8}$$

From (23.2) and (23.8), the transfer function of the transformer can be obtained and it is represented in (23.9):

$$\frac{v_2}{v_1} = \frac{1/n \cdot \left\{ s \cdot C_{12}' \left[(R_2' + s \cdot L_{lk2}') \left(\frac{1}{R_1 + s \cdot L_{lk1}} + \frac{1}{s \cdot L_m} \right) + 1 \right] + \frac{1}{R_1 + s \cdot L_{lk1}} \right\}}{\left[(R_2' + s \cdot L_{lk2}') \left(\frac{1}{R_1 + s \cdot L_{lk1}} + \frac{1}{s \cdot L_m} \right) + 1 \right]}$$

$$\cdot \frac{1}{\left(\frac{1}{R_2' + s \cdot L_{lk2}'} + s \cdot C_{12}' + s \cdot \left(C_2' + \frac{C_L}{n^2} \right) + \frac{1}{n^2 \cdot R_L} \right) - \frac{1}{R_2' + s \cdot L_{lk2}'}} \tag{23.9}$$

23.4.2 Input Impedance

From Figure 23.7 and from (23.9), the input impedance of the transformer referred to the primary side can be calculated by using the following equation:

$$Z_{in} = \frac{1}{s \cdot C_1' + s \cdot C_{12}' \cdot (1 - n \cdot (v_2 / v_1)) + ((1 - W) / (R_1 + s \cdot L_{lk1}))} \tag{23.10}$$

where

$$W = \frac{s\cdot C'_{12} + \dfrac{R'_2 + s\cdot L'_{lk2}}{R_1 + s\cdot L_{lk1}}\cdot\left(\dfrac{1}{R'_2 + s\cdot L'_{lk2}} + s\cdot C'_{12} + s\cdot\left(C'_2 + \dfrac{C_L}{n^2}\right) + \dfrac{1}{n^2\cdot R_L}\right)}{\left[(R'_2 + s\cdot L'_{lk2})\left(\dfrac{1}{R_1 + s\cdot L_{lk1}} + \dfrac{1}{s\cdot L_m}\right) + 1\right]} \cdot \frac{1}{\left(\dfrac{1}{R'_2 + s\cdot L'_{lk2}} + s\cdot C'_{12} + s\cdot\left(C'_2 + \dfrac{C_L}{n^2}\right) + \dfrac{1}{n^2\cdot R_L}\right) - \dfrac{1}{R'_2 + s\cdot L'_{lk2}}}$$

23.4.3 Resonant Frequency

The resonant frequency depends on the equivalent inductance and capacitance of the transformer model and is given by the following equation:

$$f_o = \frac{1}{2\cdot\pi\cdot\sqrt{L_{eq}\cdot C_{eq}}} \qquad \begin{cases} C_{eq} = C'_{12} + C'_2 + \dfrac{C_L}{n^2} \\ L_{eq} = L'_{lk2} + \dfrac{L_{lk1}\cdot L_m}{L_{lk1} + L_m} \end{cases} \tag{23.11}$$

Figure 23.8 presents the voltage gain versus frequency for the transformer "1A." Table 23.3 lists the electrical parameters of the transformer measured at about 2 MHz with an impedance analyzer and the resistive-capacitive load.

Comparing Figures 23.5 and 23.8, it can be observed that a higher value of the load capacitance would lower the resonant frequency of the transformer in accordance with Equation 23.11. From Figure 23.8, the maximum voltage gain is 10.25 at the frequency of 4.93 MHz.

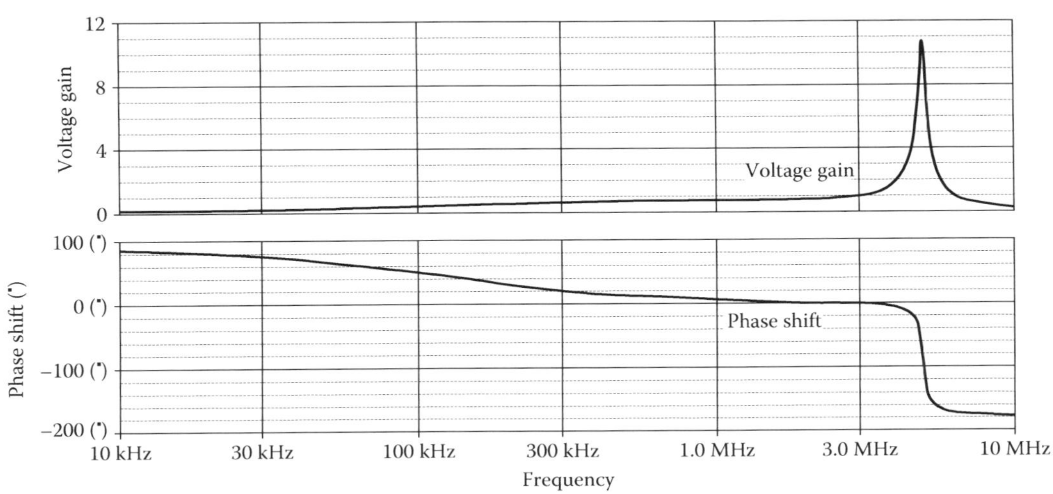

FIGURE 23.8
Transfer function for the coreless transformer with capacitive load.

TABLE 23.3

Parameters for the Transformer A1 at about 2 MHz (Resistive and Capacitive Load)

C_1 (pF)	C_2 (pF)	C_{12} (pF)	R_1 (Ω)	R_2 (Ω)	L_{lk1} (μH)	L_{lk2} (μH)	L_{m1} (μH)	R_L (kΩ)	C_L (pF)	n
3.27	3.27	9.127	1.32	1.32	0.63	0.63	1.1085	10	1000	15:15

Figure 23.9 shows the transfer function with capacitive load and for three values of R_L (100 Ω, 1 kΩ, 10 kΩ). Higher values for the voltage gain are obtained as R_L increases.

Figure 23.10 shows how the transfer function is affected when the load capacitance takes different values. As the load capacitance is smaller, the resonant frequency and the voltage gain are reduced.

From Figures 23.9 and 23.10, it can be concluded that the voltage gain can be greater than 1 in a transformer with the same number of turns in the primary and secondary windings. This feature clears the misunderstanding that coreless PCB transformer has low voltage gain due to low coupling factor.

The resonant effect is due to the partial resonant effect of the load capacitance and the leakage inductance. From these features, it can be concluded that the traditional undesirable leakage inductance can be desirable for its resonant effects.

Figure 23.11a and b show the input impedance versus frequency for the transformer with capacitive load. The maximum input impedance is 475 Ω for a value of the frequency of 3.8 MHz.

Figures 23.12 and 23.13 show the effects on the input impedance if the load resistance and the load capacitance are changed. From Figure 23.12, it can be observed that if R_L is changed from 100 to 10 kΩ, the maximum input impedance at the frequency of 3.8 MHz increases from 50 to 473 Ω. From Figure 23.13, it can be observed that if C_L is changed from 470 pF to 1 nF, the maximum input impedance and resonant frequency are reduced as shown in the following expression:

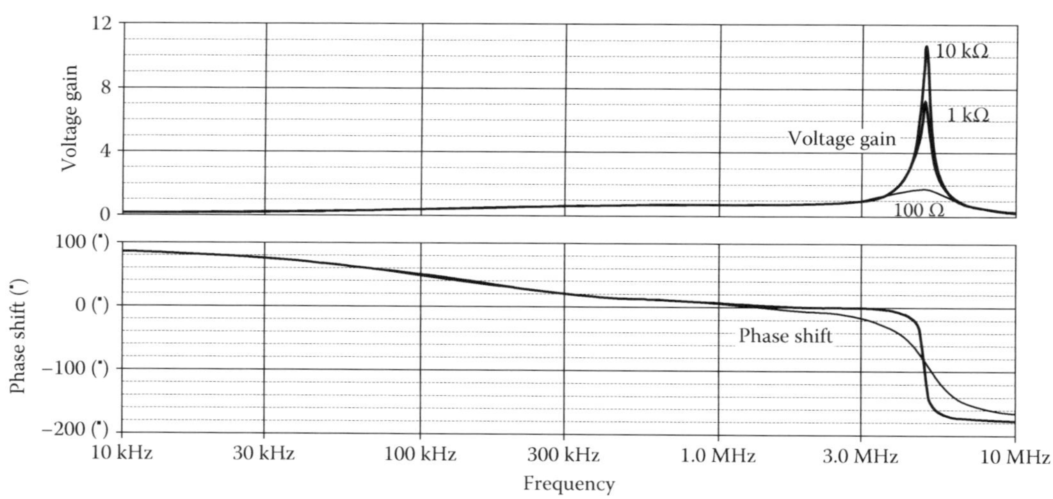

FIGURE 23.9
Transfer function for the coreless transformer: C_L = 1 nF, R_L = (100 Ω, 1 kΩ, 10 kΩ).

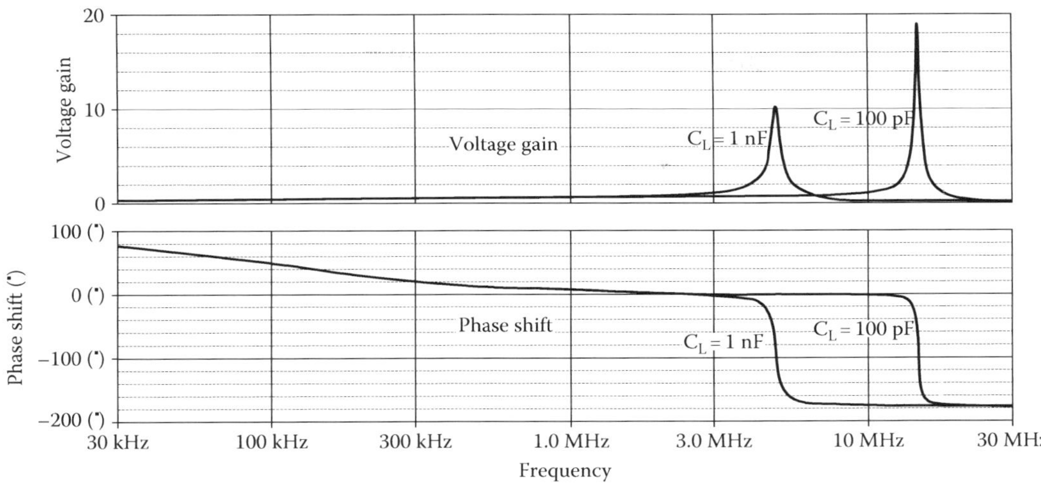

FIGURE 23.10
Effects of varying the load capacitance over the transfer function, being $R_L = 10\,k\Omega$ and $C_L = 100\,pF\ /\ 1\,nF$.

$$C_L = 470\text{ pF} \Rightarrow Z_{in(MAX)} = 878.78\ \Omega / f_o = 5.5\text{ MHz}$$

$$C_L = 1\text{ nF} \Rightarrow Z_{in(MAX)} = 473\ \Omega / f_o = 3.8\text{ MHz} \tag{23.12}$$

Figures 23.12 and 23.13 and the earlier expression indicate that the PCB coreless transformer despite its short conducting tracks can have high input impedance and they do not behave like short-circuit paths.

23.4.4 Efficiency

In coreless transformers, there is no magnetic core loss; power dissipation of the transformer is due to conductor loss.

Energy efficiency of the transformer is given by Equation 23.13, where Z_{in} is given in Equation 23.10:

$$\eta = \frac{P_{out}}{P_{in}} = \frac{|V_2|^2 / R_L}{|V_1|^2 \cdot \Re\{1/Z_{in}\}} = \frac{1}{R_L \cdot \Re\{1/Z_{in}\}} \cdot \left|\frac{V_2}{V_1}\right|^2 \tag{23.13}$$

Figures 23.14 and 23.15 show the efficiency of the transformer versus frequency when the load resistance and the load capacitance are changed. From Figure 23.14, it can be observed that as the load resistance increases, the maximum transformer efficiency decreases from 90% for $R_L = 100\ \Omega$ to 12.5% for $R_L = 10\,k\Omega$. From Figure 23.15, it can be observed that if C_L is changed from 100 pF to 1 nF, the maximum efficiency and the resonant frequency are reduced from 57.7%, 10.75 MHz, to 12.5%, 3.5 MHz. Efficiency can be increased by means of decreasing load resistance and load inductance.

As summary of this paragraph, it is concluded that the choice of the load capacitance and load resistance can determine the gain, input impedance, efficiency, and resonant frequency of the transformer.

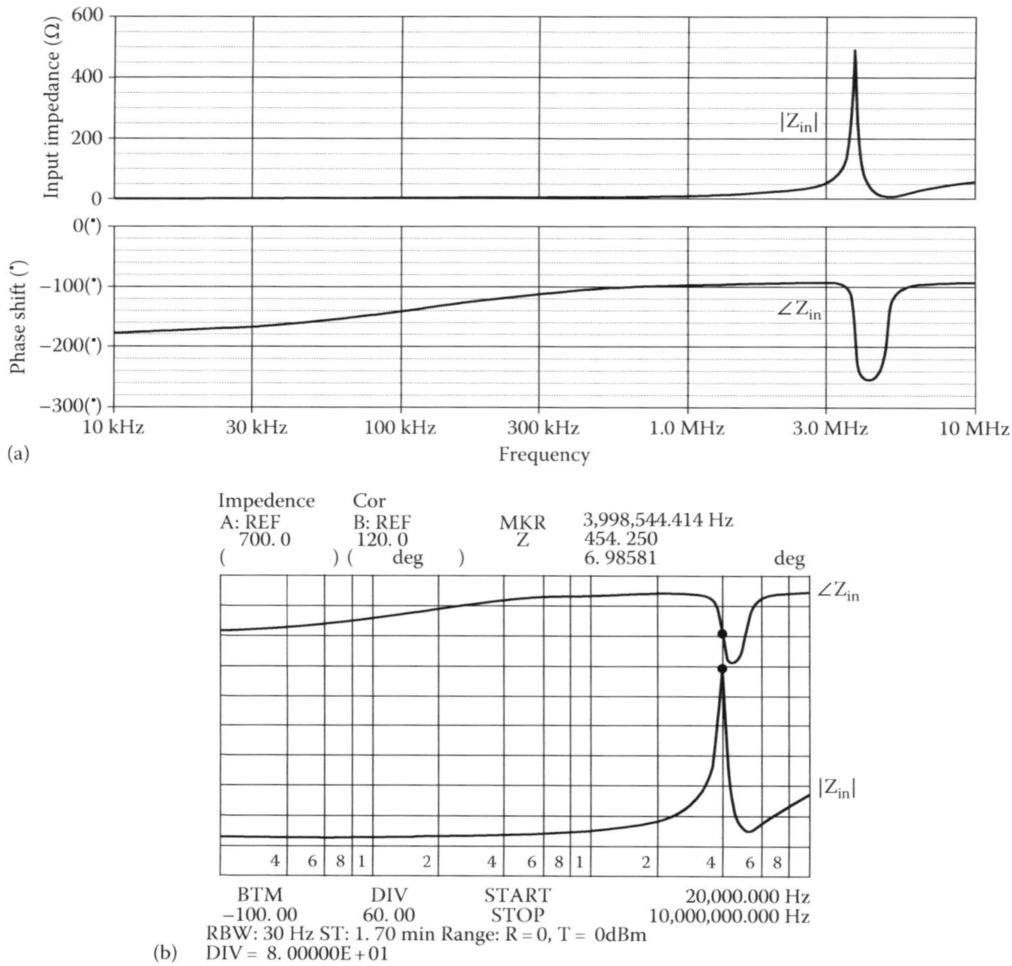

FIGURE 23.11
(a) Input impedance for the coreless transformer with capacitive load, being $R_L = 10\,k\Omega$ and $C_L = 1\,nF$. (b) Input impedance for the coreless transformer with capacitive load. $R_L = 10\,k\Omega$, $C_L = 1\,nF$. Graphical results recorded with a network analyzer (x-axis: frequency/y-axis: input impedance).

From Figures 23.9 and 23.10, it can be observed that maximum voltage gain and the frequency for this maximum gain can be adjusted by selecting a combination of R_L and C_L.

From Figures 23.12 and 23.13, it can be observed that maximum input impedance and the frequency for this maximum impedance can be adjusted by selecting a combination of R_L and C_L.

In Figure 23.14 it is shown that for a fixed value of the load capacitor, efficiency increases as the load resistance decreases, and finally in Figure 23.15 it is shown that for a fixed value of the load resistance, efficiency and resonant frequency decrease as the load capacitance increases.

Once the behavior of the transformer under different load conditions is known, the question is, how to choose the optimal operating conditions for the transformer? This question is answered in the next paragraph.

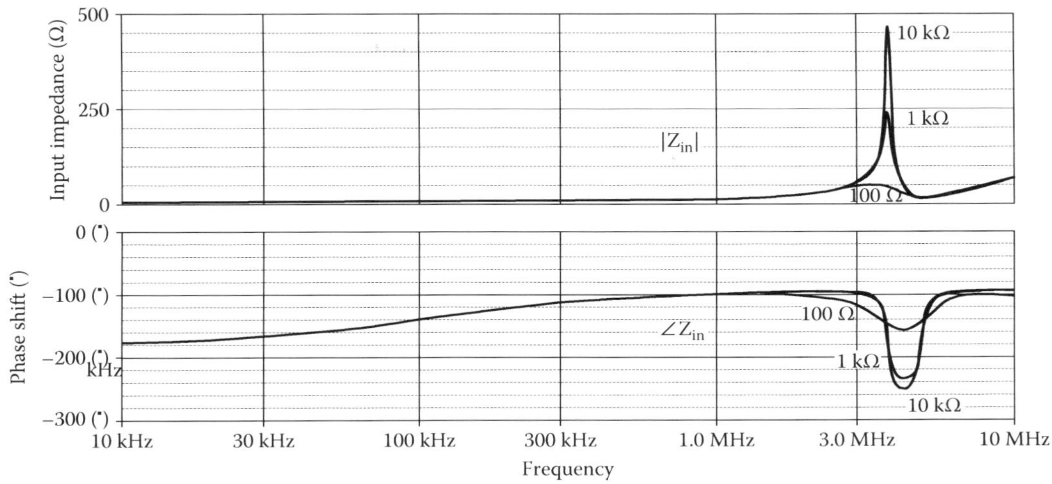

FIGURE 23.12
Variation of input impedance with load resistance: $C_L = 1\,nF$, $R_L = (100\,\Omega, 1\,k\Omega, 10\,k\Omega)$.

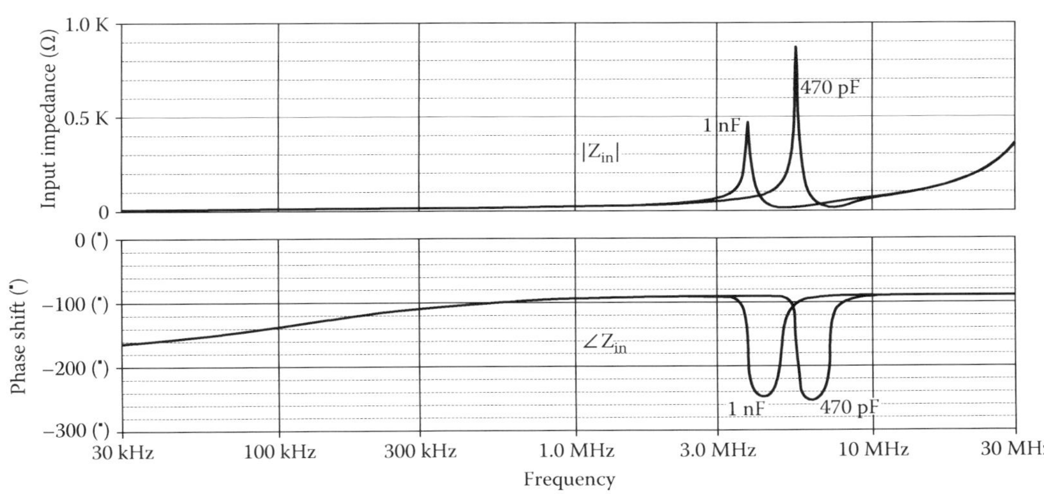

FIGURE 23.13
Variation of input impedance with load capacitance: $R_L = 10\,k\Omega$, $C_L = (470\,pF, 1\,nF)$.

23.5 Optimal Operating Conditions for Coreless PCB Transformers

From results obtained in the last paragraph, it can be concluded that with the coreless transformer, it is possible to obtain high voltage gain at frequencies under 5 MHz by selecting a capacitor of about 1 nF.

Maximum input impedance and the frequency for this maximum impedance can be adjusted by selecting a combination of R_L and C_L. The frequency at which the input impedance reaches its maximum is named Maximum Impedance Frequency (MIF).

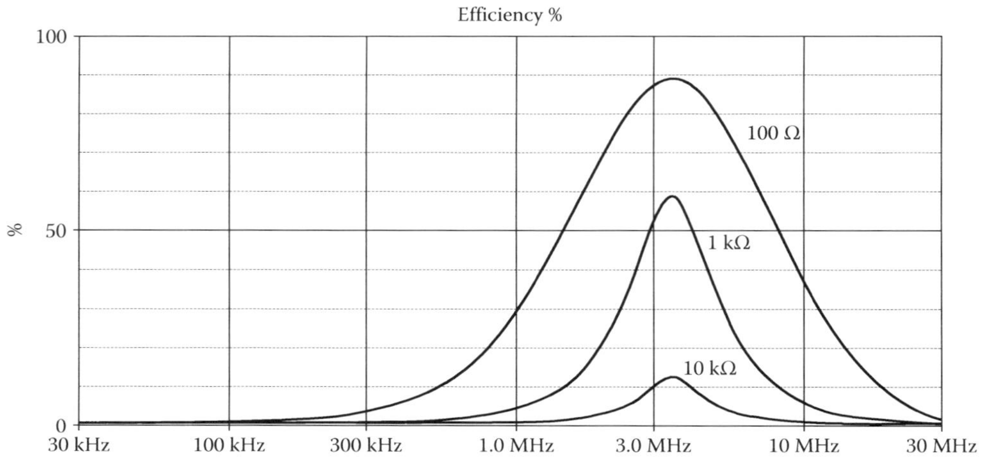

FIGURE 23.14
Variation of efficiency with load resistance: $C_L = 1\,nF$, $R_L = (100\ \Omega, 1\,k\Omega, 10\,k\Omega)$.

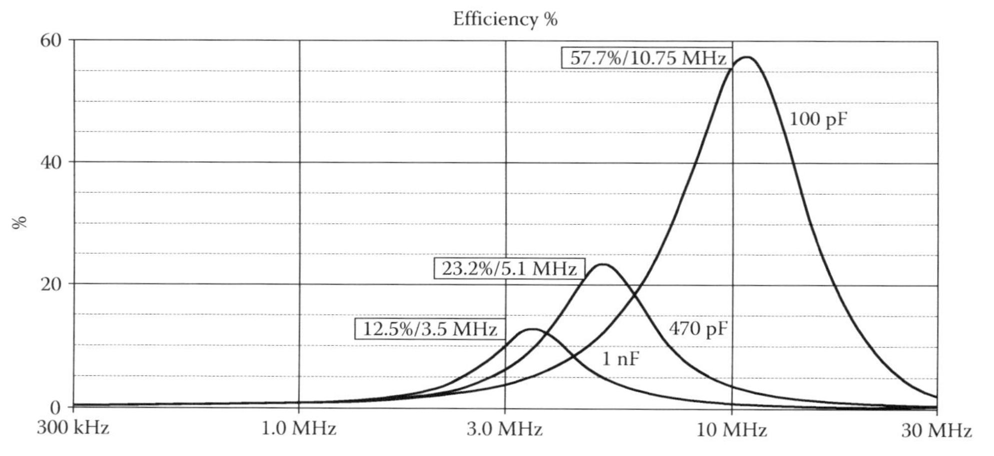

FIGURE 23.15
Variation of efficiency with load capacitance: $R_L = 10\,k\Omega$, $C_L = (100\,pF, 470\,pF, 1\,nF)$.

Figure 23.16 shows the voltage gain and input impedance versus frequency. MIF is 3.8 MHz, for this value of frequency corresponds with an input impedance of 473 Ω and a voltage gain of 1.55. MIF is a suitable operating frequency if minimum input power is required; this can be useful in signal transfer applications where minimum power requirement is preferred or in drive circuits for power MOSFET and IGBT.

Efficiency curves versus frequency for various load resistances and capacitances were plotted in Figures 23.14 and 23.15. The frequency at which the efficiency reaches its maximum is named Maximum Efficiency Frequency (MEF).

Efficiency, input impedance, and voltage gain versus frequency are represented in Figure 23.17. It can be seen that the maximum efficiency MEF is slightly less than MIF.

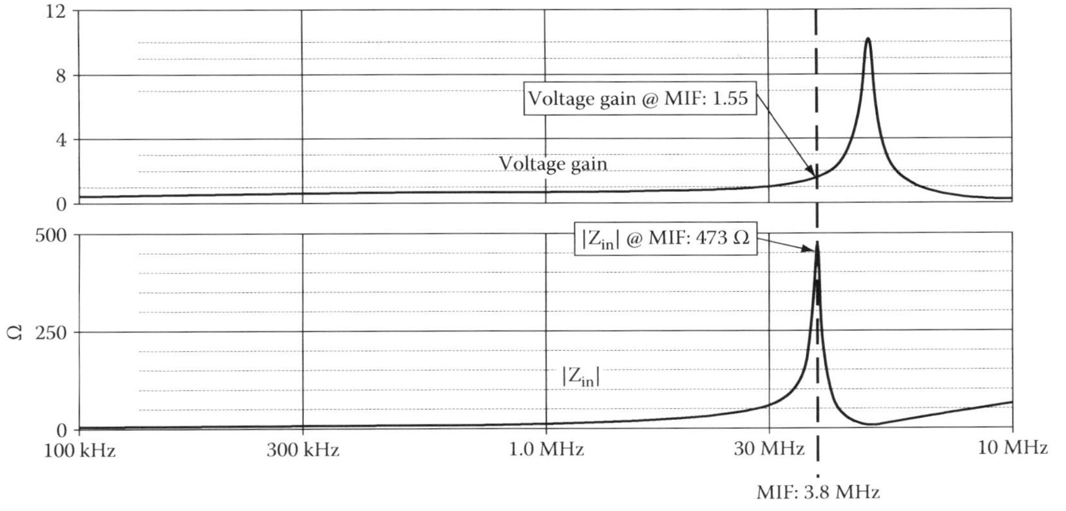

FIGURE 23.16
Voltage gain and input impedance at MIF: $R_L = 10\,k\Omega$, $C_L = 1\,nF$.

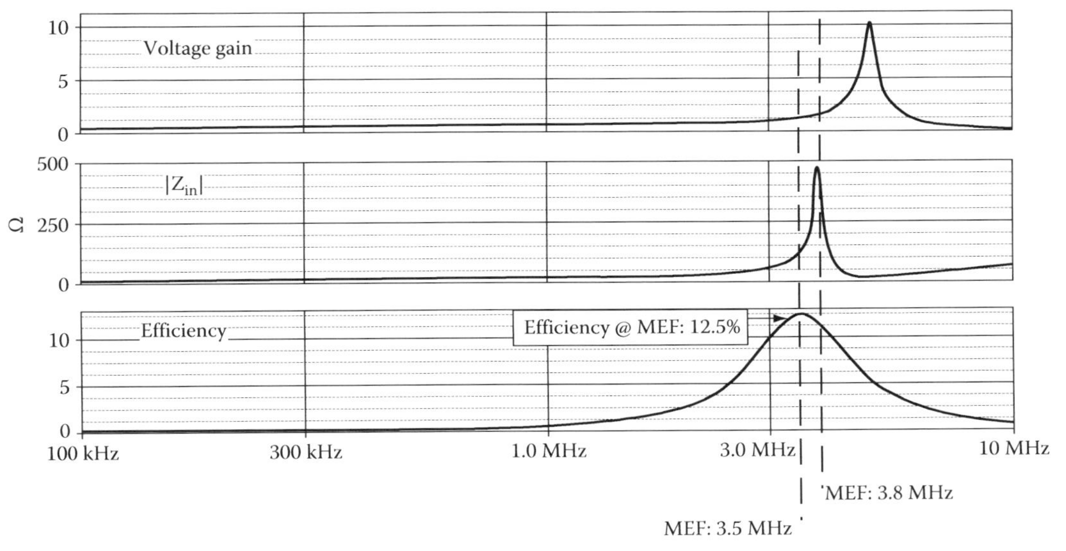

FIGURE 23.17
Efficiency at MEF: $R_L = 10\,k\Omega$, $C_L = 1\,nF$.

It is important to note that when load resistance is large, that is, load power is low, load current and current in secondary winding are small; in this situation, power dissipation of the transformer is dominated by resistive losses in the primary winding. On the other hand, increasing the transformer input impedance reduces the primary winding current. Thus, the MEF tends to approach the MIF as the load resistance increases.

When the load power increases, that is, load resistance decreases, in this situation, resistive power losses due to secondary winding current are increased. Because of proximity and skin effects, winding resistance increases as operating frequency increases. As a result, operating the transformer in relatively low frequency can reduce the power loss of the transformer when secondary current is significant.

23.6 Influence of Geometric Parameters on Coreless Transformers

Electrical parameters of coreless PCB transformers and mainly the inductances depend on the geometry of the coreless transformer. The geometry of the transformer is controlled with the following parameters (see Figure 23.18):

- Outermost radius R
- Number of turns N
- Conductor width w
- Dielectric thickness z
- Conductor thickness h

Circular spiral windings are studied in this analysis because they have less resistance and inter-winding capacitance than square or polygonal spiral windings because circular windings have shorter tracks and less overlap area than the others.

23.6.1 Influence of the Outermost Radius on the Inductance

In order to analyze the influence of the radius on the inductance, some transformers with different track separation, but with the same number of turns, are going to be analyzed. The geometry of the primary winding is the same as that of the secondary winding.

Figure 23.19 represents the inductance as a function of the transformer radius. It is found that self-inductance L_1 increases linearly with radius. The self-inductance of the primary winding is given by

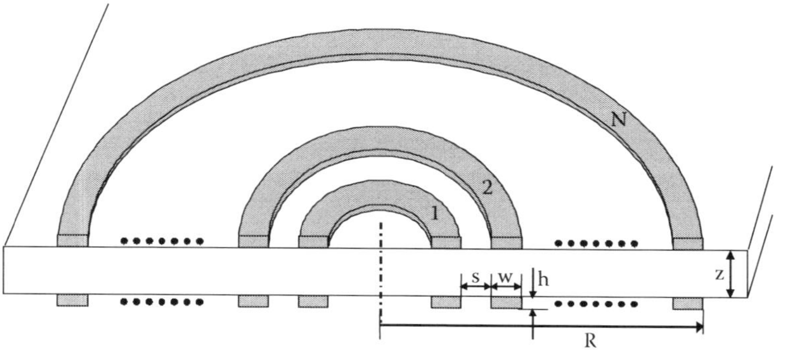

FIGURE 23.18
Geometric parameters of the coreless transformer.

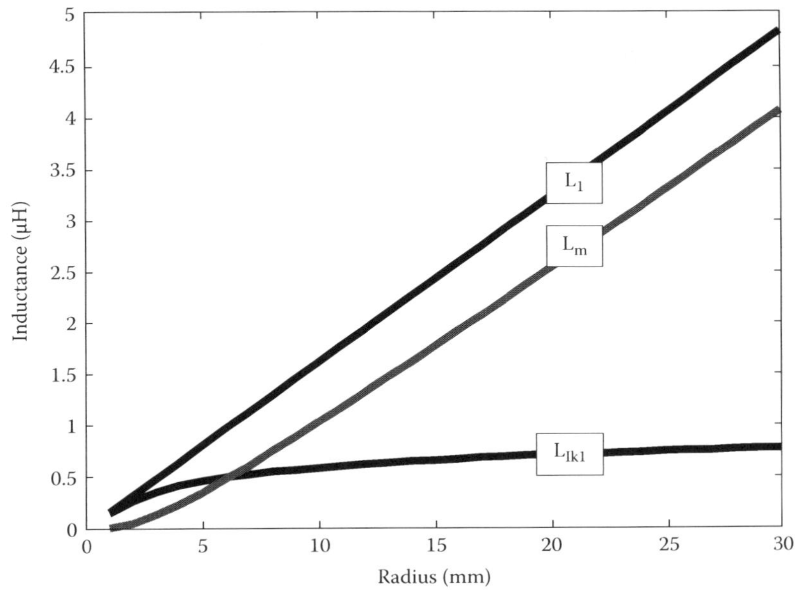

FIGURE 23.19
Inductances of the transformer versus radius.

$$L_1 = k_n \cdot R \tag{23.14}$$

where

- k_n is a constant that depends on number of turns and geometry of the primary winding
- R is the radius of the transformer

If the transformer radius is much greater than the dielectric thickness, the mutual inductance L_m and the leakage inductance L_{lk1} increase linearly.

From Figure 23.19, it can be observed that the slope of the mutual inductance is much greater than that of the leakage inductance, and then, the increase of mutual inductance is greater than that of leakage inductance. Thus, the coupling coefficient of the transformer can be improved by increasing the transformer diameter.

The mutual inductance and the leakage inductance can be represented by the following equations [21]:

$$\begin{aligned} L_m &= -k_0 + (k_n - k_1) \cdot R + (k_1 \cdot R + k_0) \cdot e^{(-R/k)} \\ L_{lk1} &= (k_1 \cdot R + k_0) \cdot (1 - e^{(-R/k)}) \end{aligned} \tag{23.15}$$

where k, k_0, and k_1 depend on the number of turns, geometry of the transformer windings, and the laminate thickness, respectively.

Graphical results presented in Figure 23.19 were obtained by testing a transformer series with the following parameters:

$$R = 1,\ldots,30 \text{ mm} \quad N = 15 \quad s = w = \frac{R}{N}$$
$$z = 1.66 \text{ mm} \quad h = 35\,\mu\text{m} \tag{23.16}$$

The polynomial approximation of the curves gives us the following solution:

$$L_1 = 0.1604 \cdot R\ \mu\text{H}$$
$$L_m = 0.0035 \cdot R^2 + 0.0859 \cdot R\ \mu\text{H}$$
$$L_{lk1} = -0.0034 \cdot R^2 + 0.0745 \cdot R\ \mu\text{H} \tag{23.17}$$

23.6.2 Influence of the Number of Turns on the Inductance

In order to see the effects of the number of turns on the inductance of coreless PCB transformers, the transformer radius is kept constant so that the track separation decreases as the number of turns increases.

Figure 23.20 represents the variation of the mutual inductance, self-inductance, and leakage inductance as a function of the number of turns. It can be observed in Figure 23.20 that the inductances follow a second-order polynomial of number of turns as given by

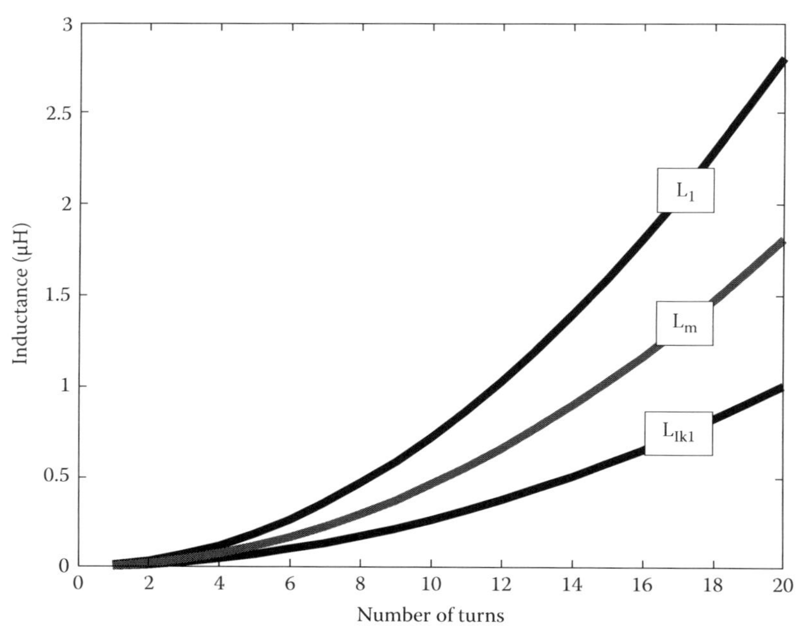

FIGURE 23.20
Inductances of the transformer versus number of turns.

$$L_1 = A_1 \cdot N^2 + A_2 \cdot N$$

$$L_m = A_3 \cdot N^2 + A_4 \cdot N$$

$$L_{lk1} = A_5 \cdot N^2 + A_6 \cdot N \tag{23.18}$$

where

A1, …, A6 are constants that depend on the radius of the transformer and the geometry of the transformer windings

N is the number of turns

Graphical results presented in Figure 23.20 were obtained by testing a transformer series with the following parameters:

$$R = 10\,\text{mm} \quad N = 1,\ldots,20 \quad w = \frac{6}{N}$$

$$z = 1.66\,\text{mm} \quad h = 35\,\mu\text{m} \quad s = \frac{R}{N} - w \tag{23.19}$$

The polynomial approximation of the curves gives the following solution:

$$L_1 = 0.006935 \cdot N^2 + 0.001775 \cdot N\,\mu\text{H}$$

$$L_m = 0.004598 \cdot N^2 - 0.000872 \cdot N\,\mu\text{H}$$

$$L_{lk1} = 0.002337 \cdot N^2 + 0.002648 \cdot N\,\mu\text{H} \tag{23.20}$$

23.6.3 Influence of the Thickness of PCB on the Inductance

The dielectric laminate plays an important role in coreless PCB transformers. The smaller the separation of printed windings is, the greater the magnetic flux coupling becomes. The greater the separation of printed windings, the smaller the flux coupling is. Figure 23.21 represents the influence of the laminate thickness on the inductances.

Graphical results presented in Figure 23.21 were obtained by testing a transformer series with the following parameters:

$$R = 20\,\text{mm} \quad N = 15 \quad s = 0.36\,\text{mm}$$

$$z = 0.2,\ldots,4\,\text{mm} \quad h = 35\,\mu\text{m} \quad w = 0.3\,\text{mm} \tag{23.21}$$

The polynomial approximation of the curves gives us the following solution:

$$L_1 = 3.34\,\mu\text{H}$$

$$L_m = -0.015 \cdot z^3 + 0.131 \cdot z^2 - 0.640 \cdot z - 3.267\,\mu\text{H}$$

$$L_{lk1} = +0.0149 \cdot z^3 - 0.1312 \cdot z^2 + 0.6396 \cdot z\,\mu\text{H} \tag{23.22}$$

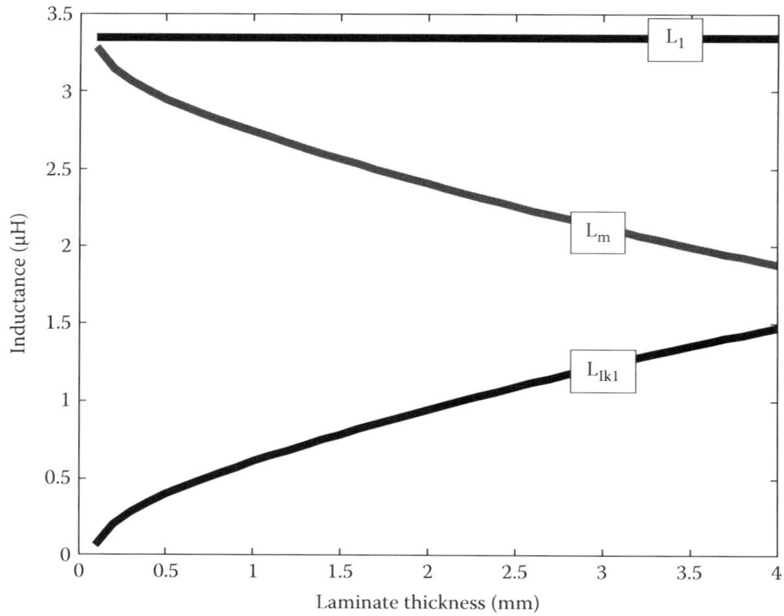

FIGURE 23.21
Inductances of the transformer versus laminate thickness.

23.6.4 Influence of Number of Turns with the Same Track Separation on the Inductance

In this analysis, the winding separation is constant and the transformer radius increases as the number of turns increases. The characteristics of the transformers tested are the following:

$$R = 1, \ldots, 20\,\text{mm} \quad N = 1 \ldots 20 \quad s = w = 0.5\,\text{mm}$$

$$z = 1.66\,\text{mm} \quad h = 35\,\mu\text{m} \tag{23.23}$$

Figure 23.22 represents the inductance versus number of turns with same track separation.

The following equation represents the polynomial approximation of the curves shown earlier:

$$L_1 = 0.0006968 \cdot N^3 + 0.0002335 \cdot N^2 - 0.0001283 \cdot N\,\mu\text{H}$$

$$L_m = 0.000691 \cdot N^3 - 0.002952 \cdot N^2 + 0.006022 \cdot N\,\mu\text{H}$$

$$L_{lk1} = 0.000006 \cdot N^3 + 0.003185 \cdot N^2 - 0.006150 \cdot N\,\mu\text{H} \tag{23.24}$$

23.6.5 Influence of Conductor Width on the Inductance

The number of turns is fixed, the track separation varies with the conductor width, and the track width ranges from 0.02 to 0.6 mm. The parameters for the transformers series are the following:

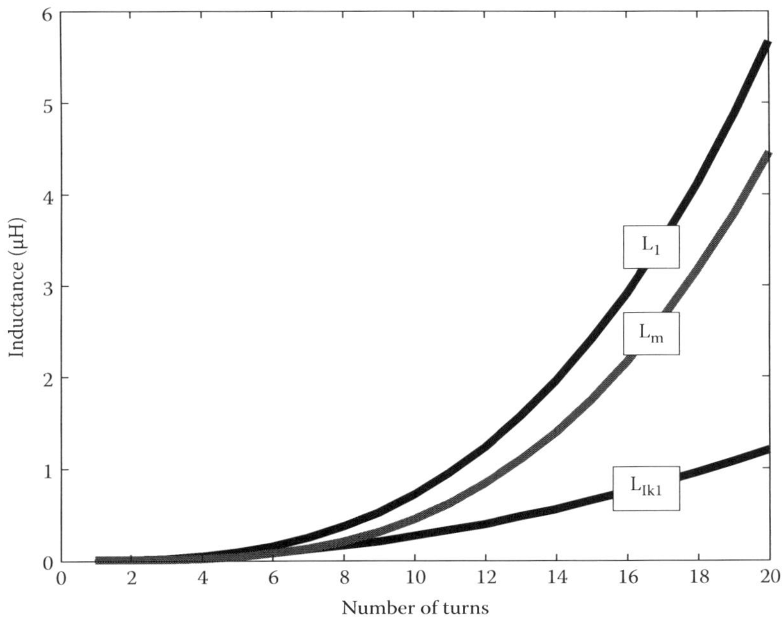

FIGURE 23.22
Inductances of the transformer versus number of turns with same track separation.

$$R = 10\,\text{mm} \quad N = 15 \quad w = 0.02, \ldots, 0.6\,\text{mm}$$

$$z = 1.66\,\text{mm} \quad h = 35\,\mu\text{m} \quad s = \frac{R}{N} - w \tag{23.25}$$

Figure 23.23 shows that leakage, self, and mutual inductances do not vary significantly with the track width.

The polynomial approximation of the curves presented in Figure 23.23 is the following:

$$L_1 = -1.573 \cdot w^3 + 2.133 \cdot w^2 - 1.104 \cdot w + 1.786\,\mu\text{H}$$

$$L_m = -0.009 \cdot w + 1.011\,\mu\text{H}$$

$$L_{lk1} = -1.573 \cdot w^3 + 2.148 \cdot w^2 - 1.104 \cdot w + 0.776\,\mu\text{H} \tag{23.26}$$

23.6.6 Influence of Conductor Thickness on the Inductance

Figure 23.24 shows that the variation of inductances is negligible when the conductor thickness varies from 1 to 40 μm and the other parameters of the transformer are the following:

$$R = 10\,\text{mm} \quad N = 15 \quad w = 0.3\,\text{mm}$$

$$z = 1.66\,\text{mm} \quad h = 1, \ldots, 40\,\mu\text{m} \quad s = 0.36\,\text{mm} \tag{23.27}$$

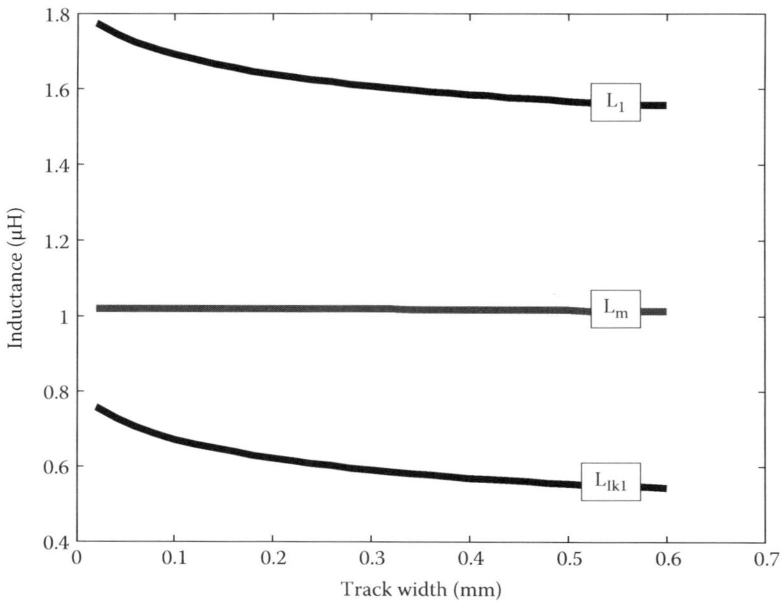

FIGURE 23.23
Inductances of the transformer versus conductor width.

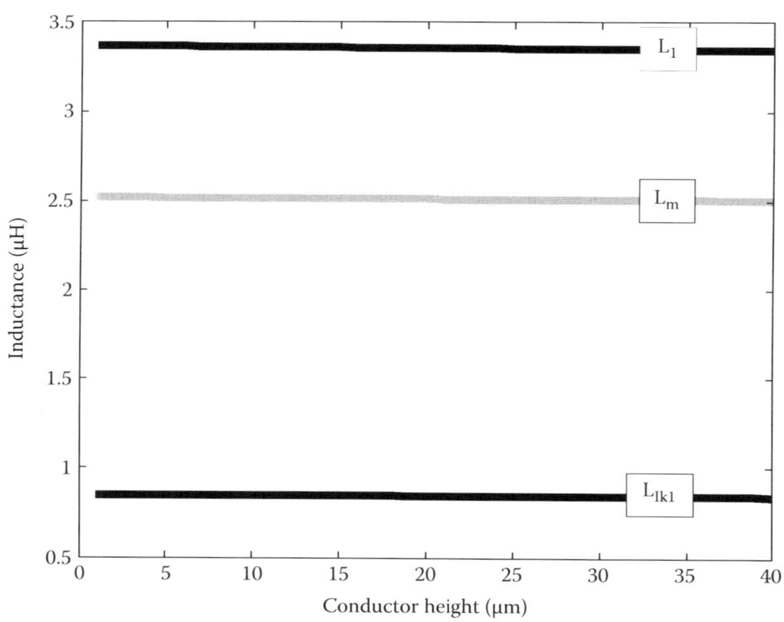

FIGURE 23.24
Inductances of the transformer versus conductor thickness.

The relationship between the inductive parameters and the conductor thickness (h [μm]) of the coreless PCB transformer series can be expressed as

$$\left.\begin{aligned} L_1 &= -0.0003 \cdot h + 0.8492\,\mu H \\ L_m &= 2.512\,\mu H \\ L_{lk1} &= -0.001 \cdot h + 3.361\,\mu H \end{aligned}\right\} h[\mu m] \tag{23.28}$$

23.7 Calculation of Coreless PCB Transformer Parameters

This paragraph gives some analytical solutions for calculating inductances, capacitances, and resistances of coreless PCB transformers. Computation using the analytical solution is more time efficient than that using finite element analysis (FEA).

23.7.1 Calculation of Winding Resistance

The resistance of the windings increases with operating frequency due to skin effects; the winding resistance is given in the following expression [23]:

$$R = R_{DC} \cdot R_{AC} \tag{23.29}$$

where

$$\left. R_{DC} = \rho \cdot \frac{L}{S} \right\} \Rightarrow \begin{cases} \rho = \text{resistivity}\,(\Omega \cdot m^2/m) \\ L = \text{conductor length}\,(m) \\ S = \text{conductor section}\,(m^2) \end{cases}$$

and

$$\left.\begin{aligned} R_{AC} &= \frac{S}{w \cdot \delta(1 - e^{(-t/\delta)})} \\ \delta &= \sqrt{\frac{2}{2 \cdot \pi \cdot f \cdot \mu_0 \cdot \sigma}} \end{aligned}\right\} \Rightarrow \begin{cases} w = \text{conductor width}\,(m) \\ \sigma : \text{conductivity}\,(S/m) \\ \mu_0 = 4 \cdot \pi \cdot 10^{-7}\,(H/m)\text{: magnetic permeability of free space} \\ t : \text{turn thickness}\,(m) \\ S : \text{conductor section}\,(m^2) \\ f : \text{frequency}\,(Hz) \end{cases}$$

23.7.2 Calculation of Winding Capacitances

The capacitance between the windings can be obtained by using the following equation and if the spiral windings are approximated by a solid disk of the same outermost radius and if the inner radius is approximated by zero [18]:

$$C_{12} = \varepsilon_0 \cdot \varepsilon_r \cdot \frac{S}{z} \Bigg\} \Rightarrow \begin{cases} S\text{: conductor section}\,(m^2) \\ z:\text{PCB thickness}\,(m) \\ \varepsilon_0 = 8.8544 \cdot 10^{-12} N^{-1} \cdot m^{-2} \cdot C^2\text{: vacuum permittivity} \\ \varepsilon_r:\text{relative permittivity of PCB material} \end{cases} \tag{23.30}$$

Exact expressions for the inter-winding capacitances are complex, and in general the value is obtained from some experimental measurements.

In order to obtain these experimental measurements, several groups of transformers with the same outermost radius have been tested. There are five groups, transformers in each group have the same track width w and different number of turns N. Figure 23.25 shows the variation of inter-winding capacitance with track width and number of turns.

For coreless PCB transformer with primary and secondary windings printed directly on the opposite sides of a double-sided PCB, the intra-winding capacitance is negligible (typically in the order of a few picofarads). Thus, the intra-winding capacitance of the secondary winding can be ignored in the high-frequency model because it is much smaller than the externally added capacitance.

23.7.3 Calculation of Winding Inductances

The spiral windings can be approximated as concentric circular windings connected in series with infinitesimal connections as shown in Figure 23.26.

If the spiral winding has N-*turns*, the total self-inductance is the summation of each mutual inductance between successive concentric circular coils M_{ij}. Mutual inductance

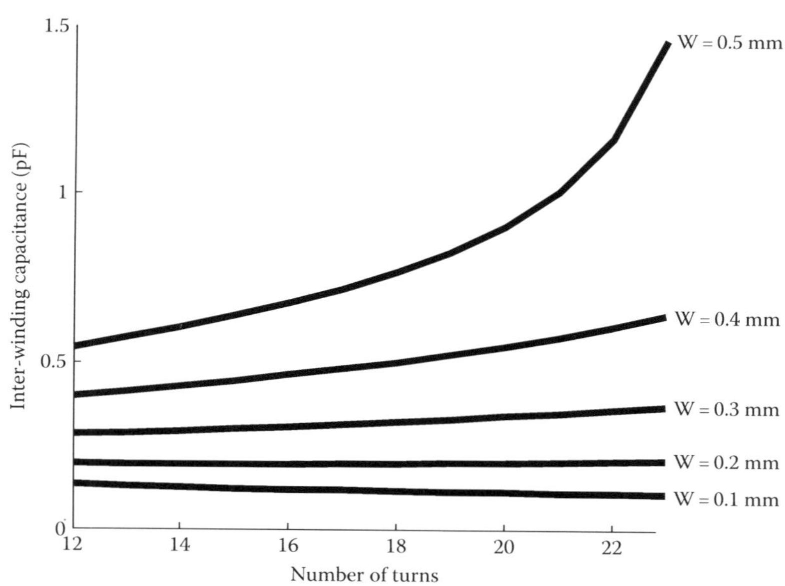

FIGURE 23.25
Variation of inter-winding capacitance with number of turns and track width.

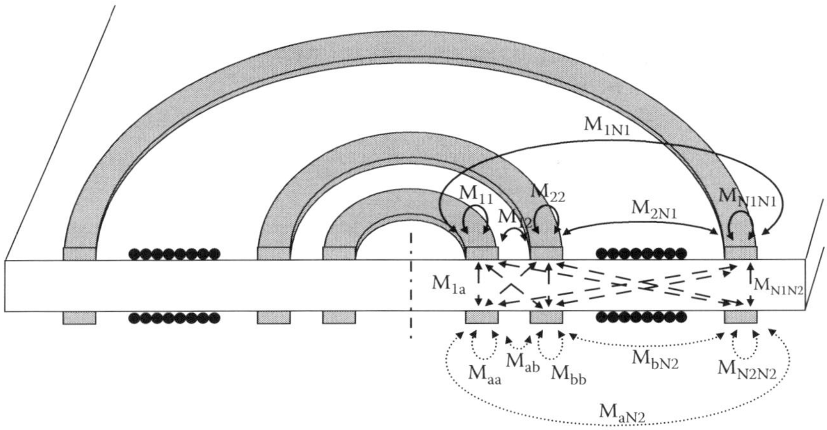

FIGURE 23.26
Approximation of spiral winding as concentric circular winding and mutual inductances between tracks.

between primary and secondary windings is the summation of mutual magnetic coupling pairs between primary and secondary coils M_{xy}. The self-inductance of the primary and secondary windings and the mutual inductance between the primary and the secondary windings are given by the following equation:

$$L_1 = \sum_{i}^{N_1} \sum_{j}^{N_1} M_{ij} \qquad L_2 = \sum_{i}^{N_2} \sum_{j}^{N_2} M_{ij} \qquad L_m = \sum_{i}^{N_1} \sum_{j}^{N_2} M_{ij} \qquad \begin{aligned} L_{lk1} &= L_1 - L_m \\ L_{lk2} &= L_2 - L_m \end{aligned} \tag{23.31}$$

where
  M_{ij} is the mutual inductance between two concentric circular coils
  L_1 is the self-inductance of the primary winding
  L_2 is the self-inductance of the secondary winding
  L_m is the mutual inductance between primary and secondary windings
  L_{lk1} is the leakage inductance of the primary winding
  L_{lk2} is the leakage inductance of the secondary winding

There are different methods in the literature for calculating mutual inductance M_{ij} between two circular tracks: Maxwell equations, method presented by Hurley and Duffy [10], method presented by Mohan [23], etc.

The starting point for inductance calculations is the equation for the mutual inductance between two filaments given by Maxwell. The mutual inductance between two filaments shown in Figure 23.27 is given by the following equation [28]:

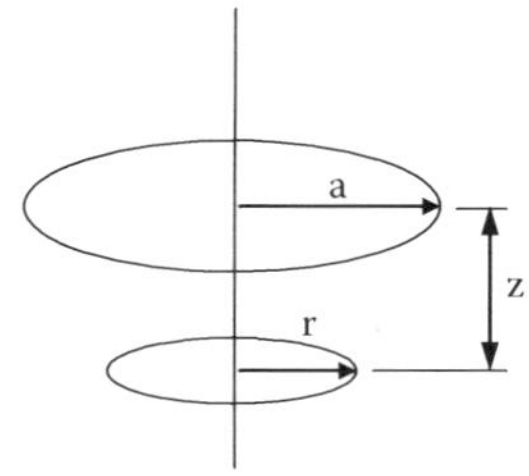

FIGURE 23.27
Circular concentric filaments in air.

$$M = \mu_0 \cdot \sqrt{a \cdot r} \cdot \frac{2}{g}\left[\left(1 - \frac{g^2}{2}\right) \cdot K(g) - E(g)\right] \tag{23.32}$$

where

$$g = \sqrt{\frac{4 \cdot a \cdot r}{z^2 + (a + r)^2}}$$

K(g), E(g) are the complete elliptic integrals of the first and second kind, respectively
a, r are the filament radius
z is the distance between circular filaments
$\mu_0 = 4 \cdot \pi \cdot 10^{-7}$ H/m is the permeability of vacuum

Coils have a finite cross section, and the standard technique is to integrate the filamentary equation over the cross section, assuming a constant current density. Alternatively, an approximate result can be obtained by placing a filament at the center of each coil, and the mutual inductance is calculated directly from the filament equation (23.32) and the distance between the coils, z, is replaced by the geometric mean distance (GMD) between the coils.

In order to approximate the circular spiral winding as concentric circular winding, the radius of the circular track can be taken as the mean value of the maximum and minimum radius of the spiral winding at the cross section. This approximation is shown in Figure 23.28.

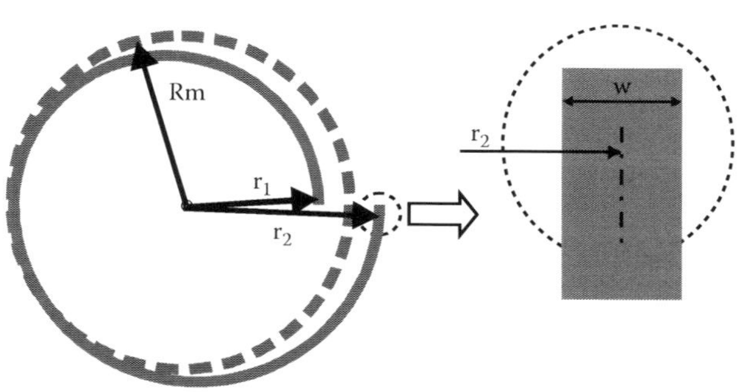

FIGURE 23.28
Approximation of circular spiral winding as concentric circular winding.

The radius of each i track Rm_i is given by the following equation:

$$Rm_i = r_1 + \frac{(r_2 - r_1)\cdot(i-(1/2))}{N}, \quad i = 1,2,3,\ldots,N \tag{23.33}$$

where
r_1 is the inner radius of the coil
r_2 is the outer radius of the coil
N is the number of turns
Rm_i is the radius of the i track

The GMD is a means of representing the total effect with regard to the inductance of two conductor cross sections on each other such that the two conductor cross sections can be replaced by two filamentary conductors. The system of two filaments with their center points separated by this GMD will have the same mutual inductance as the original two conductors. Figure 23.29a represents this statement.

The equation used to calculate the GMD between the two conductors with cross sections S1 and S2 is [29]

$$\ln(GMD_{12}) = \frac{1}{S_1\cdot S_2}\int_{S_1}\int_{S_2}\ln(d)\cdot dS_2\cdot dS_1 \tag{23.34}$$

For a two-conductor transmission line consisting of rectangular conductors, the GMD is calculated by means of integrating and also by discretizing the conductors and treating them as composite conductors (Figure 23.29b) [24]. The obtained values for the GMD (23.21) are then used to calculate the inductance of each conductor:

$$\ln(GMD_{ij}) = \frac{1}{p\cdot q}\left[\sum_{i=1}^{p}\sum_{j=1}^{q}\ln(d_{ij})\right] \tag{23.35}$$

where
d is the distance between points of areas
d_{ij} is the distance between discrete points of areas
p,q is the number of points of discrete areas

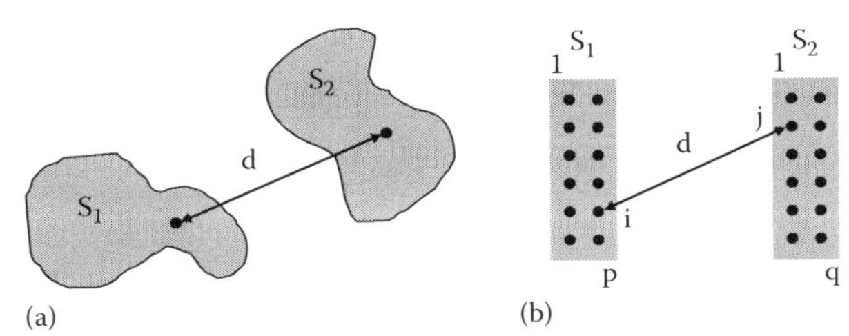

FIGURE 23.29
(a) GMD between two cross sections. (b) Discretization of two rectangular conductors for calculating the GMD.

In the case of planar magnetic components, the aspect ratio of height to width of a section is usually very high; in this situation, the current density is not constant and if this factor is taken into account, accuracy is greatly improved.

As a further improvement, it seems reasonable that an equivalent filament could be obtained provided that the filament is so placed that the correct current density is taken into account. As shown in Ref. [10], equal current division occurs at the radius given by the geometric mean (GM) of the inside and outside radii:

$$r_0 = \sqrt{r_1 \cdot r_2} \tag{23.36}$$

There are four cases to analyze (see Figure 23.30):

1. Self-inductance (MBi,Bi, MTi,Ti). Replace z by GMD of the coil from itself. Place the filament at the center or at the GM of the cross section.
2. Mutual inductance (MBi,Bj, MTi,Tj), ($i \neq j$; $z=0$). Mutual inductance between tracks of the same layer, $z=0$ (Lyle's method).
3. Mutual inductance (MBi,Ti, MTi,Bi), ($i=j$; $z \neq 0$). Replace z by GMD between sections.
4. Mutual inductance (MBi,Tj, MTi,Bj) ($i \neq j$; $z \neq 0$). Replace z by GMD. In this case, it is sufficiently accurate to take GMD = z.

In order to obtain the mutual inductance between coplanar tracks, the single filament method explained earlier is not sufficiently accurate; in this case, Lyle's method is used with two filaments replacing each section as shown in Figure 23.31. The radial dimensions are given in (23.37):

$$r_{1,2} = R_{gm} \cdot \left(1 + \frac{h^2}{24 \cdot R_{gm}^2}\right) \pm \sqrt{\frac{w^2 - h^2}{12}} \tag{23.37}$$

where

h is the height of rectangular conductor
w is the width of rectangular conductor
R_{gm} is the radius for center of GM section

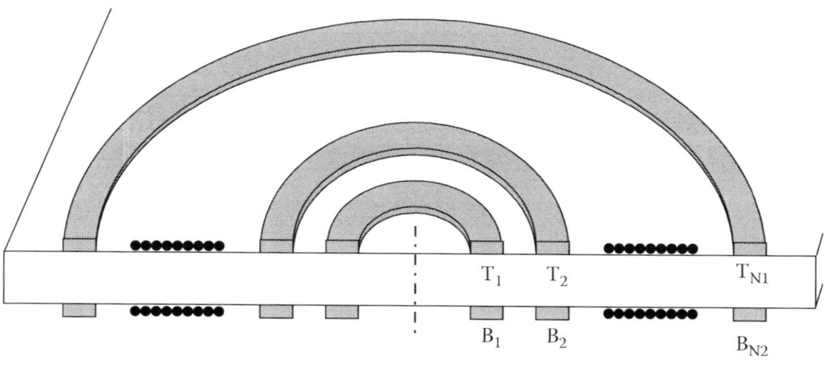

FIGURE 23.30
Determination of the vertical distance between the center of the tracks.

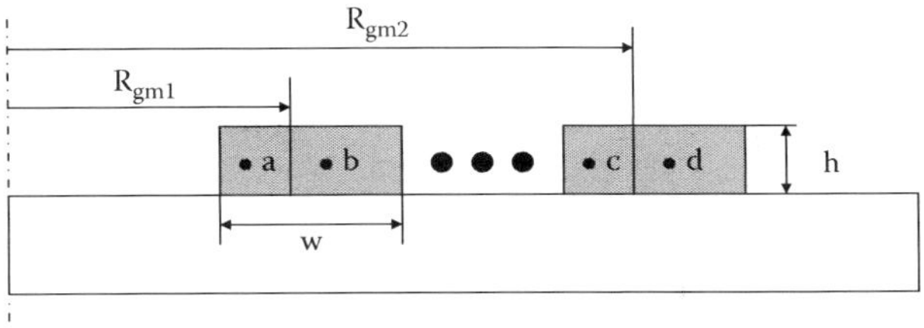

FIGURE 23.31
Lyle's method for calculating mutual inductance in concentric planar tracks.

The total mutual inductance between the two sections is the sum of the individual mutual inductances between the equivalent filaments of each section shown in Figure 23.31, each carrying half the total current:

$$M_{ij} = \frac{M_{ac} + M_{ad} + M_{bc} + M_{bd}}{4} \tag{23.38}$$

where
  a and b represent the filaments in one cross section
  c and d represent the filaments in the other cross section

Hurley and Duffy [10] presented a method for derivation of mutual inductance between two circular tracks with rectangular cross section, as shown in Figure 23.32.

The equation proposed by Hurley and Duffy for calculating the mutual inductance between the circular tracks represented in Figure 23.30 is given in (23.39):

$$M_{ij} = \frac{\mu_0 \cdot \pi}{h_1 \ln(r_2 / r_1) h_2 \ln(a_2 / a_1)} \int_0^{\infty} S(kr_2, kr_1) \cdot S(ka_2, ka_1) \cdot Q(kh_2, kh_1) \cdot e^{-k \cdot |z|} dk \tag{23.39}$$

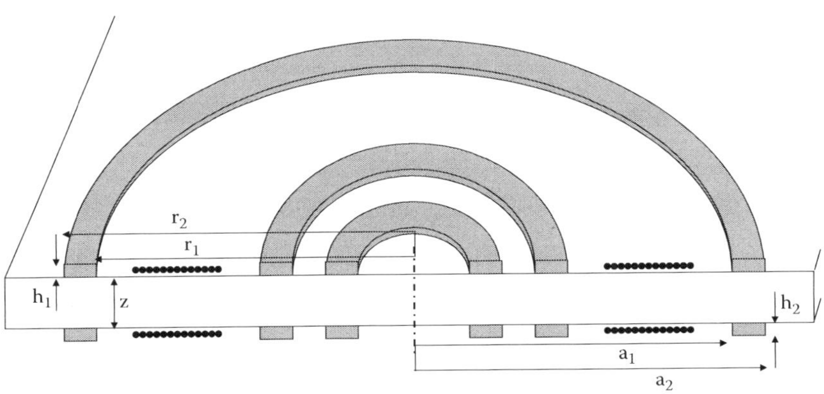

FIGURE 23.32
Geometric parameters used in the method of Hurley and Duffy.

where

$$Q(kh_2,kh_1)=\begin{cases}\dfrac{2}{k}\cdot\left(h-\dfrac{e^{-k\cdot h}-1}{k}\right), & \text{when } z=0,\quad h_1-h_2=h\\[2ex]\dfrac{2}{k^2}\cdot\left(\cosh\left(k\cdot\dfrac{h_1+h_2}{2}\right)-\cosh\left(k\cdot\dfrac{h_1-h_2}{2}\right)\right), & \text{when } z>\dfrac{h_1+h_2}{2}\end{cases}$$

$$S(kr_2,kr_1)=\frac{J_0(kr_2)-J_0(kr_1)}{k}$$

$$S(ka_2,ka_1)=\frac{J_0(ka_2)-J_0(ka_1)}{k}$$

where
J_0 is the first kind Bessel function of order zero
r_1 is the inner radius of the jth circular track
r_2 is the outer radius of the jth circular track
h_1 is the height of the ith circular track
a_1 is the inner radius of the ith circular track
a_2 is the outer radius of the ith circular track
h_2 is the height of the jth circular track
z is the separation between the circular tracks
$\mu_0=4\cdot\pi\cdot10^{-7}$ H/m is the permeability of vacuum

Mohan [23] presented a current-sheet-based approach that provides simple expressions of mutual inductance for a variety of geometries. The expressions developed in Ref. [23] are valid to calculate the inductance of geometries that approximate square and circular spirals with an acceptable degree of accuracy.

Figure 23.33 shows two circular concentric current sheets of average diameters d1 and d2. The sheets have equal width, w, and are separated by a center to center distance ρd, where d is the mean of the two average diameters (d = 0:5(d1 + d2)). The mutual inductance between the sheets is given by

$$M_{ij}\approx\frac{\mu_0\cdot d}{2}\cdot\left[\ln\left(\frac{1}{\rho}\right)-0.6+0.7\cdot\rho^2+\left(0.2+\frac{1}{12\cdot\rho^2}\right)\cdot\frac{w^2}{d^2}\right]\qquad(23.40)$$

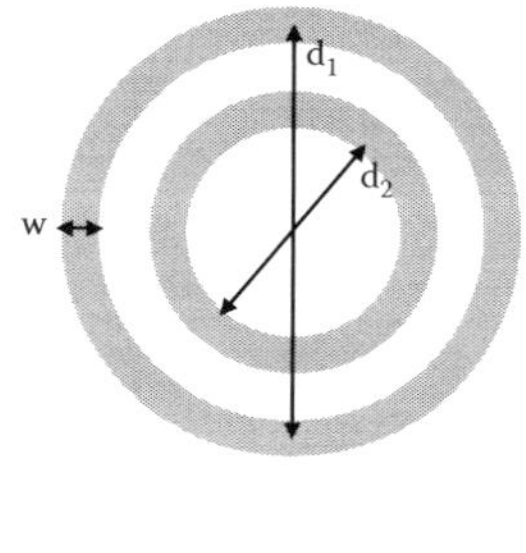

FIGURE 23.33
Geometric parameters of two coplanar windings used in approximation presented in Ref. [23].

where

$\rho = (d_1 - d_2)/(d_1 + d_2)$ is the ratio of the separation to the average diameter
$d = 0.5 \cdot (d_1 + d_2)$ is the average diameter
w is the width of conductor
$\mu_0 = 4 \cdot \pi \cdot 10^{-7}$ H/m is the permeability of vacuum

23.8 Obtaining Electrical Parameters of PCB Coreless Transformers from Actual Measurements

Inductances, resistances, and capacitances of coreless PCB transformers are usually measured by using an impedance analyzer combined with an impedance kit as shown in Figure 23.34.

Figure 23.35 contains a photograph of PCB coreless transformer connected to the impedance kit.

Electrical parameters of coreless transformers, capacitance, inductance, and resistance can be measured with the impedance analyzer. Figure 23.36 represents the transformer with the electrical components to be measured.

The following paragraphs explain the measuring process of the transformer parameters.

23.8.1 Measuring Primary or Secondary Winding Components (R_1, C_1, L_1), (R_2, C_2, L_2)

Most of impedance analyzers include the *Equivalent Circuit Mode*; this option permits the measurement of the electrical components of an electric model.

Figure 23.37 presents five possible equivalent circuits with their electrical components. In order to measure the primary/secondary winding capacitance/inductance, circuit B is selected as an example.

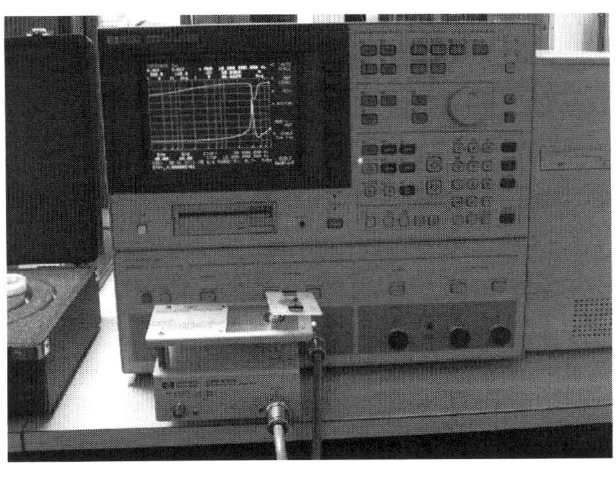

FIGURE 23.34
Network analyzer with an impedance kit connected.

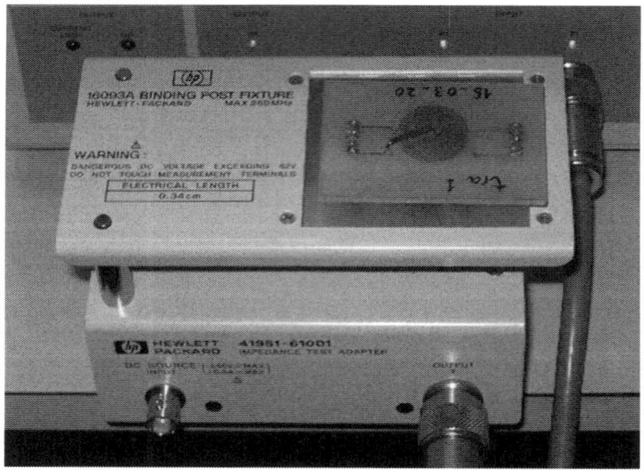

FIGURE 23.35
PCB coreless transformer connected to the impedance kit.

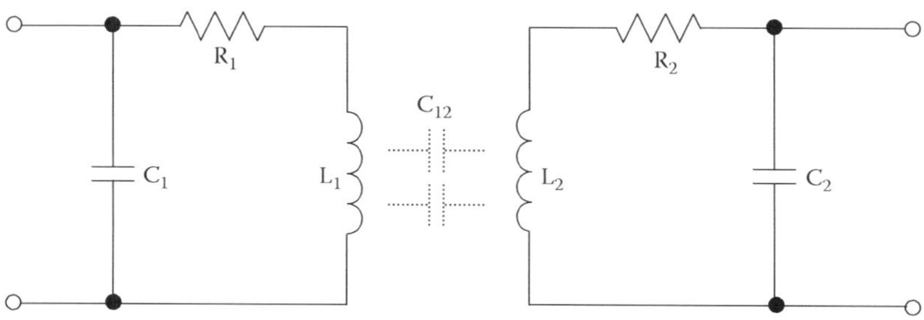

FIGURE 23.36
Electrical parameters to be measured with an impedance analyzer.

The resistance value given by the impedance analyzer is not valid because it averages the values of the parameters in the range of measurement, that is, from the start frequency to the stop frequency. Because winding resistance varies with frequency, the resistance value given by using the equivalent circuit mode is not valid. In order to measure the winding resistance, the following method can be used. The impedance of the circuit that includes the electrical components of the primary winding as shown in Figure 23.38 is given by the following expression:

$$Z = \frac{R + j \cdot \omega(L - R^2 \cdot C + \omega^2 \cdot L^2 \cdot C)}{1 + \omega^2 \cdot C \cdot (\omega^2 \cdot L^2 \cdot C - 2 \cdot L + R^2 \cdot C)} \tag{23.41}$$

From (23.41), the resistance can be obtained as

$$R = \mathrm{Re}[Z] \cdot (1 + \omega^2 \cdot C \cdot (\omega^2 \cdot L^2 \cdot C - 2 \cdot L + R^2 \cdot C)) \tag{23.42}$$

If $R^2 \cdot C \ll 2 \cdot L$, then

$$R = \mathrm{Re}[Z] \cdot (1 + \omega^2 \cdot C \cdot (\omega^2 \cdot L^2 \cdot C - 2 \cdot L))$$

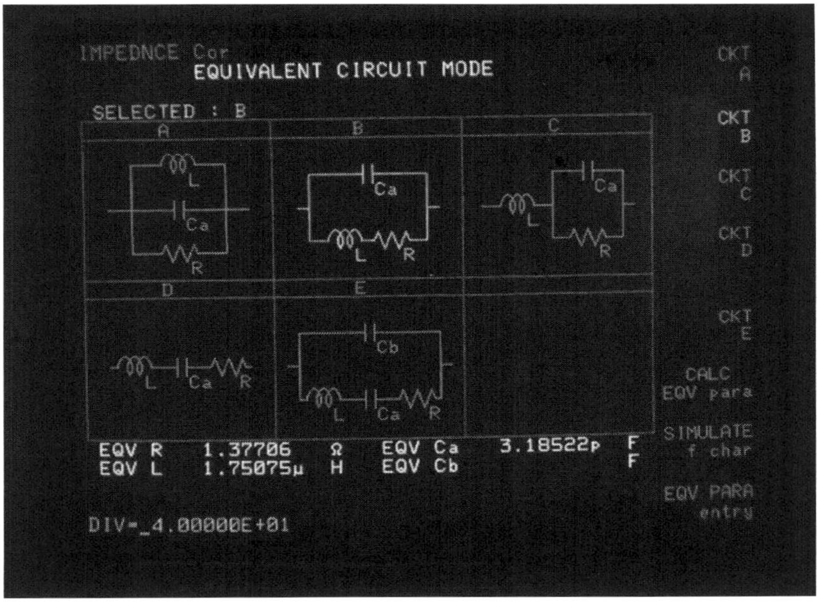

FIGURE 23.37
Impedance analyzer. Equivalent circuit mode.

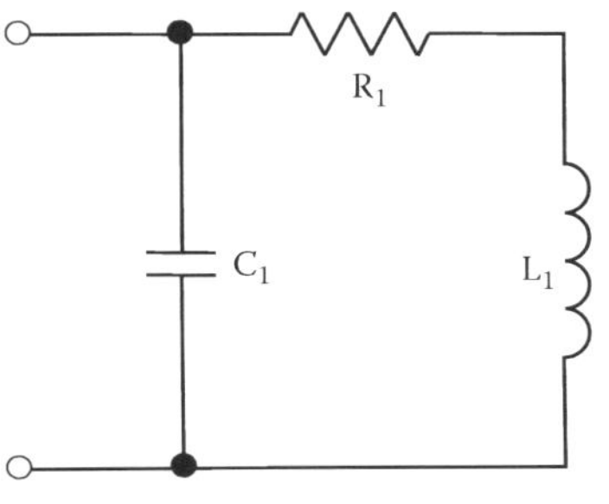

FIGURE 23.38
High-frequency model for the primary winding.

The resistance can be calculated by using the previous equation once the real part of the impedance is measured with the impedance analyzer as shown in Figure 23.39.

The real part and imaginary part of the impedance are measured for each frequency; if the value of the capacitance and the value of the inductance are considered constant in the range of variation, the resistance for each frequency can be calculated. The variation of winding resistance with frequency is represented in Figure 23.40. The variation of the resistance can be expressed as

$$R = a_0 + a_1 \cdot f + a_2 \cdot f^2 + a_3 \cdot f^3$$

$$R = 1.4442 - 7.5 \cdot 10^{-8} \cdot f + 6.4 \cdot 10^{-12} \cdot f^2 - 4 \cdot 10^{-16} \cdot f^3 \, [\Omega] \, (f : [Hz]) \quad (23.43)$$

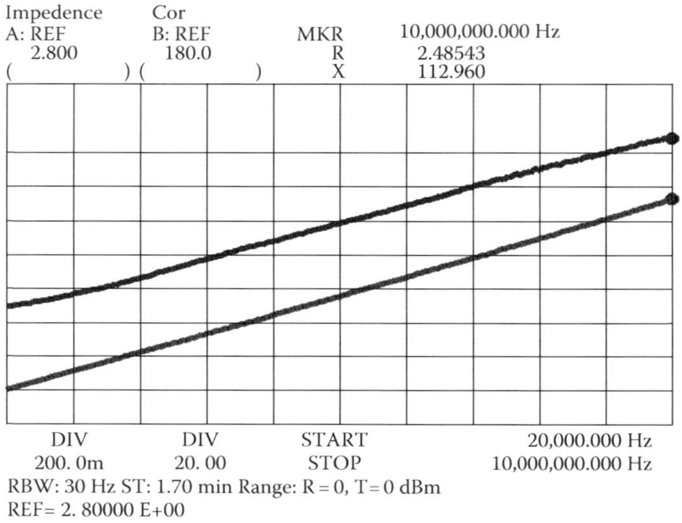

FIGURE 23.39
Measuring the real and imaginary components of the impedance with the impedance analyzer.

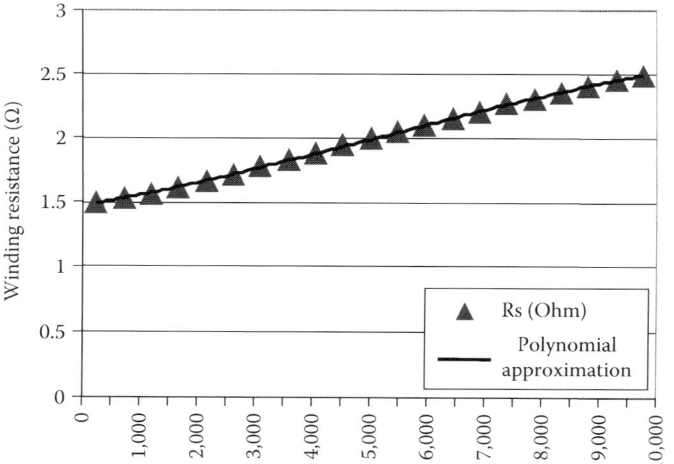

FIGURE 23.40
Variation of winding resistance with frequency and polynomial approximation.

23.8.2 Measuring the Capacitance between Windings (C_{12})

Measurement of capacitance between primary and secondary winding can be achieved by using the following disposition of the transformer windings.

As shown in Figure 23.41, the primary and secondary windings are short-circuited in order to minimize the distributed winding components. The short-circuited terminals of primary and secondary winding are connected to the impedance analyzer configured in the *Equivalent Circuit Mode*, as shown in Figure 23.42.

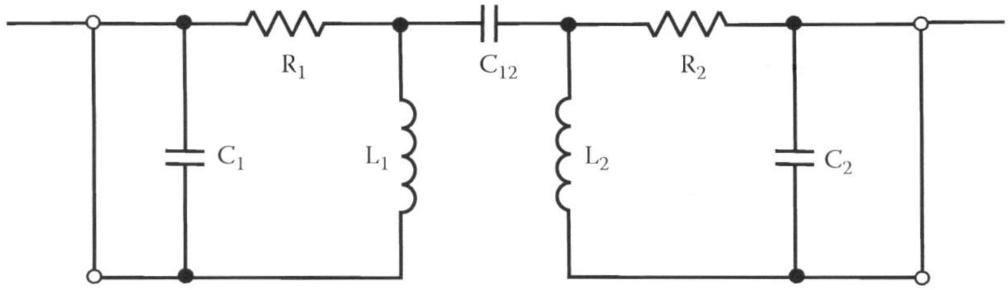

FIGURE 23.41
Disposition of transformer windings for measuring the capacitance between windings.

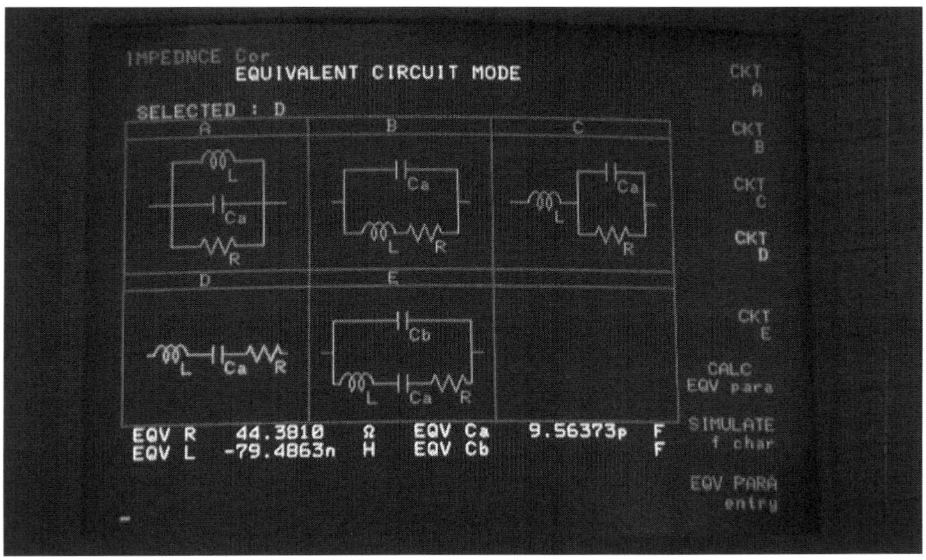

FIGURE 23.42
Measuring the capacitance between windings. Equivalent circuit mode.

23.8.3 Measuring the Mutual Inductance

Mutual inductance M of two windings with self-inductances L_1 and L_2 can be determined as explained in Ref. [22]. For this, two inductance measurements have to be made for two possible combinations of the series connection of both coupled windings: one of them for the corresponding directions of the windings, and one for the opposite directions, as shown in Figure 23.43.

Two values of inductances L_{ab} and L_{cd} are obtained as a result of the measurements:

$$L_{ab} = L_1 + L_2 + 2 \cdot L_m \quad L_{cd} = L_1 + L_2 - 2 \cdot L_m \tag{23.44}$$

Mutual inductance and leakage inductances are calculated from

$$L_m = \frac{L_{ab} - L_{cd}}{4}$$
$$L_{lk1} = L_1 - L_m \quad L_{lk2} = L_2 - L_m \tag{23.45}$$

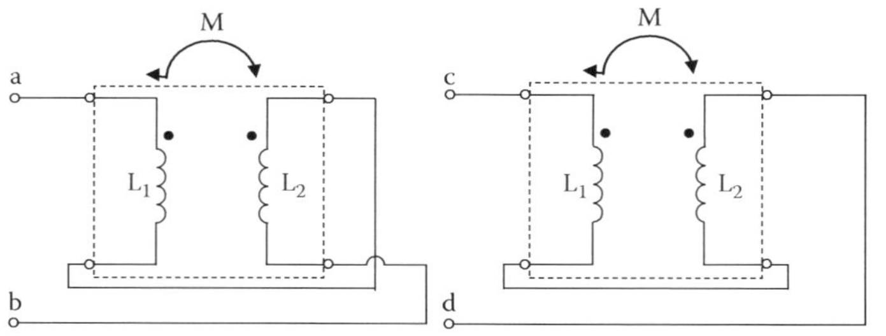

FIGURE 23.43
Transformer connection for measuring the mutual inductance.

23.9 Example of a Coreless PCB Transformer Design (I)

Design a coreless PCB transformer with the following characteristics:

- Output voltage: 12 V
- Maximum impedance frequency (MIF): 3 MHz
- Output power: 0.5 W
- Load: resistive
- Maximum dimensions: 25 mm × 25 mm
- PCB thickness: 1.66 mm
- Track width: 0.4 mm
- Track height: 35 μm
- ε_0:$10^{-9}/36\pi$ [F/m]
- ε_r: 4.7

In order to maximize the coupling coefficient of the transformer, the maximum radius is chosen. Once the radius is selected, it is necessary to follow an iterative process that starts with the selection of the number of turns and the transfer ratio. As first iteration, it is going to be selected a unity transfer ratio in order to work with a primary voltage of 12 V and the number of turns 20.

Track separation can be calculated using (23.47):

$$\text{Outermost radius} = \frac{25\,\text{mm}}{2} = 12.5\,\text{mm}$$

$$N_1 = N_2 = 20 \tag{23.46}$$

$$\text{Track separation} = \frac{\text{Outermost radius}}{\text{Number turns}} - \text{Track width}$$

$$= \frac{12.5\ \text{mm}}{20} - 0.4\,\text{mm} = 0.225\,\text{mm} \tag{23.47}$$

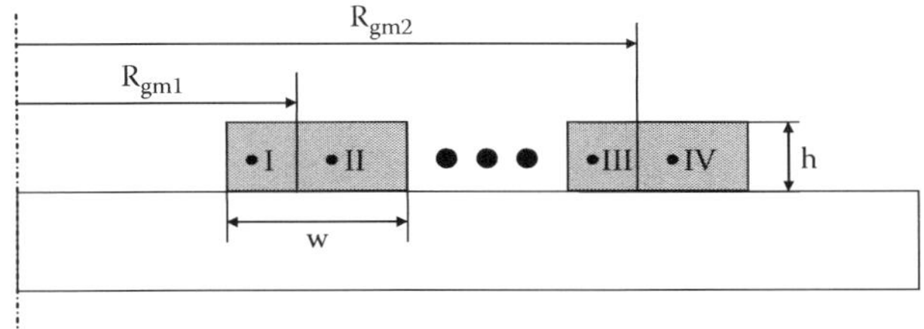

FIGURE 23.44
Lyle's method for calculating mutual inductance in concentric planar tracks.

Transformer inductances can be calculated by using the following equations, which are the solution to Maxwell equations written in terms of elliptic integrals for this particular case:

$$L_1 = \sum_i^{N_1}\sum_j^{N_1} M_{ij} \quad L_2 = \sum_i^{N_2}\sum_j^{N_2} M_{ij} \quad L_m = \sum_i^{N_1}\sum_j^{N_2} M_{ij}$$

$$L_{lk1} = L_1 - L_m \quad L_{lk2} = L_2 - L_m$$

$$M = \mu_0 \cdot \sqrt{a \cdot r} \cdot \frac{2}{g}\left[\left(1 - \frac{g^2}{2}\right) \cdot K(g) - E(g)\right] \quad g = \sqrt{\frac{4 \cdot a \cdot r}{z^2 + (a + r)^2}} \tag{23.48}$$

Mutual inductance is calculated for all possible combinations between two concentric circular coils; the coefficients of the following matrix must be obtained by using the Lyle's method Figure 23.44:

$$\begin{bmatrix} M_{11} & M_{12} & M_{13} & \dots & M_{1-20} \\ M_{21} & M_{22} & M_{23} & \dots & M_{2-20} \\ \dots & \dots & \dots & \dots & \dots \\ \dots & \dots & \dots & \dots & \dots \\ M_{20-1} & M_{20-2} & M_{20-3} & \dots & M_{20-20} \end{bmatrix} \tag{23.49}$$

The following equation shows the process of calculus for three values of the previous matrix:

M_{11}

$$a = 0.375\,\text{mm}; \quad r = 0.375\,\text{mm}; \quad z = 92.7\,\mu\text{m}; \quad g = \sqrt{\frac{4 \cdot a \cdot r}{z^2 + (a + r)^2}} = 0.99$$

$$K(g) = 3.49 \quad E(g) = 1.02$$

$$M_{11} = 4 \cdot \pi \cdot 10^{-7} \cdot \sqrt{a \cdot r} \cdot \frac{2}{g}\left[\left(1 - \frac{g^2}{2}\right) \cdot K(g) - E(g)\right] = 71.26\,\text{nH}$$

...

...

M_{23}

$$a_{I-III} = 0.915\,mm; \quad r_{I-III} = 1.5\,mm; \quad z = 0; \quad g_{I-III} = \sqrt{\frac{4 \cdot a \cdot r}{z^2 + (a + r)^2}} = 0.966$$

$K_{I-III}(g) = 2.776 \quad E_{I-III}(g) = 1.075$

$$M_{23}^{(I-III)} = 4 \cdot \pi \cdot 10^{-7} \cdot \sqrt{a \cdot r} \cdot \frac{2}{g}\left[\left(1 - \frac{g^2}{2}\right) \cdot K(g) - E(g)\right] = 1.25\,nH$$

$$a_{I-IV} = 0.915\,mm; \quad r_{I-IV} = 1.8\,mm; \quad z = 0; \quad g_{I-IV} = \sqrt{\frac{4 \cdot a \cdot r}{z^2 + (a + r)^2}} = 0.947 \qquad (23.50)$$

$K_{I-IV}(g) = 2.565 \quad E_{I-IV}(g) = 1.107$

$$M_{23}^{(I-IV)} = 4 \cdot \pi \cdot 10^{-7} \cdot \sqrt{a \cdot r} \cdot \frac{2}{g}\left[\left(1 - \frac{g^2}{2}\right) \cdot K(g) - E(g)\right] = 1.04\,nH$$

$$a_{Ii-iiI} = 1.1\,mm; \quad r_{I-IV} = 1.5\,mm; \quad z = 0; \quad g_{II-III} = \sqrt{\frac{4 \cdot a \cdot r}{z^2 + (a + r)^2}} = 0.988$$

$K_{II-III}(g) = 3.303 \quad E_{II-III}(g) = 1.03$

$$M_{23}^{(II-III)} = 4 \cdot \pi \cdot 10^{-7} \cdot \sqrt{a \cdot r} \cdot \frac{2}{g}\left[\left(1 - \frac{g^2}{2}\right) \cdot K(g) - E(g)\right] = 2.223\,nH$$

$$a_{II-IV} = 1.1\,mm; \quad r_{II-IV} = 1.8\,mm; \quad z = 0; \quad g_{II-IV} = \sqrt{\frac{4 \cdot a \cdot r}{z^2 + (a + r)^2}} = 0.976$$

$K_{II-IV}(g) = 2.94 \quad E_{II-IV}(g) = 1.05$

$$M_{23}^{(II-IV)} = 4 \cdot \pi \cdot 10^{-7} \cdot \sqrt{a \cdot r} \cdot \frac{2}{g}\left[\left(1 - \frac{g^2}{2}\right) \cdot K(g) - E(g)\right] = 1.77\,nH$$

$$M_{23} = \frac{M_{23}^{(I-III)} + M_{23}^{(I-IV)} + M_{23}^{(II-III)} + M_{23}^{(II-IV)}}{4} = 1.5717\,nH$$

...

...

M_{20-20}

$$a = 12.3\,mm \quad r = 12.3\,mm \quad z = 92.7\,\mu m \quad g = \sqrt{\frac{4 \cdot a \cdot r}{z^2 + (a + r)^2}} = 1$$

$K(g) = 6.96 \quad E(g) = 1$

$$M_{20-20} = 4\cdot\pi\cdot 10^{-7}\cdot\sqrt{a\cdot r}\cdot\frac{2}{g}\left[\left(1-\frac{g^2}{2}\right)\cdot K(g) - E(g)\right] = 76.76\,\text{nH}$$

Once all values of the mutual inductance are obtained, leakage and self-inductances are calculated as

$$L_1 = \sum_{i}^{20}\sum_{j}^{20} M_{ij} = 3.5\,\mu\text{H} \quad L_2 = \sum_{i}^{20}\sum_{j}^{20} M_{ij} = 3.5\,\mu\text{H} \quad L_m = \sum_{i}^{20_1}\sum_{j}^{20} M_{ij} = 2.4\,\mu\text{H}$$

$$L_{lk1} = L_1 - L_m = 1.1\,\mu\text{H} \quad L_{lk2} = L_2 - L_m = 1.1\,\mu\text{H} \tag{23.51}$$

As the first step in the calculation of winding resistance, the track length must be obtained as

$$l_i = 2\cdot\pi\cdot\left(\frac{R^{\bullet}}{N}\cdot i - \frac{R}{2\cdot N}\right)\Bigg\}\begin{matrix} R:\text{Outermost radius} \\ N:\text{Number of turns}\end{matrix}$$

$$l_1 = 2\cdot\pi\cdot\left(\frac{R^{\bullet}}{N}\cdot 1 - \frac{R}{2\cdot N}\right) = 2\cdot\pi\cdot\left(\frac{12.5\cdot 10^{-3}}{20}\cdot 1 - \frac{12.5\cdot 10^{-3}}{2\cdot 20}\right) = 0.002\,\text{m}$$

$$\ldots$$

$$l_{20} = 2\cdot\pi\cdot\left(\frac{R^{\bullet}}{N}\cdot 20 - \frac{R^{\bullet}}{2\cdot N}\right) = 0.0766\,\text{m}$$

$$\text{Winding length} = \sum_{i=1}^{n} l_i = l_1 + l_2 + \cdots + l_N = 0.002 + 0.0059 + \cdots + 0.0766 = 0.7854\,\text{m} \tag{23.52}$$

where l_i is the length of each spiral winding turn that is approximated by the perimeter of the circle with a radius equal to mean of the minimum and maximum radius of the spiral turn (see Figure 23.28).

Once the length is known, the winding resistance is calculated as follows:

$$R_{DC} = \rho_{CU}\cdot\frac{L}{S} = 17.24\cdot 10^{-9}[\Omega\text{m}]\cdot\frac{0.7854[\text{m}]}{35\cdot 10^{-6}[\text{m}]\cdot 0.4\cdot 10^{-3}[\text{m}]} = 0.9672\,\Omega$$

$$\delta = \sqrt{\frac{2}{2\cdot\pi\cdot f\cdot\mu_0\cdot\sigma}} = 38.15\cdot 10^{-6} \qquad f = 3\cdot 10^{6}\,\text{Hz}$$

$$w = 0.4\cdot 10^{-3}\,\text{m}$$

$$R_{AC} = \frac{S}{w\cdot\delta\left(1-e^{-t/\delta}\right)} = 1.5278 \qquad \mu_0 = 4\cdot\pi\cdot 10^{-7}\,\text{H/m}$$

$$\sigma = \frac{1}{\rho} = 58\cdot 10^{6}\,\text{S}\cdot\text{m}$$

$$R_1 = R_2 = R_{DC}\cdot R_{AC} = 0.9672\cdot 1.5278 = 1.4778\,\Omega \tag{23.53}$$

Capacitance between primary and secondary windings can be obtained with the following expression:

$$C_{12} = \varepsilon_0 \cdot \varepsilon_r \cdot \frac{S}{z} = \varepsilon_0 \cdot \varepsilon_r \cdot \frac{\pi \cdot r^2}{z} = \frac{10^{-9}}{36 \cdot \pi}[F/m] \cdot 4.7 \cdot \frac{\pi \cdot \left(12.5 \cdot 10^{-3}[m]\right)^2}{1.66 \cdot 10^{-3}[m]} = 12.3\,pF \quad (23.54)$$

In order to calculate the primary and secondary inter-winding capacitances, the graphical method explained in the paragraph "Calculation of Coreless PCB Transformer Parameters" is going to be used.

From Figure 23.45, for a transformer with outermost radius of 12.5 mm, 20 turns, and track width 0.4 mm, the inter-winding capacitance is 0.55 pF.

The value of the resistive load is given by

$$R_L = \frac{V^2}{P} = \frac{12^2}{0.5} = 288\,\Omega \quad (23.55)$$

Once the electrical parameters of the transformer are known, the following step is the calculation of the load capacitance for a given MIF of 3 MHz. As the resonant frequency is slightly higher than MIF, for f_o a slightly higher value than 3 MHz can be selected, and after various iterations the final value for C_L can be obtained:

$$f_o = \frac{1}{2 \cdot \pi \cdot \sqrt{L_{eq} \cdot C_{eq}}} \begin{cases} C_{eq} = C'_{12} + C'_2 + \dfrac{C_L}{n^2} \\ L_{eq} = L'_{lk2} + \dfrac{L_{lk1} \cdot L_m}{L_{lk1} + L_m} \end{cases}$$

$$f_o = 3.4\,MHz \Rightarrow C_L = 1.2\,nF \quad (23.56)$$

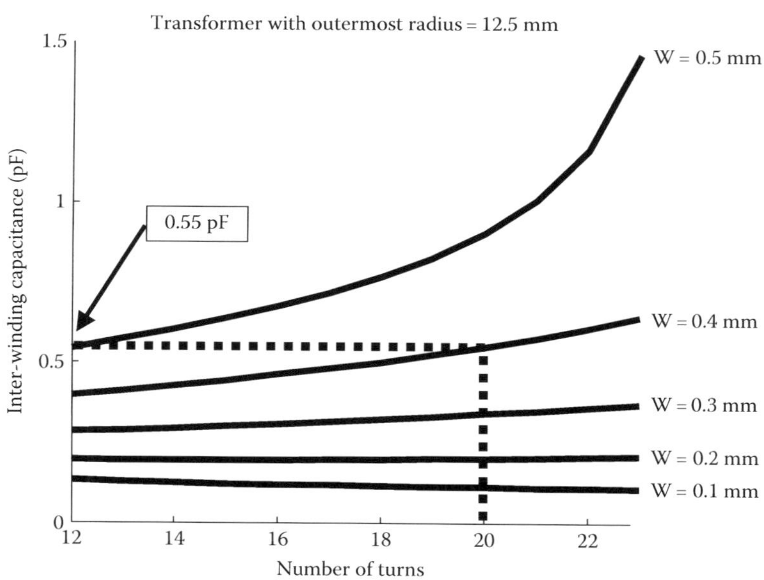

FIGURE 23.45
Obtaining inter-winding capacitance. Number of turns, 20; track width: 0.4 mm.

TABLE 23.4

Electrical Parameters for the Transformer at about 3 MHz (First Iteration)

C_1 (pF)	C_2 (pF)	C_{12} (pF)	R_1 (Ω)	R_2 (Ω)	L_{lk1} (μH)	L_{lk2} (μH)	L_{m1} (μH)	R_L (kΩ)	C_L (pF)
0.55	0.55	12.23	1.477	1.477	1.0664	1.0664	2.437	288	1200

TABLE 23.5

Electrical Parameters for the Transformer Referred to the Primary Winding (First Iteration)

C_1' (pF)	C_2' (pF)	C_{12}' (pF)	R_1 (Ω)	R_2' (Ω)	L_{lk1} (μH)	L_{lk2}' (μH)	L_{m1} (μH)	R_L' (kΩ)	C_L' (pF)
0.55	0.55	12.23	1.477	1.477	1.0664	1.0664	2.437	288	1200

Table 23.4 shows the electrical parameters for the transformer after the first iteration, and Table 23.5 shows these electrical parameters referred to the primary winding.

Figure 23.46 shows the voltage gain and the input impedance of the transformer. From Figure 23.46 the MIF is 2.4 MHz and the input impedance is 129.6 Ω and the voltage gain for MIF is 1.31.

From the results given earlier, the MIF is lower than the specified value; if the value of C_L is decreased, MIF increases but voltage gain increases as well.

In order to design a transformer with MIF of 3 MHz and unity voltage gain, the secondary number of turns is going to be reduced.

As second iteration, the following parameters are chosen:

$$N_1 = 20; \quad N_2 = 15$$
$$w_1 = 0.4\ldots\text{mm}; \quad w_2 = 0.6\,\text{mm} \tag{23.57}$$

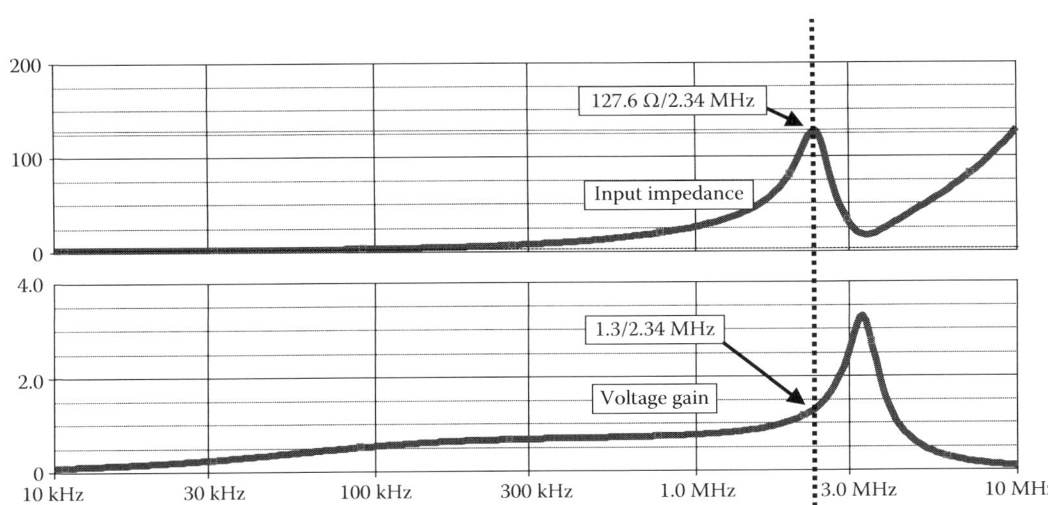

FIGURE 23.46
Input impedance and voltage gain after first iteration.

TABLE 23.6
Electrical Parameters for the Transformer at about 3 MHz (Second Iteration)

C_1 (pF)	C_2 (pF)	C_{12} (pF)	R_1 (Ω)	R_2 (Ω)	L_{lk1} (μH)	L_{lk2} (μH)	L_{m1} (μH)	R_L (kΩ)	C_L (pF)
0.55	0.13	12.23	1.477	0.7389	1.677	0.1384	1.826	288	2300

TABLE 23.7
Electrical Parameters for the Transformer Referred to the Primary Winding (Second Iteration)

C_1' (pF)	C_2' (pF)	C_{12}' (pF)	R_1 (Ω)	R_2' (Ω)	L_{lk1} (μH)	L_{lk2}' (μH)	L_{m1} (μH)	R_l' (kΩ)	C_L' (pF)
3.607	−2.22	9.17	1.477	1.313	1.677	0.246	1.826	512	1300

With these new geometric parameters and following the steps presented earlier, the new electrical parameters for the transformer are obtained as shown in Table 23.6. Table 23.7 includes the electrical parameters referred to the primary winding.

Figure 23.47 shows the voltage gain and the input impedance of the transformer after the second iteration. Figure 23.47 shows that the transformer with the new parameters accomplishes with the initial specifications.

23.10 Example of a Coreless PCB Transformer Design (II)

Design a coreless PCB transformer with the following characteristics:

- Output voltage: 15 V
- Input voltage: 5 V

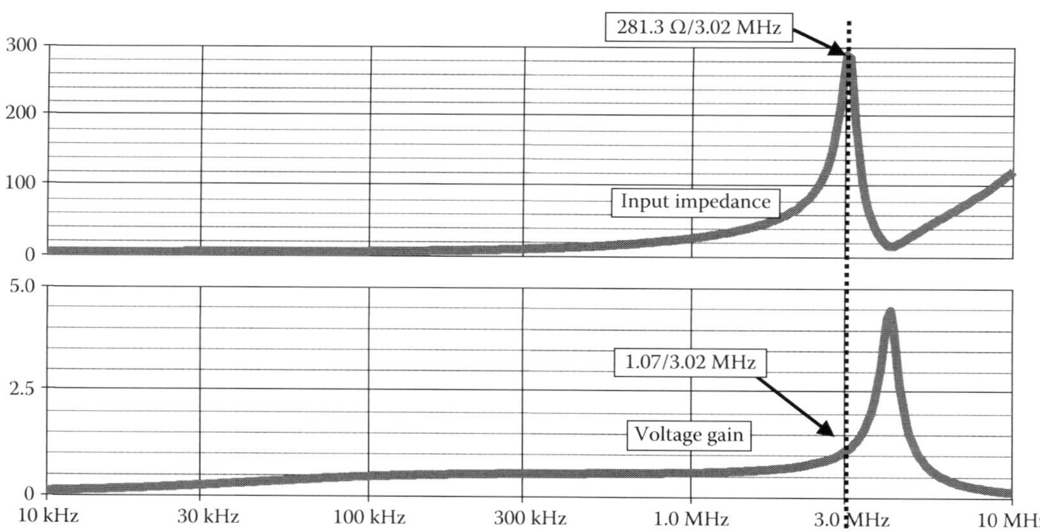

FIGURE 23.47
Input impedance and voltage gain after second iteration.

- Load: 500 Ω resistive
- PCB thickness: 1 mm
- Track width: 0.2 mm
- Track separation: 0.25 mm
- Track height: 35 μm
- ε_0: 10−9/36π [F/m]
- ε_r: 4.7

In order to design the transformer with the minimum diameter and to maximize the coupling coefficient, it is necessary to follow an iterative process that starts with the selection of the number of turns for the primary and secondary windings. As first iteration, the number of turns selected for the primary winding is 10 turns and for the secondary winding are 15 turns. The following equation shows the calculation of the radius for the primary and secondary windings:

$$N_1 = 10 \quad N_2 = 15$$
$$\text{Outermost primary radius} = 10 \cdot (0.2 + 0.25) = 4.5\,\text{mm}$$
$$\text{Outermost secondary radius} = 15 \cdot (0.2 + 0.25) = 6.75\,\text{mm} \tag{23.58}$$

Transformer inductances can be calculated by using the following based Maxwell equations:

$$L_1 = \sum_i^{N_1}\sum_j^{N_1} M_{ij} \quad L_2 = \sum_i^{N_2}\sum_j^{N_2} M_{ij} \quad L_m = \sum_i^{N_1}\sum_j^{N_2} M_{ij}$$
$$L_{lk1} = L_1 - L_m \quad L_{lk2} = L_2 - L_m$$
$$M = \mu_0 \cdot \sqrt{a \cdot r} \cdot \frac{2}{g}\left[\left(1 - \frac{g^2}{2}\right) \cdot K(g) - E(g)\right] \quad g = \sqrt{\frac{4 \cdot a \cdot r}{z^2 + (a+r)^2}} \tag{23.59}$$

Mutual inductance is calculated for all possible combinations between two concentric circular coils; the coefficients of the following matrix must be obtained by using Lyle's method (Figure 23.48). Equation 23.61 shows the process of calculus for three values of the following matrix:

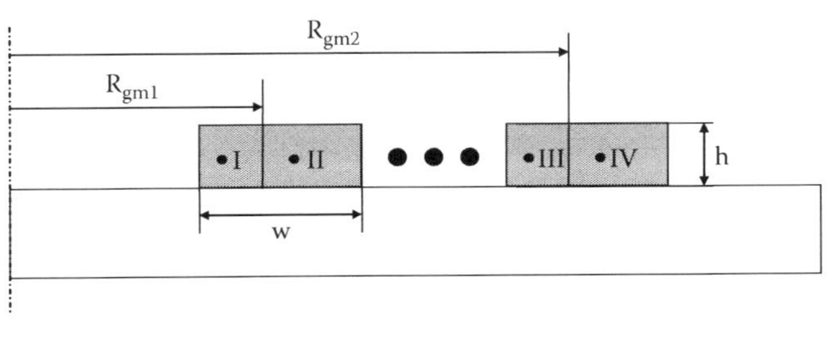

FIGURE 23.48
Lyle's method for calculating mutual inductance in concentric planar tracks.

$$\begin{bmatrix} M_{11} & M_{12} & M_{13} & \ldots & M_{1-20} \\ M_{21} & M_{22} & M_{23} & \ldots & M_{2-20} \\ \ldots & \ldots & \ldots & \ldots & \ldots \\ \ldots & \ldots & \ldots & \ldots & \ldots \\ M_{10-1} & M_{10-2} & M_{10-3} & \ldots & M_{10-10} \end{bmatrix} \tag{23.60}$$

M_{11}

$$a = 0.201\,\text{mm}; \quad r = 0.201\,\text{mm}; \quad z = 52.5\,\mu\text{m}; \quad g = \sqrt{\frac{4 \cdot a \cdot r}{z^2 + (a+r)^2}} = 0.99$$

$$K(g) = 3.44 \quad E(g) = 1.02$$

$$M_{11} = 4 \cdot \pi \cdot 10^{-7} \cdot \sqrt{a \cdot r} \cdot \frac{2}{g}\left[\left(1 - \frac{g^2}{2}\right) \cdot K(g) - E(g)\right] = 370.6\,\text{nH}$$

...

M_{23}

$$a_{I-III} = 1.2\,\text{mm}; \quad r_{I-III} = 0.724\,\text{mm}; \quad z = 0; \quad g_{I-III} = \sqrt{\frac{4 \cdot a \cdot r}{z^2 + (a+r)^2}} = 0.9712$$

$$K_{I-III}(g) = 2.847 \quad E_{I-III}(g) = 1.066$$

$$M_{23}^{(I-III)} = 4 \cdot \pi \cdot 10^{-7} \cdot \sqrt{a \cdot r} \cdot \frac{2}{g}\left[\left(1 - \frac{g^2}{2}\right) \cdot K(g) - E(g)\right] = 1.046\,\text{nH}$$

$$a_{I-IV} = 1.2\,\text{mm}; \quad r_{I-IV} = 0.6707\,\text{mm}; \quad z = 0; \quad g_{I-IV} = \sqrt{\frac{4 \cdot a \cdot r}{z^2 + (a+r)^2}} = 0.9485$$

$$K_{I-IV}(g) = 2.57 \quad E_{I-IV}(g) = 1.105$$

$$M_{23}^{(I-IV)} = 4 \cdot \pi \cdot 10^{-7} \cdot \sqrt{a \cdot r} \cdot \frac{2}{g}\left[\left(1 - \frac{g^2}{2}\right) \cdot K(g) - E(g)\right] = 0.7071\,\text{nH}$$

$$a_{Ii-iiI} = 1.1\,\text{mm}; \quad r_{I-IV} = 0.724\,\text{mm}; \quad z = 0; \quad g_{II-III} = \sqrt{\frac{4 \cdot a \cdot r}{z^2 + (a+r)^2}} = 0.9818$$

$$K_{II-III}(g) = 3.06 \quad E_{II-III}(g) = 1.04$$

$$M_{23}^{(II-III)} = 4 \cdot \pi \cdot 10^{-7} \cdot \sqrt{a \cdot r} \cdot \frac{2}{g}\left[\left(1 - \frac{g^2}{2}\right) \cdot K(g) - E(g)\right] = 1.219\,\text{nH}$$

$$M_{23}^{(II-III)} = 4 \cdot \pi \cdot 10^{-7} \cdot \sqrt{a \cdot r} \cdot \frac{2}{g}\left[\left(1 - \frac{g^2}{2}\right) \cdot K(g) - E(g)\right] = 1.219\,\text{nH}$$

$$a_{II-IV} = 1.1\,\text{mm}; \quad r_{II-IV} = 0.607\,\text{mm}; \quad z = 0; \quad g_{II-IV} = \sqrt{\frac{4 \cdot a \cdot r}{z^2 + (a+r)^2}} = 0.962$$

$$K_{II-IV}(g) = 2.72 \quad E_{II-IV}(g) = 1.08$$

$$M_{23}^{(II-IV)} = 4 \cdot \pi \cdot 10^{-7} \cdot \sqrt{a \cdot r} \cdot \frac{2}{g}\left[\left(1 - \frac{g^2}{2}\right) \cdot K(g) - E(g)\right] = 0.801\,\text{nH}$$

$$M_{23} = \frac{M_{23}^{(I-III)} + M_{23}^{(I-IV)} + M_{23}^{(II-III)} + M_{23}^{(II-IV)}}{4} = 0.942\,\text{nH}$$

…

M_{10-10}

$$a = 4.3\,\text{mm} \quad r = 4.3\,\text{mm} \quad z = 52.52\,\mu\text{m} \quad g = \sqrt{\frac{4 \cdot a \cdot r}{z^2 + (a+r)^2}} = 1$$

$$K(g) = 6.478 \quad E(g) = 1$$

$$M_{10-10} = 4 \cdot \pi \cdot 10^{-7} \cdot \sqrt{a \cdot r} \cdot \frac{2}{g}\left[\left(1 - \frac{g^2}{2}\right) \cdot K(g) - E(g)\right] = 24.05\,\text{nH} \tag{23.61}$$

Once all values of the mutual inductance are obtained, leakage and self-inductances are calculated as

$$L_1 = \sum_i^{10}\sum_j^{10} M_{ij} = 0.326\,\mu\text{H} \quad L_2 = \sum_i^{10}\sum_j^{10} M_{ij} = 1.087\,\mu\text{H} \quad L_m = \sum_i^{10_1}\sum_j^{10} M_{ij} = 0.3049\,\mu\text{H}$$

$$L_{lk1} = L_1 - L_m = 0.021\,\mu\text{H} \quad L_{lk2} = L_2 - L_m = 0.0782\,\mu\text{H} \tag{23.62}$$

As the first step in the calculation of winding resistance, the track length must be obtained as

$$l_i = 2 \cdot \pi \cdot \left(\frac{R^\bullet}{N} \cdot i - \frac{R}{2 \cdot N}\right)\Bigg\} \quad \begin{array}{l} R: \text{Outermost radius} \\ N: \text{Number of turns} \end{array}$$

$$l_1 = 2 \cdot \pi \cdot \left(\frac{R^\bullet}{N} \cdot 1 - \frac{R}{2 \cdot N}\right) = 2 \cdot \pi \cdot \left(\frac{4.5 \cdot 10^{-3}}{10} \cdot 1 - \frac{4.5 \cdot 10^{-3}}{2 \cdot 10}\right) = 0.00141\,\text{m}$$

…

$$l_{10} = 2 \cdot \pi \cdot \left(\frac{R^\bullet}{N} \cdot 20 - \frac{R^\bullet}{2 \cdot N}\right) = 0.0269\,\text{m}$$

$$\text{Winding length} = \sum_{i=1}^{n} l_i = l_1 + l_2 + \cdots + l_N = 0.00141 + 0.0043 + \cdots + 0.0269 = 0.1414\,\text{m} \tag{23.63}$$

where l_i is the length of each spiral winding turn that is calculated as the perimeter of the circle with a radius equal to the mean of the minimum and maximum radius of the spiral turn (see Figure 23.28).

Once the length is known, the winding resistance is calculated as follows:

$$R_{DC} = \rho_{CU} \cdot \frac{L}{S} = 17.24 \cdot 10^{-9}[\Omega m] \cdot \frac{0.1414[m]}{35 \cdot 10^{-6}[m] \cdot 0.2 \cdot 10^{-3}[m]} = 0.8125\,\Omega$$

$$\delta = \sqrt{\frac{2}{2 \cdot \pi \cdot f \cdot \mu_0 \cdot \sigma}} = 38.15 \cdot 10^{-5} \qquad f = 3 \cdot 10^6\ \text{Hz}$$

$$w = 0.4 \cdot 10^{-3}\ \text{m}$$

$$R_{AC} = \frac{S}{w \cdot \delta\left(1 - e^{-t/\delta}\right)} = 1.2904 \qquad \mu_0 = 4 \cdot \pi \cdot 10^{-7}\ \text{H/m}$$

$$\sigma = \frac{1}{\rho} = 58 \cdot 10^6\ \text{S} \cdot \text{m}$$

$$R_1 = R_2 = R_{DC} \cdot R_{AC} = 0.8125 \cdot 1.2904 = 0.9826\,\Omega \tag{23.64}$$

Capacitance between primary and secondary windings can be obtained with the following equation:

$$C_{12} = \varepsilon_0 \cdot \varepsilon_r \cdot \frac{S}{z} = \varepsilon_0 \cdot \varepsilon_r \cdot \frac{\pi \cdot r^2}{z} = \frac{10^{-9}}{36 \cdot \pi}[F/m] \cdot 4.7 \cdot \frac{\pi \cdot (5 \cdot 10^{-3}[m])^2}{1 \cdot 10^{-3}[m]} = 2.64\,\text{pF} \tag{23.65}$$

In order to calculate the primary and secondary inter-winding capacitances, the graphical method explained in the paragraph "Calculation of Coreless PCB Transformer Parameters" is going to be used.

From Figure 23.49, for a transformer with separation between tracks 0.25 mm, $N_1 = 10$, $N_2 = 15$, and track width 0.2 mm, the inter-winding capacitance for primary is 0.34 pF and for secondary is 0.36 pF.

Once the electrical parameters of the transformer are known, the following step is the selection of the resonant frequency and the load capacitance for this frequency; in this exercise, a resonant frequency of 2 MHz is selected:

$$f_o = \frac{1}{2 \cdot \pi \cdot \sqrt{L_{eq} \cdot C_{eq}}} \quad \begin{cases} C_{eq} = C'_{12} + C'_2 + \dfrac{C_L}{n^2} \\ L_{eq} = L'_{lk2} + \dfrac{L_{lk1} \cdot L_m}{L_{lk1} + L_m} \end{cases}$$

$$f_o = 2\ \text{MHz} \Rightarrow C_L = 10\,\text{nF} \tag{23.66}$$

Table 23.8 shows the electrical parameters for the transformer after the first iteration, and Table 23.9 displays these electrical parameters referred to the primary winding.

Figure 23.50 shows the voltage gain and the input impedance of the transformer. From Figure 23.50, the MIF is 1.95 MHz, the input impedance is 13.1 Ω, and the voltage gain for MIF is 2.77.

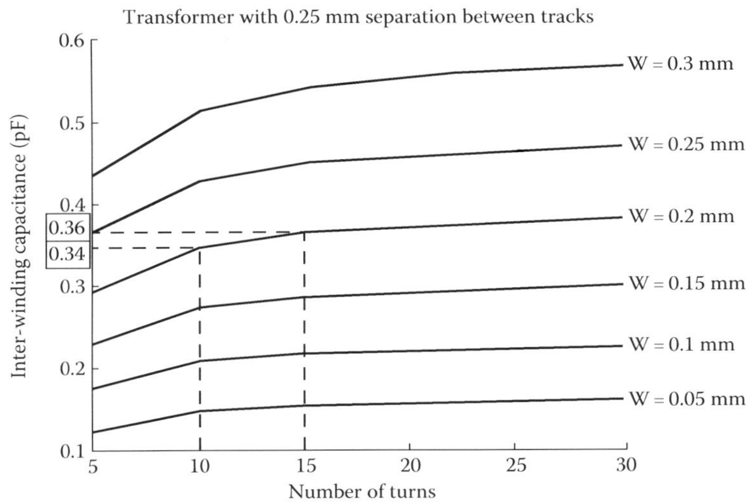

FIGURE 23.49
Obtaining inter-winding capacitance. Number of turns: $N_1 = 10$, $N_2 = 15$. Track width: 0.2 mm.

TABLE 23.8
Electrical Parameters for the Transformer at about 2 MHz (First Iteration)

C_1 (pF)	C_2 (pF)	C_{12} (pF)	R_1 (Ω)	R_2 (Ω)	L_{lk1} (μH)	L_{lk2} (μH)	L_{m1} (μH)	R_L (kΩ)	C_L (pF)
0.34	0.36	2.643	0.98	2.21	0.0162	0.782	0.305	500	10000

TABLE 23.9
Electrical Parameters for the Transformer Referred to the Primary Winding (First Iteration)

C_1' (pF)	C_2' (pF)	C_{12}' (pF)	R_1 (Ω)	R_2' (Ω)	L_{lk1} (μH)	L_{lk2}' (μH)	L_{m1} (μH)	R_L' (kΩ)	C_L' (pF)
−0.981	−0.979	3.96	0.98	0.982	0.0162	0.347	0.305	222.22	22500

In order to accomplish the requirements of this exercise, it is necessary to increase the voltage gain. A higher voltage gain can be achieved by selecting a lower load capacitance which implies higher frequency of operation; see (23.66). If we repeat the previous iteration process with several resonant frequencies higher than 2 MHz, we obtain a voltage gain of 2.97 at 8 MHz. If this frequency is too high, we have another way to increase the voltage gain by selecting a higher number of secondary turns.

As a second iteration in the design process, the secondary number of turns is going to be selected (23.67). The primary number of turns is changed in order to have similar diameter for the two windings:

$$N_1 = 12; \quad N_2 = 20$$

$$R_1 = 5.3\,\text{mm}; \quad R_2 = 9\,\text{mm} \tag{23.67}$$

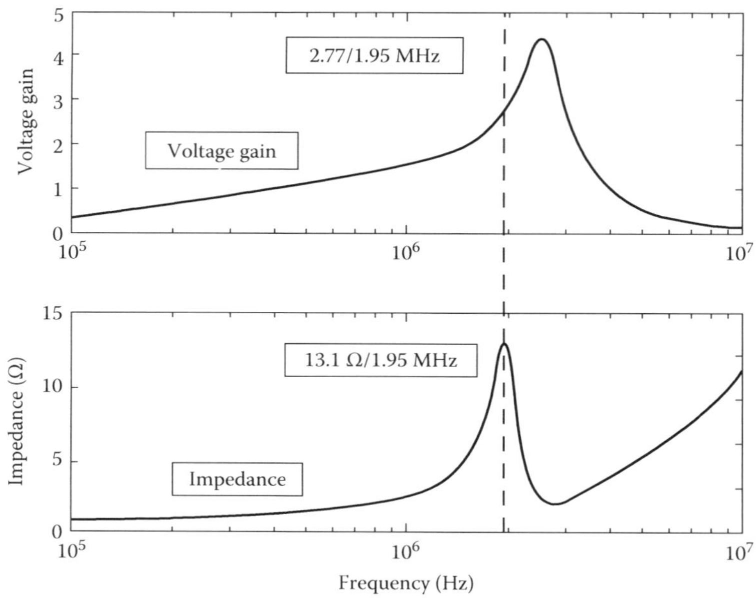

FIGURE 23.50
Input impedance and voltage gain after first iteration.

TABLE 23.10

Electrical Parameters for the Transformer at about 6 MHz (Second Iteration)

C_1 (pF)	C_2 (pF)	C_{12} (pF)	R_1 (Ω)	R_2 (Ω)	L_{lk1} (μH)	L_{lk2} (μH)	L_{m1} (μH)	R_L (kΩ)	C_L (pF)
0.35	0.37	3.80	1.41	3.93	0.0616	1.93	0.622	500	500

TABLE 23.11

Electrical Parameters for the Transformer Referred to the Primary Winding (Second Iteration)

C'_1 (pF)	C'_2 (pF)	C'_{12} (pF)	R_1 (Ω)	R'_2 (Ω)	L_{lk1} (μH)	L'_{lk2} (μH)	L_{m1} (μH)	R'_L (kΩ)	C'_L (pF)
−2.18	496.14	6.34	1.41	1.415	0.0616	0.697	0.622	55.55	4500

With these new geometric parameters and following the steps presented earlier, the new electrical parameters for the transformer are obtained and they are shown in Table 23.10. Table 23.11 includes the electrical parameters referred to the primary winding.

Figure 23.51 shows the voltage gain and the input impedance of the transformer after the second iteration. Figure 23.51 shows that the transformer with the new parameters accomplishes the initial specifications.

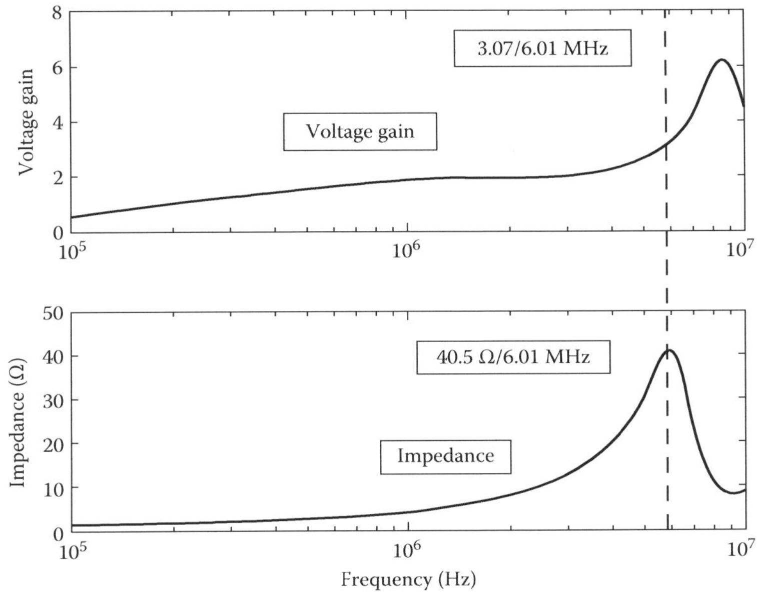

FIGURE 23.51
Input impedance and voltage gain after second iteration.

References

1. Hayano S, Nakajima Y, Saotome H, Saito Y (1991) A new type high frequency transformers. *IEEE Trans Magn*, 27, 5205–5207.
2. Linde D, Boon CA, Klaassens JB (1991) Design of a high frequency planar power transformer in multilayer technology. *IEEE Trans Ind Electron*, 38(2), 135–141.
3. Dai N, Lofti AW, Skutt G, Tabisz WA, Lee FC (1994) A comparative study of high frequency low-profile planar transformer technologies. *Proceedings of IEEE Applied Power Electronic Conference and Exposition, APEC'94*, vol. 1, pp. 226–232.
4. Skutt G, Lee FC, Ridley R, Nicol D (1994) Leakage inductance and termination effects in a high power planar magnetic structure. *Proceedings of IEEE Applied Power Electronic Conference and Exposition, APEC'94*, vol. 1, pp. 295–301.
5. Roshen W (1990) Effect of finite thickness of magnetic substrate on planar inductors. *IEEE Trans Magn*, 26(1), 270–275.
6. Mino M, Yachi T, Tago A, Yanagisawa K, Sakakibara K (1992) A new planar microtransformer for use in micro-switching converters. *IEEE Trans Magn*, 28, 1969–1973.
7. Yamaguchi K, Ohnuma S, Imagawa T, Toriu J, Matsuki H, Murakami K (1993) Characteristic of a thin-film microtransformer with circular spiral coils. *IEEE Trans Magn*, 29, 2232–2237.
8. Ahn CH, Allen MG (1998) Micromachined planar inductors on silicon wafers for MEMS applications. *IEEE Trans Ind Electron*, 45, 866–875.
9. Sullivan CR, Sanders SR (1996) Design of microfabricated transformers and inductors for high-frequency power conversion. *IEEE Trans Power Electron*, 11(2), 228–238.
10. Hurley WG, Duffy MC, O'Reilly S, O'Mathuna SC (1997) Impedance formulas for planar magnetic structures with spiral windings. *Proceedings of IEEE Power Electronics Specialists Conference, PESC' 97*, pp. 627–633.

11. Balakrishnan, Palmer WD, Joines W, Wilson TG (1993) The inductance of planar structures. *Proceedings of IEEE Power Electron, Specialists Conference*, pp. 912–921.
12. Sullivan CR, Sanders SR (1997) *Measured Performance of a High-Power-Density Micro-Fabricated Transformer in a DC–DC Converter*. IEEE Technology Update Series: Power Electronics and Applications II, New York: IEEE Press, pp. 104–111.
13. Marinova I, Midorrikawa Y, Hayano S, Saito Y (1995) Thin film transformer and its analysis by integral equation method. *IEEE Trans Magn*, 31, 2432–2437.
14. Hui SYR, Tang SC, Chung H (1998) Coreless printed-circuit board transformers for signal and energy transfer. *Electron Lett*, 34(11), 1052–1054.
15. Hui SYR, Chung H, Tang SC (1999) Coreless PCB-based transformers for power MOSFETs/IGBT's gate drive circuits. *IEEE Trans Power Electron*, 14, 422–430.
16. Tang SC, Hui SYR, Chung H (1999) Coreless PCB transformer with multiple secondary windings for complementary gate drive circuits. *IEEE Trans Power Electron*, 14, 431–437.
17. Chung H, Hui SYR, Tang SC (2000) Design and analysis of multistage switched capacitor based step-down DC–DC converters. *IEEE Trans Circuits Syst I*, 47, 1017–1025.
18. Hui SYR, Tang SC, Chung H (1999) Optimal operation of coreless PCB transformer-isolated gate drive circuits with wide switching frequency range. *IEEE Trans Power Electron*, 14, 506–514.
19. Hui SYR, Tang SC, Chung H (2000) Some electromagnetic aspects of coreless PCB transformers. *IEEE Trans Power Electron*, 15(4), 805–810.
20. Hui SYR, Tang SC, Chung H (1997) An accurate circuit model for coreless PCB-based transformers. *Proceedings of the European Power Electronics Conference, EPE'97*, pp. 4.123–4.128.
21. Tang SC, Hui SYR, Chung H (2000) Characterization of coreless printed circuit board (PCB) transformers. *IEEE Trans Power Electron*, 15, 1275–1282.
22. Szyper M (2000) *Inductance Measurement*. CRC Press LLC, Boca Raton, FL.
23. Mohan SS (1999) The design, modeling and optimisation of on-chip inductor and transformer circuits. PhD thesis.
24. Sinclair AJ, Ferreira JA (1996) Analysis and design of transmission-line structures by means of the geometric mean distance. *Proceedings of AFRICON, 4th IEEE AFRICON*, Vol. 2, pp. 1062–1065.
25. Niknejad AM (2000) Analysis, simulation, and applications of passive devices on conductive substrates. University of California, Berkeley. CA.
26. Kawabe K, Koyama H, Shirae K (1984) Planar inductor. *IEEE Trans Magn*, MAG-20, 1804–1806.
27. Rautio JC (2000) An investigation of microstrip conductor loss. *IEEE Microwave Magn*, 1(4), 60–67.
28. Hurley WG, Duffy MC (1995) Calculation of self and mutual impedances in planar magnetic structures. *IEEE Trans Magn*, 31(4), 2416–2422.
29. Higgins TJ (1947) Theory and application of complex logarithms and geometrical mean distances. *Trans AIEE*, 66, 12–16.

24

Planar Transformers

Frederick E. Bott

CONTENTS

24.1 Introduction

By planar transformers (Figure 24.1) in this chapter, we mean inductive transformers based on substrate(s) incorporating flat windings and ferrite core.*

Substrates most often take the form of printed circuit boards (PCBs). In some cases, leaded subframes are used in conjunction with PCBs in an effort to obtain greater copper cross-sectional area for higher current windings.

Historically, planar wound substrates were tailored so as to fit on standard ferrite cores intended for wirewound applications. In the past 15 years or so (since mid- to late 1990s),

* Conceptually, piezoelectric transformers also surely fall into the planar category, since these are very flat laminated structures with no core. However, in the author's experience to date, the development and applicability of piezoelectric transformers remains a very limited specialist area, and the term planar transformer is used throughout the industry to exclusively mean planar inductive transformers.

FIGURE 24.1
Standard discrete OTS planar transformer.

a range of standard low-profile ferrite cores has emerged, with corresponding industry guideline standards such as IEC62317-9 [1].

The newer low-profile planar core dimensions are designed specifically to (1) increase the flat winding areas required by substrates rather than wirewound bobbins and (2) to increase the ferrite magnetic path cross-sectional area, so as to obtain higher values of inductance from fewer turns, as is restricted by planar windings compared with wire windings. A core set may comprise a pair of E-cores (EE pair), or one E-core and one I-core (EI pair). An I-core is simply a flat ferrite plate which matches the footprint and cross-sectional area of the E-core, which can be used to close the magnetic path of the E-core instead of using a second E-core. An EE-core pair winding window has twice the height of that in an EI pair and thus can accommodate twice as many planes of a given thickness.

24.1.1 Motivation for Higher Frequency Transformers

The core set can only store a particular amount of magnetic energy for each pulse of electrical power. It can be thought of as a bucket of energy which is filled by the power source and emptied into the power load. It follows that the faster we can transfer the bucket back and forth, the more energy per second (power) we can supply to the load. Thus, we can see the main reason for wishing to maximize the operating frequency of the magnetic energy pulses applied to the ferrite. At higher frequencies, we can get away with a smaller bucket (core) while still supplying the same power to the load. We can also see that on a practical level, as the frequency increases, the amount by which the bucket can be usefully filled and emptied becomes less and less, until a point is reached where the process is no longer practical. In planar transformers, this point is largely dependent on the ferrite material.

The main advantages of planar transformers over wirewound units are the winding coil manufacturability, high-frequency conductivity performance, thermal performance, and coupling performance of the overall low-profile planar assembly with cores. Once designed and refined for production, planar transformers are ideally suited to large-scale manufacture using mass production techniques, enabling high yields of consistently manufactured high-quality, high-performance devices.

The main disadvantage of planar transformers compared with wirewound units is the development effort, cost, and time scales usually associated with getting one into production.

There are four main variations of planar transformer to consider: integrated, standard discrete, custom discrete, and discrete stackable element (DSE) (Figure 24.2).

A relative comparison of the features of each of these types, together with those for wirewound transformers, is shown in Table 24.1.

FIGURE 24.2
DSE planar transformer.

TABLE 24.1
Relative Benefits of the High-Frequency Transformer Types

	Development Cost	Applicability	Production Cost	Production Rate	Production Yield	Performance
Integrated planar	Highest	Highest	Lowest	Highest	Highest	Highest
Custom discrete planar	High	Highest	Low	Moderate	High	Highest
Standard discrete planar	Nil	Low	Low	Moderate	High	Varies
DSE planar	Low	High	Mid	High	High	High
Wirewound	Low	High	High	Low	Low	Mid

24.2 Integrated Planar Transformers

We say that the planar transformer is "integrated" if it is manufactured such that part or all of the windings are formed from the copper layers of the main power supply or system PCB. The EE-core pair is then inserted through the main PCB. Such PCB-embedded windings are sometimes also complemented by supplementary layer(s) of further windings, mounted on the board via pins or SMT pads, in which case the planar transformer type is termed "hybrid integrated."

Design and development of the integrated planar transformer is the most expensive and time consuming of all of the types considered but potentially yields the best performance results.

The design of the integrated planar transformer must be carried out as part of the main system PCB design process. Testing and optimization of this PCB will often be driven mainly by the requirements to optimize the performance of the integrated planar transformer(s). Several iterations of prototyping/design refinement production runs of the board manufacture will normally be necessary in order to achieve a reasonable level of confidence that the intended performance of the planar device is as expected and meets all applicable safety and performance standards [2]. Production runs to produce prototype boards typically involve 50–100 boards.

The development phase of a typical integrated planar transformer is shown in Figure 24.3.

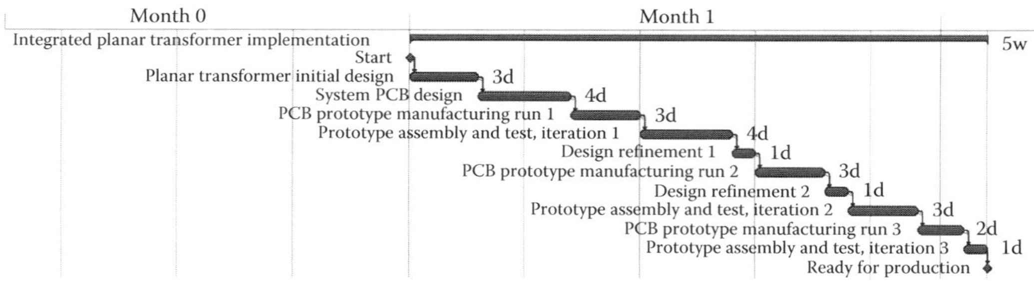

FIGURE 24.3
Integrated planar transformer implementation.

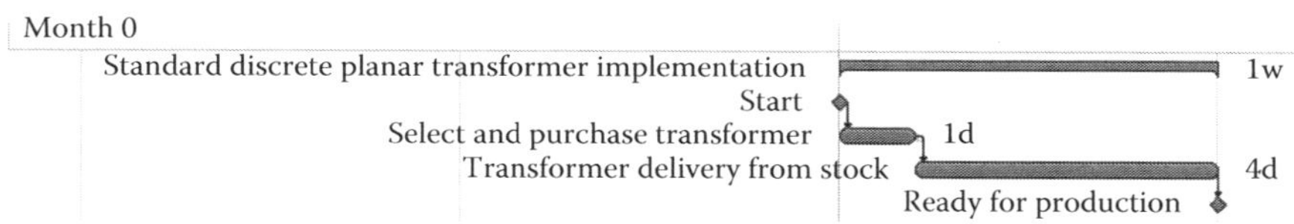

FIGURE 24.4
Standard discrete planar transformer implementation.

24.3 Discrete Planar Transformers

This is the most common form of the planar transformer, comprising readily assembled stand-alone assemblies, such as shown in Figure 24.1. Design effort and production of this form of planar transformer are normally carried out by specialist planar transformer suppliers. Such manufacturers are set up for and accustomed to rapidly carrying out all of the planar transformer development work that is necessary to meet particular equipment manufacturer's requirements, usually at a significant cost. The vast majority of discrete planar transformers in use today are custom designed in this way for specific applications. In some cases, where a transformer for a particular application may be recognized as being more widely applicable, the transformer supplier may volunteer to bear some or all of the development cost themselves in return for rights to manufacture, stock, and market the transformer as part of their own standard off-the-shelf (OTS) range (Figure 24.4).

It should be recognized that in practice, the number of applications which can be satisfied by ready-made standard OTS discrete planar transformers is relatively limited.

24.4 Discrete Stackable Element Planar Transformers

DSE planar transformers are assembled from a set of standard stackable elements [3], as shown in Figures 24.2, 24.5, and 24.6. These elements are designed and manufactured such that (1) a small set of variants can be combined in a large number of ways, so as to obtain a large number of possible devices, and (2) the elements can be assembled to core and to main system board using automatic pick and place manufacturing processes. By mass

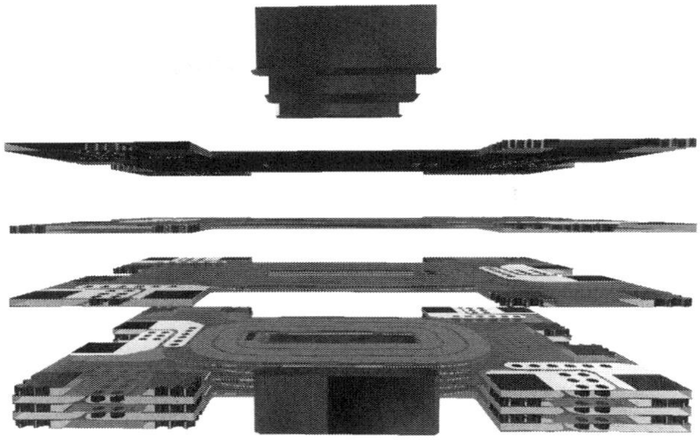

FIGURE 24.5
Exploded view of DSE elements stacking on core.

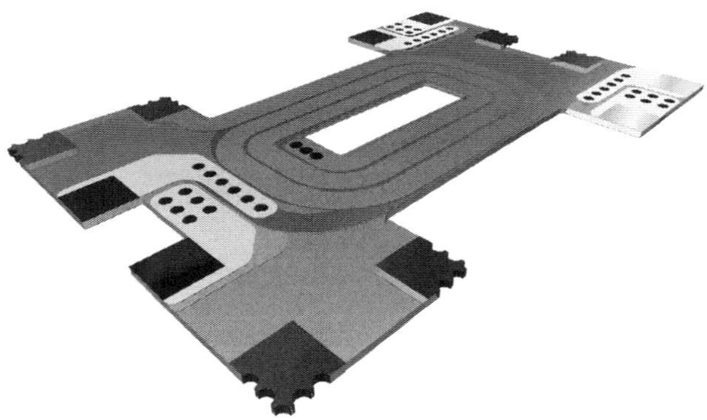

FIGURE 24.6
Single DSE plane.

producing the set of basic elements, large economies of scale and production yields can be realized. By stocking moderate quantities of basic elements, suppliers can offer a very large range of planar transformers, meeting the majority of applications with custom performance in OTS time scales.

A further feature of DSE planar transformer development is inherent suitability to SPICE and other prebuild simulation methods. Because each DSE element variant can be exhaustively characterized, the SPICE model for any DSE assembly is easily computed from the combination of the known characteristics of the individual elements.

The development and implementation of a DSE planar transformer are shown in Figure 24.7.

24.5 Planar Transformer Design Process

In order to minimize the power lost in power transformers, the ferrite supplier endeavors to minimize the core losses, while the winding supplier endeavors to minimize the

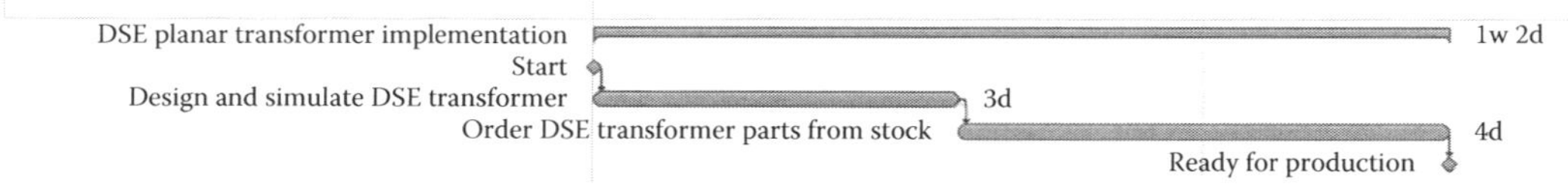

FIGURE 24.7
DSE transformer implementation.

copper losses. In the author's experience, this is often a subject of some debate, as the transformer winding supplier is most often a different company from the ferrite manufacturer. Measurement of the overall dissipated power from the transformer is fairly straightforward. However, determining the individual proportions due to core loss and copper losses on the bench can be difficult if not impossible.

24.5.1 High-Frequency Heating Effects

In wirewound high-frequency transformers, skin effects are a real concern, requiring an application-specific choice of Litz wire to be used in most examples. This is not a concern in standard PCB foil–based planar transformers. We trust readers will recall that all closed formulas describing the world of classical electromagnetics are fundamentally derived from closed solutions of Maxwell's equations [4]. We remind that these can only be solved analytically to give the effects within a conductor in the case that the conductor has continuous boundaries (i.e., circular or elliptical) [5]. We know from experience that attempts to predict skin effects in foils using closed formulas invariably produce incorrect results, always overestimating the effects, usually grossly so.

Substantiated by pragmatic references [6], we know that at frequencies of less than 1 MHz, there are no appreciable skin effects, eddy currents, proximity effects, or other alternating field phenomena affecting the resistance within standard 35 μm (1 oz), 70 μm (2 oz) copper PCB tracks of lengths much less than the wavelength of the frequency in question. As a rule of thumb, at frequencies less than 1 MHz, the current distributions within the conductors of PCB-based planar transformers can be considered as if at DC.

This covers the vast majority of power transformer applications in use today and in the near future. There are still very few power ferrites on the market today which are specified for use at frequencies beyond 1 MHz, so that is our maximum frequency bound for the purposes of this chapter.

24.5.2 Design Process

The design process for a full custom integrated or discrete planar transformer to meet the requirements of a particular application comprises two stages: (1) the electrical design and (2) the mechanical design. We will illustrate this by an example:

- 230–12 V, 400 W step-down converter.
- Transformer temperature must not exceed 100°C.

The electrical design process follows the same principles as that for wirewound ferrite transformers, with the simplification that (for the reasons explained earlier) we can assume Rdc/Rac = 1, at least up to 1 MHz.

Our task is first to determine an appropriate core size and material for the application.

TABLE 24.2

Core Figures of Interest

Core Type	EE18	EE22	EE32	EE38	EE64
Thermal resistance (°C/W)	56	35	24	18	9
Volume (cc)	0.82	2.25	4.56	10.2	36.00

We start by selecting an appropriate converter topology. Different converter topologies have different capabilities of power transmission. This has been characterized in terms of a constant [7], C:

- C = 1: push–pull converter
- C = 0.71: single-ended converter
- C = 0.62: flyback converter

As our application is fairly high powered, we select the push–pull configuration so as to obtain maximum throughput power capability from the transformer. We see the main figures of interest for the most likely popular cores in Table 24.2.

We note from general ferrite material operating characteristics that the power dissipated in the core of the transformer follows an opposite temperature characteristic to that of the winding toward the expected steady-state operating temperature. The ferrite material power loss drops toward the operating temperature, while the winding power loss increases. In the foil-based planar transformer at less than 1 MHz, we can say that the winding power loss increase is due solely to the positive temperature dependence of the copper winding resistance. By considering the transformer equivalent circuit in terms of the maximum power transfer theorem, we know that the copper and the ferrite losses need to be balanced with one another in order to maximize the power transferred between them so as to maximize the throughput power capability at the desired operating frequency. Intuitively, we can see that there is a certain self-regulatory quality about this behavior. The transformer temperature operating point in a fixed frequency converter should naturally tend toward a state of equilibrium, where the copper and the ferrite losses are balanced with one another.

Using the earlier analysis in the special case of the planar transformer at frequencies up to 1 MHz, it follows that we can treat the winding losses at the steady-state operating point as if they were being generated by a DC current independently from the power dissipated by the core and that these losses will contribute 50% of the core heating effect.

We start our calculations by choosing the smallest core in an appropriate material which satisfies our required operating temperature while delivering the required power at a manageable frequency.

We choose a ferrite material specified for higher frequency operation, up to 1 MHz. Our choice here is F47. The main material characteristics of interest for our purposes are plotted in Figures 24.8 and 24.9 (checked against manufacturer datasheets) [8].

24.6 Planar Transformer Electrical Design

Application required power capability (W):

$$P_{app} = 400 \tag{24.1}$$

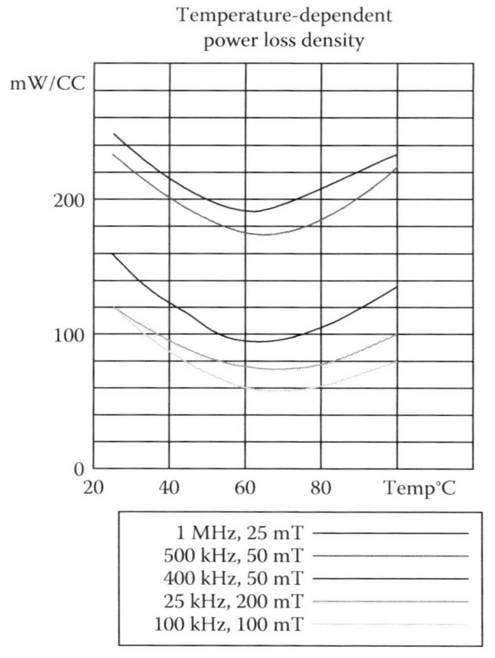

FIGURE 24.8
F47 temperature-dependent P.L.D.

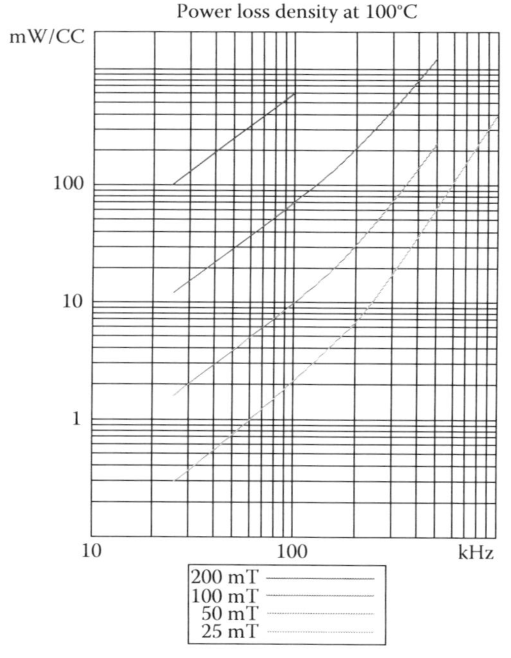

FIGURE 24.9
F47 frequency-dependent P.L.D.

Application ambient temperature (°C):

$$T_{amb} = 30 \tag{24.2}$$

Application maximum transformer temperature:

$$T_{des} = 90 \tag{24.3}$$

Transformer design allowable temperature rise:

$$\Delta t_{allowed} = T_{des} - T_{amb} \tag{24.4}$$

Expected converter operating efficiency:

$$Eff = 0.95 \tag{24.5}$$

Expected converter power dissipation:

$$P_{diss} = (1 - Eff) \cdot P_{app} \tag{24.6}$$

$$P_{diss} = 20 \tag{24.7}$$

Approximate switching to transformer loss ratio (from experience):

$$P_{sw\_tx\_ratio} = 8 \tag{24.8}$$

Expected power dissipated in transformer:

$$P_{disstx} = \frac{P_{diss}}{P_{sw\_tx\_ratio}}$$

$$P_{disstx} = 2.5 \tag{24.9}$$

Thermal resistivity of the transformer assemblies under consideration EE18, EE22, EE32, EE38, and EE64 (°C/W, from Table 24.2):

$$\theta_{th} = \begin{bmatrix} 56 \\ 35 \\ 24 \\ 18 \\ 9 \end{bmatrix} \tag{24.10}$$

Expected temperature rise in transformers under consideration:

$$\Delta t_{exp} = \theta_{th} \cdot P_{disstx} \tag{24.11}$$

Thus,

$$\Delta t_{exp} = \begin{bmatrix} 140 \\ 87.5 \\ 60 \\ 45 \\ 22.5 \end{bmatrix} \tag{24.12}$$

comparing with

$$\Delta t_{allowed} = 60 \tag{24.13}$$

Thus, we select EE32 with dt = 60 as the smallest core that might suit the application.

Allowable power dissipation allowed per core type:

$$P_c = \overrightarrow{\left[\frac{(\Delta t_{allowed})}{\theta_{th}}\right]} \tag{24.14}$$

Thus,

$$P_c = \begin{bmatrix} 1.071 \\ 1.714 \\ 2.5 \\ 3.333 \\ 6.6667 \end{bmatrix} \tag{24.15}$$

The expected transformer dissipation of 2.5 here is the result of the sum of the copper losses and the core losses.

Assuming winding losses equal to core losses, we need to obtain a core loss of roughly half of the figure given earlier, that is, 1.25 W or 1250 mW.

Volumes of cores EE18, EE22, EE32, EE38, and EE64 (cc):

$$V_c = \begin{bmatrix} 0.82 \\ 2.25 \\ 4.56 \\ 10.2 \\ 36 \end{bmatrix} \tag{24.16}$$

Required power loss density (mW/cc):

$$\varphi_{cp} = \overrightarrow{\left(\frac{1000 \cdot P_c}{2 \cdot V_c}\right)} \tag{24.17}$$

Thus,

$$\varphi_{cp} = \begin{bmatrix} 653.31 \\ 380.952 \\ 274.123 \\ 163.399 \\ 92.593 \end{bmatrix} \tag{24.18}$$

We can see from the characteristics for the chosen material that this power loss density corresponds to 50 mT at 500 kHz, or 25 mT at 900 kHz. We choose 750 kHz as an intermediate design frequency between 500 and 900 kHz. As EMC noise emissions are heavily related to switching frequency, we do not wish to switch at a higher frequency than is necessary. This helps to minimize the suppression design outlay and component count often required later to satisfy the EMC requirements of the finished system.

Our remaining task for this example is to configure the windings on the primary and secondary to obtain the appropriate voltage transformation. We endeavor to minimize the number of turns per winding, so as to maximize the proportion of area in the winding window which can be filled by paralleled conductive foils, thus shortening the trace lengths and reducing the power lost/radiated from the windings.

We choose the nominal operating point of the SMPS such that at normal working load, it will be working at 75% of maximum output capability. This leaves plenty of adjustment to increase or decrease the SMPS duty cycle so as to compensate for changes in load and/or source voltages.

Minimum number of turns we can have on the secondary:

$$N2 = 1 \tag{24.19}$$

Required output voltage:

$$V_{out} = 12 \tag{24.20}$$

Given input voltage:

$$V_{in} = 230 \tag{24.21}$$

Desired duty cycle:

$$D = 0.75 \tag{24.22}$$

Expected regulation (amount by which the transformer open circuit voltage would be expected to drop when placed on load, with no adjustment of duty cycle):

$$Reg = 20\% \tag{24.23}$$

Required number of primary turns (on each side of center tap):

$$N1 = D \cdot \frac{V_{in}}{(1 + Reg) \cdot V_{out}} \tag{24.24}$$

That is,

$$N1 = 11.979 \quad (24.25)$$

The closest practical number of turns is 12; thus, we set the primary number of turns to 12.

A rule of thumb in planar transformer design is that the winding window foil fillage should be approximately divided 50/50 between primary and secondary windings. This is again to satisfy the maximum power transfer theorem; we need the input winding dissipated power to be approximately equal to the output winding dissipated power so as to maximize the throughput capacity of the transformer.

A further rule of thumb is that in order to maximize the primary to secondary coupling (minimizing the leakage inductance), we should interleave the primary and secondary windings as far as practicable.

We know from experience that we can fit eight PCB planes of 0.75 mm thickness in the height of an EE32 planar core assembly winding window. We also know that each 0.75 mm plane can comfortably accommodate four inner layers of 70 μm ("2 oz") foil. It turns out that the easiest to design (and, in our opinion, the most practical) configuration for a winding in a plane is for half of the winding to spiral inward from the winding start toward the central leg on 50% of the available inner layers, and then the other half of the winding to spiral back outward again on the other 50% of available inner layers toward the winding finish. In order to minimize interwinding capacitance within each plane, this should be done in two-halves, one-half containing all of the layers of one spiral and the other containing all of the layers of the other spiral. The winding is then completed by connecting all of the inner layers together at the ends nearest the central leg by one or more buried vias (connecting the inner layers but not appearing on either surface of the plane, see Figure 24.10). The DSE plane of Figure 24.6 shows this type of winding. Vias are simply holes which are drilled in the PCB then plated, so as to form a tubular foil connection between adjacent layers in the PCB. Buried vias are constructed by drilling and plating, when only the inner PCB layers have been manufactured, before addition of the final outer and inner PCB layers. The vias and inner layers are then buried and thus insulated by addition of the final outer layers of substrate. This type of construction is used in the CAD/CAM file Gerber [9] graphic of Figure 24.13.

A configuration meeting all of our requirements, constructed from two types of 0.75 mm plane, one with 12 turns and the other with 1 turn, is shown in Figure 24.11.

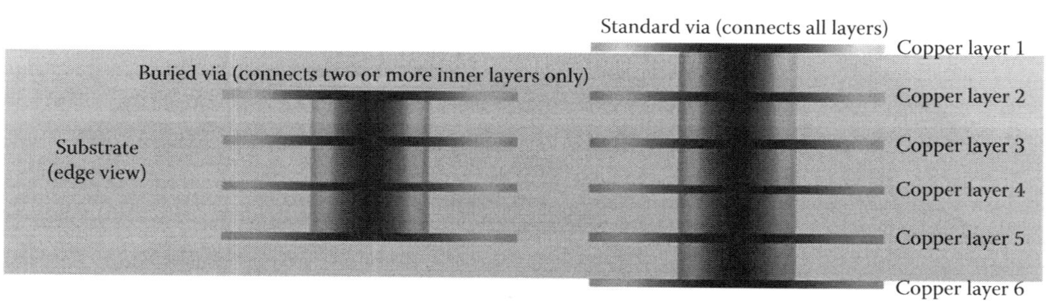

FIGURE 24.10
Standard and buried vias.

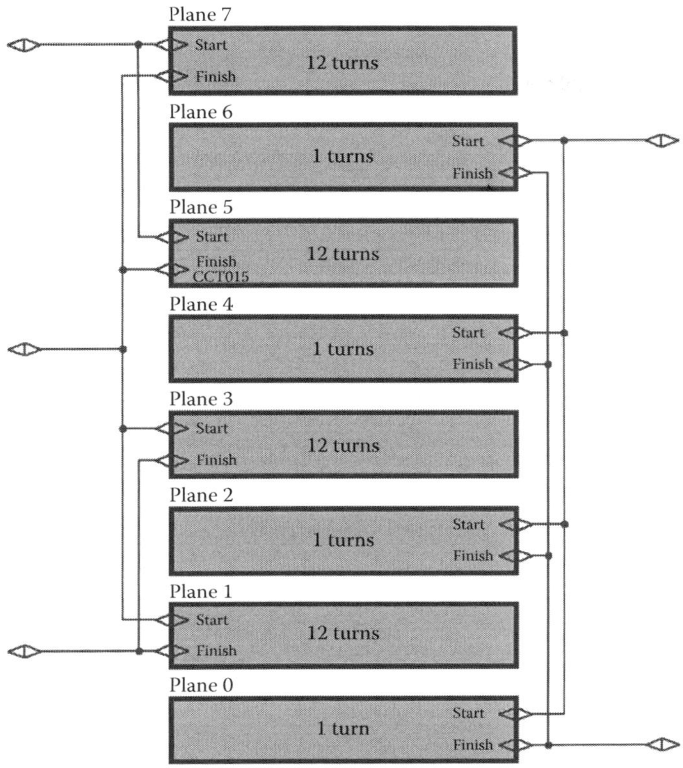

FIGURE 24.11
Example build.

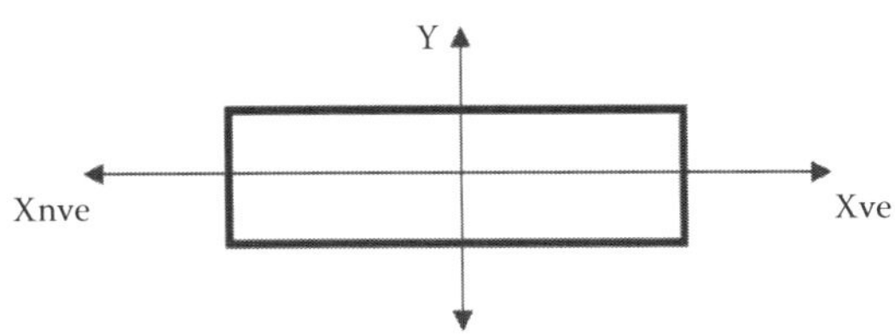

FIGURE 24.12
Reference axes relative to plane central hole.

In the build of Figure 24.11, we have connected the primary planes in two sets of pairs in such a way as to form a center tap. Plane 7 paralleled with plane 5 is the pair on one side of the center tap, and plane 3 paralleled with plane 1 is on the other side. Thus, our primary comprises a center-tapped winding of 24 turns. Our secondary is formed from four single-turn planes connected in parallel, thus has only one turn. As can be seen, there is a similar amount of copper in the primary compared with the secondary windings; thus, we have approximately balanced our primary and secondary power dissipations. In addition, the primary and secondary windings are closely interleaved, so we obtain good primary to secondary coupling.

24.6.1 Planar Transformer Mechanical Design

We show here the main processes of the mechanical design of the E32 DSE plane set. This is typical to the process which needs to be followed in the mechanical design of any custom integrated or discrete transformer. Not shown here is the design process of the vertical multilayer foil laminations within each substrate (Table 24.3).

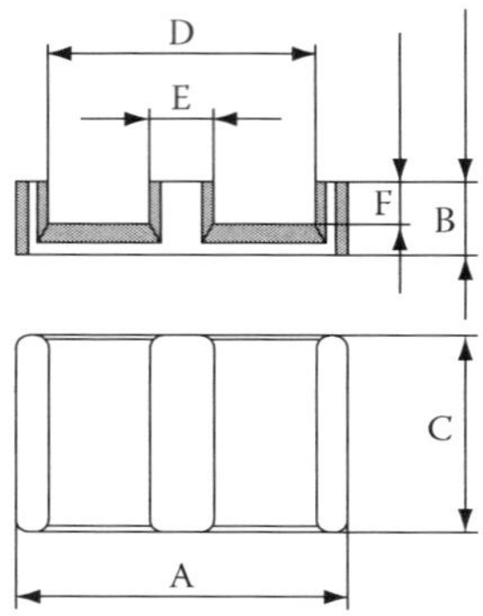

Worst-case core dimensions of interest:

$$C_{max} = 20.73 \quad D_{min} = 24.9 \quad E_{max} = 6.48 \tag{24.26}$$

Core winding channel width:

$$W_{ch} = \frac{(D_{min} - E_{max})}{2} \tag{24.27}$$

$$W_{ch} = 9.21 \tag{24.28}$$

Fit clearance assigned between plane and core:

$$FC = 0.1 \tag{24.29}$$

PCB width:

$$W_{pcb} = W_{ch} - 2 \cdot FC \tag{24.30}$$

Minimum dielectric thickness required between core and tracking:

$$MDT = 0.55 \tag{24.31}$$

TABLE 24.3

IEC62317-9 Compliant E32/6/20 Core Dimensions (mm)

A	B	C	D	E	F
31.75 ± 0.65	6.35 ± 0.15	20.33 ± 0.4	25.4 ± 0.5	6.33 ± 0.15	3.2 ± 0.15

Effective tracking width:

$$W_{tr} = W_{pcb} - 2 \cdot MDT \tag{24.32}$$

$$W_{tr} = 7.91 \tag{24.33}$$

Assigned minimum track/gap clearance:

$$MTG = 0.2 \tag{24.34}$$

Assigned numbers of turn variations in plane set (excluding one turn):

$$n = \begin{bmatrix} 2 \\ 4 \\ 6 \\ 8 \\ 10 \\ 12 \\ 16 \end{bmatrix} \tag{24.35}$$

$$Track\_widths = \left[\overrightarrow{\frac{W_{tr} - MTG \cdot ((n/2) - 1)}{n/2}} \right] \tag{24.36}$$

$$Track\_spacings = Track\_widths + MTG \tag{24.37}$$

Thus,

$$Track\_widths = \begin{bmatrix} 7.91 \\ 3.855 \\ 2.503 \\ 1.828 \\ 1.422 \\ 1.152 \\ 0.814 \end{bmatrix} \tag{24.38}$$

and

$$Track\_spacings = \begin{bmatrix} 8.11 \\ 4.055 \\ 2.703 \\ 2.028 \\ 1.622 \\ 1.352 \\ 1.014 \end{bmatrix} \tag{24.39}$$

PCB manufacturer drill sizes are still customarily in thousandths of an inch, so we need to convert track sizes in mm to thou so as to compute appropriate via sizes, which are based on drill hole sizes:

$$\text{mm\_per\_inch} = 25.4 \tag{24.40}$$

$$\text{Track\_widths\_in\_thou} = 1000 \cdot \frac{\text{Track\_widths}}{\text{mm\_per\_inch}} \tag{24.41}$$

Thus,

$$\text{Track\_widths\_in\_thou} = \begin{bmatrix} 311.417 \\ 151.772 \\ 98.556 \\ 71.949 \\ 55.984 \\ 45.341 \\ 32.037 \end{bmatrix} \tag{24.42}$$

Equivalent current capacity via sizes for tracks in the set n:

$$\frac{\text{Track\_widths} \cdot 2 \cdot 1000}{\pi \cdot \text{mm\_per\_inch}} = \begin{bmatrix} 198.254 \\ 96.621 \\ 62.743 \\ 45.804 \\ 35.641 \\ 28.865 \\ 20.396 \end{bmatrix} \tag{24.43}$$

We choose a single convenient drill size which can be used to drill all of the vias on the board, containing all of the plane variations in the set n (multiple vias can be drilled several times to provide via conductivity of single larger vias, that is 198.254 thou vias given earlier can be met by drilling 6 or 7 of 30 thou vias); this saves on manufacturing cost by alleviating need to change drills during CNC drilling of boards:

$$\text{Via\_diameter\_in\_thou} = 30 \tag{24.44}$$

With reference to the plane core central limb hole position, first track positions on y axis:

$$\text{T1Pos}_y = \frac{E_{max} + \text{Track\_widths}}{2} + \text{MDT} + \text{FC} \tag{24.45}$$

Non–via end first track positions on x axis:

$$T1Pos_{xnve} = \frac{C_{max}}{2} + FC + MDT + \frac{Track\_widths}{2} \tag{24.46}$$

$$Via\_diameter\_in\_mm = \frac{Via\_diameter\_in\_thou \times 25 \times 4}{1000} \tag{24.47}$$

$$Via\_diameter\_in\_mm = 0.762 \tag{24.48}$$

$$Via\_position\_on\_x\_axis = \frac{C_{max}}{2} + FC + MDT + \frac{Via\_diameter\_in\_mm}{2} \tag{24.49}$$

$$Via\_position\_on\_x\_axis = 11.4 \tag{24.50}$$

Via end first track positions on x axis:

$$T1Pos_{xve} = Via\_position\_on\_x\_axis + MTG + \frac{Via\_diameter\_in\_mm}{2} + \frac{Track\_widths}{2} \tag{24.51}$$

$$T1Pos_{xve} = \begin{bmatrix} 15.932 \\ 13.905 \\ 13.229 \\ 12.891 \\ 12.688 \\ 12.553 \\ 12.384 \end{bmatrix} \quad T1Pos_{xnve} = \begin{bmatrix} 14.97 \\ 12.943 \\ 12.267 \\ 11.929 \\ 11.726 \\ 11.591 \\ 11.422 \end{bmatrix} \tag{24.52}$$

$$T1Pos_{y} = \begin{bmatrix} 7.845 \\ 5.817 \\ 5.142 \\ 4.804 \\ 4.601 \\ 4.466 \\ 4.297 \end{bmatrix} \tag{24.53}$$

Core center limb area:

$$A_{cl} = E_{max} \cdot C_{max} \tag{24.54}$$

$$A_{cl} = 134.33 \tag{24.55}$$

Assigned pad/center limb area ratio:

$$K_{cl\_p} = \frac{1}{6} \tag{24.56}$$

Pad area:

$$A_{pad} = K_{cl\_p} \cdot A_{cl} \tag{24.57}$$

$$A_{pad} = 22.388 \tag{24.58}$$

Square pad dimensions:

$$L_{pad} = \sqrt{A_{pad}} \tag{24.59}$$

$$L_{pad} = 4.732 \tag{24.60}$$

Assigned number of vias per interconnect:

$$N_{int\_vias} = 6 \tag{24.61}$$

Assigned number of layers in design:

$$N_{layers} = 4 \tag{24.62}$$

Assigned copper weight (ounce per layer):

$$Weight_{copper} = 2 \tag{24.63}$$

Index of widest track in track widths array:

$$Widest\_track\_index = 0 \tag{24.64}$$

Effective ideal interconnect via the hole diameter required:

$$\phi_{thp} = Track\_widths_{Widest\_track\_index} \cdot \frac{(N_{layers} - 2) \cdot Weight_{copper}}{\pi \cdot N_{int\_vias}} \tag{24.65}$$

$$\phi_{thp} = 1.679 \tag{24.66}$$

$$\phi_{thp\_thou} = \phi_{thp} \cdot \frac{1000}{mm\_per\_inch} \tag{24.67}$$

$$\phi_{thp\_thou} = 66.085 \tag{24.68}$$

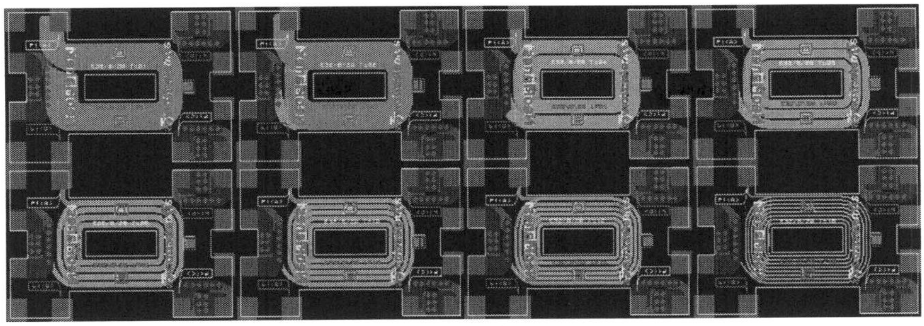

FIGURE 24.13
CAD/CAM file output Gerber [9] graphic for DSE plane set.

With the earlier calculations carried out, in addition to the design of the foil vertical lamination structure, we are ready to begin drawing the PCB layouts using appropriate PCB design CAD software. On completion, the CAD software is used to output the Gerber format computer-aided manufacturing (CAM) files normally required by the PCB manufacturer to manufacture the PCB per the design.

Figure 24.13 shows a set of DSE planes designed using the calculations given earlier.

24.7 Discrete Stackable Element Planar Transformer Design Process

The basic design of a DSE planar transformer follows a similar process to that for the discrete OTS transformer. A particular core set and turns ratio is simply selected to suit the application from the range of parts offered as stock from suppliers.

For this task, the only CAD tool that is required is the DSE transformer stack design tool. This requires no particular expertise and is available freely on the internet, downloadable from manufacturers in minutes [10]. The stack designer tool provides a simple method of selecting parts and visualizing and manipulating the planes in the stack in such a way that the primary and secondary copper fillages are maintained approximately balanced and fully interleaved. This enables rapid design of consistent high-performance planar devices with appropriate turns ratios and power capacities for the vast majority of applications and topologies to be carried out by engineers and technicians with little or no direct experience of detailed planar transformer design.

During the stack design process, the stack design tool automatically produces a stack design file. This file, when issued by a registered copy of the design tool, is automatically in the form of a protected assembly file. The protected file can thus be directly forwarded to the intended manufacturers and/or OTS parts suppliers as the transformer assembly instructions. The recipients of such files may open them using further free registered or unregistered copies of the stack design tool. This allows read-only viewing and/or printing of protected assembly designs. If the source file was generated by an unregistered copy of the stack designer tool, then the design is unprotected and modifiable by any registered or unregistered copy of the design tool. Printing protected assembly designs from the stack designer provides a means of instantly distributing assembly instructions for recipient stores and production staff.

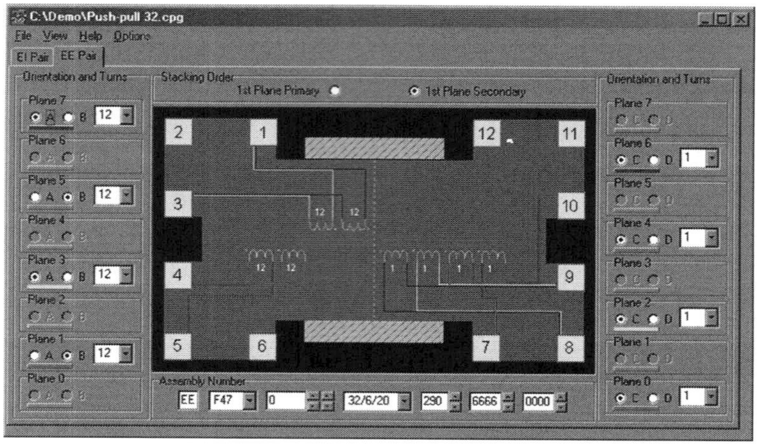

FIGURE 24.14
Example DSE planar transformer under design.

The contents of the text boxes and dropdowns visible at the bottom of the DSE stack designer tool main window in sequence are used to form the unique assembly number that specifies the assembly. From Figure 24.14, the assembly number for the example of our chapter is EE-F47-0-32/6/20-290-6666-0000.

24.7.1 Simulation Capability

Where the planar transformer design process is likely to be carried out by the main system electronic designer, or a transformer designer who is already accustomed to the use of CAD tools for the main circuit design, the DSE planar transformer is likely to be of particular interest.

The stack design tool output file is in the form of a Berkeley SPICE 3f5 [11] simulation model which includes winding resistances, leakage inductances, saturation, interwinding capacitances, pri-sec parasitic capacitances, and has B and H voltage outputs for the transformer assembly. Thus, the dynamic electrical and magnetic behavior of the device under design in the stack design tool is fully simulatable in circuit using CAD circuit design and SPICE simulation software, together with the usual SPICE models for switching transistors, controllers, etc.

This can be of considerable value in the design of new SMPS systems by helping to de-risk system designs prior to investing in physical hardware development and manufacture, per the recommendations of systems life cycle standards such as IEC15288. An example simulation showing the classic B–H hysteresis loop, simulated in push–pull mode on assembly EE-F44-0-38/8/25-290-1111-1111 at 50 kHz, is shown in Figure 24.15.

The stack design tool updates the transformer model file for every change that is made in the stack designer. Linking the CAD circuit design SPICE simulator to the stack designer output file allows the designer to interactively carry out an iterative process of adjusting the transformer model in the stack designer and resimulating in order to get the desired circuit behavior. The transformer temperature rise can also be estimated for the particular operating conditions by measuring the difference between the transformer input and output power and calculating the temperature rise which would occur from the known thermal resistance of the device.

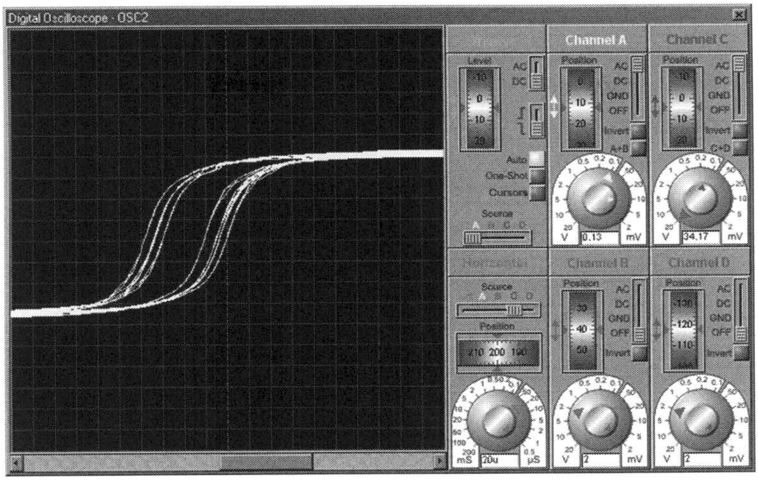

FIGURE 24.15
Classic B–H loop.

This simulation capability is unique to the DSE form of planar transformers in that a high level of confidence in the accuracy of the generated SPICE model can be obtained from being able to exhaustively characterize every element type. The characteristic of a particular build is then simply computed from the combination of the known characteristics of the individual elements. We will elaborate on how the elements are combined for simulation in the final part of this chapter.

For circuit and PCB design and SPICE simulation software, our own favorite that we would recommend is Labcenter ISIS™—part of the Proteus™ software suite [12].

Our final demo is to show how this is applied to the example 400 W 12 V converter design of this chapter.

24.8 DSE Planar Transformer Design and Simulation Example

Applying the DSE-enabled CAD simulation methodology involves the following steps:

1. Open (or construct) an appropriate circuit design file for the intended application.
2. Insert (or create) the in-circuit instance(s) of the transformer SPICE model as required by the application.
3. Set up an initial possible transformer design in the stack designer.
4. Simulate to check that the circuit performance is as expected.
5. If the circuit performance is not as expected, make changes in the stack designer and/or in the circuit design file and resimulate as necessary to obtain the desired performance.

The push–pull converter simulation schematic for the example is shown in Figure 24.16. Central to this is the transformer instance (TR1). This is set up to be linked to the stack

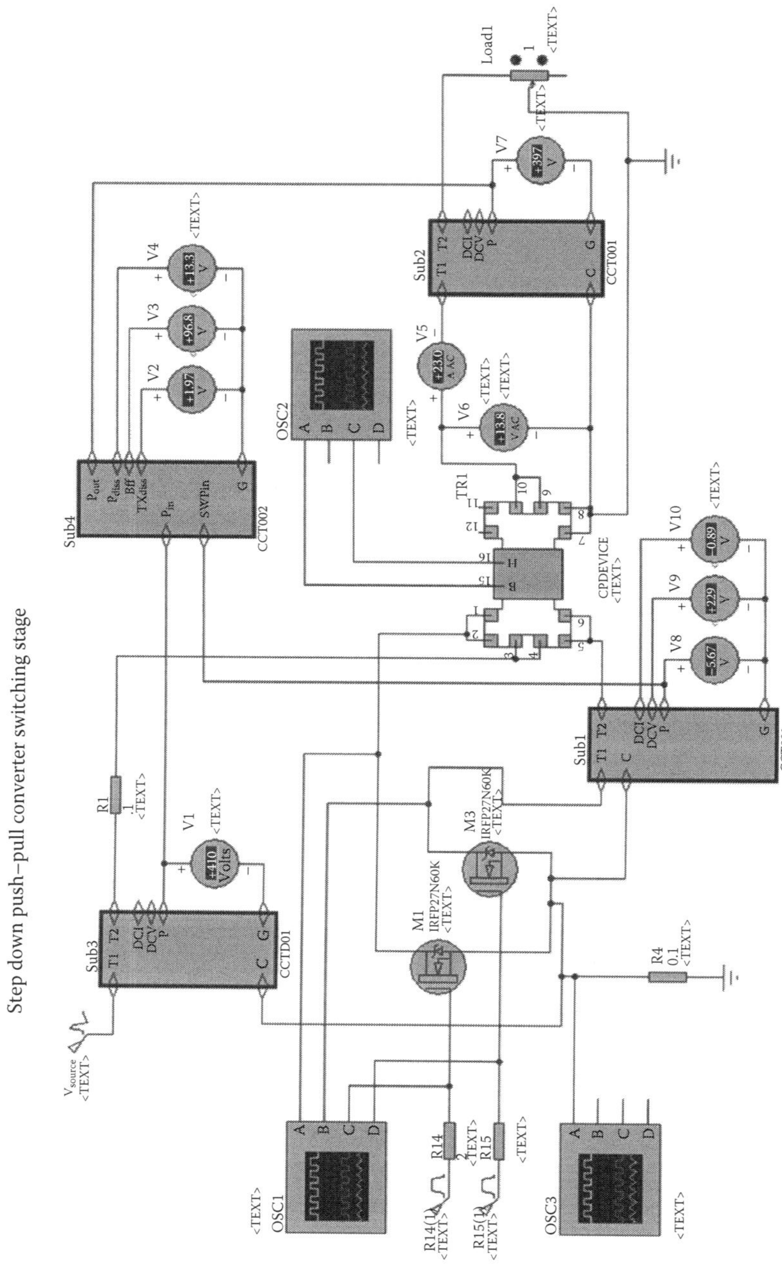

FIGURE 24.16
Push–pull converter simulation schematic.

designer.cpg output file as the SPICE source file. The transistors M1 and M2 are the MOSFET switching transistor devices we have selected to work together with the transformer for the design. These are chosen from an initial inspection of the voltage and current switching characteristics required for the design and can be changed later if necessary.

Also in the main push–pull converter simulation schematic of Figure 24.16, we can see four blocks named SUB1 to SUB4.

SUB1, SUB2, and SUB3 use a common measurement subcircuit (CCT001), shown in Figure 24.17.

In Figure 24.17, E2 outputs a voltage corresponding to that measured across the terminals T1 and C. H1 outputs a voltage corresponding to the current through the terminals T1 and T2. MULT1 produces a voltage corresponding to the product of the voltages output by E2 and H1. The numerical value of the voltage output by MULT1 thus gives the value of the instantaneous power transferred between (1) the device connected to the terminals T1 and C and (2) the device connected to T2 and C.

The resistor–capacitor pairs in CCT001, that is, R2/C1, R6/C3, and R5/C4, serve as low-pass filters which give steady average DC values for the instantaneous voltages on H1, E2, and MULT1. Thus, by connecting DC digital voltage meters to the terminals DCV, DCI, and P, we obtain real-time average values for the current, voltage, and power for convenient display during indefinite time–based simulations. From experience to date, this proves to be the most effective way to simulate switched mode power supply behavior over a period of time long enough for the system to settle on steady-state average values.

Further in the main push–pull converter simulation schematic of Figure 24.16, we see SUB4. This uses the dissipated power measurement subcircuit of CCT002, shown in Figure 24.18.

SUB4 produces DC voltages with values corresponding to the power dissipated by the converter (V4), the converter efficiency in percentage (V3), and the transformer power dissipation (V2). The converter power dissipation and efficiency are computed from P_{in}-P_{out} and P_{out}/P_{in}, respectively, where the DC values for P_{out} and P_{in} are again produced by separate instances of CCT001 (the voltage, power, and averaging subcircuit). The value for the transformer power dissipation is obtained by first obtaining the DC power dissipated across a single switching transistor. The transformer dissipated power is then obtained by computing the difference between the converter dissipated power and the power dissipated by the two switching transistors (obtained as twice the power dissipated in one

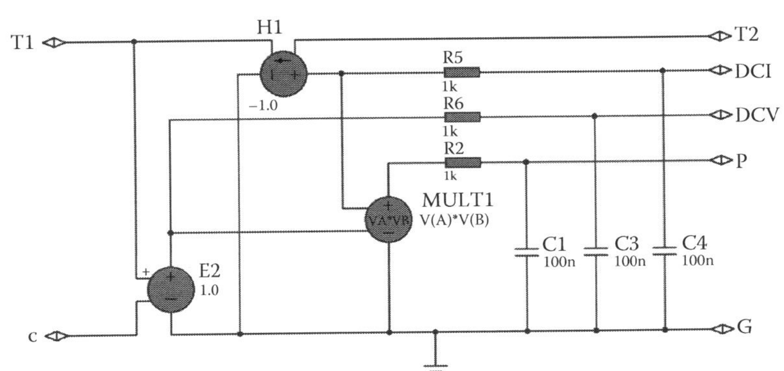

FIGURE 24.17
CCT001 V, I, and P averaging subcircuit.

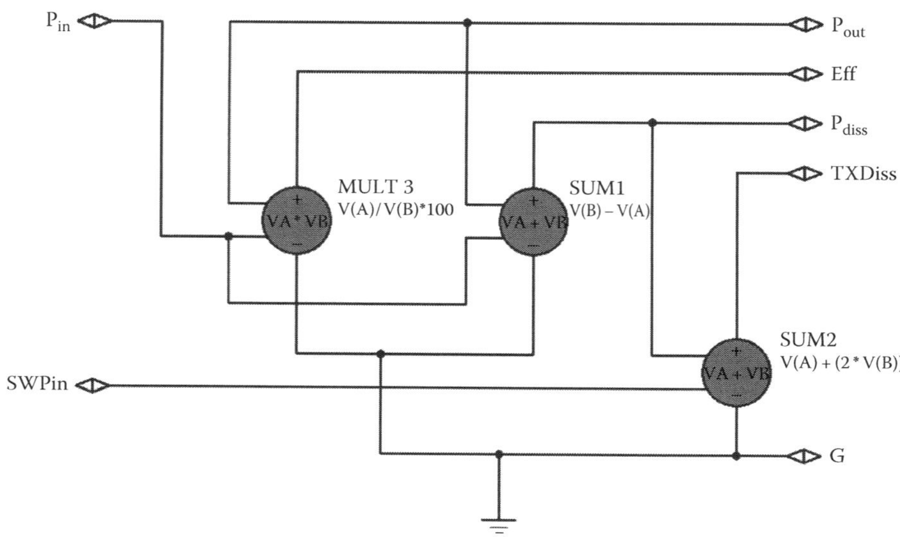

FIGURE 24.18
CCT002 dissipated power calculation subcircuit.

transistor). The remainder of the converter power dissipated is the power dissipated by the transformer. Measurement of the transistor power by an instance of CCT001 produces a negative value, indicating power dissipated (as opposed to generated), so we obtain the difference by adding in SUM2 instead of subtracting.

OSC1, OSC2, and OSC3 in the main circuit of Figure 24.16 are simulator oscilloscopes. OSC1 is used to display the voltage and current waveforms of the switching transistors. OSC2 is used to probe the B and H voltages output by the transistor model. By setting up OSC2 to be in X-Y mode, we obtain a real-time display of the familiar B–H loop in the transformer while the simulation is running.

OSC3 is used to display the current waveforms drawn from the DC supply via the switching transistors and transformer windings.

Further in the main circuit of Figure 24.16, R1 and R4 are simulator resistances placed in series with the main current path to simulate conductor resistances external to the transformer. The voltage across R4 also provides a convenient probing point for OSC3. LOAD1 is a load resistor which can be adjusted during run time, allowing us to see the effect of load changes made during simulation. R14 and R15 are the gate resistors normally fitted in the circuit to limit the capacitive switching currents to the MOSFET gates. VSOURCE is the DC supply voltage which is switched by the converter. The pulsed sources marked R14(1) and R15(1) are the pulse source generators controlling the MOSFET switches. These are set up such that they are 180° out of phase with one another as shown in the setup screens of Figures 24.19 and 24.20. A point to note is that duty cycles of 50% or more set up on these switches result in the ends of the transformer coils on either side of the center tap being effectively shorted to ground at the same time. At such instances, the transformer turns are shorted out and the transformer appears like a short circuit, providing a very-low-inductance path for current to flow from the source to ground. In the simulation, if we deliberately set the phase offset to zero, we get a current of 750 A through the combined switches at 20% duty cycle. This would surely blow the transistors and probably also the

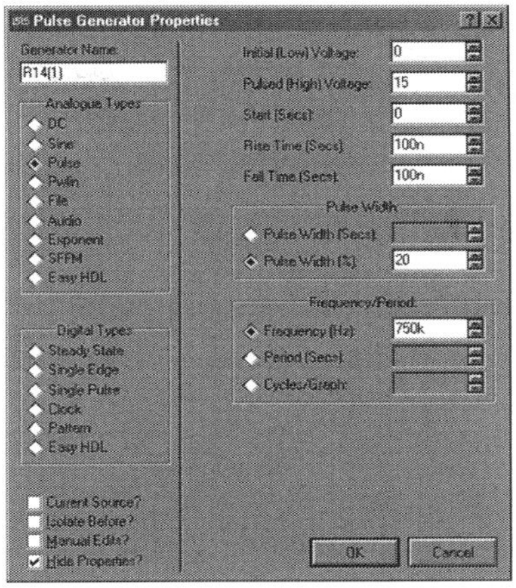

FIGURE 24.19
Switch pulse 1 setup.

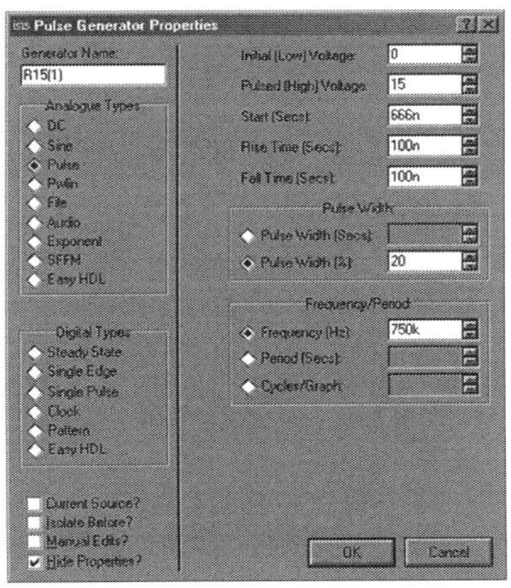

FIGURE 24.20
Switch pulse 2 setup.

transformer if it happened in real hardware*; however, we can safely simulate this and many other catastrophic scenarios in the software without the concern of having to replace prototype hardware when things go wrong.

Figure 24.21 shows the DC source voltage setup (with soft-start ramp to smooth simulation at start-up).

---

* We know from experience that the usual failure mode of switching transistor devices in switched mode power systems is catastrophic fusing of the silicon junctions due to overcurrents (i.e., they turn instantly to short circuits), often resulting in spectacular destruction of the transformer and anything else in the conduction path from the main high-current voltage source.

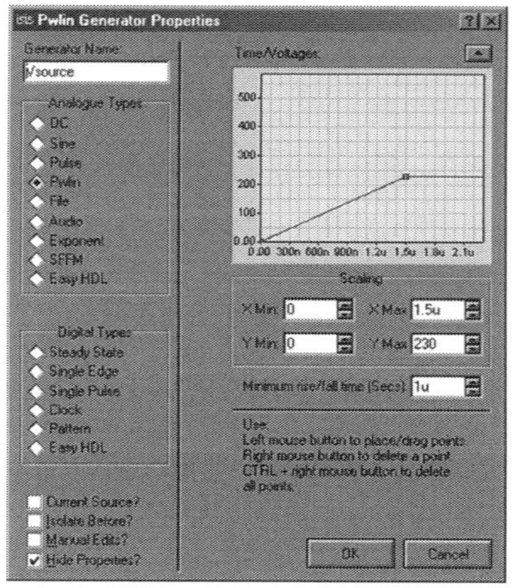

FIGURE 24.21
DC source voltage setup.

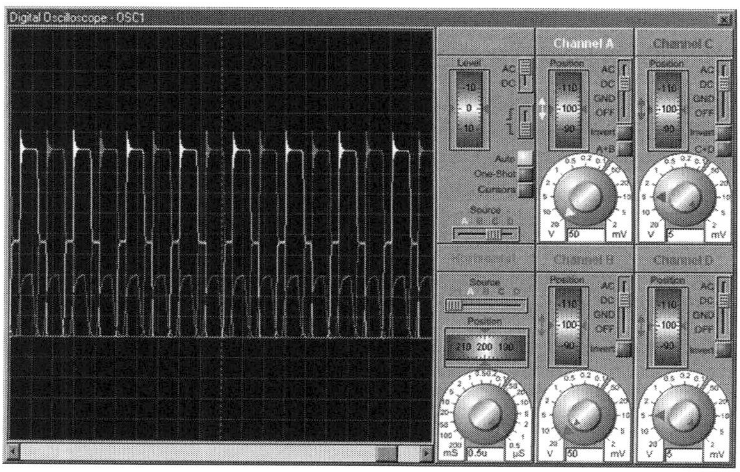

FIGURE 24.22
Transistor switching waveforms final simulation result.

24.8.1 Simulation Results

The final adjusted simulation results are shown in Figures 24.22 through 24.24. The voltage values apparent in the main circuit of Figure 24.16 are also from the final simulation, taken at the same time as the traces in OSC1, OSC2, and OSC3 in Figures 24.22 through 24.24 were recorded.

24.9 Bench Measurement of DC and Thermal Resistance

In the case of DSE and discrete OTS planar transformers, the thermal and DC resistance characteristics of the assembly are already thoroughly tested and documented, and used

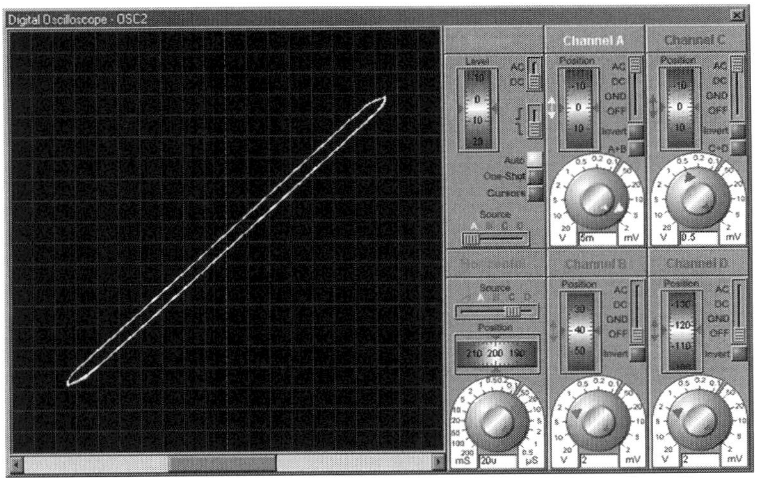

FIGURE 24.23
B–H loop final simulation result.

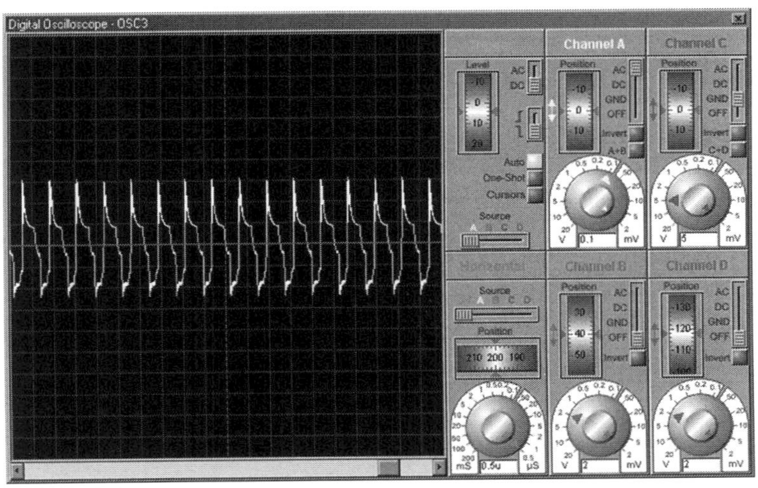

FIGURE 24.24
Current waveforms final simulation result.

as selection criteria for the chosen solution as shown earlier. However, if we have just prototyped a new planar transformer, custom designed from first principles including windings, then we would be most interested in checking the finished assembly on the bench to ensure that our prototype is likely to perform as expected.

The test setup for the thermal and DC resistance characteristics is shown in the CAD drawings of Figures 24.25 and 24.26. Figure 24.25 shows the circuit diagram of an appropriate test setup, while Figure 24.26 shows a DSE assembly setup for test.

The test procedure is to start by measuring the initial temperature of the core. The core needs to have been in the test environment for several hours beforehand so as to give a reasonable value for the test environment ambient temperature. We then roughly set

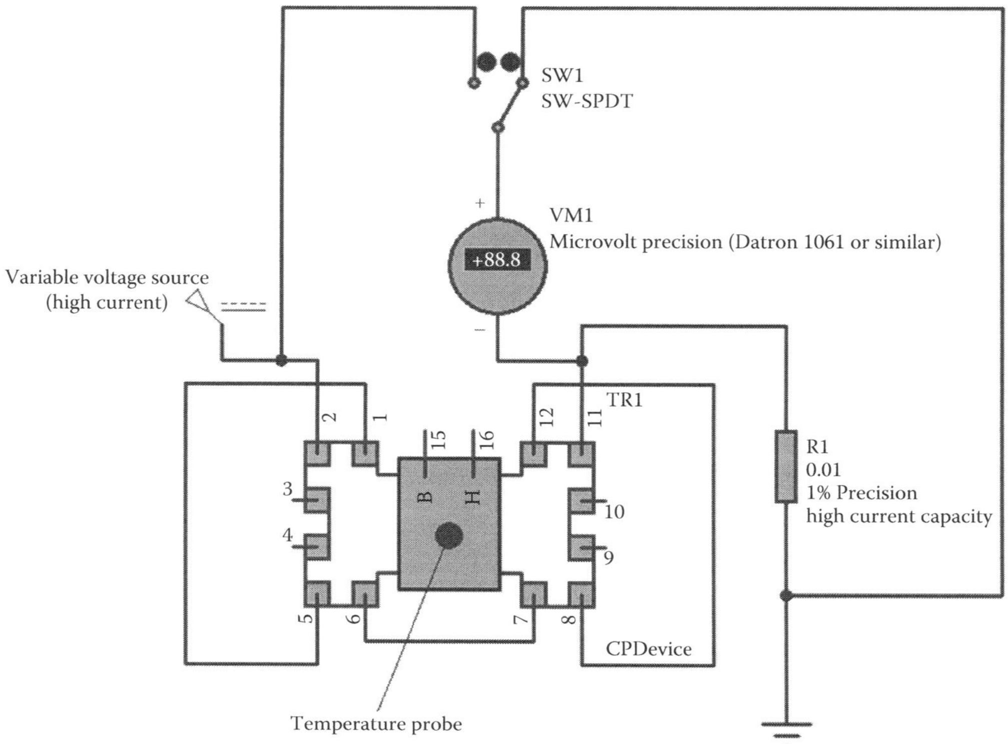

FIGURE 24.25
DC and thermal resistance test setup.

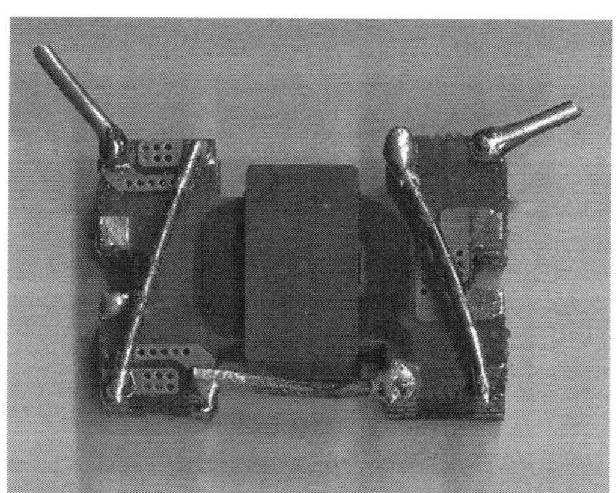

FIGURE 24.26
Unit setup for thermal testing.

a safe operating current for prolonged operation through the unit under test using the controls on our adjustable power supply. Next we measure (1) the voltage across R1 to accurately establish the value of the operating current we have set and then (2) measure the voltage across the unit under test to establish the winding resistance of the unit under test, calculated from the operating current. The voltage value on the digital voltmeter (displaying results with microvolt accuracy) can be seen steadily climbing as the temperature of the unit under test increases. When this settles at a final value (up to an hour later), the core has reached an equilibrium temperature, and the temperature rise of the core is established. A final measurement of the voltage across R1 is then taken in order to calculate the final value current. To obtain the final value power dissipated by the assembly, we multiply the final value voltage across the unit under test with the final value current. The thermal resistance is then obtained by dividing the core temperature rise by the final value power. A further useful parameter which is obtained from this exercise by recording the time at each measurement is the assembly thermal time constant (τ), which can be used to estimate the effects and recovery times of temporary overloads.

24.10 DSE Transformer Spice Model

The DSE-based transformer SPICE model takes the form of multiple subcircuits as shown in Figure 24.27, the DSE transformer SPICE model block diagram. The stack designer software manipulates the connections and numerical parameters to these and incorporates/removes subcircuits per the elements and core set comprising the device assembly.

The blocks PLANE 0 to PLANE 7 in Figure 24.25 represent the plane wound elements in the transformer. The model which appears within each of these is shown in Figure 24.28.

The core block in the model block diagram of Figure 24.27 contains the core model illustrated in Figure 24.29.

A full description of the core mathematics or how the planes and core models operate with one another is beyond the scope of this text. However, we show here the basics of how the model fits together so as to provide readers with some insight into the model operation and to perhaps provide a basis for reconstruction and/or further research.

An output file from the stack designer is given in the appendix to show the format of the SPICE model. The stack designer produces this uniquely for the assembly in design by converting the set of interconnected plane and core models, as illustrated here, into SPICE format, with the addition of interwinding/interplane parasitic capacitances, leakage resistances, and adjustments for core gapping and other effects.

24.11 Conclusion

For ultimate custom performance regardless of development cost or time scales, the choice of planar transformer development strategy has to be full custom design of each integrated planar transformer, together with iterative physical prototyping.

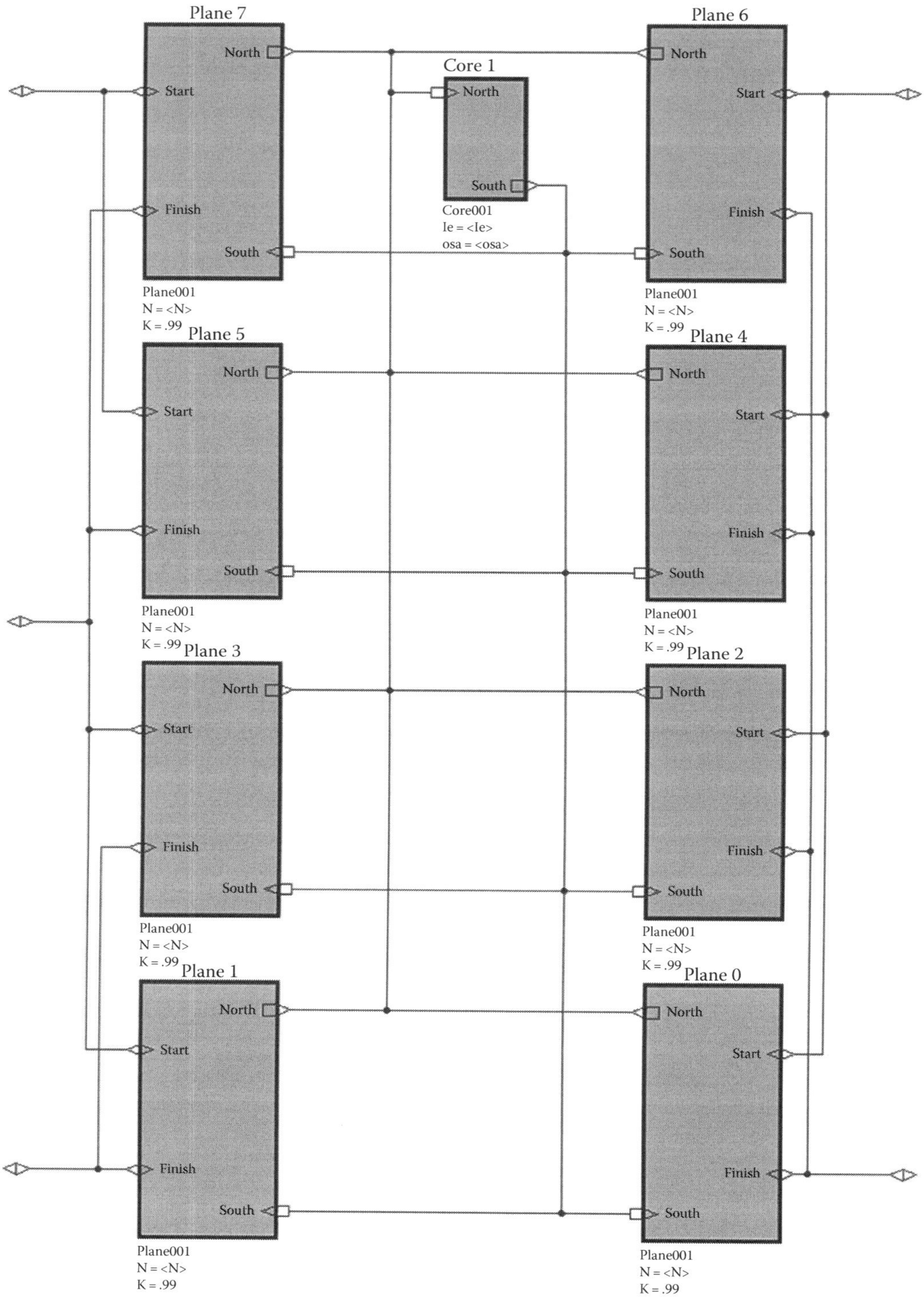

FIGURE 24.27
General DSE transformer SPICE model block diagram.

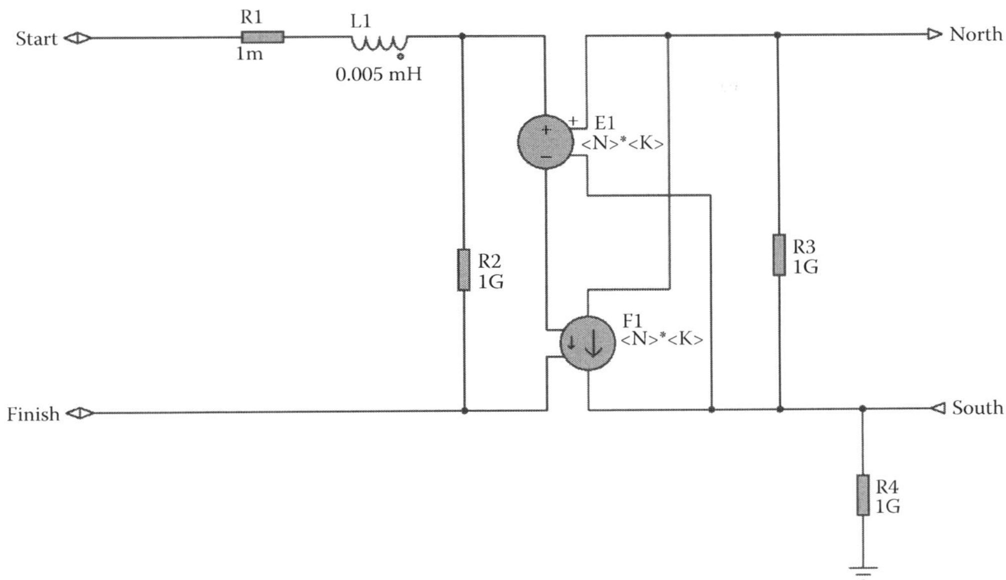

FIGURE 24.28
DSE transformer plane model.

The DSE-based approach enables the rapid development of SMT planar transformers with custom performance in OTS time scales. Also, the DSE approach uniquely offers electronic designers with little or no experience of transformer design a means of routinely devising, simulating, and optimizing SMT planar transformer designs in-circuit using standard electronic CAD design tools with SPICE modeling capability.

A very small proportion of applications may appear to be suited to ready-made standard OTS discrete transformers. However, development of systems in anticipation of such, on only paper specifications, carries a high risk of failure. This is due to the ultimate suitability of the system being unknown until after prototyping or manufacture of the physical system. It will normally be far more expensive to redesign a system than to redesign a transformer.

Appendix 24.A: DSE Transformer Configuration/SPICE Model

```
*******************************************************
* Planar Transformer/Choke Spice 3f5 compatible Model
* for Noctiluca CurlPlanes(TM) device assemblies.
* Generated by CPgo(TM) CAD software v3.0
*******************************************************
* Characteristics modelled: hysteresis, saturation,
* stray capacitances, stray inductances, copper
* resistances, core gapping, device loading.
* Valid frequency range:  100Hz - 1.5MHz.
```

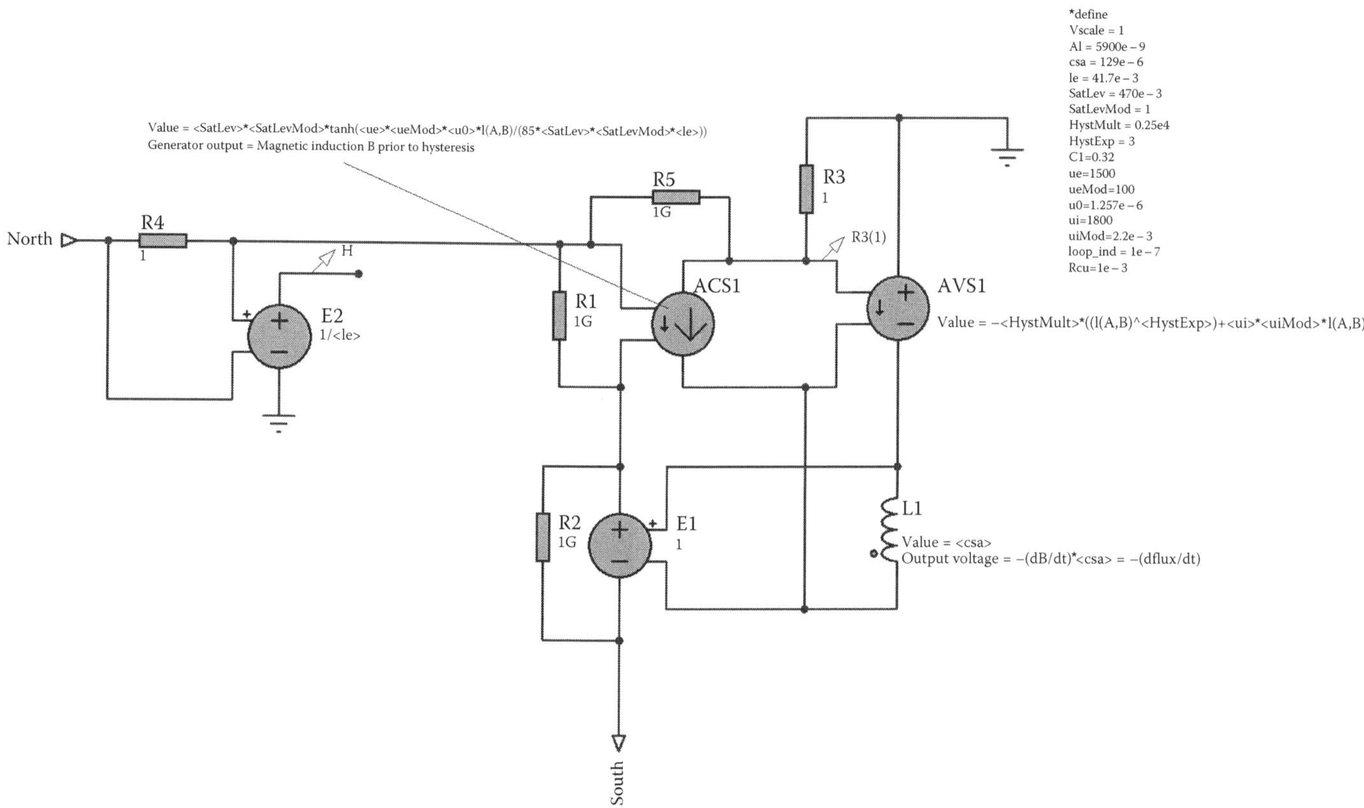

FIGURE 24.29
DSE transformer core model.

```
* Modelled temperature:    25 Degrees C
* (Effects at other temperatures not addressed yet)
*****************************************************
*===Author===
* Device Designer:  Frederick Bott
* Company:   Noctiluca Ltd.
* Address:
* E-mail:
* Website:
* Phone Number:
* Fax Number:
*===Assembly===
* Device: EE-F47-0-32/6/20-290-6666-0000
*===Assembly===
*
*
*                  Device SMT Footprint Pads   B  H
*                  | | | | | | | | | |   |   |  |  |
.SUBCKT CPDevice 1 2 3 4 5 6 7 8 9 10 11 12 15 16
*EE pair
*Plane 0 Orientation=C
XPL0 7 10 13 14 PL32/6/201
*Plane 1 Orientation=B
XPL1 4 5 13 14 PL32/6/2012
*Plane 2 Orientation=C
XPL2 8 9 13 14 PL32/6/201
*Plane 3 Orientation=A
XPL3 1 3 13 14 PL32/6/2012
*Plane 4 Orientation=C
XPL4 7 10 13 14 PL32/6/201
*Plane 5 Orientation=B
XPL5 4 5 13 14 PL32/6/2012
*Plane 6 Orientation=C
XPL6 8 9 13 14 PL32/6/201
*Plane 7 Orientation=A
XPL7 1 3 13 14 PL32/6/2012
C0145 4 10 24e-12
C1256 3 10 24e-12
C2367 3 9 24e-12
C0034 4 9 12e-12
*XCR0 13 14 15 16 CR32/6/20F47EE
R101 1 0 1e9
R102 2 0 1e9
R103 3 0 1e9
R104 4 0 1e9
R105 5 0 1e9
R106 6 0 1e9
R107 7 0 1e9
R108 8 0 1e9
R109 9 0 1e9
R110 10 0 1e9
R111 11 0 1e9
R112 12 0 1e9
*Pins: Start Finish North South
```

```
.SUBCKT PL32/6/201 1006 1005 1004 1003
R6 1006 1002 2.9000E-003
C4 1001 1000 0.0000E+000
E2 1001 1000 1004 1003 9.8000E-001
V1 1005 1000 0
F2 1003 1004 V1 9.8000E-001
L2 1001 1002 2.3561E-009
.ENDS PL32/6/201
*Pins: Start Finish North South
.SUBCKT PL32/6/202 1006 1005 1004 1003
R6 1006 1002 8.3000E-003
C4 1001 1000 1.2000E-011
E2 1001 1000 1004 1003 1.9600E+000
V1 1005 1000 0
F2 1003 1004 V1 1.9600E+000
L2 1001 1002 9.4245E-009
.ENDS PL32/6/202
*Pins: Start Finish North South
.SUBCKT PL32/6/204 1006 1005 1004 1003
R6 1006 1002 2.9700E-002
C4 1001 1000 1.2000E-011
E2 1001 1000 1004 1003 3.9200E+000
V1 1005 1000 0
F2 1003 1004 V1 3.9200E+000
L2 1001 1002 3.7698E-008
.ENDS PL32/6/204
*Pins: Start Finish North South
.SUBCKT PL32/6/206 1006 1005 1004 1003
R6 1006 1002 6.9900E-002
C4 1001 1000 1.2000E-011
E2 1001 1000 1004 1003 5.8800E+000
V1 1005 1000 0
F2 1003 1004 V1 5.8800E+000
L2 1001 1002 8.4821E-008
.ENDS PL32/6/206
*Pins: Start Finish North South
.SUBCKT PL32/6/208 1006 1005 1004 1003
R6 1006 1002 1.3030E-001
C4 1001 1000 1.2000E-011
E2 1001 1000 1004 1003 7.8400E+000
V1 1005 1000 0
F2 1003 1004 V1 7.8400E+000
L2 1001 1002 1.5079E-007
.ENDS PL32/6/208
*Pins: Start Finish North South
.SUBCKT PL32/6/2010 1006 1005 1004 1003
R6 1006 1002 2.1410E-001
C4 1001 1000 1.2000E-011
E2 1001 1000 1004 1003 9.8000E+000
V1 1005 1000 0
F2 1003 1004 V1 9.8000E+000
L2 1001 1002 2.3561E-007
.ENDS PL32/6/2010
```

```
*Pins: Start Finish North South
.SUBCKT PL32/6/2012 1006 1005 1004 1003
R6 1006 1002 3.9180E-001
C4 1001 1000 1.2000E-011
E2 1001 1000 1004 1003 1.1760E+001
V1 1005 1000 0
F2 1003 1004 V1 1.1760E+001
L2 1001 1002 3.3928E-007
.ENDS PL32/6/2012
*Pins: Start Finish North South
.SUBCKT PL32/6/2016 1006 1005 1004 1003
R6 1006 1002 6.0850E-001
C4 1001 1000 1.2000E-011
E2 1001 1000 1004 1003 1.5680E+001
V1 1005 1000 0
F2 1003 1004 V1 1.5680E+001
L2 1001 1002 6.0317E-007
.ENDS PL32/6/2016
*Pins: North South B H
*.SUBCKT CR32/6/20F47EE 1009 1008 1002 1007
V1 13 105 0
V2 15 103 0
BACS 15 103 I=4.7000E-001*6.6000E-001*ATAN(1.5000E+003*3.2500E+002*1.25
70E-006*I(V1)/(85*4.7000E-001*6.6000E-001*3.2500E-002))
BAVS 100 101 V=-2.5000E+003*((I(V2)^3.0000E+000)+1.8000E+003*2.2000E-
003*I(V2))
C1 103 104 1e-6
E1 105 14 101 104 1
H1 16 0 V1 -2.3981E+001
L1 104 101 1.2900E-004
L2 0 100 1e-7
R2 14 105 1e9
R3 15 0 1
R5 15 13 1E9
*.ENDS CR32/6/20F47EE
.ENDS
*********************************************************
* Disclaimer: Models are supplied for evaluation only,
* and as such may or may not reflect typical baseline
* specifications. While great care has been taken in
* the preparation of the CPgo Spice model engine, some
* performance aspects may not be modelled fully. Models
* are supplied "As is", with no direct or implied
* responsibility on the part of Noctiluca Ltd.
* The right to change the model engine for
* improvement purposes without prior notice is
* reserved. In all cases, the current Manufacturers
* data sheets for Ferrites and CurlPlanes are the final
* design guidelines, and are the only actual
* performance guarantees.
*********************************************************
*CS:
*12G53
```

Acknowledgments

The author would like to thank Ian Wilkinson and Andy Hayling, formerly of TT group MMG Neosid Letchworth, and former TT group Company Prestwick Circuits for their joint efforts in working toward the production of the TT MMG versions of the DSE planar products featured here.

Thanks to John Jameson and Iain Cliffe of Labcenter Electronics Ltd. for kindly supplying the latest version of Proteus for the purposes of this chapter. All of the author's historical work on the schematics, PCB, and simulation development (including development of the SPICE model engine in Cpgo) to date has been carried out using Proteus. Figures 24.10, 24.11, 24.13 through 24.23, and 24.25 through 24.27 were all produced from Proteus.

Thanks to Professor Bulent Ertan and Dr. Paul Lefley of Leicester University for the invitation to write this chapter.

References

1. International IEC Standard 62317–9 (2007) ferrite Cores-Dimensions-Part 9: Planar Edition 1:2006 consolidated with amendment 1:2007, Edition 1.1.
2. International IEC Standard 60950–1 (2005) Information technology equipment-Safety-Part 9: Planar Cores, Second edition.
3. Bott F, Method and means of forming a desired coil configuration. U.K. Patent No. GB2337863.
4. Maxwell JC (1873) *A Treatise on Electricity and Magnetism*. Clarendon Press, Oxford, U.K.
5. Belevitch V (1971) The lateral skin effect in a flat conductor. *Philips Tech Rev* 32(6/7,18), 221–231.
6. Johnson H Dr., The proximity effect. High speed digital design newsletter. 4(1). http://www.sigcon.com/Pubs/news/4_1.htm
7. Roespel G (1978) Effect of the magnetic material on the shape and dimensions of transformers and chokes in switched mode power supplies. *J Magn Magn Mater* 9, 145–149.
8. MMG Neosid Ferrite Material Data Sheets. http://www.mmgca.com/catalogue/MMG-Neosid.pdf
9. Barco Graphics N.V. Gerber RS-274X Format.1998.
10. Noctiluca Ltd. CPgo™Software. https://www.noctiluca.com/Support/Downloads/downloads.html
11. Berkeley SPICE 3f5, Appendix D, SPICE3 User's Manual. http://bwrc.eecs.berkeley.edu/classes/icbook/spice/.
12. Labcenter Electronics Ltd. Proteus™Software http://www.labcenter.co.uk
13. International Standard ISO/IEC 15288 (2008) Systems and software engineering—System life cycle processes (Ingénierie des systèmes et du logiciel—Processus du cycle de vie du système), Second edition 2008-02-01.

Index